Adaptation, Diversity, and Ecology
Mammalogy

George A. Feldhamer
Southern Illinois University at Carbondale

Lee C. Drickamer
Southern Illinois University at Carbondale

Stephen H. Vessey
Bowling Green State University

Joseph F. Merritt
Carnegie Museum of Natural History

WCB
McGraw-Hill

Boston   Burr Ridge, IL   Dubuque, IA   Madison, WI   New York   San Francisco   St. Louis
Bangkok   Bogotá   Caracas   Lisbon   London   Madrid
Mexico City   Milan   New Delhi   Seoul   Singapore   Sydney   Taipei   Toronto

WCB/McGraw-Hill
A Division of The McGraw-Hill Companies

MAMMALOGY: ADAPTATION, DIVERSITY, AND ECOLOGY

 This book is printed on recycled, acid-free paper containing 10% postconsumer waste.

1 2 3 4 5 6 7 8 9 0 KGP/KGP 9 3 2 1 0 9 8

ISBN 0–697–16733–X

Vice president and editorial director: *Kevin T. Kane*
Publisher: *Michael D. Lange*
Sponsoring Editor: *Margaret J. Kemp*
Senior developmental editor: *Kathleen R. Loewenberg*
Marketing manager: *Michelle Watnick*
Senior project manager: *Gloria G. Schiesl*
Production supervisor: *Mary E. Haas*
Designer: *K. Wayne Harms*
Photo research coordinator: *John C. Leland*
Art editor: *Jodi K. Banowetz*
Compositor: *Precision Graphics*
Typeface: *10/12 ACaslon Regular*
Printer: *Quebecor Printing Book Group/Kingsport*

Cover/interior design: *Kristyn Kalnes*
Cover image: *Kathy Bushue/Tony Stone Images*
Illustrator: *ISIS/Sandra R. Sevigny*

The credits section for this book begins on page 544 and is considered an extension of the copyright page.

Library of Congress Cataloging-in-Publication Data

Mammalogy : adaptation, diversity, and ecology / George A. Feldhamer . . . [et al.].
    p.   cm.
    Includes bibliographical references and index.
    ISBN 0–697–16733–X
    1. Mammalogy.   I. Feldhamer, George A.
QL703.M36   1999
599—dc21   98–12994
                                                       CIP

www.mhhe.com

For Carla, Andy, and Carrie.

George

For my parents, Mae Elizabeth Drickamer and Harry G. Drickamer.

Lee

In memory of David E. Davis

Steve

With love and appreciation to my wife, Colleen, and big brother, Bob.
Love to my twin brother, Rich (the least-published brother and Mom's third favorite).

Joe

BRIEF CONTENTS

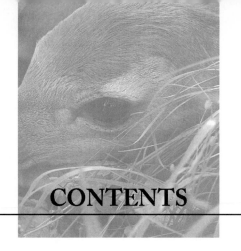

CONTENTS

PREFACE

The amount of information available on mammals has increased dramatically in the last 25 years. In *Mammalogy: Adaptation, Diversity, and Ecology*, we have attempted to "make sense" of this information explosion. Any such attempt is a balancing act between breadth and depth—enough breadth to include numerous subject areas within mammalogy and sufficient depth to avoid superficiality. In terms of form and function, the approximately 4600 species of mammals represent the most diverse class of vertebrates. Mammals are terrestrial, arboreal, or marine; they burrow, run, or fly; and they feed on meat, nectar, blood, pollen, leaves, or a variety of other things. They range in size from 3-gram pygmy shrews and hog-nosed bats to 160-million-gram blue whales. Mammalogy can be approached from a variety of directions and subdisciplines—anatomy and physiology, behavioral ecology, molecular genetics, systematics, conservation, zoogeography, and paleontology, to list a few—all of which ultimately are complementary and interrelated. We explore the diversity and complexity of mammalian form and function in this textbook, as well as phylogenetic and ecological relationships.

Mammalogy: Adaptation, Diversity, and Ecology is intended for use in upper-division undergraduate and graduate courses with students who have a basic background in vertebrate biology. The book's length is tailored to a one-semester mammalogy course.

The text consists of 29 chapters arranged in 5 parts. In part I (chapters 1 through 4), we introduce the subject of mammalogy, the history of the discipline, methods (including current molecular techniques important in systematics and population analyses), and the evolution of mammals. In part II (chapters 5 through 9), we discuss biological functions and physical structure of mammals. Adaptive radiation in form and structure among the currently recognized orders is covered in part III (chapters 10 through 19). Morphology, fossil history, conservation and economics, and a brief synopsis of all extant families are covered for each order. Behavioral, ecological, and zoogeographical considerations are the focus of part IV (chapters 20 through 26). Finally, in part V (chapters 27

through 29), we explore mammalian parasites and diseases, including zoonoses, domestication of mammals, and conservation issues. **Suggested readings** and **questions** designed to help generate critical thinking and discussion are found at the conclusion of each chapter. **Literature cited** within a chapter is collected at the end of the text to avoid redundancy. **Technical terms** are in boldfaced type the first time they appear and are defined in both the text and the **glossary**. Although there is continuity between the part divisions and chapters of the text, instructors can select certain chapters based on individual interest, emphasis, or time constraints without sacrificing clarity and understanding.

All four of us bring several years of field and laboratory experience with mammals in a variety of settings—as well as our individual specializations, viewpoints, and biases—to the endeavor of writing this book. We have benefited from this collaboration and hope it is reflected in the text. More important, we have profited through the years from the suggestions, ideas, and constructive criticism of many teachers, colleagues, students, and friends.

We want to express our appreciation to the many reviewers who read drafts of this text in part or in whole.

Reviewers

David M. Armstrong *University of Colorado, Boulder*

Richard Buchholz *Northeast Louisiana University*

Jack A. Cranford *Virginia Polytechnic Institute and State University*

Jim R. Goetze *Laredo Community College*

Dalton R. Gossett *Louisiana State University*

Kay E. Holekamp *Michigan State University*

Carey Krajewski *Southern Illinois University*

Thomas H. Kunz *Boston University*

Peter L. Meserve *Northern Illinois University*

Christopher J. Norment   *SUNY College at Brockport*

Larry S. Roberts   *University of Miami*

Robert K. Rose   *Old Dominion University*

Michael D. Stuart   *University of North Carolina, Asheville*

John A. Vucetich   *Michigan Technological University*

Wm. David Webster   *University of North Carolina, Wilmington*

John O. Whitaker, Jr.   *Indiana State University*

Bruce A. Wunder   *Colorado State University*

Many other persons helped significantly in the preparation of the text, including Marge Kemp, Kathy Loewenberg, Gloria Schiesl, Mary Reeg, Jodi Banowetz, Sandra Sevigny, LouAnn Wilson, and Linda Davoli. We also thank science librarian Kathy Fahey and her staff at Southern Illinois University at Carbondale for their assistance throughout this project. Thanks also to Marjorie K. Laughrey, Colleen Hannaken, and Kristin Vessey for their editorial assistance.

We hope this book does justice to past and present mammalogists upon whose research and teaching efforts it is largely based. Just as important, we hope that it will prove useful to students who will be the mammalogists of the future and that through it, students will better appreciate and explore the mysteries of mammals—those "fabulous furballs."

About the Authors

George A. Feldhamer is an associate professor of zoology, and coordinator of the Environmental Studies Program, at Southern Illinois University at Carbondale. His research has focused exclusively on mammalian populations, ecology, and management; introduced cervid biology; and threatened and endangered species. He is a former associate editor of the *Wildlife Society Bulletin*, and coeditor of *Wild Mammals of North America: Biology, Management, and Economics*. He is curator of the mammal collection at SIUC and has 20 years of experience teaching an upper division mammalogy course.

Lee C. Drickamer has been a professor of zoology at Southern Illinois University at Carbondale for 11 years. He is a past-president of the Animal Behaviour Society, past-secretary-general of the International Council of Ethnologists, past chair of the Division of Animal Behavior of what is now the Society for Integrative and Comparative Biology, and former editor of *Animal Behaviour*. His research emphases have included social factors affecting development and reproduction in house mice and swine, behavioral ecology of house mice and deer mice, social biology of primates, intrauterine position effects on behavior and reproduction of mice and swine, and the consequences of mate selection for offspring viability in house mice.

Stephen H. Vessey is professor emeritus of biological sciences at Bowling Green State University. His research interests include the behavioral ecology of mammals, especially primates and rodents. He has been studying a population of white-footed mice in northwestern Ohio for more than 25 years. He is a former associate editor of the *Journal of Mammalogy* and is a Fellow of the Animal Behavior Society. He has taught mammalogy and animal behavior at Bowling Green for 28 years, coauthoring a textbook in animal behavior with Lee Drickamer.

Joseph F. Merritt is the resident director of Powdermill Biological Station, the field station of the Carnegie Museum of Natural History. He is a physiological ecologist specializing in adaptations of small mammals to cold. Dr. Merritt is the author of *Guide to Mammals of Pennsylvania,* published by the University of Pittsburgh Press, and editor of several technical monographs on specific taxa of mammals. He has served on the Editorial Committee of the American Society of Mammalogists since 1990, and is currently the managing editor of the *Journal of Mammalogy*. Dr. Merritt teaches mammalogy at the University of Pittsburgh's Pymatuning Laboratory of Ecology and courses in mammalian ecology at Antioch New England Graduate School and at the Adirondack Ecological Center, SUNY College of Environmental Science and Forestry.

Adaptation, Diversity, and Ecology

Mammalogy

PART ONE

Introduction

Although widely distributed throughout much of North America, bobcats (*Lynx* [*Felis*] *rufus*) are rarely seen.

CHAPTER

1

THE STUDY OF MAMMALOGY

WHAT IS MAMMALOGY?

Mammalogy is the study of the animals that constitute the Class Mammalia, a taxonomic group of vertebrates (Phylum Chordata, Subphylum Vertebrata) within the Kingdom Animalia. Humans (Homo sapiens) are mammals, as are many domesticated species of pets and livestock, as well as wildlife, such as deer and squirrels, with whom we share our natural surroundings (figure 1.1). Many of the species of animals, such as elephants, whales, large cats, and the giant panda, that have aroused public concern for their survival are mammals.

Mammals share a number of common features, including (1) the capacity for internal temperature control, often aided by a coat of fur; (2) the possession of mammary glands, which, in females, provide nourishment for the young during early development; and (3) with a few exceptions, the ability to give birth to live young. These and many other features of mammals are discussed in detail in chapter 4 and in parts II and III.

Animal biology can be studied from a taxonomic perspective, that is, by concentrating on groups of organisms, such as mammals (mammalogy) or birds (ornithology). Or, the functional perspective can be used, concentrating on processes, as in physiology and ecology. In this book, we combine both approaches. The disciplines of biochemistry, physiology, animal behavior, and ecology, among many others, all contribute to mammalogy. Our goal is to explore and integrate discoveries from all these disciplines to provide the most enlightening and productive approach to the study of mammals.

Throughout the book, we weave together at least four major themes: evolution, methods for investigating mammals, diversity, and the interrelationships of form and func-

tion. A basic underlying theme for all of biology is evolution by natural selection. Beginning with chapter 4, we take up the thread of evolutionary thought, giving particular emphasis to both speciation and adaptations of mammals. Chapter 3 begins the second thread, scientific methods, which deals with how mammalogists formulate questions (hypotheses) for investigation and what methods they use to answer these questions. The third thread, which is covered in parts II and IV, involves how form, function, and behavior are tightly interwoven and shaped by natural selection to provide solutions to the key problems of survival and reproduction in mammals. Our fourth thread, mammalian diversity, is emphasized in part III, but examples offered throughout the text further underscore this theme.

WHY STUDY MAMMALS?

Most of us have at least a passing interest in mammals, but we seldom stop to think why the formal study of mammalogy is important. **Mammalogy** can be approached from a variety of directions and for diverse reasons (Wilson and Eisenberg 1990). Mammals were a resource for early humans. Knowledge about them was important if humans were to successfully hunt or trap them. Some mammals, such as the sabre-toothed cats that coexisted with early humans, were potential predators on humans. Knowledge of their habits was important for survival. Indeed, there are still locations throughout the world where wild animals, including grizzly bears (*Ursus arctos*) in western North America, may attack and kill humans. Mammals continue to be important to humans as food. People with a subsistence way of life may depend on capturing or killing free-ranging mammals. More industrialized cultures depend on domesticated livestock for food. In addition, humans have a long tradition of using in

A   **B**   **C**

Figure 1.1   Mammals with whom we share the world.   In addition to our own species, mammals with whom we share our world can be grouped roughly into (A) domestic pets and livestock, such as a house cat (*Felis silvestris*), (B) wildlife in our familiar environment, which we may see often or in other cases rarely, such as a fox squirrel (*Sciurus niger*), and (C) wildlife from other lands, particularly endangered or threatened species, such as an African elephant (*Loxodonta africana*).

numerous ways mammals, including hides, bones, fur, or blubber from whales and seals.

Mammals serve the needs of humans as pets and for recreational hunting. Humans keep many types of mammals as pets, ranging from cats, dogs, and mice to more exotic species, such as large cats, primates, and even skunks. Much of the practice of veterinary medicine, which developed originally to serve the needs of agriculture, is now devoted to the diagnosis and treatment of illnesses and injuries affecting our mammalian pets. Many species are hunted for sport in North America, including the cottontail rabbit (*Sylvilagus floridanus*), the white-tailed deer (*Odocoileus virginianus*), and elk (*Cervus elaphus*). Exotic forms of mammalian wildlife, including free-ranging populations of fallow deer (*Dama dama*), sika deer (*Cervus nippon*), and feral hogs (*Sus scrofa*), have been introduced in several states, most notably Texas, to provide additional game species. Some exotics have become major pests after introduction for sport or trade because their interactions with native species were unforeseen. An example is the introduction of the Indian mongoose (*Herpestes javanicus*) on many islands in the Caribbean Sea and on the Hawaiian Islands. Mongooses were introduced to control rodents that had, in turn, been brought to the islands by humans. Mongooses consume the eggs and young of many native bird species, however, as well as compete with other native animals.

Some mammals pose risks for humans and other animals because they serve as reservoirs or vectors for a variety of diseases and parasites (e.g., the black rat [*Rattus rattus*] or black-tailed prairie dog [*Cynomys ludovicianus*] are vectors for plague). Knowledge of the life cycles of parasites and the symptoms of various mammal-borne diseases is necessary for avoiding and treating these health hazards.

Some mammals can damage portions of our environment or negatively affect other mammals. Rats, mice, and occasionally other small mammals with whom we share our living areas do great harm to both our property and food stores. Some rodents exhibit explosive population growth and overrun large areas of planted cropland. A better understanding of the reproductive and population biology of such agricultural pests can lead to means for controlling them. Moles or gophers may damage our lawns, and beavers can cause flooding of forests and croplands. It is sometimes difficult to realize that these mammals are just carrying out their normal activities, which, unfortunately, often lead them into conflict with humans. Our anthropocentric (human-centered) perspective of life leads us to view many "normal" activities of nonhuman mammals as being in conflict with our goals.

Another currently important reason for studying mammals is conservation. After driving many species into or close to extinction, some effort is being made to reverse the trend. Toward that goal, some people work to understand and protect the habitats of endangered or threatened species. Others study social and reproductive biology under natural conditions or to establish captive breeding programs designed to eventually reintroduce species into their natural habitats. Good examples are current efforts involving the black-footed ferret (*Mustela nigripes*) and red wolf (*Canis rufus*).

Because we are mammals, we can learn much about ourselves by studying similar processes that occur in other mammals. Some animals serve as models for various diseases or as subjects for developing or testing vaccines for eventual use on humans. We also maintain large colonies of some mammals in captivity to better study a whole variety of physiological, behavioral, and related medical phenomena. Work on particular species broadens and enhances our knowledge about such basic processes as developmental biology, immunology, endocrinology, and reproduction.

RESOURCES FOR MAMMALOGISTS

A variety of resources is available to help us learn about mammals. Those who study mammals over many years develop personal libraries of pertinent materials, including general reference works and guidebooks containing keys for identifying mammals and providing basic information on the habits of particular species.

A great deal of literature is available on all aspects of mammals. Some volumes encompass worldwide coverage, such as *Walker's Mammals of the World* (Nowak 1991); Macdonald's (1984) *Encyclopedia of Mammals*; and Wilson and Reeder's (1993) *Mammal Species of the World*. Other books provide coverage of a particular continent or faunal region, such as Hall and Kelson's *The Mammals of North America* (1959), Hall's second edition of the same work (1981), and Burt and Grossenheider's *A Field Guide to the Mammals* (1980), covering North America; *The Mammals of Southern Africa* (Stuart and Stuart 1988), *Mammals of Australia* (Strahan 1995), *The Mammals of the Palearctic Region* (Corbet 1988), and *Wild Mammals of North America* (Chapman and Feldhamer 1982). In the United States, many books cover the mammals of various regions, for example, *Wild Mammals of New England* (Godin 1977), *Mammals of the Intermountain West* (Zeveloff 1988), *Mammals of the Great Lakes Region* (Kurta 1995), and *Guide to the Mammals of the Plains States* (Jones et al. 1985), as well as of practically every state; for example, *Mammals of Wyoming* (Clark and Stromberg 1987), *Mammals of Indiana* (Mumford and Whitaker 1982), and *Mammals of Pennsylvania* (Merritt 1987). Other works are specialized treatises on particular taxonomic groups (e.g., *The Natural History of Badgers* by Neal [1986]) or even monographs on particular species (e.g., *White-tailed Deer Ecology and Management*, edited by Halls [1984]). There are books on practically every mammalian order, particular families, and on many individual species. A variety of journals also are devoted strictly to mammals, such as *Journal of Mammalogy, Acta Theriologica, Mammalia, Mammal Review,* and others published by national or international professional societies devoted to the study of mammals.

Chapter 1 *The Study of Mammalogy* **7**

Mammalogists rely on good university or college libraries for extensive collections of books and journals containing information on mammals. The *Zoological Record*, which began in 1848 and is issued annually, is the best overall source for literature on mammals. It provides information by species, subject, author, and geographic area. Computerized bibliographic databases are very useful for locating literature on a particular subject or a specific mammalian species, or works by a particular author. A note of caution is in order; most computerized bibliographic databases are limited to information going back only 10 to 15 years. Works published earlier than that are not recorded. Because mammalogy has a history of significant work dating back more than a century, additional sources beyond databases should be consulted. One method uses reference lists in books or papers to compile a retrospective list of articles that encompasses a broad time period.

Mammals are studied as both living organisms and preserved specimens. Living mammals are studied in various situations, including in the wild in their natural habitats. Observing mammals directly through such means as trapping or radiotelemetry can provide particular insights and may involve travel to exotic places or simply being "out in the field." Other types of research are conducted in a captive setting, usually a laboratory facility, zoological park, or aquarium. Some species are either rare enough in the wild or are small, nocturnal, or secretive enough to necessitate the use of a zoological park or laboratory setting to conduct research. Many people who visit zoos are unaware that they also function as places of scientific study. In some exhibit buildings or enclosed areas not open to the public, species conservation

and related investigations are taking place (figure 1.2). Today, a number of zoos have separate facilities for the study of species that may be endangered or threatened. An excellent example is the Smithsonian Institution (National Zoological Park) facility at Front Royal, Virginia, which was formerly a major horse-breeding station for the U.S. Army cavalry. Investigations of domestic animals, both livestock and pets, also are carried out under more controlled conditions in a variety of laboratory research settings.

The field of mammalogy is fortunate that, beginning several centuries ago, museum collections of preserved mammals were started, and public and private menageries, the forerunners of the modern zoological parks, came into existence. For a short but thorough summary of this topic, with particular emphasis on North America, consult Wilson and Eisenberg (1990). Today, a large network of museums and related collections of mammals provides study skins, whole mounts, skeletal remains, and, in some instances, preserved soft anatomy and tissues for genetic studies (Hafner et al. 1997). These collections also include large numbers of fossils discovered by paleontologists. Without such materials, we would not be able to discern very much about the evolutionary history of mammals. Although we generally see such museums in their role as educational institutions, they also contain vast storage and work areas where professionals curate and study the preserved materials (figure 1.3). In several countries (e.g., Canada), computerized databases exist of all of the preserved and fossil materials from most museums, permitting investigators throughout the country to access the location of and information about particular specimens. A global network of this sort may be developed in the coming decades.

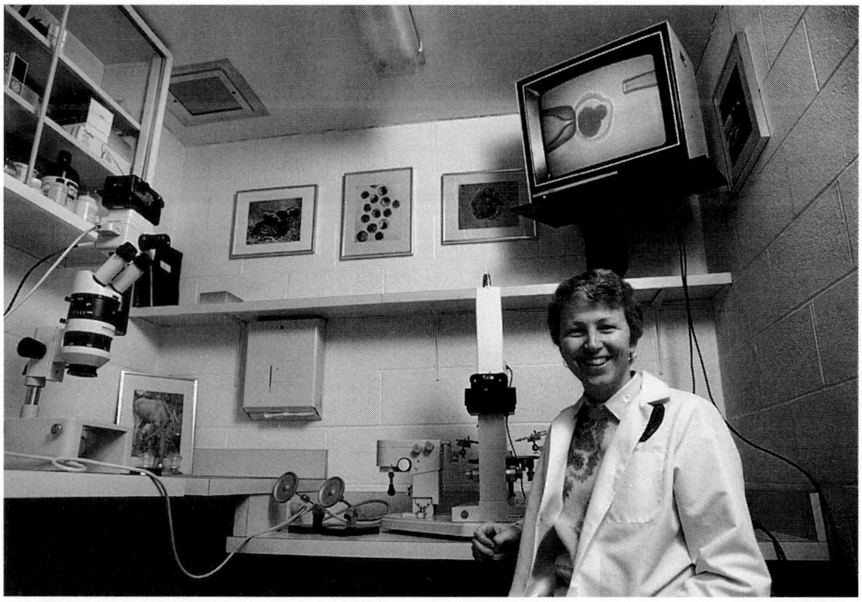

ORGANIZATION OF THE BOOK

This book is divided into five parts. In the first part, after these introductory remarks, we examine the history of mammalogy (chapter 2). Chapter 3 explores how mammalogists proceed with their investigations; what questions they ask, and how they answer them. Chapter 4 deals with the evolution of mammals and the diagnostic characteristics that define them.

Part II integrates the morphological features of mammals with their physiological functions and behavior. This material and the examples provide a foundation for structure and function, and illustrate variation among mammals, including specialized adaptations that serve as solutions to particular problems.

Figure 1.2 National Zoological Park, Washington, D. C. Exhibit areas of many zoological parks that are not open to the public are important in terms of the studies taking place there on breeding and social biology. Often these areas involve breeding programs of endangered species, such as that shown here in a behind-the-scenes view from the Cincinnati Zoo.

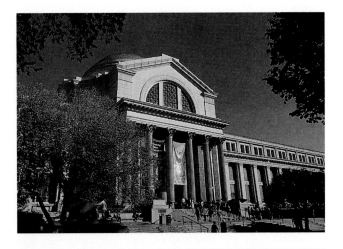

Figure 1.3   National Museum of Natural History, Washington, D.C.   Museums are important repositories for large collections of preserved and fossilized specimens of mammals and other animals. The more familiar function of such facilities, serving to educate and entertain the public, may also be viewed as an important reason for the study of mammals.

Part III is a taxonomic examination of biodiversity among the 26 orders of mammals, emphasizing traits that characterize each mammalian family. Methods for discerning relationships among mammals at various levels of taxonomic classification are covered, including such recent developments as the use of protein allozymes and DNA.

Part IV examines the interactions of mammals from behavioral and ecological perspectives. Vaughan (1986) commented that behavior and related topics were slow in developing. In earlier textbooks by Cockrum (1962) and Gunderson (1976), as well as earlier editions of Vaughan's text, we find scant coverage of behavior and only modest coverage of ecology. As is true for ornithology and herpetology, a shift in emphasis has taken place in the last quarter century: Behavior and ecology are now receiving considerably more attention. An examination of recent issues of the *Journal of Mammalogy* further illustrates our point. Fifty percent or more of the papers deal with these topics. Some earlier books on mammals, including the text by Davis and Golley (1963) and the volume on natural history by Bourliére (1970), provided substantial coverage of ecology and behavior. In part IV, we build on the foundations of classical mammalogy and combine description, systematics, form, and function to examine the ecology and behavior of mammals.

The final section (part V) covers several specialized topics. Animal diseases and parasites (chapter 27) are important from both biological and practical perspectives. Domestication of mammals (chapter 28) has long been a part of human life. Recently, emphasis on management of animals in zoological parks and on game farms and ranches worldwide has been renewed. Conservation biology (chapter 29) takes on renewed importance each time we read about the threat to an endangered species. This last chapter attempts to apply what we have learned throughout the book to the issues of habitat conservation, reproductive biology, and related species preservation efforts.

Summary

Mammals are one of the classes of vertebrates or animals with backbones. All mammals share a series of common characteristics, including internal control of body temperature, (often aided by an insulating layer of fur), mammary glands, and (with a few exceptions) live birth of young. Mammals are studied for a variety of reasons, including their use for food and other products; as subjects of recreational hunting; as pets; as pests that cause damage; their conservation; their role in disease and related health considerations; because we are mammals ourselves; and for aesthetic interest.

Our study of mammals

1. Examines the history of the discipline, the methods used by mammalogists, and the evolution and characteristics of mammals

2. Explores the details of relationships between structure and function in the morphological and physiological systems of mammals

3. Reviews the taxonomic subdivisions of mammals

4. Explores the behavior and ecology of mammals

5. Provides special chapters on diseases, parasites, domesticated mammals and those kept on game ranches, and conservation.

Our approach weaves four major themes together: the process of evolution by natural selection as it has shaped mammals, in terms of both adaptation and speciation; how mammalogists ask and go about answering questions; the interrelationships of morphology, physiology, and behavior; and the diversity of mammals.

A variety of helpful resources are available for students and professionals interested in mammalogy. These include places to study mammals, such as zoological parks, aquaria, laboratories, and the natural setting. Excellent collections of preserved material from mammals are located in museums. Finally, modern libraries, with their collections of books, journals, and computerized databases, contain vast quantities of readily accessible information.

Discussion Questions

1. Make a list of all of the mammals (use common names for now) that you have encountered in the past month. Note also where you saw them and any key features you used to distinguish them first as mammals and then as individual species.
2. After reading the section on reasons for studying mammals, compile your own list of reasons for investigating this group of animals. You can use these reasons as a starting point and provide some more detailed purposes, or you can start from scratch to come up with some reasons for the study of mammals and see how your list compares with the one provided. For each

    reason on your list, provide a brief specific example of how that rationale for the study of mammals has already affected you.
3. At your library, locate the computer terminals that access bibliographic databases. Select four topics related to mammalogy. Search these topics and the several permutations of those topics that occur to you as you peruse the search output. This exercise should familiarize you with the use of such databases and can provide references for topics that may become part of a required paper in the course you are taking.

Suggested Readings

Bourliére, F. 1970. The natural history of mammals. Alfred A. Knopf, New York.

Macdonald, D. (ed.) 1984. The encyclopedia of mammals. Facts on File Publications, New York.

Wilson, D. E., and J. F. Eisenberg. 1990. Origin and applications of mammalogy in North America. Pp. 1–35 *in* Current mammalogy, vol. 2. (H. H. Genoways, ed.). Plenum, New York.

CHAPTER

2

History of Mammalogy

arly in human evolution, our ancestors became aware of other mammals with whom they shared habitat. In some cases, this knowledge about mammals served to determine possible prey that could be food. In other instances, information about potential predators was necessary for humans to avoid becoming food for larger carnivores. Other mammals have always been competitors with humans for food and shelter.

Early recorded indications of human knowledge about mammals come from cave paintings and petroglyphs (figure 2.1). Other evidence comes from sites where humans either drove herds of mammals over cliffs to their death or forced them through a narrow passage to be trapped and clubbed or otherwise killed. Prior knowledge of the behavior and movement patterns of mammals was crucial to the development of an effective hunting strategy. Some relationships between early humans and mammals were even closer. Dogs were domesticated from wolves (*Canis lupus*), and sheep (*Ovis aries*), goats (*Capra hircus*), and cattle (*Bos taurus*) were domesticated from wild ungulates (chapter 28). Religious icons, such as small statues (figure 2.2), indicate that humans ascribed certain mystical powers to the animals with whom they shared their world. Throughout much of human history, many mammalian species have also been used as beasts of burden. It has only been within the last several hundred years that other major modes of transportation have replaced mammals. Even today in many locations around the world, mammals are a major means of transport.

In the sections that follow, we trace the development of mammalogy from the days of classical Greece and Rome through the period of exploration and the dawn of natural history and ending with a brief examination of the discipline of mammalogy today.

Figure 2.2   **Early mammalian carving.**   The mammal shown in this statue from a Meso-American archeological site is a cat, and is represented because it shared the environment with the people who carved it. The fact that it is depicted in this small carving suggests that it also may have been the subject of some form of religious importance or it may have served as a clan totem.

Mammalogy and the study of mammals are important to humans for at least three reasons. First, as humans, we are mammals, and thus we share similar physiological, behavioral, social, and ecological traits. Gaining knowledge about nonhuman mammals aids in understanding ourselves and human evolution. Second, some species of mammals, perhaps a disproportionately high number relative to other vertebrates, are endangered in our world today. If we are to conduct effective programs to preserve these species, we need an enhanced understanding of mammals. Third, mammals have been most useful to scientists in the search for general principles of evolution, ecology, and behavior. This last point is covered in greater detail in part IV.

FIRST INTEREST IN MAMMALS

From the days when most humans were hunters and gatherers through the beginnings of agriculture about 9000 to 10,000 years ago, a body of knowledge about mammals developed and was passed from generation to generation. With the advent of written language, some of this knowledge was recorded in glyphs, replacing or augmenting earlier depictions in art and oral stories. Particularly important in these early times was knowledge about mammals as food sources and work animals. Mammals were first domesticated in the Middle East and Asia. These early rudiments of formal interest in mammals, including scholarly writings, were later fostered in Egypt, Greece, and Rome as part of a growing body of information about the natural world.

Figure 2.1   **Cave painting of a mammal.**   Early humans daily dealt with other living mammals in their environment. The large ungulate mammal represented here in a cave painting from Alpera, Spain, was likely hunted by the people who painted it. The animal was eaten for food, and its bones, fur, and other products were also put to good use as tools and clothing.

Interest in mammals involved both curiosity about living forms and attention to various fossils that were discovered, collected, and passed among what were then called natural philosophers, a group that included Hippocrates (460–377 BC) and Aristotle (384–322 BC). Although Aristotle did not actually generate a classification scheme for living organisms, he did group animal forms as he saw them (Singer 1959). The category he labeled as having red blood and viviparous reproduction (mammals) included three major groups: (1) viviparous quadrupeds, subdivided into the ruminants with cutting teeth in the lower jaw only and having cloven hoofs (e.g., sheep, oxen), the solid-hoofed animals (e.g., horses), and other viviparous quadrupeds; (2) cetaceans (whales and their relatives); and (3) humans.

Fossilized bones and teeth were discovered from time to time in the ancient world. These remains often raised questions about the origin of these animals and their relationships to existing animal life. Even early in the history of mammalogy as a science, fossils were important in attempts to understand mammals, their history, and their distribution around the globe (Miller and Gidley 1934).

Aristotle and Pliny the Elder (AD 23–79) made extensive records of what they observed and heard about various mammals. Although these writings are primarily anecdotal, they are interesting, as the following quote from Pliny illustrates:

In the mountains of Mauretania it is said that the herds of elephants move at the new moon down to a river by the name of Amilo, ceremoniously cleanse themselves there by spraying one another with water, and after having thus paid their respect to the heavenly light return to the forests bearing their weary calves with them. It is also said that when they are to be transported overseas, they refuse to go onboard until the master of the ship has given them a promise under oath to convey them home again. Further, they are so modest that they never mate except in secluded spots, while adultery never occurs amongst them. Towards weaker animals they show compassion, so that an elephant when passing through a flock of sheep will with his trunk lift out of the way those he meets, for fear of trampling on them. (Nordenskiöld 1928:55)

Other early natural historians, such as the Roman anatomist Galen (AD 130–201), performed dissections, thereby generating new knowledge about the structure and function of different organ systems of animals, including mammals. Over the next several centuries, various expeditions, undertaken primarily for trade and warfare, brought the Europeans into greater contact with areas of North Africa and the Middle East and introduced them to new varieties of animals. Between the fifth or sixth century and the sixteenth century, very little new knowledge was gained. Natural philosophy and the various scientific disciplines waited almost 1000 years before reemerging.

SEVENTEENTH- AND EIGHTEENTH-CENTURY NATURAL HISTORY

Naturalists

By the 1600s, interest in natural history had reawakened. European explorers were traveling to many parts of the world during this time. The materials they discovered, described, and brought back from various countries for further study served to stimulate interest in the life sciences. Although none of these early natural historians was strictly a mammalogist, their broad interests included mammals. They undertook the early study of mammals around the world and the development of a number of concepts that are still applied broadly to all living organisms.

Notable among the early naturalists to visit North America was Mark Catesby (1683–1749). He made two lengthy trips from England to the colonies, one from 1712 to 1719 and a second from 1722 to 1726. He was fortunate to have the financial backing of several wealthy English patrons, enabling him to travel and explore freely, particularly in what is now the southeastern United States. In 1748, Catesby completed a three-volume treatise entitled, *The History of Carolina, Florida and the Bahama Islands*, in which he provided original descriptions and illustrations of a number of North American mammals (figure 2.3).

Among the Europeans who contributed to the growing body of information on mammals was Georges Buffon (1707–1788), a Frenchman who compiled and wrote a 44-volume *Histoire Naturelle*. The following passages dealing with mammals illustrate both the state of knowledge at the time and the style of writing that characterized science dur-

Figure 2.3 **Illustration from Mark Catesby.** During his two visits to the southeastern portion of North America, Catesby painted and described a number of mammals, including the chipmunk (*Tamias striatus*) shown here.

ing this period (Buffon 1858; a later translation from the French). The first concerns the lion (*Panthera leo*):

> *Both the ancients and the moderns allow that the Lion, when newly born, is in size hardly superior to a weasel; in other words, that he is not more than six or seven inches long; and if so, some years at least must necessarily elapse before he can increase to eight or nine feet. They likewise mention that he is not in condition to walk till two months after he is brought forth; but, without giving entire credit to these assertions, we may, with great appearance of truth, conclude that the Lion, from the largeness of his size, is at least three or four years in growing, and that, consequently, he must live seven times three or four years, that is, about twenty-five years. (Vol. II:28)*

This second passage from Buffon concerns the European badger (*Meles meles*), and although the exact terms are not used, he describes aspects of imprinting and omnivory:

> *The young badgers are easily tamed; they will play with young dogs, and, like them, will follow any person whom they know, and from whom they receive their food; but the old ones, in spite of every effort, still remain wild. They are neither mischievous nor voracious, as the fox and wolf are, yet they are carnivorous; and though raw meat is their favourite food, yet they will eat anything that comes in their way, as flesh, eggs, cheese, butter, bread, fish, fruit, nuts, roots, &xc. They sleep the greatest part of their time, without, however, being subject, like the mountain rat or the dormouse, to a torpor during the winter; and thus it is that, though they feed moderately, they yet are always fat. (Vol. I:255)*

Several key developments in Europe contributed to the renewed interest in biology (Gunderson 1976). John Ray (1627–1705) first attempted to define a species as a group of organisms that can interbreed. As more and more distinct types of organisms were discovered and described, the need became apparent for some system of **taxonomy;** a way of classifying and organizing information on all living plants and animals. From the 1500s onward, a number of individuals proposed schemes for classification, although none proved entirely satisfactory. A Swedish botanist, Carl von Linné (known generally by his Latinized name, Carolus Linnaeus; 1707–1778; figure 2.4) published a series of editions of a scheme of classification. His tenth edition of *Systema Naturae,* published in 1758, is generally cited as the basis for modern taxonomy of living animals. The Linnaean system considered species to be fixed, discrete, individually created entities. A key feature of this system was that it involved a hierarchical arrangement of the various levels of classification. Thus, species were grouped into genera, genera into families, and families into classes. Together, the genus and species constituted the **binomial nomenclature** (two names) that scientists use today. After classification

Figure 2.4   Carolus Linnaeus.   Linnaeus is considered by many as a founder of modern systematics. He formalized the system of binomial nomenclature by which all distinct species are given two names. He also developed an hierarchical scheme for classifying living and fossil organisms.

systems like that proposed by Linnaeus were adopted, patterns became evident with regard to the grouping together of organisms that shared similar traits. These groupings reflected possible relationships between organisms. Linnaeus and some of his contemporaries were the first true systematists.

In addition to those who traveled and those who worked in Europe, there were some notable naturalists in North America. Thomas Jefferson (1743–1826) had a keen interest in all of the sciences, including natural history. As vice president of the United States, he published a paper in the *Transactions of the American Philosophical Society* on fossil ground sloths (*Megalonyx*) (Jefferson was president of the society at the time). The story is often told that when he was president of the United States, Jefferson spread out the bones of several fossil mammals on the floor of the East Room of the White House. Included in this array were mastodons, bison, deer, and many smaller mammals. These materials were part of the trove of materials recovered from Big Bone Lick (now in Kentucky), excavated by William Clark (of Lewis and Clark fame). Besides his scientific contributions, Jefferson sponsored expeditions of exploration. He was responsible, as president, for sending Meriwether Lewis and William Clark on their journey westward to the Pacific in 1804–1806 and arranged for Clark to explore Big Bone Lick in 1807.

Concepts and Ideas

The numerous discoveries of the seventeenth and eighteenth centuries led to the formulation of key concepts and ideas that continue to influence mammalogy today. The Linnaean system of classification eventually made apparent the relationships between various groups of animals. New attention was turned to the origins of these animals and to the idea of evolution. Three individuals who contributed to the development of the theory of evolution deserve special mention: Erasmus Darwin, the grandfather of Charles Darwin; Thomas Malthus; and Charles Lyell. Their work helped provide the bases for development of the theory of evolution by natural selection. Erasmus Darwin (1731–1802), an Englishman and a physician by profession, published a number of papers on scientific topics. His comprehensive treatise, *Zoonomia*, explored all the laws of organic life. Although he did not attempt to explain the origin of species, Erasmus Darwin proposed that the diversity of living organisms resulted from influences of the various environments in which they lived.

Thomas Malthus (1766–1834), also an Englishman, was a university professor in London. His primary contribution, *The Principle of Population*, argued that the human population had the potential to grow beyond its limits. He reasoned that self-control and restraint were necessary if humans were to avoid the problems stemming from overpopulation. The notion of overpopulation was part of the rationale that Charles Darwin used to formulate the theory of evolution by natural selection.

Charles Lyell (1797–1875; figure 2.5), often considered to be the founder of modern geology, proposed that processes that influenced the physical world in the past are still active in the present. He called this process "uniformatarianism." Lyell also contended that these changes occurred gradually rather than catastrophically, meaning that time was necessary for such changes. These thoughts had a major influence on Charles Darwin.

NINETEENTH-CENTURY MAMMALOGY

Explorations and Expeditions

Although this section concentrates primarily on events that occurred in North America, expeditions to other continents and to islands of the Pacific Ocean also contributed greatly to existing knowledge. Discoveries from expeditions that were important to mammalogy in North America include those by (1) Lewis and Clark and, soon thereafter, Zebulon Pike; (2) trappers and fur traders who moved into the West in the early 1800s, as well as whalers and other seafaring explorers; and (3) naturalists who accompanied U.S. Army troops in the West or who were members of survey parties working to find routes for railroads. Together, these various expeditions provided an opportunity to discover, collect, and describe new species of mammals, their habitats, and their habits.

Figure 2.5   Charles Lyell.   Lyell proposed that to understand the present state of things, particularly with respect to geology, we needed to examine the progression of changes that lead to the present state. This idea of change through time was relatively novel and served to further the thinking that lead to the theory of evolution by natural selection.

Lewis and Clark, and Pike

Shortly after the Louisiana Purchase in 1803, President Jefferson authorized an expedition to explore this new territory. The party, led by Meriwether Lewis (1774–1809) and William Clark (1770–1838; figure 2.6), traveled from St. Louis to the Pacific Ocean and back between May 1804 and September 1806. Their route took them through much of the northern Great Plains and the northwestern United States. No other single exploration of North America added as much information about natural history and ethnology. It is noteworthy that this and many subsequent expeditions of discovery were funded, at least in part, by the federal government. In the case of the Lewis and Clark expedition, the total cost to the U.S. government was reported to be $2500.

Lewis, Clark, and the members of their party were the first to report on and describe a number of small and large mammals. Among these were pronghorn antelope (*Antilocapra americana*), grizzly bear (*Ursus arctos*, formerly *Ursus horribilis*), eastern wood rat (*Neotoma floridana*), and black-tailed prairie dog (*Cynomys ludovicianus*). The following passage from the diary of Meriwether Lewis for 14 May 1806, when the party was in Idaho on its return journey eastward, illustrates the types of observations made of mammals (figure 2.7):

Figure 2.6   Meriwether Lewis and William Clark.   These two explorers were responsible for leading the nearly 2.5-year expedition from St. Louis to the Northwest in the first decade of the 1800s. Although they were not the first, nor the last, to engage in such exploratory travel, their efforts were more extensive than those of any who had preceded them. Their narratives and the materials they brought back stimulated considerable work on the diversity of mammals and other living organisms.

The hunters killed some pheasants, two squirrels, and a male and female bear, the first of which was large, fat, and of a bay color; the second meager, grizzly, and of smaller size. They were of the species [Ursus horribilis] common to the upper part of the Missouri, and might well be termed the variegated bear for they were found occasionally of a black, grizzly, brown, or red color. There is every reason to believe them to be of precisely the same species. Those of different colors are killed together, as in the case of these two, and as we found the white and bay associated together on the Missouri; and some nearly white were seen in this neighborhood by the hunters. Indeed, it is not common to find any two bears of the same color; and if the difference in color were to constitute a distinction of species, the number would increase to almost twenty. Soon afterward the hunters killed a female bear with two cubs. The mother was black, with a considerable intermixture of white hairs and a white spot on the breast. One of the cubs was jet black and the other of a light reddish-brown color. (Lewis and Clark 1979, vol. II:1010–1011)

Sadly, few of the mammal specimens gathered during the Lewis and Clark expedition survive. Materials the expedition brought back were deposited in Peale's Museum at Philadelphia. When that museum was dissolved in 1846, the contents were sold at public auction. Half the collection, containing most of the mammals, went to the showman, P. T. Barnum, and subsequently were destroyed in a fire in 1865. The Library of the American Philosophical Society in Philadelphia still has the diaries of Lewis and Clark. These have been published in several editions and are a rich source of information on the mammals of the American West.

Shortly after Lewis and Clark had begun their travels, President Jefferson sent a 26-year-old army officer, Lieutenant Zebulon Pike (1779–1813), to find the source of the Mississippi River. That expedition lasted approximately 6 months from the fall of 1805 into spring of 1806, during which Pike and his associates thought that they had found the source. They had not, although this mistake was not known for some years. Pike, after whom Pike's Peak in Colorado is named, was sent almost immediately on a second journey to explore the region south of the Missouri River and west of the Mississippi. This area included parts of what are now the Great Plains, the American Southwest, and northern Mexico. These travels added to the growing body of knowledge on mammals and other natural history of the western United States.

Trappers, Fur Traders, and Whalers

Beginning in the late 1700s and continuing until the mid-1800s, the beaver (*Castor canadensis*) became the center of an

Figure 2.7   Grizzly bears.   These two photos illustrate the sort of color differences that exist within the North American species of grizzly bear. It is easy to see that the passage quoted in the text has a real basis in terms of variation in color pattern of these bears.

entire industry (Gunderson 1976). Initially, the fur trade in North America involved bartering with the native peoples. As animal numbers declined in the eastern United States, the fur trade shifted westward. Eventually, individual trappers braved the rigors of life alone in the mountains and valleys of the West. Their life was uncluttered by "modern" conveniences, and, in good years, the payoff was excellent. During the early nineteenth century, John Jacob Astor's Pacific Fur Company, the Hudson Bay Company, the Missouri Fur Company, and a number of smaller firms turned the beaver pelt operations into a well-organized economic boom (Chittenden 1954). The records from the fur-trading companies for beaver and other mammals (e.g., lynx [*Lynx canadensis*] and snowshoe hare [*Lepus americanus*]) brought in by the mountain men who trapped and hunted them, are a valuable data set for historical population estimates of these mammals. By the 1850s, the fur trade was almost nonexistent because the supply of animals had been so depleted by overtrapping. At about this same time, garments made of beaver fur went out of style in Europe, eliminating the demand for the pelts.

Whales, particularly sperm whales (*Physeter catodon*), were the basis for a nineteenth-century industry on the high seas (figure 2.8). Again the primary focus was economic, but significant data were gathered on populations of sperm whales and other related species. The wide-ranging voyages of the whaling ships resulted in considerable knowledge about the oceans and the movements and habits of sea mammals. Sadly, with the advent of more efficient hunting methods in the twentieth century, populations of many species of whales were badly overexploited. This led to the current bans on whaling, observed by all but a very few countries in the world (see chapter 29).

Army and Railroad Survey Expeditions

In the nineteenth century, a number of U.S. Army expeditions traveled the western states on reconnaissance or in search of suitable locations for forts. Many of these expeditions benefited from the presence of medical personnel who often were also natural historians. These medical officer/naturalists numbered more than 100 and included names familiar to students of mammalogy, such as Say, Baird, and Mearns. Thomas Say (1787–1843) was one of the first of these surgeon/naturalists. He accompanied Major Stephen Long on expeditions to the Rocky Mountains and up the Mississippi and Minnesota

Figure 2.8   Sperm whale.   Sperm whales, among the largest of all mammals, were hunted heavily during the nineteenth century, and, with modern technology, their populations were decimated during the first half of the twentieth century. Due to current restrictions on whaling, adhered to by almost all countries, populations of this species appear to be making a comeback.

Rivers. Although he is rightly known more for his contributions to entomology, he supplied descriptions of and data on living mammals and participated in fossil finds important in describing the phylogenies of several mammalian groups.

Spencer Fullerton Baird (1823–1887) helped to found the U.S. National Museum (now the National Museum of Natural History) within the Smithsonian Institution. He published a monograph entitled *General Report on North American Mammals* in 1859, with descriptions of more than 730 species of mammals. Many of the mammals described in the monograph were discovered during railroad surveys searching for the best route to the Pacific Ocean.

Edgar Alexander Mearns (1856–1916) served as the medical officer and naturalist for the Mexico–United States International Boundary Commission. Mearns published *The Mammals of the Mexican Boundary of the United States* (1907) as a result of this service and collected over 7000 mammal specimens. A substantial portion of the early collection of mammals at the American Museum of Natural History in New York City is the result of his work.

Another young physician, who was influenced by Baird and deserves special mention for his contributions to mammalogy, was C. Hart Merriam (1855–1942; Brown and Wilson 1994). In 1889, Merriam initiated a new publication series, *North American Fauna*, which continues today, covering a wide range of topics related to aspects of systematics, taxonomy, and natural history of North American mammals. He is probably best known for an 1890 paper on the effects of changes in elevation and latitude on the presence of certain plants and animals, which served as the basis for the concept of life zones. He developed and refined a number of the techniques used in systematic mammalogy, including an emphasis on cranial traits and dentition. It was under Merriam's auspices that the Division of Economic Ornithology and Mammalogy of the U.S. Department of Agriculture became the Bureau of Biological Survey (in 1905). His efforts did much to foster the rapid development of mammalogy as a distinct science, and he served as the first president of the American Society of Mammalogists.

Other explorers also made contributions, including those who spent decades attempting to find the elusive Northwest Passage through the Arctic to connect the Atlantic and Pacific Oceans. They encountered animals such as the polar bear (*Ursus maritimus*), arctic fox (*Alopex lagopus*), arctic ground squirrel (*Spermophilus parryii*), and numerous other small mammals, some of which had been known to science, but others that were recorded for the first time. Other explorers penetrated interior regions of Africa, Asia, Australia, and South America, finding and describing numerous mammals that were unknown to scientists in western Europe and North America.

There were also enterprising, private, individual collectors. One of whom, Martha A. Maxwell of Colorado, may be considered a pioneer female in mammalogy (Schantz 1943). She spent many years constructing what we now would call a diorama (in this instance, a very large one with 100 mammals and 400 birds). At the request of the Colorado Legislature, her work was featured at the Centennial Exhibition in Washington, D.C., in 1876.

Museums

The great expeditions of discovery and exploration of the nineteenth century resulted in the collection of literally thousands of mammalian specimens, many of which were deposited in museums. Many of these repositories initially were in the eastern United States, but over time, major museums were established west of the Mississippi, such as those at Lawrence, Kansas, and Berkeley, California. Financed by individuals, governments, or universities, museums served as repositories for specimens and for developing collections of written materials on mammals and other fauna and flora. Books and papers were important sources of information for those who worked to describe and document new species; avoiding renaming a type of mammal already described was a common problem. The collections and accompanying libraries were modeled after those developed in Europe in the eighteenth century. Resources for such endeavors were usually only available in or near larger population centers. Hence, Boston, New York, Philadelphia, Pittsburgh, Washington, D. C., and several smaller centers became the initial focal points for museum activity in the United States.

Additional information on these museums and how they began can be found in the edited volume on the history of the American Society of Mammalogists (Birney and Choate 1994). One example of this process should be instructive. The founding of a major center for the study of mammalogy and training of mammalogists at the University of California at Berkeley was due, in large measure, to the efforts of Annie M. Alexander (1867–1950). An early interest in travel and natural science resulted in her support for and leadership of three expeditions to Alaska in 1906, 1907, and 1908. She became friends with C. Hart Merriam and discussed with him her idea of a museum for the study and preservation of the rapidly disappearing wildlife of the western states. Her proposal to the University of California at Berkeley was accepted, and in 1908, the museum was established with Joseph Grinnell as its first director. Its early growth was fostered by Alexander's substantial financial contributions and the hundreds of specimens she helped to collect (Stein 1996).

Museums have traditionally had two major functions. One has been to educate the public by exhibiting their collections. Recently, this role has been expanded, and conservation education is becoming a key focus for many museums. A second major function has been to provide a base for research on mammals and for training new generations of mammalogists. These two functions are still priorities for many major museums. Our ability to study various mammalian systems or to work on the systematics and phylogeny of a particular group of mammals depends on the continued existence and maintenance of these museum collections. Mammal collections at universities range from small to modest holdings used primarily for teaching in such courses as mammalogy and vertebrate natural history to large, specialized research collections of mammals from an entire region of the world

New Theories and Approaches

Biology was revolutionized during the second half of the nineteenth century. Charles Darwin (1809–1882), Alfred Russell Wallace (1823–1913), and other naturalists accompanied ships from European countries on journeys to many regions of the world. In the process, these ship's scientists were able to observe and collect specimens of plants and animals. From those collections, extensive observations made of native fauna and flora in many distant lands, observations of domestic animals and plants, and knowledge of earlier work (particularly the ideas of Malthus, Lyell, and others), Darwin and Wallace independently arrived at the theory of evolution by natural selection. Their joint paper, presented to the Linnean Society in London in 1858, began a process of debate and acceptance concerning the ways in which animal life adapts to changing conditions and the manner by which new species arise. The resulting theory of evolution by natural selection has become the unifying principle for all of life science.

The plant-breeding trials of the Augustinian monk Gregor Mendel (1822–1884) on the inheritance of traits in peas helped set the stage for modern genetics. He described dominant and recessive characters in plants and formulated the laws of segregation and independent assortment. The subfield of population genetics has become very important in modern mammalogy, both for understanding the interplay of ecology and evolution and as a foundation for conservation biology (chapter 29).

The process of conducting science was also undergoing dramatic changes. Experimental manipulations to test specific hypotheses became standard procedure. This augmented, but did not replace, reliance on observations as the basis for understanding natural phenomena. In addition, more refined equipment and technology, such as better microscopes, provided new insights that changed the way experimental science was conducted, thus paving the way for modern mammalogy. In chapter 3, we will examine in more detail the variety of research methods and techniques that are used by mammalogists. Because the older techniques are a valuable part of today's research tools in mammalogy, we also review much of the history of methods in mammalogy.

Early Written Works on Mammals

During the nineteenth century, a number of monographs were written on the mammalian fauna of North America. Some of these were portions of multivolume sets covering all of the plant and animal life on the continent. Sir John Richardson's (1787–1865) series entitled *Fauna Boreali Americana* (1829) contained a volume on mammals. Examples of more specialized books dealing only with quadrupeds are Thomas Bewick's (1753–1828) *General History of*

Quadrupeds (1804), the first truly American book on mammalogy, and *The Viviparous Quadrupeds of North America*, produced by John James Audubon (1785–1851) and John Bachman (1790–1874) between 1846 and 1854.

The first textbooks of mammalogy appeared during the second half of the nineteenth century. Most of these were compendia, some several volumes in length. They provided descriptions of all of the mammals of North America, including those known for several centuries from the eastern portion of the country, to which were added all of the new forms discovered during western explorations. Many of these books follow a typical pattern, true even of some, but not all, subsequent textbooks of mammalogy. For instance, the first 75 pages of the 750-page book by William Henry Flower (1831–1899) and Richard Lydekker (1849–1915), *Introduction to the Study of Mammals Living and Extinct* (1891), provide an introduction to the structure and function of mammals, with an emphasis on skeletal and dental traits. The remainder of this text is devoted to accounts of the various orders of mammals. With the vast increase in knowledge during the twentieth century, some authors (e.g., D. E. Davis and F. B. Golley [1963], *Principles in Mammalogy*), as well as our own treatment of the subject, have taken a different approach, focusing on comparisons of physiology, anatomy, ecology, and behavior across the orders of mammals.

In addition to more general works, some treatises dealt with smaller groups of mammals or even single species. An example of the first type is the work of Elliott Coues (1842–1899) on a rodent family, the Muridae (1877). Coues later became the first curator of mammals at the U.S. National Museum in Washington, D. C. A prime example of the single-species treatment is *The American Beaver and His Works* (1868) by Lewis H. Morgan (1818–1881). Morgan summarized what was known about the beaver (*Castor canadensis*), exemplified in this short passage about beaver dams and lodges (figure 2.9):

> *The dam is the principal structure of the beaver. It is also the most important of his erections as it is the most extensive and because its production and preservation could only be accomplished by patient and long-continued labor. In point of time, also, it precedes the lodge since the floor of the latter and the entrances to its chamber are constructed with reference to the level of water in the pond. The object of the dam is the formation of an artificial pond, the principal use of which is the refuge it affords to them when assailed, and the water connection it gives to their lodges and to their burrows in the banks. Hence, as the level of the pond must, in all cases, rise from one to two feet above these entrances for the protection of the animal from pursuit and capture, the surface level of the pond must, to a greater or lesser extent, be subject to their immediate control. (Morgan, 1868:82–83)*

Volumes like those noted are the forerunners of specialized monographs that abound today in mammalogy. Any mammalogy student conducting literature research will now

Figure 2.9 **Beaver.** The beaver was the center of a major industry for much of the first half of the nineteenth century. Trappers and fur traders, both individuals and companies, constituted a key segment of the economy of the Rocky Mountains and northwestern United States, declining quickly as the number of beaver dwindled to isolated populations.

find entire books on particular topics, for example on reproductive physiology or foraging behavior. As noted in chapter 1, books are available for almost every taxonomic group, and a mammal guide exists for just about every state or region of North America, as well as most geographical and faunal regions of the world.

EMERGENCE OF MAMMALOGY AS A SCIENCE

The extensive survey work and exploration that occurred in the 1800s and just after the turn of the century resulted in a critical mass of specimens, records, and other information on mammals. That combined with increasing numbers of professional scientists working almost exclusively on mammals lead to the

emergence of mammalogy as a distinct discipline. These developments helped foster the formation of the American Society of Mammalogists in 1919. The history of the society and its role in furthering the study of mammals has been thoroughly covered in a recent volume edited by Birney and Choate (1994) to commemorate the seventy-fifth anniversary of its founding. Other histories of North American mammalogy have been written by Storer (1969), Hoffmeister (1969), and Hamilton (1955). The American Society of Mammalogists has become the largest professional group in the world whose main focus is the biology of mammals.

The collections of mammals in museums, particularly those at universities, played a pivotal role in the establishment of mammalogy as an academic discipline within the life sciences. One of the most distinguished of those early mammologists was Joseph Grinnell (1877–1939). The son of a physician, Grinnell grew up on and around Indian reservations in Oklahoma, Nebraska, South Dakota, and North Dakota (figure 2.10). He spent most of his adult life in an academic career at the University of California at Berkeley. One of his key scientific contributions was the concept of the **niche:** the idea that organisms have functional roles within the framework of the community. While at Berkeley, Grinnell began the museum collection, introduced courses in vertebrate zoology to the curriculum, and trained graduate students. Among Grinnell's graduate students were William H. Burt (1903–1987), who played a vital role in developing the notions of home range and territory; Lee R. Dice (1887–1977), whose efforts contributed to our knowledge about interspecific competition and its effects on the structure of communities; and E. Raymond Hall (1902–1986), who conducted extensive research on the taxonomy and distribution of mammals (figure 2.10). Together, the academic descendants of the Grinnell group constitute the single largest branch of the genealogical tree within mammalogy (Whitaker 1994). Another major lineage was formed from the graduate students of W. J. Hamilton, Jr., (1902–1990) at Cornell University (figure 2.10). Hamilton's major emphasis was on life history traits and ecology.

A   **B**   **C**   **D**   **E**

Figure 2.10 **Distinguished mammalogists.** The five men shown here were among a number of professionals who played key roles in establishing mammalogy as a distinct subdiscipline within vertebrate zoology. Each of them was associated with a particular museum and each trained a number of graduate students; their academic descendants are numerous among today's mammalogists. (A) Joseph B. Grinnell; (B) William H. Burt; (C) Lee R. Dice; (D) E. Raymond Hall; (E) William J. Hamilton, Jr.

The presence and role of women in mammalogy has changed rather dramatically in the past several decades. As noted earlier, several women played instrumental roles in the early days of the discipline; however, few women were trained in the major academic and professional centers for mammalogy before the late 1960s. The history of women in mammalogy and the shift toward greater involvement of women has been well documented by several recent articles in the *Journal of Mammalogy* (Horner, et al. 1996; Kaufman, et al. 1996; Smith and Kaufman 1996, Stein, 1996). These changes are reflected in the membership composition of the American Society of Mammalogists, as well as in authorship of papers published in the *Journal of Mammalogy* and presented at annual meetings.

Excellent museums at Harvard, Yale, Michigan, Cornell, Kansas, Texas Tech, and many other universities contributed to the growth of mammalogy, both through their collections and the training of students by faculty associated with the museums. At various times during the first half of this century, courses in mammalogy were started, often growing out of courses in vertebrate zoology. After World War II, the number of mammalogy courses grew rapidly until the 1980s, after which the number declined slightly.

The emergence and growth of mammalogy as a distinct subdiscipline within zoology can be characterized by a progression of trends in the accumulation of knowledge. Initially, most information came from observation and description. With time, approaches involving experimental manipulations contributed greatly to our knowledge about mammals. Ecology and life history research were major foci for mammalogists before and for several decades after World War II. This was followed by an upsurge of interest in physiological processes during the 1960s and 1970s. At about the same time, and continuing into the 1980s, research began on models of various types, aided by the introduction of computers. Processes that use computer modeling ranged from population biology to energetics. Most recently, considerable attention has been devoted to molecular genetics. Fostered by the development of a variety of DNA-based techniques, molecular genetics has aided investigations in areas ranging from population genetics to reproductive success. No approach has been dropped from the mammalogist's repertoire in this research progression. Because of the cumulative nature of the process, mammalogists need to be trained in a wide variety of techniques and approaches. Mammalogy today involves a broad range of scientists who study systematics, paleontology, behavior, physiology, ecology, anatomy, biochemistry, and other biological topics, although the degree of individual specialization varies. Modern mammalogy integrates knowledge across all of these disciplines.

Throughout the twentieth century, the importance of mammalogy has grown as scientists have found new areas in which they can apply their knowledge about mammals (Wilson and Eisenberg 1990). Wild mammals have been food for humans for millennia. Domesticated livestock and their products have been critical to human cultural and eco-

nomic development for 10,000 years (chapter 28). Other mammals, such the beaver and the mink (*Mustela vison*), provided the basis for fur trapping and fur farming, although the demand for these products has declined in recent years. Mammals also provide us with oils (whales, seals) and ivory (elephants, walruses), although suitable substitutes are now available. Interest in mammals extends to their actions as pests in terms of livestock predation, crop depredation, and destruction of the environment. Rodents of various species cause billions of dollars in damage annually to cereal crops and grain stores in many parts of the world. Many home owners have had to deal with burrows made by animals such as gophers and moles. On the other hand, many rodents have vital ecological roles in terms of insect and seed control.

Mammals are also of interest to us in several ways in connection with medical science. A number of species serve as useful laboratory models for various human diseases or physiological functions. Several species of rodents, most notably house mice (*Mus musculus*) and Norway rats (*Rattus norvegicus*), have been the primary subjects for many laboratory investigations of general biological principles and more specifically of studies in human medicine. Many procedures used to treat coronary ailments were first tested on domestic dogs (*Canis [familiaris] lupus*) or pigs (*Sus scrofa*). Some mammals are known to be reservoirs for or vectors of diseases. A good example of this is the rat (*Rattus* spp.), which carries the fleas responsible for transmitting bubonic plague (chapter 27). Rabies, which is carried by mammals, also poses a serious threat to humans. The recent outbreak of a hantavirus, involving several species of *Peromyscus*, is another example of a disease carried by a mammal (figure 2.11). White-tailed deer (*Odocoileus virginianus*) and white-footed mice (*Peromyscus leucopus*) are primary reservoirs of the larval and adult ticks that carry Lyme disease, which has become a major health concern as it spreads in the eastern United States. The virus that leads to AIDS in humans likely had its origin as a similar virus in African primates.

Lastly, mammals have become valuable to humans for recreation. Sport hunting is a major industry in many regions of the United States and throughout the world. In the past 20 years, particularly in such places as Africa and South America, mammals have become a valuable resource in ecotourism (figure 2.12). Mammals are also prized for their aesthetic value; most people enjoy seeing a herd of deer feeding, a female bear and her cubs ambling across a mountain meadow, or a cat silently stalking its prey. Television offers considerable numbers of programs dealing with the nonhuman mammals with whom we share the world.

Mammalogists work in a variety of settings today. Many are at universities and colleges, and a large number are employed by state and federal agencies. On the state or provincial level, agencies that deal with wildlife and conservation have mammalogists on their staffs, as do those that deal with parks and recreation. Positions for natural heritage biologists have expanded in recent decades. At the federal level, programs in agriculture and forestry, in which mam-

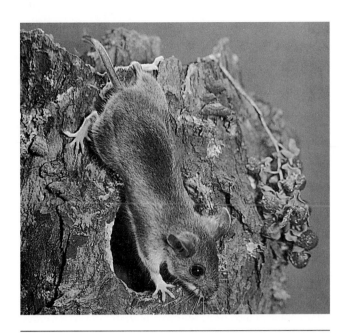

Figure 2.11 **White-footed mouse.** The white-footed mouse shown here, along with the white-tailed deer, are distributed throughout much of the eastern United States. Both species are vectors of Lyme disease, carried to humans by ticks.

Figure 2.12 **Herd of zebras and wildebeests.** The mixed herd of zebras (*Equus burchelli*) and wildebeests (*Connochaetes taurinus*) are a major feature of the Masai Mara– Serengeti ecosystem. They are among the many East African mammals that have become a valuable resource as tourists have flooded the region in recent decades.

mals are studied with regard to pest management, and wildlife management of nature preserves, forests, and other federally owned lands, use the talents of mammalogists. With the advent of laws to protect the environment and the need to rehabilitate polluted land, a variety of private environmental and consulting firms have hired mammalogists. Finally, many nongovernmental organizations, such as the Nature Conservancy and the National Audubon Society, employ scientists trained in mammalogy. These individuals work in local, national, or international operations primarily surveying mammalian fauna and striving to protect and conserve mammalian species.

Among recent technological advances used by mammalogists to push the frontiers of the discipline forward are night–vision scopes and radiotelemetry to watch and monitor, respectively, the movements and behavior of animals. Beginning in the 1950s, molecular techniques, such as pro-

tein electrophoresis, were applied to the study of mammals. Current DNA-based techniques are used to assign parentage for individual animals, and both nuclear and mitochondrial DNA sequencing techniques aid in studying phylogenies of various mammalian groups. These and related topics pertaining to the methods we use to study mammals are the subjects of the next chapter.

In closing, we take note of the fact that although the discovery and recording of new species of mammals reached its peak during the nineteenth century and early years of the twentieth century, new mammals are still being discovered. More than 100 previously undescribed mammals have been recorded since 1980 (Morell 1996). Among these are rodents from Madagascar and the Philippines, as well as several primates from South America. Thus, work still remains to be done to complete our inventory of mammalian diversity.

Summary

The earliest humans were aware of and interested in other mammals that shared their environment. They needed knowledge about mammals to obtain meat and avoid becoming prey. The ancient Greeks and Romans (e.g., Aristotle and Pliny) recorded their observations of the known mammals of their time. After a hiatus in discovery during the Dark Ages, extensive interest in natural history was revived in the 1600s. Expeditions by Europeans resulted in the

collection of plant and animal specimens from previously unknown places. The diversity of mammals and other animals represented in these new collections stimulated interest in general zoological principles. Among those whose contributions are singled out in this chapter were Mark Catesby, who visited America in the first half of the eighteenth century; Georges Buffon, who wrote an extensive series of volumes summarizing what was known of living organisms, including mammals; and Carolus Linnaeus, who developed a hierarchical system of classification based on binomial nomenclature. Several ideas developed in the late eighteenth and early nineteenth centuries by individuals such as Malthus and Lyell served to stimulate thinking about the origins of biological diversity.

The nineteenth century was an age of exploration in North America. The century began with the travels of Lewis and Clark, supported by President Jefferson. Others who contributed to the progress of natural history during this century included trappers and fur traders, whalers, surveyors for the army, and those who searched for the Northwest Passage in the Arctic. Preserved materials and fossil specimens from these expeditions were often deposited in museums. The museums and their associated libraries became key centers of learning about mammals.

The cornerstone for all of modern biology, the theory of evolution by natural selection, was first presented in 1858 by Darwin and Wallace. Other scientific developments, including the beginnings of modern genetics and the use of experimental manipulations in the course of conducting research, characterized natural science during the last half of the nineteenth century. It was during this period that the first written works devoted solely to mammals were published; some of these were summary volumes, covering all known mammals from North America or the world, whereas others dealt with specific groups of mammals or even individual species.

The twentieth century has seen mammalogy emerge as a distinct discipline within the life sciences. The American Society of Mammalogists, founded in 1919, has served as the primary scholarly organization for mammalogists in North America. Current mammalogists are a diverse array of scientists working on research ranging from molecular genetics to ecology and systematics, but sharing the common subject of mammals. University museums in particular have become important foci for mammalogy, in terms of both their education and conservation functions and as fertile grounds for training new mammalogists and conducting research.

Mammals are of interest to humans for a variety of reasons. They are used for food and other products. In medical science, nonhuman mammals are models for some human diseases, and they are reservoirs for or vectors of disease. Mammals are also pests, destroying grain and other crops, and they have an aesthetic value in terms of recreation and sport hunting. New techniques have been developed in recent decades for monitoring mammal movements and activities and for studying molecular phylogeny to understand the evolutionary relationships among mammals, both living and extinct.

Discussion Questions

1. Suppose that you were living in the nineteenth century and were about to travel across the country from St. Louis to the area that is now Los Angeles via a southern route through Texas, New Mexico, and Arizona. Using any of several mammal guides that you may have available, generate a list of the mammals that you might expect to encounter.

2. Select any three of the nineteenth-century naturalists mentioned in this chapter and, using available reference books, write one-page synopses of their contributions to biology and to mammalogy in particular.

3. In the manner of the early naturalists, spend some time watching members of one or more mammal species, preferably in a field setting. Write down your observations in a journal format. Using your observations, what sort of questions can you generate about various aspects of the biology of these mammals?

4. The most important theory for biology is evolution by natural selection, first postulated by Darwin and Wallace. From your previous biology background, write down what you believe are the major components of the thinking that resulted in this theory. Check your answer by referring to an introductory biology textbook or other reference that contains a section on the theory of evolution and how it developed.

5. Using a good historical atlas and a guide to living mammals, construct a list of the mammals that Lewis and Clark might have encountered on their western expedition. For each mammal, list the state or states where it might have been found.

Suggested Readings

Birney, E. C. and J. R. Choate (eds.). 1994. Seventy-five years of mammalogy (1919–1994). Special Pub. No. 11, Am. Soc. of Mammal.

Gunderson, H. I. 1976. The evolution of mammalogy, a history of the science. Pp. 3–38 *in* Mammalogy. McGraw-Hill, New York.

Wilson, D. E. and J. F. Eisenberg. 1990. Origin and applications of mammalogy in North America. Pp. 1–35 *in* Current mammalogy vol. 2. (H. H. Genoways, ed.). Plenum, New York.

CHAPTER 3

Methods and Techniques for Studying Mammals

FIELD METHODS

Trapping and Marking
Monitoring
Geographic Information and Global Positioning Systems
Observational Methods

LABORATORY METHODS

Morphometrics
Physiological Measures
Genetics and Molecular Techniques

Some of the variety of current techniques used to study mammals were developed specifically for mammalogy, but others originated from other fields and have been applied to mammals. This chapter provides a brief overview of some of the most commonly used methods in mammalogy. Because most courses in mammalogy also have a laboratory component and because no overview of techniques can replace a laboratory manual, this chapter seeks to provide, (1) a general understanding of the methods used by mammalogists; (2) general explanations for some relatively complex methods, such as molecular techniques; and (3) references for those who wish to pursue some of these methods in greater detail. Most sections of the chapter have references that can serve as the basis for exploration of the more detailed aspects of the various techniques. In many instances, these references also provide information on vendors of specific types of equipment and supplies needed for using the techniques.

Although considerable overlap exists between the methods employed in field and laboratory settings, we will take the practical expedient of dividing the discussion of methods into these two basic categories. Bear in mind, however, that laboratory procedures have also been used in the field and vice versa.

In all studies of mammals, some training in statistics and experimental design is necessary. These aspects of methods are not covered here, but a number of textbooks on statistics are available (Sokal and Rohlf 1981; Siegel and Castellan 1988; Glantz 1992; Zar 1996).

FIELD METHODS

Mammalogists often ask such questions as: What is the population size of squirrels in a particular forest or woodlot? What are the sizes and shapes of their home ranges? Does the social system include dominance hierarchies or territories? When are the squirrels most active? Answering questions like these requires capturing, marking, and monitoring individual animals.

Trapping and Marking

Trapping

Methods for capturing wild mammals include a variety of trapping and netting techniques (Eltringham 1978; Bush 1996; Lehner 1996; Schemnitz, 1996). Whether the animals are live-trapped or killed depends on the nature of the study and the reason for their capture. Although many mammals trapped today are captured alive, in some instances kill-trapping is necessary and justified. For instance, museum collections need reference specimens to permit identification of species, and the skeletal or tissue materials needed for analyses cannot be obtained from live animals. Trapping to remove animals from an area, particularly an abundant or

pest species, can often best be accomplished by using kill traps, guns, or poison. If trapped specimens are to be used for a museum collection, it is necessary to ensure that the specimen is not damaged by the capture procedure.

Mammal live traps are available in a variety of types and sizes. For small rodents, the most widely used types are the Longworth and Sherman traps. For somewhat larger mammals, such as raccoons (*Procyon lotor*), traps with wire mesh sides are available, such as the Havahart or National traps. Large box-type traps are constructed for capturing coyotes (*Canis latrans*), bears, or other carnivores, or for white-tailed deer (*Odocoileus virginianus*). The dimensions and operation of such traps fit the subject species (figure 3.1). Larger box-type traps are also used, for example, for African carnivores such as brown hyenas (*Hyaena brunnea*) (Mills 1996). Box traps can also be used, in some instances, to capture groups of smaller animals. Rood (1975) used large box traps to capture banded mongoose groups (*Mungos mungo*) in East Africa. In the same manner, large enclosures have been used to capture groups of ungulates (Taber and Cowan 1969) and primates (Rawlins et al. 1984). Pitfall traps, consisting of a can or bucket buried in the ground, are used to capture small mammals, particularly shrews (Mengak and Guynn 1987; Cawthorn 1994), which may also be captured in bottles (Gerard and Feldhamer 1990). Padded leghold traps of various types are used to capture canids and the smaller felids. Mist nets, most commonly used for capturing birds, are often used to catch bats, particularly when the bats use a regular flight path on entering or leaving their roost (Kunz 1988). Larger nets, fired by guns or rockets, have been used to capture white-tailed deer (Hawkins et al. 1968).

The types of kill traps used with smaller mammals include Museum Special, Victor, and McGill traps. These come in various sizes, depending on the rodent they are designed to

Figure 3.1   Stephenson box trap. Large box traps of the type shown here are used to capture larger mammals, such as white-tailed deer. Animals captured in this manner can be measured, tagged, or dyed for individual identification, or fitted with collars for tracking by radiotelemetry.

catch. Several types of traps are available for capturing burrowing mammals, including the harpoon mole trap and the Macabee-type gopher trap (DeBlase and Martin 1981).

Capture guns, involving a tranquilizer drug or some form of anesthetic, may be used for larger mammals (reviewed by Harthoorn 1976; Bush 1996). This technique is quite useful when the animal needs to be held for a brief time only. For example, lions (*Panthera leo*) and leopards (*Panthera pardus*) can be immobilized with darts, examined, measured, and given an antagonist drug to reverse the effects of the tranquilizer (Bertram and King 1976). Capture guns can also be used to catch animals for translocation or for bringing them into captivity for further study. Finally, specially modified guns have been used to shoot marking devices into animals, particularly whales (Kemp et al. 1929; Brown 1978).

The care of wild mammals in captivity has been important since the advent of zoological parks and, more recently, since mammals have been used as laboratory subjects. Such mammals may be held in cages or enclosures. In a landmark volume, *Management of Wild Animals in Captivity*, Crandall (1964) spelled out many of the procedures to be followed in caring for wild animals, including mammals. Caring for wild mammals in captivity has become increasingly significant in the past 20 years as zoos and related organizations have shifted some of their emphasis from exhibition to conservation. Working with captive populations of mammals has enabled us to gain a better understanding of their reproductive biology, thereby aiding in breeding programs (Tudge 1992). A major recent volume, *Wild Mammals in Captivity, Principles and Techniques* (Kleiman et al. 1996), provides an up-to-date assessment of the procedures used to care for a wide range of mammals.

Marking

Animal marking techniques vary with the type of mammal and whether they are living free, held in zoos, or used in laboratory investigations (Stonehouse 1978). For techniques used with zoo animals and mammals used as laboratory stock, see Ashton (1978), Lane-Petter (1978), or Rice and Kalk (1996). We are concerned here only with free-ranging mammals or those maintained in seminatural conditions.

In some instances, physical features of the individuals can be used for identification. These natural marks include, for example, the color patterns on the flukes of whales of several species (Clapham and Mayo 1987; Hammond et al. 1990) or the variations in shape and coloration of dorsal fins of bottle-nosed dolphins (*Tursiops truncatus;* Würsig and Würsig 1977) or killer whales (*Orcinus orca;* Balcomb and Bigg 1986); such markings can be photographed to generate a library or catalog. These photos are circulated among investigators and serve as a permanent, readily used method for repeated identification of the same individuals. In a similar fashion, vibrissae spot patterns can be used to identify individual lions (Rudnai 1973). Some members of most species of larger mammals, such as hoofed stock or primates,

can be identified, with a bit of experience, by their individual markings. These individual characteristics may include natural differences in physical appearance, scars or other similar marks, or differences in walking or climbing patterns resulting from a physical injury or deformity.

We can use a variety of marking devices to tag mammals. Some of these, including individually numbered metal or plastic tags placed in the ears or webbing of the feet, can be used to identify the animal only when it is captured. Bird bands have been used on the legs of small mammals, particularly bats (Stebbings 1978; Kunz and Robson 1996).

For field identification at a distance, investigators have used a variety of dyes, tags, and flags (figure 3.2; Giles 1971; Stonehouse 1978; Bookhout 1996; Lehner 1996). Freeze-branding with liquid nitrogen results in white hairs growing in where the mark was applied. Kummer (1968) used both plastic ear tags and sprayed-on dyes to mark hamadryas baboons (*Papio hamadryas*). Loss of tags used for marking or of hair due to molt can reduce the effectiveness of these techniques, but marks made with freeze branding are permanent.

Radioisotopes placed in collars on animals or as subcutaneous implants have been used for marking individuals (Linn 1978). Sheridan and Tamarin (1988) developed the use of radionuclides for individual identification of meadow voles (*Microtus pennsylvanicus*) to study longevity and reproductive success. Radionuclides are safer than radioisotopes, do not harm the animals tagged with them, and pose less danger to the environment than radioisotopes.

Figure 3.2 **Animal marking.** The cougar (*Felis concolor*) shown here has a neck collar.

Some investigators remove or modify tissue as a means of marking mammals (Twigg 1978). Toe clipping is one such method (Golley 1961), as is ear notching (Berry 1970); the latter marks are readily visible from some distance in many mammals, particularly with the use of binoculars. Clipping fur in patterns and using depilatories to remove patches of hair have been done with Norway rats (*Rattus norvegicus*; Chitty and Shorten 1946) and numerous other mammals. Various branding and tattooing techniques have been used on some mammals (figure 3.3). Care must be taken with any of these marking methods, and experimental controls must be established to ensure that the mark itself does not alter either the behavior or survivorship of the animal, or the reactions to it by conspecifics or potential predators.

Small transmitters, called passive integrated transponders (PIT), can be implanted or injected subcutaneously. The device emits a radio signal that is different for each individual (Prentice et al. 1990). When the animal is near the decoding device or antenna, its number appears on a lighted display and can be recorded by a data-logging system. Harper and Batzli (1996) put PIT tags into voles (*Microtus pennsylvanicus* and *M. ochrogaster*) and placed antennae along runways used by the animals. They found that several individuals used the runway system each day, and voles were most active near sunset and sunrise. The transponders can

also be employed with pets, particularly dogs, as a means of protecting against theft.

Mammals captured with traps can be saved as preserved specimens, particularly as study skins and skeletons, and used for future reference. These materials can be important for investigating taxonomy and phylogeny, including the critical process of species identification. Techniques for accomplishing these tasks can be found in Anderson (1965) and DeBlase and Martin (1981). Animal signs and constructions, such as houses, burrows, or lodges, are important adjuncts to observations and records (Murie 1954).

Monitoring

Several procedures have been developed for following the activities of mammals and for recording various physiological functions in free-ranging individuals. The most widely used of these techniques is **radiotelemetry** (Kenward 1987; White and Garrott 1990). A telemetry system includes a transmitter, usually battery-powered, that emits a radio signal. This signal is monitored using a receiver that changes the signal from high frequency to within the range of human hearing and amplifies it. Transmitters can be placed in a collar, usually around the animal's neck, or they may be implanted in the animal's abdominal cavity or subcutaneously (figure 3.4). There are four major

Figure 3.3   Tattoo marking.   The rhesus macaque has been given a tattoo consisting of a letter and a number. These sorts of markings can be used to identify the individual animal, both for observational purposes and for record keeping. For example, the young born to this female can be noted, and their survival monitored in addition to her own group affiliation, interactions, and survival.

Figure 3.4   Radio collars.   A variety of sizes of radio collars have been developed to use with mammals of different sizes. The collar, battery, and transmitter should not exceed about 10% of the weight of the animal. Thus, with constraints on collar size in smaller mammals, there are often limitations on the size of the battery and consequently on the duration of time the transmitter will be functional.

techniques for montoring signals from radiotransmitters attached to mammals: hand-held antennae of various types, antennae mounted on ground vehicles or aircraft, antennae mounted on towers, and satellites (Samuel and Fuller 1996).

In addition to pinpointing locations of specific animals, radiotelemetry can be used to monitor a number of other functions. Information on rates of movement can be obtained either by calculating time intervals between location fixes or using an accelerometer (Dyhrepoulsen et al. 1994; Bradshaw et al. 1996). Radiotelemetry can used for recording dispersal, home range area, habitat use, and homing behavior (figure 3.5). The use of space and interactions of feral swine (*Sus scrofa*) and collared peccaries (*Pecari tajacu*) were studied in south Texas using radiotransmitters (Ilse and Hellgren 1995). Radiotelemetry can also be used to estimate survival rates (White and Garrott 1990). Techniques for estimating home range, survival, and other parameters obtained from radiotelemetry can be found in Krebs (1989), Lehner (1996), White and Garrott (1990), and Wilson and colleagues (1996). Various physiological functions, including body core temperature and heart rate, can be relayed via radiotelemetry (Folk and Folk 1980; Samuel and Fuller, 1996). This information is used to assess animal activity and to estimate energy expenditure.

Numerous problems, both methodological and statistical, can confound or complicate the collection and analysis of data using radiotelemetry. To obtain unbiased information, the collar or other device by which the monitoring system is attached to the animal should not interfere with the animal's normal activities (White and Garrott 1990). The animal's behavior may be affected by the capture, handling, and attachment procedure (Fuller 1987). For example, Mikesic and Drickamer (1992a) recorded a reduction in wheel-running activity for at least 96 hours after radiotransmitter collars were placed on house mice (*Mus musculus*).

The system used for obtaining animal locations must be adequately designed. Triangulation, the procedure most often used for estimating the location of a transmitter, uses two or more directional bearings obtained from known positions that are remote from the transmitter's location. Problems with accuracy of the recording system and possible signal interference from electrical lines or objects that cause a directional bias in the recordings must be noted. Such objects can be either near the receiving antenna or at some point between the subject animal and the antenna. Valid estimates of parameters such as home range entail consideration of issues such as the independence of the data.

Autocorrelation is the obtaining of locations or fixes that are too close together in time; that is, dependent on the previous location fix. This results in a lack of independence of the data points used, for example, to calculate a home range. If the radiotelemetry locations are gathered using a time interval that is too short, the home range estimate obtained will be too small. The length of time required between radiotelemetry locations varies with the species, time of day, season, and nature of its activities. It is therefore necessary, when designing a proper radiotelemetry study, to know something about all these aspects of the mammal's biology. Here again, the need for thorough training in statistics and experimental design is apparent (Kenward 1987; White and Garrott 1990).

Another technique, powdertracking, has been developed for determining the locations of individuals (Lemen and Freeman 1985). Powdertracking involves coating a small mammal with a fluorescent powder, usually by shaking it with the powder in a small bag. The trail produced after the individual is released can then be traced the next night using an ultraviolet light source. This technique is used to measure home ranges of several species of small mammals (Jike et al. 1988; Mullican 1988; Mikesic and Drickamer 1992b) and to compare values obtained with powdertracking versus radiotelemetry. Powdertracking can also be used to study foraging behavior and seed caching by small rodents (Hovland and Andreassen 1995; Longland and Clements 1995).

All of these methods for capturing, marking, and monitoring mammals can be used to gather data for a variety of purposes. The presence and abundance of species of mammals in each trapping location can provide estimates of habitat use and various indices of diversity (Krebs 1989; Lancia et al. 1996; Wilson et al. 1996). One of the most common procedures involves trapping mammals, marking them in some manner and releasing them, and then recapturing subsamples from the population. Different population parameters can be estimated from such data, including estimates of population size, life history traits such as survival, and patterns of reproduction that may vary seasonally (Krebs 1989; Sutherland 1996). A variety of procedures

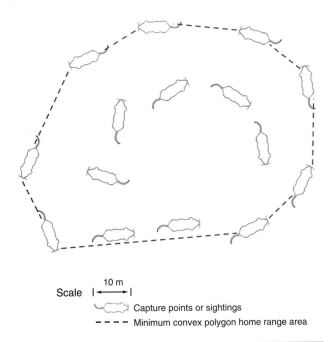

Scale |←— 10 m —→|

⌒⌒⌒ Capture points or sightings

– – – – Minimum convex polygon home range area

Figure 3.5   Home range.   The home range of a house mouse (*Mus musculus*) can be plotted from a series of radiotransmitter locations. Home range area can vary with the age, sex, season, population density, and reproductive condition of the animal.

have been employed to estimate population size, such as the Lincoln-Petersen (Krebs 1989), Schnabel (Schnabel 1938), and Jolly-Seber methods (Seber 1965, 1982). In addition, estimates of areas of activity, home range size, and home range overlap can be obtained from multiple captures of the same animals (White and Garrott 1990; Lehner 1996). For most of these procedures, the problem of members of a population not being equally catchable can lead to erroneous estimates or calculations. If, for instance, animals of different ages or sexes are more or less likely to be captured by the method employed, then the conclusions may be flawed. In addition, individuals of some species may become either prone to enter traps regularly if they have been captured previously ("trap-happy"), or they may avoid traps after an initial capture ("trap-shy") (Nichols and Hines 1984). We must always examine and test the assumptions of the particular method being used.

Geographic Information and Global Positioning Systems

Geographic information systems (GIS) are computerized software programs used to store and manipulate geographic information (Aranoff 1993; Koeln et al. 1996; Demers 1997). First developed in the 1970s (Tomlinson et al. 1976), the procedures involve taking geographic data obtained from maps, aerial photographs, satellite images, ground surveys, and other methods and entering them into computer programs as digitized bits of information. The data then can be viewed in a variety of forms, generating images of the area being studied. GIS databases are now available for much of North America and substantial portions of the rest of the world. The **global positioning system (GPS)** involves the use of satellite-based radio signals. Using a receiver held in the hand or mounted on a vehicle, the exact latitude and longitude at any location can be determined. GPS information can be stored for subsequent analysis and has been used to aid in obtaining geographic coordinates in connection with collection of data for GIS.

The GIS technique has a number of potential applications in mammalogy (McLaren and Braun 1993); for example, a regional landscape analysis was used for the northern Great Lakes region to predict the locations of suitable habitat for the gray wolf (*Canis lupus*). Interrelationships between habitat fragmentation and development in Minnesota, Michigan, and Wisconsin indicate that wolves can move across unfavorable areas from source populations to other available habitats (Mladenoff et al. 1995). For another example of the use of GIS, involving caribou (*Rangifer tarandus*), see chapter 23. A number of problems must still be resolved concerning the use of GIS (Aranoff 1993), but it promises to be a valuable tool for future studies of mammals.

Observational Methods

Many mammals can be observed directly. This is particularly true for those that are large, live in open habitats, and are diurnal; some nocturnal mammals have been viewed using night–vision scopes (Vessey, 1973; Waser 1975; Arletazz 1996). Thermal infrared imaging has been used for estimating population size, locating animals, and noting habitat locations (Boonstra et al. 1994; Sabol and Hudson 1995). Watching mammals during daylight hours provides data on a variety of life history characteristics, such as movements, habitat use, predator–prey relations, and foraging. Behavior, both individual and social, is perhaps most commonly recorded using observational methods.

Basic techniques for making observational records have been summarized by Altmann (1974), Crockett (1996), and Lehner (1996). Whatever the observation technique used, many aspects of the activities that we wish to record vary. There are differences within the same individual over time, among individuals, among populations, and seasonally. Thus, an appropriate experimental design and adequate knowledge of various statistical tests and the assumptions that underlie them are important with all of these methods. Observations can be made with the unaided eye; by using binoculars or spotting scopes of various types; and with various photographic methods, including videotaping, still cameras, and aerial photography. Videotaping has proven especially valuable because it allows the same sequence to be viewed many times to capture smaller nuances of, for example, facial expressions or postures. Videotaping lets several people watch the same sequence, thereby often reducing judgment errors in recording data.

The two most common methods of recording data are recording focal animal samples and scan samples. **Focal animal sampling** is the recording of all occurrences of specified behavior patterns or interactions of a selected individual or individuals during a period of prescribed length, such as 10 minutes. Only the behavior related to the hypothesis being tested is recorded. For example, focal animal sampling has been used to study the foraging behavior of ungulates (Focardi et al. 1996). Mammalogists stationed on high perches recorded the movements of fallow deer (*Dama dama*) with regard to vegetation types and where they stopped to forage. Numerous studies of primates have also used this technique. Silk and colleagues (1996) recently examined postconflict interactions among female baboons (*Papio cynocephalus*) in Botswana (figure 3.6). They recorded higher rates of interactions between opponents immediately after a conflict than at other times; probably a form of reconciliation.

The second technique, **scan sampling,** involves recording the current activity of all or selected members of a group at predetermined intervals. These intervals may be as varied as once every 15 seconds or every minute, or at much greater intervals, such as once every half hour. The current activity of each animal is noted in terms of a predetermined set of categories. These categories may be quite general, for example, resting, moving, or grooming, if the general behavior of the animals is being studied. They may, however, be quite specific, for instance, mounts, intromissions, and ejaculations, if a particular behavior sequence such as mating is being examined. Scan sampling has been used to investigate

Figure 3.6   **Social interactions among female baboons.**
Focal animal sampling can be used to examine the rates of interactions among animals in a group. Watching a single individual for a prescribed time period provides an accurate picture of her activities, in the case shown here, with two females exhibiting aggression toward one another.

relationships between space availability, breeding behavior, and pup survival in southern elephant seals (*Mirounga leonina*; Baldi et al. 1996). The spaciousness of beaches at Peninsula Valdés, Argentina, results in fewer aggressive interactions among females than at other, more crowded, breeding sites for this species. Furthermore, pups are much more likely to survive, because fewer interactions result in fewer pups being harmed or starved due to separation from their mothers. Scan sampling and focal animal sampling are often used together. The long-term studies of red deer (*Cervus elaphus*) on the island of Rhum (Clutton-Brock et al. 1982) are a good example.

In addition to these two basic forms for recording observational data, the following are among the other available methods: recording all occurrences of a particular behavior pattern or patterns in a group of animals; sampling sequences of behavior, such as those involved in communication; and ad libitum sampling in which all the activities of all organisms in view are recorded. Details of these methods and some of the problems associated with them can be found in Altmann (1974) and Lehner (1996). These references also cover additional, more specialized methods used for behavioral observation work, such as sound playback. For this method, calls or songs that have been recorded are played back to animals in particular situations to assess their behavioral reaction, providing information on the functions of these vocalizations.

One key issue that arises whatever type of sampling method is used is the observability of the animals under study. The term **observability** as used here has two meanings. In the first sense, it refers to whether the habitat permits regular, direct observation. With some mammals living in dense vegetation or with those that are arboreal, foliage can interfere with an investigator's ability to record all their

actions at all the times prescribed by the method in use. In such cases, it may be necessary to make some form of correction for the "missing" observations. A second meaning for the term *observability* refers to the fact that when mammals are watched, some of them may be seen for differential portions of the time period depending on age, sex, or dominance status. For example, male rhesus macaques (*Macaca mulatta*) of high social status are seen more frequently than those of low social status in the central area of a group (Drickamer 1974). If the relationship between dominance status and mating activity is being studied, differential observability could lead to potentially erroneous conclusions. If these higher-ranking males were seen more often engaging in mating activities with females, we might decide that social rank conferred an advantage in terms of mating success. Knowledge of the fact that these higher-ranking individuals were observed much more frequently than lower-ranking males, however, necessitates some form of correction to account for differential observability. When such a correction factor is applied to the data on rhesus macaque males, no significant differential mating frequency between high- and low-ranking males was apparent (Drickamer 1974). It is therefore important in all situations in which differential observability by age, sex, dominance status, or some other variable may occur to record how frequently animals of these different classes are seen in the observation area. In this way, a correction can be made if needed.

LABORATORY METHODS

Some of the techniques described in the following sections are used primarily in the laboratory, although many of them also are used in the field. Many of these laboratory procedures provide key information that can be combined with field data and can be used to answer such questions as: What sort of variation in size and pelage occurs within and between populations of raccoons? How can the diet of raccoons be investigated to determine that they are carnivores? What other species are most closely related to raccoons? How do the endocrine systems and energetics of raccoons relate to such life history characteristics as survival and rates of reproduction?

Morphometrics

Museum collections of mammals consist of study skins, mounts, axial and cranial skeletons, teeth, and, for some mammals, horns or antlers. **Morphometrics** is the measurement of the characteristics of these specimens. In some cases, ectoparasites are collected from the fur and skin of mammals, or the animals are dissected and internal parasites are removed for preservation (see chapter 27). General body measurements are recorded for each specimen first. These usually involve body mass, total length, tail length, hind foot length, and ear length. Some mammals, such as bats and cetaceans, require additional or specialized measurements. For mammals the size of raccoons or smaller, a study skin is often made. With larger mammals, the skin may be tanned for preservation. Details of study skin preparation and parasite collection can be found in DeBlase and Martin (1981) and Anderson (1965).

From study skins, we can assess the thickness of the fur and the relative presence and distribution of different types of hairs (see chapter 5), as well as the color of the pelage (figure 3.7). After the study skin has been completed and the internal organs examined, the skeleton, or often just the skull since that is used for species identification, can be cleaned using one of the following methods: using a colony of beetles of the genus *Dermestes* that eat the remaining flesh, boiling the carcass and then cleaning off the flesh, maceration using bacteria (Hildebrand 1968), or enzyme digestion (Harris, 1959; Mahoney 1966). Additionally, some smaller mammals can be preserved in a 10 percent formalin solution. Some museum specimens, particularly those for display, are prepared as whole mounts. This can be done with expert taxidermy using the skin of the animal, or, with some smaller mammals, by freeze-drying. Today, some tissues and organs are preserved using ultra-low-temperature freezers so that they can be sampled or examined using histological techniques at a later date. Freezing can include materials preserved for DNA analyses (see under "Genetics and Molecular Techniques" in this chapter).

After the skeleton has been prepared using one of these methods, its features can be assessed. Such measures include examination of the dentition, often using a dissecting microscope to ascertain the cusp patterns and other key features (see chapter 4). The cranium and remainder of the skeleton are measured using rulers and micrometers. The detail with which these measurements are recorded depends on the nature of the investigation. Properly preserved museum

specimens can be used many times for different studies. Stored in air-tight museum cases, such specimens can last for hundreds of years.

These specimens, both as originally found and after proper preparation, can be used as the basis for at least two types of studies. First, the age, health, and reproductive condition of the animal can be assessed, as well as any intraspecific variation. Age can be determined via a variety of methods involving tooth eruption patterns, tooth wear, microscopic examination of cross sections of tooth roots for cementum annuli, and, in some instances, the eye lens weight (Bloemendal 1977; Bruns Stockrahm et al. 1996). Health can be determined by observations made during the dissection of the carcass and preparation of the study skin, including the numbers and types of ectoparasites and internal parasites. Reproductive condition is determined by the presence of scrotal testes (males) in most terrestrial mammals, or the presence of fetuses in utero or evidence of lactation (females). Female ovaries and uteri can be examined histologically for the presence of corpora lutea or placental scars (Laws 1967). Male testes and epididymydes can be examined for sperm production and presence.

Prepared mammal material also can be used to study intraspecies variation. The thickness of the pelage may vary, for example, along a latitudinal gradient. Examination of museum specimens gathered at different points along this gradient permits the testing of this hypothesis. Another hypothesis that could be tested similarly is that pelage color varies in relation to the soil color in different areas of the range of the same species. Benson (1933) found this sort of relationship for small mammals living in New Mexico, and Dice (1939) reported on intraspecific variation in pelage color that was related to habitat for the cactus mouse (*Peromyscus eremicus*).

Second, prepared material can be used for identifying unknown mammal specimens. Taxonomy is the theoretical study of classification, including its bases, principles, procedures, and rules (Simpson 1961). **Biological classification** involves grouping items, mammals in this case, into ordered categories according to their attributes, reflecting their similarities and consistent with their evolutionary descent. Various methods of classification have been used over the years. The oldest and still most common form of classification involves the use of physical features—the traits that are derived from morphometrics. Classification keys have been developed through **phylogenetics,** the study of evolutionary histories for various groups of living organisms (Simpson 1961). A primary goal of such studies is to develop a treelike diagram of the relationships in a phylogeny, called a **dendrogram.** (Several examples of dendrograms are examined in subsequent chapters discussing orders of mammals.) The nodes in such a tree represent either extant derived or extinct ancestral species; the branches are the pathways that connect the nodes and illustrate evolutionary relationships among the species. A simple representation of the branching pattern or phyletic lineage that does not attempt to represent rates of

Figure 3.7 **Study skins.** The sequence of study skins shown here, of white-footed deermice, can be used to examine differences in pelage thickness or color, which might vary with habitat or geographical location.

evolutionary divergence is generally called a **cladogram.** A **phylogram** goes one step further by attempting to represent the degree of genetic divergence among the taxa represented by the lengths of the branches and the angles between them.

A **taxonomic key** consists of an arrangement of the traits of a group of organisms into a series of hierarchical, dichotomous choices. By choosing one or the other of a pair of traits based on examining the specimen, we can identify the unknown animal. Many such keys have been generated for mammals as a group and for specific subgroups (see DeBlase and Martin 1981). Prepared specimens of mammals served as the primary basis for these sorts of investigations for centuries. They are still of principle importance today, but as we shall learn in the final section of this chapter, molecular techniques now form a valuable adjunct to these older methods. Molecular methods are, in fact, simply another "morphological" data set and are fallible just like other older methods.

Several decades ago, another method for determining phylogenies, **numerical taxonomy,** became quite popular (Cole 1969; Felsenstein 1983; Mayr and Ashlock 1991). Numerical taxonomy organizes individuals into taxa based on unweighted or weighted estimates of overall similarity. Although not entirely discarded, this technique is not used as much today.

Cladistics analyzes unweighted, nonmetric characters (traits) to organize organisms exclusively into taxa on the basis of joint descent from a common ancestor (Hennig 1965, 1966). Consult Duncan and Stuessy (1984), Mayr and Ashlock (1991), or Weir (1996) for additional information on methods used in phylogenetics. Bronner (1995) used variations in chromosomes (see "Genetics and Molecular Techniques" in this chapter) as the basis for a cladistic analysis of the relationships among seven species of golden moles (Chrysochloridae). The three primary characters were the numbers of chromosomes, the lengths of chromosomes, and the positions of the centromeres. Mites inhabiting the fur and hair follicles served as the basis for a cladistic analysis of the relationships among seven genera of New World primates (O'Connor 1988).

Because cladograms are used in part III, some comments on how they are constructed should be helpful. The characters for the group of organisms for being described with the cladogram are divided into two types. **Plesiomorphic** characters are primitive, that is, they are ancestral or appeared earlier. **Apomorphic** characters are derived, or of more recent origin. The sharing of derived traits among organisms is defined as **synaptomorphy,** and such organisms are more closely related in terms of their evolutionary history than organisms that do not share the derived character (figure 3.8A). **Convergence** involves the presence of a similar character for two taxa whose common ancestor lacked that character (see figure 3.8A). Although convergence could result in the same derived characters evolving independently in two or more distantly related groups, it is not very probable. To construct a cladogram representing the proposed sequence of evolutionary derivation for the different taxa in our group of organisms, sister groups are used. **Sister groups**

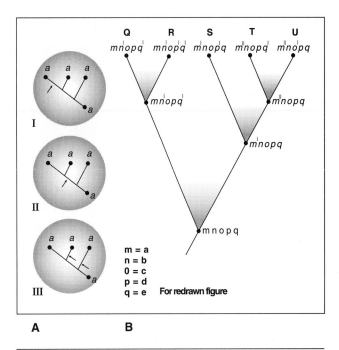

Figure 3.8 **Cladograms.** (A) The types of similarities between traits used in cladistics. The letters *p* and *p′* indicate two forms of the same character, with *p* representing the more primitive state and *p′* the derived state. Arrows (→) indicate the points in evolutionary history where the change from *p* to *p′* occurred. Part I is an example in which two of the three taxa share the primitive character. Part II exemplifies a situation in which two of the three taxa share the derived character. In part III, convergence has occurred. (B) The cladogram illustrates putative phylogenies for five taxa (*Q–U*). Characters a–e can occur as plain letters, meaning their most primitive form, and as single- (′) or double-prime (″) letters, indicating, in sequence, more derived forms of the character. Using this cladogram, taxa *Q* and *R* are a sister group to taxa *S, T,* and *U,* and *S* is a sister group to taxa to *T* and *U.*

are defined as the two groups resulting from a split, or bifurcation, in the evolutionary history of the group of organisms being studied (figure 3.8B).

Physiological Measures

Metabolism

Two different procedures are available for assessing metabolic rate in mammals. **Metabolic rate** is energy expenditure measured in kilojoules per day. Oxygen consumption as a measure of metabolic rate has been used for more than half a century, although the technology used for making such measurements has improved through time (Morrison 1948; Morrison et al. 1959; McNab 1995). The test subject is placed in a chamber where the rate of oxygen uptake can be determined. Gas pressure in the chamber is held constant by neutralizing the carbon dioxide respired by the animal and adding oxygen as needed. Data gathered using this technique reveal distinct differences in metabolic rate across mammal taxa that reflect variations in body size, diet, and climate (McNab 1974, 1995). Oxygen consumption as a measure of metabolic rate has also been used to study individual animal activities, such

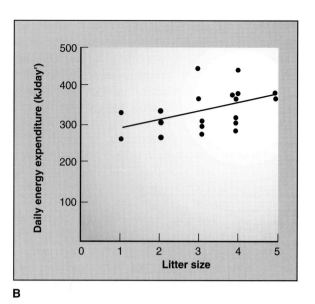

A

B

Figure 3.9   Metabolic activity.   (*A*) Oxygen consumption was used as a measure of metabolic rate for golden-mantled ground squirrels. Data from laboratory determinations of metabolic rate were used in connection with field observations of the activities of the ground squirrels to determine energy budgets. (*B*) Total daily energy expenditure of 20 female golden-mantled ground squirrels on peak days of lactation (days 31–35 after parturition) in relation to litter size (measured as the number of young that emerged from the burrow on days 35–38 after parturition). The regression line equation is $y = 19.2x + 274$, where x is the litter size and y is the daily energy expenditure.

Source: Data from Kenagy, et al. "Energy Expenditure During Lactation in Relation to Litter Size in Free-Living Golden-Mantled Ground Squirrels" in Journal of Animal Ecology, *59:73-88, 1990.*

as locomotion in golden-mantled ground squirrels (*Spermophilus saturatus*) (Kenagy and Hoyt 1989; figure 3.9A). The increase in daily energy expenditure was significant with increasing litter size (figure 3.9B).

A second method for measuring metabolic rate, developed for use under field conditions, involves injection of doubly labeled water in which both the oxygen and hydrogen are isotopes (Nagy 1987, 1989; Kenagy et al. 1989a, 1989b, 1990). Repeated blood samples taken at intervals of from several hours to several days after the injection are analyzed. An assumption of this method is that the animal is in a steady state, not gaining or losing body mass. Then, the proportional turnover rates of oxygen and hydrogen isotopes measure the rate of carbon dioxide production. Carbon dioxide production, in turn, can be equated with metabolic rate (see Nagy 1987 for further details). This method has been used, for example, to study daily energy budgets in lactating golden-mantled ground squirrels (Kenagy et al. 1990) and annual patterns of energy expenditure in the same species (Kenagy et al. 1989). Metabolic rates were determined for a small Australian marsupial, the honey possum (*Tarsipes rostratus*), using this doubly labeled water method (Nagy et al. 1995). The technique has also been used to study larger mammals, such as the springbok antelope (*Antidorcas marsupialis*; Nagy and Knight 1994).

Total body electrical conductivity (TOBEC) provides information on the nutritional condition of the animal with respect to its fat-free lean body mass and fat reserves, rather than supplying a value for metabolic rate. The method relies on the difference in conductivity between lipids and other body constituents to estimate lean body mass. The animal is placed in a chamber surrounded by a solenoid, which measures the animal's effect on the electromagnetic inductance of the coil. The value obtained by subtracting the instrument's reading with an empty chamber from that when a test subject is present can be converted to an estimate of fat-free body mass (see Walsberg 1988 or Roby 1991 for details). Fat reserves can be computed by subtracting the TOBEC value for lean body mass from the animal's overall mass. The device has been used for smaller animals, for example, Belding's ground squirrels (*Spermophilus beldingi*) (Bachmann 1994), guinea pigs (*Cavia porcellus*; Rafael et al. 1996), and house mice (*Mus musculus*; Koteja 1996), as well as for larger animals, including sheep (*Ovis aries*; Wishmeyer et al. 1996).

Body temperature measurements can be used as an indication of thermoregulation and in connection with assessment of metabolic rates. Both manual measurements and radiotelemetry can be used to determine body temperatures in different species (Morrison 1965; Morrison and McNab 1967; Osgood 1980; Vogt et al. 1983). Such data provide information on the strategies mammals use to regulate body temperature in relation to ambient conditions (see chapter 8) and on relationships between diet, activity period, and changes in body core temperature.

Finally, we can use stomach contents, urine, and fecal materials to evaluate various aspects of the diet and nutrition of mammals. Examination of scats for remaining parts of plants or animals that have been consumed tells us about the current diet of a species. Stomach contents also provide indications of the foods eaten (Whitaker 1963, 1966). Both scats and stomach contents are examined for parts of seeds, portions of vegetation, hair, feathers, and bones to help identify the diet. For example, the diets of grizzly bears (*Ursus arctos*) inhabiting areas in and around Yellowstone National Park (Mattson et al. 1991) and of rodents living in a California coastal sage scrub community (Meserve 1976) have been analyzed by examining fecal and stomach contents.

Chemical analyses of urine and feces can also provide data on the energy and nutritional composition of the diet. For example, in penned gray wolves (*Canis lupus*), DelGiudice and colleagues (1987) found that fasting produced changes in urine levels of several compounds, including the ratios of urea, potassium, and sodium with respect to urinary creatinine. Similar measures were made on free-ranging wolves (Mech et al. 1987). These sorts of measures are taken to help determine the nutritional status of wild mammals. For example, the mineral and nitrogen content of feces of some neotropical bats is a direct consequence of the fish, crustaceans, beetles, and moths they have consumed (Studier et al. 1994).

Analysis of mammalian feces can also supply information concerning parasites. We examine feces for remains and ova of parasites. Parasite load can be related to dominance and reproduction in baboons (*Papio cynocephalus;* Hausfater and Watson 1976). Parasite numbers are used as a measure of habitat disturbance in muriquis (*Brachyteles arachnoides*) in Brazil (Stuart et al. 1993).

Stress

Several measures of stress have been developed for mammals. Adrenocorticotrophic hormone (ACTH) is produced in and released by the adrenal cortex in response to stress. Adrenal weight is therefore a convenient way to assess prolonged levels of increased stress and ACTH production. Adrenals collected from laboratory or field specimens can be fixed in formalin solution, cleaned, and weighed. We then examine the tissue histologically. These techniques have been used with a variety of small mammals, including house mice (Bronson and Eleftheriou 1964), meadow voles (Christian 1963), and California ground squirrels (*Spermphilus beecheyi*) (Adams and Hane 1972). The findings for small mammals have been related to population growth and regulation (Christian 1950, 1963; Christian and Davis 1964; see chapter 8).

Blood levels of corticosteroids have been used to measure adrenal activity and stress. **Radioimmunoassay (RIA)** is a method for assessing cortisol levels on small blood samples. RIA involves using radioactively labeled antibody mixed with blood samples from the animal (Abraham et al. 1977) in a competitive binding assay. Results from these samples are compared with a standard curve to determine the level of circulating hormone (see Nelson 1995 for details). Because the adrenal response is relatively rapid, samples must be taken very soon after an animal is restrained. Examples of the use of RIA include measuring increases in cortisol levels in rhesus macaque (*Macaca mulatta*) mothers

separated from their infants (Champoux and Suomi 1994) and the effects of various stressors related to cage size and movement in long-tailed macaques (*M. fascicularis;* Crockett et al. 1993). Other hormone assays use enzyme-linked immunosorbent assay (ELISA) or enzyme immunology (EIA) binding steroid hormones in a similar antigen-antibody reaction to produce a quantifiable color change (Nelson 1995).

Reproduction

Because production of progeny is a basic measure of individual fitness and is important to understanding the health of populations, considerable attention has been given to various aspects of mammalian reproductive biology. At the most direct level, measurements of the proportion of females anually producing progeny, litter size, and survival of adults and young give us useful information. Many techniques have been developed for exploring the underlying mechanisms in males and females.

For many seasonally breeding male mammals, the testes are scrotal for only a portion of the year; assessing whether the testes are scrotal or abdominal is thus an important field observation. Testis size and weight and the weights of various sex accessory glands are useful indices of reproductive activity (Gilmore 1969; Inns 1982). Sperm counts, obtained by examining ejaculates with a microscope, provide data on the relative number of sperm and can be used to note the proportion of defective or abnormal sperm (Berndtson 1977). RIA has been used to measure changes in androgen levels (Kantongol et al. 1971; Baldwin et al. 1985). Testosterone levels vary seasonally, but also exhibit changes with respect to stress, aggression and dominance status, sexual stimulation, and the general health of the animal.

Most female mammals also breed seasonally. A first step in assessing female reproduction is detection of **estrus,** the manifestation of the internal processes surrounding the maturation of a follicle and release of ova. Behaviorally, many female mammals in estrus exhibit greater tolerance to mating approaches and solicitation by conspecific males. They may remain stationary, adopt a lordosis posture with the head and hindquarters elevated and back depressed to permit intromission, or move their tail to one side when courted by a male or when pressure is applied to the back region by an investigator. The contents of the vagina can be used in some mammals as an indication of estrus. One secondary action of the increase in circulating estrogens is the cornification of the cells that line the vaginal walls as ovulation approaches. Flushing the vagina with a small aliquot of water and examining it under a microscope, sometimes with appropriate histological staining procedures, results in a rapid indication of whether the animal is or has recently been in estrus (Vandenbergh 1967, 1969; Kennelly and Johns 1976; Turner and Bagnara 1976).

Ovulation rate, or number of ova released, can be determined in live animals by flushing the reproductive tract at a time immediately after ovulation. A somewhat easier procedure involves histological examination of the ovaries to study

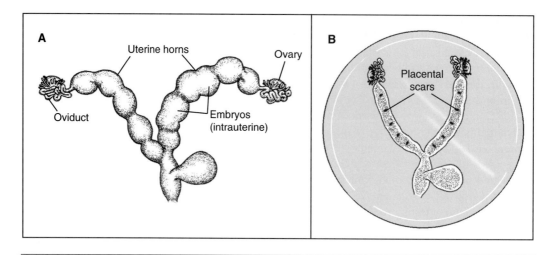

Figure 3.10   Placental scars.   (A) The swellings in the two uterine horns from a pregnant white-footed mouse; each swelling represents a fetus and associated placenta. (B) The uterine horns, dissected from the abdominal cavity are compressed between the bottom and inverted top of a Petri dish, revealing the placental scars along both horns. This information can be used to determine the number of young a female has had implanted in her uterus.

follicular development and the presence of corpora lutea ("yellow bodies")—locations from which ova have been released (Cheatum 1949; Cowles et al. 1977). This technique requires that the animal be sacrificed. Counting placental scars (figure 3.10) indicates the number of fetuses that implanted (Hensel et al. 1969; Oleyar and McGinnes 1974). In some instances, however, fetuses are resorbed after they implant.

In living female mammals that must be examined repeatedly, several other procedures can be employed to assess reproduction. **Laparotomy** is an internal examination, performed by trained personnel, of the reproductive tract of the animal to explore the condition of the ovaries and uterus (Adams et al. 1989). With the invention of fiber optic techniques, laparotomy can be replaced by **laparoscopy.** For some larger mammals, particularly in captive or zoo settings, ultrasound is now used to determine the morphology of the reproductive tract and the existence and condition of a developing fetus (Adams et al. 1989, 1991).

We can perform RIA and EIA assays on blood samples to determine circulating levels of hormones related to female reproduction, including estrogens, progesterone, and prolactin (Harder and Woolf 1976; Austin and Short 1984). More recently, it has become possible to make these hormone determinations using urine and feces (Wasser et al. 1988; Wasser et al. 1991; Safar-Hermann et al. 1987). This highly useful, noninvasive technique has been applied to both male and female mammals and holds great promise for obtaining hormone measurements under free-ranging field conditions.

Genetics and Molecular Techniques

Molecular techniques, both protein assays and deoxyribonucleic acid (DNA) assays, contribute to our understanding of chromosomal variation, taxonomy, behavior, and conservation biology. One basic type of analysis involves an examina-

tion of the chromosome number and their cytogenetic structure. For more detailed explanations of the techniques noted in the following sections and their uses, consult Avise (1994) or Ferraris and Palumbi (1996).

Chromosomes

A **karyotype** is a depiction of the numbers and shapes of the chromosomes characteristic of a species (figure 3.11). We can compare numbers of chromosomes, their relative sizes, and the structural organization within different chromosomes to explore taxonomic relationships among mammalian species. Variation in karyotypes has been used, for example, to explore evolutionary relationships among nine species from five different genera of golden moles (Insectivora: Chrysochloridae; Bronner 1995). In a similar manner, chromosomes of two species, *Octodontomys gliroides* and *Octomys mimax,* have been used to understand their phylogenetic history with respect to other members of the South American octodontid rodents (Contreras et al. 1994). Variation in chromosome numbers occurs within some species of mammals, including sika deer (*Cervus nippon;* van Tulnen et al. 1983). Some wood mice (*Apodemus*) and harvest mice (*Reithrodontomys*) have what are known as supernumerary, or B, chromosomes in additional to their normal chromosome complement (Zima and Macholán 1995). Studies of these extra chromosomes in *Apodemus peninsulae* and *A. flavicollis* indicate that there is no clear pattern to the presence of these B chromosomes. They may be regulated by stochastic effects and not by any biological or ecological variables.

Protein Assays

Protein assays were developed before the more recent DNA-based analyses. One technique, called **protein immunology,** involves a procedure not unlike that for RIA discussed ear-

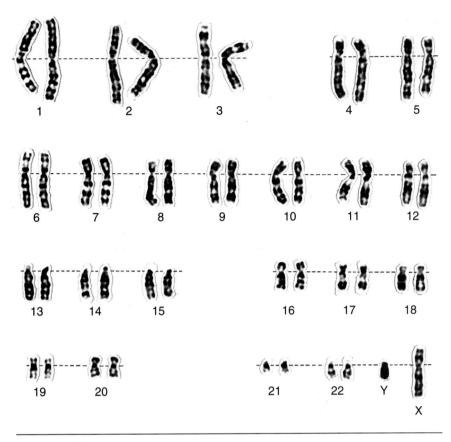

Figure 3.11 **Karyotype.** The numbers and structural characteristics of the chromosomes of a mammal, as in the human shown here, form its karyotype. The lengths and cytogenetic structural features of these chromosomes can be used for comparative studies exploring phylogenetic relationships. Differences in chromosome numbers and their structure may prevent interbreeding between animals of different species.

lier. Rabbits, goats, and other mammalian species are used to produce antibodies to a specific protein from a test species. The cross-reactivity of the homologous (original) antigen (protein) and a heterologous antigen (protein from a different, related species) provide an estimate of the degree of genetic relationship between the two species. The results are usually expressed as differences in immunological units and are assumed to be a measure of the rate of amino acid substitution in the proteins of the two species. Protein immunology has been used to examine phylogenetic relationships among some mammalian taxa (Prager and Wilson 1978; Prager et al. 1980). Several of the DNA-based techniques discussed in the following section have generally replaced these methods.

A widespread technique, **protein electrophoresis,** is sometimes called allozyme analysis. Nondenatured proteins migrate through a gel made of agarose, acrylamide, or other material at different rates when an electric charge is applied to the gel (figure 3.12). This procedure uses the characteristic migration distance of various proteins to identify and compare individuals. Most proteins move from the cathode (negatively charged) pole toward the anode (positively charged pole) according to their charge, size, and shape. After the

proteins have migrated, the gel is is stained to make the protein "bands" visible. Samples of known protein of the desired types can be run as standards against which unknown samples can be judged. This technique is relatively simple and fast, and thus many genotypes can be determined with only a day's work in the laboratory.

Protein electrophoresis information can be used to assess phylogenetic relationships. Because alleles at loci within the genome code for protein production, differences in proteins, as measured by their electrophoretic patterns, can be informative about relationships between taxa. This method has been used to investigate the systematics of deermice (Genus *Peromyscus;* Selander et al. 1971; Smith et al. 1973; Selander 1982). Allozyme analysis was also used to explore the systematics of Bolivian tuco-tucos of the Genus *Ctenomys* (Cook and Yates 1994). Another use for protein electrophoresis involves measuring genetic heterozygosity in populations of mammals. Protein polymorphism of house mice (*Mus musculus*) living on several farms revealed that genetic similarity followed closely the patterns of spatial distribution (Selander 1970). Polymorphism here refers to the presence of several different alleles in the population. The pattern of spatial and genetic similarity was true even within individual barns, where clusters of animals with similar genetic patterns ("demes") lived. A comparable situation often exists with respect to demes for larger species, such as white-tailed deer (Sheffield et al. 1985). Finally, allozyme analyses have been used for parental assignment by exclusion. Hanken and Sherman (1981) used the exclusion method to determine parentage in Belding's ground squirrels. They also found that some litters involved multiple sires. Several complications and problems with interpretations of data obtained by protein electrophoresis are discussed by Avise (1994).

DNA Assays

Recent decades have witnessed the development of a variety of new techniques based on DNA analysis. The method of **DNA–DNA hybridization** was developed in the 1970s (Britten et al. 1974; Avise 1994). DNA is double-stranded with pairs of nucleotide bases held together by hydrogen bonds. When DNA is heated, these strands separate but remain intact. This is called the melting point of DNA. When the strands are allowed to cool, they collide with one another by chance, and complementary base pairs again link together. When DNA from a single species has been used, cooling

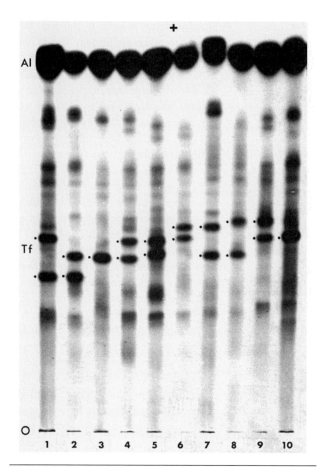

Figure 3.12   Protein electrophoresis.   Evidence from protein electrophoresis of albumin (Al) and transferrin (Tf) in white-footed mice. These are two of the most common types of proteins found in mammalian blood. Note here the transferrin allozyme variation, indicating multiple alleles involved in this trait. Data of this sort can be used to assign parentage using an exclusion method. They can also be used to determine the degree of inbreeding.

various bacterial strains have provided a new set of DNA-based techniques, among which is the assaying of **restriction fragment-length polymorphisms (RFLPs).** The process involves several steps. First, DNA is isolated from cells of the animal being studied. It is then cut with one or more restriction enzymes, and the resulting fragments are placed on a gel (usually agarose or acrylamide) for electrophoresis. Larger fragments of DNA migrate more slowly than smaller ones on these gels. Staining the gel permits the viewing of the fragments sorted by size. A number of variables that are important to such processes are beyond the scope of our coverage here (see Avise 1994). The source of the DNA for RFLP analyses can be either nuclear DNA or mitochondrial DNA (mtDNA). Nuclear DNA usually consists of all of the chromosomal DNA from that particular species. Mitochondrial DNA, inherited only along maternal lines, consists of a single loop of DNA (figure 3.13). RNA genes are also associated with ribosomes and have been used, in some instances, for studies similar to those conducted on DNA.

RFLPs have proven useful in several types of studies in mammalogy. For humpback whales (*Megaptera novaeangliae*), analyses of mtDNA revealed that females are quite site-specific (philopatric), returning to specific locations within their summer feeding grounds in subpolar and temperate oceans and to winter breeding grounds in and near the tropics (Baker et al. 1990). The relationships among the various subspecies of the leopard (*Panthera pardus*) have been worked out in some detail using restriction enzymes (Miththapala et al. 1996). These findings have implications

results in homoduplexes. When DNA samples from two species are heated and cooled together, heteroduplexes are formed. The homoduplexes and heteroduplexes are separated and tested for thermal stability by slowly raising the temperature and assessing the mixture for DNA strand disassociation. Because the homoduplexes involve DNA strands from the same species, this second disassociation should be at a higher temperature than in heteroduplexes from the same mixture. A measured difference in thermal stability of the homoduplexes and heteroduplexes provides a quantitative estimate of the genetic distance or divergence between the two species. This technique has been used, for example, to examine the phylogeny of the Order Cetacea (Milinkovitch 1992). The data support an ungulate origin for this order. This technique was also used to investigate the intergeneric relationships of bandicoots (Family Peramielidae) in Australia (Kirsch et al. 1990).

A second DNA technique involves enzymes called restriction endonucleases. These enzymes cleave, or cut, DNA at specific locations, leaving fragments of varying lengths. The discovery of these enzymes and their isolation from

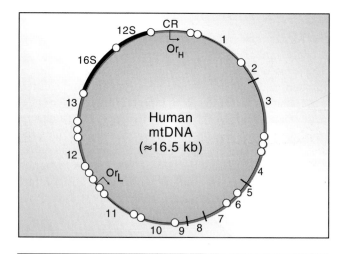

Figure 3.13   Mitochondrial DNA.   DNA associated with mitochondria (mtDNA) occurs in a single double-stranded loop. A number of specific loci on the mitochondrial DNA of various mammals have been mapped and are used for phylogenetic studies. These loci have been sequenced and serve as the basis for cladograms. The human mtDNA loop is depicted here. The loop has a control region (CR), genes coding for two different RNAs (12s and 16s), and loci coding for 13 different polypeptides (1–13). Comparisons of mtDNA loops can be used, for example, to determine evolutionary relationships between humans and the other great apes.

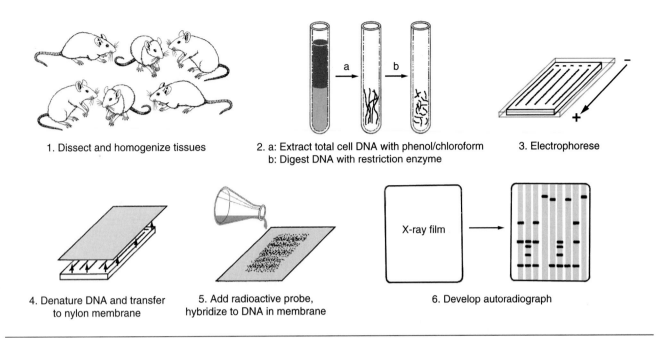

1. Dissect and homogenize tissues

2. a: Extract total cell DNA with phenol/chloroform
   b: Digest DNA with restriction enzyme

3. Electrophorese

a   b

X-ray film

4. Denature DNA and transfer to nylon membrane

5. Add radioactive probe, hybridize to DNA in membrane

6. Develop autoradiograph

Figure 3.14   **DNA** fingerprinting.   The sequence of events involved in using variable number tandem repeats (VNTRs) to obtain a DNA fingerprint proceeds from extraction of the DNA from tissue to the visualization of the results via an autoradiograph.

for conservation biology and the recognition of different subgroups with respect to programs for endangered species. Mitochondrial ribosomal genes were mapped with the RFLP technique for bats of the Genus *Lasiurus* (Morales and Bickham 1995). As with some similar studies, RFLP results agree with some and disagree with other phylogenies based on more traditional morphological traits for these same species.

The process known as **DNA fingerprinting,** sometimes referred to as minisatellite DNA (Avise 1994), is based on the fact that DNA possesses, along various regions of the chromosome, **variable numbers of tandem repeats (VNTRs)** of nucleotide base sequences. When DNA is cleaved by restriction enzymes, repetitive units of 16 to 64 base pairs in length are produced (figure 3.14). The fragments are run on gels and then visualized by a process called the Southern blot method (Jeffreys 1987; Avise 1994). The band patterns produced in this manner are unique to each individual. This technique permits us to make exact parental assignments, which, in turn, can provide estimates of individual reproductive success, a key to understanding evolutionary processes. DNA fingerprinting also provides information on mate selection, that is, which females mated with which males. If a tissue sample is not available, even hair or feces can be used. DNA fingerprinting can provide evidence concerning the exact identity of that individual.

Examples of the use of DNA fingerprinting with mammals come most readily from human forensics. Crimes such as rape and homicide have involved DNA evidence used in the courts for convictions (Kirby 1990; Chakraborty and Kidd 1991) Because young mammals are often observed with their birth mother, the question becomes assignment of paternity.

DNA fingerprinting assists with paternity assignment in several taxa of primates, including macaques, baboons, and lemurs (Martin et al. 1992; Turner et al. 1992). DNA fingerprinting has been used to demonstrate monogamy as the mating system resulting in litters of California mice (*Peromyscus californicus;* Ribble 1991), although data from allozyme analyses indicate that multiple paternity occurs in the related white-footed mouse (*P. leucopus;* Xia and Millar 1991).

The next key development in molecular techniques was the **polymerase chain reaction (PCR;** Mullis et al. 1986). PCR can be used to multiply particular segments of DNA literally millions of times, starting from very tiny amounts (figure 3.15). The process has several steps (Avise 1994): (1) DNA is heated to denature the double-stranded structure to single strands; (2) primers are annealed to either end of the specific region of the DNA that is to be amplified; and (3) during cooling, new strands of DNA complementary to the region selected by the primers flanking either end are generated using a DNA polymerase that is stable throughout the temperature range used for these procedures (see figure 3. 15). A primer consists of a short sequence of 20 to 30 nucleotide bases that are highly similar to the DNA sequences on either end of the target section of DNA that is to be amplified. Multiple repetitions of the heating and cooling cycle result in a large number of multiples of the original DNA sequence located between the selective primers. The ability to use the PCR procedure has made three final molecular techniques very important in current research on mammals: RAPD, microsatellite DNA, and gene sequencing.

Random amplified polymorphic DNA (RAPD) involves the use of restriction enzymes with short primers in

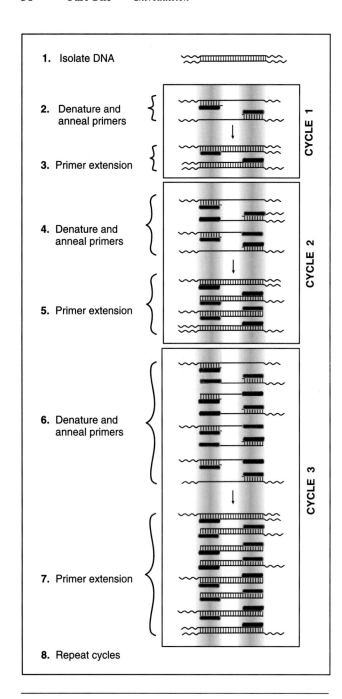

1. Isolate DNA

2. Denature and anneal primers

3. Primer extension

CYCLE 1

4. Denature and anneal primers

5. Primer extension

CYCLE 2

6. Denature and anneal primers

7. Primer extension

CYCLE 3

8. Repeat cycles

Figure 3.15   Polymerase chain reaction. PCR involves a sequence of events through which the DNA is cycled multiple times, with a significant amplification at each cycle. The process begins with DNA from the nucleus or from mitochondria and a pair of selected primers that target a specific segment of the DNA for amplification.

conjunction with PCR (Ferraris and Palumbi 1996). The resulting DNA fragments from a DNA sample are 200 to 2000 base pairs in length and appear as a series of bands on a gel. The primers are not locus-specific, and thus the amplified loci are said to be anonymous. The resulting gel banding pattern is called a RAPD marker. The presence and absence of bands can be used to compare samples from different animals for parental assignment by exclusion (Lewis and Snow

1992) and to compare populations with regard to genetic structure (Gibbs et al. 1990; Lynch and Milligan 1994). Because of some problems of variability in RAPD results and differences in interpretation of findings from this procedure, it has not been widely used (see Ferraris and Palumbi 1996).

Microsatellite DNA is a procedure involving tandem repeats of short sequences, usually as multiples of two to four of the original DNA. Specific portions of the DNA are amplified selectively, using pairs of primer sequences usually developed uniquely for the subject species. When the products from the PCR reaction for a microsatellite locus are run on an electrophoresis gel, a single band should be evident if the animal is homozygous at that locus. Two bands are visible if the organism is heterozygous. There are extensive libraries of these primer sequences, for example, for mice (Tautz 1989), bovines (Fries et al. 1990), and humans (Valdés et al. 1993). Constructing additional libraries and extending existing compilations of primers are ongoing processes (Stallings et al. 1991). Each locus that is amplified may have several, up to 5 or even 10, alleles facilitating the use of the technique. The microsatellite procedure has a number of applications, including parental assignment, estimating gene flow between populations, and estimating heterozygosity in mammalian populations (Ellegren 1991).

The final approach that has come into widespread use involves **DNA sequencing**—the exact listing of the string of the four nucleotide bases (A [adenine], G [guanine], T [thymine], C [cytosine]) within a section of DNA. This approach has been employed for both nuclear and mitochondrial DNA. Several different methods are employed for DNA sequencing (see Avise 1994), but the most widely used procedure (Sanger et al. 1977) involves five steps:

1. Denaturing the double-stranded DNA into single strands.

2. Annealing a primer with a DNA sequence complementary to the target sequence to the DNA.

3. Dividing the mixture into four subsamples (one for each of the bases in DNA) and subjecting these to a series of reactions to radioactively label the different nucleotide bases with different tags.

4. Running the mixtures through the PCR procedure.

5. Placing the products on large polyacrylimide gels, after which the banding patterns are visualized by autoradioagraphy (figure 3.16).

The DNA sequence can then be read from the gels.

DNA sequencing has been used, for example, to assess phylogenetic relationships for northern hair seals (Otariidae; Mouchaty et al. 1995). Mitochondrial DNA from muscle obtained from seven species of northern hair seals was analyzed with respect to a 458-base-pair sequence in the cytochrome b gene region. As with many of the current molecular techniques, some revisions in the phylogenies of these seven species are proposed based on the mtDNA analyses. Primate

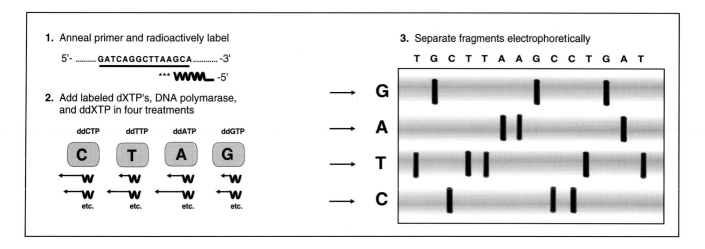

Figure 3.16   DNA sequencing.   The Sanger (1977) method for producing a sequencing gel for DNA. The DNA sequence can be determined by reading the gel from top to bottom. For example, the short sequence depicted here would start TGCTTAA. . . .

phylogeny has also been the subject of a number of investigations using DNA sequencing (Koop et al. 1986, 1989). Several decades of research lie ahead with respect to working out the molecular phylogenies of mammals and resolving the differences between findings obtained from the DNA-based procedures and earlier findings based on morphological traits.

A wide variety of approaches is available for constructing phylogenetic trees using DNA-based molecular techniques, some details of which are beyond the scope of our coverage here (see Avise 1994; Ferraris and Palumbi 1996; Weir 1996). As noted earlier, cladograms involve a pattern of nodes and branches, representing taxa and the pathways that connect them in terms of phylogenetic relationships. We will use cladograms in several later chapters of the book to enhance discussions of various orders of mammals. Because of the size of the databases used for such procedures, computer programs have been developed to construct the cladograms (summarized by Swofford and Olsen 1990). Each of these programs uses a different sequence of procedures (algorithm) to generate a variety of possible cladograms to represent the phylogeny of the particular group being investigated. The rule of **parsimony** is used to determine which of these computer-generated cladograms best represents the evolution of the groups involved. The tree with the fewest steps or branching points is generally accepted as the best representation of the phylogeny (Felsenstein 1983).

One way to summarize the uses of these molecular techniques is to examine how the various procedures are used for different tasks (table 3.1). Certain techniques are

Table 3.1   Levels of evolutionary divergence at which various molecular genetic methods normally provide informative phyogenetic markers

Hierarchical Level	Protein Immunology	Protein Electrophoresis	DNA–DNA Hybridization	RFLP Analyses of			DNA Sequencing
				mtDNA	scnDNA	VNTR Loci	
Genetic identity or nonidentity	—	*	—	*	*	**	*
Parentage	—	*	—	*	**	**	*
Conspecific populations	—	**	—	**	**	*	*
Closely related species	*	**	*	*	*	—	*
Intermediate taxonomic levels	**	*	**	—	—	—	**
Deep seperations (>50mya)[†]	*	—	*	—	—	—	**

From D.M. Hills and C. Moritz. Molecular Systematics, *1990. Copyright ©1990 Sinauer Associates, Sunderland, MA.*

(**) highly informative; (*) marginally informative, but not an ideal approach for reasons of cost-effectiveness or other difficulties; (—) inappropriate use of method. Not all categorizations are absolute. For example, some isozyme characters, such as presence or absence of duplicate gene products, can be useful at higher taxonomic levels. ([†]) mya = millions of years ago

best for specific tasks, and others can provide marginal answers. DNA fingerprinting (VNTRs) is the best technique for establishing genetic identity, whereas parentage can be assigned equally well using any of three techniques (see table 3.1). When in search of a particular type of answer, select the technique that provides the strongest data for the conclusions you are seeking.

Clocks

An underlying assumption of most of the molecular methods used to explore phylogenetics is that the rate of change for nucleotide bases in DNA has been a constant. But, are rates of change of base pairs in DNA, and hence of amino acid sequences in proteins, constant (Avise 1994; Ferraris and Palumbi 1996)? If we wish to assign dates to the branch points of our tree, we need to know something about the rates of mutation that produce changes in DNA. It is then possible to calibrate a clock and use it to estimate, for example, the time since divergence for species, genera, or families in a cladistic tree. The validity of the assumption of constant change is called into question, at least in some systems, because molecular evolution seems to proceed at variable rates (Ayala 1986; Britten 1986; summarized in Avise 1994). Among the key factors that may influence perceived differences in rates of evolution are generation time, exposure to mutagens leading to more rapid introduction of new genes, and the efficiency with which DNA repair mechanisms work in different lineages. Some caution must be used when interpreting phylogenetic trees constructed using various molecular techniques.

Investigators working in this field today, however, generally are confident that the procedures provide, in a cumulative sense and over long periods of geological time, a sound picture of both the evolutionary relationships among various taxa and the time frame during which various divergences occurred. Molecular-based clocks can be used in both a relative and an absolute sense. The relative time frame for taxonomic events derived using molecular clocks is probably quite accurate. The agreement between molecular clocks and what is called sidereal time (geologic, or "real," time) is somewhat less certain and subject to the variations noted earlier. A serious difficulty with this technique, and one that scientists are working to resolve, is the lack of fossil material available for many taxa to use as a check against the molecular clock.

Summary

Many techniques are used to study mammals under both field and laboratory conditions, and some are used in both settings. Applying of techniques from laboratory settings to field situations and vice versa can be a most productive approach for generating ideas and testing hypotheses.

A primary set of field techniques includes trapping and marking animals. Various traps have been employed, depending on the size of the mammal to be captured. Darts, fired from modified guns, with tranquilizers to anesthetize animals have been used for many species. Marking techniques include natural features such as color patterns or physical attributes, ear tags, toe clipping and ear punching, and fur dyes. Animals have also been identified using radionuclides and transmitters. Individually identified animals can be used for a variety of studies, including population biology, movement patterns, and behavior.

Radiotelemetry is widely used for monitoring mammals to obtain information on position and various physiological functions. Powdertracking can provide information on animal locations, movements, and foraging habits. Geographic information systems (GIS) and the global positioning system (GPS), based on satellites, have added new dimensions to the study of mammals.

Observational methods used to study mammals, particularly their behavior, include focal animal sampling and scan sampling. The possible differential observability of animals is a key concern when conducting behavioral studies.

One of the earliest laboratory procedures, and still an important technique for the study of mammals, is morphometrics. These measurements of the physical characteristics of mammals have served as the basis for much of mammalian taxonomy. Special techniques have been developed for preparing study skins and skeletons, which form permanent collections for reference and continued study. Phylogenetics, the study of the evolutionary histories of the organisms, uses characteristics to generate cladograms. These tree diagrams contain nodes, which represent extant and extinct species, and branches, which represent the evolutionary pathways connecting them. Taxonomic keys, developed from characteristics of each species, can be used to identify unknown specimens.

Various physiological measures have been made on mammals using laboratory-based techniques that are now extended to field applications. These include determinations of metabolic rate via gas analysis or the doubly labeled water technique, examination of the composition of urine and feces to assess diet and nutritional condition, and the TOBEC method for determining lean body mass and fat reserves. Analyses of hormones related to stress and reproduction, carried out by radioimmunoassay (RIA), are performed on blood, feces, and urine samples. Other measures of reproduction include assessment of reproductive condition, weights of sex organs and accessory glands, and determinations of litter size and the proportion of females in a population that produce progeny.

Molecular techniques have become quite important in the recent history of mammalogy. Analyses of chromosome numbers and structure provide insights into phylogenetic history. Allozyme analyses, via protein electrophoresis, have been used for several decades for systematics, assessing genetic heterozygosity, and parental assignment via exclusion procedures. DNA–DNA hybridization techniques provide information on the cross-reactivity of DNA from different mammals and thus about phylogenetic relationships.

Restriction endonucleases, to generate fragments of varying lengths (RFLPs), have been used with both nuclear and mitochondrial DNA. Data from this procedure have provided better information on philopatry and systematics. One of the most widely uses procedures employs variable length tandem repeats (VNTRs) and is commonly called DNA fingerprinting. Hypervariable regions of DNA are cleaved with enzymes and then visualized on gels as a series of bands. These banding patterns can be analyzed to obtain information for parental assignment, mate selection, and determinations of multiple paternity.

The development of the polymerase chain reaction (PCR) for amplifiying segments of DNA opened new avenues for the use of molecular techniques. Microsatellite DNA consists of short sequences of nucleotides that are targeted by site-specific primers and then amplified using PCR. When these amplified fragments are visualized using gel electrophoresis, the single band or pair of bands, can be used for parental assignment and for assessing genetic heterozygosity. Finally, DNA sequencing—determining the exact string of nucleotides in a segment of nuclear or mitochondrial DNA—has seen wide application to phylogenetics and the construction of cladograms.

This suite of new molecular techniques provides exciting possibilities for the future study of mammals. The use of particular techniques depends on the exact nature of the question being tested. Some care should be exercised in using any of these techniques in terms of the underlying assumptions. For those techniques that rely on DNA as a molecular clock, some evidence exists that the rate of change in base pair sequences over long periods of time may vary—reason for some caution in interpreting times of divergence in phylogenetic trees.

Discussion Questions

1. Suppose you wish to study the relationship between geographic location and metabolic rate within a species of small mammal. Which of the techniques covered in this chapter would you use? List the general procedures you would follow. What are some of the problems you might expect to encounter in conducting such a study? How would you try to resolve those problems?

2. On an expedition to Central Africa, you discover several individuals of a bat species that seems unfamiliar and is not easily identified. What procedures would you use to attempt to identify the bats?

3. You are asked to develop a general activity time budget for pronghorn antelope (*Antilocapra americana*) living at several sites in northern Colorado. Describe the methods you might use to conduct such a study. What problems might be encountered? What measures could you adopt to potentially resolve these difficulties?

4. In the course of using a taxonomic key for the Genus *Microtus*, you become concerned that it is not complete and has some inaccuracies. Describe how you would use the older procedures for taxonomic classification to update the key. In addition, which modern molecular techniques would you use to aid in this process and why? How might you go about resolving any differences between the keys developed by these two different methods?

5. Without knowing the details of how GIS works, but with the knowledge that it provides a rather complete picture of the nature of the habitats in your location, describe how you would determine the relationships between the presence and abundance of the mammals in your locale in relation to the various types of habitats present.

Suggested Readings

Animal Diversity Web (University of Michigan Museum of Zoology): www.oit.itd.umich.edu/ projects/ADW/.

Aranoff, S. 1993. Geographic information systems: a management perspective. WDL Publications, Ottawa, Ontario.

Bookhout, T. A. (ed.). 1996. Research and management techniques for wildlife and habitats, 5th ed., rev. Wildlife Society, Bethesda, MD.

Kleiman, D. G., M. E. Allen, K. V. Thompson, and S. Lumpkin. 1996. Wild animals in captivity. Univ. of Chicago Press, Chicago.

Krebs, C. J. 1989. Ecological methods. HarperCollins, New York.

Lehner, P. N. 1996. Handbook of ethological methods, 2d ed. Cambridge Univ. Press, New York.

White, G. C. and R. A. Garrott. 1990. Analysis of wildlife radio-tracking data. Academic Press, San Diego.

CHAPTER

4

Evolution and Dental Characteristics

Mammals evolved from reptiles during the 100-million-year period from the late Paleozoic to the early Mesozoic era. During this time, pronounced morphological changes took place from reptilian traits to those that characterize mammals today. The distribution, adaptive radiation, and resulting diversity seen in today's mammalian fauna are products of the evolutionary process operating throughout hundreds of millions of years—a process that continues today. In this chapter, we describe evolutionary morphological changes from reptilian to mammalian structural organization, including development of mammal-like traits in synapsid reptiles and the emergence of early mammals. We continue with changes in mammals throughout the Mesozoic era and the explosive adaptive radiation of mammals beginning in the early Cenozoic era. We also describe dental characteristics of early mammals and how these teeth developed into the dentition seen in modern mammals.

SYNAPSID REPTILES

Mammals arose from reptiles, specifically from reptiles in the subclass **Synapsida.** The synapsids were one of four subclasses of early reptiles, distinguished by the number, size, and position of lateral temporal openings (fossa) in the skull (figure 4.1) for the attachment of jaw muscles. Synapsids first appeared in the late Paleozoic era, about 320 million years ago (mya) in North America (table 4.1). They were the dominant land animals for 70 million years, but had passed their evolutionary peak by the time of emergence of the dinosaurs. Two orders within the synapsids are separated both spatially and temporally. The more primitive of the two orders, the **Pely-**

cosauria, is known from fossil remains in North America, and more recently South Africa (Dilkes and Reisz 1996). The more advanced order, the **Therapsida,** is known from later fossil remains in Russia, South Africa (Carroll 1988), and China (Rubidge 1994; Jinling et al. 1996). Therapsids often are referred to as "mammal-like reptiles."

PELYCOSAURS AND THERAPSIDS

Pelycosaurs, like all synapsids, were distinguished by a single lateral temporal opening, with the postorbital and squamosal bones meeting above. They were common by the end of the Pennsylvanian epoch (figure 4.2). By that time, they had radiated into three suborders. The Ophiacodontia, with several known families, were the most primitive. They were semi-aquatic and ate fish. The Edaphosauria were terrestrial herbivores and probably were preyed on by the third group—the carnivorous Sphenacodontia (Kermack and Kermack 1984). The sphenacodonts, such as *Dimetrodon* (figures 4.3A and 4.4), were the dominant carnivores throughout the early Permian period. All sphenacodonts shared a unique feature—a reflected lamina of the angular bone in the lower jaw. This feature was to become part of the development of the middle ear in later synapsids and mammals. The sphenacodonts eventually gave rise to the Therapsida. By the middle to late Permian period, all pelycosaurs were replaced by the more advanced therapsids (figure 4.3B), which were smaller and much more like mammals in their general morphology.

The oldest known therapsids date from the late Permian period (see table 4.1). Therapsids may be divided into two suborders: **Anomodontia** and **Theriodontia** (see figure 4.2). The

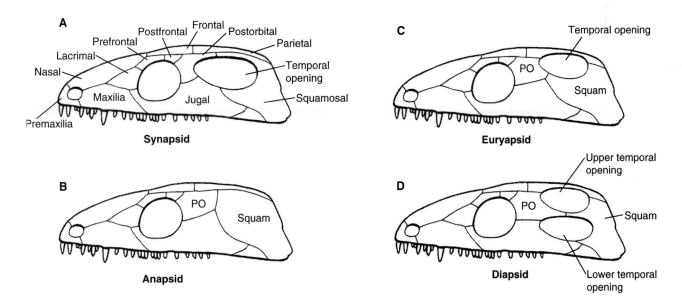

Figure 4.1   **Temporal openings of skulls in four subclasses of reptiles.**   (A) Synapsida, the group from which mammals arose. Note the articulation of the postorbital (PO) and squamosal (Squam) bones above the single temporal opening. (B) Anapsida, with no temporal opening. This group led to turtles. (C) Euryapsida, with the postorbital and squamosal bones meeting below the temporal opening, led to plesiosaurs and other forms of extinct marine reptiles. (D) Diapsida, named for the two temporal openings on each side of the skull. Lizards and snakes arose from the diapsids.

Table 4.1.   Geologic time divisions

Era	Period	Epoch	Myr BP (approx.)	Biological Events
Cenozoic	Quaternary	Recent	0.01	
		Pleistocene	1.8	
	Neogene / Tertiary	Pliocene	5	Most mammalian families in evidence
		Miocene	24	
	Paleogene	Oligocene	37	
		Eocene	54	Origin of most mammalian orders
		Paleocene	65	Mammalian radiation Dinosaurs extinct
Mesozoic	Cretaceous		144	Dinosaurs abundant
	Jurassic		213	
	Triassic		248	First mammals
Paleozoic	Permian		286	Mammal-like reptiles
	Carboniferous	Pennsylvanian	320	First reptiles
		Mississippian	360	
	Devonian		408	First amphibians
	Silurian		438	First jawed fishes
	Ordovician		505	First vertebrates and land plants
	Cambrian		590	Invertebrates

anomodonts included two lineages, the Dinocephalia and the Eotitanosuchia, neither of which left Triassic descendents. The largest and most successful group of anomodonts was the Dicynodontia. They enjoyed a worldwide distribution (although the continents were not in their current positions) and were the dominant terrestrial herbivores for 60 million years from the mid-Permian until the late Triassic period, when the last of the various lines of anomodonts became extinct.

The other suborder of therapsids was the Theriodontia. They were primarily carnivorous and much more diverse and successful than the herbivorous anomodonts. Several different theriodont lines are recognized in the fossil record (see figure 4.2). The gorgonopsians were the prevalent theriodonts throughout the late Permian period, but they did not survive into the Triassic period. The therocephalians were a much more advanced and diverse group. They paralleled the other advanced theriodont group, the **Cynodontia,** in some of their mammal-like characteristics, including a secondary palate and complex **cheekteeth** (postcanine teeth; the premolars and

molars). Therocephalians did not show changes in position of the jaw muscles, however, as did cynodonts. The therocephalians were extinct by the early Triassic period, and only the cynodonts possessed the specialized cranial and skeletal features that eventually led to the evolution of the mammals.

Cynodontia

Cynodonts existed for 70 million years, throughout the Triassic to the middle Jurassic period. During this time, the several recognized families of cynodonts developed many of the transitional anatomical features leading from mammal-like reptiles to the earliest mammals (Hotton et al. 1986). Later cynodonts included diverse herbivores (gomphodonts and tritylodonts), as well as carnivores (cynognathids and tritheledonts). Several cynodont characteristics approached mammalian grade, that is, a level of organization similar to mammals. These characteristics included changes in dentition to tricuspid (**cusps** are the projections, or "bumps," on the

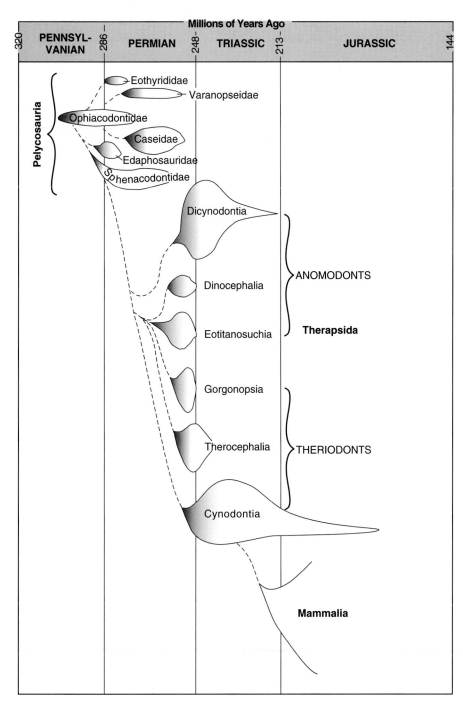

Millions of Years Ago

| PENNSYL-VANIA | PERMIAN | TRIASSIC | JURASSIC |

320 — 286 — 248 — 213 — 144

Pelycosauria

Eothyrididae
Varanopseidae
Ophiacodontidae
Caseidae
Edaphosauridae
Sphenacodontidae

Dicynodontia

Dinocephalia — ANOMODONTS

Eotitanosuchia — **Therapsida**

Gorgonopsia

Therocephalia — THERIODONTS

Cynodontia

Mammalia

Figure 4.2 **The major groups of mammal-like reptiles.** The Pelycosauria were early synapsid reptiles. The more advanced Therapsida arose from the carnivorous group of pelycosaurs called sphenacodonts. Therapsids included lineages of both herbivores (the anomodonts) and carnivores (the theriodonts). It was from a branch of the theriodonts, specifically the cynodonts, that mammals arose over 200 mya. Widths of each lineage suggest relative abundance.

toral and pelvic girdles, and thoracic ribs); and their phalangeal formula (Crompton and Jenkins 1979; Dawson and Krishtalka 1984). Several of these changes are discussed in more detail later. Changes in cynodonts and early Mesozoic mammals also included more erect posture and efficient movement, as well as increased adaptability in feeding. For example, development of the masseter muscles was associated with changes in the jaw, especially enlargement of the dentary bone and reduction in the number and size of postdentary bones (see figure 4.4), and development of the zygomatic arch, temporal openings of the skull, and the eventual development of the lateral wall of the braincase. These developments reduced stress on the jaw joint, increased the force of the bite, and protected the brain.

The evolution of reptiles to mammals was a continuum, with a *mammal* necessarily (if somewhat arbitrarily) defined by the articulation of the squamosal and dentary bones. As noted previously, mammal-like reptiles had several bones in the lower jaw in addition to the dentary (see figure 4.4; figure 4.5A). The joint between the lower jaw and the cranium was formed by the quadrate and articular bones. In transitional forms, as cynodonts became more mammal-like, postdentary bones continued to decrease in size. The reptilian quadrate-articular joint remained, and an additional joint formed between the cranial squamosal and the surangular bones of the jaw. This occurred because of the progressive enlargement of the dentary bone and concurrent reduction in size of the postdentary bones. Certain lineages of cynodonts had a double-hinged jaw joint; the dentary bone articulated with the squamosal laterally, and the quadrate and articular bones also formed a medial jaw articulation (Crompton 1972).

This later joint served not only as a hinge but also to transmit sound to the tympanic membrane. This membrane was supported by the reflected lamina of the angular bone (figure 4.6). Only one relatively large middle ear bone, the **stapes,** conducted sound from the tympanic membrane to the inner ear in mammal-like reptiles. Eventually, sound

occlusal [chewing] surface of a tooth) and double-rooted cheekteeth; jaw structure and masseter muscles (increased dentary size with reduction in postdentary bones and development of a glenoid fossa on the squamosal bone); hearing; postcranial skeleton (differentiated vertebrae, including modification of the first two vertebrae—the atlas/axis complex, modified pec-

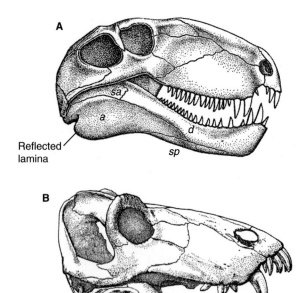

Reflected lamina

Figure 4.3   **Contrast between pelycosaurs and therapsids.**
(A) Skull of *Dimetrodon,* a common carnivorous sphenacodont
Pelycosaur. Relatively primitive features include the small temporal opening and large angular bone (*a*). Other postdentary (*d*) bones include the
surangular (*sa*) and the splenial (*sp*). These bones are larger than those
in (B) *Titanophoneus,* a more advanced therapsid, with a large temporal
opening and smaller postdentary bones. *Note:* Not to the same scale.

transmission became the only function of the quadrate and articular bones as the dentary became the only bone in the lower jaw. The articulation between the dentary and squamosal bones is the characteristic used to define a mammal. The position and reduction in size of the quadrate-articular joint was associated with the transformation of these bones into **ossicles** in the mammalian middle ear (see figure 4.5B). The long attachment arm of the **malleus** to the tympanic membrane in modern mammals is called the **manubrium** and is derived from the former retroarticular process of the articular bone. In conjunction with the large tympanic membrane and the much smaller fenestra ovalis, this lever system (enhanced by the long "lever" arm of the manubrium) not only transmits sound waves but also amplifies them. Low-pressure sound waves carried by air are increased to the higher pressure necessary for conduction through the fluid of the inner ear—the cochlear endolymph. This results in increased auditory acuity. As with many other changes from reptilian to mammalian organization, these two features are interrelated. That is, the reduction in size of the postdentary jaw bones increased auditory acuity, especially of high-frequency sound, as well as the efficiency of chewing.

By the time the cynodonts became extinct in the mid-Jurassic period, several well-defined groups of mammals already existed. Thus, from mammal-like reptiles that existed 320 mya to emergence of the earliest identifiable mammals about 100 million years later, evidence in the fossil record indicates several anatomical trends in organization that resulted in a mammalian grade (see the next section). Associated changes must also have occurred in the soft

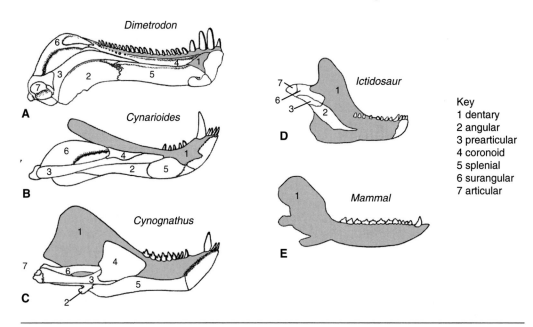

Key
1 dentary
2 angular
3 prearticular
4 coronoid
5 splenial
6 surangular
7 articular

Figure 4.4   **Enlargement of the dentary bone.**   The progressive enlargement of the dentary bone (shaded)
and reduction in postdentary bones is evident when comparing jaws of primitive mammal-like reptiles:
(A) *Dimetrodon,* an early Permian pelycosaur; (B) Cynarioides, a late Permian therapsid; (C) *Cynognathus,* an early
Triassic cynodont; and (D) *Ictidosaur,* a late Triassic–early Jurassic cynodont. The dentary is the sole bone in the jaw
of mammals (E).

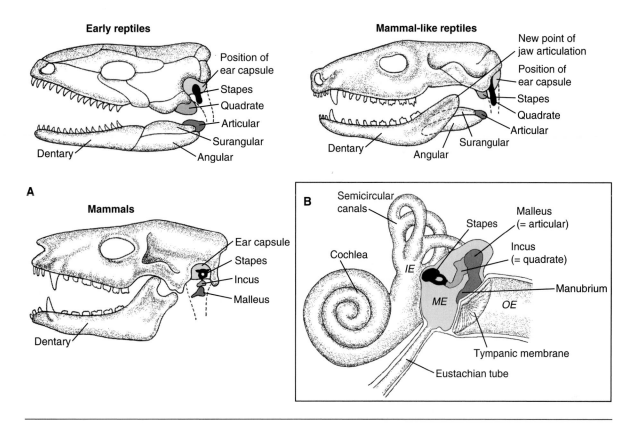

Figure 4.5   Transition of the jaw and history of the ear ossicles.   (A) Simplified transition of the jaw structure from reptiles through mammal-like reptiles to mammals, showing the increase in size of the dentary bone and decrease in postdentary bones. The quadrate and articular bones of mammal-like reptiles eventually changed from their dual role of jaw joint and sound transmission to solely sound transmission in mammals. (B) Outer ear (*OE*), middle ear (*ME*), and inner ear (*IE*) of modern mammals. The tympanic membrane is now supported by the tympanic bone, derived from the former reflected lamina of the angular bone (see figure 4.6). The articular bone has become the first of the three small bones (ossicles) in the middle ear, specifically, the malleus. The second ossicle, the **incus,** is derived from the quadrate bone. The mammalian stapes, much reduced in size from the reptilian stapes, connects the incus to the inner ear through the fenestra ovalis (the "oval window"). Thus, in mammals, the stapes is not connected directly to the tympanic membrane as in reptiles, but instead is connected through a lever system of two small bones—the malleus (former articular bone) and the incus (former quadrate bone)—the familiar "hammer, anvil, and stirrup."

anatomy, physiology, metabolism, and related features of mammal-like reptiles that are not evident from fossil remains. These features were interrelated in terms of increased efficiency of metabolism needed for endothermy, better food gathering and processing methods, increased auditory acuity, and other adaptations to maintain internal homeostasis and enhance survival.

Monophyletic Versus Polyphyletic Origins of Mammals

During the early Jurassic period, one of two things happened. Either a single phyletic line of therapsid reptiles gave rise to early mammals, or two or more therapsid lines independently achieved the mammalian grade of organization (figure 4.7). In the first case, early mammals split into two divergent lines and gave rise to prototherians (which today include the monotremes—see chapter 10), and the other line, after many adaptive radiations, gave rise to therians (the marsupials and

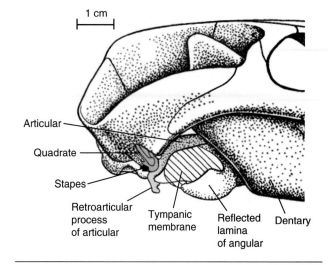

Figure 4.6   Tympanic membrane in a mammal-like reptile.
Posterior part of the cranium in an advanced mammal-like reptile, *Thrinaxodon liorhinus*. Note the jaw joint, the location of the tympanic membrane, and its relationship to the postdentary bones. The quadrate and articular bones form the jaw joint as well as transmit sound from the tympanic membrane to the stapes.

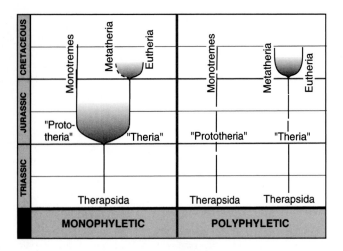

Figure 4.7   Diagram of alternative hypotheses of the origin of mammals.   In the monophyletic scheme, the major groups of mammals alive today diverged from a common ancestral line of reptiles. Conversely, the three major groups may have converged from at least two distinct, separate lines of reptiles. Most authorities suggest mammals had a monophyletic origin.

eutherians, or "placental," mammals). This scenario describes a monophyletic origin for mammals. In the second case, the characteristics evident in these three major mammalian groups (infraclasses) are strictly convergent, and mammals have a polyphyletic origin. The question persists because of different interpretations of a fragmentary and incomplete fossil record. This necessitates "grade-level" designation, that is, grouping by similar levels of organization. If the reptile-to-mammal decision is based on independently evolved characteristics, such as the three bones in the middle ear that were derived from the jaw joint, then by definition mammals are polyphyletic. As noted, however, most authorities consider mammals to be defined by the single dentary bone with squamosal-dentary articulation. This character is shared by common ancestry, meaning mammals are monophyletic in origin. Regardless of the exact reptilian lineage of mammals, what can we say about the earliest Mesozoic mammals that appeared about 200 mya?

THE FIRST MAMMALS

Early Mesozoic Mammals

As we have seen, therapsid reptiles achieved many mammalian features; however, only mammals had the primary jaw joint composed of the dentary and squamosal bones. Also, they had **diphyodont** dentition (only two sets of teeth during an individual's lifetime) with complex occlusion. The mammalian fauna in the Mesozoic era and for the first 100 million years of the existence of mammals, however, was strikingly different from what we see today. Until the early Cenozoic era, mammals were a relatively rare and insignificant part of the fauna compared with the larger, widespread, specialized,

and well-adapted lineages of reptiles. Mammals were small (mouse-sized), relatively uncommon, and probably nocturnal. Because the majority of the terrestrial reptiles were diurnal, mammals could better avoid predation by being nocturnal. Throughout this era, mammals appeared to be a narrowly restricted reptilian offshoot, confined to a few phyletic lines. It is interesting that the fossil record shows little overlap in size between the smallest dinosaurs and the largest mammals for the 140 million years these two groups shared the terrestrial environment. In general, the smallest dinosaurs were many times larger than the largest mammals. Mammals did not attain large body sizes until after the extinction of the dinosaurs (Lillegraven 1979). Mammalian phylogeny may be depicted as in figure 4.8, with early radiation into two major groups: prototherians (sometimes referred to as atherians) and therians. The information available on relationships among groups is minimal and uncertain. As noted by Lillegraven (1979:5) ". . . the formalization of Prototheria and Theria cannot be adequately defended. . . . The terms are, however, deeply entrenched in the literature, remain generally useful in communication, and thus are retained. . . ." The classification of archaic mammalian taxa discussed in the next section is given in table 4.2.

Early Prototherians

Members of the Family Morganucodontidae are among the most primitive known mammals, with the Genus *Morganucodon* abundant in the fossil history. Morganucodontids may have been ancestral to later major groups, including the unknown ancestors of monotremes (see figure 4.8). Relationships among lines of Mesozoic mammals are unclear, however, and affinities of morganucodontids to later groups remain uncertain. They may be considered "prototherian" only in that they did not form the ancestry of the marsupials and placental mammals. Although monotremes today are numerically an insignificant part of the modern mammalian fauna, prototherian mammals during the Mesozoic era were numerous and diverse.

Several distinct mammalian structural features are present in morganucodontids and evolved among several different lineages of reptiles (Crompton and Jenkins 1979; Carroll 1988). These features included dentary–squamosal articulation, although involvement of the quadrate and articular bones remained. Besides incisors and canines, the cheekteeth were differentiated into premolars and molars. The occlusal surfaces (the portion of the crowns that contact each other when an animal chews) of the upper and lower molars were clearly mammalian. The cochlear region was large relative to skull size. The first two vertebrae were similar to those seen in later mammals, and two occipital condyles were present. Thoracic and lumbar vertebrae and the pelvic region were distinct from the reptilian pattern. Morganucodontids had a mammalian posture with the legs beneath the body, not splayed out as in reptiles. Also, the vertebrae allowed flexion and extension of the spine during locomotion. Thus, many

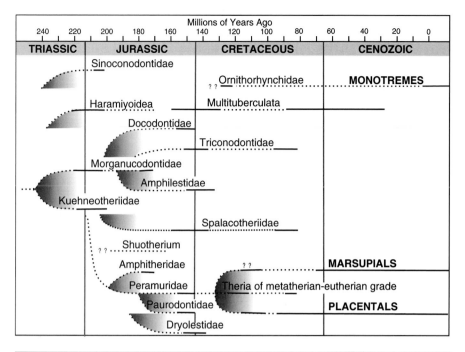

Figure 4.8   **Early evolution and relationships of mammals, assuming a monophyletic origin.**   Dashed lines represent uncertainty of lineages due to a fragmented and incomplete fossil record. The morganucodontids were the most primitive of the lineages that eventually produced the Prototheria, but many of these lineages no doubt remain undiscovered, and the transitions among the major phyletic groups shown are not clear. Similarly, the Kuehneotheriidae are the most primitive "therian" mammals, but again, many gaps exist in the fossil history until the transition to a true metatherian-eutherian grade.

jaw articulation in this group. This demonstrates an important point. The suite of characteristics seen in mammals evolved at different rates within and among a number of lineages, and often in association with retained primitive reptilian characters.

A third order, the **Multituberculata,** was another successful mammalian line that extended about 120 million years from the late Jurassic period to the late Eocene epoch, concurrent with the emergence and radiation of flowering plants. They were primarily herbivorous and had a single pair of large, procumbent lower incisors, as do modern rodents. They also had as many as three pairs of upper incisors, however. The order is named for the molariform teeth, which had up to eight large, conical cusps (figure 4.9C). These cusps were arranged in triangles in anterior molars but in longitudinal rows in the posterior teeth. The posterior lower premolar often was very large and was used for shearing (figure 4.9D). Multituberculates probably were replaced by true rodents, primates, and other eutherian herbivores.

interrelated features of the skull and postcranial skeleton that define mammals were evident in morganucodontids (as well as in late cynodonts). These features continued to be refined in later groups of late Jurassic and early Cretaceous lineages of mammals. The most prominent of these lines were the triconodonts, amphilestids, docodonts, and multituberculates—groups defined on the basis of tooth structure and associated adaptive feeding types.

The **Triconodonta** were a successful lineage that extended from the late Triassic to the late Cretaceous period—120 million years. They included the early Morganucodontidae. The triconodonts were small, carnivorous mammals named for their molars, which had three cusps arranged in a row (figure 4.9A). The **amphilestids** occurred from the mid-Jurassic to the early Cretaceous periods. They had a linear row of cusps much like morganucodontids. The amphilestid Genus *Gobiconodon*, from the early Cretaceous period, is noteworthy because it had deciduous molars (Carroll 1988). The **Docodonta** are known only from the late Jurassic period and may have arisen from the triconodonts. Based only on teeth and jaws from which they have been described, they appear to have been omnivores. The lower molars were rectangular with prominent cusps. These are the most complex teeth seen in fossils from the Jurassic period, and achieve the same complexity as later Cretaceous therians (figure 4.9B). The advanced teeth of docodonts were in contrast to the retention of reptilian

Early Therians

Remains of the earliest therians (pantotheres) occur in rock strata that also contain the prototherian morganucodontids. Two orders are known only from teeth and jaw fragments: the **Symmetrodonta** and the **Eupantotheria.** The earliest

Table 4.2.   Classification of archaic (Mesozoic) mammalian taxa discussed in the text.*

Class Mammalia
  Subclass Prototheria
    Order Triconodonta
      Morganucodontidae
      Amphilestidae
    Order Docodonta
      Docodontidae
  Subclass Allotheria
    Order Multituberculata
  Subclass Theria
    Order Symmetrodonta
      Kuehnoetheriidae
    Order Eupantotheria
      Dryolestidae
      Peramuridae

Source: Reprinted, with permission, from Carroll, R.L. 1988. Vertebrate paleontology and evolution. W.H. Freeman, New York.
*Also refer to figure 4.8 for other lineages not mentioned in the text.

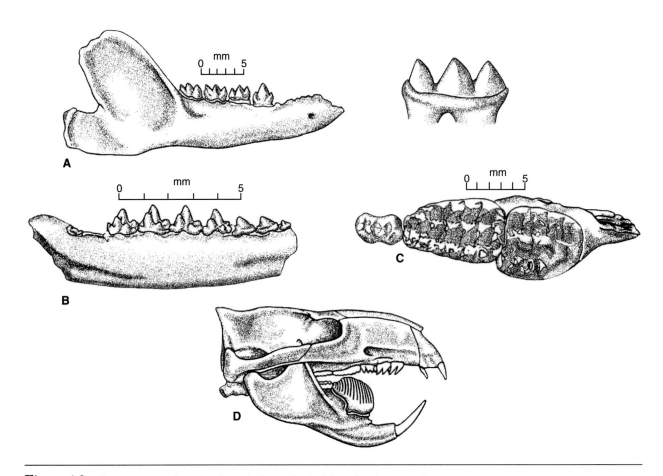

Figure 4.9   **Representative dentition from the "prototherian" line of early mammals.**   (A) Lateral view of the right jaw and medial view of a lower molar from *Triconodon,* a triconodont; (B) Lingual view of the left mandible from the docodont *Borealestes serendipitus;* (C) Occlusal view of the upper premolar and molars of the multituberculate *Meniscoessus robustus;* and (D) lateral view of the skull of *Ptilodus,* showing the enlarged, shearing posterior lower premolar. Note the completely formed braincase for muscle attachment and the large dentary with coronoid process. The actual length of the skull is approximately 8 cm.

known symmetrodonts, within the Family Kuehnoetheriidae, are the Genera *Kuehneotherium* and *Kuhneon*—very small carnivores or insectivores from the late Triassic peroid. Later Jurassic pantotheres radiated into numerous different lines and adaptive feeding niches during the Cretaceous perod. A significant feature of pantotheres was their molars, which had three principal cusps in triangular arrangement. This tribosphenic tooth pattern (the basic pattern for later mammals; see the next section) allowed for both shearing and grinding food. The most diverse family of eupantotheres, the **Dryolestidae,** may have been omnivorous and survived into the early Cretaceous period. Based on derived dental characteristics, advanced therians, that is, distinguishable metatherians and eutherians, probably originated within the eupantothere Family **Peramuridae** by the middle to late Cretaceous period, if not before (see figure 4.8). Peramurids are known only from the late Jurassic Genus *Peramus.*

Tribosphenic Molars

As noted previously, early mammals had tooth cusps arranged longitudinally (see figure 4.9A and C). Metatheri-

ans and eutherians, and their immediate ancestors in the Cretaceous period (referred to generally as "theria of metatherian-eutherian grade"), had more advanced **tribosphenic** (or tritubercular) molars (Butler 1992; Smith and Tchernov 1992). These are named for the three large cusps arranged in a triangular pattern. A tribosphenic upper molar (figure 4.10) consists of a **trigon** with three cusps, a **protocone** that is lingual (the apex of the triangle points inward toward the tongue), and an anterior, labial (outward toward the cheek) **paracone** and posterior **metacone**. A lower molar, or **trigonid** (an *-id* suffix always denotes mandibular dentition), consists of these three cusps and a "heel," or talonid basin. In lower molars, the protoconid is labial (not lingual, as in the upper molars), whereas the paraconid and metaconid are lingual. The talonid of the lower molars also has smaller accessory cusps. These often include a labial **hypoconid,** a posterior **hypoconulid,** and a lingual **entoconid** (see figure 4.10). Thus, the occlusal view of the trigon(id) of tribosphenic molars is a somewhat asymmetrical, three-cusped triangle (see figure 4.10A and C). The apex of the triangle points lingually (inward toward the tongue) in upper molars, and labially (outward toward the cheek) in

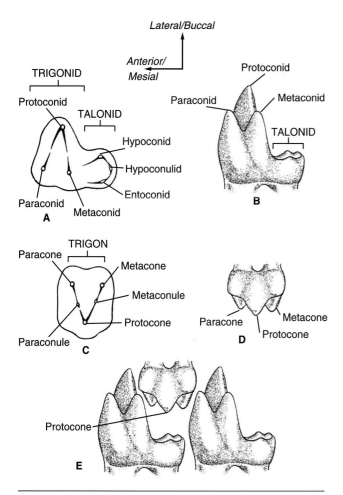

Figure 4.10 Nomenclature of the cusps of tribosphenic dentition.   A lower molar in (A) occlusal view and (B) lingual view, and an upper molar in (C) occlusal and (D) lingual views. The upper and lower dentition is shown in occlusion in (E). The addition of a fourth cusp, the hypocone, forms a quadrituberculate upper molar.

lower molars. During occlusion, a crushing or grinding action occurs as the protocone of an upper molar contacts the talonid basin of the opposite lower molar. Food is not only crushed but also sheared. Shearing results from several facets of the upper and lower molars coming together (see Bown and Kraus 1979), for example, the anterior face of a paracone and the posterior face of a protoconid and metaconid.

The basic pattern of the tribosphenic molar in early mammals was very important because it is believed to be ancestral to modern therian mammals. It is seen today in lineages such as marsupials and insectivores and has been modified in other modern mammals. For example, molars have become square (**euthemorphic**) with the addition of another main cusp (the **hypocone**) posterior to the protocone. Such four-cusped (**quadritubercular**) molars occur in many species of modern mammals, including humans. Cusps are often connected by a series of crests or ridges, as in many insectivores (see figure 11.1). Hershkovitz (1971) provided an exhaustive treatment of cusp patterns, homologies, and associated terminology.

CENOZOIC MAMMALS AND MAMMALIAN RADIATION

The different mammalian lineages seen today began with diversification of mammals during the early Cenozoic era. This radiation resulted from two major events that occurred worldwide. The first was the extinction of the dominant terrestrial vertebrate fauna, the dinosaurs, at the end of the Cretaceous period. There are several hypotheses as to why dinosaurs died out so quickly. Nobody knows for sure, although evidence exists for the hypothesis that extinction of the dinosaurs was caused by a large asteroid that struck the earth, resulting in major climate and vegetation changes. Whatever the reason, disappearance of the dominant Mesozoic reptiles opened new adaptive opportunities and resulted in a worldwide mammalian radiation. Rapid expansion and divergence of the mammals was also facilitated by the breakup of the large continental land mass (**Pangaea**) that had been in place during much of the time of the dinosaurs (Fooden 1972). Continental drift throughout the early Cenozoic era (figure 4.11) allowed major genetic differentiation of the various phyletic lines to proceed in relative isolation. These two factors, in addition to ever-expanding faunal and floral diversity worldwide, allowed mammals to occupy increasingly specialized ecological roles. As a result, for most modern mammals, ordinal differentiation was underway by the early Cenozoic era, and for many groups, probably since the late Mesozoic era. Most extant orders are recognized in the fossil record by the beginning of the Eocene epoch, and most families date from before the Miocene. Mammals have been the dominant terrestrial vertebrates ever since— for the last 65 million years.

Interrelationship of Characteristics and Increased Metabolism

The changes in skeletal features noted in the following section and in table 4.3 occurred in association with metabolism, physiology, and reproduction—all of which were related to and developed concurrently with maintenance of endothermy. Features of the soft anatomy related to endothermy are not visible in the fossil record because these organ systems do not fossilize; however, many of these features can be inferred. Evolutionary changes from reptiles to mammals can be related to increased metabolic demands of mammals. Mammals need approximately 10 times the amount of food and oxygen as do reptiles of similar size to maintain their high body temperature. Endothermy demands an efficient supply of oxygen to the lungs for aerobic metabolism, a widespread and constant food supply, and the ability to obtain and process that food quickly and efficiently. Thus, from reptilian to mammalian organization, most of the trends summarized in the following section relate directly to efficient homeostasis. All these trends are no doubt interrelated

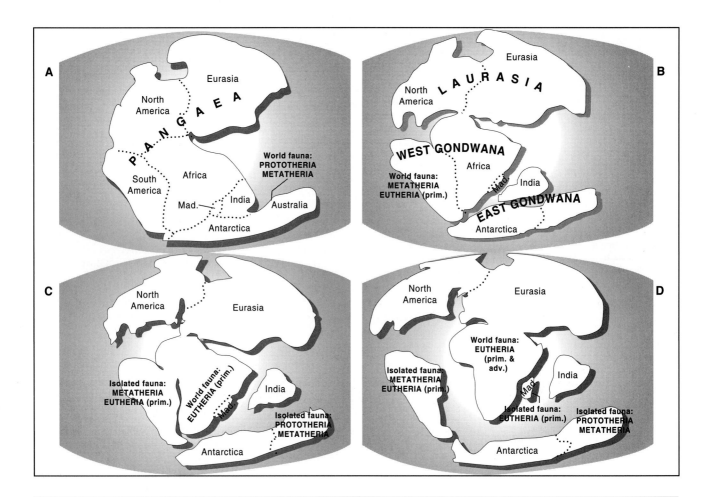

Figure 4.11   Early continental land masses.   The breakup of the single large land mass (Pangaea) beginning about 200 mya and eventual isolation of the continents that promoted differentiation of the various mammalian phyletic lines following the Cretaceous period. (A) 2.00 • 10^8 years ago; (B) 1.8 • 10^8 years ago; (C) 1.35 • 10^8 years ago; (D) 6.5 • 10^7 years ago.

Sources: Data from J. Fooden, "Breakup of Pangea & Isolation of Relict Mammals in Australia, South America, and Madagascar" in Science, *175:894–898, 1972; and U.S. Petroleum Institute.*

in a much more complicated and sophisticated manner than can be appreciated from a simple reconstruction from fossil history. The adaptive significance and interrelationship in postdentary bones and increase in the size of the dentary bone offer an excellent example of this. These interrelated changes in anatomy not only increased efficiency of chewing and digestion but also directly enhanced auditory acuity through greater efficiency of vibrations from the tympanic membrane. Enhanced hearing can help an individual avoid predators or capture prey more efficiently.

SUMMARY OF ANATOMICAL TRENDS IN ORGANIZATION FROM MAMMAL-LIKE REPTILES TO MAMMALS

Several morphological trends were evident in the evolution of mammals from their mammal-like reptilian ancestors (figure 4.12). Many of these trends have been noted, as have

some of the interrelationships among them. Also, remember that different characters appeared at different times and in different phyletic lines. The process of change from reptile to mammal certainly did not proceed in orderly, progressive, or easily defined steps. The following trends are evident, however, in the evolution of mammals from reptiles:

1. The temporal opening of the skull of therapsids is enlarged (see figure 4.3). This was associated with eventual movement of the origin of the jaw muscles from the inner surface of the temporal region in mammal-like reptiles to the outer surface of the braincase and newly developed zygomatic arch in mammals (figure 4.13).

2. These changes parallel development of a larger, heavier dentary bone for processing the food necessary for higher metabolic activity and maintenance of homeostasis. Thus, the dentary bone became progressively larger as the postdentary bones were reduced in size (see figure 4.4). As noted earlier, the articular and quadrate bones diminished in size and became part of the middle ear. Remember, however, that several other

Table 4.3 Different characteristics of reptiles and mammals

Reptiles	Mammals
More than one bone in mandible; with quadrate-articular articulation of jaw joint	Single bone in mandible; with squamosal-dentary articulation
One occipital condyle	Two occipital condyles
Long bones without epiphyses (indeterminant growth)	Long bones with epiphyses (determinant growth)
Unfused pelvic bones	Fused pelvic bones
Secondary palate usually absent	Secondary palate present
Middle ear with one ossicle (stapes-columella)	Middle ear with three ossicles (malleus, incus, and stapes)
Phalangeal formula 2-3-4-5-3 (4)	Phalangeal formula usually 2-3-3-3-3
Dentition homodont and polyphyodont	Dentition often heterodont and diphyodont
Epidermis with scales	Epidermis with hair
Oviparous or ovoviviparous	Viviparous (except for the monotremes)
Three-chambered heart in most	Four-chambered heart with left aortic arch
Ectothermic with low metabolic rate	Endothermic with high metabolic rate
Nonmuscular diaphram	Muscular diaphragm
No mammary glands	Mammary glands present
Relatively small, simple brain	Relatively large, complex brain

bones in the lower jaw were retained for a long time, even after the emergence of the dentary-squamosal articulation.

3. The maxillary and palatine bones extended posteriorly and medially, forming a secondary palate (figure 4.14). This resulted in more efficient airflow, allowing a constant supply of oxygen to the lungs while permitting chewing and thus enhancing metabolism. It also may have affected suckling in neonates (Maier et al. 1996).

4. Dentition changed from **homodont** (uniform, peglike tooth structure with little occlusion) to strongly **heterodont** (teeth differentiated on the basis of form and function) in association with obtaining and processing foods more efficiently. Chewing efficiency also was enhanced by the development of tribosphenic molars.

5. A change occurred from one occipital condyle in reptiles to two in advanced mammal-like reptiles and mammals (see figure 4.14). This reduced tension on the spinal cord when the head is moved up and down and allowed finer control of head movements, but decreased lateral movement.

6. Limbs rotated 90 degrees from the "splayed" reptilian stance (i.e., horizontal from the body and parallel to the ground) to directly beneath the body (perpendicular to the ground) (figure 4.15). Additional changes resulted in the pectoral and pelvic girdles, including loss of the coracoid, precoracoid, and interclavicle bones in the

pectoral girdle, although monotremes still retain them. In the pelvic girdle, the separate bones found in reptiles fused in mammals and moved to a more anteriodorsal orientation. Mammals can therefore move with less energy expenditure than reptiles.

7. Cervical and lumbar ribs were lost completely, and the number and size of thoracic ribs were reduced. In association with changes in the vertebrae and scapula (figure 4.16), as well as others, this again allowed for more flexibility in movement, especially dorsoventral flexion of the spine.

8. The number of carpal and tarsal bones was reduced, and the phalangeal formula was reduced from the reptilian 2-3-4-5-3 (forefeet) and 2-3-4-5-4 (hind feet) to the typical 2-3-3-3-3 found in most mammals (Hopson 1995).

Many of these skeletal transformations, as well as differences in the soft anatomy, are evident between reptiles and mammals today (see table 4.3). The generally accepted feature used in the recognition of early mammals is a jaw joint with squamosal-dentary articulation. This is first seen in the Mesozoic era about 220 mya and was the result of a 100-million-year process of change. Any single criterion separating mammal-like reptiles and early mammals becomes more arbitrary, however, as fossil history becomes more complete.

CHARACTERISTICS OF MODERN MAMMALS

We noted several characteristic skeletal trends associated with the evolution of mammals. The single dentary bone and three ossicles of the middle ear are unique to mammals. Two occipital condyles, epiphyses on many long bones (which result in determinant growth, unlike reptiles), and a tympanic bone are other mammalian skeletal characteristics. Among the vertebrates, several aspects of the soft anatomy of mammals are also unique.

Probably the most obvious mammalian feature is hair, or fur. These terms are synonymous—structurally, no difference exists between hair and fur. There are several types of hair, and one or more types make up the **pelage** (coat) of mammals. Mammals have a four-chambered heart, with a functional left aortic arch. Birds also have a four-chambered heart but with a functional right aortic arch. The biconcave **erythrocytes** (red blood cells) of mammals are **enucleate** (without a nucleus). Not having a nucleus enhances the oxygen-carrying capacity of these cells. Female mammals have milk-producing **mammary glands (mammae).** This character is, of course, the basis for the name of the Class Mammalia. Finally, mammals have a muscular **diaphragm** separating the thoracic and abdominal cavities. Other aspects of the soft anatomy can be used to characterize mammals. These either are not unique to mammals or are not found in all mammals. For example, the **corpus callosum,**

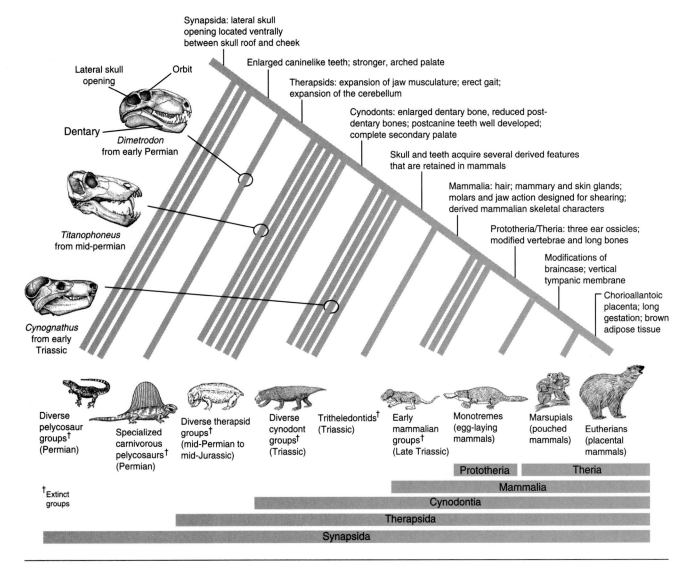

Figure 4.12   Mammalian evolutionary trends.   General summary of mammalian evolutionary trends from ancestral synapsid reptiles to modern mammals.

a bundle of nerve fibers that integrates the two cerebral hemispheres of the brain in eutherians, does not occur in monotremes and marsupials. Likewise, a true **placenta** occurs only in eutherians (except for marsupial bandicoots—see chapter 10). Aspects of mammalian dentition are discussed in the following sections. Other general mammalian characteristics, including **locomotion** (movement), feeding, hair, and reproduction, are examined in greater detail in part II.

DENTITION

Teeth are one of the most important aspects of living mammals. Also, many fossil lineages are described only on the basis of their teeth. Although all mammals begin life on a diet of milk, they eventually enter into one of a variety of adaptive feeding modes. An individual's teeth reflect its trophic level

and feeding specialization. A number of different feeding niches are available, and as a result, mammalian dentition shows a number of different modifications. These modifications are derived in large part from the basic tribosphenic pattern, which allowed much more efficient processing of food necessary for endothermy and is retained in more primitive groups, such as insectivores, tree shrews, elephant shrews (chapter 11), and some marsupials (chapter 10). Besides their role in feeding, teeth also may function secondarily in burrowing, grooming of fur, and defense. Whereas mammals show little skeletal variation, except in their limbs, a great deal of variation occurs in their dental patterns.

Teeth may occur in three bones in mammals: the premaxilla and maxilla of the cranium and the mandible (dentary bones). Most species have teeth in all three of these bones; others have a much reduced dentition in only one or two of these bones. Still other species are **edentate,** that is, they have no permanent teeth at all.

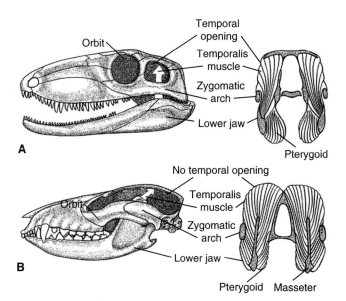

Figure 4.13   **Muscle attachment and the temporal opening.**
(A) Lateral view and cross section of the skulls of a mammal-like reptile
and (B) a mammal showing movement of the origin of the muscle
attachment to the lower jaw from inside the cranium to outside the
cranium. Muscle attachment was around the edge of the temporal
openings in mammal-like reptiles. Muscle attachment moved to the
outside of the cranium with complete ossification of the braincase
and formation of the zygomatic arch in mammals.

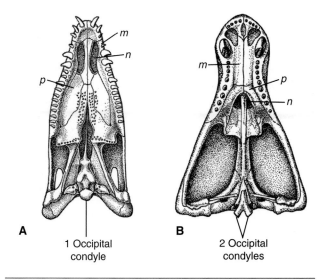

Figure 4.14   **Formation of the secondary palate.**
Ventral views of the cranium of *Dimetrodon*, (A) a pelycosaur and
(B) *Cynognathus*, a more advanced cynodont. Note that the internal
nares (*n*) open immediately to the front of the mouth in the primitive
form, but in the cynodont, the air enters the back of the mouth because
of the medial extension of the maxillary (*m*) and palatine (*p*) bones
that form a secondary palate. Also notice the one occipital condyle
in *Dimetrodon* and the development of two occipital condyles
in *Cynognathus*.

Tooth Structure

The portion of the tooth above the gum line is the **crown,**
and the **roots** are below the gum line (figure 4.17). In most
species, **enamel** overlays **dentine** in the crown of the tooth.
Enamel is harder, heavier, and more resistant to friction than
any other vertebrate tissue. It is acellular, cannot regenerate,
and is made up of crystalized calcium phosphate (hydroxya-
patite). Enamel is ectodermal in origin, whereas dentine is of
dermal origin and makes up most of the tooth. In some
species, such as the aardvark and xenarthrans (chapter 14),
the teeth have no enamel. Rodent incisors have enamel only
on the anterior surface, causing differential wear and continuous sharpening of the incisors for gnawing. Within the
dentine is the **pulp cavity,** in which blood vessels and nerves
maintain the dentine. In **open-rooted** teeth, growth is continual, and such teeth are termed "ever-growing." Incisors of
rodents are a prime example of ever-growing, open-rooted
teeth (see figure 17.1). Alternatively, teeth stop growing and
begin to wear down with age when the opening to the pulp
cavity closes. Wear on teeth that are **closed-rooted** may be
used to estimate an individual's age. Just as the crown of a
tooth usually is covered with enamel, the root is covered with
cementum. This is a modified bony material deposited
throughout an individual's life. Cementum annuli often are
deposited much like the rings of a tree, and as with tree
rings, the annuli can sometimes be used to determine an individual's age. A socket in a bone containing the tooth roots
is an **alveolus.** Connective tissue between the cementum
and the alveolar bone holds the tooth in place.

One of the anatomical trends noted previously was the
change from homodont dentition of reptiles to heterodont
dentition in mammals. Certain modern mammalian lines
have homodont dentition, for example, the toothed whales
(odontocetes) and armadillos (Xenarthra: Dasypodidae).
Still other groups, such as the platypus and spiny echidnas
(monotremes), are edentate. So are several unrelated mammalian lineages that feed on ants and termites, for example,
true anteaters (Xenarthra: Myrmecophagidae) and pangolins, or scaly anteaters (Pholidota: Manidae). A special

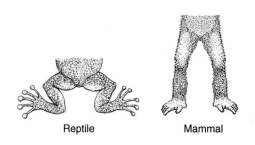

Figure 4.15   **Conformation of limbs.**   General body form in a
modern reptile, with the limbs horizontal to the main axis of the body
and parallel to the ground, compared with a mammal, with limbs vertical
to the main axis of the body and perpendicular to the ground.

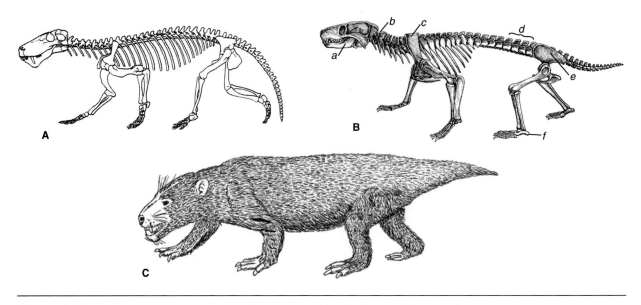

Figure 4.16   **Evolutionary trends in therapsid reptiles toward the development of mammals.**   (A) *Lycaenops,* an early theriodont from the late Permian period and (B) an early Triassic cynodont, *Thrinaxodon,* showing formation of distinct mammalian characteristics, including enlargement of the dentary bone, with the coronoid process extending above the zygomatic arch (*a*); the second cervical vertebra (axis) with a spine (*b*); enlargement of the scapula (*c*); formation of distinct lumbar vertebrae and associated reduction in the number of ribs (*d*); enlargement of the pelvic bones (*e*); and formation of a heel bone (*tuber calcanei*) and distinct plantigrade feet (*f*). (C) Lifelike reconstruction (hair is hypothetical) of *Thrinaxodon,* which was about the size of a weasel.

case of edentate mammals is the mysticete whales, in which teeth have been replaced with **baleen** in the upper jaw (see chapter 16). The edentate condition is secondarily derived; that is, teeth develop and sometimes emerge in embryos but are resorbed or lost prior to parturition.

Most mammals have heterodont dentition, with well-defined incisors, canines, premolars, and molars. **Incisors** are the anteriormost teeth, with the upper incisors rooted in the premaxilla. All lower teeth are rooted in the dentary bones.

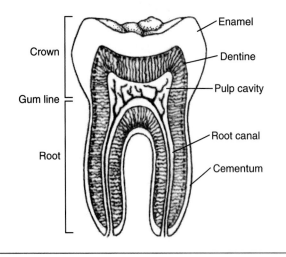

Figure 4.17   **Longitudinal section of a mammalian molar.**   The tooth is seated in an alveolus (socket). With increased age, the enamel wears away, and progressively more dentine is exposed in most species. Also, an additional layer of cementum often is deposited each year.

The incisors often function to cut or gnaw, as in rodents and lagomorphs. Incisors usually are structurally simple with a single root. Sometimes, though, they are highly modified and serve a variety of purposes. In shrews (Insectivora: Soricidae), the first pair of incisors are long and curved. They appear to function as forceps in seizing insect prey. Vampire bats have bladelike upper incisors for making incisions. Incisors are sometimes modified as tusks, as in elephants and male narwhals (*Monodon monoceros*). Some groups, such as deer, have lost the upper incisors but have retained the lower ones. They clip vegetation by cutting against a tough, pad-like tissue in place of the upper incisors.

Canines are posterior to the incisors. There is never more than one pair of upper and lower canines in modern mammals. These teeth generally are **unicuspid** (i.e., they have one cusp) with a single root. In carnivores and some other groups, canines are often enlarged and elongated for piercing and tearing prey. They may even form tusks in certain species, such as the walrus (*Odobenus rosmarus*) and pigs (Artiodactyla: Suidae). In species of deer without antlers, musk deer (Genus *Moschus*) and the Chinese water deer (*Hydropotes inermis*), males have elongated, tusklike upper canines.

Premolars are posterior to the canines and anterior to the molars. Generally, teeth in the posterior part of a dental arcade are structurally more complex than anterior teeth. Premolars are generally smaller than molars and have two roots, whereas molars usually have three. Premolars may be unicuspid, or they may look the same as molars, but premolars have deciduous counterparts ("milk teeth"). **Molars** have multicusps and no deciduous counterparts; that is, they are not replaced. The premolars and molars are often considered together as "cheekteeth," "postcanine," or "molariform" teeth, especially in species in which they are difficult to differentiate. Molariform dentition is used for grinding food. As such, these teeth usually have the greatest degree of specialization in cusp patterns and ridges associated with a particular feeding niche. The height of the crown varies among species. Teeth with low crowns are termed **brachyodont** and often are found among omnivores. Herbivores consume forage that is often highly abrasive and contains large amounts of silica. This wears teeth down more rapidly than does a carnivorous diet, and it is therefore adaptive for an herbivore to have high-crowned or **hypsodont** cheekteeth.

In addition to crown height, occlusal surfaces are quite variable. Specifically, the cusp patterns are often highly modified. Brachyodont cheekteeth are often **bunodont,** with rounded cusps for crushing and grinding, as in most monkeys and pigs. Alternatively, the cusps may form continuous ridges, or **lophs,** such as occur in elephants, a pattern termed **lophodont**. Sometimes the lophs are isolated and crescent-shaped, as in deer, in which case they are called **selenodont** (figure 4.18A–C). Loph patterns may become so complex that it is difficult to discern the original cusp pattern. A great deal of diversity in cusp patterns is evident among individual families or within an order such as the rodents (e.g., see figure 17.2). Another example of specialization is found in many modern carnivores, which have **carnassial** or **sectorial** teeth for shearing. Carnassial teeth in modern carnivores are always the last upper premolar and the first lower molar. These teeth are particularly well developed in the cat (Felidae) and the dog families (Canidae). They still occur but are less evident in more omnivorous groups, including many of the mustelids (see figure 15.4). When carnassials are used on one side of the mouth, they are not aligned on the other side and cannot be used. Likewise, in most species, when the anterior dentition (incisors or canines) is used, cheekteeth do not occlude, and when the animal chews with the cheekteeth, incisors do not come together.

Many species have lost teeth through evolutionary time so that a gap, or **diastema,** occurs in the tooth row. All rodents and lagomorphs have lost their canines and have a diastema between their incisors and the anteriormost cheekteeth (see figure 17.1). Deer (Family Cervidae) also have a diastema between the lower incisiform teeth and the cheekteeth. Deer also can be used to illustrate that teeth in a given position may resemble the teeth next to them. What appears to be the last lower "incisor" in a deer jaw is actually a canine. Because it functions as an incisor, however, its form has changed through time to accomodate its function. It has become "incisiform," that is, indistinguishable from the three true incisors on each side of the midline (see figure 4.18D).

The structure of the lower jaw and primary use of different muscle groups differ between herbivores and carnivores. In herbivores, the mandibular condyle and its articulation with the fossa of the cranium is elevated above the mandibular dentition. This gives maximum advantage to

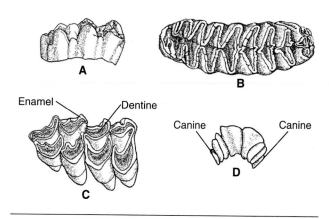

Figure 4.18 Occlusal surfaces. Teeth showing general types of occlusal surfaces: (*A*) a pig, with a bunodont surface; (*B*) the lophodont dentition of an African elephant, with cusps in the form of transverse ridges; and (*C*) a deer, with selenodont teeth forming crescent-shaped ridges. The enamel appears lighter than the dentine. (*D*) Dorsal view of the lower incisors of a white-tailed deer, showing the lateral "incisiform" canines. *Note:* The teeth are not to the same scale.

the masseter muscles in closing the jaw. In carnivores, the temporal muscles are the primary muscle group closing the jaw, and the mandibular articulation is level with the dentition (figure 4.19). Also, carnivore jaws close in a shearing manner like scissors. Conversely, when the jaw of an herbivore closes, all the opposing teeth occlude together.

Tooth Replacement

Generally, mammals have two sets of teeth during their lifetime; that is, they are diphyodont. The deciduous, or "milk," teeth, are replaced by permanent dentition later in life. In many species, the pattern of tooth replacement is used to estimate the age of individuals. In eutherian mammals, all the teeth except the molars have deciduous counterparts. It is unclear, however, whether molars are permanent teeth without preceding milk teeth or "late" milk teeth without succeeding permanent teeth. In metatherians, only the last premolar is deciduous; the remaining postcanine teeth are not replaced. Deciduous incisors and canines do not erupt, and only the permanent anterior dentition is seen (Luckett and Woolley 1996). Certain eutherians, such as shrews, appear to have only permanent teeth because the deciduous teeth are resorbed during fetal development. With few exceptions, replacement of deciduous teeth by their permanent counterparts is vertical, the permanent tooth erupting from below and pushing out the worn-down deciduous tooth. In elephants (Order Proboscidea) and manatees (Sirenia: Trichechidae), however, tooth replacement is horizontal. Teeth in the posterior part of the mandible slowly move forward and replace anterior cheekteeth as they wear down and fall out (see figure 18.4).

Dental Formulae

A dental formula provides a useful shorthand description of the total number and position of teeth in a given species. Dental formulae are always given in the following order:

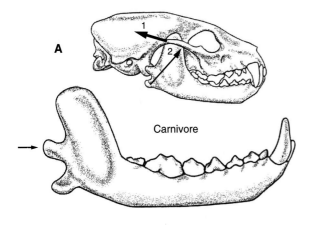

Carnivore

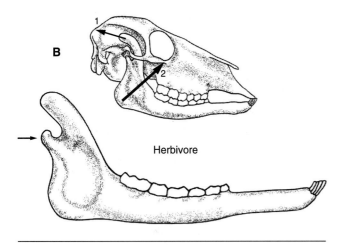

Herbivore

Figure 4.19   Mandibular condyle and muscle groups.
The position of the mandibular condyle (*arrow*) relative to the plane of
the teeth differs between (A) carnivores and (B) herbivores. Thus, the
temporalis muscles (1) are the primary group of chewing muscles in
carnivores, whereas the masseters (2) are the primary group in herbivores.

incisors, canines, premolars, and molars. Thus, the dental
formula for the brown hyena (*Parahyaena brunnea*) is 3/3,
1/1, 4/3, 1/1. This means there are three upper (numbers
above the line) and three lower incisors (numbers below the
line), one upper and lower canine, four upper and three
lower premolars, and one upper and lower molar. Because
dentition is bilaterally symmetrical, dental formulae are
given for one side of the mouth only and may be multiplied
by two to arrive at the total number of teeth in the mouth.
The hyena has 17 teeth on each side of the mouth (9 upper
and 8 lower), for a total of 34. Many species have lost teeth
in a particular position through evolutionary time. For example, white-tailed deer (*Odocoileus virginianus*) have lost
the upper incisors and canines. Thus, their dental formula
is 0/3, 0/1, 3/3, 3/3 = 32 (figure 4.20). Again, because the
tooth order given in a dental formula is always the same,
words or abbreviations for incisors, canines, premolars, and
molars are not necessary. If there are no teeth in a position,
as in the white-tailed deer, a zero is shown, but the position

is not deleted. Thus rodents, which have no upper or lower
canines, have a canine formula of 0/0.

Abbreviations often are used when describing particular teeth. Superscripts and subscripts may be used with abbreviations for tooth type. For example, P^2 refers to the
second upper premolar, whereas P_2 means the second lower
premolar. Alternatively, you may see capital letters used to
refer to upper teeth (e.g., P2), and lowercase letters for lower
teeth (p2). Care must be taken to avoid confusion when referring to particular teeth.

Primitive Dental Formulae

For most species of extant mammals, there is a maximum
number of each type of tooth. Presumably, these maxima
represent the ancestral condition. Thus, although species
tend to lose teeth (see the next section), very few exceed the
primitive number of incisors, canines, premolars, and molars.
In eutherians, the primitive dental formula is 3/3, 1/1, 4/4,
3/3 = 44. This means ancestral eutherian mammals had 44
teeth. Thus, most modern eutherians will not have more
than three upper or lower incisors per quadrant, more than
one upper and lower canine, and so forth. For metatherians,
the primitive dental formula is 5/4, 1/1, 3/3, 4/4 = 50. Although very few species exceed the primitive number, the
toothed whales often have more than 44 teeth; some species
actually have over 200. Among terrestrial species, only the

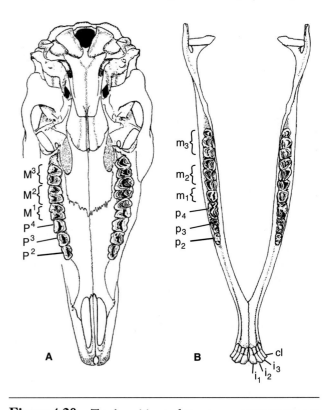

Figure 4.20   Tooth position and structure.   (A) Ventral view of
the upper dentition and (B) dorsal view of the lower dentition of the
white-tailed deer, showing the number and structure of teeth in each
position. For further discussion of the abbreviations, see text.

giant anteater (*Priodontes giganteus*), bat-eared fox (*Otocyon megalotis*), and the marsupial numbat (*Myrmecobius fasciatus*) exceed the primitive numbers of teeth.

The evolutionary trend is toward reduction from the primitive dental formula. In the white-tailed deer noted earlier, three upper and lower premolars occur in each quadrant (side of the mouth). One premolar (the first position) has been lost over evolutionary time. The most anterior upper and lower premolars actually are P^2 and P_2, although they may be described as the "first" premolars in the arcade (row). Similar examples could be cited for most other species.

Dental Anomalies

Occasionally, an individual's dental complement is different from that normally seen in the species. Such congenital anomalies or abnormalities may involve **supernumerary** dentition (extra teeth in a position) or, conversely, **agenesis** (reduced number of teeth in a position). Anomalies may be unilateral, occurring on one side of the jaw, or bilateral, occurring on both sides. Although they are rare, dental anomalies have been reported in representative species from most orders of mammals (Miles and Grigson 1990).

Summary

The evolution of mammals from therapsid reptiles occurred during a 100-million-year period from the late Paleozoic to the early Mesozoic era, with mammals appearing about 220 mya. During this time, numerous changes occurred in the skull, dentition, and skeleton from the reptilian to the mammalian form. These skeletal changes, and concurrent changes in soft anatomy, adapted mammals for improved ability to maintain homeostasis. Mammals became more efficient at gathering and processing food than reptiles and developed a much higher metabolic rate (although some dinosaurs may also have had high metabolic rates). This set the stage for the explosive adaptive radiation of mammals during the 100-million-year period after their initial appearance. The radiation of numerous phylogenetic lines of mammals, from shrews to elephants and rodents to whales, occurred after the extinction of the dominant terrestrial vertebrates—the dinosaurs. Mammalian radiation was further enhanced by genetic isolation of phylogenetic lines resulting from continental drift and the separation of the continental land masses, as well as increased diversity of flowering plants throughout the world. Most mammalian orders today are recognized from the Eocene epoch, whereas most families are evident by the Miocene.

Diversity of form and function is manifested in the highly interrelated characteristics of modern mammals. The broad diversity of modern mammals in terms of their dentition, locomotion, pelage, feeding, and reproduction enables them to adapt to the wide range of biomes and habitats. Many of the general characteristics noted in this chapter are discussed in detail in part II. The extent of mammalian diversity is seen in part III, where the orders and families of extant mammals are examined.

Discussion Questions

1. What reasons can you give for mammals being so much smaller than even the smallest dinosaurs for the 140 million years they were on earth together? What might have been the adaptive advantages to mammals of having been so small?
2. Because they were so small, what morphological and physiological characteristics were necessary for early mammals to survive?
3. Why does heterodont dentition of a mammal allow an individual a much broader range of feeding possibilities than the homodont dentition of reptiles?
4. Before reading chapter 6, how many different mammalian feeding adaptations can you think of?
5. How did the concurrent rise in the diversity of other fauna and flora early in the Cenozoic era affect the potential for early mammals to radiate into different lineages?

Suggested Readings

Eisenberg, J. F. 1981. The mammalian radiations: An analysis of trends in evolution, adaptation, and behavior. Univ. of Chicago Press, Chicago.

Lillegraven, J. A., Z. Kielan-Jaworowska, and W. A. Clemens (eds.). 1979. Mesozoic mammals: the first two-thirds of mammalian history. Univ. of California Press, Berkeley.

Lucas, S. G. and Z. Luo. 1993. *Adelobasileus* from the Upper Triassic of West Texas: the oldest mammal. J. Vert. Paleontol. 13:309–334.

Kemp, T. S. 1982. Mammal-like reptiles and the origin of mammals. Academic Press, New York.

Rowe, T. 1988. Definition, diagnosis and origin of Mammalia. J. Vert. Paleontol. 8:241–264.

PART TWO

Structure and Function

In part II, we examine the relationships between structure and function by investigating mammalian organ systems. Our goal is to understand how mammalian anatomy and physiology have been shaped by natural selection to accommodate varied environmental circumstances. We give special attention to ways in which natural selection has molded mammalian anatomy. In part IV, we will explore the interconnections between morphological and physiological systems and mammalian behavior.

With chapter 5, we begin by exploring the body's outer covering (the skin, or integument), the internal support structures that make up the cranial, axial, and appendicular skeleton, and the muscles that operate these skeletal structures. Chapter 5 concludes with an examination of locomotor adaptations among mammals. Chapter 6 covers the ways in which mammals obtain and process food, examining variations in teeth and how these different forms affect mastication. The chapter then explores the absorption of nutrients in the intestines and storage of reserve energy, and concludes by examining foraging strategies. Chapter 7 is devoted to control and regulatory systems. The nervous system receives information from sensory organs, processes and stores information, and mediates responses. Endocrine organs and their products, hormones, also mediate responses to external and internal stimuli. In chapter 8, we examine the concept of homeostasis and the series of systems and processes by which homeostasis is maintained, including the heart and circulatory system; the lungs and respiration; the kidney and osmoregulation; temperature regulation; and hibernation, torpor, and estivation. In chapter 9 we explore the processes by which new life is generated (reproduction) and events that occur during development of the adult body form and function.

The cryptic color pattern of fawn white-tailed deer (*Odocoileus virginianus*) helps keep them hidden from predators.

CHAPTER

5

Integument, Support, and Movement

Imagine a cheetah (*Acinonyx jubatus*) slowly stalking a Thomson's gazelle (*Gazella thomsonii*) on the African savanna: The cheetah hides in the dappled shade under an acacia, then warily follows the gazelle at a distance. Suddenly, it sprints after the gazelle in a final, all-out dash to capture its prey (figure 5.1). We are amazed by the cheetah's incredible burst of speed, by the graceful, powerful movements of its legs as it strides toward the gazelle. What body structures enable the cheetah to make such a swift attack? What body processes are at work in the smooth, forceful run of this predator?

The cheetah, like other mammals, has complex structures and processes that control its appearance and movement. Its fur protects and camouflages it. Its legs move by the intricate internal coordination of muscles and bones. Each stride involves particular skeletal and muscle movements as the cheetah's forepaws and then its hind paws contact the ground to propel it forward. To understand the dynamics of this locomotor process, we must first examine the arrangement of skin, bones, muscles, and other structures in the cheetah's body. This chapter starts with the outside of the animal—the integument—and then moves under the skin to explore skeletal features and the muscles that provide the force to make it move. In the final section, we put the pieces together and consider the particulars of locomotor patterns in mammals.

INTEGUMENT

An animal's characteristic shape and colors are the result of its skin, or integument. The **integument** is the outer, boundary layer between an animal and its environment. The integumental features of mammals evolved for two primary

Figure 5.1 Cheetah and gazelle. This scene of a cheetah attempting to run down a gazelle shows us the general body form of each animal covered by its integument and the color patterns of the fur on each species. Information in this chapter should enable you to better understand the internal support system (skeleton) and the muscle system that operates that skeleton during the patterns of running locomotion being exhibited here.

reasons. First was the need to retain water. An impervious integument became necessary when vertebrates first invaded the terrestrial environment. Second was the need to retain body heat. As we already noted in chapter 4, endothermy is a key feature of mammals. Their skin and fur trap escaping heat and insulate their internal structures from extremes of temperature. The integument evolved to serve a variety of additional functions in mammals, including heat dissipation through evaporative water loss, communication via hair color patterns, and secretion of scents from specialized glands, to name a few.

Integument Structure and Function

Skin

Skin consists of an outer epidermis and inner dermis, below which is the hypodermis or subcutaneous region (figure 5.2; see Romer and Parsons 1977). The **epidermis** consists of three layers. The outer stratum corneum is composed of dead, keratinized (cornified) cells, which wear off with time and are replaced from below. The thickness of this layer varies over regions of the body surface and from one kind of animal to another. This thickness variation relates to the function of the particular part of the body. For example, the foot pads of many mammals have a thicker stratum corneum. The soles of the feet regularly make contact with rough and sometimes irritating surfaces in the environment, such as stones, twigs, and thorns. This exposure leads to more wear in those areas and to the need for greater protection.

In the middle layer of epidermis, the stratum granulosum, the cells become keratinized. The innermost layer is the stratum germanitavium. In this region, cells are produced by mitosis and eventually move outward. The epidermis has no blood supply, and metabolic needs of cells in the stratum germanitavium are met via diffusion of nutrients from the underlying dermis. Located at the boundary between the epidermis and dermis are cells carrying melanin pigments called melanocytes. The pigments function to absorb harmful ultraviolet light from the sun that might otherwise injure the underlying dermal tissue. For example, the tongues of giraffes (*Giraffa camelopardalis*) are black or purple to protect the tissue from sunburn as the animals lift their head into the top branches of trees to forage for leaves. In some cases, pigmented skin also provides patches of color that may be used for communication. For example, blue and red coloration of the scrotum and perineal region of vervets (*Chlorocebus aethiops*) is used in dominance displays. Derivatives of the skin, such as hair, claws, hooves, nails, horns, and antlers, are discussed in the following subsections.

The **dermis** consists of connective tissue and is vascularized (see figure 5.2). Embedded within the dermis and, in some cases, extending into the hypodermis are a number of structures, some of which are epidermal in origin; for example, hair follicles, sweat glands, and sebaceous glands. Other dermal structures include blood vessels and sensory endings

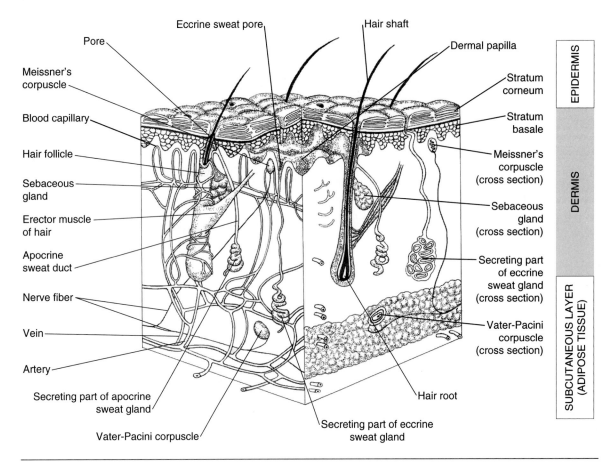

Figure 5.2   Mammalian skin.   The structure of the mammalian skin, including the three major layers and the various structures and tissue types included within each layer. The skin is the largest organ of the body and serves protective, sensory, insulative, and other functions.

of nerves. The blood vessels provide nourishment for both the dermis and epidermis and remove waste. Nerve endings involve receptors for temperature (Pacinian corpuscles) and pressure (Meissner's corpuscles), which send messages to the central nervous system about objects with which the skin comes into contact.

The innermost layer of the integument, the **hypodermis,** consists of fatty tissue. Fat storage cells serve as insulation, a cushion against external blows, and energy reserves. The base of each hair follicle is located in this layer, along with the vascular tissues, parts of sweat glands, and portions of many of the dermal sensory receptors mentioned earlier.

Hair

Hair (pelage) is one of the unique distinguishing features of mammals (Ryder 1973). It evolved primarily as an insulating outer covering, permitting mammals to venture into cooler climates and to be active throughout the year and at night. The predecessors of mammals, reptiles, are restricted to more temperate and tropical latitudes, to times of the year when the ambient conditions are sufficiently warm, and primarily, though not exclusively, to periods of the day when

temperatures are above a certain threshold. Hair, or fur, also serves to protect the animal from ambient conditions.

Although they may exhibit some superficial similarities, the hair of mammals and the feathers of birds are not homologous traits. That is, they have different evolutionary origins. Feathers are homologous to the scales of some reptiles, possibly developing from the fringes found on some scales. Hair and feathers also have different developmental pathways. Feathers and scales share the same form of germinal bud during development. The hair follicles of mammals develop from within the epidermis.

A hair follicle begins its development in the stratum germanitivum layer of the epidermis (Butcher 1951), and as it grows, it projects downward into and through the dermis to form a root, or papilla, protruding into the subcutaneous fat layer (figure 5.3). Several structures are associated with these follicles. An erector muscle (erector pili) can stand the hair up, which may function to increase insulating capacity or, in some instances, such as in an aggressive encounter, to make the animal appear larger. Nerve endings found near the papilla send information to the central nervous system when the hair comes into contact with some external object.

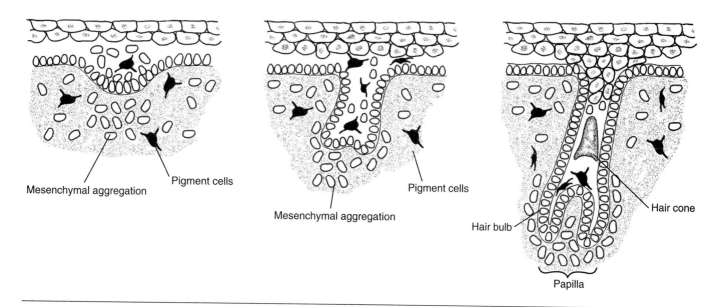

Mesenchymal aggregation — Pigment cells

Mesenchymal aggregation — Pigment cells

Hair bulb — Hair cone

Papilla

Figure 5.3 **Hair follicle.** In the developing hair follicle, cells invaginate as they migrate inward. Eventually, a complete hair bulb and cone develop in the papilla, resulting in a follicle.

A hair has three well-defined structural layers. The **medulla** extends throughout the center, or **shaft,** of the hair. The medulla may have a variety of different patterns, depending on the species. In some species, including deer, the medulla is absent, leaving a hollow center. The **cortex,** composed of tightly packed cells, surrounds the medulla and makes up most of the hair shaft. A thin, transparent **cuticle** makes up the outer layer and forms a scalelike pattern on the surface of the hair. These patterns vary greatly, depending on the species, although totally unrelated species may have quite similar cuticular patterns. Conversely, closely related species may have totally different patterns.

Although mammalian hair has been classified in several ways (de Meijere 1894; Toldt 1935), we use categories based on growth patterns and functions (DeBlase and Martin, 1981). Hair can be categorized as either **angora,** which refers to continuous growth resulting in long, flowing hair that may be shed or not (e.g., a horse's mane) or **definitive,** referring to hairs that attain a particular length and are shed and replaced periodically (e.g., body hairs of a dog). The two main functional types of hair are vibrissae and body hair. (1) **Vibrissae** are long, stiff hairs, with extensive enervation at the base of the follicle that are found on all mammals except humans. They exhibit definitive growth. Vibrissae are usually referred to as "whiskers" and function to warn the animal when its head or eyes are about to make contact with objects that might result in injury. Vibrissae are also found on the tails of fossorial species (mammals that dig) and on the legs of some mammals. Vibrissae may be either active, meaning they can be erected voluntarily, or passive, meaning that they are not voluntarily controlled but respond to external stimuli. (2) **Body hair,** the outer layer of hair or fur we most often see, is termed **guard hair,** or overhair, and comes in three types. **Spines** are stiff, enlarged guard hairs that

exhibit definitive growth. Porcupines, echidnas, and hedgehogs have spines, which protect them from potential predators or other enemies. **Bristles** are firm, generally long hairs that exhibit angora growth, such as in the manes of horses or lions. Bristles most likely function in communication, augmenting or accentuating facial expressions (lions) or body postures (horses). **Awns** are the most common guard hairs we see on mammals. They have an expanded distal end with firm tips and a weak base and exhibit definitive growth. They usually lie in one direction, giving pelage a distinctive nap. There are three types of awns. **Wool** is long, soft, and usually curly, such as in sheep. The most common underhair, simply called **fur,** consists of closely spaced, fine, short hairs. Finally, **velli** is the term used for very short, very fine hairs. Velli are sometimes referred to as "down," or "fuzz," and include embryonic hair found on newborn mammals.

Color is another important feature of mammalian hair. Two types of melanin pigments occur in mammalian hair. **Pheomelanin** (or **xanthophylls**) mixtures produce various shades of red and yellow. **Eumelanin** provides shades of black and brown. Some hairs, called **agouti** hairs, show mixtures or banding from these two types of pigments. The permanent absence of any pigment produces a white hair and results from a genetic mutation called **albinism.** Albinism is different from seasonal color change to a white pelage and can occur over either the entire body of the animal or, in some instances, particular portions of the body. Animals that are completely or partially white are likely to be at a selective disadvantage because they will be more easily detected by predators. Although albinism is uncommon, a few populations of mammals are known in which albinism is prominent. These are often, as with the albinistic gray squirrels (*Sciurus carolinensis*) in Olney, Illinois, under some form of legal protection. The opposite of albinism is **melanism**—the condition in which the animal is generally all black, due again to a genetic mutation. Examples of populations exhibiting melanism include fox squirrels (*Sciurus niger*) and red foxes (*Vulpes vulpes*). Because these animals are less likely to encounter problems with predator detection, more melanistic than albinistic forms of mammals survive to reproduce; thus, there are more populations with melanistic animals than with albinistic animals.

Skin and hair color may be influenced by several factors. We have already mentioned the need for protection from ultraviolet radiation. Mammals from polar regions generally have lighter skin and hair color than those from the tropics, due in part to the less intense incident sunlight in polar locales. Human populations provide an excellent example of the manner in which pigmentation works: people of all races who inhabit the tropics have darker skin than their counterparts from higher latitudes. Curiously, small terrestrial mammals from arid (desert) regions are often paler than those from more humid areas—a phenomenon called Gloger's rule. This phenomenon probably relates to the color of the substrates in these different types of habitats. The desert floor is often more lightly colored than forest or grassland habitats, and potential prey species need to blend in with the background (Benson 1933). The notion that pelage coloration is a potential antipredator adaptation was recently tested for oldfield mice (*Peromyscus polionotus*) in the southeastern United States by Belk and Smith (1996). Their data generally support the idea of predation as a selective force, although some of their findings suggest that further tests of the hypothesis are needed.

Some specific color patterns relate to an animal's habitat and behavior (Cott 1966). A ventral body area that is more lightly colored than the dorsal surface of the body, a phenomenon called **countershading,** is often found in semi-arboreal and scansorial species. It apparently results from selection for animals that look white (or like the sky) to any animal seeing it from below and dark (or like the ground and vegetation) to any animal seeing it from above. Two types of color patterns aid mammals that either are seeking to avoid predators or are themselves predators and are attempting to remain concealed from possible prey. **Cryptic coloration** is a pelage pattern that matches the general background color of the animal's habitat. White-tailed deer fawns (*Odocoileus virginianus*) are an example; the light-colored spots on their brown fur allow them to "disappear" when sitting in dappled shade. Other mammals have patterns of stripes or colors that stand out from the basic background fur color—a property called **disruptive coloration.** These patterns allow the animal to blend in with the background mixture of sunlight and vegetation. Body shapes and outlines are not as readily distinguished when animals have disruptive coloration patterns. Some mammals may appear larger through optical illusion, as is the case with the vertical black-and-white banding patterns of zebras (*Equus* spp.). Many color patterns, particularly those in the facial region and around the eyes, relate to communication functions (see chapter 20).

Many mammals periodically replace their definitive hair and sometimes their angora hair in a process called **molting,** which occurs in three principal molt types, or patterns. Young mammals, particularly rodents, undergo a **postjuvenal molt** that starts after weaning (figure 5.4). Such molts have been well described for many species; for example, meadow voles (*Microtus pennsylvanicus*), Ecke and Kinney 1956) and several species of deer mice (*Peromyscus* spp.; Osgood 1909; Storer, et al. 1944).

Most mammals living at temperate and northern latitudes undergo a rather rapid process of **annual molt** each year during which most hairs are replaced. This process is important for animals in these cooler environments because daily activities wear down and damage hairs. This would result in less effective insulation during the next winter if the hair were not replaced. Sunlight also bleaches out the color if hair is not replaced. Some mammals undergo **seasonal molts,** changing their pelage more than once each year. Mammals such as the ermine (*Mustela erminea*) and arctic fox (*Alopex lagopus*) have a white coat during the winter, matching the snow-covered landscape they inhabit, and change to a dark coat for the summer (figure 5.5). In this way, their pelage always matches the predominant color of the habitat, enabling these predators to hunt more effectively. Alternatively, some mammals probably blend in with seasonally changing backgrounds as a mechanism for predator avoidance (see figure 17.27). They also gain the advantage of having their hair replaced twice each year, keeping it fresh and maintaining its effectiveness as insulation.

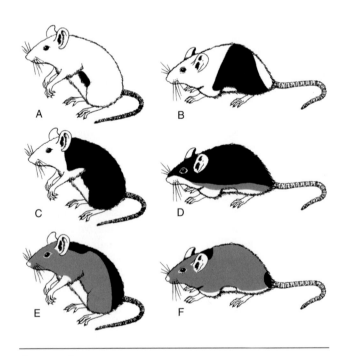

Figure. 5.4   Molting.   The postjuvenal molt of the golden mouse (*Ochrotomys nuttalli*) proceeds in a regular manner from the underside of the flanks dorsally and finally to the back of the neck and the base of the tail. The letters refer to the chronological sequence, A–F, during the molt.

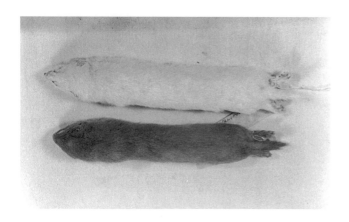

Figure 5.5 **Seasonal changes of coat color.** Weasels of several species (*Mustela* spp.) change their coat color seasonally. They molt to a mostly white pelage for the winter months, which camouflages them against the snowy background. They molt again in the spring to a pelage with brown on the dorsal surface and flanks and a white ventrum.

Animals in molt can be recognized by the short or partially elongated hairs, which are dark to the base. These grow up among the fully grown hairs, which are light to the base. Also, fully developed molt lines can sometimes be seen, especially in short-haired species such as moles, shrews, and pocket gophers. On the underside, the skin of a mammal undergoing molt is dark or dark-speckled as a result of the partially elongated hairs in which pigment is just being laid down at the skin surface. Fully grown hairs are white at their bases, thus making the skin appear white. In such condition, when the animal is not molting, the pelt is referred to as "prime" in the fur industry.

Glands

Two types of glands are located in or near the skin of mammals—sweat glands and sebaceous glands—each of which has several subtypes and particular functions. Sweat glands (see figure 5.2) evolved in mammals as a means of dissipating heat. Mammals eliminate excess heat in several ways, which are detailed in chapter 8. **Eccrine sweat glands** have their own separate ducts leading to the body surface, where water is forced outward, resulting in evaporation and cooling. **Apocrine** (or sudoriferous) **sweat glands** are highly coiled structures located near hair follicles. The distribution of sweat glands over the body surface varies. In mammals such as cats and dogs, sweat glands are located primarily on the pads of the feet, while in primates, sweat glands can be found over major areas of the body. Primate apocrine sweat glands are found primarily in the axillae (armpits), the anogenital region, the naval area, the nipples, and the ears. Eccrine glands are distributed over the remainder of the body. In the ears, the apocrine glands are modified to produce a waxy substance (cerumen) that helps to keep the tympanum pliable. Some mammals, such as short-beaked echidnas (*Tachyglossus aculeatus*) and rodents (Order Rodentia), lack sweat glands entirely.

An important distinguishing feature of mammals, the trait from which the group takes its name, are the mammary glands, or mammae. These are similar to apocrine glands in development and structure. The number of mammae and their location varies: primates have 2 thoracic mammae, perissodactyls have a pair of inguinal mammae, many rodents have up to 5 pairs of mammae distributed from the axillae to the inguinal region, and some marsupials may have as many as 20 mammae in their pouches. The North American opossum (*Didelphis virginiana*) has 13 mammae: 12 in a circle and 1 in the center. Mammary glands consist of elaborate, elongated duct systems with hollow sacs at the ends. During pregnancy, with hormonal stimulation, these sacs increase in number and enlarge. The milk they produce is then carried via the duct system to the nipples, or teats. The composition of milk from mammals varies (table 5.1). In general, the young of mammals whose milk has a high proportion of protein gain weight more rapidly (table 5.2). Composition also varies significantly during lactation, within an individual female, and over the period of milk production.

Sebaceous glands are generally associated with hair follicles and secrete oils to keep the hair moist and waterproof. The glands are usually situated so that when the erector pili muscle moves the hair, the gland is squeezed and some of its product is forced into the space around the hair shaft. The lanolin of sheep is produced by such specialized sebaceous glands. Modified sebaceous glands, called Hartner's glands, provide lubrication for the eyes and nictitating membranes.

Scent glands may be either modified sweat glands or modified sebaceous glands. The nature of the scents produced by mammals and their functions vary widely. Scents may be used for defense, such as in skunks (*Mephitis* spp.); territorial marking, as in all species of deer (Family Cervidae); marking used for orientation, as in weasels and badgers (Famly Mustelidae); and a variety of other social interactions (Ralls 1971; Brown and Macdonald 1985). Some mammals spread glandular-scented secretions onto their fur, possibly to enhance the effectiveness of any communication or to make the scent last longer. The yellowish color often seen on the fur of opossums is an example of such behavior.

Claws, Nails, Hooves, Horns, and Antlers

Claws, hooves, and nails are keratinized structures derived from the stratum germanitivum layer of the epidermis. The basic, most primitive structure, the **claw,** occurs at the ends of the digits and grows continuously. The claw consists of two parts (figure 5.6A): a lower or ventral **subunguis,** which is continuous with the pad at the end of the digit, and an upper or dorsal **unguis,** which is a scalelike plate that surrounds the subunguis. Generalized claws found in many mammals are useful for climbing, digging, aggressive intraspecific interactions, and defending against a variety of enemies. In some mammals, such as felids, the sharp claws are retractable. In other groups, such as canids, the claws are usually not as sharp and are nonretractable. Having retractable claws (figure 5.7),

Table 5.1 Milk composition The composition of milk in various mammals varies in terms of the proportions that are comprised of protein, fat, and sugar. Notice in particular the high fat content for pinnepeds and cetaceans, as well as for lagomorphs. In the last case, the mother returns usually only once a day to feed her young.

	Water	Protein	Fat	Sugar	Ash
Marsupials					
Kangaroo (wallaroo)	73.5	9.7	8.1	3.1	1.5
Primates					
Rhesus Monkey	88.4	2.2	2.7	6.4	0.2
Orangutan	88.5	1.4	3.5	6.0	0.2
Human	88.0	1.2	3.8	7.0	0.2
Edentates					
Giant Anteater	63.0	11.0	20.0	0.3	0.8
Lagomorphs					
Rabbit	71.3	12.3	13.1	1.9	2.3
Rodents					
Guinea pig	81.9	7.4	7.2	2.7	0.8
Rat	72.9	9.2	12.6	3.3	1.4
Carnivores					
Cat	81.6	10.1	6.3	4.4	0.7
Dog	76.3	9.3	9.5	3.0	1.2
European red fox	81.6	6.6	5.9	4.9	0.9
Pinnipeds					
California sea lion	47.3	13.5	35	0	0.6
Harp seal	43.8	11.9	42.8	0	0.9
Hooded seal	49.9	6.7	40.4	0	0.9
Cetaceans					
Bottle-nosed dolphin	44.9	10.6	34.9	0.9	0.5
Blue whale	47.2	12.8	38.1	?	1.4
Fin whale	54.1	13.1	30.6	?	1.4
Ungulates					
Indian elephant	70.7	3.6	17.6	5.6	0.6
Zebra	86.2	3.0	4.8	5.3	0.7
Black rhinoceros		1.5	0.3	6.5	0.3
Collared peccary		5.8	3.5	6.5	0.6
Hippopotamus	90.4	0.6	4.5	4.4	0.1
Camel	87.7	3.5	3.4	4.8	0.7
White-tailed deer	65.9	10.4	19.7	2.6	1.4
Reindeer	64.8	10.7	20.3	2.5	1.4
Giraffe	77.1	5.8	12.5	3.4	0.9
American bison	86.9	4.8	1.7	5.7	0.9
Cow	87.0	3.3	3.7	4.8	0.7

From *F. Boulieré*, The Natural History of Mammals, *3rd ed. Revised, 1964, Alfred A. Knopf. Reprinted by permission.*

Table 5.2 Protein and body growth There is a strong positive relationship between the proportion of protein in mother's milk and the rate of growth for young in mammals.

	Average Number of Days Needed to Double Birth Weight	Protein Content of Milk (g/1000)
Human	180	12
Horse	60	26
Cow	47	33
Pig	18	37
Sheep	10	51
Dog	8	93
Cat	9	101
Rabbit	6	123
Harp seal	5	119

Source: Data from F. Boulieré, The Natural History of Mammals, *3rd ed. revised, 1964, Alfred A. Knopf, New York, NY.*

particularly ones that are sharp, may be advantageous in terms of peaceful interactions with conspecifics and young. It is interesting to note that felids generally do much more object manipulation with their forepaws than canids, an action probably facilitated by the ability to retract the claws and use the sensory organs on the footpads to a greater extent.

Nails are a specialized variation of claws that evolved in primates to facilitate better gripping and more precise in object manipulation by the hands and feet. Only the dorsal surface of the end of each digit is covered by the nail (see figure 5.6B). The number of sensory endings in the dermis at the ends of the digits of primates increased greatly in conjunction with the development of nails.

Hooves, found in the ungulates (perissodactyls and artiodactyls), are another specialized variation of claws. They consist of a large mass of keratin (the unguis) completely surrounding the subunguis (see figure 5.6C). This adaptation is combined with a reduction in the number of digits in ungulates. Most ungulates have either one (perissodactyls such as horses) or two (artiodacytls such as cattle or deer) hooves at the end of each leg. The keratinized tissue wears away slowly at the outer surface as constant contact is made with the substrate.

Head ornamentation in extant mammals, which occurs only in the artiodactyls and perissodactyls, is of several types, including horns and antlers, as well as minor variations (Geist 1966). **True horns** occur only in the Family Bovidae. Horns are formed from the combination of an inner core of bone, which derives from the frontal bone of the cranium and a sheath of keratinized material, which derives from the epidermis (see figure 19.19). True horns grow continuously through the life of the animal, are unbranched, and are never shed. Growth rings at the base of the horns of some species (e.g., gazelles, *Gazella* spp.) can be used to determine the age of individuals. True horns usually occur on males, but sometimes also on females.

Pronghorn antelope (*Antilocapra americana*), which are endemic to the American West, have horns with a structure like that of true horns, consisting of a bony core and outer keratinized sheath derived from the epidermis. The sheath is shed annually, however, and a new sheath develops as the old one is lost (O'Gara and Matson 1975) . They also differ in having a small secondary branch, or "prong," which has no bony core. Pronghorns occur in most males and some females in this species. Giraffes and okapi (Family Giraffidae) have two bony projections on the skull, which derive from separate bone centers and then fuse to the skull near the junction of the parietal and frontal bones. These are not true

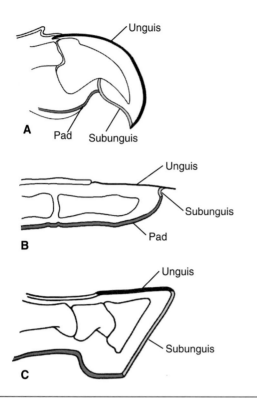

Figure 5.6 **Claws, nails and hooves.** Similarities in the general structure but distinct variation with regard to specialized functions are evident for (A) claws, (B) nails, and (C) hooves. Top views of each part are lateral sections; bottom views are ventral views. Unguis is solid dark-colored; subunguis, stippled; and pad, dotted.

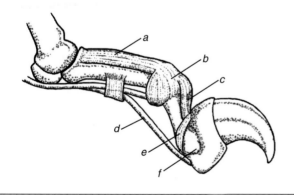

Figure 5.7 **Retractable claws.** The retractile claw mechanism of the mountain lion (*Felis concolor*) involves the (a) extensor expansion, (b) middle interphalangeal joint, (c) extensor tendon, (d) flexor digitorum profundus tendon, (e) lateral dorsal elastic ligament, and (f) distal interphalangeal joint.

horns. They are permanently covered by skin and hair. They occur in both sexes. Rhinoceroses (Family Rhinocerotidae) have "horns" that are composed of large masses of elongated, keratinized epidermal cells, shaped like hair. The cells grow from dermal papillae on the head of the rhinoceros and fuse to form the horn.

Antlers are made entirely of bone and, unlike true horns, are branched (see fig. 19.17). During the growth period, the antler is entirely covered by a layer of "velvet"—a highly vascularized, enervated skin that supplies the growing tissue with nutrients. When bone growth is complete, the velvet is shed. After the annual **rut,** or mating season, the antlers are lost. A new set is grown each year. Antlers are characteristic of members of the Family Cervidae and are found only on males, except for the caribou (*Rangifer tarandus*), in which they may occur on both sexes.

Although head ornamentations can have a variety of functions, they are all adapted for intraspecific competition among males. This is especially evident in Rocky Mountain bighorns (*Ovis canadensis*): males of this species compete during the rut by running toward each other at great speed and butting heads. In many instances, horns may be used defensively to ward off potential predators. Antlers or horns may also be important with respect to social rank in mammals, such as wildebeest (*Connochaetes taurinus;* Walther 1984). Some of these functions are considered in greater detail in part IV.

BASIC SKELETAL PATTERNS

Full treatment of the varied skeletal patterns of mammals would require a separate course in comparative anatomy. Our goal here is to offer an overview of the general patterns of mammalian cranial, axial, and appendicular skeletons with an emphasis on the relationships between structure and function. The skeleton serves as the major framework of the body on which various organs, muscles, and other tissues are suspended. It is also a basis for movement, through lever action. The mammalian skeleton can be partitioned into three subdivisions. The skull includes the **cranium,** which surrounds the brain, and the facial area, including the mandible. The postcranial skeleton can be divided into the **axial skeleton,** consisting of the spinal column and rib cage, and the **appendicular skeleton,** consisting of the pectoral and pelvic girdles, arms (forelimbs), and legs (hind limbs).

Cranial Skeleton

The head consists of two major regions: the cranium and the facial, or rostral, area, including the eyes, nose, and jaws (figure 5.8). The cranium includes a series of bones (or plates): the parietals, frontals, squamosals, and occipitals. These bones are connected to one another at sutures. The primary function of the cranium is to protect and support the brain within. Cranium size is positively related to the

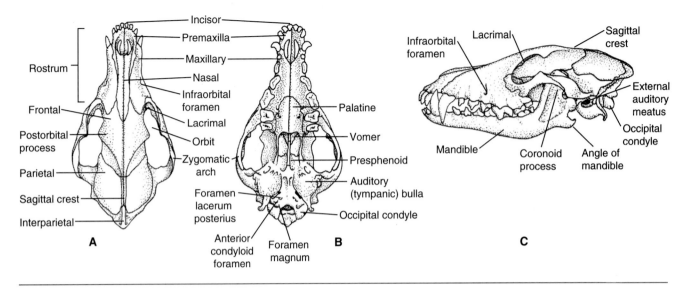

Figure 5.8   Divisions of the skull.   The skull of a coyote (*Canis latrans*) exhibiting the two major divisions of the cranium and the front of the skull, the bones that comprise the upper and lower jaws, and the various foramina discussed in the text. (A) Dorsal view, (B) ventral view, (C) lateral view.

size of the brain, although many other factors contribute to the size of the skull. For example, many mammalian crania exhibit a **sagittal crest,** the bony midline ridge on the top, and **lamboidal crest,** at the rear of the cranium. These ridges have developed, through evolutionary selection pressures, as places of attachment for muscles that help hold the head upright at the front of the vertebral column and operate the jaws. The **zygomatic arch** surrounds and protects the eye but also serves as a place of attachment for jaw muscles. The head of most mammals is connected to the main trunk of the body by a neck, or cervical portion of the vertebral column, which allows for head movement independent of the body.

A key characteristic of the skull is the number of openings, or **foramina,** located on several of the surfaces, notably in the facial region and where the skull joins the axial skeleton (see figure 5.8). The larger of these include paired openings for the eyes and ears. A complex series of foramina also make up the nasal passages. Contained within the nasal area are the **turbinal bones,** which increase the surface area for reception of chemical cues and secrete mucus to aid in filtering small particles from incoming air. A pair of small openings in the palate of most mammals lead to Jacobson's organ, which contains receptors for nonvolatile olfactory cues. The **foramen magnum** (meaning "large opening") is located at the base of the skull and provides the pathway for the spinal cord leaving the brain and entering the vertebral column. Many smaller openings in the braincase provide passage for blood vessels and nerves.

The front of the face contains two major sense organs—the eyes and the nose. A third sensory mode, the ears, are generally located at the top perimeter or sides of the face. The ears exhibit a remarkable evolutionary transition: two of the small bones involved in transmitting sound from the tympanum to the inner ear in mammals—the **incus** and

malleus—are derived from the reptilian **quadrate** and **articular bones,** respectively. These bones were part of the jaw and its articulation with the skull in reptiles. These two bones have moved from the jaw into the ear during the evolution of mammals. The third bone of the middle ear—the stapes—is analogous and possibly homologous to the columella found in amphibians, reptiles, and birds.

The protrusion of the jaw and nasal region varies considerably in mammals, ranging from the nearly flattened face of the primates to the extended snout of carnivores such as the canids (figure 5.9). These variations relate to differences in diet, the importance of odor cues, the position of the skull relative to the vertebral column, and the locations of muscles involved in holding the skull in position and operating the jaws. Probably the most important change is the flattening of the face in many primates to accommodate stereoscopic, binocular vision with the consequent diminution of the rostrum and olfactory senses.

The upper jaw, or **maxilla,** is a fixed part of the skull composed of a series of bones: the premaxillary, maxillary, and palatine (see figure 5.8). The lower jaw, or **mandible,** is mobile and consists of a single **dentary bone.** It contains a depressed area called the coronoid fossa, where the masseter muscle inserts, and, in many groups of mammals, a coronoid process, where the temporalis muscle inserts. The mandible articulates with the skull via the mandibular condyles fitting into the fossae of the squamosal bones, which are part of the skull. Specifically, mammals have a dentary-squamosal articulation. Teeth are one of the most important features of mammals and a key to their overall evolutionary success (see chapter 4).

The **hyoid apparatus,** located in the throat region, consists of a series of bones that are modified from ancestral gill arches in fish. These structures have evolved to protect and support the base of the tongue, trachea, larynx, and esophagus.

A

B

C

D

Figure 5.9   **Facial region.**   The facial region of the head varies widely in mammals, depending on such factors as diet, particularly as it affects the jaws and jaw musculature; the positioning and importance of the sensory systems concentrated in the facial region; and the position of the head relative to the axial skeleton of the animal. This variation is exemplified by views of the facial region of (A) gorilla (*Gorilla gorilla*), (B) coyote, (C) Indian tiger (*Panthera tigris*), and (D) white-footed mouse (*Peromyscus leucopus*).

Axial Skeleton

Mammals that live in water have that medium to help support their weight, but for terrestrial mammals, the skeleton and muscles must perform this task. Thus, the backbone of mammals, suspended above the ground by the limbs, provides a basic framework. This framework could not function as a support structure without the ligaments and cartilage that connect the bones and the tendons that connect bones and muscles. The skeleton, muscles, and their associated structures must be thought of as a single unit. The vertebral column can be thought of as the deck or girders of a bridge, with the legs as the pillars (Thompson 1942). Many of the tissues and organs of the body cavity are attached to or suspended from elements of the axial skeleton, particularly the vertebral column. A major distinguishing feature in mammals is that the pillars (legs) must move as levers, via muscle action, to provide locomotion.

The vertebral column of mammals consists of five regions. From anterior to posterior, these are the **cervical, thoracic, lumbar, sacral,** and **caudal** vertebrae. All mammals (except sloths and manatees) have seven cervical vertebrae. The thoracic vertebrae articulate with the ribs and number from 12 to 15. The lumbar vertebrae number from 4 to 7, though these are sometimes partially or entirely fused. In most mammals, the sacral vertebrae are fused to form the **os sacrum,** to which the pelvic girdle attaches. The number of caudal, or tail, vertebrae varies with tail length.

Two key vertebrae are located at the junction of the vertebral column and the skull: the **atlas** and **axis.** The nature of their articulation surfaces with each other and with the base of the skull, in combination with the neck musculature, allows for the movement of the skull. This ability to move the skull in many directions lets mammals position the sensory receptors located on the skull to take in information. Fish, amphibians, and some reptiles have much less flexibility. Usually, they rotate their heads only by turning the entire body.

The remainder of the axial skeleton consists of the **ribs,** which are attached to the thoracic vertebrae on the dorsal surface and, in most cases, to the sternum on the ventral surface. The rib cage, or thoracic basket, surrounds and protects the vital internal organs—the heart and lungs. The ribs also provide locations for attachment of muscles, including those involved in posture and movement of the forelimbs. Lastly, suspensory ligaments, connected to the ribs, support many internal organs of the abdominal cavity associated with the digestive system.

Appendicular Skeleton

One portion of the appendicular skeleton consists of regions of attachment for the forelimbs (shoulder, or **pectoral girdle**) and hind limbs (hip, or **pelvic girdle**) to the axial skeleton (figure 5.10). A general mammalian pectoral girdle consists of a clavicle and scapula. The pelvic girdle is made up of three pairs of bones fused together: the ilium, ischium, and pubis. Forelimbs consist of a proximal bone connected to the pectoral girdle (the humerus), two middle elements

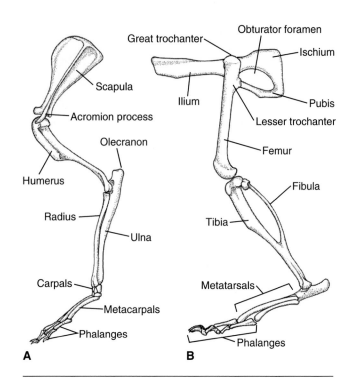

Figure 5.10   Pectoral and pelvic girdles.   The bone patterns of (A) the pectoral and (B) pelvic girdles and the forelimbs and hind limbs for the Norway rat (*Rattus norvegicus*). Each involves bone elements for connection to the axial skeleton, a single proximal bone, usually two distal bones, a terminal joint comprising a number of bones, and digits (or phalanges).

(the radius and ulna), and the carpals, metacarpals, and phalanges (digits). Early mammals had five digits, but in many mammals, such as ungulates, varying degrees of loss and fusion occurred during the evolution of specialized forms of locomotion. Rotation of the forelimb, at the shoulder or elbow joints or both, varies among mammals, depending on whether the limb is used strictly for locomotion or climbing, object manipulation, or other functions.

The proximal hind limb element (the femur) articulates proximally with the pelvic girdle in a ball-and-socket joint, and distally with a pair of lower leg bones (the fibula and tibia); there are variable numbers of tarsals, metatarsals, and phalanges (see figure 5.10). The fibula and tibia are separate bones in some mammals but are fused to varying degrees in others. In some instances, bones of the ankle and hind foot are also fused. Rotational movement of the hind limbs of mammals varies considerably less than for the forelimbs, with hind limbs being generally restricted to use in locomotion and climbing.

MUSCLES OF MAMMALS

Our presentation of the basic muscle patterns of mammals is, of necessity, much briefer than might be justified by the enormous body of comparative knowledge that exists on the subject (Young 1957; Romer and Parsons 1977; Hildebrand et al. 1985). Muscles and the tendons that connect them to the bones provide the force to move the levers, resulting in general body movements, locomotion, and other actions. Mammals have the same three basic types of muscles found in other vertebrates: **striated** (voluntary muscle), **smooth** (involuntary muscle), and **cardiac** (heart muscle). The structure and physiology of these muscle types are generally similar to those of other vertebrates and are not covered in further detail here.

The distribution of muscles of mammals follows a pattern coincident with the skeleton (figure 5.11), with the bulk of the musculature relating to locomotion. The figure illustrates well the fact that the striated muscles are arranged to provide for movement, particularly of the limbs. Various muscle groupings of the rabbit, such as the adductors, rectus femoris, and psoas major, provide for movement of the hind limbs so as to propel the animal forward. A similar set of muscles, including the biceps and triceps, provides for the forward and backward actions of the forelimbs. Yet another set of muscles, including the obliquus capitas, splenius capitas, semispinalis capitas, basioclavicularis, levator scapulae, and sternomastoid, both support and rotate the head. Still other striated muscles are involved in movements of the ears and tail. Many mammals, such as primates, have considerable musculature for the refined control of the hands and feet used in climbing or manipulating objects. Many mammals evolved musculature for producing a variety of facial expressions or the ability to make hair on the back of the neck stand up. These postures and expressions are often used for communication.

MODES OF LOCOMOTION

We can now use the information on the skeleton and muscles to understand the manner in which different mammals move through their environment. Mammals exhibit various combinations of adaptations of bone and muscle patterns that fit their particular life-style. Most mammals use walking and running as the primary means of locomotion. Mammals such as deer can make graceful jumps over obstacles, but the jumps are really an extension of their running gait. A second form of locomotion—hopping, jumping, and leaping—may be best exemplified by kangaroos and related marsupials, but some rodents and lagomorphs also use this mode of travel.

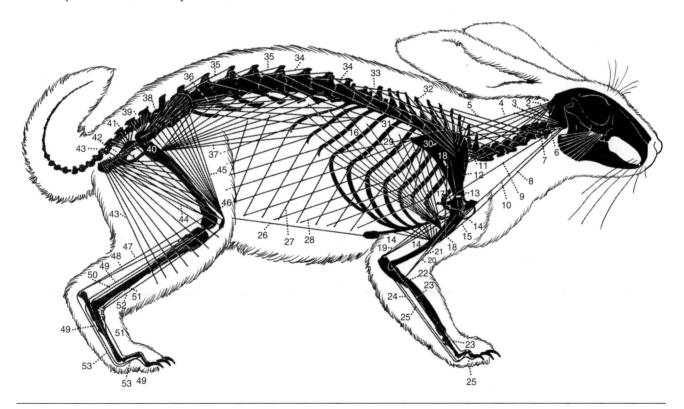

Figure 5.11 **Muscle groups.** Schematic representation of the muscle groups for the European rabbit (*Oryctolagus cuniculus*). Notice in particular the groups of muscles that operate the forelimbs and hind limbs and the manner in which these muscle groups are designed to work in opposition to one another. 1. masseter; 2. obliquus capitis; 3. splenius capitis; 4. semispinalis capitis; 5. longissimus cervicis; 6. longissimus capitis; 7. obliquus capitis inferior; 8. basioclavicularis; 9. levator scapulae; 10. sternomastoid; 11. scalenus; 12. supraspinatus; 13. infraspinatus; 14. pectoralis; 15. cleido humeralis; 16 latissimus dorsi; 17. subscapularis (displaced caudally); 18. deltoid; 19. triceps; 20. biceps brachii; 21. brachialis; 22. extensor carpi ulnaris; 23. extensor digitorum communis; 24. flexor digitorum sublimis; 25. flexor digitorum profundus; 26. rectus abdominis; 27. transversus abdominis; 28. external oblique; 29. serratus anterior; 30. trapezius; 31. ilio-costalis; 32. longissimus; 33. semispinalis dorsi; 34. longissimus dorsi; 35. multifidus; 36. sacro-spinalis; 37. psoas major; 38. gluteus medius; 39. piriformis; 40. gluteus maximus; 41. abductor caudae; 42. gemellus inferior; 43. biceps; 44. adductors; 45. rectus femoris; 46. vastus intermedius; 47. gastrocnemius and plantaris; 48. soleus; 49. flexor digitorum longus; 50. peroneal muscles; 51. extensor digitorum; 52. tibialis anterior; 53. plantaris

Reprinted from The Life of Mammals *by J. Z. Young and M. J. Hobbs (1975) by permission of Oxford University Press.*

Certain mammals use water for part or all of their daily activities, and for them swimming is an important mode of movement. This group includes the sea mammals (whales, dolphins, walruses, and seals), as well as many freshwater mammals (e.g., beavers and muskrats) that use the water either as a permanent home or a place to find food or shelter. Several types of mammals, including some rodents and marsupials, have evolved gliding modes of locomotion. Bats are the only mammalian group in which true flight has evolved.

A number of terrestrial mammals climb trees or rocky cliffs. Climbing is a primary mode of locomotion for arboreal species, including many primates, sloths, and numerous other mammals. Many mammals exhibit more than one general form of locomotion, depending on the circumstance and the activity. A beaver may therefore amble along in a somewhat cumbersome quadrupedal walk on land but swim gracefully in its aquatic habitat. Some mammals are specialized for burrowing, particularly those of the Order Insectivora and some rodents. They spend much of their lives digging either through the leaf and detritus layer on the top of the ground or underground.

Walking and Running

Most animals (except humans) move quadrupedally. In Australia, some marsupials move about on two feet, and a few Old World primates use some bipedal locomotion. Humans, however, are the only habitually bipedal mammals. Those mammals that spend at least a portion of their time running are called **cursorial**; those that locomote almost exclusively by walking are called **ambulatory**. Many, but not all, cursorial mammals, such as carnivores and dasyurid marsupials, run on one or more toes and are termed **digitigrade**. Ungulates run with only their hooves on the ground, a form of locomotion termed **unguligrade**. Most ambulatory mammals walk on the soles of their hands and feet and are called **plantigrade**. Speeds achieved by mammals vary considerably with the different forms of locomotion (table 5.3).

Hildebrand (1985a) lists four requirements for animals that walk or run, including (1) support and stability even though the feet only contact the substrate intermittently, (2) propulsion to move the body forward, (3) maneuverability, and (4) endurance (see also Rewcastle 1981). The first challenge is a postural one. Large, heavy species, such as elephants and hippopotamuses, have their legs directly under the body. They sacrifice agility for the support necessary to bear their weight and are termed **graviportal** (very heavy-bodied). Other, lighter mammals, such as deer, have their limbs positioned slightly outside the main body axis. In terms of forward movement, **gait** is a pattern of regular oscillations of the legs in the course of forward movement (figure 5.12). For ambulatory mammals, each foot is on the ground at least half of the time, but in cursorial mammals, each foot is in contact with the substrate less than half the time. Walking, pacing, and trotting all involve equal spacing of the feet making contact with the ground at even time intervals; we label this a symmetrical gait. Contact with the ground at uneven intervals, as in galloping or

Table 5.3   Running speeds of land mammals.
Some of these speeds were measured over varying distances, and for some animals, this is a maximum speed, maintainable only for short distances.

Species	Speed (km/h)
Cheetah	110
Pronghorn antelope	98
Thomson's gazelle	80
Wildebeest	80
Elk (wapiti)	72
Coyote	70
European hare	65
Spotted hyena	64
Cape buffalo	55
Giraffe	50
Grizzly bear	50
Human	45
African elephant	40
Tree squirrel	20
Three-toed sloth	1

bounding, is an asymmetrical gait. Walking is the most stable gait because of the foot's lengthy contact time with the ground. Pacing, seen often in canids, involves moving the two legs on the same side of the body together. This pattern greatly reduces the possibility of the front and hind limbs of the same side coming into contact (figure 5.12C). Trotting involves moving the two legs that are diagonally opposite one another in synchrony and is a more stable pattern than pacing in its support of the body (figure 5.12B). Stability is also affected by foot size; larger animals tend to have larger feet, providing for more contact with the substrate.

Models of movement (Hildebrand 1980, 1985a; Lanyon 1981) are based on the idea that the legs swing below the body like a pendulum. The joints and their associated muscles, tendons, and ligaments work like levers (Maynard Smith and Savage 1956). Several predictions coming from such models are generally supported by the evidence from mammals. Faster mammals have relatively longer legs than other related species. Cursorial mammals tend to have limb joints that restrict movement to a single plane, forward and backward, parallel to the body. The forward movement of such a lever system is, in a sense, the sum of the forward movements of the component levers. The insertion points of many muscles used for running have shifted so that they are nearer to or in the same plane as the lever joint they operate. Reducing the weight in the distal portions of the limbs also adds to speed. This is accomplished by bone fusion in the lower leg, (e.g., the cannon bone of horses) and, in some instances, by moving the muscle mass that operates the levers nearer to the center of mass of the body, but with longer tendons to operate the joint. Tendons and ligaments can function like springs when they are flexed, storing energy (Alexander and Bennet-Clark 1977) that is released in the next extension; for example, in the long "springing" ligaments of the distal leg section of ungulates.

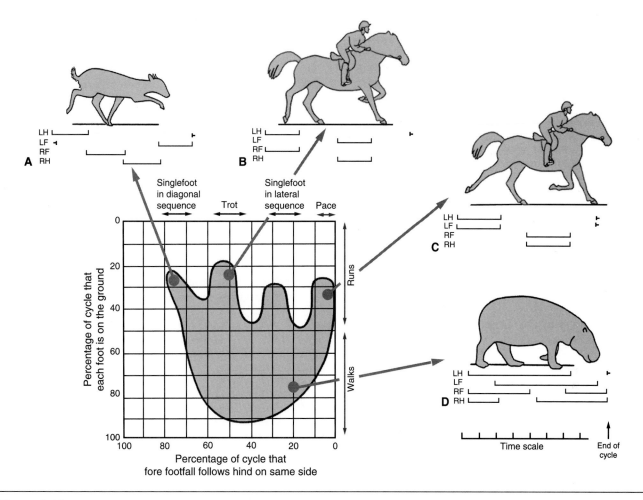

Figure 5.12   Walking and running gaits.   Diagrammatic representation of the gaits of mammals that walk or run as a mode of locomotion. The graph shows the percentage of the time that each foot is in contact with the ground as a function of the percentage of the cycle for which forefoot contact follows contact by the hind limb on the same side. Patterns are shown for (A) hippopotamus (*Hippopotamus amphibious*), (B) a trotting horse (*Equus caballus*), (C) a pacing horse, and (D) a galloping cheetah (*Acinonyx jubatus*).

Maneuverability, involving a change of direction or speed, is required by both predators and prey. Two patterns have been observed for change of direction. First, most cursorial mammals can alter their gait momentarily, such that both legs on the same side or both hind legs strike the ground at the same time, providing for an angular shift in the forward direction. In some cases, mammals can turn their bodies, (e.g., spinal flexion), while they are in the air and achieve some alteration of the direction of forward motion. Just as when we round a curve in an automobile, mammals making a turn lean their bodies in the direction of the turn to maintain balance and control. The larger the mammal, the more maneuverability becomes a problem; small mammals are generally more agile than large ones. Having very flexible limbs, such as primates do, is a function of the joints at the shoulder and hip, as well as the limb bones.

Endurance is a function of the integration of the animal's entire musculoskeletal system. Some of the adaptations just discussed contribute to efficient motion of the limbs. The nature of the fibers in muscles used for running and the man-ner in which these are employed during running also contribute to endurance (Armstrong 1981; Goldspink 1981). The study of joint mechanics, models for ambulatory and cursorial movement, and research into the energetics of these processes in mammals, as well as other vertebrates, have become quite sophisticated and complex in recent years (Hildebrand 1985a).

Consider again the example of the cheetah and gazelle from the beginning of the chapter to examine cursorial locomotion further. The gazelle uses unguligrade movement as it flees from the predator, and the cheetah uses digitigrade locomotion, though it may, at times, appear to be bounding. The top speeds for these animals (see table 5.3) might predict that the cheetah will win the race; however, the distance over which the animal can travel at the prescribed speed must also be considered. Cheetahs can move swiftly only for relatively short distances, say a few hundred meters at most, whereas the endurance of the gazelle at top or near top speeds is considerably longer. Thus, the cheetah needs to be close enough to the gazelle when it begins its dash so it can overtake the gazelle before the prey outdistances the predator.

Jumping and Ricocheting

Jumping and ricocheting are called **saltatorial** locomotion. Jumping, or springing, involves the use of all four feet, as in the case of lagomorphs. Ricocheting involves propulsion provided only by the two hind limbs, such as in kangaroos, kangaroo rats, and jumping mice (figure 5.13; also see figures 17.10 and 17.12). Most mammals that employ a ricochetal form of movement spend much of their time in a bipedal position, using the forepaws only occasionally for locomotion. The forelimbs of such animals are shorter than the hind limbs and are often used for manipulating objects in the environment, principally food. It is noteworthy that most mammals (indeed, most vertebrates) that use saltatorial movement are relatively small, with the kangaroos and wallabies (up to 90 kg body mass) being exceptions.

Hind limbs of saltatorial mammals exhibit convergent evolution (Emerson 1985). The principal adaptation is a lengthening of one or more segments of the hind limb. In mammals that use jumping or ricochetal locomotion, the tibia is lengthened to varying degrees. By relying on the elasticity of the tendons in their leg joints, such mammals gain efficiency in their locomotion. As their limbs, joints, muscles, and tendons recover from one jump, energy is stored, particularly in the tendons of the lower leg and ankle. This energy is then released for the next propulsive thrust forward (Biewener et al. 1981). Other adaptations for saltatorial locomotion involve

1. A posterior shift in the center of body mass to avoid the problem of tumbling forward (or backward) when propulsive thrust from the hind limbs occurs

2. An enlargement of muscles in the hip region, with lengthy tendons

3. Changes in the size and arrangement of bones in the pelvic girdle to accompany the shifts in musculature and center of body mass

4. Larger hind feet for take-off and landing

5. A longer tail for balance

One group of animals that uses a variation of the ricochet method deserves special mention—the indrid lemurs (figure 5.14). These primates are found on the island of Madagascar. Indrid lemurs use a locomotory style termed "vertical clinging and leaping." They cling to the sides of tree trunks with all four limbs and bound downward to the ground using only their hind feet, off the ground, and back to a similar clinging position on another tree. Their tarsi are lengthened rather than either of the two more proximal segments of the hind limb, as is the case in most other mammals. These animals also exhibit some awkward bipedal and quadrupedal movement and move inefficiently on the ground.

Swimming

All mammals that spend a significant portion of their time in water have evolved from terrestrial ancestors. **Amphibious** mammals, including beavers (*Castor* spp.) and otters (*Lutra* spp.), spend their time in both terrestrial and aquatic habitats. With few exceptions, amphibious mammals live in a medium that is cooler than their average body temperature. They have a thick coat of fur and sometimes more body fat, which act as insulation and protect them against excessive heat loss to the surrounding medium, as well as provide buoyancy. The feet and tail are often modified (figure 5.15; see figure 17.11). Muskrats possess laterally flattened tails; beavers have tails that are dorsoventrally flattened. Such tails can be used for either control of movement, like a rudder, or propulsion. Webbing between the toes in many aquatic species increases surface area for contact with water during locomotion. Water shrews (*Sorex palustris*) have stiff hairs between the toes (**fimbriation**) that serve the same function. Amphibious mammals use oscillatory propulsion, swimming in the water by using their limbs much as they would to move on land (Webb and Blake 1985).

Mammals that live most of the time in water but come onto land periodically for certain activities are called **aquatic.** These primarily include the seals. Some aquatic mammals, including the eared seals and sea lions (Family Otariidae) and the walrus (Family Odobenidae), use the oscillatory form of locomotion similar to that observed for amphibious mammals (Scheffer 1958). Evolutionary adaptations of these animals include alterations of the forelimbs and hind limbs so that the bones now are within a flipper that acts like a paddle, propelling the animal through the water. The earless seals (Family Phocidae), in contrast, use an undulating form of locomotion through lateral movements of their hind quarters and rear flippers. Both kinds of seals are excellent divers, descending regularly to depths of 300 to 400 m and occasionally even deeper (Kooyman et al. 1976).

Figure 5.13   Saltatorial locomotion.   Kangaroos, such as the eastern gray kangaroo (*Macropus giganteus*) shown here, move about using a saltatorial form of locomotion. Notice the large lower portion of the hind leg and foot, used in propelling the animal forward with each bound.

A

B

Figure 5.14   **Richochetal locomotion.**   Lemurs such as the ring-tailed lemur use a specialized form of saltatorial locomotion involving ricochetal movement from (A) one tree, (B) to the ground, and back to another tree.

As adaptations to their aquatic medium, seals are fully furred and have a layer of insulating fat that is much greater than that of amphibious mammals. The buoyancy provided by water changes some demands on the structural support system, resulting in a general simplification of the axial skeleton. Tails are absent or rudimentary in seals and thus do not play any role in locomotion. Movement on land can be awkward or clumsy and involves limited use of the flippers to pull and push the animal over the ground.

The evolution of particular adaptations for swimming reaches its pinnacle among **marine** mammals—those that never come onto land. These include baleen (Order Mysticeti) and toothed whales (Order Odonotoceti), and dugongs and manatees (Order Sirenia). The axial skeleton of marine mammals is further simplified, with the cervical vertebrae being partially fused. The hind limbs are completely eliminated, except for rudiments of the pelvic girdle, and the sacrum is absent. Tails of whales and sirenians are modified into a horizontal fluke, used for propulsion, in this instance, in the form of an undulating oscillator. Corresponding modifications of the muscle mass operate the tail fluke. As the fluke goes up and down, water is pushed backward and the animal is propelled forward. The buoyancy provided by the water has removed some of the restrictions on body size that would accompany a terrestrial existence. Thus, we find that whales are among the largest mammals. Whales and sirenians have few hairs on their skin but instead have an excellent layer of insulating body fat, called blubber, just beneath the skin. We discuss other noteworthy adaptations of marine mammals in chapters 16 and 18.

Figure 5.15   **Swimming.**   Muskrats are amphibious mammals, spending a portion of their time in the water. They have a tail that is laterally flattened (from side to side), which can be used as a propeller for some movement and a rudder when the animal swims with its front and hind limbs.

Flying and Gliding

Among mammals, only the bats (Order Chiroptera) have evolved true powered **flight,** or **volant** locomotion. With the exception of swimming, flying is the most energetically efficient means of moving a given body mass between two points (Norberg 1985, 1990).

Bats have a number of morphological adaptations for flight. The forearm and hand have elongated. The thumb is represented by a claw on the edge of the wing and is used to crawl on the ground or up a surface. The fingers are greatly elongated, supporting much of the membrane of the wing. A thin sheet of skin, called the flight membrane **(patagium),** extends from the tip of the fifth digit of the forelimb to the hind limb and connects all along the inner edge to the side of the body and includes the tail (figure 5.16). Although some bats attain body masses of hundreds of grams, most weigh less than 100 g, and many species are 15 to 20 g or less. The small body mass is due, at least in part, to the constraints imposed by flying.

Three major types of problems are encountered by bats and any animal that attempts to fly: (1) lift, (2) drag, and (3) power. Lift is generated when an airstream passes over a wing surface, because air moves faster across the upper surface than the lower. This generates a pressure differential, in which lower pressure on the upper surface lifts the wing upward. The wing structure of both bats and birds includes a dorsal surface that is curved upward and a concave ventral surface, called the camber of the wing. Camber contributes to a relatively longer but faster movement of air passing over the top of the wing, which helps generate the pressure differential between the upper and lower surfaces. Bats can change the configuration of their wings relative to the air by changing the angle of attack. Up to the point of stalling, increasing the angle of attack results in greater lift. Bats generally have broad wings with a high **aspect ratio**—the surface area of the wing divided by its length. Short, deep wings are very important to the maneuverability that bats need to negotiate around the variety of obstacles in their environment (e.g., in caves). Similarly, high aspect ratios are

found in birds that live in forests, where a high degree of flight control is necessary.

Wing loading is the mass in grams of the bat divided by the surface area of its wings in square centimeters. As a rule, animals with low wing loadings can fly at slower speeds without stalling than those with higher wing loadings. In general, the surface area of the wings of bats is proportional to body mass (Norberg 1985). That is, smaller, lighter bats have smaller wing surface areas, and larger, heavier bats have larger wing surface areas. Bats have lower wing loadings than birds, they can fly more slowly, and they are more maneuverable (Vaughan 1970). A recent study of little brown bats (*Myotis lucifugus*) revealed a relationship between wing loading and habitat partitioning. Bats with high wing loading and less maneuverability were more likely to forage in more open habitat (Kalcounis and Brigham 1995). Conversely, smaller individuals of the same species were found foraging more frequently in habitats containing more obstacles, such as trees.

Drag refers to anything that retards forward motion. Three sources of drag affect flight: friction from air along the body surface, friction generated at the leading edge of the wing as it passes through the air, and turbulence generated by air slipping from underneath the wing. Bats have relatively long, slender bodies that aid in reducing surface drag. Less body surface area is therefore present at right angles to the direction of movement, and the form of the body allows air to pass with less friction. Bats exert a degree of control over the leading edge of their wing by shifting the breadth of three membranes (propatagium, dactylopatagium brevis, and dactylopatagium minus) that come into forward contact with the air (see figure 5.16). These three membranes can be used like the flaps on the fronts of the wings of aircraft, increasing drag but also adding to lift. The turbulence created by air spilling out from underneath the wings is most pronounced near the wing tip. Part of this problem can be solved, particularly at slow flight speeds, by shifting the positions of the membranes to permit air to move from below the wing without generating as much turbulence. Having wings that are sharply tapered at the tip is another adaptation for minimizing air loss and turbulence.

Power is needed to move the wings through the air to generate the lift needed for flight and to move the animal forward. The major power for locomotion comes from the adductor and abductor muscles of the forelimb. Bats typically fly rather slowly and are very maneuverable compared with birds or insects. Wing motion patterns analyzed using slow-motion photography and strobe lights reveal that in normal flight, bats have a powerful downstroke with the limb fully extended, during which air is forced backward as well as downward. The tip of the wing bends during this process and completes an almost propellerlike movement. On the upstroke, the limb is flexed and moved upward, backward along the body, and returns to the original extended position in preparation for the subsequent downstroke. Bats generate the recovery stroke during flight using muscles located in the back region. This contrasts with birds,

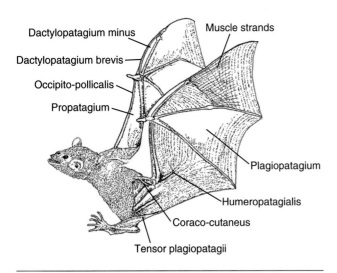

Figure 5.16   Flight.   (A) A view of a fruit-eating bat (*Artibeus lituratus*) skeleton reveals the general size, the disproportionately elongated bones of the forearms and hands, and fingers. (B) The flight membrane (patagium) of bats is made up of several parts, that play different roles in the process of flight.

in which the return stroke is powered by muscles attached to the sternum in the thoracic region.

Gliding as a means of movement has evolved independently in three separate groups of mammals: the Orders Rodentia and Dermoptera and some marsupials. Species such as North American flying squirrels (*Glaucomys* spp.), Australian honey gliders (*Petaurus breviceps*), and colugos (Order Dermoptera) have developed the ability to glide over relatively short distances. Mammals that glide have a patagium that extends along the lateral surface of the body connecting the forelimb and hind limb as well as the head and tail (see figures 17.8 and 17.13). This membrane is much thicker than in bats, and its size is controlled by positioning the limbs. Increases in speed are achieved by decreasing the surface area of the membrane and the angle of attack. Conversely, speed is decreased, until the stall speed is attained, by increasing membrane surface area or angle of attack or both.

Climbing

Climbing mammals use their forelimbs and hind limbs to move about their habitat, which is often arboreal. We do not include species such as the mountain goat (*Oreamnos americanus*) that climb about on rocky mountain surfaces in this catagory; they use some form of walking or running locomotion. Climbing mammalian species face the dual problems of avoiding falling to the ground, where they could be injured or subject to predation, and making their way within the discontinuous pattern of tree branches and limbs where they live (Cartmill 1985).

Several adaptations enable different groups of mammals to solve the problems of climbing vertically and clinging to and moving about in trees. Many mammals have claws that enable them to attach to trees; several rodent species are excellent examples. Smaller mammals can climb directly up vegetation using only their claws. Other clawed species climb trees by a combination of using claws and grasping the tree from the side (figure 5.17) with lateral pressure from the forelimbs or hind limbs. In other instances, particularly among primates, we see the evolution of **prehensile** (grasping) hands and feet. These species have friction pads to aid in gripping tree limbs securely and increased numbers of sensory receptors on the hands and feet. In species such as sloths, a third solution to survival in the arboreal habitat has evolved—hanging underneath tree branches. In all three instances, animals generally locomote with an alternating forefoot and hind foot pattern, propelling themselves upward or moving laterally among the tree branches. As a further adaptation, primates such as gibbons have evolved longer and stronger forelimbs that they use for **brachiation,** swinging from branch to branch using the arms. Alternatively, species such as tree squirrels use leaping forms of locomotion to traverse large gaps between branches or to go from one tree to another.

Most climbing mammals have long tails, which in many instances are used for balance. In some lemurs and Old

Figure 5.17   **Tree climbing.**   When a black bear (*Ursus americanus*) climbs a tree, it uses a combination of grasping the tree trunk with its forepaws and hind feet, aided by lateral pressure from the arms and legs, and the claws on the ends of its digits.

World monkeys, tails are used as a brace against the tree during upward and downward climbing. Many rodents that climb, such as harvest mice (*Reithrodontomys* spp.), use their tails for balance when climbing. In a few species of South American primates, the tail has become a prehensile appendage; such animals can use their tails to grasp branches. The distal portion of the tail in these monkeys has developed friction pads and increased sensory receptors similar to the gripping hands and feet of other climbing mammals.

Digging and Burrowing

Mammals that dig in the soil to find food or create shelter are called **fossorial** (Hildebrand 1985b). Mammals have evolved a variety of means of digging, including use of the limbs and feet, teeth, and head. Digging and scratching with the forelimbs, often aided by the hind limbs, may be the most familiar form of digging. Increased size and strength in the claws and the structure and musculature of the limbs and pectoral girdle are associated with digging. Pocket gophers and armadillos are good examples of mammals that scratch and dig to create burrows. Digging activity often consists of a series of rapid alternate strokes by the forelimbs to loosen the soil and move it backward, followed by similar actions by the hind limbs to move the soil farther from the digging site or out of the burrow, or to compact the soil behind the animal. Some mammals turn around and use their forelimbs and head to push dirt forward bulldozer style along a tunnel or out of the burrow.

Mammals that primarily use their teeth for digging include bamboo rats (Genus *Rhizomys*) and mole rats (Genus *Spalax*). Characteristic adaptations of these animals include enlarged heads with strong rostrum and zygomatic arches for muscle attachment involved in using the large incisors to cut through soil (Orcutt 1940; Krapp 1965). Other mammals, such as golden moles (family Chrysochloridae), use their heads either to force the surface layer of soil upward or as a means of compacting soil (Puttick and Jarvis 1977). The musculature used to raise the head and neck is highly developed in these species. Bateman (1959) describes a golden mole (*Amblysomus hottentotus*) weighing less than 60 g that moved an iron plate weighing more than 9 kg with its head in order to escape from a fishbowl filled with soil! Mole rats actually use a combination of their teeth, feet, and head to establish and maintain their elaborate systems of tunnels. Other fossorial mammals use various combinations of these digging methods.

Mammals such as moles spend virtually their entire lives underground. Moles have evolved a general body form and sleek pelage commensurate with this mode of existence. They have forefeet and hind feet that are highly modified for digging and earth-moving functions (see figure 11.4), extensive vibrissae in the facial region, noses developed as touch receptors, and extremely small eyes (figure 5.18). Many shrews have some of these same adaptations for living either in the soil or in the detritus layer just above it.

Many other mammals, ranging from rodents to canids, dig burrows for several functions, including nesting, giving birth to and caring for dependent young, food storage, hibernation, and as refuges from the elements or predation (Reichman and Smith 1990). Burrows may be used on a

Figure 5.18   Burrowing.   Moles, such as this eastern American mole (*Scalopus aquaticus*), live their entire lives underground. A key adaptation for this fossorial existence is the modification of the hands and feet for digging. Moles build extensive tunnel systems just below the surface of the ground, primarily as a means of finding food.

relatively permanent basis over many months or years, or they may be used briefly and abandoned. Hibernating mammals often dig a separate, deeper burrow for the winter. For example, hamsters (Family Cricetidae) make separate chambers for sleeping, storing food for midwinter feeding bouts, and excreting. Woodchucks (Genus *Marmota*) usually have a summer burrow in an open area and a winter burrow in more forested habitat.

It is important to note that most mammals use two or more means of movement. The principal form of locomotion in seals involves the use of flippers, the tail, and pelvic undulations for propulsion in water, but most seals can also move themselves forward on land. So too can tree squirrels and other climbing mammals use one means of locomotion on the ground and another to move about in the trees.

Summary

In this chapter, we addressed four major topics: (1) integument, (2) skeleton, (3) muscles, and (4) locomotion. Together, these topics tell us a great deal about the integration of form and function in the evolutionary development of the diversity of mammalian forms. We also discussed some specialized adaptations for dealing with problems of movement. The integument consists of the epidermis and dermis. Together, they contain glands of several types, hair follicles, fatty tissue, sensory endings, claws, hooves, nails, and other structures. The major functions of the integument are insulation, protection, sweating as a form of cooling, and reception of certain types of sensory information. In addition,

several structures (e.g., horns and antlers) are derivatives of the integument and underlying bones.

The basic mammalian internal framework consists of a cranium, an axial skeleton, and an appendicular skeleton. The cranial skeleton includes the skull (itself having numerous openings for particular functions, such as the nasal passages and foramen magnum), the face with locations for the major sense organs, and the bones of the jaws. The axial skeleton comprises the vertebral column and rib cage. The axis and atlas vertebrae are key to the high degree of head mobility found among mammals. The appendicular skeleton has a pelvic girdle, a pectoral girdle, and the forelimbs and hind limbs. Each limb consists of two segments of long bones and a third distal segment with ankle or wrist bones and digits.

Mammals have three types of muscles: striated, cardiac, and smooth muscle. Locations of striated muscles, including their points of origin and insertion, closely parallel the skeleton and the functional life-style of every mammalian species.

There are six major categories of locomotion in mammals: (1) walking and running, (2) jumping and ricocheting, (3) swimming, (4) flying and gliding, (5) climbing, and (6) digging and burrowing. Each of these major modes of movement is characterized by key adaptations of limb structure and associated musculature. Many additional modifications have occurred in the sensory and motor systems of mammals that live all or part of their lives in water and of bats, the only mammals to exhibit true flight.

Discussion Questions

1. Spend some time watching your own dog or cat or those living in your neighborhood. Note the pattern of leg movements and the order in which the feet strike the ground. What form of walking and running locomotion are these animals using? As you observe an animal, describe the nature of the lever actions and the underlying sequences of muscle movements. Where is the animal's heel in relation to the ground as it walks? How does this compare with your heel when you walk?

2. If possible, visit a zoo or wild animal park and take careful notes on the nature and distributions of different kinds of hair you can see on the various mammals. Can you relate the patterns of hair types to the locations and climates in which these mammals originated?

If any distinctive color patterns are visible, can you guess what functions they serve?

3. Many mammals have a variety of processes, or ridges, that protrude from their bones. Examine some bones and speculate on why and how these small but significant alterations of the basic bone pattern occurred. Knowing the location of the bones in the mammal's skeleton should help you decide which muscles may have attachments on a particular bone.

4. After examining tables 5.1 and 5.2 in some detail, what additional conclusions can you draw concerning the relationships between mammalian milk composition and growth? What other data might you need to expand your analysis?

Suggested Readings

Cott, H. B. 1966. Adaptive coloration in animals. Methuen, London.
Day, M. H. (ed.). 1981. Vertebrate locomotion. Academic Press, New York.
Hildebrand, M., D. M. Bramble, K. F. Liem, and D. B. Wake (eds.). 1985. Functional vertebrate morphology. Harvard Univ. Press, Cambridge.
Macdonald, D. 1987. The encyclopedia of mammals. Facts on File Publications, New York.
Young, J. Z. 1957. The life of mammals. Oxford Univ. Press, New York.

CHAPTER 6

Foods and Feeding

MODES OF FEEDING

Insectivorous
Carnivorous
Herbivorous
Specializations in Herbivory
Omnivorous

FORAGING STRATEGIES

Optimal Foraging
The Marginal Value Theorem
Food Hoarding

ammals, like all organisms, require energy and nutrients for maintenance, growth, activity, and reproduction, that is, for survival. Maintaining a high body temperature, however, which is a key feature of Class Mammalia, requires regular acquisition of food. The food of mammals ranges from microscopic forms such as diatoms and crustaceans—a staple in the diet of the largest mammals, the baleen whales—to sedentary forms such as plants used by the most abundant mammals, the rodents. Mammals consume food of high energy content (blood of vertebrates and insects) as well as of low energy value (grasses and stems). The food of mammals may be highly specialized and restricted (nectar of localized plants) or rather general and readily available (grasses and herbs). To meet their high energy needs, mammals have evolved a diverse array of trophic, or nutritional, specializations. The adaptive radiation in food-gathering morphologies are diverse and reflect the diversity of available food.

In this chapter, we detail the feeding apparatus of mammals, focusing on the capturing (teeth, tongue, and jaw musculature) and processing (alimentary canal) of food. Feeding integrates the sense organs and locomotor adaptations (see chapters 5 and 7). Although different orders of mammals are sometimes grouped according to their modes of feeding (i.e., Carnivora), food habits cannot be employed as a systematic criterion, because many members of an order may depart from these feeding generalizations. Thus, to enhance understanding of nutritional adaptations, we suggest consulting specific chapters to unite anatomical specializations of different groups with their dietary habits. At the end of the chapter we will examine briefly some general principles regarding mammalian foraging strategies.

MODES OF FEEDING

We understand the life-history traits and food habits of extant mammals by examining their teeth. As you learned in chapter 4, all mammals, except certain whales, monotremes, and anteaters, have teeth, and these structures are inextricably linked with food habits. As mammals evolved in the Mesozoic era, major changes occurred in their dentition and jaw musculature; teeth became differentiated to perform specialized functions. Within extant species, several trophic groups can be recognized, namely insectivorous, carnivorous, herbivorous, and omnivorous mammals. Other specialized modes of feeding have evolved from these four basic plans (figure 6.1).

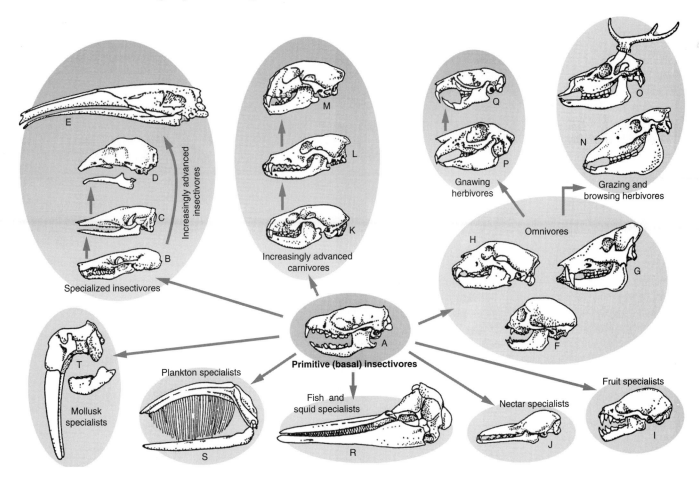

Figure 6.1   Skull and dentition specialization.   Feeding specializations in the dentition and skulls of mammals relate to their dietary habits: (A) hedgehog, (B) mole, (C) armadillo, (D) anteater, (E) giant anteater, (F) marmoset, (G) peccary, (H) bear, (I) fruit-eating bat, (J) nectar-eating bat, (K) raccoon, (L) coyote, (M) mountain lion, (N) horse, (O) deer, (P) jackrabbit, (Q) woodrat, (R) porpoise, (S) right whale, and (T) walrus.

Insectivorous

Insectivory

Mammals that consume insects, other small arthropods, or worms are referred to as **insectivorous** (meaning "insect-eating"). We know from examination of Triassic mammals that the insectivorous feeding niche represented the primitive, or basal, condition of eutherian mammals. Today, this feeding niche is exploited by members of seven groups of mammals: echidnas and platypuses (Monotremata); hedgehogs, shrews, and moles (Insectivora); most bats (Chiroptera); anteaters and armadillos (Xenarthra); pangolins (Pholidota); aardvarks (Tubulidentata); and the aardwolf (Carnivora; see figure 6.1). Many other orders of mammals also contain members that exhibit insectivorous habits. The dentition of hedgehogs, shrews, moles, and most bats is typified by numerous sharp teeth with sharp cones and blades for piercing, shearing, and ultimately crushing the tough chitinous exoskeletons of insects. In many forms, the lower incisors are slightly **procumbent** (pointing forward and upward) to aid in grasping prey (see figure 11.11). Because insectivorous mammals consume minimal amounts of fibrous vegetative material, prolonged fermentation is not required;

their alimentary canals are short, and most insectivores and chiropterans lack a cecum (figure 6.2).

Aerial Insectivores

The most abundant foods are plants and insects; it is therefore not surprising that the most abundant mammals are rodents and bats. Chiropterans occupy ecological niches in almost all habitats of the world; the diversity of their diets is unparalleled among extant mammals (chapter 12). The majority (70%) of microchiropterans are insectivorous (Black 1974; Whitaker et al. 1996). All bats residing north of 38°N and south of 40°S latitude are insectivorous. Throughout their range, insectivorous bats feed on a diverse array of arthropods, ranging from scorpions, spiders, and crustaceans to soft-bodied and hard-bodied insects. Insects are either captured in the mouth or trapped by a wing tip or the uropatagium (see figure 12.1). Foraging styles vary depending on the species. Insectivorous bats are voracious eaters: Mexican free-tailed bats (*Tadarida brasiliensis*) in the southwestern United States, totaling some 50 million individuals, may consume up to 250,000 kg of insects each night (Kunz et al. 1995; Whitaker et al. 1996). Lactating female big brown bats (*Eptesicus fuscus*) nightly consume a quantity of

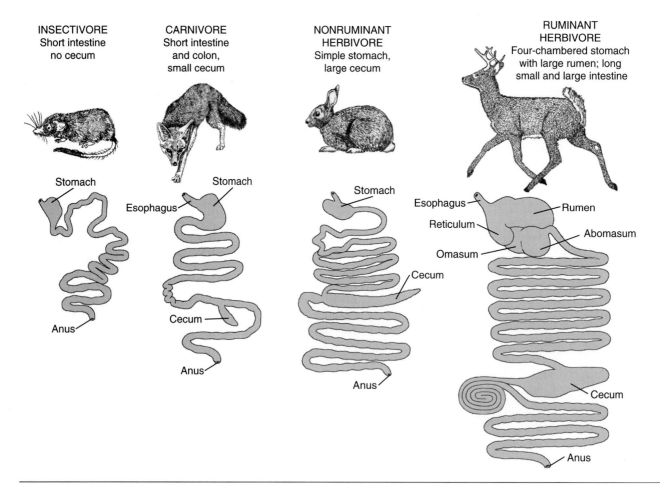

Figure 6.2   Digestive system.   The digestive systems of mammals, illustrating the differences in morphology that correspond to different diets.

insects equivalent to more than their body mass (Kurta et al. 1990). Within a given habitat, feeding assemblages of bats can be quite diverse and may be divided into different guilds, such as species that glean insects or those that forage within forest openings, over water, and in open-air zones above the forest canopy (Findley 1993; Nowak 1994).

The trophic niche of insectivorous bats may be assessed by determining the morphological attributes of a species, including its wing and jaw morphology, brain size, and external dimensions (Findley and Wilson 1982). The size of prey varies in relation to the predator's jaw morphology, from very small midges and mosquitoes to large beetles. For example, Freeman (1979, 1981, 1988) predicted food habits of molossids by assessing jaw structure and mechanics; beetle-eaters were characterized by more robust skulls and fewer but larger teeth, whereas moth-eaters had delicate skulls and numerous yet smaller teeth. Most insectivorous bats are generalists and opportunistic feeders, but remarkable specialists do occur (Whitaker 1994). Pallid bats (*Antrozous pallidus*) of the southwestern United States, for example, feed on beetles, Jerusalem crickets, sphinx moths, scorpions, and small vertebrates gleaned from the ground. Golden-tipped bats (*Kerivoula papuensis*) of southeastern Australia feed by gleaning, flying slowly in dense vegetation and hovering and plucking orb spiders from their webs (Richards 1990; Strahan 1995).

Terrestrial Insectivores

The platypus is a semiaquatic insectivore that feeds on benthic worms, insects, mollusks, and small invertebrates—those creatures that live at the bottom of a body of water (chapter 10). Food obtained during a dive is stored in large cheek pouches that open to the rear of the bill. When the cheek pouches are full, the platypus rests on the surface of the water, and the food is transferred to the rear of the mouth and masticated by horny pads. As in other insectivores, the alimentary canal of platypuses is simple and lacks gastric glands. Cheek pouches are thought to replace the stomach as a food storage area (Harrop and Hume 1980).

Three species of mammal produce a venomous saliva: the northern short-tailed shrew (*Blarina brevicauda*) of North America, the European water shrew (*Neomys fodiens*), and the solenodons (Genus *Solenodon*) from Haiti and the Dominican Republic. Research by Tomasi (1978) and Martin (1981) demonstrated the importance of venom in the hoarding behavior of *Blarina*. In both *Blarina* and *Neomys*, the toxin is stored in submaxillary glands and is administered to the prey through a concave medial surface in the first lower incisors. Extracts of this neurotoxin affect the nervous, respiratory, and vascular systems, causing paralysis and death (Lawrence 1945). *Blarina* bites its prey, immobilizing it, and caches it below ground in a comatose state. Caching sites are marked by defecation and urination and provide shrews with a source of fresh food for some time. The ability to cache unused prey ensures that a predictable

quick energy source is accessible and readily available if prey is scarce (Churchfield 1990). The toxin produced by the submaxillary glands in two of the species of *Solenodon* appears to be similar to that of *Blarina* and *Neomys* (Rabb 1959).

Several groups of insectivorous mammals are **myrmecophagous** (meaning "ant-eaters"). Representatives include the armadillo (*Dasypus*), silky anteater (*Cyclopes*), giant anteater (*Myrmecophaga*), pangolin (*Manis*), aardvark (*Orycteropus afer*), and numbat (*Myrmecobius fasciatus*), which feed on colonial insects, such as ants and termites. Reduction of teeth is common among myrmecophagous mammals, and their dentition departs from the "insectivorous" design of the hedgehogs, shrews, and moles. They possess numerous peg-like teeth (armadillos) or no teeth at all (echidnas, anteaters, and pangolins). The marsupial numbat, the sole member of the Family Myrmecobiidae, possesses numerous, small, delicate teeth—the total number may be as high as 52. The aardvark (Order Tubulidentata) is a special case, characterized by columnar cheekteeth composed of vertical tubes of dentine within a matrix of pulp (see figure 14.11). Mammals that consume colonial insects such as termites and ants possess long, extendible, wormlike tongues. Elongated snouts and strong front feet used as digging tools enable anteaters and aardvarks to burrow rapidly into and tear apart termite hills. Their highly maneuverable, sticky tongues are effective in reaching the inner recesses of ant and termite nests. The tongue may be three times the length of the head; in several groups of anteaters, the tongue is anchored at the posterior end of the sternum rather than the throat (Hildebrand 1995; figure 6.3). Greatly enlarged salivary glands situated in the neck produce a viscous, sticky secretion that coats the tongue and is important in the breakdown of chitin.

Echidnas consume ants, termites, and earthworms, but they do not have teeth or even horny grinding plates on the rear of their jaw, as do platypuses. Rather, a pad of horny

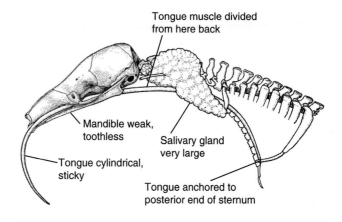

Figure 6.3 **Anteater tongue specialization.** The long, wormlike tongue of the collared anteater (*Tamandua*) is anchored to the posterior end of the sternum and can be protruded extensively to assist in capturing ants and termites.

spines on the back of the tongue grinds against similar spines on the palate to crush the exoskeletons of arthropods. The mouth of the echidna is positioned at the very tip of its elongated snout and can be opened only enough to permit passage of the long, sticky, protrusile tongue. Because there are no glands in the stomach of echidnas, digestive enzymes are not present. The amylase present in the saliva therefore assists in the breakdown of insect chitin within the stomach. Like the insectivores, the monotremes have a simple alimentary canal with a tiny, nonfunctional cecum (Harrop and Hume 1980).

Insects represent a staple in the diet of many other mammals; for example, several species of nasute harvester termites (Genus *Trinervitermes*) form the chief food of the aardwolf (*Proteles cristatus*) of southern Africa (Koehler and Richardson 1990). A single aardwolf was estimated to consume about 105 million termites a year (Kruuk and Sands 1972). The bat-eared fox (*Otocyon megalotis*) of eastern and southern Africa consumes primarily termites and beetles; close to 70% of its diet consists of harvester termites (Genus *Hodotermes*) and dung beetles (Family Scarabaeidae). A combination of extremely long ears and small, numerous teeth enhances the bat-eared fox's ability to detect, capture, and consume its prey.

Members of Order Rodentia are notably omnivorous, but one member, the grasshopper mouse (Genus *Onychomys*) of North America, is unique in having a diet composed almost entirely of grasshoppers, crickets, and ground-dwelling beetles (McCarty 1975, 1978). Grasshopper mice have evolved specialized attack strategies to avoid the defensive secretions of insect prey such as beetles (Genera *Elodes* and *Chlaenius*). When attacking a whip-scorpion, *Onychomys* first immobilizes the tail and then attacks the head. A marsupial "equivalent" of the grasshopper mouse, the mulgara (*Dasycercus cristicauda*), resides in the arid sandy regions of central Australia and specializes in consuming large insects, spiders, and scorpions.

Insectivorous mammals are broadly distributed throughout the class and exhibit remarkable adaptations for locating food. Like platypuses, which rely on tactile receptors on their bill to locate food under water, certain moles employ a similar system underground. Talpids have poor vision but acute hearing and touch. The snouts of moles and desmans are equipped with several thousand sensitive tactile organs, known as **Eimer's organs,** located on the nose (Quilliam 1966; Gorman and Stone 1990). In the star-nosed mole (*Condylura cristata*) of North America, touch receptors are distributed among 22 fleshy, tentaclelike appendages around the tip of the nose (figure 6.4) See chapter 20 for a discussion of electroreception in mammals.

Like the long protrusile tongues of anteaters, some arboreal primates and marsupials employ elongated digits to secure well-hidden insect prey. The third finger of the aye-aye (*Daubentonia madagascariensis*) of Madagascar and the fourth finger of two species of striped possums (*Dactylopsila*) from Australasia are uniquely adapted as probes for removing insects from the crevices of trees (figure 6.5). For example, using

A

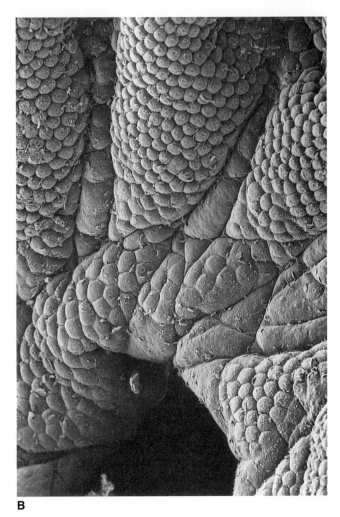

B

Figure 6.4   **Mole specializations.**   (A) The star-nosed mole of North America is unique among mammals in possessing 22 fleshy, tentaclelike appendages surrounding the tip of its nose. (B) Eimer's organs on the nose of a European mole (*Talpa europaea*), as shown in a scanning electron micrograph.

A
B

Figure 6.5   Insect eaters.   (A) The aye-aye (*Daubentoria madagascariensis*) feeds primarily on the tree-burrowing larvae of beetles. (B) It bites into the bark with its powerful incisors and crushes and extracts insect larvae with its elongated third fiinger.

their keen hearing aye-ayes detect larval insects hidden under the bark of dead branches. An aye-aye exposes prey by first gnawing off the overlying bark with its incisors, then inserting its third finger to crush and extract larvae, which it transfers to its mouth. Aye-ayes may have filled this insect-eating niche on Madagascar, which elsewhere is occupied by woodpeckers (Macdonald 1984). Tarsiers (Genus *Tarsius*) from islands of southeast Asia appear to be exclusively insectivorous and carnivorous. Equipped with long legs (the name *tarsier* refers to the elongated tarsal, or ankle), exceedingly large eyes that face forward to permit stereoscopic vision, and a keen sense of hearing, tarsiers are exquisite predators. They capture arthropods, such as ants, beetles, and cockroaches, in trees or on the ground by leaping and pinning the prey down with both hands and quickly dispatching the victim with several bites.

Carnivorous

Carnivory

Carnivorous (meaning "meat-eating") mammals feed primarily on animal material. Members of this group comprise the flesh- or meat-eating members of the Order Carnivora (chapter 15) and the marsupial dasyurids (chapter 10). The Carnivora include the canids, mustelids, felids, and allies. The Order Carnivora is represented by a diverse array of feeding types and dental morphologies, ranging from obligatory meat-eaters with large carnassial teeth, such as felids and hyaenids, to members such as the giant pandas (*Ailuropoda melanoleuca*) with crushing molars that feed exclusively on bamboo shoots (see figure 6.1). During their evolution, carnivores retained a versatile dentition, with different teeth adapted for cutting meat, crushing bone, and grinding insects and fruits (Van Valkenburgh 1989). Animal material is mostly protein and is converted to energy more efficiently than plant material. Thus, like the insectivorous

mammals, the alimentary canal of carnivorous mammals is short and the cecum small or absent (see figure 6.2).

Most carnivores are predators typified by strong skulls, jaws, and teeth—namely sharp incisors and canines—designed to kill and dismember prey. Killing techniques of predators differ; felids and mustelids kill with a single, penetrating bite, whereas hyaenids and canids may kill with several shallower bites. Once the prey is subdued, carnivores rely on their large, strong, pointed canine teeth to tear and shear flesh into hunks, which are then swallowed without being finely divided in the mouth. Also, most carnivores have a pair of carnassial teeth—a combination in which the last upper premolar and the first lower molar teeth form a powerful shearing mechanism when the mouth is closed (figure 6.6). The carnassials are most highly developed in the felids and canids and least developed in the more omnivorous families of ursids and procyonids (figure 15.4). Members of the Order Carnivora bite by employing a chopping motion. Because they have large, crushing molars in addition to carnassials, dogs can crush bones, whereas cats cannot.

Terrestrial Carnivores

With the exception of the otters (Subfamily Lutrinae), which are efficient marine and aquatic carnivores, most mustelids hunt on land. Mustelids are active, fierce hunters, many with specialized methods of killing prey. For example, the long-tailed weasel (*Mustela frenata*) of North America kills its victim by inflicting a rapid bite to the base of the skull or by severing the jugular vein with its sharp teeth. It first consumes the brain, then the heart, lungs, and ultimately the entire body, including bones and fur.

Felids are highly adapted for capturing and consuming vertebrate prey. Their senses of smell and hearing are acute. Their eyes, larger than those of most carnivores, face forward, thus providing the binocular vision and depth perception vital

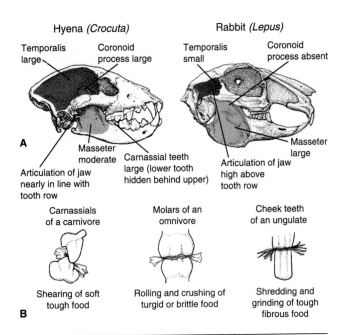

Figure 6.6   Kinds of teeth. (A) Comparison of the jaw mechanics of a carnivore (*left*) and a herbivore (*right*). Note that the carnivore possesses a large temporalis muscle and a moderate masseter muscle attached to a large coronoid process. Herbivores possess a large masseter muscle and a small temporalis, and the coronoid process is absent. (B) The occlusal surface of teeth are adapted for processing three principal kinds of food: (*left to right*) shearing of soft, tough food (e.g., the carnassials of a carnivore); rolling and crushing of brittle food (e.g., molars of an omnivore); and shredding and grinding of tough fibrous food (e.g., cheekteeth of an ungulate).

to locating prey. Their long, stiff, highly sensitive vibrissae are especially useful for foraging at night. Long, sharp, usually re-tractile claws serve as effective meat hooks for capturing, slashing, and manipulating prey. Cats use their long, sharp canines for grasping prey and their well-developed carnassials for shearing food. The tongue of felids is covered with many sharply pointed papillae, which are well suited for holding prey and scraping meat from a carcass.

Canids are opportunistic hunters that rely on high intelligence, social organization, and superb behavioral adaptability (see chapter 22). Small canids hunt singly or in pairs, whereas larger canids, such as gray wolves (*Canis lupus*), hunt in packs of up to 30 members seeking prey that are far larger than themselves (Paradiso and Nowak 1982).

The jaw muscles of carnivorous mammals differ from those of herbivores (see figure 6.6) with regard to the relative importance of the three major adductor muscles of the mandible: the **temporalis, masseter,** and **pterygoideus.** Carnivores must first seize and hold their prey with the canines, which requires a large force at the front of the jaws. The very large temporalis muscles function in holding the jaws closed and aid in the vertical chewing action. In carnivores, the masseter muscle is comparatively small and serves to stabilize the articulation of the jaw, and the pterygoideus muscle helps position the carnassials. In contrast, herbivores rely heavily on a large masseter muscle to maintain a horizontal movement of molars for grinding fibrous food.

Aerial Carnivores

Most chiropterans feed on insects taken on the wing. Certain species of bats are quite specialized, however, feeding on small vertebrates such as rodents, birds, frogs, lizards, small fish, and even other bats. The "carnivorous" bats comprise five families: Megadermatidae, Nycteridae, Phyllostomidae, Noctilionidae, and Verspertilionidae. As noted earlier, the shape of the skull and teeth in bats is a good indicator of diet. For example, the carnivorous false vampire bat (*Vampyrum spectrum*), the largest bat in the New World, has a massive skull equipped with strong, sharp canine teeth and shearing molars adapted for crushing bones and cutting flesh. *Vampyrum* was once thought to be a true vampire bat, but it does not consume blood. Its diet consists of birds, bats, rodents, and some insects and fruit (Gardner 1977). Asian false vampire bats (*Megaderma lyra*) prey on small vertebrates, such as mice, baby birds, and frogs, which are carried to the roost to be eaten. The frog-eating bat (*Trachops cirrhosus*), a phyllostomid, primarily consumes insects and small vertebrates such as lizards. Able to locate and distinguish between different species of frogs by listening for and analyzing their unique calls, *Trachops* can discriminate between poisonous and palatable species (Ryan and Tuttle 1983).

The morphology of teeth, alimentary canal, and limbs is strongly correlated with food habits, a fact that is well illustrated by the only **sanguinivorous** (meaning "blood-eating") mammals—the vampire bats. Three species consume blood, and all are phyllostomids confined to the New World from Mexico to northern Argentina. The vampire bat (*Desmodus rotundus*) preys exclusively on mammals, but the white-winged vampire bat (*Diaemus youngi*) and the hairy-legged vampire bat (*Diphylla ecaudata*) prefer avian prey (Greenhall et al. 1983; Greenhall et al. 1984; Greenhall and Schutt 1996). *Desmodus rotundus* are medium-sized bats weighing between 25 and 40 g (figure 6.7). Morphologically, they are adapted for a diet of blood:

1. The rostrum is reduced, supporting upper incisors and canines that are unusually large and knifelike, with the sharp points of the incisors fitting into pits in the lower jaw.

2. Cheekteeth are tiny.

3. The tongue possesses a pair of grooves at each border that function like drinking straws.

4. The stomach is long and tubular, highly distensible, and well vascularized to enhance the storage of blood and absorption of water.

5. The small intestine is thin-walled and twice as long as the stomach.

6. The kidneys have a unique excretory ability linked with feeding and roosting behavior (McFarland and Wimsatt 1969).

A

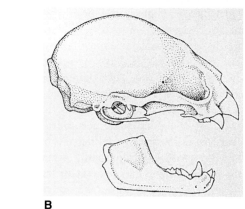

B

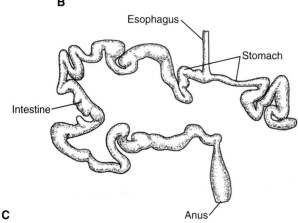

C

Figure 6.7   Vampire bats.   (A) The vampire bat (*Desmodus rotundus*) occurs only in the New World, ranging from northern Mexico south to southern South America. (B) The skull of the vampire bat showing the bladelike upper incisors and canines. The sharp points of the upper incisors fit into distinct pits in the lower jaw behind the incisors. (C) The alimentary canal of the vampire bat. The stomach serves to store large amounts of blood and absorb water, rather than to digest protein, as with most mammals.

Also, the humerus is strong and well-developed. It supports a thumb that is unusually long and equipped with three pads that function like a sole. The forelimbs are unique and greatly aid terrestrial locomotion (Altenbach 1979).

Desmodus rotundus feeds chiefly on domestic livestock. They detect vascular areas of the victim by use of specialized heat-sensitive pits surrounding their nose (Kurten and Schmidt 1982). Bats typically land on the ground near the leg of the host and climb or jump to a feeding site, usually the legs, shoulders, or neck. The upper incisors and canines are used to remove a small piece of skin from the victim or make an incision several millimeters deep. Movement of blood is facilitated by several anticoagulants in the bat's saliva. The flow of blood is maintained by peristaltic waves of the tongue as the bat rapidly licks and continuously abrades the wound. Up to 13 *Desmodus* at a time have been observed feeding on the neck of a cow, with a feeding time of 9 to 40 minutes. Within a 3-hour period, seven bats may feed from the same wound, one after another.

The long and highly vascularized stomach of *Desmodus* does not function for protein digestion as in most mammals. Rather, it is important in the storage of blood and the absorption of water to concentrate the blood. Average consumption of blood in the wild is about 20 mL/day (Wimsatt and Guerriere 1962). Bats may consume up to 50% of their body mass in blood per night, which might impose serious constraints on their ability to fly. Vampire bats cope with this potential problem by employing a unique "two-phase" renal function: in the first phase, which occurs at the feeding site, water is excreted; during the second phase, which takes place at the roost, urine is concentrated. About an hour after feeding, bats rapidly lose much of the water taken in with the blood meal—about 25% of the ingested blood is excreted as urine. This weight loss is essential to enabling the bats to fly back to the roost. At the roost, digestion of the partially dehydrated blood continues. Bats concentrate wastes and thus excrete a highly concentrated urine. The kidney of vampire bats may surpass that of many desert mammals in its ability to concentrate urine and thus conserve water. In addition to relying on superb kidney function for water conservation, vampire bats practice blood sharing at the roosting site by regurgitating blood into the mouth of another bat (Wilkinson 1985, 1987; Altringham 1996).

Aquatic Carnivores

Odontocete cetaceans that feed on fish show a second type of specialized dentition represented by the **baleen**, or **whalebone**, whales. Baleen whales (the mysticetes) are filter, or suspension, feeders—that is, they strain small organisms, known collectively as plankton, from the water by use of the baleen sieves in the front of their mouths (Voelker 1986). The structure and function of baleen and feeding habits of the mysticetes are detailed in chapter 16.

Although baleen whales are touted as the filter-feeders par excellence, they are equaled by one of the pinnipeds. The crabeater seal (*Lobodon carcinophagus*) is distributed along the leading edge of the Antarctic pack ice and is a major consumer of **krill**. The population consumes up to 160 million tons per year (Øritsland 1977; Bonner 1990). The teeth of crabeater seals are well adapted for straining krill from the

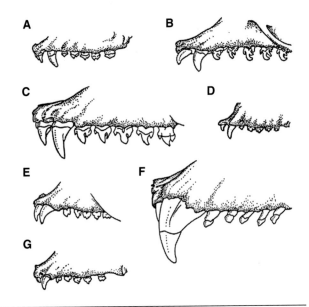

Figure 6.8   Pinniped dentition.   Comparison of the dentition of seals: (A) harbor seal (*Phoca vitulina*), (B) crabeater seal (*Lobodon carcinophagus*), (C) leopard seal (*Hydrurga leptonyx*), (D) ringed seal (*P. hispida*), (E) Ross seal (*Ommatophoca rossii*), (F) elephant seal (Genus *Mirounga*), and (G) bearded seal (*Erignathus barbatus*). The teeth of the upper and lower jaws of the crabeater seal (B) intermesh to form sieves adaptive in straining krill from the sea.

sea. Their cheekteeth possess elaborate cusps (figure 6.8; see also figure 15.8) such that when the jaws close, the cusps intermesh to form an effective sieve for separating krill from the water being forced out of the mouth. Unlike baleen whales, crabeater seals are selective feeders, and foraging is directed at individual prey rather than large masses. This eliminates the amount of water taken in, and thus a less extensive filtering system is required. Seals locate a swarm of krill and direct their snout to the prey. By depressing the floor of their mouth, they effectively suck the prey into their mouth. Once inside, the jaws are closed, and the tongue is raised, expelling excess water and filtering the krill through the matrix formed by the interlocking lobulate cheekteeth. When a sufficient bolus of food is collected in the mouth, it is swallowed. The average meal of krill for a crabeating seal is about 8 kg (Øritsland 1977).

Predators that eat fish are **piscivorous** (meaning "fish-eating"). At least three species of bats are known to capture and eat fish: *Noctilio leporinus, Myotis vivesi,* and *Myotis adversus* (Nowak 1994). Bulldog, or fisherman, bats (*N. leporinus*) are remarkable for their structural and behavioral modifications for capturing and consuming fish. Bulldog bats (the name stems from their large, jowllike upper lip resembling that of a bulldog—see figure 12.17) are characterized by unusually long hind limbs and large feet equipped with sharp, recurved, and laterally flattened claws. Bulldog bats employ echolocation to detect ripples caused by fish swimming near the surface of the water (Brown et al. 1983; Wenstrup and Suthers 1984; Altringham 1996). They skim low, dragging their feet through the water with limbs and

hooklike claws rotated forward, thus acting as a gaff (fishing spear). Once gaffed, fish are quickly transferred to the mouth, where long, thin canines combine with large upper lips and elastic cheeks to form a sort of internal pouch to secure the slippery fish. Fish up to 8 cm long may be captured, and from 30 to 40 fish may be taken per night. *Noctilio* usually forages over small pools, slow-moving rivers, or sheltered lagoons (Findley 1993; Nowak 1994).

Among mammals, piscivory is common in seals, sea lions, and dolphins. Toothed whales, porpoises, and dolphins (Odontocetes) are fish and squid specialists (see figures 6.1 and 16.7). They have numerous, small, simple, teeth that are all alike (homodont). To optimize prey capture, the mouth of a porpoise or dolphin forms a fish trap similar to that used by other fish-eating vertebrates, such as gars, crocodiles, and mergansers. The adaptive value of this morphology is clear—fish are active and slippery and must be trapped and swallowed quickly to prevent their escape. Killer whales (*Orcinus orca*) are the largest predator among "warm-blooded" animals. These opportunistic feeders consume fish, squid, baleen whales and smaller cetaceans, pinnipeds, penguins and other aquatic birds, and marine invertebrates. Foraging success is optimized by cooperation. Pods vary from 4 to 40 individuals, with adults and older juveniles, aided by underwater vocalizations, cooperatively herding the fish. Sperm whales (Family Physeteridae) feed primarily on squid. When consumed, the beak of a squid may act as an irritant to stimulate the production of ambergris from the stomach and intestines of the whale (Macdonald 1984; Voelker 1986). **Ambergris,** once used as a fixative in the cosmetic industry, is a form of excrement from sperm whales. One lump taken from a sperm whale in the Antarctic weighed 421 kg (926 lb; Slijper 1979).

Like the odontocetes, the jaws and teeth of pinnipeds are adapted for grasping prey, not chewing it, and most prey are swallowed whole. Because meat requires only a short period for digestion compared with vegetable matter, one would expect pinnipeds to have a short alimentary canal (see figure 6.2). This is not the case with some seals, however: an adult male southern elephant seal, (*Mirounga leonina*) possessed a small intestine 202 m (660 ft) in length—42 times its body length! The reason for this long gut is unknown. Like other carnivorous mammals, the cecum, colon, and rectum of seals are relatively short.

Herbivorous

Herbivory

Herbivorous (meaning "plant-eating") mammals consume green plants and thus constitute the base of the consumer food web. Plant food is far more abundant than animal food, but its energy content is lower. Gaining access to the protein within leaves and stems is difficult due to the tough fibrous cell walls of plants. We can divide herbivores into two main groups: (1) browsers and grazers, such as the hooved mammals—the Perissodactyla and Artiodactyla (chapter 19), and (2) the gnawers—the Rodentia, and Lagomorpha (chapter 17). Other important herbivores are the kangaroos, wallabies, wombats, langurs, sloths, elephants and hyraxes, and the aquatic grazers such as manatees and dugongs. Herbivores feed on a great diversity of foods, including grasses, leaves, fruit, seeds, nectar, pollen, and even the sap, resins, or gums of plants. Herbivores share unifying characteristics in the design of the skull, teeth, and alimentary canal, which are adapted for feeding on cellulose-rich herbs and grasses for which mammals lack digestive enzymes. In general, herbivorous mammals are typified by skulls in which canines are reduced or absent (see figure 6.1) and broad molars are adapted for crushing, shredding, and grinding fibrous plant tissue (see figure 6.6). Rodents are characterized by the presence of a single pair of ever-growing, chisel-like incisors on both the upper and lower jaws. Lagomorphs have an additional pair of "secondary" upper incisors that are located immediately behind the first pair. Because canine teeth are absent, a wide gap (diastema) occurs between the incisors and cheekteeth. Plant-eaters typically possess a long intestine with either a simple stomach (nonruminant herbivores) or one that has internal folds and is divided into several functionally different chambers (ruminant herbivores; see figure 6.2).

Omnivorous mammals, such as primates, bears, pigs, and some rodents, possess teeth with low crowns (brachyodont), well-developed roots and root canals, and rounded cusps (bunodont) adaptive for a generalized diet (see figure 6.6). These teeth are not very effective in shredding and grinding tough fibrous plant tissue, and several modifications of this basic design have evolved among herbivores. In response to dietary needs, elephants and some rodents have cheekteeth characterized by transverse ridges, or lophs, on their grinding surface (lophodont). In the ungulates, wear on the surfaces of the teeth from chewing on course plant tissue produces and maintains the transverse ridges. Because enamel, dentine, and cement have different hardnesses, continuous grinding maintains the rough surface. In horses, these teeth are hypsodont (high-crowned) with complex, folded ridges on the surface. In **ruminant artiodactyls,** such as goats, cows, and deer, the grinding surfaces of the teeth have longitudinal crescents, or "half-moons" (selenodont), adaptive for breaking down tough plant material.

The jaw muscles of herbivores differ from those of carnivores. The movement in mastication is from side to side (not up and down, as in carnivores), and the upper cheekteeth slide across the complementary surfaces of the lower teeth in a sweeping motion. For herbivores, the major muscles involved in mastication are the masseter and pterygoideus, whereas the temporalis muscle is smaller than that found in carnivores (see figures 6.6 and 15.5). Nonruminant grazers such as horses use their large incisors to snip and cut tough, fibrous stems. They consume large quantities of fibrous food and have robust lower jaws supporting a large masseter muscle used primarily for closing

the jaw. Because leaves require less mastication than grasses due to their lower fiber content, the lower jaw and masseter complex of deer and other browsers are not as pronounced as in horses. Unlike horses and other perissodactyls, ruminant artiodactyls have lost their upper incisors and crop foliage by use of the lower incisors biting against a callous pad on their upper gum, which acts as a sort of cutting board. Cows are well adapted for eating grasses, which they pull free by twisting them around their mobile tongue held against their lower incisors. After biting off the foliage, the cow holds it within the diastema before moving it back to the cheekteeth for grinding. A distance between the lower incisors and the cheekteeth offers the advantage of allowing for a narrow snout that can penetrate into small spaces to crop food, as seen in smaller deer, antelopes, and rodents (see figure 6.1).

Although the specialized teeth of herbivores effectively shred and grind the cell walls of plant tissue and release their contents, only certain enzymes can digest cellulose. Mammals, however, do not produce these **cellulolytic** (cellulose-splitting) enzymes, so they rely on symbiotic microorganisms residing in their alimentary canal. These microorganisms break down and metabolize the cellulose of plants and release fatty acids and sugars that can be absorbed and used by the mammal host. Rodents and lagomorphs become inoculated with the appropriate anaerobic protozoans and bacteria by eating maternal feces, whereas young ungulates commonly consume soil to acquire their microorganisms. Ungulates have evolved two different systems for breaking down cellulose: foregut (rumination) and hindgut fermentation (Putman 1988; Robbins 1993).

Foregut Fermentation

Rumination (digastric digestive system), also called **foregut fermentation,** is typified by artiodactyls, such as camelids, giraffids, antilocaprids, cervids and bovids, as well as by kangaroos and colobus monkeys. Foregut fermenters possess a complex, multichambered stomach with cellulose-digesting microorganisms. After food is procured by cropping or grazing, it immediately passes to the first and largest chamber of the network, the **rumen** (see figure 6.2; figure 6.9). Here, the food is moistened and kneaded, mixing it thoroughly with microorganisms that ferment the food. Large particles of food float on top of the rumen fluid and pass to the second chamber, the **reticulum**—a blind-end sac with honeycomb partitions in its walls. The reticulum is where a softened mass called the "cud" is formed. Fermentation occurs in both the rumen and reticulum, and both absorb the main products of fermentation, short-chain fatty acids. When the animal is at rest, this softened mass is regurgitated, allowing the animal to "chew its cud," or "ruminate." At this time, the mass is further broken down by a potent enzyme, **salivary amylase.** The food is then swallowed a second time; and enters the third chamber, the **omasum,** where muscular walls knead it further. The fourth,

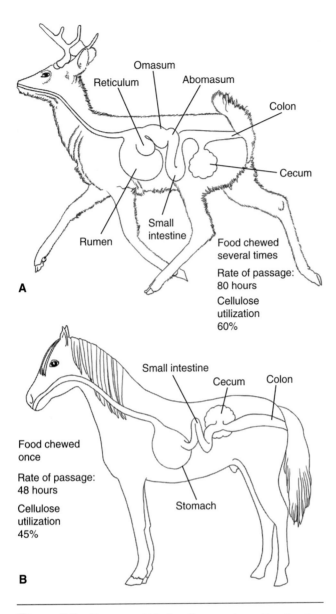

A

Omasum

Reticulum
Abomasum

Colon

Cecum

Small
intestine

Rumen

Food chewed
several times

Rate of passage:
80 hours

Cellulose
utilization
60%

B

Small intestine
Cecum   Colon

Food chewed
once

Rate of passage:
48 hours

Cellulose
utilization
45%

Stomach

Figure 6.9   **Foregut and hindgut fermentation.**   Two digestive systems of ungulates: (A) Foregut fermentation (digastric digestive system) is typified by artiodactyls, such as cervids and bovids, and by kangaroos and colobus monkeys; (B) Hindgut fermentation (monogastric digestive system) is characteristic of perissodactyls, such as horses, zebras, asses, tapirs, and rhinoceroses, in addition to other herbivores such as elephants, lagomorphs, and rodents.

and final, chamber, the **abomasum**, is the true stomach. Here, digestive enzymes are secreted that kill any escaping microorganisms, and protein digestion is completed. Digested material then passes into the small intestine, where the products of microbial digestion and acid digestion are absorbed. Some additional fermentation and absorption occur in the cecum.

Hindgut Fermentation

The **monogastric system**, also called **hindgut fermentation**, is characteristic of horses, tapirs, rhinoceroses, elephants, lagomorphs, hyraxes and rodents. Hindgut fermenters masticate food as they eat, initiating digestion with salivary enzymes. Digestion continues by enzymatic activity within the simple stomach, and food then moves rapidly into the small intestine as new food is eaten. Unlike ruminant artiodactyls, hindgut fermenters do not regurgitate their food. Nutrients are absorbed in the small intestine. Finely ground particles of food pass from the small intestine into the cecum, and larger food particles move through the large intestine and are passed as feces. Fermentation of ingested cellulose by microorganisms occurs in the cecum and large intestine (colon).

The two kinds of fermentation processes that take place in herbivores have clear advantages and disadvantages (Montgomery 1978; see figure 6.9). Foregut fermentation tends to be very efficient because microorganisms begin to break down the plant material before it reaches the small intestine, where it is absorbed. Furthermore, in foregut fermenters, microorganisms from the rumen are themselves broken down by acids in the true stomach (abomasum). The resulting material, which contains the carbohydrates and protein synthesized by the microorganisms, as well as the products of fermentation, moves into the small intestine and colon. Lastly, the microorganisms in the rumen can detoxify many harmful alkaloids in the plants that foregut fermenters consume.

In contrast, food passes rapidly into the small intestine in hindgut fermenters and is then mixed with microorganisms in the cecum. These animals do not digest the microorganisms present in the cecum and thus cannot exploit this potential source of nutrients. In addition, hindgut fermenters must absorb toxic plant chemicals into the bloodstream and transport them to the liver for detoxification or sequestration.

Efficiency may indeed be the trademark of foregut fermenters; however, hindgut fermenters are able to process material much more rapidly. For example, food moves through the gut of a horse in about 30 to 45 hours, whereas it may take a cow from 70 to 100 hours to process food. Hindgut fermenters efficiently digest food high in protein, because large volumes of food can be processed rapidly. Furthermore, hindgut fermentation is effective when forage is dominated by indigestible materials, such as silica and resins, because these compounds move quickly through the alimentary canal by bypassing the cecum. In sum, due to their lowered efficiency, hindgut fermenters must eat large volumes of food in a short time. The foregut system is comparatively slow, because food cannot pass out of the rumen until it has been ground into very fine particles. Thus, ruminants do poorly on forage containing high levels of resins and tannins because these compounds inhibit the function of microorganisms in the rumen. Furthermore, plants with high silica content break down slowly and thus impede movement of food out of the rumen.

The digestive physiology of herbivores influences their ecology and distribution in several ways. Ruminants benefit most from foods that require an optimally efficient digestive system, whereas the best forage for hindgut fermenters is that which facilitates speed of digestion. Each strategy has advantages for survival in particular ecological niches. Speed of digestion may not be important to ruminant artiodactyls. When food is available in the form of tender, short herbage with high protein content, ruminant digestive efficiency pays off. The process of rumination also permits animals to feed quickly and then move to safe cover to chew the cud at leisure. For environments such as the arctic tundra, where food is limited but of high quality, ruminants such as the musk ox (*Ovibos moschatus*) and caribou (*Rangifer tarandus*) have an advantage. As we will learn in chapter 8, some ruminants, such as elands and oryx, also survive well in the deserts of Africa. Although absorption of the products of protein digestion is similar in ruminants and hindgut fermenters, ruminants have the advantage of being able to recycle urea. This allows these mammals to survive with very little water, whereas perissodactyls residing in xeric areas must drink daily to balance the urea in their urine.

When food is of low quality, with high fiber content, yet is not limited in quantity, a premium is placed on the ability to process large quantities quickly. Perissodactyls can survive in regions typified by seasonal drought and poor-quality food—places where ruminants could not process food fast enough to survive.

On the Serengeti Plains of East Africa, dense migrating herds of ungulates influence plant succession and finely partition available resources. They respond to growth of grasses in a predictable sequence (figure 6.10; Gwynne and Bell 1968; Bell 1971). First, perissodactyls such as zebras (*Equus burchellii*) enter the long-grass communities of the Plains and consume many of the longer stems of grasses. Next, come large herds of wildebeests (*Connochaetes taurinus*), trampling and grazing the grasses to short heights. The last invasion of ungulates is Thomson's gazelles (*Gazella thomsonii*), which feed on short grass during the dry season. In addition to different spatial and temporal division of resources, these ungulates sort out available food according to different parts of plants. Zebras consume mostly stems and sheaths of grasses and almost no leaves. Wildebeests eat great numbers of sheaths and leaves, and gazelles eat grass sheaths and herbs not consumed by the other two species. Because grass stems and sheaths are low in protein and high in lignin, and leaves are high in protein and low in lignin, it appears that zebras fare poorly and gazelles do quite well. As we learned earlier, however, perissodactyls, such as zebras, process twice the volume of plant material that ruminant artiodactyls can and therefore compensate with quantity (volume) for what their forage lacks in quality. In addition,

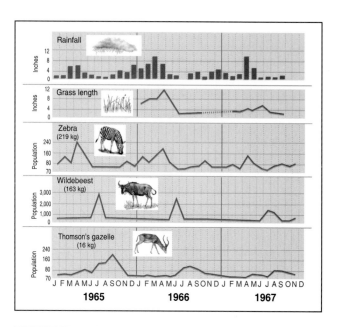

Figure 6.10   **Ungulates of the Serengeti.**   Populations of migrating zebras, wildebeests, and Thomson's gazelles in relation to rainfall and length of grass on the Serengeti Plains of Africa.

Source: Data from R.H.V. Bell, "A Grazing Ecosystem in the Serengeti" in Scientific American, *225:86-93, 1971.*

because zebras are larger than wildebeests and gazelles, they require less energy per unit mass than the smaller mammals.

There are more species of ruminant artiodactyls than perissodactyl hindgut fermenters among the ungulates. This may be due in part to the ability of ruminants to recycle nitrogen and digest protein-rich bacteria. As a result of this ability, they do not have to gain all of their amino acids in forage and can focus feeding on specific, preferred species of plants. This independence of forage selection permits ruminants the freedom to partition resources in finer fashion than most perissodactyls can.

Gnawing Mammals

One of the most successful groups of herbivores are the gnawing mammals—namely the rodents and lagomorphs. Like the ungulates, rodents and lagomorphs cannot produce the enzyme cellulase, so they facilitate the fermentation of fibrous forage with the aid of bacteria and protozoa. As with the perissodactyls, rodents and lagomorphs do not ruminate, and hindgut fermentation occurs in the colon and cecum (see figure 6.2). The only rodents that lack a cecum are dormice (Family Gliridae), indicating that their diet possesses little cellulose. Compared with ruminants, the stomach of rodents is simple but possesses from one to three chambers (Carleton 1973, 1985; Hume 1994). The small intestine is comparatively short, and the hind gut (colon and cecum) is complex, with the cecum having many spiral folds, recesses, and saclike expansions (Bjornhag 1994). Within the Order Rodentia, variation in morphology of digestive systems is correlated with diet. For example, the sciurids (squirrels, chipmunks, and marmots), which feed on a variety of seeds, nuts, fruits, and herbs, have a much simpler digestive system than grass-eating arvicolines (voles and lemmings; Batzli 1985; Batzli and Hume 1994). Variation in the morphology and function of the gastrointestinal tracts of voles correlates with diet and thermal stress. For example, high-fiber diets coupled with cold acclimation of arvicoline rodents such as prairie voles (*Microtus ochrogaster*) may result in higher rates of food intake and increases in the size of the hind gut (Hammond and Wunder 1991).

Rodent skulls are characterized by their large gnawing incisors (see figure 17.1), but they also show features common to all herbivores. The incisors (see figure 6.1) are used to gnaw through hard plant coverings to reach the tender material inside; as well as for nibbling grasses and shrubs. The lips can be folded in behind the incisors to prevent chips of bark or soil from entering the mouth during gnawing. As with other herbivores, a diastema posterior to the incisors results from the absence of canine and premolar teeth. Most mouselike rodents lack premolars, but jumping mice (Genus *Zapus*) have one on both sides of the upper jaw and squirrel- and cavy-like rodents have one or two premolars on either side of the jaw. Food can also be held in the diastema before it is passed back to the cheekteeth for processing—this is most obvious in grass-eating herbivores. Among rodents, members of at least four families (hamsters, pocket gophers, pocket mice, and squirrels) have either internal or external cheek pouches that open near the angle of the mouth. External cheek pouches can be everted for cleaning. Cheek pouches are well adapted for carrying food; an early biologist discovered a total of 32 beechnuts in the cheek pouches of an eastern chipmunk (*Tamias striatus;* Allen 1938). In herbivores, three main masticatory muscles—the masseter, pterygoideus, and digastricus—regulate how these animals shred and grind tough, fibrous food (see figure 6.6). The arrangement and function of these muscles are responsible for the forward and backward jaw movements of rodents, in contrast to the lateral chewing movements of lagomorphs.

Coprophagy

Digestion of cellulose in hindgut fermenters, such as rodents and lagomorphs, occurs in the cecum. Because there is no regurgitation and the rate of passage of forage is rapid, these mammals process a minimal amount of the fiber when they first ingest plants. As a result, **coprophagy** (refection), the feeding on feces, has evolved in lagomorphs, rodents, and shrews (McBee 1971; Kenagy and Hoyt 1980; Macdonald 1984). The cecum, located between the small and large intestines, houses bacteria that aid in digestion of cellulose (figure 6.11). Most products of digestion, except for certain nutrients such as essential B vitamins produced by microbial fermentation, pass through the gut into the bloodstream. Such minerals and vitamins would be lost if lagomorphs did not eat some of their feces and so pass them through the gut

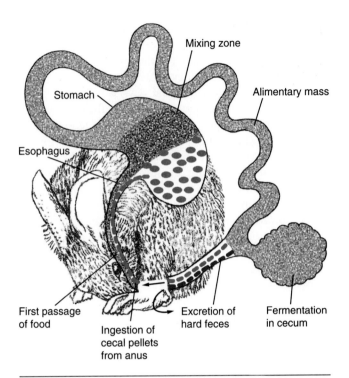

Figure 6.11   Coprophagy. Coprophagy occurs in shrews, rodents, and lagomorphs. The digestive tract of lagomorphs is highly modified for coping with large quantities of vegetation. The alimentary canal has a large cecum, which contains bacterial flora to aid in the digestion of cellulose.

twice. To optimize the uptake of essential vitamins and minerals and enhance assimilation of energy, lagomorphs produce two kinds of feces. The first are moist, mucus-coated, black cecal pellets excreted and promptly eaten directly from the anus. These are stored in the stomach and mixed with food derived from the alimentary mass. Second, lagomorphs produce hard, round feces that are passed normally. The frequency of coprophagy in rabbits is usually twice daily. Prevention of coprophagy in laboratory rats resulted in a 15% to 25% reduction in growth. Another coprophagous mammal is the mountain beaver (*Aplodontia rufa*), which extract fecal pellets from their anus individually with their incisors. They cache these in underground fecal chambers, reingesting them at a later time. Coprophagy has been documented in about nine species of shrews, including the northern short-tailed shrew (*Blarina brevicauda;* figure 6.12; Merritt and Vessey, in press). The adaptive significance of coprophagy for shrews is unknown, but it may represent a technique of reducing daily food intake and extracting certain essential nutrients and vitamins from available food.

Herbivory is not confined to ungulates, rodents, and lagomorphs. Orders such as Carnivora and Chiroptera have species with strong herbivorous habits. The Orders Rodentia and Chiroptera comprise the greatest number of species of mammals; their success is due in part to adaptive radiation in feeding techniques, including insectivory, granivory, folivory, frugivory, nectarivory, and gumivory (Eisenberg 1981).

Figure 6.12 **Short-tailed shrew.** The northern short-tailed shrew (*Blarina brevicauda*), one of the rare venomous mammals, is a common inhabitant of moist habitats of eastern North America. Its poison, produced by submaxillary glands, is administered to its victim though a medial groove in the lower incisors. The poison quickly immobilizes small prey, which may be cached in a comatose state to be available as a source of fresh food for some time after capture.

Specializations in Herbivory

Granivory

Herbivorous mammals that consume primarily fruits, nuts, and seeds are referred to as **granivorous** (meaning, seed-eating). Equipped with large, external, fur-lined cheek pouches and a keen sense of smell, heteromyid rodents represent the most specialized seed-eaters. Kangaroo rats (Genus *Dipodomys*), kangaroo mice (Genus *Microdipodops*), and pocket mice (Genus *Perognathus*) of North American deserts are primarily granivorous. Seeds are also the mainstay for tropical and subtropical species of heteromyids (Genera *Heteromys* and *Liomys*) that harvest fruits, nuts, and seeds from shrubs and trees and cache these propagules in underground burrows (Fleming 1970, 1974; Sanchez-Cordero and Fleming 1993). Several species of the Family Muridae also feed heavily on seeds. The diversity and availability of seeds in desert ecosystems is a key to the evolutionary success of the heteromyids. In terms of the biomass of seeds harvested, heteromyids are rivaled only by ants as important granivores inhabiting North American deserts (Brown and Davidson 1977). Rodents are reported to use over 75% of all seeds produced at certain Mohave and Chihuahuan Desert sites (Brown et al. 1979). In the Mohave Desert of California, *Dipodomys merriami* consumed over 95% of the seeds produced by the annual *Erodium cicutarium* (Soholt 1973). Maximum numbers of seeds produced in desert habitats of North America range from 80 to 1480

kg/ha (Tevis 1958; French et al. 1974; M'Closkey 1978). Minimum densities of seeds remaining in the soil years after the last seed crop are rarely below 1000 seeds/m^2 (Nelson and Chew 1977; Reichman and Oberstein 1977).

As a result of the abundant seed resources and competition with ants, birds, and other rodents, heteromyids have evolved fascinating morphological and behavioral adaptations to optimize their foraging success. All heteromyids cache seeds. They collect large quantities of seeds and store them either in larders within their burrows or scatter hoard them in small buried caches outside the burrow. They employ their large cheek pouches to collect many seeds in single foraging bouts (Brown et al. 1979). Seeds used by heteromyids are derived primarily from grasses and forbs and are quite small, usually less than 3 mm long and weighing less than 25 mg. Kangaroo rats collect most of their seeds directly from plants by clipping fruiting stalks and removing seeds from felled seed heads or by plucking seeds from fruit located close to the ground. Heteromyids may also collect seeds, primarily located in clumps, from the surface of the soil or strain them from the soil (Eisenberg 1963; Randall 1993; Reichman and Price 1993). Seeds are relocated by olfactory cues coupled with memory (Jacobs 1992; Rebar 1995). *Dipodomys merriami* and *Perognathus amplus* may locate seeds below the surface of the soil by detecting concentrated odor characteristics of buried seeds (Reichman and Oberstein 1977; Reichman 1981; Smith and Reichman 1984).

Folivory

Animals that exhibit adaptations for consuming leaves and stems are referred to as **folivorous** (meaning "leaf-eater"). Within the Class Mammalia, about 42 genera, or 4%, of mammals specialize in the consumption of leaves and stems (Eisenberg 1978). As in the grazing and browsing ungulates, consumption of leaves and stems requires considerable morphological adjustment in dentition, jaw musculature, and gut morphology (Eisenberg 1978). In response to predation of leaves by herbivores, however, plants have evolved diverse chemical defenses (Freeland and Janzen 1974; Belovsky and Schmitz 1994; Foley and McArthur 1994). A detailed discussion of plant defenses to herbivory is not within the scope of this chapter. Instead, we choose to highlight specific mammalian practitioners that exemplify mechanisms adaptive in folivory. We encourage you to examine recent reviews of the ecological and evolutionary consequences of plant-herbivore interactions (Batzli 1985; Belsky 1986; Palo and Robins 1991; Batzli 1994; Coley and Barone 1996).

Leaves are difficult to digest and have poor nutritional value. In addition, many leaves contain potentially toxic phenolics and terpines. In spite of these obstacles, three species of marsupials subsist on seemingly unpalatable leaves of *Eucalyptus*: koalas (*Phascolarctos cinereus*), greater gliders (*Petauroides volans*) and common ring-tailed possums (*Pseudocheirus peregrinus*) of eastern Australia. These species have evolved a remarkable suite of anatomical, physiological, and behavioral adaptations to consume *Eucalyptus*. Of special interest among

Figure 6.13   Koala.   The koala (*Phascolarctos cinereus*) occurs in *Eucalyptus* forests and woodlands of Australia. It lives alone or in small groups and is well adapted for arboreal life. Climbing is enhanced by its strong limbs, sharp claws, and use of two opposable digits on its forepaw and the first toe of the foot.

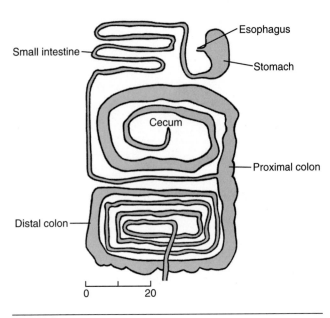

Figure 6.14   Koala digestion.   The digestive tract of the koala is well adapted for digesting fibrous leaves of *Eucalyptus*, the staple in its diet. The caecum, measuring up to 2.5 m (8 ft), is the site of microbial fermentation.

these is the koala (figure 6.13). Only about five species of *Eucalyptus* comprise the bulk of the diet of koalas. The koala's jaw is very powerful and is equipped with sharply ridged, high-cusped molars that finely grind *Eucalyptus* leaves. Koalas consume about 500 g of leaves each day. The stomach is small and the small intestine is of intermediate length; the colon and cecum, however, are extremely long and wide (figure 6.14). The cecum, which is the site of microbial fermentation, is the most capacious of any mammal, measuring four times the koala's body length! Hind gut microflora are thought to detoxify certain essential oils of *Eucalyptus*—about 15% of which pass through the alimentary canal without transformation or absorption (Eberhard et al. 1975). Toxic compounds are inactivated in the liver through the action of glucuronic acid and then are excreted. The cecum also plays a vital role in absorption of water from *Eucalyptus* leaves: The water content of fecal pellets of koalas is low (about 48%), similar to that of camels and kangaroo rats (Schmidt-Nielsen 1964). The frugal water economy of koalas is essential, because they do not drink free water and must rely on water derived from their folivorous diet.

Folivory is represented in about 12% of the genera of primates. With this group, notable folivores include the in-

dris, howlers, langurs, gorillas, and colobus and leaf monkeys of Africa and Asia The diet of gorillas (*Gorilla gorilla*), the largest of all primates, consists of about 86% leaves, shoots, and stems (Fossey and Harcourt 1977), yet their close relatives, the chimpanzees, feed mainly on fruit. Leaves are the staple in the diet of all species of colobus monkeys except Brelich's snub-nosed monkey (*Pygathrix brelichi*), which feeds on fruits such as wild cherries, pears, and cucumbers. Colobus monkeys have flexible food preferences. They consume fruits, flowers, buds, seeds, shoots, sap, and arthropods, in addition to leaves. These monkeys are unusual because, like ruminant artiodactyls and kangaroos, they possess a greatly modified forestomach and thus are able to ruminate. Colobus monkeys have large stomachs with a large upper, sacculated region where fermentation by microorganisms occurs and a lower region typified by high concentrations of digestive enzymes. Digestion is further enhanced by enzymes produced by large salivary glands augmented by mastication with high, pointed cusps on their cheekteeth.

Two-toed and three-toed sloths of South America (Genera *Choloepus* and *Bradypus,* respectively) feed almost exclusively on leaves, stems, and fruit. Like other leaf-eating mammals, sloths possess an extremely large, compartmentalized stomach containing cellulose-digesting bacteria. As in colobus monkeys, the sloth's stomach may be one-third its body mass. The colon and cecum of sloths are relatively simple. Feces and urine are passed only once per week, and thus the rectum is quite expanded, an adaptation for storing feces during prolonged periods between defecations.

Contrary to our stereotyped view of Order Carnivora, at least two species of "carnivores" practice folivory. The red panda (*Ailurus fulgens*) consumes mostly bamboo sprouts, roots, and fruit. The giant panda (*Ailuropoda melanoleuca*) is well known for its consumption of bamboo shoots but feeds on only about 5 species out of the 20 available. Pandas are unique in possessing an extra "digit" on their forepaws (Schaller et al. 1989). This enlargement of one of the wrist bones acts as sort of a "thumb" to oppose the rest of the digits, enabling pandas to grip and manipulate slender pieces of bamboo with great dexterity. The carnassial teeth of *Ailuropoda* are well adapted for crushing and slicing fibrous plants. Because bamboo is low in nutritional value, giant pandas spend about 12 hours a day consuming up to 40 kg of bamboo, yet digest less than 20% of what they eat. Much of the stem is passed through the gut relatively unchanged, resulting in many large feces.

Other notable arboreal folivores include the dermopterans (colugos) of southeast Asia, the prehensile and South American porcupines, tree hyraxes, and about 18 species of rodents. Leaves may represent an important part of the diet of megachiropterans (Kunz and Ingalls 1994; Kunz and Diaz 1995).

Frugivory

Animals that exhibit adaptations to consume a diet of fruit, the reproductive part of plants, are referred to as **frugivorous** (meaning "fruit-eating"). Mammals from several orders are known to specialize in the consumption of fruit: pteropodid and phyllostomid bats; phalangerids; and primates such as indrids, lorisids, cercopithecids, colobine; and the pongids. Because fruit may have a hard outer covering, the teeth of some frugivores are adapted for piercing and crushing. Mammals that subsist on softer fruits typically possess a reduced number of cheekteeth with a bunodont occlusal pattern.

Approximately 29% of all species of bats depend on plants as a source of food. They are distributed within two families: the Pteropodidae of the Old World tropics and Phyllostomidae of the New World tropics (Fleming 1982, 1993; Fleming and Sosa 1994; Racey and Swift 1995). No fruit- or nectar-feeding microchiropterans occur in the Old World (Nowak 1994). Bats play an essential role in the pollination of flowers **(chiropterogamy)** and dispersal of seeds. Because some seeds exhibit higher rates of germination after passing through the gut of a mammal, this method of dispersal is termed **chiropterochory**. Frugivorous bats comprise either principal or partial pollinators and dispersers of close to 130 genera of tropical and subtropical plants. Bats are particularly important to those species of plants that blossom only at night (e.g., avocados [Genus *Persea*], balsa [*Ochroma lagopus*], durian [*Durio zibethinus*], *Eucalyptus*, figs [Genus *Ficus*], guava [*Psidium guajava*], kopok [*Ceiba pentandra*], mangoes [*Mangifera indica*], papaya [*Carica papaya*], and wild bananas [*Musa paradisiaca*]).

Megachiropterans (Old World flying foxes and fruit bats) are restricted to the tropical forests of Africa, Asia, and the Australian region, where succulent fruits are plentiful. Marshall (1985) reported that megachiropterans feed on fruits of at least 145 plant genera in 50 families. Most species locate fruit by smell. Old World fruit bats possess comparatively few teeth, and the lower molars are reduced in number and possess large, flat grinding surfaces. The canines are the principal piercing teeth (see figure 6.1). Rather than biting off and swallowing mouthfuls of fruit, megachiropterans crush the pulp of ripe fruit in their mouth, swallow the juice, and spit out most of the pulp and seeds. They can bite into fruit while hovering, or they may hang onto a branch with one foot and press the fruit to their chest with the other foot and bite into it. If the fruit is small, they may carry it with them to a branch and hang their head downward while they consume it (Nowak 1994). Most megachiropterans crush pulp from soft fruits and may extract only the juice and reject the pulp and seeds or digest the pulp and excrete intact seeds. Because this activity often occurs at some distance from the harvesting site, seeds are dispersed great distances. Several species of Old World fruit bats (*Rousettus aegyptiacus, Epomophorus wahlbergi,* and *Eidolon helvum*) are principal agents of dispersal of the baobab (*Adansonia digitata*), an important tree in the African savanna (Start 1972; Nowak 1994).

Although phyllostomids may be viewed as generalists (chapter 12), over half of the species consume some fruit. Most fruit-eating phyllostomids are in the Subfamilies Carolliinae or Stenodermatinae of the neotropics or the Brachyphyllinae of the Antilles (Gardner 1977; Findley 1993). Members of the Subfamily Carolliinae possess reduced molars and consume ripe, soft fruits such as bananas and figs. In contrast, members of the Subfamilies Stenodermatinae and Brachyphyllinae have more robust molars adapted for crushing fruit (see figure 6.1). *Artibeus,* a neotropical fruit bat, possesses a short rostrum and a high coronoid process on the dentary, which supports a strong temporalis muscle and enhances a vertical chewing motion. Like many of the megachiropterans, *Artibeus* bites chunks from fruit and crushes the pulp for its juices with its broad, flat posterior teeth. They eat their own weight in fruit each night, and food passes through their alimentary tract rapidly, in about 15 to 20 minutes. *Artibeus* is important in the dispersal of seeds of tropical fruits. *Artibeus jamaicensis* is a generalist frugivore, consuming up to 92 taxa of plants, with figs forming a staple in its diet throughout its range (Gardner 1977; Fleming 1982; Handley et al. 1991).

Nectarivory

Insects and hummingbirds are not the only animals that feed on flowers. Some mammals are also exquisitely adapted to feed on nectar and in the process transfer pollen. **Nectarivorous** (meaning "nectar-eating") mammals are represented by about six genera of bats (both mega- and microchiropterans) and the marsupial honey possums. Their skulls are characterized by elongated snouts; small, weak teeth (bats); reduced numbers of teeth (honey possums); and poorly developed jaw musculature (see figure 6.1). The tongue is

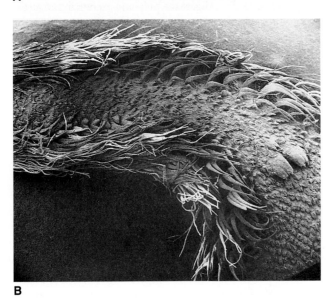

A

B

Figure 6.15   Bat tongues.   (A) The long, slender tongue of a long-nosed bat (Genus *Leptonycteris*) is used to transport nectar from plants to its mouth. The tongue may be extended up to 76 mm, almost the length of the bat's body. (B) Dorsal surface of the tongue of *Leptonycteris* showing the brushlike tip consisting of rows, or tufts, of hairlike projections that slope toward the throat.

long, slender, and protrusile and typically has a brush tip (see figures 6.15, 10.28, and 10.30) consisting of many rows of hairlike papillae pointing toward the throat (Greenbaum and Phillips 1974; Hildebrand 1995). Long-nosed bats of the Family Phyllostomidae are well adapted to feed on fruits, pollen, nectar, and insects. Flowers of the *Agave* and the saguaro cactus (*Carnegia gigantea*) are a staple in the diet of the big long-nosed bat (*Leptonycteris nivalis;* Hensley and Wilkins 1988), a bat (20–25 g) with high energy demands. During feeding, *L. nivalis* crawls down the stalk of an *Agave,* thrusting its long snout into the corolla of the flower. It licks nectar with its tongue, which can be extended up to 76 mm.

Big long-nosed bats emerge from feeding covered with pollen. They are agile flyers and form foraging flocks containing at least 25 bats that feed at successive plants. While feeding, bats circle the plants and take turns feeding on the flowers. Flocks show a cohesiveness typified by minimal antagonistic behavior. Flocking seems to confer the adaptive advantage of an increased foraging efficiency, which is critical for minimizing energy expenditure.

Another phyllostomid regarded as an obligate pollen feeder is the Mexican long-tongued bat (*Choeronycteris mexicana*). Analysis of stomach contents of bats from Central America showed a majority of pollen grains from pitahaya (Genus *Lemaireocereus*), cazahuate (Genus *Ipomoea*), *Ceiba, Agave,* and garambulla (Genus *Myrtillocactus;* Arroyo-Cabrales et al. 1987). Parallel adaptations for nectarivory are well known in megachiropterans such as the long-tongued fruit bats (Genus *Macroglossus*) and blossom bats (Genus *Syconycteris*). Common blossom bats (*S. australis*) of eastern Australia feed on nectar and pollen of a variety of rainforest plants, such as cauliflorous (Genus *Syzygium*), *Banksia, Melaleuca, Callistemon,* and certain *Eucalyptus.* Blossom bats locate nectar and pollen with their large eyes and keen sense of smell. *Syconycteris* typically lands on an inflorescence and gathers pollen and nectar by use of its long snout and brush-tipped tongue—it does not hover while feeding, as do many other nectarivores. These bats are unique in that they do not consume pollen directly from the flower. Their body hairs are covered with small, scalelike projections in which pollen lodges. This pollen is "consumed" while the animal grooms its fur and wings after a foraging episode. As with other nectarivores and frugivores, pollen rapidly passes through the gut and appears in the feces 45 minutes after ingestion. The majority of protein in the diet of these bats is provided by the pollen, whereas sugars in the nectar help to meet energy demands (Law 1992, 1993).

A species of marsupial is the premier terrestrial nectarivore. The honey possum (*Tarsipes rostratus*), the only species within the marsupial Family Tarsipedidae, is found only on the sand plain heaths of southwestern Australia (chapter 10). It weighs only 7 to 12 g (figure 6.16). Unlike most mammals, this mouse-sized marsupial does not climb by the aid of claws but has digits that are expanded at the tip with short, nail-like structures adaptive in gripping branches. The honey possum locates food by smell, inserting its long snout into a flower and using its protrusile tongue (reaching some 25 mm beyond the nose), which has bristles at the tip (see figure 10.30), to lick pollen from protruding anthers. The stomach is small and may act as a temporary storage compartment; the intestine is short; and there is no cecum. Pollen passes though the gut in about 6 hours.

In addition to chiropterans and other previously mentioned mammals, the following groups are important seed dispersers: primates, tapirs (Genus *Tapirus*), African elephants (*Loxodonta africana*), one-horned rhinoceroses (*Rhinoceros unicornis*), European badgers (*Meles meles*), and foxes (Genus *Dusicyon*).

Figure 6.16 **Nectarivore.** The honey possums or noolbengers inhabit the coastal sand plain heaths of southwestern Australia. With their thin tongue, bearing bristles and a tuft at the tip, honey possums lick nectar and pollen from the anthers of native plants, such as bottle-brush and *Banksia*. Honey possums are a significant pollinator of plants of southwestern Australia.

Gumivory

Animals that primarily consume plant exudates, such as resins, sap, or gums, are termed **gumivorous** (meaning "gum-eating"). This peculiar dietary specialization occurs in eight species of marmosets (Genus *Cebuella*), mouse lemurs (Genus *Microcebus* and *Phaner*), four species of petaurid gliders (Genus *Petaurus*), and Leadbeater's possum (*Gymnobelideus leadbeateri*). The diet of the fork-marked mouse lemur (*Phaner furcifer*) of Madagascar consists of close to 90% gum exudates from the trunks and branches of trees. Lemurs use a "tooth comb" (formed by their procumbent lower incisors and canines) to scrape off gum released from the surface of a plant. During feeding, lemurs easily cling to the surface of the trunk by use of needle-sharp claws. They digest gums within the enlarged cecum, which houses symbiotic bacteria. Although lemurs are able to scrape off saps and gums exuded as a result of damage from wood-boring insects, marmosets are the only primates that actually gouge holes to liberate plant juices. The incisors of marmosets are composed of thickened enamel on the outer surface and lack enamel on the inner surface, thus producing chisel-like instruments. By anchoring their upper incisors in the bark, marmosets use their lower incisors to gouge oval holes in the trunks of trees. These holes may measure 2 to 3 cm across, and certain trees may be riddled with channels of holes 10 to 15 cm in length (Tatersall 1982). Like the lemurs, the claw-like nails of marmosets are essential adaptations for clinging to vertical trunks while feeding on sap, gums, and resins. Plant and insect exudates make up the bulk of the diet of petaurid gliders of Australia. Yellow-bellied gliders (*Petaurus australis*) obtain sap from *Eucalyptus* by biting out small patches of bark of the trunk or main branches. After the flow of sap dries up, they move on to a new area. As a result, some trees become heavily scarred after several years of feeding (Goldingay and Kavangah 1991).

Mycophagy

Animals that consume fungi are referred to as **mycophagous** (meaning "fungus-eating"). Fungi of various types are an important component in the diet of a diverse array of mammals, representing insectivores, herbivores, carnivores, and omnivores. Fogel and Trappe (1978) provide a thorough review of mycophagy in mammals, detailing both the specific taxa of fungi consumed by mammals and the mammalian groups known to exhibit mycophagy. Fungi preferred by mammals include the higher Basidiomycetes, Ascomycetes, and Phycomycetes (Endogonaceae) and lichens (Maser et al. 1978). These groups of fungi are especially well represented in the diets of sciurids, murids, and members of the marsupial Family Potoroidae, the potoroos, bettongs, and rat kangaroos (Maser et al. 1978; Ure and Maser 1982; Maser et al. 1985; Maser and Maser 1987; Taylor 1992; Johnson 1994; Pastor et al. 1996). Fungi constitute a principal component of the diet of mammals on a year-round basis (Maser et al. 1986), with seasonal peaks in consumption reflecting availability (Merritt and Merritt 1978). When ingested by mammals, sporocarps pass through the digestive tract and are excreted without morphological change or loss of viability (Trappe and Maser 1976), while the other tissues are digested. The rate of passage of spores varies from 12 to 24 hours in the Cascade golden-mantled ground squirrel (*Spermophilus saturatus*) and the deer mouse (*Peromyscus maniculatus*), respectively (Cork and Kenagy 1989a). Fleshy fungi are 70% to 90% water and provide protein and phosphorous to the consumer. Sporocarps of hypogeous (below ground) fungi are reported to contain high concentrations of nitrogen, vitamins, and minerals (Cork and Kenagy 1989b). Fungi often contain complex carbohydrates associated with cell walls, however, and thus many small mammals are unable to efficiently digest and access the nutritious material of fungi. Some mammals possess modifications of the digestive tract that enable them to use fungi as a primary food. Potoroids, however, possess an enlarged foregut supporting fermentation of food by microbial symbionts. Long-nosed potoroos (*Potorous tridactylus*) are able to digest much of the cell wall and sporocarps of certain hypogeous fungi, thus accessing more of the fungi's available energy (Claridge and Cork 1994).

Certain coniferous and deciduous trees possess a relationship between their root systems and the mycelia of certain fungi. The combination of host and fungus is called a **mycorrhiza**—a mutualistic relationship; trees rely on filaments of the fungi to extract water and nutrients from the soil, and fungi derive nutrition from the sugars of the trees (Trappe and Maser 1976). Mycorrhizal fungi do not produce aboveground fruiting bodies however, and thus rely on small mammals to disperse their spores. Many forest-dwelling small mammals, such as shrews, mice, voles, and squirrels, consume large quantities of spores of mycorrhizal fungi that pass unchanged through their digestive tracts. The dispersal of sporocarps by small mammals serves the essential function of reinoculating habitats for reestablishment of forests following natural catastrophes or deforestation.

Omnivorous

Most mammals are **omnivorous** (meaning "everything-eating") and notably opportunistic. Each order of mammals contains omnivorous species; however, omnivory is best illustrated in opossums, primates such as humans and many monkeys, pigs, bears, and raccoons (see figure 6.1). The dentition is versatile, adapted to process a variety of foods. Omnivorous mammals retain piercing and ripping cusps in the anterior teeth but typically have flat, broad cheekteeth with low cusps (bunodont) adapted for crushing food (see figure 6.6). The stomachs of omnivores such as pigs (*Sus scrofa*) are comparatively simple. The small intestine is elongated, and the colon is large with many folds and bands of longitudinal muscle. The cecum of most omnivores is poorly developed due to the lack of fibrous plant material in the diet. The success of raccoons (*Procyon lotor*) in North America is a good example of a mammal that combines omnivory with opportunism to enhance survival. Although they exhibit distinct food preferences, availability largely dictates selection. During spring and summer in eastern North America, *P. lotor* feeds mainly on animal matter, including insects, earthworms, snails, bird eggs, and small mammals. They also feed on carrion and commonly visit creek edges to search for crayfish, frogs, fish, and other aquatic prey. During late summer, autumn, and winter, fleshy fruits and seeds, such as wild grapes, acorns, beechnuts, and berries, constitute the bulk of the raccoon's diet.

FORAGING STRATEGIES

During the preceeding discussion of dietary types we learned many aspects of how mammals obtain their food. In that discussion, we developed some general rules to describe the processes that occur when an animal, such as a mammal, forages for food. These rules apply to all foraging behavior, regardless of the type of diet—to nectarivorous bats searching for flowers and to coyotes looking for small mammals or fruit. In addition, when some mammals discover a food source, they hoard the food they find. We conclude the chapter with an examination of patterns of hoarding behavior. Predictions derived from many of the general rules concerning foraging behavior have been tested on a variety of birds, on some other vertebrates, and on insects such as bees. To date, although mammalogists maintain a consistent interest in foraging theory, fewer aspects of this theory have been thoroughly tested on mammals. Part of the problem may be that when the various foraging models are applied to some mammals (ungulates in particular), the models have proved to be relatively poor predictors (Belovsky 1984; Owen-Smith 1993).

Optimal Foraging

Models of foraging behavior are created to generate testable predictions about such behavior that can be evaluated by empirical studies. One general theoretical framework for analyzing the foraging behavior of mammals is the notion of **optimal foraging;** these models attempt to predict the combination of costs and benefits that will ultimately maximize the animal's fitness (Charnov 1976; Pyke et al. 1977). Optimal foraging models involve three types of issues (Stephens and Krebs 1986): (1) The decisions an animal makes while it forages. Should it eat a particular type of food or prey? Should it remain in one place to capture its prey (sit-and-wait behavior) or actively seek out prey (stalking)? (2) The currency involved. What is being maximized? (3) How the constraints or limits of the animal affect its foraging pattern. How large a prey item can it handle? Does the fruit have a hard or soft covering?

For mammals, foraging decisions depend a great deal on its past evolutionary history: What particular morphological, physiological, and behavioral traits adapt it for a particular type of diet? Much of this chapter has dealt with the adaptations of the dentition, skulls, and alimentary canals of mammals with respect to various types of diets. Some mammals, such as those we labeled omnivores, are food generalists, consuming a wide variety of food types. Others, such as koalas, are food specialists, being highly adapted to exploit a particular type of food.

Two types of currencies are generally considered when examining foraging behavior: energy (calories) and time. Should the animal minimize its time hunting to gain a fixed quantity of food, or should it maximize its energy intake in a fixed amount of time? A lion spots a potential prey. What "decision rules" does it use to determine its actions? How close is the prey, and how much energy will be needed to pursue and capture it? How big of a meal does it represent? What are the dangers in pursuing this particular prey? Given the time that it will take to pursue and subdue this prey, would the lion be better off if, instead, it went off in search of some other prey? The lion does not actually compute a cost-benefit analysis before acting, but we can use the economic model as a means of generating testable predictions about its behavior.

Several key constraints influence a mammal's foraging strategy. The first are constraints related to the time spent traveling to find food—search time. Consider the differences between an insectivorous bat that must find swarms or patches with insects versus a nectarivorous bat searching for groups of flowers (Pleasants 1989; Jones 1990; Barclay and Brigham 1994). For smaller mammals, the danger of starvation may be greater than for large mammals, and thus the constraint imposed by time and the need to regularly find food sources affects foraging decisions. Second, foods vary in how easily their energy "packets" can be extracted by a mammal—handling time. Some foods, such as grasses, require almost no manipulation; they are snipped or chewed off and swallowed. On the other hand, some fruits and seeds may require the animal to spend considerable time and energy to extract the nutrition. In an experiment that combined field and laboratory testing, *Peromyscus polionotus* made their food selections among millet, sunflower seeds, and peanuts based, in part, on handling time (Phelan and Baker 1992).

Third, mammals that are themselves potential prey must always balance the need for finding and consuming food with some degree of vigilance. Individual mammals may do this by keeping a wary eye or ear attuned to the surroundings as they feed. Group-living mammals may have various systems for detecting predators and alerting other members of the group. Meerkats (*Suricata suricatta*) feed in groups, and individual members take turns being the sentinel, watching for predators such as snakes or hawks. The sentinel, of course, is sacrificing valuable feeding time, and thus the meerkats rotate this duty to ensure that each individual has sufficient time for energy intake (Ewer 1968). Another way to reduce predation risk is to hoard food; by having food stores in a protected location, the mammal does not have to venture forth, exposing itself to potential predators. Hoarding behavior is discussed in more detail in the last section of this chapter. Fourth, some mammals have nutrient constraints. The contrasts between foregut and hindgut fermenters particularly exemplify this point. Moose (*Alces alces*) must consume both sufficient energy and sufficient sodium to meet their daily needs (Belovsky 1978), but the capacity of the rumen also must be considered (figure 6.17).

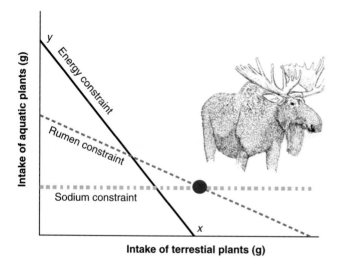

Figure 6.17 Moose feeding. The foraging constraints affecting daily feeding by moose involve (1) the need for sufficient energy (solid line), (2) the need for sufficient sodium intake (dashed line), and (3) the size of its rumen (broken line). Aquatic plants are bulkier but contain more sodium than terrestrial plants. The moose's average intake lies within the triangle defined by the three constraints, near the circle.

Source: Data from G.E. Belovsky, "Diet Optimization in a Generalist Herbivore: the Moose" in Theoret. Pop. Biol., 14:105-134, 1978.

Marginal Value Theorem

A second foraging behavior model examines how animals make decisions about their feeding sites; these are often called **patch models** (Wiens 1976; Bell 1991). Many food items consumed by mammals are found in patches, or clumps. Some questions are associated with locating patches, devising search patterns within patches, and deciding when to leave a patch and search for another and when to return to a patch (if the resource is renewable, such as nectar). For example, if a wild pig (*Sus scrofa*) locates some acorns, it initially moves from one acorn to the next rather rapidly. Soon, however, the patch becomes depleted and the acorns are more difficult to find. Mapping the path used to find acorns can provide insights concerning the pig's foraging decisions. Finally, at what point should the pig give up on that patch and attempt to locate another, perhaps more profitable, patch?

The **marginal value theorem** deals with the decisions about when to leave patches (Charnov 1976). The model relates the energy an animal spends moving between resource patches to the energy that can be obtained in that patch. At some point, the gain in energy relative to the costs of continued foraging in that patch becomes marginal; this is often referred to as the giving-up time (McNair 1982). Leaving a patch that is becoming depleted of food means the animal is confronted with the costs of searching for another patch and of traveling to another patch. The basic question is, based on diminishing returns within a patch, when should the animal leave?

Our discussion of foraging behavior can best be summarized by the following formula:

$$\text{Net rate of energy intake} = \frac{\text{Energy from food} - \left(\text{Search energy} + \text{Pursuit energy} + \text{Handling energy} + \text{Eating energy}\right)}{\text{Search time} + \text{Pursuit time} + \text{Handling time} + \text{Eating time}}$$

Exploring and testing the various components of this formula have been and will continue to be significant research problems for mammalogists. Feeding is, after all, of primary importance for fitness, measured as survival and reproductive success.

Food Hoarding

During autumn in seasonal environments, certain mammals, such as squirrels, are commonly observed harvesting fallen nuts and establishing food stores in preparation for winter. This habit, called **food hoarding**, or **caching**, has been reported for members of 6 orders and 30 families of mammals (Vander Wall 1990). Caching techniques, location, composition, season, and how food is used vary at both the individual and species levels. Caching is performed by employing either **larder hoarding** (concentration of all food at one site) or **scatter hoarding** (one food item stored at each cache site). Vander Wall (1990) documents the mammalian taxa that store food and summarizes their food-storing behavior (table 6.1).

Table 6.1 Mammalian taxa known to store food with a summary of their food-storing behavior

Order Family	Dispersion*	Food Type†	Substrate/ Location††	Storage Duration§
Marsupials				
Pygmy opossums (Burramydae)	L	S	N	L
Insectivores				
Shrews (Soricidae)	L	I,SM,A,Fi,S	N,B	L
Moles (Talpidae)	L	I	B,N	L
Primates				
Squirrel monkey (Cebidae)	S?	Mi	?	S
Green monkey (Ceropithecidae)	L?	Fr	C	S
Chimpanzee (Hominidae)	S?	Me	?	S?
Carnivores				
Foxes, wolves (Canidae)	S,L	SM,MM,Bi,E,Re	S,Sn,C	S,L
Bears (Ursidae)	S,L	LM	S,L	S
Weasles, mink (Mustelidae)	L	SM	B,N,S,C	S,L
Hyenas (Hyaenidae)	S	Ca	W	S
Tigers, bobcats (Felidae)	S	MM,LM	S,L,Sn,T	S
Rodents				
Mountain beaver (Aplodontidae)	L	V	B	S,L
Squirrels and chipmunks (Sciuridae)				
Chipmunks	L,S	S,Nu,Bu	B,N,S	L
Red squirrels	L	Co,Fu	L,F,T	L
Tree squirrels	S	Nu,Fr,Fu	S,T	L
Ground squirrels	L,S	S,Nu,V	B,S	L
Flying squirrels	S,L	Nu	S,L,T	L
Pocket gophers (Geomyidae)	L	Bu,Ro	B,Sn	L
Kangaroo rats (Heteromyidae)	L,S	S	B,S	L
Beavers (Castoridae)	L	WV	W	L
Mice, hamsters (Muridae)				
New World mice	L,S	S,Nu,I	B,S	L
Woodrats	L	V,S,Nu,Fr	N	L
Hamsters	L	S	B	L
Gerbils	L	S,Nu,Ro,V	B	L
Mole-rats	L	Bu,Ro,V	B	L
Mole-rats	L	Bu,Ro,V	N	?
Voles and muskrats	L	V,Bu,Ro,WV	B,G,Sn,F	L
Old World rats and mice	L,S	S,Nu	B,S	L
Dormice (Myoxidae)	L	Nu,Fr,Mi	C	L
Jerboas (Dipodidae)	L?	?	B	L?
Old World porcupines (Hystricidae)	S	?	S	?
Agoutis and acouchis (Dasyproctidae)	S	Nu,Fr	S	L
Octodonts (Octodontidae)	L	Bu	B	L
Tuco-tucos (Ctenomyidae)	L	Bu	B	L?
African mole-rats (Bathyergidae)	L	Ro,Bu	B	L
Lagomorpha				
Pikas (Ochotonidae)	L	V	G,B	L

From S.B. Vander Wall, Food Hoarding in Animals, *1990. Copyright ©1990 University of Chicago Press. Reprinted by permission.*

*Dispersion patterns: L=larder; S=scattered.

†Food types: A=amphibians; Bi=birds; Bu=bulbs; Ca=carrion; Co=cones; E=eggs; Fi=fishes; Fr=fruit; Fu=fungi; I=invertebrates; LM=large mammals; Me=meat; Mi=miscellaneous; MM=medium mammals; Nu=nuts; Re=reptiles; Ro=roots; S=seeds; SM=small mammals; V=green vegetation; WV=woody vegetation.

††Substrates and locations: B=burrow chambers; C=cavity or chamber (not in burrow); F=foliage; G=ground surface; L=litter; N=nest; Sn=snow; S=soil; T=tree trunk and branches; W=water.

§Storage duration: S=short term (generally <10 days); L=long term (generally > 10 days).

Most mammals that cache food are members of the Orders Rodentia or Carnivora. Only one marsupial is reported to store food, the pygmy possum (*Burramys parvus*). The pygmy possum is the only mammal restricted to the alpine and subalpine zones of Australia. It is a true hibernator, and in autumn, it caches seeds and fruits of heathland shrubs as a food reserve during its winter dormancy.

Of the 312 species of shrews, only 7 cache food. This behavior has been documented primarily in the laboratory with food provided ad libitum (Merritt and Vessey, in press). Among shrews, the premier scatter hoarder is the northern short-tailed shrew (*Blarina brevicauda;* see chapter 8). European moles (*Talpa europaea*) collect and cache earthworms and insect larvae. After mutilating the head segments of earthworms, moles cache the worms in chambers and walls of galleries near the nest as food for the winter. When soil temperature increases in the spring, some of the remaining worms, trapped in the galleries, may regenerate a head and burrow to escape (Gorman and Stone 1990).

Members of five families of carnivores cache food; the canids, ursids, mustelids, hyaenids, and felids (Vander Wall 1990). Groups of predators follow several distinctive caching patterns: (1) canids bury prey in shallow surface depressions; (2) felids and ursids do not excavate a hole but rather rake soil and leaf litter over the prey; (3) mustelids either cache the prey in their dens or, like canids, bury it in shallow surface pits; and (4) hyenas may submerge prey in water (Macdonald 1976; Elgmork 1982). Because the food items captured by these predators are usually single carcasses, these predators practice larder hoarding.

The most common hoarders are rodents. They store food in many different locations, usually for periods of more than 10 days. Larder and scatter hoarding techniques are employed to cache foods ranging from seeds and nuts to woody vegetation, roots, and invertebrates. As winter approaches, eastern chipmunks (*Tamias striatus*) carry large amounts of food in their cheek pouches and cache the food in their burrows for winter use. Preferred items in their winter diet include hickory nuts (Genus *Carya*), beechnuts (Genus *Betula*), maple seeds (Genus *Acer*), acorns (Genus *Quercus*), and a long list of seeds of woody and herbaceous plants. Flying squirrels (*Glaucomys volans*) are more selective than eastern chipmunks; hickory nuts may comprise up to 90% of the nuts they store for the winter. Unlike chipmunks (Genus *Tamias*) and tree squirrels (Genus *Sciurus*), red (*Tamiasciurus hudsonicus*) and Douglas (*T. douglasii*) squirrels establish large surface middens to cache conifer cones and mushrooms. Heteromyids primarily cache seeds, but some species also store fruit, dried vegetation, and even fungi. Foods are cached in burrows or shallow pits on the surface of the ground near entrances to the burrow, or possibly grouped in small haystacks on the ground.

Some small mammals store food in scattered surface caches and underground chambers, whereas others concentrate their cache in a single or just a few larder sites. These foods are consumed during the winter when food supplies are scarce. Eastern woodrats (*Neotoma floridana*) gather and store large quantities of fruit, seeds, leaves, and twigs in their large surface dens constructed of sticks (Wiley 1980). Perishability and nutrient content of food influence caching decisions of woodrats (Post and Reichman 1991; Post et al. 1993). Mole-rats (*Spalax microphthalmus*) of the steppes of Eurasia are noteworthy for establishing many storerooms up to 3.5 m in length, which they pack with rhizomes, roots, and bulbs. These large caches are essential for meeting the energy demands of mole-rats during the long winter on the steppes, when the ground surface is frozen and foraging is restricted.

Among lagomorphs, only the pikas (Genus *Ochotona*) establish food caches, called hay piles. Hay-gathering behavior is common to most species of pikas. *Ochotona princeps* of North America collect green vegetation during late summer and autumn and establish hay piles under overhanging rocks within talus (rubble or scree) deposits (chapter 17). Hay piles function as a source of food during winter, as well as serve as a safeguard against an unusually harsh or prolonged winter (Conner 1983; Dearing 1997). Size and placement of hay piles vary greatly among different species of pikas. *Ochotona princeps* establishes comparatively large hay piles, weighing up to 6 kg (Millar and Zwickel 1972), and up to 30 species of plants may be found in one hay pile (Beidleman and Weber 1958). Daurian pikas (*O. dauurica*) inhabiting the steppes of northern Manchuria establish hay piles weighing 1 to 2.5 kg (Vander Wall 1990). Pallas's pikas (*O. pallasi*) make hay piles measuring up to 100 cm in height placed on the ground over burrow entrances. They carry pebbles (up to 5 cm in diameter) in their mouth, placing them near the burrow entrance to prevent the hay from being scattered by the wind.

Beavers (*Castor*) stockpile woody vegetation in submerged caches below the ice near their dens, which remains a fresh and handy food supply through the winter. The phrase "busy as a beaver" refers to the energetic activity of beavers generating these food stockpiles.

Summary

Just as mammals range in size from shrews to elephants, their food ranges from microscopic organisms to prey that is larger than the predator. Types of food range from those low in calorie content, such as grasses and nectar, to proteinaceous foods, such as insects, blood, and pollen. Food may be highly restricted, such as underground fungi available only during summer, or ubiquitous and available year-round, such as insects in tropical environments.

Mammals have evolved many fascinating adaptations to procure food. Evolutionarily, the precursor of all mammalian feeding groups was represented by insectivores of the Triassic. Insectivores today include aerial (bats) and terrestrial (platypuses, shrews, moles, anteaters, echidnas, aardwolves, etc.) forms. These mammals have one thing in common—

teeth equipped with sharp cones and blades for crushing insect exoskeletons. Insectivorous mammals also possess additional adaptations that optimize their ability to capture and process prey, such as the toxic bites of shrews for immobilizing prey and the sensitive tactile organs (Eimer's organs) of some talpids for efficiently locating insects below ground.

Mammals that feed primarily on flesh are called carnivores and include the canids, mustelids, felids, hyaenids, and dasyurid marsupials. Carnivorous mammals have sharp incisors and canines, as well as a pair of carnassial teeth—a characteristic of meat-eaters. Carnivores possess a keen sense of smell and acute vision and hearing. Some carnivores such as canids rely on intelligence, social organization, and behavioral adaptability to optimize their hunting success. Carnivores hunt prey on land, in the air, and in the water. Three species of "carnivorous" mammal specialize in the consumption of blood—the vampire bats. The largest of all mammals, the baleen whales, consume the smallest prey—diatoms and krill.

Herbivorous mammals consume plants and include the ungulates and gnawing mammals. The artiodactyls and perissodactyls possess high-crowned cheekteeth with transverse ridges, which are well adapted for shredding and grinding coarse plant material. Ungulates, rodents, and lagomorphs have microorganisms in their alimentary canals that break down and metabolize cellulose. The ruminant artiodactyls (foregut fermenters) have a very efficient but slow digestive process. The perissodactyls (hindgut fermenters) have a faster but less efficient digestion. Gnawing mammals (lagomorphs and rodents) and the ungulates are unable to produce the enzyme cellulase. Thus, they ferment fibrous plant material using bacteria and protozoa in the alimentary canal. Rodents and lagomorphs possess large gnawing incisors followed by a conspicuous gap (diastema) and a battery of cheekteeth characterized by many peculiar patterns adapted to consuming a diverse array of foods. Herbivorous mammals that consume seeds are called granivores and include mostly the desert mammals in the Family Heteromyidae. Leaf-eaters, or folivores, must work very hard for minimal nutritional reward. The koala of Australia, with its elongated cecum, and many species of primates, sloths, and the pandas are noteworthy folivores. Fruit eating (frugivory) is characteristic of pteropodid and phyllostomid bats, phalangerids, and certain primates. Megachiropterans feed on fruits of some 145 genera of plants and are important pollinators of flowers and dispersers of plant seeds. This frugivorous niche is filled in the New World by the phyllostomids, notably the Genus *Artibeus*. Nectarivorous mammals are represented in about six genera of mega- and microchiropterans and the diminutive honey possums of Australia. Adaptations for consuming nectar and pollen include greatly elongated snouts and long, slender, protrusile tongues equipped with a brush tip for gleaning pollen and nectar from flowers. Mammalian gumivores consume the exudates of plants, such as resins, sap, or gums. This unusual dietary specialization is represented by various marmosets, mouse lemurs, petaurid gliders, and Leadbeater's possum.

Fungi are important in the diet of many insectivores, herbivores, carnivores, and omnivores. Many species of rodents, called mycophagists, act as important consumers and dispersers of mycorrhizal fungal spores and thus contribute to forest succession.

Various models developed to explain foraging behavior have been tested for mammals. These include the optimal foraging theory, which models dietary decisions, the currency that is maximized (energy or time), and the constraints that affect feeding behavior. Patch models are used where food is distributed in clumps. The marginal value theorem makes predictions about when a mammal should give up feeding in one location and find another patch. These models have been applied to mammalian foraging with varying degrees of success.

Food hoarding, or caching, occurs in 6 orders and 30 families of mammals. The techniques, location, and composition of the cache and how the food is used vary greatly depending on the individual, species, and season.

Discussion Questions

1. What would the diet of primitive mammals likely have been? What modern mammals possess teeth like those of primitive mammals?
2. Compare the digestive systems of Artiodactyla and Perissodactyla. Comment on their digestive efficiency and speed of processing food. How do the differences in processing food of foregut and hindgut fermenters relate to the ecology of ungulates?
3. How can you distinguish between the skull of a herbivore and that of a carnivore?
4. Of a group of mammals consisting of a bovid, mustelid, primate, and pinniped, which do you predict would have the shortest gut in proportion to its body length? Why?
5. Compare the feeding behavior of right whales to that of rorquals.
6. Explain the mechanism and adaptive significance of the toxic bite of the northern short-tailed shrew (*Blarina brevicauda*).

Suggested Readings

Chivers, D. J., and P. Langer. 1994. The digestive system in mammals: Food, form and function. Cambridge Univ. Press, New York.

Clutton-Brock, T. H. (ed.) 1977. Primate ecology: studies of feeding and ranging behaviour in lemurs, monkeys and apes. Academic Press, London.

Dawson, T. J. 1995. Kangaroos: biology of the largest marsupials. Comstock Publ. Assoc., Ithaca, NY.

Gittleman, J. L.(ed.). 1989. Carnivore behavior, ecology, and evolution. Comstock Publ. Assoc., Ithaca, NY.

Gittleman, J. L. (ed.). 1996. Carnivore behavior, ecology, and evolution, vol. 2. Comstock Publ. Assoc., Ithaca, NY.

Owen, J. 1980. Feeding strategy. Univ. of Chicago Press, Chicago.

Rogers, E. 1986. Looking at vertebrates: a practical guide to vertebrate adaptations. Essex, England.

Stevens, C. E., and I. D. Hume. 1996. Comparative physiology of the vertebrate digestive system, 2d ed. Cambridge Univ. Press, New York.

Vander Wall, S. B. 1995. Influence of substrate water on the ability of rodents to find buried seeds. J. Mammal, 76:851–856.

CHAPTER 7

The Nervous and Endocrine Systems, and Biological Rhythms

CONTROL SYSTEMS

Two major control processes regulate internal events in mammals: the nervous and endocrine systems. Part III, on mammalian diversity, covers a number of specialized aspects of these systems, as do several chapters within this part. This chapter examines some general aspects of these two body control systems that are unique to mammals. We begin by exploring the sensory mechanisms of mammals and the features of their central nervous system. We then treat the key aspects of the mammalian endocrine system and some of the functional processes regulated by hormones. The activity patterns of mammals are governed by a series of biological rhythms. These rhythms are, in turn, regulated by activities in the nervous and endocrine systems. We conclude the chapter with an examination of both the mechanisms that underlie these rhythms in mammals and their functional consequences. For background in the general biology of the nervous and endocrine systems, consult Hickman and colleagues (1995).

NERVOUS SYSTEM

Sensory Systems

Each of the major sensory systems in mammals has undergone adaptations that help to define the group. The importance of a particular sensory modality varies among mammals; vision, olfaction, and hearing predominate to varying degrees in most groups.

Vision

Detection and use of radiant energy in the form of light of particular wavelengths is important to most but not all mammals. The mammalian eye is generally similar to the eye of other amniote vertebrates. Because humans use vision perhaps more than any other sense, we may tend to overemphasize its importance. Among some insectivores that are primarily fossorial and cetaceans living in the aquatic environment, the eyes are greatly reduced and can really only differentiate light from dark. At the other extreme, primates, felids, and some other mammals have binocular, stereoscopic vision. Many nocturnal mammals possess a **tapetum lucidum,** usually in the choroid; this reflective layer lying outside the receptor layer of the retina. The tapetum lucidum provides the typical eye shine when light, for example from car headlights, strikes the retina at night. This structure aids in night vision by reflecting light that has passed through the receptor layer back toward the retina.

The proportions of the eye vary among mammals depending on environmental light intensity (figure 7.1). The cornea and lens are positioned farther from the retina in diurnal mammals than in nocturnal mammals and are also less sharply curved to provide the appropriate focal length. Together, these adaptations provide for a larger image in diurnal mammals. Mammals possess both **rods** for black and white vision and **cones** for color vision. Diurnal mammals have a greater proportion of cones, and nocturnal mammals have a greater proportion of rods. Color vision is not very useful at low light levels, and the predominance of rods provides a sharper black-and-white image for nocturnal mammals.

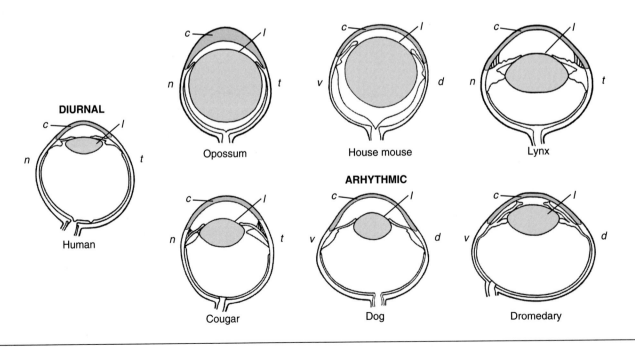

Figure 7.1   Eye proportions.   The eyes of vertebrates vary in the intraocular proportions depending on normal light intensity at the time of activity. *d* = dorsal side of eyeball; *n* = nasal side of eyeball; *t* = temporal side of eyeball; *v* = ventral side of eyeball; *c* = cornea; *l* = lens.

Hearing

Reception and use of auditory cues, along with olfaction, are the two sensory systems most prominent in the majority of mammals. Auditory signals are used for a variety of functions, including predator detection and communication. Mammals are unique in possessing pinnae (external ears), which function to focus and direct sound, and thus aid in their acute sense of hearing. Pinnae have been lost secondarily in some insectivores, fossorial mammals, cetaceans, and phocid seals. For some mammals, as in many ungulates and particularly bats, the ears can be rotated to facilitate reception of sound waves in order to determine the direction of a sound source. Mammals can detect a wide range of wavelengths and intensities of sound. Hearing ranges extend from **infrasound,** frequencies of less than 20 Hz, in elephants (Payne et al. 1986; Langbauer et al. 1991) to **ultrasound,** frequencies of greater than 20,000 kHz, in bats and cetaceans (Gould 1983). Echolocation in bats and cetaceans is discussed in chapters 12 and 16.

The internal ear in mammals has several specialized functions (figure 7.2). The semicircular canals are the body's balancing mechanism. Three ossicles, or bones, are present in the ears of mammals: the malleus, incus, and stapes. The incus and malleus are unique to mammals. The stapes is homologous with the hyomandibular bone of fishes and the columella of reptiles and amphibians. The ossicles transmit sound pressure information impinging on the **tympanum** (ear drum) at the interior end of the external auditory meatus through the middle ear to the inner ear. The coiled cochlea of the inner ear contains the organ of Corti, which transduces (changes) pressure information into neural impulses.

Taste

Although the sense of taste is quite important for some mammals in dietary selection, it is secondary in most groups. Varying types of sensory receptors are located on different portions of the tongue, usually in discrete patches. For some mammals, including humans, these receptors detect four basic classes of substances: sweet, sour, salty, and bitter.

Olfaction

Chemoreception via olfactory receptors located in the nasal cavity is well developed in most mammals. Smell is important for a variety of activities, including locating food and detecting danger, as well as in social relations. Mammals have evolved a variety of glands that produce odorous substances, such as sebaceous glands in hamsters and male goats; preorbital glands in deer; interdigital glands in sheep; preputial glands in deer, beaver, and mice; anal glands in mustelids; and perineal glands in guinea pigs and rabbits (figure 7.3). Many mammals also obtain chemosensory information via the vomeronasal organ (Jacobson's organ), which is located near the floor of the nasal cavity on each side of the nasal septum and attached to the vomer. In whales and some higher primates, the sense of smell is

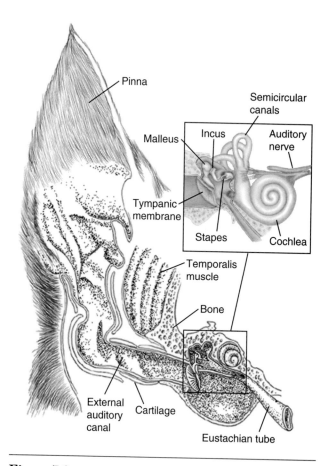

Figure 7.2   **Mammalian ear.**   The features of the ear include an external pinna for collecting sound, the ossicles of the middle ear for transmitting the sound pressure, and the organ of Corti in the inner ear for transducing the pressure signal to neural impulses. The semicircular canals, important for balance, are located in the middle ear.

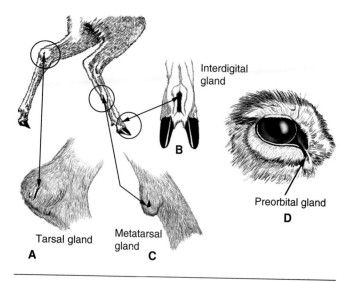

Figure 7.3   **Scent glands of white-tailed deer.**   (A) Tarsal glands are located on the inner surfaces of the hind legs. (B) Interdigital glands are located between the hooves of each foot, (C) Metatarsal glands are found on the outer surface of the hind legs, (D) Preorbital glands are located at the front of each eye.

reduced, and it is absent in dolphins and porpoises. Although some aspects of the olfactory epithelium, where chemical information is transduced to neural impulses, are modified or specialized in mammals, the general morphology is similar to that in reptiles.

Touch

In addition to pressure receptors associated with hairs, most mammals possess vibrissae. Vibrissae (whiskers) are elongated hairs located on the muzzle and around the eyes and ears. Some fossorial mammals also have vibrissae on their tails. These tactile organs enable the animal to detect nearby objects and may also function to protect the face and eyes. Many mammals, ranging from moles to elephants, have specialized touch sensors in the snout and lips. Tactile sensory systems are most highly developed, however, in smaller mammals that burrow and tunnel.

Other mammals have highly evolved tactile receptors in their digits. The skin in these regions has lost its hairy cover, and receptors are concentrated near the surface. This is particularly true for primates, but it also characterizes other groups, including carnivores. This enhanced sense of touch is useful in locomotion, processing food, and social relations.

Central Nervous System

The central nervous system consists of the brain and spinal cord. Although some generally minor aspects of the mammalian spinal cord are specialized and help to differentiate mammals from other vertebrates, major changes have occurred in brain morphology and organization. The mammalian brain can be subdivided into five major regions (figure 7.4). A central, unique feature of mammalian evolution has been the expansion of the cerebral hemispheres (neocortex) of the telencephalon. All of the other brain re-

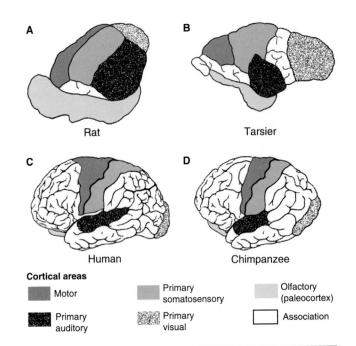

Cortical areas

■ Motor	▨ Primary somatosensory	□ Olfactory (paleocortex)
■ Primary auditory	▨ Primary visual	□ Association

Figure 7.5   **Brain areas of neocortex.**   Considerable differences occur among the four mammals shown here with respect to both (A) the relative areas devoted to different sensory systems in terms of storage areas in the cerebrum and (B) the amount of association cortex.

gions have been virtually surrounded by this new gray matter. Cells in the cerebral hemispheres are arranged in layers, with all of the cells in each particular region of the cortex devoted to processing the same type of information.

The amount of cerebral tissue in each major region varies considerably among different mammals (figure 7.5). The general structures depicted in figure 7.4 are all underneath the cortex shown in figure 7.5, as the cortex wraps around these more ancestral brain regions. In the rat (*Rattus norvegicus*) and even the tarsier (*Tarsius syrichta*), large proportions of the cerebral cells are in regions pertaining to the different sensory modalities. Very little **association cortex**—regions of the cortex that are not specifically identifiable as sensory or motor cortex—occurs where information is processed and integrated across the various sensory modalities. The rat has considerably more olfactory cortex than the tarsier, but the tarsier has more visual cortex. These patterns reflect the relative use of these sensory modalities by rats (rodents) and tarsiers (primates). In the chimpanzee (*Pan troglodytes*) and human (*Homo sapiens*), the area devoted to association cortex has increased enormously (Camhi 1984). In addition, of course, the total size of the brain has increased from a volume of about 20 cm^3 in the rat to about 1200 cm^3 in the human. Both absolute and relative changes must be considered. There is a significant trend toward increase in the ratio of brain size to body mass among mammals over evolutionary time (Jerison 1973; Eisenberg 1981). Although only partially visible in figure 7.5, brain surfaces in rats and even in tarsiers are rather smooth, which contrasts with the multiple infoldings and overall rough surface ap-

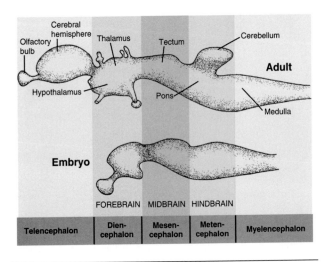

Figure 7.4   **Mammalian brain.**   The five major regions of the mammalian brain. A rather primitive mammal is depicted here; considerable additional material is found in the cerebral hemispheres of higher mammals.

pearance of the brains of chimpanzees and humans. These differences in texture reflect a vast increase in the overall surface area of the brain, much like the difference between traveling along a straight coastline and one with many bays and coves. The increased surface area reflects the presence of many more neurons and more complex interconnections between various brain regions.

Because of the increased association areas and modest but significant changes in the underlying neural structures in mammalian brains, mammalian sensory pathways are more complex than those in most other vertebrates (Camhi 1984). An impulse arriving from the visual or auditory nerves splits and goes to a number of areas of the brain, rather than to one or a few locations, as happens in lower vertebrates. These multiple pathways provide for at least two major characteristics of mammalian central nervous systems. First, considerably more storage area is available for retaining stimulus information, permitting easier recall and later use. Second, incoming information is better integrated, both across modalities and with respect to relating current and past stimuli.

In mammals, both the mode of impulse transmission and the manner in which the effector neurons convey information to the muscle at the motor end-plates are generally similar to these mechanisms in other vertebrates. Results from research in the past several decades indicate that more types of neurotransmitters are involved in nerve impulse transmission across synapses in mammals than in other vertebrate groups. To date, more than 20 different neurotransmitters have been identified in mammalian central nervous systems. Nerve regeneration occurs in the peripheral nervous system of mammals, much as it does in other vertebrates, but it does not occur readily in the central nervous systems of mammals. Because of the difficulties associated with spinal cord injuries in humans, a great deal of research is currently devoted to studying this problem, using mammals and other animals as models.

ENDOCRINE SYSTEM

Hypothalamic-Pituitary System

The master control mechanisms for the endocrine system in mammals are the hypothalamus and the pituitary gland (figure 7.6). The **hypothalamus,** part of the midbrain, is located

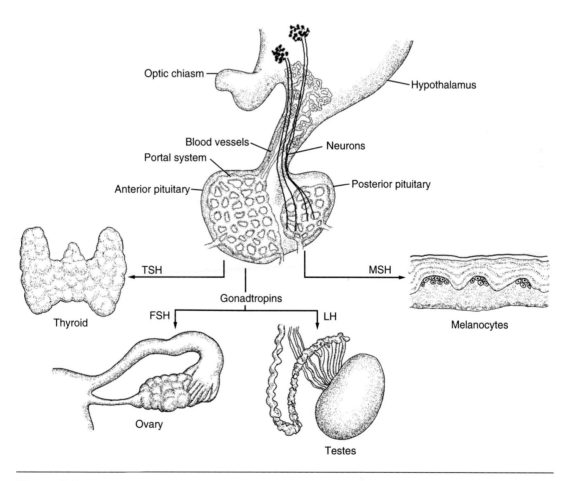

Figure 7.6 **Mammalian endocrine system.** Shown here are a portion of the endocrine glands and the influences on them from the two major regions of the pituitary (anterior and posterior). In particular, the systems shown are either unique to mammals or specialized in mammals, for example, the thyroid gland and its control of homeothermic metabolism. *Abbreviations:* TSH = thyroid stimulating hormone; FSH = follicle stimulating hormone; LH = luteinizing hormones; and MSH = melanocyte-stimulating hormone.

below the thalamus and contains collections of neuron cell bodies (nuclei). Sensory input to the hypothalamus comes from other brain regions and from cells within the hypothalamus that monitor conditions in the blood that passes through the region. The **pituitary gland** is located directly below the hypothalamus and is connected to it by both blood vessels and neurons. The vascular connection is known as the **hypothalamic-pituitary portal system.** The pituitary gland is divided into anterior and posterior regions. Endocrine functions of the anterior pituitary are regulated by releasing neurohormones that are carried in the blood from the hypothalamus (see figure 7.6). Neurons that pass from the hypothalamus to the posterior pituitary release hormones (neurosecretions) directly into that region. Thus, the hypothalamus controls the pituitary gland, and in turn, the pituitary gland and the hormones it releases play critical roles in regulating the endocrine processes of the various tissues and organs of the body (see figure 7.6).

Glands and Hormones

Endocrine glands are specialized groups of cells that produce chemical substances that are released into the bloodstream. These substances, called **hormones,** act as messengers as they travel to cells throughout the body. Mammals possess most of the same types of peripheral endocrine glands as other vertebrates, and they have both steroid and protein hormones (see figure 7.6). The mammalian endocrine system has some unique aspects, however. Some hormones that mammals share with other vertebrates have taken on in mammals specialized functions associated with reproduction, including ovulation, pregnancy, and lactation (see chapter 9). These hormones include follicle-stimulating hormone, luteinizing hormone, estrogen, testosterone, prolactin, progesterone, and oxytocin. The reaction to stress, involving adrenal gland hormones, has been studied extensively in some mammals, with particular regard to the effects of stress on reproduction and the immune system (see chapter 24). Hibernation and kidney function also involve specializations of the endocrine system that are unique to mammals (see chapter 8). In addition to the processes covered in other chapters, the endocrine system significantly affects mammalian physiology and behavior in other ways. Several examples in the following sections provide insight into these relationships.

Functional Aspects

Because mammals are endothermic homeotherms, hormones are important for regulation of metabolic processes. The thyroid gland is the key modulator of metabolic activity. Through its hormone, thyroxin, and related compounds, the thyroid triggers increased cellular respiration in target tissues, resulting in additional heat production. Cold-adapted individuals have higher rates of thyroid output than those that have not been exposed to cold. The insulin-glucagon hormone system is responsible for maintaining the homeostasis of the blood carbohydrate levels. Insulin acts to lower blood glucose by promoting the conversion of glucose to glycogen and lipids. Glucagon has **antagonistic** (opposite) **actions** to insulin. It raises blood glucose levels by promoting mobilization of fat reserves. An additional hormone that may be unique to mammals is calcitonin, which was not discovered until 1961. This fast-acting hormone, released from the parathyroid-thyroid complex, results in lower blood levels of calcium. This action is the opposite of that for parathyroid hormone, which raises blood calcium levels.

Hormones play key roles in many mammalian behavior systems (Nelson 1995). Some mammalian chemical communications systems (see chapter 20) rely on hormones. **Priming pheromones** (chemical substances that trigger generalized internal physiological events, such as the production and release of hormones) can function to regulate puberty in some rodents and are mediated by hormonal conditions in the donors (Vandenbergh 1983; Vandenbergh and Coppola 1986). The puberty-accelerating pheromone in male mouse urine depends on the testosterone level (Lombardi and Vandenbergh 1977). The puberty-delaying pheromone present in the urine of grouped females depends on adrenal gluccocorticoids (Drickamer and Shiro 1984). Chemosignals also influence reproduction in larger mammals; for example, the urinary and vaginal compounds released by white-tailed deer (*Odocoileus virginianus*) (Murphy et al. 1994) can act as sexual attractants and signal reproductive condition. Aggression in both male and female mammals is mediated, in part, by hormones. Higher levels of aggression in males are sometimes associated with higher levels of testosterone in some mammals, although this relationship is not true for all mammals under all conditions (Brain and Poole 1976; Barkley and Goldman 1977; Sapolsky 1997). The maternal aggression exhibited by female mammals around the time of parturition is mediated in part by prolactin (Svare and Mann 1983). Various aspects of behavior, including reproduction, are influenced by prenatal exposure to hormones and are called **organizational effects** (vom Saal 1979, 1989). In particular, in species with multiple births, female embryos positioned in utero between two male embryos are masculinized, which affects their behavior and reproductive success (Drickamer 1996). Other behavior systems are also under varying degrees of hormonal regulation.

BIOLOGICAL RHYTHMS

Biological rhythms are patterns of activity that occur at regular intervals during the life of an animal. **Circadian rhythms** have a period of about 24 hours. **Circannual rhythms** have a period of about a year. **Ultradian rhythms** are those with a period of less than 24 hours. Because these cyclical patterns govern many of the activities of mammals, we examine them both in terms of their functional significance and the underlying mechanisms.

Circadian

Probably our most common perceptions about activity patterns in mammals relate to their cycle of daily activity. Some mammals are **diurnal,** meaning they are active primarily during daylight hours and quiescent at night; others are **nocturnal,** exhibiting peak activity during hours of darkness and resting when there is daylight. Some mammals, including types that are primarily diurnal or nocturnal, exhibit **crepuscular** activity, moving about more and feeding near sunrise and sunset. The primary cue for entrainment, or establishment and modification, of circadian rhythms (called a ***Zeitgeber*** or time-giver) is photoperiod. The photoperiod sets and changes the biological clock.

Activity patterns of about a day govern many aspects of the lives of mammals, just as they do for other animals. Members of the same mammalian species feed, sleep, mate, give birth, and disperse at about the same time of day. Various factors influence the timing of the activity phase. For mammals on which predation pressure is high, being nocturnal may reduce their chance of being detected. Alternatively, for some predatory mammals, such as insectivorous bats,

peak activity may occur near dusk and just after nightfall, when large numbers of insects are active. Most mammals tend to give birth during the inactive phase of the daily cycle. Being able to move is necessary for avoiding predation (figure 7.7). For mammals that live in migratory herds, such as wildebeest (*Connochaetus gnou*), giving birth during the inactive phase means that newborn calves have sufficient time within several hours after birth to become mobile enough to travel with the group. Nocturnal rodents that give birth during daylight hours in their burrows or nests can still forage during their normal nighttime activity phase. Obtaining sufficient resources just prior to parturition and at the onset of lactation may be critical for reproductive success.

Early mammals probably were nocturnal, and many of those living today are still nocturnal. Others have adopted a diurnal activity rhythm. Being active at different times of the day has both advantages and disadvantages. Being active at night decreases the chances of heat stress and of being seen and preyed on. In addition, because the air is more humid, olfactory communication is enhanced. Social interactions involving visual communication are reduced at

Figure 7.7   Timing of parturition.   Many mammals give birth to their young during the resting or sleep portion of their daily cycle. In this instance, a newly born wildebeest has a few hours to strengthen its wobbly legs so it can stand and move with the herd.

night. Competition for food may also be reduced with, for example, birds, which are primarily diurnal. Diurnal mammals, on the other hand, are better able to use vision to forage, and they can use complex visual communication. Heat stress, being seen by potential predators, and competition for food can pose problems for diurnal mammals. They therefore tend to be larger than nocturnal mammals and live in larger, more complex social groups, on average, than mammals that are active primarily at night.

At least one mammalian evolutionary sequence may have revolved around circadian rhythms. Fossil history suggests that rodents evolved nearly concurrently with early primates from insectivorelike precursors and soon became dominant in the terrestrial habitat. Competition for food and nesting sites between rodents and early primates may have brought about two of the key changes that characterize primate evolution: a shift to a diurnal activity phase and adoption of a more arboreal existence (Simons 1972). Shifting activity to the trees may also have led to selective advantages for a more diurnal activity pattern because locomotion and foraging in the trees are safer during daylight. This is only one of several explanations for the evolution of primates as arboreal, diurnal animals, but it illustrates the possible importance of circadian activity patterns over the course of evolutionary time.

Although they are not completely known, considerable progress has been made in the last decade in deciphering the mechanisms that regulate circadian rhythms in mammals. The timing mechanisms appears to have three major components located in (1) the retina, (2) a portion of the thalamus in the brain called the intergeniculate nucleus, and (3) the suprachiasmatic nucleus (SCN) of the hypothalamus. Light striking the retina follows several pathways, two of which lead to the SCN and intergeniculate nucleus (Moore 1973; Harrington et al. 1985). One pathway from the intergeniculate nucleus then leads to the SCN. Animals with **ablation** (destruction by electrical or chemical techniques) of the SCN exhibit irregular cycles, implicating the SCN as the putative primary oscillator in mammals (Rusak and Zucker 1979; Moore 1982). Additional investigations that reveal spontaneous neural activity in the SCN, as well as the activity of neurotransmitters in this brain region, provide further evidence for its role as the primary pacemaker (Gillette et al. 1993; Glass et al. 1993). Although the pineal gland does not appear to be a primary component of the timing mechanism, its removal (pinealectomy) in rats alters some aspects of circadian rhythms. These findings suggest that the pineal gland, via its secretion of melatonin, may connect various circadian processes based on the clock located in the SCN (Warren and Cassone 1995).

Circannual

Circannual rhythms regulate a number of processes in mammals. As we will see in the next chapter, the annual process of hibernation is a key adaptation for some species that inhabit colder climates. The annual pattern of breeding in many mammals, particularly in temperate climates and at more northern latitudes, is also governed by a circannual rhythm. In tropical climates, the reproductive cycles of some mammals respond to rainfall and the resulting changes in vegetation, as well as to population numbers of other animals. These resources provide the food necessary to reproduce and to support progeny when they are weaned. Some species of bats migrate each year between a summer breeding location and a winter hibernation location. The ultimate mechanisms behind circannual rhythms in mammals likely can be tied primarily to food resources and secondarily to climatic conditions.

Like circadian rhythms, the primary cue for annual patterns is probably photoperiod. Hormonal changes are cued by photoperiod, resulting in alterations of physiology, morphology, and behavior. For some mammals, including most small mammals in North America, an increasing photoperiod is correlated with these events and reproduction begins. In other instances a decline in photoperiod correlates with the fall mating season, with young born the following spring, such as occurs with white-tailed deer or Rocky Mountain bighorn sheep (*Ovis canadensis*).

Seasonal changes in coat color (pelage dimorphism) are also triggered by circannual cycles. Shifts between a white winter coat that matches the snow-covered landscape, for example, in weasels (*Mustela* spp.) and snowshoe hares (*Lepus americanus;* figure 7.8) and a darker coat in summer are critical to these animals' survival as predators or prey. The weasel's darker pelage of summer and white coat of winter each provide the camouflage necessary for stalking and preying on small mammals that constitute its diet. The snowshoe hare, on the other hand, needs its white winter coat as a means of preventing detection by predators. That same winter coat would stand out against a darker background of vegetation during summer months.

For some mammals, factors other than photoperiod can act as the primary cue triggering events during an annual cycle. In the hotter, more humid climates of the southern United States, some species of small mammals, such as the beach mouse (*Peromyscus polionotus*), breed in the fall, when temperatures have declined from summer highs (Blair 1951; Drickamer and Vestal 1973). In other mammals, annual patterns of reproduction may be linked to certain plant compounds. For example, reproduction in montane voles (*Microtus montanus*) is triggered by a compound found in some fresh green plants (Negus and Berger 1977; Berger et al. 1987). The compound 6-methoxybenzoxazolinone (6-MBOA) may cue the start of reproduction in the spring, when fresh plant shoots begin to grow. Both the number and size of litters produced are influenced by 6-MBOA (Berger et al. 1987; table 7.1). Finally, social cues can be important with regard to hormones and seasonal patterns of reproduction. The social activities of conspecifics, often regulated by hormonal events, may, for example, synchronize reproductive activity among members of a population of rhesus

A

B

Figure 7.8 **Pelage color changes.** Mammals that live in more northern latitudes often change their pelage color during the course of an annual molt. The snowshoe hare is white in winter and darker in summer, providing some degree of camouflage by nearly matching the color of the background. This annual shift in pelage color is regulated by external photoperiod and mediated by internal hormonal events.

macaques (*Macaca mulatta*) (Vandenbergh 1969; Vandenbergh and Drickamer 1974).

Photoperiod is thought to be the primary cue for circannual rhythms in most mammals because over many thousands of years, the annual cycle of changes in photoperiods is more stable and consistent than any other environmental cue. Changes in climatic conditions, for example, are more variable, and thus cueing on such events could lead to wasted energy. Through evolutionary time, photoperiod may have become the primary cue, and climate and changes in food resources (e.g., vegetation or the supply of insects) may have come to be secondary cues. Some evidence for the effect of photoperiod on annual cycles comes from the work of Davis and Finnie (1975). Woodchucks (*Marmota monax*) from the Northern Hemisphere were sent to Australia. There, they entrained on some feature of the environment and eventually reversed their annual pattern, exhibiting hibernation at the appropriate time in the Southern Hemisphere, possibly cued by changes in photoperiod. Additional evidence for the importance of changing photoperiod affecting circannual cycles comes from work by Heller and Poulson (1970) on two species of ground squirrels (Genus *Spermophilus*) and four species of chipmunks (Genus *Eutamias*). Individuals of several species of each genus were held under constant ambient and photoperiod conditions in the laboratory. They maintained circannual rhythms under these conditions even though various proximate cues (e.g., diet) were sometimes manipulated. The evidence is correla-

tional, and a causal link between photoperiod and circannual rhythms in mammals has not yet been fully demonstrated. Davis (1991) tested the possible effects of irradiance (light intensity) on the entrainment of circannual rhythms in California ground squirrels (*S. beecheyi*). Light intensity varies with the angle of the sun, and irradiance varies with latitude. Irradiance therefore could be an excellent predictor

Table 7.1. **6-MBOA Affects Reproduction**

Pairs of montane voles (*Microtus montanus*) given either a control treatment or one of three doses of 6-MBOA differed significantly with respect to both the total number of pups produced and the sex ratio. Sex ratio differences were computed using chi-square analyses.

Treatment (μg)	Total Number of Offspring	Sex Ratio (Male:Female)
Control	218	1.00:0.70
0.1	262	1.00:0.97*
1.0	272	1.00:1.08[†]
10.0	285	1.00:1.02[‡]

Source: Data from P.J. Berger et al., "Effects of MBOA on Sex Ratio and Breeding Performance in Microtus montanus*" in Biol. Reprod., 36:255-260, 1987.*

*Value differs from the control at p = 0.08.
[†]Value differs from the control at p = 0.02.
[‡]Value differs from the control at p = 0.04.

of entrainment of annual cycles. Davis held California ground squirrels in captivity inside a building but exposed them to the seasonal cycle of irradiance characteristic for Santa Barbara, California. The squirrels subsequently exhibited typical circannual patterns of weight gain and torpor. The possibility that irradiance is a key *Zeitgeber* for entrainment of circannual rhythms in mammals deserves further attention.

Hibernation is the primary phenomenon for which the underlying neural and hormonal correlates for circannual rhythms have been studied (Kayser 1965). The pineal gland plays a central role as a photoendocrine transducer in mammals (Reiter 1980; Gwinner 1986). Research has yet to reveal a critical role for this gland in circannual rhythms, however. Pinealectomies in ferrets (*Mustela putorius*) and golden-mantled ground squirrels (*Spermophilus lateralis*) did not eliminate circannual cycles (Herbert 1972; Zucker 1985), but the annual cycles of these animals were not synchronized with changes in photoperiod. Similarly, **lesions** (destroying an area of brain tissue via electrical or chemical means) in the central nervous system have not demonstrated that any particular structures are essential for circannual rhythms, although such treatments produce variations in the annual pattern. These studies have included lesions in such brain areas as the SCN, which are essential to circadian rhythms (Zucker et al. 1983). Findings to date suggest couplings between the mechanisms mentioned earlier for circadian periodicities and those involved in annual patterns, but additional studies are needed to further elucidate the proximate mechanisms underlying circannual rhythms. One problem associated with research on these rhythms is the long time needed for data collection; waiting several years to obtain results from tests designed to determine whether a particular treatment has influenced rhythmic patterns is difficult and costly.

Ultradian

Some mammals have regular cycles of activity with periods that are shorter than a day in length. Many microtine rodents exhibit 6 to 12 evenly spaced, short bouts of activity and rest during a 24-hour day (Daan and Slopsema 1978; Madison 1985; figure 7.9). Halle (1995) studied this phenomenon in tundra voles (*Microtus oeconomus*). Until recently, the primary methods for exploring ultradian rhythms involved either live-trapping with trap checks made at very short intervals or radiotelemetry. Both of these methods, however, can possibly interfere with the ongoing activities of the animals. Halle's method used passage counters (Mossing 1975)—small tubes (7 cm wide by 5 cm long) placed in the runways used by these animals. A small gate in the center of the tube is attached to a microswitch. When a vole passes through the tube on its way along a runway, the switch closes, tripping an event recorder. The method does not per-

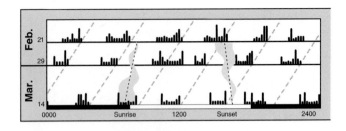

Figure 7.9   Ultradian rhythms.   The activity rhythm of a female meadow vole (*Microtus pennsylvanicus*) weighing 30 g and fitted with a radiotransmitter collar. The animal was free-ranging in a field covered with 1 m of snow in Quebec, Canada. The vertical axis represents movement (in meters) at 15-minute intervals of detection. Spacing between the 3 days depicted reflects the time intervals between those days. The pair of dotted lines represent time of sunrise and sunset. Dashed lines represent ultradian periods of activity of about 4 hours duration and with an underlying circadian period of 23.85 hours/day.

mit individual identification but gives a better overall picture of undisturbed activity rhythms. The method also depends on the animals using the runways, something done very regularly by voles but less often by other small mammals.

Halle (1995) found that root voles have about seven activity periods per day. Diurnal periods are 3.0 to 3.5 hours in duration, whereas nocturnal periods are 3.5 to 4.0 hours long. The voles were more active during daylight than at night but showed the highest levels of activity near sunrise and sunset. Changing seasons (shifts in daylength) resulted in changes in the ultradian patterns as the animals' internal clocks were reset by sunrise and sunset. The significance of ultradian rhythms in rodents most likely involves their metabolic needs and related foraging activities. Curiously, these ultradian activity bouts are synchronous within family groups (Mossing 1975; Webster and Brooks 1981), a phenomenon that has yet to be explained.

Ultradian rhythms are not restricted to rodents. Rhesus macaques exhibit regular short bouts of social activity between periods of more self-directed behavior (Maxim et al. 1976). Because these mammals are diurnal, the ultradian pattern was observed only during the active daylight portion of the daily rhythm. Periods for these ultradian patterns were generally 40 to 45 minutes in duration. The underlying mechanism for this ultradian rhythm in these monkeys may relate to an oscillator that affects cycles of rapid eye movement (REM) sleep (Kripke 1974).

With ongoing research on circadian rhythms rapidly producing results, it seems likely that attention may focus on ultradian rhythms in the future. Not only are new technologies likely to result in better detection of such rhythms, but we are also likely to find that such patterns have heretofore unrecognized functional significance. With an increased understanding of the proximate mechanisms for circadian rhythms, it may soon be possible to explore the underlying evolutionary causes of ultradian rhythms.

Summary

Like other vertebrates, internal processes in mammals are regulated by the nervous and endocrine systems. For many mammals, the senses of hearing and olfaction are most important. For others, vision is also key. Mammals hear and use sound over a wide range of wavelengths from 20 Hz to more than 20,000 kHz. Taste is of relatively minor importance for most mammals. Touch receptors are most highly developed in fossorial species.

The most obvious feature of the mammalian brain is the relative increase in the neocortex. In general, brain size has increased progressively both in absolute terms and relative to body mass for larger versus smaller mammals. Nerve cells in the cerebrum are associated with two properties that help to define the uniqueness of mammals: storage of information and associations made between events, both past and present. Mammals have more types of neurotransmitters than other vertebrates.

The mammalian endocrine system is characterized by control from the pituitary gland and its connections with the hypothalamus. Specialized functions have evolved in mammals for hormone systems associated with reproduction, hibernation, kidney function, metabolic processes, and a number of behavioral systems. Chemosignals excreted in urine that affect reproductive physiology, aggression, and developmental effects involving intrauterine position are all examples of processes that are regulated, in part, by hormones.

Biological rhythms, including circannual, circadian, and ultradian patterns, govern a number of life history features of mammals. Daily rhythms include cycles of activity and sleep, feeding behavior, mating, and the timing of births. The control of these rhythms involves the retina, the suprachiasmatic nucleus, and part of the thalamus. Circannual patterns are evident in hibernation, seasonal mating, and, for a few species of mammals—particularly bats and some cetaceans—migration. Changes in photoperiod are the most consistent cues for the establishment and maintenance of circannual patterns in mammals, although light intensity may be a good *Zeitgeber*. Other factors that affect annual cycles include climate, chemicals in plants, and social influences. Ultradian rhythms have received less attention than other mammalian rhythms. These short cycles, occurring within each daily pattern, are characteristic of some small mammals, as they awake, become active, feed, and return to rest at intervals of several hours.

Discussion Questions

1. Expansion of the cerebral hemispheres in mammals has resulted in more complex learning processes than those found in most other vertebrates. After consulting an appropriate book on the subject (see "Suggested Readings" in this chapter), summarize the differences in learning capacities between mammals and other vertebrates.
2. Using information from this and other chapters, summarize the integration of neural and hormonal mechanisms underlying these adaptational systems in mammals: (a) hibernation; (b) response to stress; and (c) female reproduction.
3. Although daylength is a primary cue for mammals with respect to seasonal activities and breeding in particular, other more proximate environmental cues can act as secondary triggers. Select any five species of mammals from different regions of the world and, after learning a little about their environmental conditions, suggest which environmental factors might be important secondary cues for each species.
4. What is the importance of the comparative method with regard to studies of mammalian nervous systems, endocrine systems, and biological rhythms? What are some of the possible advantages and disadvantages of using this method? (Excellent references for this method are Gittleman 1989 and Harvey and Pagel 1991.)

Suggested Readings

Brown, R. E. 1994. An introduction to neuroendocrinology. Cambridge Univ. Press, New York.

Camhi, J. M. 1984. Neuroethology. Sinauer Assoc., Sunderland, MA.

Drickamer, L. C., S. H. Vessey, and D. B. Meikle. 1996. Animal behavior: mechanisms, ecology and evolution, 4th ed. Wm. C. Brown, Dubuque, IA.

Nelson, R. J. 1995. An introduction to behavioral endocrinology. Sinauer Assoc., Sunderland, MA.

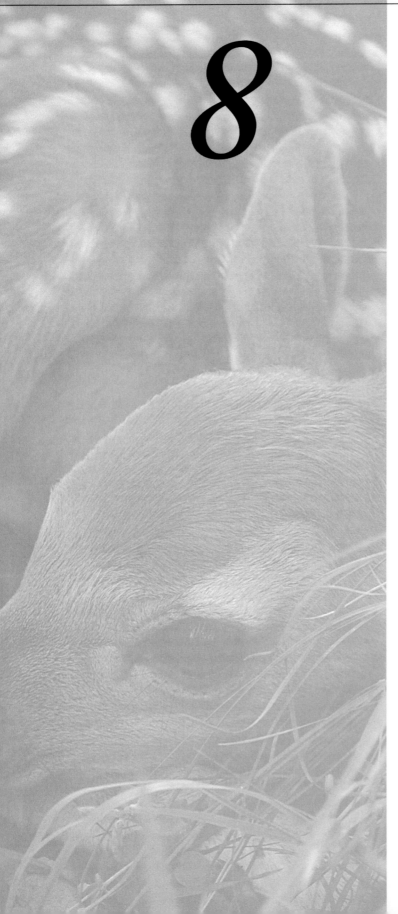

CHAPTER

8

Environmental Adaptations

Land mammals reside in environments characterized by extremes in temperature, precipitation, and altitude. Marine environments also vary greatly in terms of water pressure at different depths. Different terrestrial mammals may be faced with temperatures ranging from close to –65°C in the Arctic to 55°C in Death Valley, California. Daily fluctuations during winter can be extreme, for example, in Montana, the temperature in one day can plummet from 6°C to –49°C. Summer is no relief for mammals, and a hot spell in western Australia may reach 38°C and last for 162 continuous days! Although marine mammals live in water, they also experience temperature variation ranging from –2°C near the poles to 30°C near the equator. So how do mammals cope with such extremes? This chapter focuses on the diversity of mammalian adaptations that enhance survival in variable environments.

For many years, the terms **warm-blooded** and **cold-blooded** were used to conveniently divide animals into two groups: the vertebrates (typified by high, constant body temperature) and the invertebrates (typified by variable body temperatures). These terms are inaccurate and unscientific, however, as well as being unclear. Thermoregulatory terminology can be tricky and ambiguous, so it should be well defined (Hainsworth 1981; Bartholomew 1982; Hill and Wyse 1989; Pough et al. 1989). The ability of mammals to colonize inhospitable environments is due largely to their ability to use **endothermy**—the maintenance of a relatively constant body temperature by means of heat produced from *inside* (*endo*) the body. The degree of endothermy varies in both space and time. No animal maintains the same body temperature over all parts of its body. The term **homeothermy,** the regulation of a constant body temperature by physiological means, is more precise than endothermy (Hill and Wyse 1989:99). The term **ectothermy** refers to the determination of body temperature primarily by sources *outside* (*ecto*) the body. This is not necessarily a passive system, and many ectotherms employ behavioral means to regulate their body temperature; for example, lizards will bask in the sun or rest on a warm rock. A similar term, **poikilothermy,** emphasizes the variation in body temperature under environmental conditions, as opposed to focusing on the mechanism by which the body temperature is maintained. Some examples of poikiotherms are clams, starfish, and many other invertebrates. Mammals and birds that are capable of dormancy—hibernation, daily torpor, and estivation—are referred to as **heterothermic.** In such cases, regulation of body temperature may vary on different parts of the body (regional heterothermy) or at different times (temporal heterothermy). Many species of bats, for example, exhibit temporal heterothermy: They maintain constant body temperature while foraging but permit their body temperature to approach ambient temperature when they rest.

Maintaining an internally regulated body temperature offers numerous benefits for mammals. Both mammals and birds have high metabolic rates—at least eight times that of ectotherms. In terms of energy expenditure, maintaining such high body temperatures plus high levels of activity is costly, but it has the advantage of enhancing coordination of biochemical systems, increasing information processing, and speeding central nervous system functions. As a result, mammals have a refined neuromuscular system, thus enhancing their efficiency at capturing prey and escaping from predators. Mammals gain independence from temperature extremes in nature, can extend activity periods over a 24-hour period, and colonize many environments and ecological niches throughout the world. Mammals can match their thermoregulatory pattern to suit a given environment and thus take advantage of nutritional resources year-round.

Endothermy is thought to have evolved from an ectothermal condition probably two to three times in the Mesozoic era. Internal heat production probably evolved in the late Triassic within the group of mammal-like reptiles in response to selective pressure favoring sustained activity and temperature regulation. The evolution of endothermy in mammals has been the subject of considerable debate (Bakker 1971; Bennett and Ruben 1979; Farlow 1987; Hayes and Garland 1995).

HEAT TRANSFER BETWEEN A MAMMAL AND THE ENVIRONMENT

Most placental mammals residing in thermally benign environments maintain their body temperature between 36° and 38°C (Cossins and Bowler 1987). In general, the body temperature of birds is several degrees higher than that of mammals. Monotremes and marsupials, on the other hand, may have a core body temperature ranging from 30° to 33°C, but they thermoregulate quite well. To keep a constant body temperature, mammals must maintain a delicate balance between heat production (energy in) and heat loss (energy out)—thermodynamic equilibrium—to survive (Porter and Gates 1969). Heat is produced through metabolism of food or fat, cellular metabolism, and muscular contraction. The factors affecting the exchange of energy between a mammal and the environment are sunlight (solar radiation), reflected light, thermal radiation, air temperature and movement, and the pressure of water vapor of the air (figure 8.1). The properties of a mammal that influence the exchange of energy are metabolic rate, rate of moisture loss, thermal conductance of fat or fur, absorptivity to radiation, and the size, shape, and orientation of the body. Most of the heat produced by mammals is lost to the environment passively by **radiation, conduction,** and **convection** (air movement) to a cooler environment and by the **evaporation** of water. Mammals must make adjust their energy balance to meet different demands in their environment.

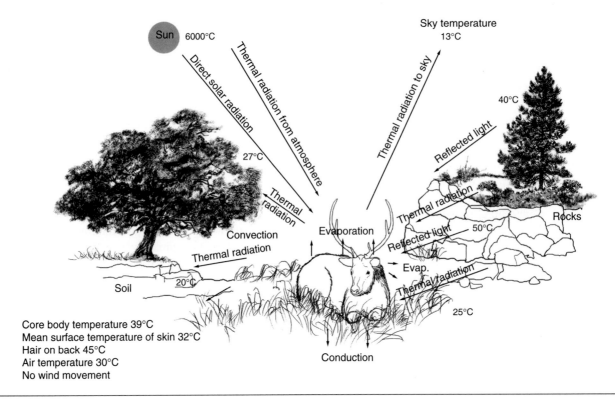

Figure 8.1   **Energy exchange.**   Illustration of the energy exchanges between an animal and its environment under warm conditions.

TEMPERATURE REGULATION

As mentioned in chapter 7, mammals regulate body temperature by continuously monitoring outside temperatures at two locations: on the surface of the skin and at the hypothalamus. The hypothalamus, or mammalian "thermostat," is located in the forebrain below the cerebrum and operates by comparing a change in body temperature with a reference, or set point, temperature. Each species may have a different **set point,** or comfort zone, which is set at the hypothalamus. For most eutherian mammals, heat is generated by muscle contraction (shivering), brown fat (nonshivering thermogenesis), and activity of the thyroid gland. In marsupials and monotremes, heat production is due primarily to the activity of skeletal muscle (Augee 1978; Nicol and Andersen 1993). For mammals residing in variable environments, body temperature is maintained around a given set point, with each mammal's set point determined by such factors as insulation, behavioral postures, activity levels, and microclimatological regimes.

ADAPTATIONS TO COLD

Most placental mammals maintain a core body temperature of about 38°C. Monotremes and marsupials tend to show lower body temperatures, with echidnas and platypuses exhibiting normal body temperatures ranging from 28° to

33°C (Augee 1978; Grigg et al. 1989, 1992). For each species, a range of environmental temperatures, referred to as the **thermoneutral zone,** occurs within which the metabolic rate is minimal and does not change as ambient temperature increases or decreases. The upper and lower limits of the thermoneutral zone are referred to as the upper and lower critical temperatures (figure 8.2). Decreasing environmental temperatures require that an animal increase its metabolic rate to balance heat loss. The temperature at which this becomes necessary is called the **lower critical temperature.** This temperature varies from species to species and is seasonally adjusted by the interplay of insulatory thickness, behavioral attributes, and integration with the hypothalamus. As environmental temperatures decrease, adjustments in **thermal conductance** (the rate at which heat is lost from the skin to the outside environment) and metabolism are necessary if a species is to maintain euthermy. If this is not possible, death due to **hypothermia** (low body temperature) results. Thermal conductance (*C*) is expressed as the metabolic cost (expressed in milliliters of oxygen per gram of body mass) for a given time interval per degree Celsius difference between body temperature (T_b) and environmental temperature (T_a)

$$C = MR/T_b - T_a$$

(McNab 1980; Hill and Wyse 1989). Some mammals and birds from polar regions are so well insulated that they can withstand the lowest environmental temperatures on earth

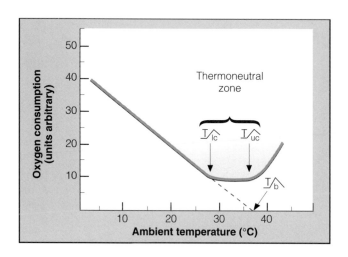

Figure 8.2   **Thermoneutral zone.**   Relationship of oxygen consumption compared with ambient temperature in a hypothetical mammal. T_{lc} = lower critical temperature; T_{uc} = upper critical temperature; T_b = core body temperature.

Source: Data from G.A. Bartholmew, "Body Temperature and Metabolism"
in Animal Physiology: Principles and Adaptations, *4th edition,* Macmillian
Company, New York.

building, and huddling in groups). By modifying the pelage, for example, an animal can reduce the gradient of heat flow. Do not forget that hair is a principal means of conserving energy (see chapter 5). Hair is a poor thermal conductor, so it greatly decreases the amount of body heat lost to the environment by decreasing thermal conductance.

At the other end of the thermoneutral zone is the **upper critical temperature,** the temperature above which animals must dissipate heat by evaporative cooling to maintain a stable internal temperature. This zone is important to mammals residing in desert environments and is less variable than the lower critical temperature. Mammals employ many different mechanisms, such as seeking shelter in cooler underground burrows and restricting surface activity to hours of darkness, to avoid upper critical temperatures. Furthermore, evaporative cooling can be facilitated by pulmonary means, panting, sweating, spreading saliva, or using a form of dormancy called estivation. If a mammal cannot combat an increase in temperature by dissipating heat through evaporation or by muscular activity of panting, its body temperature will keep increasing, ultimately resulting in death. The mechanisms used to cope with heat are examined later in the chapter.

Energy Balance in the Cold

Winter is a potentially stressful time for northern mammals (and for mammals from extreme southern latitudes), and they employ a wide array of adaptations to cope with the food shortages and cold stress that the season brings. The key to survival in the cold is to maintain energy balance (Wunder 1978, 1984). Two major features determine the energy needs of mammals in the natural environment: the *physical environment* and the *activities* of the mammal (i.e., type and level of behavior, growth, or production young; figure 8.4). To meet their energy needs, mammals must acquire

(about −70°C; figure 8.3) by simply increasing their resting metabolic rate (Scholander et al. 1950; Irving 1972). As we will learn later, however, some species can significantly decrease body temperature (undergo hibernation or torpidity) to combat seasonal decreases in food. To maintain body temperature within the thermoneutral zone when residing in cold environments, the individual must reduce the rate of heat loss to the environment. Mammals can achieve this reduction in many different ways: by being larger, possessing enhanced body insulation, changing peripheral blood flow, or modifying behavior (e.g., curling, **piloerection** [the fluffing of fur], nest

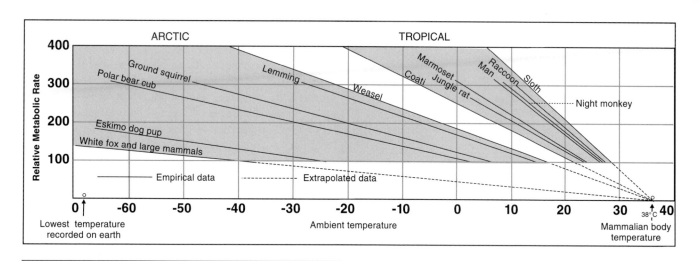

Figure 8.3   **Metabolism and temperature.**   Metabolic rates of various mammals in relation to ambient temperature. The normal resting metabolic rate for each animal, in the absence of cold stress, is given the value 100%. Any increase at lower temperature is expressed in relation to this normalized value, making it possible to compare widely different animals.

Source: Data from P.F. Scholander et al., "Body Insulation of Some Arctic and Tropical Mammals and Birds" in Biol. Bull., *99:225-235, 1950.*

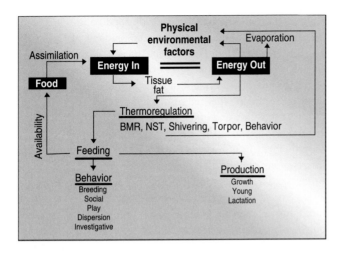

Figure 8.4   Energy balance.   Conceptual model of avenues of energy balance for a small mammal, indicating a primary cascade for energy allocation. Lines represent both total energy flow and rate functions.

Table 8.1.   Winter survival mechanisms of mammals

Avoidance

Body size
Insulation
Appendages
Coloration
Modification of microclimatic regime
 Communal nesting
 Construction of elaborate nests
 Foraging zones
Food hoarding
Reduction in level of activity
Reduction in body mass
Dormancy

Resistance

Increase in thermogenic capacity through BMR, NST, shivering

Abbreviations: BMR = basal metabolic rate; NST = nonshivering thermogenesis.

energy—they must feed. Given a finite source of food during winter, mammals allocate energy initially to thermoregulation. Temperature regulation is essential for mammals to perform normal functions. For example, a mammal may freeze to death even with abundant food if it cannot assimilate that energy and produce heat fast enough to balance heat loss **(turnover capacity).** Also, even a mammal with a high heat-generating capacity will perish if there is not enough food (energy) available in the environment to meet its needs. In sum, mammals residing in cold regions must have sufficient food available and sufficient turnover capacity to meet their thermoregulatory needs.

AVOIDANCE AND RESISTANCE

Unlike many species of passerine (perching) birds inhabiting cold regions, most mammals do not undertake long migrations. They remain as residents during winter months and have evolved mechanisms of *avoidance* and *resistance* to cold (table 8.1). Avoidance of cold entails energy conservation, whereas resistance requires energy expenditure—a debt that must be repaid. Mammals rarely rely on a single mechanism to enhance survival in the cold but instead exhibit a suite of strategies that integrate behavioral, anatomical, and physiological specializations finely tuned to a specific habitat and life-style. Thermoregulatory mechanisms vary according to taxon and the attributes of each species.

Avoidance of Cold

Body Size and Metabolism

Does a mammal's size enhance its energy conservation capacity in the cold? To answer this question we must closely examine the relationship of body size and metabolism. Resting,

or basal, metabolic rate (BMR) is the volume of oxygen consumed per unit of time at standard pressure and temperature (STP) and is known for many species of mammals. McNab (1988) provides rates of basal metabolism and body masses for 320 species constituting 22 orders of mammals. MacMillen and Garland (1989) studied metabolic rates, body masses, climatic data, and latitudes for many species of rodents. Hill (1983) provides an excellent review of the thermal physiology and energetics of the Genus *Peromyscus*. It makes intuitive sense that larger animals require more food to maintain their body temperature and level of activity. Figure 8.5 (solid line) shows that larger animals consume more oxygen (they eat more food) than small animals—this is **total metabolic rate.** To compare oxygen consumption between animals of different sizes, metabolic rate is adjusted for body size and expressed as **mass-specific metabolic rate** (rate of oxygen consumption per gram of body mass). Now it is apparent that the mass-specific metabolic rate decreases as body size increases (see figure 8.5, broken line). Given this relationship, we can estimate metabolic rate from the body mass of a species. The slope of the solid line in figure 8.5 has a value of about 0.75; that is, the metabolic rate varies as the 0.75 power of body mass (Kleiber 1932, 1961). This equation is quite useful in estimating the metabolic rate of mammals of known body size and has important ecological and evolutionary consequences. The study of the consequences of a change in size, or in scale, is important in predicting the metabolic rates of mammals. Physiologists define **scaling** as the structural and functional consequences of a change in size or scale (Schmidt-Nielsen 1964; Calder 1984; McNab 1988).

In euthermic animals, heat production must equal heat loss, and heat loss is proportional to surface area. This fact led to the formulation of **Bergmann's rule,** which states that, " . . . races from cooler climates in species of warm-

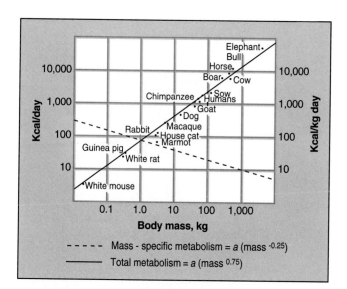

Figure 8.5   Body size and metabolism.   Linear relationship comparing the log of body size in placental mammals with the log of total metabolism (solid line) and the mass-specific metabolism (broken line). *a* = proportionality constant (3.8) for placental mammals.

Source: Data from M. Kleiber, "Body Size and Metabolism" in Hilgardia, *6:315–353, 1932.*

blooded vertebrates tend to be larger than races of the same species living in warmer climates . . ." (Mayr 1970:197). There is much debate concerning the interpretation and implications of Bergmann's research (Scholander 1955; Rosenzweig 1968; McNab 1971; Calder 1984; Lindstedt and Boyce 1985; Geist 1987). This "rule" implies that some energetic advantage may be gained through a decreased surface-area-to-volume ratio. We can further generalize that the amount of heat loss depends on both the animal's surface area and the difference in temperature between its body surface and the surroundings. Large mammals, for example, have less heat loss than small mammals because of their large body mass relative to surface area. For example, visualize a mammal as a cube. With linear dimensions of 1 cm on each side, the cube has a surface of of 6 cm² and a volume of 1 cm³. Hence, the surface-area-to-volume ratio is 6:1. If you double the linear dimension of the cube, the total surface area increases to 24 cm², and the new volume is 8 cm³. Now the surface-area-to-volume ratio is 3:1. By doubling the length, heat conservation is enhanced by reducing surface area relative to volume. Granted, mammals are not cubes, but arguments in support of an energetic interpretation of Bergmann's rule follow this logic: A larger mammal has less total surface area per unit volume and thereby benefits from a reduced rate of cooling. This interpretation sounds rather convincing, but there are problems. Small mammals know little about per-gram efficiency and are only concerned with total food requirements; a larger mammal clearly requires more food—a clear disadvantage for herbivores during winter. Furthermore, if we analyze body size of mammals distributed over a wide latitudinal range, we see mixed results.

For example, McNab (1971) indicated that mammalian body size is not inversely related to temperature. Of 47 species examined, only 32% followed the trend predicted by Bergmann's rule. Moreover, Geist (1987) contended that Bergmann's rule was invalid. Although body size of large mammals initially increased with latitude, it reversed between 53° and 63°N; small body size occurs at the lowest and highest latitudes.

Body mass of mammals may change seasonally (voles and shrews are known to decrease mass during winter; Dehnel 1949; Churchfield 1990; Merritt 1984, 1995), and body mass may also change in different species due to competition for food (McNab 1971, 1989). The long, thin shape of weasels (Genus *Mustela*) has distinct disadvantages in cold climates. Their shape exposes a large surface area to the cold air, and their unique feeding requirements dictate that they pursue prey through small crevices and a labyrinth of subterranean runways, which is energetically very expensive. Their mobility would be greatly compromised by a dense, heavy pelage; thus, a short fur is essential for optimal agility. Compared with woodrats (Genus *Neotoma*), weasels have about a 15% higher surface-area-to-mass ratio—they cannot assume a spherical form by curling as woodrats can. As a result, cold-stressed weasels have metabolic rates 50% to 100% greater than less slender mammals of comparable size. So, being long and skinny may facilitate capturing prey, but weasels pay a high energy cost in the form of a rapid rate of heat loss. As a result, weasels must consume more food to maintain energy efficiency (Brown and Lasiewski 1972; King 1989, 1990; McNab 1989).

Insulation

For mammals that reside in cold environments, the most direct method of decreasing heat loss is to increase the effectiveness of insulation (see table 8.1). For mammals and birds, the best way to reduce thermal conductance is to possess fat, fur, or feathers between the body core and the environment. Insulation value increases with the thickness of fur (figure 8.6) and is at maximum in such mammals as Dall sheep (*Ovis dalli*), wolves (*Canis lupus*), and arctic foxes (*Alopex lagopus*). The easiest way to understand the benefits of insulation is to think in terms of extending the limits of the lower critical temperature. The arctic fox, for example, has a lower critical temperature of –40°C, and at an environmental temperature of –70°C has elevated its metabolic rate only 50% above its normal metabolic rate! It is easy to understand that if an animal lowers its metabolic rate, it can get along with less food—a real plus if food is limited during winter months.

Some tradeoffs are associated with insulation. The insulating value of fur is a function of its length and density; some small mammals such as lemmings (Genera *Dicrostonyx, Lemmus*) have very long fur relative to body mass. A thick coat on predators like weasels, however, would compromise their ability to forage in narrow crevices. Scholander (1955) contended that small mammals could not accumulate

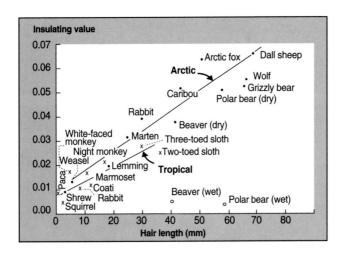

Figure 8.6 The length of fur. The insulating values of the pelts of arctic mammals (•) are proportional to the length of the hair. Pelts from tropical mammals (x) have approximately the same insulating value as those of arctic mammals at short hair lengths, but long-haired tropical mammals like sloths have less insulation than arctic mammals with hair of the same length. Immersion in water greatly reduces the insulative value of hair, even for such semiaquatic mammals as the beaver and polar bear (O).

Source: Data from P.F. Scholander et al., "Body Insulation of Some Arctic and Tropical Mammals and Birds" in Biol. Bull., *99:225–235, 1950.*

enough insulation in the form of fat or hair to cope with low winter temperatures without compromising agility. Research documenting the importance of pelage insulation in small mammals has focused primarily on the Families Muridae and Soricidae (Churchfield 1990). When these groups are faced with cold stress, heat loss was reduced by increasing pelage insulation by 11% to 19% during winter months. For small mammals, pelage changes may be useful in conserving energy and compensating for the large difference between body and environmental temperatures during winter. As we will see later, however, small mammals that do not have a thick, long pelage take advantage of stable microclimatic regimes below ground and may employ social thermoregulation to cope with cold. On the other hand, larger mammals benefit greatly from insulation in the form of fat and fur. An excellent discussion of adaptation to cold by northern cervids is provided by Marchard (1996).

Large arctic mammals such as muskox (*Ovibos moschatus*), caribou (*Rangifer tarandus*), and wolves have long, dense fur as well as thick layers of fat beneath the skin. The muskox, in particular, is equipped with an immense cloak of coarse guard hairs over 30 cm in length and an underfur of thick, dense, silky wool (figure 8.7). The hairs of the muskox are thicker at the tip than at the root so that they form an almost airtight coat. The animal's heat-losing extremities (tail and ears) are well hidden under the pelage. These heat-conserving adaptations permit the muskox to survive in the arctic tundra where winter temperatures commonly reach −40°C.

Figure 8.7 An immense fur coat. Muskox are inhabitants of the arctic tundra of North America. They possess a massive dark brown coat composed mainly of long threads of high-quality wool. The outer coat is covered with long, coarse guard hairs. Heat-losing extremities are buried in the coat, an adaptation for survival in the arctic cold.

Insulating fur is a great advantage for thermoregulation in larger terrestrial mammals (Hudson and White 1985; Marchard 1996). Aquatic mammals have a slightly different strategy. For example, muskrats (*Ondatra zibethicus*), water shrews (*Sorex palustris*), beavers (*Castor canadensis*), and northern fur seals (*Callorhinus ursinus*) can function in waters at 0°C because of structural modifications of their fur that trap air. Their skin stays dry, shielded from the water by a layer of air. This layer of air not only helps with thermoregulation but assists with buoyancy. It is therefore easy to see that a polar bear (*Ursus maritimus*) with dry fur is far better off than one with wet fur. Fur is actually of limited value to an aquatic mammal. Body warmth is maintained because air is trapped between hair and the skin; if this air is displaced by water, the fur loses most of its insulating value, which falls close to zero when the fur becomes wet.

Because the thermal conductivity of water is almost 25 times greater than that of air, marine mammals must protect themselves from heat loss, particularly when inactive (Schmidt-Nielsen 1997). Air trapped in dry underfur acts as an effective insulator for some groups. Marine mammals such as whales (chapter 16), walruses, and seals (chapter 15) possess a particularly effective insulator—**blubber** (the thick layer of white adipose tissue below the dermis of the skin that prevents

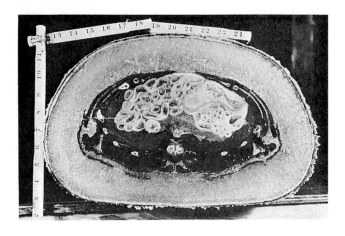

Figure 8.8 **Blubber.** Cross section of a frozen seal showing the thick layer of blubber. Of the total area in the photograph, 58% is blubber and the remaining 42% is muscle, bone, and visceral organs. The measuring stick is graduated in inches.

heat loss from the body core; figure 8.8). Blubber also acts to streamline the body, adjust buoyancy, and provide an energy reserve. Marine mammal fat is entirely blubber; marine mammals have no visceral fat like other mammals. Blubber thickness varies from 5 cm in small seals to 60 cm in the bowhead whale (*Balaena mysticetus*). In some whales, such as the gray whale (*Eschrichtius robustus*), blubber is also an energy reserve during migration. The blubber of seals is a better insulator than that of whales because it contains less fibrous connective tissue (Bonner 1990). During winter, the layer of blubber in a ringed seal (*Phoca hispida*) may account for 40% of its mass. Northern fur seals (*Callorhinus ursinus*) and other eared seals (Family Otariidae) are characterized by a dense covering of fur that traps air bubbles. Because they rely less on insulation in the form of blubber, they must remain active to maintain normal body temperature in frigid waters.

Activity on land, even with the air temperature at 12°C, can produce overheating, however, which may prove lethal. Marine mammals control overheating by remaining in the water, thus keeping their fur or skin moist; by panting; and also by waving their hind flippers, which are supplied with abundant sweat glands. Another way to control heating is by sleeping; in some species, sleeping reduces heat production by almost 25%. A southern elephant seal (*Mirounga leonina*) may reach 3500 kg. This large, dark mass basking on pack ice in the Antarctic is subject to significant warming due to absorption of solar radiation. Blubber, permeated by blood vessels, is not passive insulation. For large elephant seals and Weddell seals (*Leptonychotes weddellii*) resting on pack ice, loss of heat is essential. In response to heat stress, these seals can increase the rate of blood flow through their skin by dilating special vessels that join arterioles directly to venules, bypassing the capillary beds. Because these bypasses (called arteriovenous anastomoses) are very near the surface of the skin, the increased flow of blood permits heat to be lost to the environment.

Appendages

The fur thickness of northern mammals may increase as much as 50% during winter months. Appendages such as legs, tail, ears, and nose, however, are potentially great heat "wasters." They cannot be well insulated, but to prevent heat loss from appendages, arctic land mammals permit temperatures at appendages to decrease, often approaching the freezing point. Appendages must be supplied with oxygen and kept from freezing, however, by circulating blood through them. This is done, without constant loss of heat from the extremity, by a process called **countercurrent heat exchange,** a form of peripheral heterothermy. This physiological mechanism shunts blood through a heat exchanger (the **rete mirabile,** or **miraculous net**) that intercepts the heat on its way out and maintains the extremity at a considerably lower temperature than the core—the net result is a reduction of heat loss. As warm arterial blood passes into a leg, for example, heat is shunted directly from the artery to the vein and then carried back to the core of the body (figure 8.9). As a consequence, the appendages of many northern mammals are maintained at comparatively cold temperatures. Anatomical studies have shown that the foot pads of arctic canids possess a massive arteriovenous plexus through which blood flow to and heat loss from the foot pads is controlled (Henshaw et al. 1972). To keep the extremities soft and flexible at such low temperatures, the fats in the feet of northern mammals must have very low melting points, perhaps 30°C lower than ordinary body fats (Storey and Storey 1988). Countercurrent heat exchange is well developed in the tail of beavers. In air at low temperature, loss of heat from the tail may be reduced to less than 2% of the resting metabolic rate. Then, when faced with higher temperatures, over 25% of the heat produced at resting metabolism may be dissipated through the tail (Coles 1969).

During inhalation, heat and water are added to air before it reaches the lungs. During exhalation, heat is typically lost to the environment due to warm air loss and evaporation of water. Heat recovery (heat exchange) in the respiratory system is slightly different from the countercurrent heat exchange process discussed for the vascular system. In the vascular system, heat exchange occurs at the same time between two channels, but in the respiratory system, the exchange occurs at different times in the same channel (Schmidt-Nielsen 1997). Caribou (*Rangifer tarandus*) demonstrate the principle. When active, they produce a considerable amount of heat, which, due to their excellent insulation, may exceed their thermoregulatory requirements. Some must therefore be lost, or their body temperature would increase. When resting, however, caribou do not generate as much heat and must minimize heat loss to the environment. Caribou and other large mammals possess a flap of skin at the opening to the nasal passages. This flap of skin reduces the size of each opening to a small slit when the animal is inactive, thus minimizing heat loss from the lungs. When resting, the temperature of expired air is lower than body temperature. When

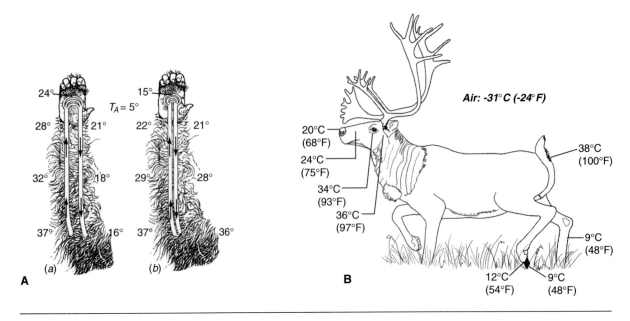

Figure 8.9 **Countercurrent heat exchange.** (A) A diagram representing the circulation in a limb of a mammal showing hypothetical temperature changes of the blood in the absence (a) and presence (b) of countercurrent heat exchange. Arrows indicate direction of blood flow. In (b), the venous blood takes up heat (thus cooling the arterial blood) all along its path of return because even as it becomes warmer and warmer, it steadily encounter arterial blood that is warmer yet. (B) Regulation of external body temperature in the caribou. Temperature regulation is accomplished in part by countercurrent heat exchange. An intricate meshwork of veins and arteries acts to keep the temperature of the legs near that of the environment so heat is not lost from the body.

(A) Source: Data from R.W. Hill and G.A. Wyse, Animal Physiology, *2nd edition, 1989, Harper & Row, New York. (B) Data from J.F. Merritt, "Animals of the Arctic" in* Arctic Life: Challenge to Survive, *(M.M. Jacobs & J.B. Richardson III, eds.), 1983, The Board of Trustees, Carnegie Institute, Pittsburgh, Pennsylvania.*

exercising, the nostrils flare open, increasing the surface area of the mucous lining of the nose for heat exchange. Due to the increased rate of respiration during activity, the volume of air in the nasal passages has less time for heat exchange to occur. The temperature of expired air during exercise is therefore higher when the animal's heat production exceeds its requirements for temperature regulation. Thus, respiratory heat exchange for large arctic species serves a dual function: minimizing heat loss while at rest and increasing heat loss during periods of activity (Hainsworth 1981:223).

Conservation of heat in species occupying cold regions can also be enhanced by reducing the length of the exposed extremity. **Allen's rule** states that the appendages of endothermic animals are shorter in colder climates than those of animals of the same species found in warmer climates. For instance, arctic foxes have small, rounded, densely furred ears; a reduced muzzle; and short, stubby legs. Foxes living in deserts, such as kit foxes (*Vulpes macrotis*) of the southwestern United States or the fennec (*V. zerda*) of North Africa, have large, elongated, thinly furred ears, and comparatively long legs (figure 8.10). Thus, by reducing unnecessary body surface area, animals can reduce heat loss. It seems logical from the standpoint of thermal physiology that a long nose and tail, as well as long ears and legs, tend to add unnecessarily to the body surface and would be great heat wasters for cold-adapted species. For species residing in hot climates,

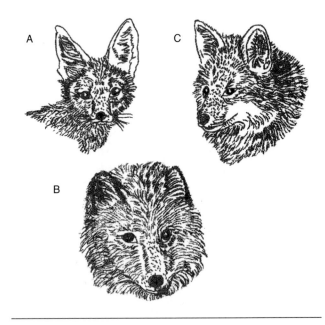

Figure 8.10 **Ears and heat loss.** The ears of foxes help to regulate body heat. (A) The kit fox of the American Southwest has relatively long ears that help to dissipate heat. (B) In contrast, the relatively short ears of the arctic fox help to conserve heat. (C) The red fox (*Vulpes vulpes*) of temperate America possesses intermediate-sized ears.

Figure 8.11 **A White fur coat.** The polar bear has a circumpolar distribution. It is one of the largest terrestrial carnivores in the world, feeding primarily on seals and fish. The thick white coat is composed of hollow hairs that trap warm insulating air and also act as an efficient solar collector for this bear when wandering from ice floe to ice floe in search of prey. Although their unique hairs convey a thermal advantage, polar bears enhance survival in cold by possessing an enormous thermal inertia (large body size) and large stores of subcutaneous fat.

long appendages are a valuable commodity because they aid in the dissipation of heat. Like Bergmann's rule, the validity of this generalization is debatable, and there are exceptions. For example, the body form of bush dogs (*Speothos venaticus*) in the tropics of South America is not unlike that of arctic

foxes. Furthermore, the length of ears and tail in hares (Genus *Lepus*) follows Allen's rule, but that for rabbits does not (Genus *Sylvilagus;* Stevenson 1986).

Coloration

Although northern mammals exhibit some fascinating anatomical specializations for coping with harsh winter climates, perhaps the most obvious trait defining them is their white pelage (figure 8.11). **Gloger's rule** states that, "Races in warm and humid areas are more heavily pigmented than those in cool and dry areas" (Mayr, 1970:200). Pigmentation in many mammals tends to be paler the closer the species habitat is to the Arctic, with northern races (subspecies) of animals generally being lighter in color than their southern counterparts (Flux 1970; Johnson 1984). Coloration of pelage of many arctic mammals either remains white year-round or changes to white during winter. Those that remain white year-round include the polar bear (actually their hairs are transparent and pigmentless), arctic hares (*Lepus arcticus*) of the far north, and northerly forms of the caribou and gray wolf. It is noteworthy that arctic hares in the far north, such as those on Ellesmere and Baffin Islands and in Greenland (figure 8.12), do not change color seasonally but remain white during the short arctic summer. Because hares are so well adapted to winter, the cost in energy of changing to brown during the short summer exceeds the benefits of permanent adaptation to winter. A seasonal color change (dimorphism) is seen in collared lemmings, ermines, and arctic foxes, however. Species that undergo seasonal dimorphism accomplish the color changes in different ways. The hairs of weasels, for example, turn white along its entire length. This

Figure 8.12 **Arctic hares in the far north.** *Lepus arcticus* reside on the arctic tundra—the most northern distribution of all leporids. They normally show seasonal changes in coat color. Arctic hares inhabiting the far north, however, keep their white coats year-round, even though there is no snow cover in midsummer.

complete molt is characterized by each hair being lost and a new, white (winter) hair replacing the old, brown (summer) hair. In contrast, the snowshoe hare (*Lepus americanus*) exhibits a different type of molt. Rather than going through a complete molt pattern, only the tips of the winter hairs of *L. americanus* are white; the bases remain gray. The timing of molt patterns for the Old World mountain hares (*L. timidus*) is most strongly influenced by length of day (photoperiod) but is also correlated with average ambient temperature and duration of snow cover (Angerbjörn and Flux 1995). Furthermore, species of mammals that show seasonal color changes do not change color over their entire geographic range. For example, the North American long-tailed weasel (*Mustela frenata*) shows seasonal color changes north of about 40°N latitude, but south of this zone, they remain brown year-round. Individuals residing in the zone of overlap between these color groups show gradations in color from white to pied or brown (Hall 1951).

The specific cue (proximate factor) that triggers the onset of molt is decreasing daylength. In autumn, the optic nerve receives the stimulus, which is then transmitted to the hypothalamus. The factor that is ultimately responsible for establishing the duration of the winter coat is probably its thermal advantage, accounting for the close correlation between molting and mean annual temperatures (Flux 1970; Johnson 1984). Temperature and photoperiod, however, are not the sole determinants of white winter coloration. Transplantation experiments have demonstrated that heredity, as well as temperature, is involved in controlling pelage changes in mammals (Kliman and Lynch 1992). The molt cycle is geared to changes in the environment by way of changes in the hormonal system (namely, the endocrine glands: thyroid, pituitary, adrenal cortex, and pineal) as mediated by the brain and the hypothalamus. The seasonal cycles of molt and reproduction are closely related and are coordinated by the neuroendocrine system. In spring, in addition to changes in pelage color, density, and length, the brain signals the pituitary to secrete gonadotropins, which stimulate the gonads to prepare for the approaching breeding season. During spring, the hair follicles and testes of male weasels enlarge simultaneously; the testes begin to manufacture testosterone and sperm, and the hair follicles accumulate melanin and manufacture hair for a complete coat replacement (King 1989).

Why northern animals turn white in winter is not fully known. We assume that if the mechanism were not maintained by natural selection (if it did not confer an adaptive advantage more often than a disadvantage), it would disappear. We assume that the color white acts to conceal both predators and prey. This adaptation may conceal polar bears from a potential victim. While actively searching for prey, however, a weasel is surely easily detected, even with a white pelage. It is possible that the weasel's white color in snowy regions may allow it to blend with its background (cryptic coloration) and thus avoid predation by hawks, owls, and foxes. Many questions concerning the adaptive nature of color changes in northern mammals have yet to be answered

(Walsberg 1983; Marchand 1996). For example, how can the blue color phase be explained that commonly occurs in arctic foxes inhabiting the Pribilof Islands and many coastal areas of Alaska and Canada? If color acts to camouflage these animals, why are they slate-blue during winter? If this blue phase is not adaptive, then how did it evolve and how is it maintained in the population? If cryptic coloration is important to conceal predator from prey, why don't gray foxes (*Urocyon cinereoargenteus*) or fishers (*Martes pennanti*) turn white in snowy regions?

The color white may also convey a thermal advantage. According to the laws of physics, black-colored animals lose heat by radiation faster than white-colored ones. Following Gloger's rule, we therefore expect to find white animals occurring in cold regions. Some investigators have misinterpreted the pertinent physical laws, however. Radiation of heat from an animal's body is in the form of infrared energy, which is unrelated to visible coloration. All animals are therefore considered to be thermodynamic "blackbodies," meaning that they absorb all incident radiation and reflect none. The color of the fur and underlying skin may be important to the amount of heat absorbed from solar radiation—its peak intensity is in the visible range. When exposed to direct solar radiation, dark-colored skin or fur absorbs more incident energy than light-colored skin or fur. The conservation of this heat energy depends on the length and thickness of the fur, not on its color. Many small mammals, such as voles and shrews, show increases in pelage density and length during winter (Khateeb and Johnson 1971; Bozinovic and Merritt 1992). Animals living in northern regions may not only possess white fur but also have thicker, denser pelages. For example, mountain hares possess a white winter coat that is twice the length and thickness of their summer coat (Flux 1970). Generalizations concerning the adaptive significance of different biogeographical "rules" or trends must take into account that the survivability of northern mammals may not result solely from forces that maximize heat conservation.

For some animals, the thermal advantage of hair is not necessarily contingent on its color, density, or length, but rather on the anatomy of the hairs and color of the skin beneath it. For many members of the Family Cervidae, insulation depends on the air contained by the highly medullated guard hairs that provide insulation (Johnson and Hornby 1980). Research with polar bears indicates that a combination of their transparent, pigmentless hair and black skin may enhance their heat-conserving abilities (Grojean et al. 1980; Walsberg 1983). Pigmentless hair traps and transmits to the skin 90% of sunlight in the invisible ultraviolet portion of the spectrum but only 10% of the visible light. Energy in the form of heat from the ultraviolet light is absorbed by the dark skin to help warm the body, while the visible light is reflected as white color. The hairs of polar bears act like optical fibers, with ultraviolet light entering at one end and bouncing along the inside of the hair shaft to reach the dark skin, where it is absorbed.

Modification of Microclimatic Regime

In northern regions, nonhibernating small mammals, particularly those of the rodent Families Muridae and Sciuridae (e.g., *Glaucomys*), construct elaborate nests and engage in communal nesting (Muridae: Madison et al. 1984a; West and Dublin 1984; Wolff 1989; Bazin and MacArthur 1992; Sciuridae: Muul 1968; Stapp et al. 1991; Stapp 1992). Not only small mammals use huddling: 23 raccoons (*Procyon lotor*) were reported occupying one winter den (Mech and Turkowski 1966). During winter, female striped skunks (*Mephitis mephitis*) formed a communal nest with other females or with a single male (Wade-Smith and Verts 1982). The greatest gain from huddling, however, should accrue to small mammals with a large surface-area-to-mass ratio and a limited capacity for increasing the insulation value of their pelage.

Both nest building and huddling conserve body heat by reducing in thermal conductance. Huddling in groups reduces each individual's exposed surface, thus reducing the cold stress and the metabolic requirement for heat production. Studies traditionally employed laboratory-based calorimetry to demonstrate energy savings of communal nesting during winter (Glaser and Lustick 1975; Martin et al. 1980; Casey 1981). More recent studies linked laboratory experimentation with field-derived data on nesting habits by employing radiotracers and radiotelemetry methods (Madison et al. 1984a; Andrews and Belknap 1986).

When muskrats huddle in a group, a major part of each individual's body surface is in contact with a neighboring animal (figure 8.13). Curling and retracting the extremities reduces heat loss, and microclimatic modification in the form of an elaborate, well-insulated nest below ground and snow cover (subnivean) adds greatly to energy savings. Sealander (1952) showed that at low temperatures, individuals of the Genus *Peromyscus* formed a "communal" group. Those at the bottom of the group enjoyed temperatures well above ambient levels, but by continually shifting position, each mouse in the huddle was periodically rewarmed and thus avoided hypothermia. Because heat loss by conduction or convection varies in direct proportion to the amount of surface area exposed, the energy savings of huddling can be easily calculated (Marchand 1996). Vickery and Millar (1984) provided a model for predicting the energy advantages and disadvantages of huddling. When applying their data on *Peromyscus,* they found that huddling confers a distinct energy savings for mice subjected to ambient and nest temperatures well below thermoneutrality.

Many species of small mammals (principally rodents of the the subfamilies Arvicolinae and Cricetinae) form aggregations during winter to conserve energy (Merritt 1984). Rodents construct elaborate nests of grasses and herbs either under the litter, on the ground, or within a hollow tree or log. During the winter in northern regions, these nests are commonly located within the subnivean environment, which aids in insulating the nest site from fluctuating supranivean (above snow) temperatures. Radiotelemetry studies demonstrated that up to six muskrats may inhabit the same winter lodge in the marshes of Manitoba, Canada (MacArthur and Aleksiuk, 1979). The resting metabolic rate of a group of four muskrats huddling in such a lodge during winter with an environmental temperature of −10°C shows up to a 13% energy savings over that of a single animal. In northern latitudes, aggregations of at least six muskrats are common. The physical structure and communal use of beaver lodges in southeastern Manitoba were assessed by Dyck and MacArthur (1993). Winter air temperatures outside the lodges reached a low of −41.4°C, but temperatures within the chambers of occupied lodges did not fall below 0°C. The mean monthly temperature of the nesting chamber consistently exceeded the mean monthly exterior air and water temperatures. The ameliorated microclimate within the lodges also facilitates periodic rewarming of foraging

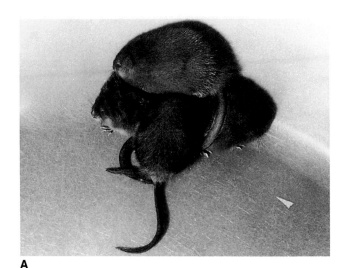

A

B

Figure 8.13   Social thermoregulation.   Examples of (A) close and (B) loose aggregation responses of muskrats exposed to a temperature of 5°C.

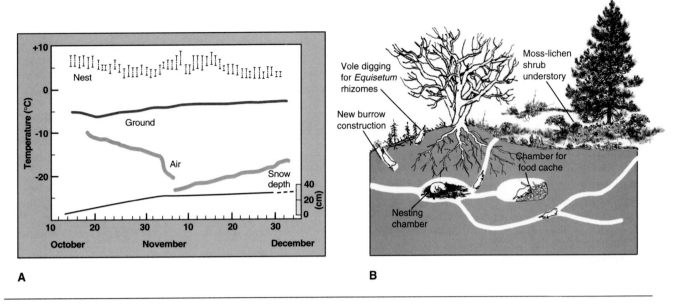

A   **B**

Figure. 8.14   **Nest temperatures of taiga voles.**   (A) Daily mean air temperature, ground temperature, and temperature in the nest of the taiga vole from 15 October to 1 December 1977, at a site 155 km northwest of Fairbanks, Alaska. The daily ranges of nest temperature are shown by the vertical bars. Increasing ground temperatures after 20 October were due to progressively deeper snow cover. (B) The communal winter midden-tunnel system and activities of a group of taiga voles.

Source: Data from J.O. Wolf and W.Z. Lidicker, Jr., "Communal Winter Nesting and Food Sharing in Taiga Voles" in Behav. Ecol. Sociobiol., 9:237–240, 1981.

beavers, thus minimizing thermoregulatory costs during rest. Nests of taiga voles (*Microtus xanthognathus*) occupied by 5 to 10 individuals remain between 7° and 12°C warmer than ground temperature within the subnivean environment and as much as 25°C warmer than supranivean temperatures (figure 8.14). Furthermore, nests were not completly vacated, so that foraging voles returned to a warm nest (Wolff 1980; Wolff and Lidicker 1981). Huddling by deer mice and voles may, for example, reduce energy requirements in the cold by as much as 16% to 36% (e.g., Gebczynska and Gebczynski 1971; Vogt and Lynch 1982; Andrews and Belknap 1986). For *P. leucopus* a combination of torpor and huddling of three individuals within a nest at an ambient temperature of 13°C resulted in a daily energy savings of 74%, compared with nontorpid, individual mice without a nest (Vogt and Lynch 1982). Aggregate nesting during winter is common for species of *Peromyscus* (Madison et al. 1984b; Wolff and Durr 1986). The average daily metabolic rate of *P. maniculatus* is also lower in winter (Merritt 1984). When *P. maniculatus* and *P. leucopus* are syntopic (live within the same locality) in the Appalachian Mountains of Virginia, radiotelemetry studies have shown that *P. maniculatus* prefers nesting high in large, hollow trees, year-round, whereas *P. leucopus* uses both ground and tree nest sites in summer but shifts to underground nest sites in winter (Wolff and Durr 1986). Radiotelemetry studies have demonstrated that both species nest together during winter months (Wolff 1989).

Most tree squirrels are solitary and euthermic during winter. Southern flying squirrels (*Glaucomys volans*), however, form "huddles" of up to 20 individuals (but groups fewer than 10 are more common) in hollow trees to conserve energy during winter. *Glaucomys volans* and *G. sabrinus* undergo periodic bouts of torpor during winter due to extended periods of food shortage and low temperature (Muul 1968). A group of six southern flying squirrels in New Hampshire, huddling within a wooden nest box and surrounded by temperatures of 6°C, reduced their energy expenditure by 36% (Stapp et al. 1991). This behavioral strategy, augmented by a nearby supply of hoarded nuts, enhanced survival during winter.

Construction of Elaborate Nests

Conservation of heat by a group of huddling animals is greatest when the nest is well insulated. Researchers have evaluated the thermal capacity of nests by calculating their shape (Wolfe 1970; Wolfe and Barnett 1977) and composition (King et al. 1964; Layne 1969). The resistance of nests to heat loss can also be measured quantitatively (Wrabetz 1980). Interspecific differences in nest-building behavior were correlated with available microhabitat and nest site preferences (Layne 1969). King and colleagues (1964) demonstrated a geographic correlation in the amount of nest material used by *Peromyscus*. Northern forms used more nesting material under constant temperatures than southern forms. Sealander (1952) demonstrated a seasonal difference in nest-building behavior: *P. leucopus* and *P. maniculatus* constructed more elaborate nests during winter than in summer, with winter nests conferring greater resistance to low temperatures. Pierce and Vogt (1993) employed outdoor enclosures and demonstrated that *P. leucopus* and *P. maniculatus* from northern New York constructed larger nests during winter compared with other times of the year. *P. maniculatus* constructed the largest nests.

Foraging Zones

Small mammals such as shrews and voles are active during midwinter and do not undergo physiological heterothermy (Wunder 1985; Merritt 1995). The thermal regime of the foraging zone, therefore, is crucial in dictating energy budgets of these winter-active mammals. The climatological regime of the foraging and nesting sites of shrews, voles, and mice has been examined during winter in Michigan, Ontario, and Pennsylvania (Pruitt 1957; Randolph 1973; Merritt 1986, respectively). In mixed deciduous forests, many small mammals forage in tunnels within soil covered by a rich layer of leaves. During winter, this foraging zone provides a stable, comparatively warm thermal regime. Snow covering the ground also provides additional insulation. Although ambient temperatures may reach –29°C in mid-

January, the minimum temperature at the soil-leaf litter interface is about –4°C and 1°C within a subsurface tunnel (figure. 8.15). Snow cover is an integral part of the life of small mammals (Merritt 1984; Marchand 1996). The presence of a sufficient depth of snow, called the **heimal threshold** (Pruitt 1957), insulates the subnivean environment against widely fluctuating environmental temperatures. High mortality rates among *Clethrionomys gapperi* and *Peronmyscus leucopus* during midwinter are attributable to a lack of snow cover to insulate the forest floor (Pruitt 1957; Beer 1961; Fuller et al. 1969). The period of autumn freeze is also crucial to the survival of small mammals due to great fluctuations in temperatures in their foraging zone. This period was found to produce increased mortality rates in small mammals studied in the Rocky Mountains of Colorado (Merritt and Merritt 1978). Snow thickness in autumn was insufficient to insulate soil against fluctuating ambient temperatures, which may reach –20°C. Within the stable foraging zone, many species of small mammals establish caches of food to ensure that a predictable, quick energy source is

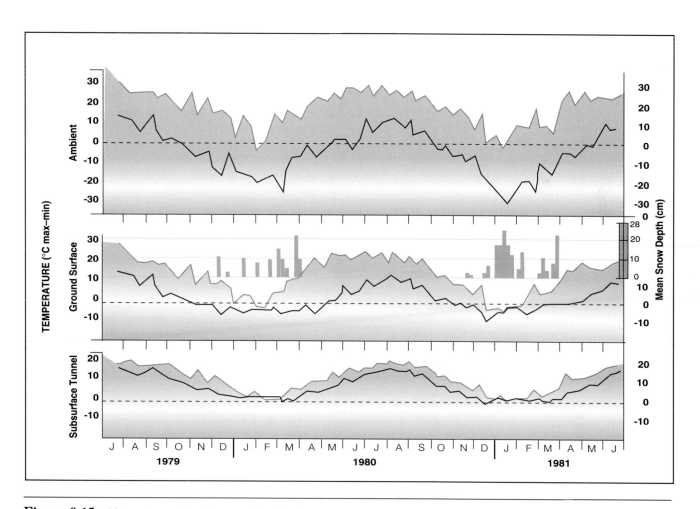

Figure. 8.15   **Thermal zone of small mammals.**   Maximum and minimum temperatures recorded on an Appalachian Mountain site from July 1979 to June 1981. Temperatures are recorded from 1.5 m above ground surface (ambient), at ground surface, and in a subsurface tunnel. Snow depth is shown by shaded bars.

Source: Data from J.F. Merritt (Ed.), "Winter Ecology of Small Mammals" in Special Publication of Carnegie Museum of Natural History, #10, 1984.

readily available during winter, when food is scarce. Food hoarding is reported for 6 orders and 30 families of mammals, with most practitioners found in the Orders Rodentia and Carnivora. Because hoarding behavior is so important in the biology of Class Mammalia, a detailed discussion of this topic is provided in chapter 6.

Reduction in Level of Activity

Daily and seasonal temperature changes influence activity patterns of winter-active small mammals. Nonhibernating species residing in seasonal environments must secure adequate nourishment during winter to maintain a constant, high body temperature. Among winter-active mammals, shrews are characterized by rapid heat loss due to their large surface-area-to-volume ratio, high metabolic rate, and resulting high caloric requirements. We would therefore intuit them to be poor candidates for enduring cold stress. The northern short-tailed shrew (*Blarina brevicauda*) of eastern North America conserves energy by using cached food reserves and avoids cold environmental temperatures by restricting foraging and nesting to zones characterized by stable microclimates. Because temperature changes influence the activity of invertebrates within the soil, foraging by soricid predators may also be influenced. Churchfield (1982) used time-lapse photography to analyze the influence of temperature on the activity and food consumption of *Sorex araneus*. Shrews were active throughout the day and night with peaks in activity every 1 to 2 hours. Activity outside the nest in summer was 28% but dropped to 19% during winter, concomitant with a decrease in food consumption. Martin (1983) showed an annual activity range for *B. brevicauda* of 7% to 31% of the day. Daily activity during winter was greatly reduced, averaging only 11.6% during the coldest months. *Blarina* spends 80% to 90% of the day resting in its nest at a low metabolic rate, sleeping for long periods, and being intermittently highly active. The reduction in caloric intake and foraging activity during winter represents a survival tactic for coping with cold. In contrast, seasonal metabolism of *B. brevicauda* increases from a low in summer to a maximum in autumn and winter. Northern short-tailed shrews may depart from the typical metabolic profile of shrews due to their proclivity for hoarding food (Merritt 1986).

Unlike the case for *Blarina*, food is severely limited for many rodents during winter months. For species of *Peromyscus* inhabiting northern latitudes, torpidity combined with communal nesting is important to conserving energy. For example, in western Kansas, *P. leucopus* spent 72% of the day in the nest during winter, compared with 28% of the day in summer (Baar and Fleharty 1976). During winter, when the cost of thermoregulation is highest and food resources are lowest, available energy must be used to maintain metabolism. Energy loss can be minimized in part by curtailing locomotor activity. For voles, cold temperatures during winter, especially at night, may select for more diurnal activity. During winter, voles demonstrate increased movements on warm days or forage below a mantle of snow. When there is no insulating blanket of snow, low temperatures stimulate the use of nests and inhibit activity in voles (Madison 1985).

Reduction of Body Mass

Small mammals inhabiting seasonal environments undergo a general decline in body mass during winter called **Dehnel's phenomenon** (Dehnel 1949). This loss of mass or slowed growth reduces caloric needs during this period when food is scarce. Seasonal declines in mass are reported for many microtine species and shrews (Churchfield 1990; Merritt 1995). Some northern species, namely collared lemmings (Nagy 1993) and northern short-tailed shrews, depart from this trend, however, and gain mass during winter. *Peromyscus* does not conserve energy in this way during winter: Their ability to undergo torpor coupled with communal nesting, food hoarding, and use of elaborate nests aids considerably in conserving energy during winter. The cues for such fluctuations in mass are complex and include photoperiod, environmental temperature, and availability of food. The common shrew in Eurasia also demonstrates body mass decreases during winter. Declines in body mass of shrews can be significant—*Sorex araneus* residing in Great Britain, Poland, and Finland lose up to 45% of body mass (Churchfield 1982; Pucek 1965; Hyvärinen 1969), and *Sorex cinereus* in North America showed a decline of 53% in body mass from early summer to winter (Merritt 1995). Churchfield (1990) indicated that a latitudinal gradient may exist for changes in body mass of shrews. Small size during winter confers an energy advantage by reducing food requirements. A small mammal eats less food and has a greater assimilation efficiency than a large one; consequently, it can reduce foraging time during cold and thus conserve energy. Although small body mass in winter decreases food requirements, it will reduce cold tolerance due to an increase in surface-area-to-volume ratio.

The survival mechanisms discussed have one thing in common, namely, they pertain to animals that are active during most of the winter and thus must forage to find food to maintain a high body temperature. Even small mammals such as *Peromyscus* that exhibit short-term torpidity are quite thermolabile—because their body temperature decline is slight, they can quickly elevate their body temperature, forage on cached food, and then rewarm in a communal nest. As depicted in our model of the energy budget (see figure 8.4), thermoregulation is the highest priority: If a mammal cannot maintain euthermy, it cannot conduct other activities. But some small and medium-sized mammals residing in northern environments are quite lethargic during winter. They solve the problem of scarcity of food and low temperature by entering a prolonged and controlled state of dormancy called torpor, or hibernation.

Dormancy

Small mammals, such as bats and rodents, maintain high body temperature when active but are able to save energy by temporarily abandoning euthermia. **Dormancy** is defined as a period of inactivity characterized by a reduced metabolic rate and lowering of body temperature. **Torpor** is a form of dormancy characterized by a lowering of body temperature, metabolic rate, respiration, and heart rate. During the winter, torpor is referred to as hibernation, and during the summer, it is called estivation. Torpor that occurs daily is logically called daily torpor. These energy-conserving responses are sometimes grouped as forms of **adaptive hypothermia** (Bartholomew 1982). Unfortunately, "hibernation biologists" have difficulty agreeing on consistent terminology. The difference between these patterns should be treated as points along a continuum because one condition may shade imperceptibly into another. To understand these concepts, we must first establish some definitions. Our terms attempt to follow Bartholomew (1982), Lyman and colleagues (1982), French (1992), Geiser and Ruf (1995), and Schmidt-Nielsen (1997).

The energy savings for mammals coping with cold depend on their drop in body temperature and length of time spent in a state of dormancy. Torpor is a type of dormancy in which body temperature, heart rate, and respiration are not lowered as drastically as in hibernation. Body temperature declines markedly but not usually below 15°C. The lowest range of tolerable body temperatures during torpor is about 10° to 22°C (Wang and Wolowyk 1988). Patterns of torpor may extend for a period of hours or several days. Daily torpor is a response of small mammals to an immediate energy emergency. Examples of animals that undergo daily torpor are species of rodents (namely, *Peromyscus*), many marsupials, insectivores, bats, and some primates. Torpor is a quite plastic condition and can provide significant energy savings for animals coping with cold stress (Bartholomew 1982). The ability of many murid rodents to undergo periodic bouts of torpor during winter is well known (Lyman et al. 1982; French 1992). This strategy aids in combating cold stress and scarcity of food during winter and is commonly accompanied by other energy-saving strategies. It is noteworthy that the ability to abandon homeothermy for torpor has never been shown in voles and most shrews, which must continue to forage under coldest conditions (Wunder 1985; Churchfield 1990). **Hibernation** is defined as a profound dormancy in which the animal remains at a body temperature ranging from 2° to 5° C for periods of weeks during winter. Recently, however, Barnes (1989) measured core body temperatures as low as −2.9°C in hibernating arctic ground squirrels (*Spermophilus parryii*)! The term *hibernation* is also referred to as seasonal torpor, deep hibernation, or true hibernation. Animals that undergo hibernation include ground squirrels, marmots, and hedgehogs. No mammal remains continuously in a dormant state during the entire period of hibernation, however, and the length of the period of hibernation varies with ambient temperature, body size, and species. We defer treatment of estivation (a form of shallow torpor typified by desert small mammals) for the later discussion of how mammals cope with heat.

Dormancy is exhibited by five orders of mammals and at least six groups of birds (swifts, goatsuckers, humming-birds, sunbirds, manakins, and colies). Representative mammals and their sizes are given by Hudson (1978), Lyman and colleagues (1982), and more recently by Geiser and Ruf (1995). All mammals exhibiting adaptive hypothermia have a hypothalamic set point below which they do not allow body temperature to fall. For example, in very cold environments, the lowest temperature for hibernation may be regulated at just a couple degrees above freezing! For species in slightly warmer environments, body temperatures range slightly higher (5° to 10°C), and for species that employ torpor only for short-term emergencies, body temperatures range from 10° to 15°C. When challenged in such a way, these animals must arouse from torpor and restore their body temperature all the way to normal before reentering the state of hibernation. In terms of energy, this can be very costly.

The largest mammals to undergo hibernation are marmots (Genus *Marmota*), which weight about 5 kg. Hibernation is not possible or even necessary in very large mammals because of the large amount of energy necessary for arousal (Morrision 1960). Also, due to their size, large mammals can store sufficient energy in the form of internal fat to meet winter demands. Contrary to popular belief, bears do not hibernate. Instead, they undergo a period of **winter lethargy** and decrease their body temperature only about 5° to 6°, to around 33°C. For example, as indicated by Hainsworth (1981:249), a grizzly bear (*Ursus arctos*) that weighs 386 kg requires 347,400 cal (calories) to raise its temperature 1°C (specific heat of animal tissue is about 0.9 cal/g • °C). If the grizzly bear were truly hibernating, we would assume that it must raise its body temperature from 5° to 37 °C for arousal. The heat required to perform this arousal would be 11,116,800 cal! Just to maintain a basal metabolic rate, this bear would require about 6,100,000 cal/day. Also, many females are pregnant during the winter. During arousal, heat production increases above basal levels, and the amount of heat required in such a short period (daily) may not be feasible. Because large mammals can store more energy internally relative to their rate of energy use, a decrease in body temperature is less critical to them.

Mammals capable of adaptive hypothermia are found in the mammalian groups Prototheria, Metatheria, and Eutheria. Hibernation has not been reported for members of the Orders Cetacea, Edentata, Carnivora, Tubulidentata, Lagomorpha, Perissodactyla, and Artiodactyla. The short-nosed echidna (*Tachyglossus aculeatus*) and 15 species of marsupials undergo adaptive hypothermia (Geiser and Ruf 1995). Marsupials capable of torpor range in size from Giles'planigale (*Planigale gilesi*, 8 g) to the western quoll (*Dasyurus geoffroii*, 1 kg). Of the insectivores, the hedgehogs (Genera *Erinaceus* and *Aethechinus*) and some tenrecs of Madagascar hibernate, as do the golden moles (Family Chrysochloridae). In Europe, hedgehogs have been a principal subject of hibernation studies. Shrews do not hibernate, but some may undergo shallow torpor. Two metabolic levels characterize the Family Soricidae (Vogel 1976; Nagel 1977; Genoud 1988). The Subfamily Crocidurinae (white-toothed shrews) is characterized by a low metabolic rate and low body temperature and undergoes torpor. In contrast, the Subfamily Soricinae (red-toothed shrews) exhibits high metabolic rates and elevated body temperatures and cannot undergo torpor (Churchfield 1990; Merritt 1995). The different metabolic levels of the two subfamilies represent evolutionary responses to different climates; white-toothed shrews are adapted to warmer, more southern latitudes, whereas the red-toothed shrews have evolved in more northerly latitudes.

Bats avoid cold by hibernating, migrating, or doing both. The picture of temperature regulation in the microchiropterans is more complicated, however (Kunz 1981; Lyman et al. 1982; Thomas 1995; Altringham 1996). Bats of the Suborder Microchiroptera (see chapter 12) are well-known hibernators, and most species occurring in the temperate zones spend the winter in caves for this purpose. Vespertilionid bats in temperate regions typically hibernate from October to April. In a hibernation site (hibernaculum), the ambient temperature may be near 5°C, and the bats are in deep hibernation, maintaining a body temperature of about 1°C above the ambient. This period is punctuated with occasional arousals during which the animal urinates, drinks, or changes location. Arousals occur every 1 to 3 weeks and last for only a few hours each time. In preparation for winter, temperate zone bats may establish body fat in autumn equal to a third of their body mass. In summer, shallow daily torpor (lasting for only a few hours) may occur during the day while bats are roosting. Body temperature rises again before feeding at dusk. Winter hibernation in bats differs from short-term torpor largely in the length of dormancy and the temperature decrease. The duration of hibernation for bats differs widely among species and within a species, depending on the geographic area. In the northeastern United States, for example, the little brown bat (*Myotis lucifugus*) hibernates for 6 to 7 months. Periods of hibernation for bats in warmer areas are considerably shorter. The larger members of Suborder Megachiroptera are euthermic and maintain body temperature between 35° to 40°C (Ransome 1990; McNab and Bonaccorso 1995). Bartholomew and colleagues (1970) however, reported torpor in smaller species of fruit bats (*Nyctimene albiventer* and *Paranyctimene raptor*), and Coburn and Geiser (1996) described torpor in blossom bats (*Syconycteris australis*). When food was withheld, blossom bats remained in torpor from 1 to 10 hours at a body temperature of about 18°C. Torpor in primates is limited to the small lemurs of Madagascar. Mouse lemurs (Genus *Microcebus*), the smallest primates, range in mass from 29 to 63 g and exhibit an average body temperature during torpor of 24.9°C (Ortmann et al. 1996).

The greatest number of hibernators are found in the Order Rodentia, specifically squirrels (Family Sciuridae). Rodents also show the greatest variation in length of dormancy. Within this group, we see a continuous integration between daily and seasonal torpor. Ground squirrels and marmots undergo periods of deep hibernation (Armitage et al. 1990; Ferron 1996). Body temperature in marmots decreases from about 39°C during summer to between 2° and 8°C during deep hibernation. The lowest body temperature reported for hibernating marmots is about 4.0°C (Ferron 1996). Typically, the heartbeat slows from 100 bpm (beats per minute) to 15 bpm, and oxygen consumption falls to a tenth the normal rate. Marmots may only breathe once every 6 minutes when in deep hibernation; hibernating animals lose 30% of their weight by mobilizing body fat during winter. It is noteworthy that, with the exception of the woodchuck (*Marmota monax*), all species of the Genus *Marmota* are social and hibernate in groups (Arnold 1988).

Richardson's ground squirrels (*Spermophilus richardsonii*) of northwestern North America also undergo periods of deep hibernation. Adult squirrels may enter hibernation as early as mid-July and emerge 8 months later, in mid-March. During much of their long period of hibernation, body temperature is about 3° to 4°C and increases to about 38°C for their 4-month active phase. Like many hibernators, torpor in ground squirrels during winter is interrupted by frequent intervals of rewarming to euthermy, and entry and arousal show a stepped progression in body temperature. Hibernation has been well studied in different ground squirrels, especially in golden-mantled (*Spermophilus lateralis*), Richardson's, arctic (*S. parryii*) and thirteen-lined (*S. tridecemlineatus*) ground squirrels.

Murids commonly survive cold by employing daily torpor coupled with communal nesting. Because their body temperatures are not as depressed as those of such deep hibernators as ground squirrels, they show a great deal of thermolability. All members of the Genus *Peromyscus* probably undergo some form of dormancy. Torpor occurs diurnally and lasts for less than 12 hours (Hudson 1978). For example, *Peromyscus* may be active on a warm day in mid-January in the north, but when ambient temperatures reach about 2° to 5°C, these small rodents may rapidly decrease body temperature to 13°C and undergo short-term torpor. The lability of body temperature in *P. leucopus* seems greater than that of other species of the genus (Hart 1971). Among the murids, hibernation (deep torpor) is reported for members of the Subfamily Zapodinae (Genera *Zapus* and *Napaeozapus*; Brower and Cade 1966; Muchlinski 1980). In the Wasatch Mountains of Utah, the period of hibernation of *Z. princeps* ranges from early September to late July depending on elevation of the hibernacula (Cranford 1978). A summer active period of about 87 days spanned the period between snow melt in early summer and the beginning of the autumn snowfall season. Mice hibernated at an average depth of 59 cm, with no food caches. The mean soil temperature during hibernation was 4.6°C, and emergence from hibernation was cued by increasing soil temperature (figure 8.16). Birch mice (*Sicista betulina*) of northern and eastern Eurasia undergo daily torpor in response to cold. They decrease body temperature to about 4°C and wake spontaneously at night to feed. Most hamsters (namely, golden or Syrian and Turkish—*Mesocricetus auratus*, *M. brandti*) do not readily hibernate or

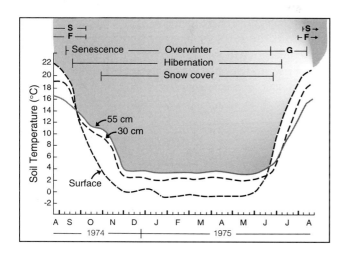

Figure 8.16   **Soil temperatures in a montane environment.** Daily average soil temperatures at the surface, 30 cm, and 55 cm depth at 2900 m in Albion Basin, Alta, Utah, from 15 August 1974 to 15 August 1975. The lines are plotted through the weekly mean. *Abbreviatons:* G = plant growth; F = flowering; S = plants setting seed.

Source: Data from J.A. Cranford, "Hibernation in the Western Jumping Mouse (Zapus princeps)" in Journal of Mammalogy, *59:496–509, 1978.*

exhibit torpor. On the other hand, Djungarian hamsters (*Phodopus sungorus*), inhabitants of Siberia, undergo periodic daily torpor with body temperature near 19°C. Color change and torpor in these hamsters are quite variable; some individuals turn white in winter and exhibit torpor, whereas others remain brown and do not enter torpor. Dormice (Genera *Glis* and *Eliomys*) are also reported to undergo hibernation in response to food deprivation.

The **cycle of dormancy** can be divided into three phases: entrance, period of dormancy, and arousal (Hudson 1973). An excellent example of the cycle of dormancy is that measured for a "classic hibernator," Richardson's ground squirrels (Wang 1978, 1979). Many mammals prepare for **entrance** to winter dormancy by putting on fat. For example, the woodchuck (*Marmota monax*) of North America begins putting on weight in midsummer. Just prior to hibernation, its weight is about 30% greater than in early summer. In autumn, after lining its hibernaculum with leaves and grasses, the obese woodchuck moves into its den, plugs the entrances, and curls up into a tight ball. The earthen plug is important for maintaining proper temperature and humidity in the den during winter. Increase in body mass among hibernators before dormancy can be impressive, reaching 80% in golden-mantled ground squirrels (*S. lateralis*), for example. During entrance, animals decrease heart rate, blood pressure, and oxygen consumption and finally exhibit a decline in body temperature. For some species, the environmental cues that signal preparation for hibernation are associated with the time of year and are induced by a combination of low temperature or lack of food. For others, entrance into dormancy may occur without an external stimulus. For such

species, the duration of the daily light and the temperature cycle may synchronize to maintain the annual rhythm of dormancy. For juvenile Richardson's ground squirrels, entrance into hibernation begins in early September with stepped periods of torpor alternating with periodic rewarming bouts reaching euthermy (figure 8.17A). The burrow temperature during the period of entrance is about 13°C.

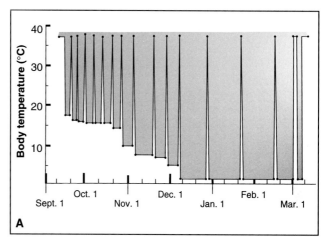

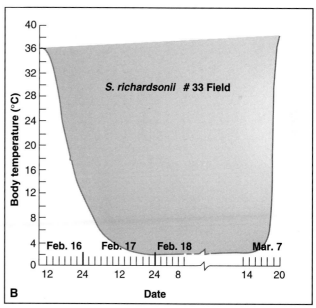

Figure 8.17   **Duration of torpor in a ground squirrel.**   (A) A *complete torpor season* in a juvenile male Richardson's ground squirrel under field conditions. Note the seasonal variations in duration of torpor and the decrease of minimum body temperature during torpor. (B) Typical dynamic changes of body temperature (telemetered) during a *single torpor episode* in Richardson's ground squirrel under field conditions. Note the relatively prolonged period during entry into torpor (24 hours or longer) and the relatively rapid rise of body temperature during arousal from torpor.

Source: Data from L.C.H. Wang, "Energetic and Field Aspects of Mammalian Torpor: the Richardson's Ground Squirrel" in Strategies in Cold: Natural Torpidity and Thermogenesis, *L.C.H. Wang and J.W. Hudson, eds., 1978, Academic Press.*

The **period of dormancy** is signified by body temperature leveling off in December. From December to March, the periods between arousal become maximum, averaging 14 to 19 days, and body temperature stabilizes at about 3°C. In late February, the periods of dormancy become progressively shorter, and in early March, the squirrel exhibits **arousal**. Changes in body temperature of a specific episode of torpor of Richardson's ground squirrel are illustrated in figure 8.17B. Entry into deep torpor began shortly after noon on 16 February and stabilized at 3°C, 24 hours later. The ground squirrel maintained this level of torpor until late afternoon on 7 March, at which time arousal began. Within only 3 hours, the squirrel increased its body temperature from 3° to 37°C. During this arousal phase, the squirrel maintained the elevated body temperature for 14 hours and then fell into torpor again. This stepped progression occurred until total arousal was accomplished. The longest time spent in deep torpor was during the months of December and January, and the shortest was in July and March—the entry and arousal phases of dormancy, respectively (table 8.2). The duration of torpor bouts ranged from 3 to 19 days (occurring in July and January, respectively). Periods of arousal account for most of the energy used during dormancy. Energy costs associated with arousal from 3° to 37°C include (1) the cost of warming from hibernation to 37°C, (2) the cost of sustaining euthermy (37°C) for several hours, and (3) the cost of maintaining metabolic rate above torpor as the body temperature slowly declines during reentry into torpor. For the entire period of dormancy, these three metabolic phases account for an average of about 83% of the total energy used by the squirrel. Energy for arousal from hibernation is provided principally by nonshivering thermogenesis in brown adipose tissue augmented by shivering. This energy expense seems rather wasteful, and the function of such periodical arousals is not

understood. Because Richardson's ground squirrels do not store food, they are not employing periods of arousal to consume underground food caches. It is possible that periodic arousal permits hibernators to assess environmental conditions conducive to emergence.

Resistance to Cold

Increase in Thermogenic Capacity

Mammals employ many different tactics of energy conservation to avoid cold stress. If cold stress persists, however, mammals must resist it by processes that generate heat and thus require energy (see table 8.1). The major ways in which endotherms increase heat production are muscular activity and exercise, involuntary muscle contractions (shivering), and nonshivering thermogenesis (see figure 8.4). The most conspicuous mechanism by which endotherms increase heat production is by muscular activity from locomotion or shivering; however, the means need not be apparent for it to increase the rate of heat production (Bartholomew 1982; Kleinebeckel and Klussman 1990; Schmidt-Nielsen 1997). Shivering is well documented for humans, canids, felids, selected rodents, lagomorphs, and ungulates, as well as many species of birds. Small mammals residing in variable environments, however, employ a means of heat production called **nonshivering thermogenesis** that does not involve muscle contraction (Smith and Horwitz 1969; Jansky 1973; Girardier and Stock 1983; Trayhurn and Nicholls 1986). **Brown adipose tissue,** the site of nonshivering thermogenesis, was first observed by Conrad Gesner in 1551 in the interscapular area of the Old World marmot (*Marmota alpina*). Brown fat, once referred to as the "hibernating gland," is found in all hibernating mammals and is the pri-

Table 8.2.   Seasonal variations in estimated time budget for torpor in Richardson's ground squirrel (*Spermophilus richardsonii*) under natural conditions

Month	*x* Days per Torpor	*x* Hours per Intertorpor Homeothermy	Total Hours per Torpor Cycle	Number of Possible Torpor Cycles	Percentage of Month Spent in Torpor
July	3	11.3*	83.3	4.46†	43.1
August	4.3	11.3*	114.5	6.49	90.1
September	6.0	11.3	155.3	4.64	92.7
October	8.3	10.1	209.3	3.55	95.2
November	11.0	8.6	272.6	2.64	96.8
December	15.0	7.5	367.5	2.02	98.0
January	19.1	7.5††	465.9	1.60	98.4
February	14.2	14.0	354.8	1.90	96.1
March	6.0	25.0	169.0	3.12§	60.4

From L.C.H. Wang in Canadian Journal of Zoology, *57:1520, 1979. Copyright ©1979 National Research Council of Canada, Research Journals. Reprinted by permission of NRC Research Press.*

*Assumed to be same as in September.
†Since torpor was not observed until July 17, only half of July was counted as available for torpor.
††Assumed to be same as in December.
§Allow 22 days in March before termination of torpor season in this species.

mary thermogenic tissue of cold-adapted small mammals (especially rodents and shrews). It is also well developed in newborn species of mammals, including humans. Within mammals, brown fat has been reported in many species spanning seven orders: Chiroptera, Insectivora, Rodentia, Lagomorpha, Artiodactyla, Carnivora, and Primates. Brown adipose tissue is unique to eutherian (placental) mammals and probably evolved early in the radiation of this group (Hayward and Lisson 1992). The wide occurrence of nonshivering thermogenesis in small mammals was reviewed by Heldmaier (1971) and Jansky (1973). Marsupials and monotremes do not possess brown adipose tissue (Hayward and Lisson 1992). Oliphant (1983) reported the presence of brown adipose tissue in ruffed grouse and chickadees; however, microanatomical studies based on multilocularity, increased vascularity, mitochondrial density, and cytochrome-c oxidase activity indicate that birds do not possess functional brown fat (Olson et al. 1988; Saarela et al. 1989, 1991; Brigham and Trayhurn 1994).

Unlike **white adipose tissue,** characterized by a single large droplet of fat with a peripheral nucleus (Pond 1978), brown fat contains many small droplets (multilocular) with a centrally located nucleus. In contrast to white adipose tissue, brown fat is highly vascular and well innervated. The cells contain many mitochondria, whereas the white fat cells have comparatively few. Brown fat is capable of a far higher rate of oxygen consumption and heat production than white fat. The reddish brown color of brown fat is derived from iron-containing cytochrome pigments in the mitochondria, the essential part of the oxidizing enzyme apparatus of brown adipose tissue. White fat serves primarily as insulation and a storage site for food and energy. Brown fat, with its rich supply of mitochondria, serves as a miniature internal "blanket" that overlies parts of the systemic vasculature and becomes an active metabolic heater applied directly to the bloodstream (Wunder and Gettinger 1996). Deposits of brown fat can be rather diffuse but are principally found in the interscapular, cervical, axillary, and inguinal regions in close proximity to blood vessels and vital organs (Hyvärinen 1994).

Temperature receptors in the skin sense cold and send impulses to the preoptic area of the hypothalamus—the "mammalian thermostat" located in the brain. Impulses are then relayed along the sympathetic nerves to the brown adipose tissue, where nerve endings release the neurohormone **norepinephrine.** At the brown adipose tissue, norepinephrine activates an enzyme (lipase) that splits triglyceride molecules into glycerol and free fatty acids. In the brown fat cell, the mitochondrial respiration is "uncoupled" from the mechanism of adenosine triphosphate (ATP) synthesis so that the energy of oxidation of the fatty acids is dissipated as heat instead of being used for ATP synthesis. A special protein, called **thermogenin,** is responsible for the uncoupling (Himms-Hagen 1985). When bats arouse from hibernation, the brown fat pad is much warmer than the rest of the body (Hayward and Lyman 1967; figure 8.18). The close proximity of Selzer's vein just beneath the interscapular brown fat

permits rapid passage of warmed venous blood directly to the heart and brain with a minimum of heat loss.

Research on northern small mammals has shown that dramatic increases in metabolic rate in cold are due to nonshivering thermogenesis. Nonshivering heat production usually tracks environmental temperatures, falling to the lowest rates in spring and summer, increasing in autumn, and peaking in winter. Typically, small mammals demonstrate a significant inverse relationship between nonshivering heat production and environmental temperature. Wunder and colleagues (1977) showed a 29% increase in oxygen consumption for prairie voles (*Microtus ochrogaster*) in winter compared with summer. Alaskan red-backed voles (*Clethrionomys rutilus*) exhibited a 96% increase in metabolism in winter. Although species of *Peromyscus* commonly employ many different survival adjustments (Merritt 1984), they are adept at increasing metabolism during winter. Maximum metabolism for *Peromyscus* from Iowa and Michigan

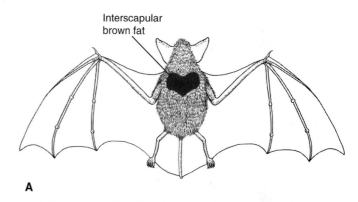

A

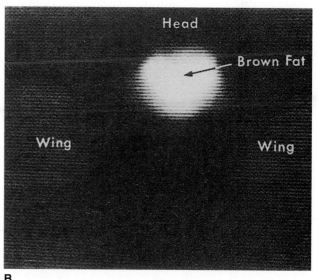

B

Figure 8.18   Brown fat.   (A) Diagram showing the position of interscapular brown fat in the big brown bat (*Eptesicus fuscus*). (B) Photograph taken with heat-sensitive film indicating that the temperature of interscapular brown fat in the big brown bat is high during arousal from hibernation.

ranged from a 24% to 70% increase over summer rates (Lynch 1973; Wickler 1980, respectively). Shrews typically show very high rates of nonshivering heat production. For example, for masked shrews (*Sorex cinereus*) residing in Pennsylvania, nonshivering thermogenesis in winter was almost twice that measured in summer—an increased capacity of 182% (Merritt 1995). Maximum metabolism, resting metabolic rate, and nonshivering thermogenesis were compared for 12 species of shrews from tropical and temperate regions (Sparti 1992). Maximum metabolism, and consequently improved cold tolerance, was pronounced in temperate species. Smith and Horwitz (1969) found a direct correlation between mass of brown fat and nonshivering thermogenesis. Like nonshivering thermogenesis, monthly changes in the mass of brown fat are inversely related to minimum environmental temperature for many species of voles, mice, and shrews.

ADAPTATIONS TO HEAT

To survive in deserts, mammals must cope with a variety of demanding environmental challenges, such as intense heat during the day, cold nights, paucity of water and cover, and a highly variable food supply. Desert ecosystems are widespread and abundant—35% of the earth is covered with deserts. Mammals have successfully colonized desert ecosystems, as evidenced, for example, by the rich and diverse fauna of heteromyid and dipodid rodents (see chapter 17) in the deserts of North America and the Old World, respectively (Schmidly et al. 1993; Wilson and Reeder 1993).

The challenges faced by mammals in desert environments are even more severe than those encountered in cold regions; water, as well as food, is scarce, and the problem in temperature regulation is reversed. When we considered cold stress to mammals, we assessed the role of thermal radiation, conduction, convection, and other environmental influences on heat transfer between an individual and the environment, and then we focused on the many mechanisms employed to reduce heat flow to the environment (see figure 8.1). Recall that small mammals in cold regions reduced heat loss by up to 19% by increasing pelage insulation. Increased insulation reduced the gradient between the warm core of the body and the outside environment. In desert environments, however, the gradient between the internal and environmental temperatures is reversed. In some deserts, mammals may have to cope with air temperatures that reach 55°C and ground temperatures that exceed 70°C. Unlike arctic mammals concerned with conserving heat, desert mammals must dissipate heat or avoid it to maintain euthermy. Our discussion now focuses on the complex anatomical, physiological, and behavioral adaptations that enhance the survival of mammals in desert ecosystems.

Water is essential for survival. It constitutes 70% of the body mass of mammals, and water loss must be balanced by water gain. In mammals, **osmoregulation**—the maintenance of proper internal salt and water concentrations—is performed principally by the kidney. In addition to producing a concentrated urine, desert mammals cope with a lack of water by producing very dry feces. Evaporation, occurring mostly from the respiratory tract, is the major avenue of water loss but is also an important device for cooling. In this section, we examine evaporation across the skin and from respiratory passages, and the ways in which mammals keep cool in xeric (very dry) environments by employing evaporative cooling. Temperature regulation is also influenced by changes in insulation, appendages, metabolic rate, and body size. Water balance is achieved by eating succulent vegetation, drinking available water, and "metabolically" converting food into water. Desert mammals also conserve water by behavioral means such as selecting saturated burrows in which they reside during the heat of the day. Lastly, we will discuss a form of dormancy called estivation, which mammals residing in hot environments employ to survive heat and reduced food availability.

Water Economy

The Mammalian Kidney

Most of the elimination of excess water and soluble salts, urea, uric acid, creatinine, and sulfates occurs in the kidney. Mammalian **kidneys** are paired, bean-shaped structures located within the dorsal part of the abdominal cavity (figure 8.19). The kidney in cross section displays the following areas and structures. The outer **cortex** contains the renal corpuscles, convoluted tubules, and blood vessels. Masses of cortical tissue fill in between the pyramids of medullary tissue. The inner **medulla** is divided into triangular wedges called **renal pyramids.** Their broad bases are directed toward the cortex, and narrow apices **(renal papillae)** are oriented toward the center of the kidney, opening into the **calyx** and expanded **pelvis.** Ducts leading into the pelvis, the **ureters,** empty into the **urinary bladder,** which functions as a storage organ for urine. Another duct, the **urethra,** drains the bladder and carries its contents to the outside.

Nephrons (about 1.5 million in each kidney) are the functional units of the kidney and consist of a closed bulb, **Bowman's capsule** (or glomerular capsule), connected to a long coiled tube. Tubules of the various nephrons empty into collecting ducts that discharge into the pelvis of the kidney and then connect to the ureter. A microscopic mass of capillaries called the **glomerulus** is enclosed within the capsule. The capsule plus the inner glomerulus is called a **renal corpuscle.** Blood reaches the kidney via the large renal artery, a branch of the descending aorta. The blood arrives at the glomeruli by afferent arterioles. The blood is collected from the glomeruli by a number of venules and leaves the kidney by way of renal veins making their way to the inferior vena cava. Blood vessels enter a convoluted network within the glomerular capsule, then move through proximal and distal convoluted tubules and the **loop of Henle,** finally emptying into a branch of the renal vein. Exchange of substances takes

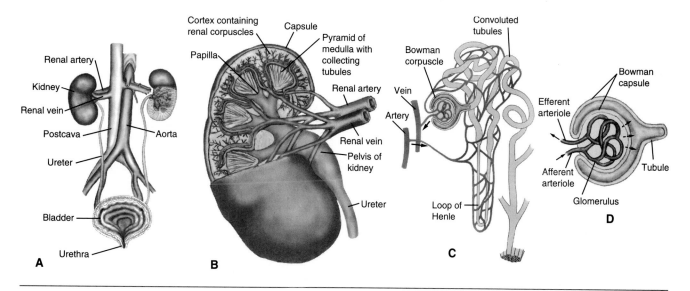

Figure 8.19   The human excretory system. (A) Ventral view of the entire system; (B) Median section of one kidney; (C) Relationship of Bowman's corpuscles, tubules, and blood vessels; (D) A single Bowman's corpuscle and adjacent tubule (shown also in cross section)—solid arrows show flow of blood, broken arrows show the excretory path. (C) and (D) are diagrammatic and much enlarged.

place through active transport and osmosis almost exclusively between blood capillaries and nephrons.

The mammalian kidney performs many different roles including glomerular filtration, tubular reabsorption, and tubular secretion. These functions are well described by Hill and Wyse (1989) and Schmidt-Nielsen (1997). For mammals residing in desert environments, the ability to concentrate urine is paramount and closely tied to the function of the kidney. Because of the urine-concentrating ability of kidneys, mammals are able to produce urine that is hyperosmotic to that of blood plasma—up to 25 times the concentration of plasma. Understandably, the highest urine concentrations are found in mammals residing in desert habitats (Yousef et al. 1972; French 1993). The concentrating ability of the mammalian kidney in different species is closely related to the respective lengths of their loops of Henle and collecting ducts that transverse the renal medulla. The prominence of the medulla is commonly expressed as the relative medullary thickness (RMT), an important index of kidney adaptation (Sperber 1944). RMTs of mammals residing in arid areas are greater than those of mammals from more mesic environments (those with more moisture). The relationship between RMT and the kidney's maximum urine-concentrating capacity was first quantified by Schmidt-Nielsen and O'Dell (1961) and has proven very useful as a means of comparing kidney function in mammals. The anatomy of the papilla of the medulla can be compared visually for different species (figure 8.20). In desert-adapted small mammals, the papilla may extend beyond the margins of the renal capsule into the ureter. This extension is pronounced in small desert rodents (MacMillen and Lee 1969; Altschuler et al. 1979), shrews (Lindstedt 1980), and

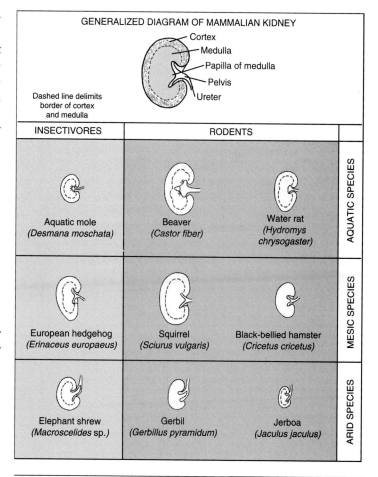

Figure 8.20   The kidneys of selected mammals. Aquatic species show little or no development of the papilla of the medulla. *Desmana* and *Hydromys* lack the papilla. *Castor* has two very shallow papillae. Mesic species have papillae. The papilla is especially well developed in arid species, so much so that it often penetrates well into the ureter (e.g., *Macroscelides, Gerbillus,* and *Jaculus*).

bats (Geluso 1978), suggesting that these species possess very powerful kidneys. In contrast, aquatic mammals, such as beavers, aquatic moles, water rats, and muskrats, have very short loops (shallow papillae) and produce a less concentrated urine.

Urine and Feces

Comparative studies of renal function and morphology in mammals indicates a direct relationship between the ecological distribution of a species and its ability to conserve urinary water. The ability to concentrate urine in mammals is associated with long loops of Henle and tubules in the kidney that enhance the countercurrent exchange function. Species that reside in arid habitats tend to possess kidneys better adapted for water conservation. Representatives are found in the rodents (Families Muridae, Heteromyidae, Sciuridae), xenarthrans (Family Dasypodidae), lagomorphs (Family Leporidae), and many chiropterans. Most mammals lose water by excretion in the urine and elimination in the feces. But the preceding groups have the ability to produce relatively dry feces and concentrated urine. For example, the average values for maximum urine concentration in desert heteromyids is superior to those of most mammals and comparable to those of other desert-adapted small mammals, such as dipodids from Asia and North Africa (see chapter 17) and murids from Australia and southern Africa (French 1993). For example, the laboratory white rat can produce urine with twice the osmotic concentration that humans can achieve. The dromedary, and even dogs and cats, have a urine-concentrating equivalent to that of the white rat. As expected, the amount of water loss from feces is quite low for desert small mammals. The feces of Merriam's kangaroo rat (*Dipodomys merriami*) are over 2.5 times as dry as those of white rats (834 versus 2246 mg of water/g of dry feces). Furthermore, heteromyids commonly decrease fecal water loss by assimilating over 90% of the food they ingest. Desert rodents (e.g., kangaroo rats, sand rats, and jerboas) produce urine concentrations of 3000 to 6000 mOsm/L; Australian hopping mice (*Notomys alexis* and *N. cervinus*), can produce urine concentrations of over 9000 mOsm/L (MacMillen and Lee 1969, 1970). Intuitively, we think of heteromyid rodents such as kangaroo rats and pocket mice as leaders in water conservation. Many other small mammals, however, such as pallid bats (*Antrozous pallidus*), canyon and house mice (*Peromyscus crinitus* and *Mus musculus*), and golden hamsters (*Mesocricetus auratus*), have evaporative water losses equivalent to that of desert heteromyids.

An additional way that desert rodents economize water loss is by producing a highly concentrated milk. The milk of *D. merriami* averages 50.4% water—a concentration comparable to that produced by seals and whales (Kooyman 1963). Furthermore, it has been demonstrated that in desert rodents, canids, and kangaroos, mothers reclaim water by consuming the dilute urine and feces of their young. This behavior may regain about a third of the water originally secreted as milk (Baverstock and Green 1975).

Diet

Since free drinking water is not available for desert mammals, they must obtain water from other sources, such as succulent plants or the body fluids of their prey or by consuming dry food. This requires subsisting on **metabolic water,** which is created in the cells by the oxidation of food, especially carbohydrates (Hill and Wyse 1989). Some desert mammals consume succulent plants for a source of water. Desert woodrats (*Neotoma lepida*) and cactus mice (*Peromyscus eremicus*) of southwestern North America consume large quantities of cactus (Genus *Opuntia*) as a source of both food and water. In addition, cactus is a staple in the diet of other xeric-adapted mammals, such as northern pocket gophers (*Thomomys talpoides*) inhabiting the dry short-grass prairies of Colorado (Vaughan 1967). Gophers and murids such as woodrats have evolved the ability to metabolize oxalic acid, an abundant compound of cactus that is toxic to other mammals. Some desert mammals depend on moist food such as cactus. Others rely primarily on dry seeds or halophytic plants (those that grow in salty soils), and their intake of water is quite minimal. Kangaroo rats (*Dipodomys merriami*), pocket mice of southwestern North America, and fat sand rats (*Psammomys obesus*) of the Saharo-Arabian deserts are able to subsist on dry food only and do not require water.

Halophytic plants of the Family Chenopodiaceae form a staple in the diet of many desert-dwelling small mammals but are, by definition, extremely high in salt concentrations. Many small mammals that consume halophytes possess kidneys that produce highly concentrated urine. Fat sand rats (*P. obesus*) obtain water from the leaves of the saltbush (*Atriplex halimus*). These gerbillid rodents are unusual in being diurnal and wholly herbivorous, whereas other members of the family are nocturnal and granivorous. *Psammomys* scrapes off the outer surface of leaves with their teeth before consuming them—negligible amounts of leaf are scraped from moist plants and substantial amounts from dry plants (Kam and Degen 1992). Leaves possess up to 90% water but have high concentrations of salts and oxalic acid. To consume this plant material, the rat must produce urine with extremely high concentrations of salt as well as be able to metabolize large concentrations of oxalic acid. A parallel development occurs in the chisel-toothed kangaroo rat, (*Dipodomys microps*), an inhabitant of shrub habitats of western North America. This heteromyid harvests leaves of the chenopod (*A. confertifolia*) rather than foraging on seeds, as do most heteromyids. The epidermis of *Atriplex* leaves is high in electrolyte concentration, but the more internal parenchyma is low in electrolytes and high in starch. *Dipodomys microps* is able to consume the inner tissue by shaving off the peripheral epidermis, thus minimizing its consumption of salt. This kangaroo rat can perform such a task while conspecifics cannot because it possesses lower incisors that are broad, flattened anteriorly, and chisel-shaped, permitting access to the plant's inner tissue. Because other sympatric kangaroo rats lack this adaptation, they cannot exploit this unique food resource and must rely on unpredictable seed crops (Kenagy 1972). Many other species exhibit tolerance for high levels of salt in their food or water (e.g., highland desert mice [*Eligmodontia typus*] of South America, fawn hopping-mice [*Notomys cervinus*] and Tammar wallabys [*Macropus eugenii*] of Australia, and western harvest mice [*Reithrodontomys megalotis*] of western North America, to mention just a few).

Desert carnivores and insectivores meet their moisture requirements by relying on their food rather than on free water. Southern grasshopper mice (*Onychomys torridus*) of hot, arid valleys and shrub deserts of southwestern North America consume primarily arthropods, including scorpions, beetles, and grasshoppers. Studies of water balance demonstrate that grasshopper mice can be maintained in the laboratory for more than 3 months on a diet of only fresh mouse carcasses. Grasshopper mice are able to survive the arid conditions of the desert because of their preference for animal foods high in water content (Schmidt-Nielsen 1964). Kit foxes (*Vulpes macrotis*), badgers (*Taxidea taxus*), coyotes (*Canis latrans*), desert hedgehogs (*Hemiechinus auritus*), fennecs (*V. zerda*), and the Australian dasyurid (*Dasycercus cristicauda*) are also able to subsist on a meat diet with minimal supplementation by free drinking water. Fennecs, xeric-adapted canids inhabiting the deserts of northern Africa, maintained water balance for a minimum of 100 days when fed only mice and no drinking water (Noll-Banholzer 1979).

In East Africa, many plains antelope are able to endure long periods of intense heat without drinking water. Two examples of nonmigratory ungulates, the eland (*Taurotragus oryx*) and oryx (*Oryx beisa*), are able to survive indefinitely without drinking water in an ecosystem typified by environmental temperatures reaching 40°C. A critical part of their ability to survive without drinking water is contingent on the fact that they can use metabolic water. Elands consume large quantities of the leaves of *Acacia*, which contain about 58% water. Oryx feed primarily on grasses and shrubs, a staple being the shrub *Diasperma*. Leaves of this shrub fluctuate in water content. During the day, when air temperature is high and humidity low, the leaves contain only 1% water; at night, they increase to 40% water due to decreased temperature and increased relative humidity. The oryx takes advantage of the variable water content of *Diasperma* by consuming it only late at night when water content is highest. During other hours, oryx opportunistically consume more succulent species of plants according to availability.

Temperature Regulation

Evaporation

Metabolic processes, such as kidney function, all require energy. Changes in metabolic processes produce heat, and in desert environments, internal heat must be lost or an individual overheats and dies. **Evaporative cooling,** the major mechanism employed by mammals to reduce body temperature, is very effective as long as an animal has an unlimited supply of water. Evaporative cooling is relatively simple. When mammals cool by evaporation, they take advantage of a physical property of water's ability to absorb a great deal of heat when it changes state from a liquid to a vapor. In desert ecosystems, however, heat is intense and water scarce, so evaporative cooling is of limited utility, except as a short-term response to a temperature crisis. In terms of thermal stress, it is clearly maladaptive for a kangaroo rat (Genus *Dipodomys*) to venture into the desert sun. For such a small mammal to maintain normal body temperature under such circumstances, it would have to evaporate 13% of its body water per hour. This would be highly taxing, as most species die when they lose 10% to 20% of their body water. As we know from discussing how mammals cope with cold, many factors can modify the direct influence of environmental stressors. Although evaporative cooling requires some trade-offs, it represents a major line of defense for mammals combating heat.

We now focus on four major mechanisms of water loss known as **insensible,** or **transpirational water loss.** This water loss occurs by diffusion through the skin and from the surfaces of the respiratory tract. It includes sweating, panting, saliva spreading, and respiratory heat exchange (Hill and Wyse 1989). Keep in mind that small mammals are limited by body size in the extent to which they can store and lose heat. As we will learn later, selection of cool and saturated microclimates represents a crucial strategy for conserving water by small mammals residing in xeric ecosystems.

Sweating

For many mammals, water loss occurs through the skin by way of sweat glands. There are two types of sweat glands: **apocrine** (found on the palms of the hands and bottom of the feet; they do not secrete for thermoregulation) and **eccrine** (distributed throughout the body; they secrete for evaporative heat loss). Water released from eccrine sweat glands evaporates from the surface of the skin, cooling it and the underlying blood. Sweating in response to overheating occurs only in primates and several species of ungulates; it does not occur in rodents and lagomorphs. For humans working in a hot, dry environment, as much as 2000 mL/ hour of water may be produced by eccrine sweat glands and lost by evaporation. Sweating appears to have evolved in mammals whose fur does not represent an appreciable barrier to surface evaporation, but the mechanism is not quite so simple. For animals that sweat, such as camels, the insulative barrier provided by the pelage takes on special significance.

Panting

Humans sweat to increase cooling by evaporation. In contrast, canids possess very few sweat glands and cool primarily by **panting**—a rapid, shallow breathing that increases evaporation of water from the upper respiratory tract. Panting is a common method of evaporative cooling for many carnivores and smaller ungulates, such as sheep, goats, and many small gazelles (Schmidt-Nielsen et al. 1970a). All mammals lose some heat as a result of evaporation of water from their respiratory passages. Inspired air is cooler and less humid than expired air, thus, heat is released from the evaporatory surface in both warming and humidifying the air. Water (and heat) is conserved during expiration when the warmed, moist expired air meets the cooler respiratory surfaces.

Sweating Versus Panting

A major difference between sweating and panting is that the panting animal provides its own air flow over the moist surfaces, thus controlling the degree of evaporative cooling. A sweating animal has minimal control over the degree of evaporation. Another shortcoming of sweating is that sweat contains large amounts of salt. A profusely sweating human may lose enough salt in the sweat to become salt deficient. This is why we are reminded to drink lots of liquid and limit strenuous exercise outside on very warm days. In contrast, panting animals do not lose any electrolytes and do not become sodium-stressed. Panting does, however, have some drawbacks. The muscular energy associated with panting generates more heat than sweating, thus adding to the heat load. Second, the increased ventilation generated by panting can result in severe respiratory alkalosis—an elevation of serum pH attributable to excess removal of carbon dioxide.

Cool Brains

Panting has the major advantage of allowing an animal under sudden heat stress (e.g., a gazelle pursued by a cheetah) to maintain a high body temperature and yet keep its brain at a lower temperature. Taylor (1972) described this fascinating adaptation in some mammals, notably carnivores and artiodactyls. The brain is kept cooler than the body by the now familiar mechanism of countercurrent heat exchange. Arteries carrying warm blood from the heart toward the brain come into intimate contact with venous blood cooled by evaporation of water from the walls of the nasal passages (figure 8.21). Within the **cavernous sinus**—a network of small vessels immersed in cool venous blood located in the floor of the cranial cavity where heat exchange occurs. The venous blood from the nasal passages cools the warmer arterial blood heading toward the brain. As a result, the brain temperature may be 2° or 3°C lower than the blood in the core of the body (Baker 1979). Taylor and Lyman (1972) found that the small Thomson's gazelle of East Africa running for 5 minutes at a speed of 40 km/h exhibited a core body temperature of 44°C, but its brain was maintained at the cooler level of 41°C. Other devices may augment cooling of the brain due to panting. Cabanac (1986) described the cooling of the brain from venous blood returning from facial skin, exchanging with warm arterial blood within the cavernous sinus and influencing the temperature of the brain.

Saliva Spreading

When faced with heat stress, many rodents and marsupials spread saliva on their limbs, tail, chest, or other body parts.

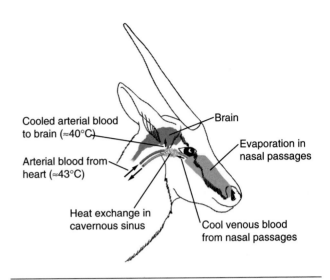

Figure 8.21   Cool brains.   Schematic representation of the anatomical arrangement that promotes cooling of the brain in certain mammals. Dark vessels symbolize arterial blood flow, and light vessels, venous blood flow. In the cavernous sinus, the two flows are in intimate juxtaposition.

Grooming saliva assists in evaporative heat loss. This technique is less effective than sweating for evaporative cooling because the fur must be soaked with saliva before heat can be lost from the underlying surface of the skin. Furthermore, this technique is only effective for a short time. Because of their small size, most rodents have limited supplies of internal water to replenish the high rates of loss. Nonetheless, many rodents rely solely on saliva spreading for evaporative cooling. This mechanism is especially useful when heat stress is relatively short, for example, to prevent excessive hyperthermia while searching for cool refuge sites.

Secretion of saliva, like sweating and panting, is controlled by the hypothalamus. An increase in body temperature from a set point between 37° and 38.5°C is detected at preoptic tissues of the anterior hypothalamus and results in the activation of a salivary control center in the brainstem. Saliva then flows from the submaxillary and parotid glands (the salivary glands); saliva spreading thereby decreases or stabilizes the rising body temperature, which, in turn, signals the hypothalamus by a negative feedback mechanism. Certain nondesert rodents have been shown to increase evaporative heat loss to more than 100% of heat production, and at least half of this total is attributable to evaporative cooling due to saliva spreading.

Respiratory Heat Exchange

Earlier, we discussed the importance of respiratory countercurrent heat exchange as a mechanism of conserving heat loss from respiratory passages in cold environments. Now, we see how water can be saved by evaporation in the lungs for animals residing in desert ecosystems. Many desert mammals, such as kangaroo rats, are able to cool expired air and thus reduce the amount of water lost to the environment. A brief discussion of the cycle of respiration helps elucidate this process of water conservation. As air is inhaled, it passes over moist tissues in the nasal passages where it is warmed and humidified. As the relatively dry air passes over the moist tissues of the nasal passages, these tissues are cooled due to evaporation, and heat is transferred from them to warm the inhaled air. During exhalation, the returning warm, saturated air from the lungs condenses on the cool walls of the nasal passages, thus conserving water. This countercurrent exchange—evaporation on inhalation and condensation on exhalation—conserves both water and energy. Kangaroo rats (Genus *Dipodomys*) are extremely efficient at cooling expired air because of the unique morphology of their nasal passageways (Schmidt-Nielsen et al. 1970b). Compared with another desert inhabitant, the cactus wren, the nasal passages of the kangaroo rat are very narrow with a large wall surface, which enhances heat exchange between the air and the nasal tissues (figure 8.22). In birds such as the cactus wren (*Campylorhynchus brunneicapillus*), the passageway for air flow is wider and shorter, and thus the surface area for contact is smaller. As a result, the nasal passageways of the cactus wren are less efficient at heat exchange than those of the kangaroo rat.

Depending on the ambient temperature and humidity, about 65% to 75% of the water vapor added to inspired air is recovered in the kangaroo rat's nasal passage during exhalation (figure 8.23). Kangaroo rats may inspire air at

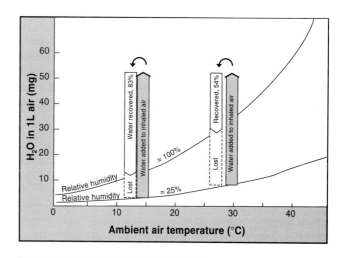

Figure 8.23   **Respiratory water exchange.**   Diagram showing the recovery of water from exhaled air in the kangaroo rat.

Source: Data from K. Schmidt-Nielsen et al., "Counter-Current Heat Exchange in the Respiratory Passages: Effect on Water and Heat Balance" in Respiratory Physiology, *9:263-276, 1970.*

about 15° and 30°C that is 25% saturated with water vapor (these are arbitrarily selected air temperatures and humidity levels). In figure 8.23, the shaded bars indicate the amount of water vapor that must be added to 1 L of outside air as it is inhaled and brought to saturation at the body temperature of 38°C. The unshaded bar next to each shaded bar is the corresponding temperature of exhaled air under the selected conditions. If air is inhaled at 15°C, for example, it will be exhaled at 13°C. In this case, the cooling of the exhaled air causes the recondensation (recovery) of about 83% of the water that was added on inhalation; note that only about 17% of the water needed to saturate the respiratory air is lost by the animal. At 30°C, less water is recovered, but even so, 54% of the evaporated water is saved compared with what would be lost had the air been exhaled at body temperature. By cooling expired air, kangaroo rats reduce the amount of water lost. As shown by the shaded bars in the figure, the amount of water lost and recovered depends on the saturation deficit between inspired (25% humidity) and expired (100% humidity) air plus the air temperature of the inspired air. Expired air is always saturated, and thus its water content is directly related to its temperature.

Evaporation of water from the respiratory passages is a major avenue of water loss for small mammals residing in dry habitats (French 1993). Evaporation is minimized in desert-adapted small mammals by a reduction in both respiratory and cutaneous water losses. As detailed in the section on "Avoidance of High Temperatures," water loss through the skin (cutaneous) is reduced in arid-adapted small mammals by changes in the microhabitat. For example, of the total water lost by *Dipodomys merriami*, 84% is due to evaporation in the respiratory tract and only 16% is lost through the skin (Chew and Dammann 1961).

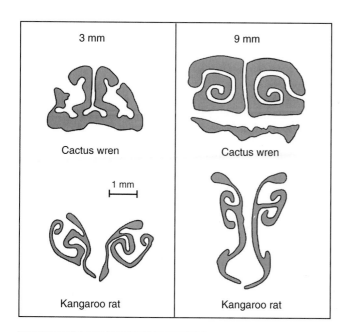

Figure 8.22   **Comparative anatomy of nasal passages.**   Cross sections of the nasal passageways of the cactus wren and kangaroo rat. The passages are wider and the wall area smaller in the bird than in the mammal of the same body size (about 35 g). In both animals, the profiles were obtained at a depth of 3 mm and 9 mm, respectively, from the external openings.

Insulation

Earlier, we discussed how animals residing in cold environments used their pelage to decrease the flow of heat from the body to the environment. In contrast, animals adapted to hot climates, use their fur to minimize the rate of heat they absorb from the environment. Dromedaries must cope with surface temperatures of 70°C and Merino sheep with temperatures as high as 85°C. Because the core body temperature of these ungulates is about 40°C when they are heat-stressed, the temperature gradient between the surface of the pelage and the body core is extremely steep. A shallower gradient in the opposite direction exists between core and skin if sweating occurs, as it does in camels. The maintenance of these two opposing temperature gradients is an essential component of the temperature regulation process of mammals residing in hot environments. Pelage insulation not only retards movement of heat from the environment to the skin but prevents large amounts of heat in the form of incident solar radiation from reaching the skin. The heat returns to the environment from the surface of the hair by convection and radiation (see figure 8.1). The reduction of surface temperature due to forced convection and air movement is significant and decreases the steepness of the thermal gradient across the pelage, thus reducing the heat load on the animal. In addition, the steepness of the gradient from the surface of the pelage to the skin is affected by the quality of the hair. For example, many mammals residing in arid regions have sleek, glossy, light-colored pelages that reflect many of the wavelengths of sunlight, reducing heating due to solar radiation.

Furthermore, the pelage of a mammal is not uniformly distributed. Guanacos (*Lama guanacoe*) reside in regions of South America characterized by intense solar radiation and high air temperatures. They possess gradations in pelage (figure 8.24) ranging from bare skin at the axilla, groin, scrotum, and mammary glands to thick pelage on the dorsum. These bare and sparsely furred areas, seen also in many desert antelopes, serve as **thermal windows** through which some of the heat gained from solar radiation can be lost by convection and conduction. Horns, found in the Family Bovidae (see chapter 19), also serve as thermal windows for species residing in many different climates. Because horns are often richly vascularized, under certain circumstances they exhibit local vasodilation and serve as sites of heat loss, similar to bare patches on the coats of desert mammals.

Data on insulation of small desert rodents indicate a trend toward increased insulation. McNab and Morrison (1963) studied *Peromyscus* from arid and mesic environments and concluded that desert subspecies showed a marked increase in insulation compared with their nondesert relatives. When the body temperature is less than the ambient temperature (e.g., during summer in desert ecosystems), it is advantageous for the mammal to have fur of a low conductance so as to slow the inward conduction of heat. The body temperature of desert rodents decreases at night; during the day, heat is a gradually stored in the fur, with a concomitant rise in body temperature.

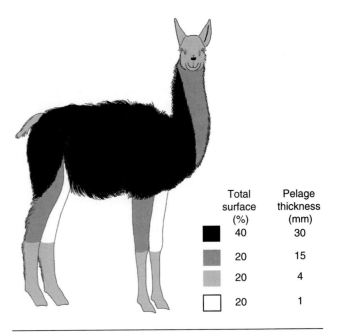

Total surface (%)	Pelage thickness (mm)
40	30
20	15
20	4
20	1

Figure 8.24   Thermal windows.   Schematic representation of the distribution of wool of differing thickness in the guanaco, a South American member of the Family Camelidae. The white areas are thermal windows of almost bare skin.

Source: Data from P.R. Morrison, "Insulative Flexibility in Guanaco" in Journal of Mammalogy, *47:18–23, 1966.*

Appendages

In the discussion of adaptations to cold, the physiological mechanism called peripheral heterothermy via countercurrent heat exchange was examined (see figure 8.9). These vascular arrangements, in which arteries and veins are closely juxtapositioned, are widely reported to occur in the appendages of species that display regional heterothermy. The extent to which such a system can work for animals coping with heat stress depends on several properties of the countercurrent exchanger. Cooling of the outgoing arterial blood in the appendages is promoted by a high degree of contact between arteries and veins and by a relatively slow rate of blood flow through the exchanger. The arms of humans, the flippers of dolphins, and the limbs of tropical mammals, such as sloths, have a unique arrangement of countercurrent exchange. They possess two sets of veins: one superficial and distant to the major arteries, and the other deep and part of the exchange system. By changing the return of blood along these two venous systems, the mammal can emphasize heat dissipation or conservation in its extremity according to its thermal needs. The large ears of jackrabbits (*Lepus californicus*) are excellent heat dissipators. Most excess heat generated during activity may be lost by dilation of the arteries in the ears (Hill et al. 1980). Thermal conductivity of the heat exchange system of jackrabbits is reported to be about 10 times greater for animals at 23°C than at 5°C. Hart (1971) provides an excellent review of studies confirming the importance of appendages as dissipators of heat for desert mammals.

Metabolic Rate

Seed-eating rodents residing in desert ecosystems tend to exhibit low basal metabolic rates. Examples include heteromyids of North America (McNab 1979; Hinds and MacMillen 1985; French 1993), and murids of Australia (MacMillen and Lee 1969, 1970) and Asia (Shkolnik and Borut 1969). French (1993) summarized the factors related to energy consumption in heteromyid rodents. Low metabolic rates may be used to reduce overheating when occupying a closed burrow system or to cut pulmonary water loss in a dry environment. Metabolic rates of heteromyids are about a third less than those of other mammals when at rest. In addition, metabolic reductions are possible by undergoing estivation during times of food scarcity.

Body Size

Earlier, we used the analogy of the mammal as a cube to help illustrate the energy implications of surface-area-to-volume ratios. Now we consider the implications of body size for mammals coping with heat rather than cold. Animals living in hot environments gain heat from two sources: (1) the environment through conduction and convection plus radiation from ground and sun (see figure 8.1) and (2) by metabolic heat gain. A mammal's **heat load** (sum of the environmental and metabolic heat gain) is therefore roughly proportional to its body surface area. Because small mammals have a much larger surface-area-to-volume ratio, they lose heat more readily than large mammals. Using the surface relationship, we can estimate the quantity of water required to dissipate a certain heat load. A small mammal such as a kangaroo rat would need to evaporate a great deal more water (relative to body size) to eliminate its heat load than a large mammal (figure 8.25). To avoid this, as we shall see, small mammals escape the heat (curtail the heat load) by retreating to underground burrows during the day. In addition, desert mammals can decrease their body temperature to conserve water and decrease heat load.

Dormancy

As we learned earlier, many mammals pass through unfavorable climatic periods by becoming inactive. A period of dormancy in reaction to cold is called torpor and hibernation. A similar period in reaction to dry or hot conditions is called **estivation.** Estivation occurs in marsupials and insectivores but is most common among the rodents. It occurs at relatively high body temperatures, and the animals seem somewhat lethargic rather than torpid. If body temperatures are not measured, estivation may go unnoticed. For example, pigmy possums (*Cercartetus nanus*), an Australian marsupial, are able to eat and exhibit normal mobility at a body temperature as low as 28°C. Little pocket mice (*Perognathus longimembris*) exhibit a wide repertoire of behaviors while in estivation, ranging from eating at 23°C to shifting postures at temperatures as low as 6°C (figure 8.26). Typically, rodents capable of dormancy have narrow thermoneutral zones

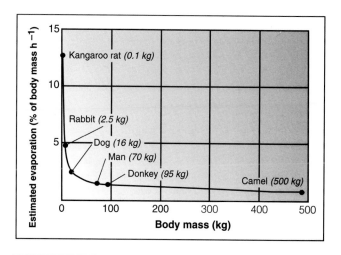

Figure 8.25   **Evaporation in relation to body mass.**
Estimated evaporation compared with body mass of different mammals. For a mammal to maintain a constant body temperature under hot desert conditions, water must be evaporated in proportion to the heat load. Because of the larger relative surface area of a small animal, the heat load, and therefore the estimated evaporation in relation to the body size, increases rapidly in the small animal. The curve is calculated on the assumption that heat load is proportional to body surface.

Source: Data from K. Schmidt-Nielsen, Desert Animals: Physiological Problems of Heat and Water, *1964, Oxford University Press, New York.*

(28°–35°C) and low basal metabolic rates. Estivation has not been as intensely studied as hibernation.

Among heteromyids, estivation is reported for all species of pocket mice and kangaroo mice (Genera *Perognathus, Chaetodipus,* and *Microdipodops*) but is poorly developed in the kangaroo rats (Genus *Dipodomys*; French 1993). Within heteromyids, two basic dormancy "profiles" occur: (1) those species that forage year-round and employ shallow torpor to cope with energy emergencies (e.g., species of *Chaetodipus* that tolerate body temperatures of 10°–12°C for less than 24 hours), and (2) those species of *Perognathus* (e.g., *P. parvus*) that use supplies of cached seeds during dormancy that may last for up to 8 days at a body temperature as low as 2°C. A thorough discussion of dormancy in the Family Heteromyidae is provided by French (1993). One of the smallest rodents in the world, Genus *Baiomys* (adults = 9 g) of southwestern North America, undergoes bouts of estivation. If food and water are limited, these small mice readily decrease body temperature from 38° to 20°C. Periods of estivation are reported for species of *Peromyscus* inhabiting hot regions. The cactus mouse (*P. eremicus*), an inhabitant of deserts of southwestern North America, undergoes torpor (estivation) in summer and winter in response to limited food and water (figure 8.27). Cactus mice remain in burrows during the driest part of the summer and enter torpor in response to food and water shortages at environmental temperatures of about 30°C. Under laboratory conditions, when deprived of food and water, *P. eremicus* enters torpor at an ambient temperature of 19.5°C, maintaining a body temperature of about

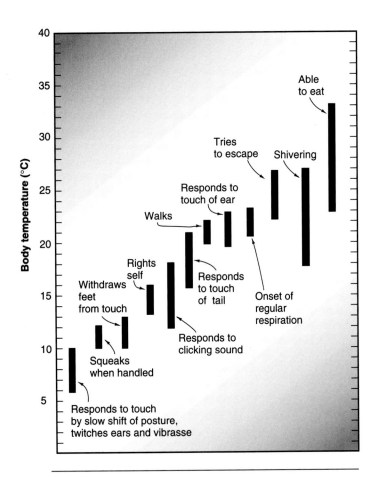

Figure 8.26   Body temperature and behavior.   Range of minimal body temperatures for various patterns of behavior exhibited by the little pocket mouse during arousal from hibernation.

Source: Data from G.A. Bartholomew and T.J. Cade, "Temperature Regulation, Hibernation, and Estivation in the Little Pocket Mouse," Perognathus longimembris" *in* Journal of Mammalogy, *38:60–72, 1957.*

22°C for several hours. Body temperature does not drop below 15°C during torpor (MacMillen 1965). From the preceding discussion, it is apparent that rodents exploit many different adaptive patterns in dormancy, with the ultimate goal of saving energy in unpredictable environments.

Case Study of Estivation

Kangaroo mice can be used to illustrate the energy savings accrued by estivation. These small heteromyid rodents (10–14 g) occur in the Upper Sonoran sagebrush desert of western North America. Seeds, a staple in their diet, are gleaned from the sand and hoarded in underground burrows. Although environmental temperatures may range from 0° to 30°C, kangaroo mice rarely experience drastic changes in temperature due to their fossorial (burrow-digging) and nocturnal habits. In most deserts, seed production is quite seasonal and limited to very brief periods. As a result, strict homeothermy for granivorous rodents such as kangaroo mice would be a waste of energy. Kangaroo mice therefore cope with such environmental unpredictability by exhibiting great thermolability; duration and frequency of estivation is

contingent on the availability of seeds and environmental temperature. For example, when food is in excess and environmental temperatures high, kangaroo mice exhibit homeothermy; if food is limited and temperatures unfavorable, they shift into heterothermy (estivation). Periods of estivation may last for only a few hours or for several consecutive days, and the more time they spend in dormancy, the less food they need. Brown and Bartholomew (1969) measured the influence of environmental (ambient) temperature on the energy needs (oxygen consumption) of homeothermic and torpid kangaroo mice (figure 8.28). At ambient temperatures of 5° to 25°C, the body temperatures of mice were 32° to 37°C when they were homeothermic, but during torpor, their body temperatures were just 1° to 3°C above ambient levels. As shown in the figure, the vertical distance between the "homeothermic" and "torpid" lines represents the energy savings per unit time of torpor at that temperature.

Avoidance of High Temperatures

Unlike large mammals, small mammals avoid extreme temperatures in desert ecosystems by adhering to fairly definite periods of activity. With the exception of ground squirrels and chipmunks, all desert rodents of North America are nocturnal and fossorial. Small mammals optimize survival through avoiding extremes in temperature by residing in

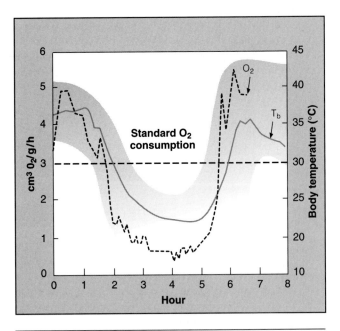

Figure 8.27   Torpor in the cactus mouse.   Pattern of changes in oxygen consumption and body temperature in the cactus mouse during entry into and arousal from torpor at an ambient temperature of 19.5°C. The standard consumption line is for active mice at an ambient temperature of 20°C. The cycle of torpor was initiated by deprivation of food and water.

Source: Data from R.E. MacMillen, "Aestivation in the Cactus Mouse" in Comp. Biochem. Physiol., *16:227–248, 1965.*

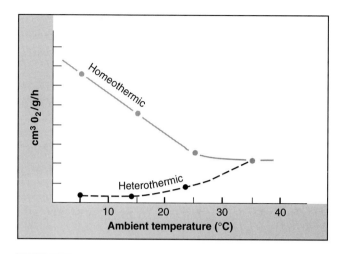

Figure 8.28   Torpor and euthermy in the kangaroo mouse.
The effects of ambient temperature on the oxygen consumption of
kangaroo mice when maintaining normally high body temperatures and
when in torpor. The points are mean values.

*Source: Data from J.H. Brown and G.A. Bartholomew, "Periodicity and Energetics
of Torpor in the Kangaroo Mouse, (*Microdipodops pallidus*)" in* Ecology,
50:705–709, 1969.

burrows below ground during the heat of the day. For a typical desert ecosystem, the temperature of the air in the burrow
is mild compared with the extremes that occur on the surface
(figure 8.29). Heteromyids search for seeds at night when it
is cool and spend the day in relatively cooler and more humid burrows. They are seldom exposed to air with high temperatures. Within burrow environments, the air is saturated
with water vapor. This is essential for granivorous rodents for

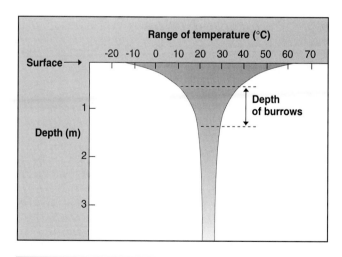

Figure 8.29   Soil temperature changes in deserts.   Diagram
showing the range of temperature compared with the depth of the soil.
Temperature fluctuations are less extreme below the surface of the
desert at depths typical of burrows of kangaroo rats. A temperature of
25°C represents an "average" burrow temperature.

Source: Data from K. Schmidt-Nelson, Desert Animals: Physiological Problems of
Heat and Water, *1964, Oxford University Press, New York.*

two reasons: first, the seeds gathered on the dry surface absorb water when stored in the burrow, and second, evaporative water loss is reduced considerably while the rodent is in
its burrow because there is no saturation deficit.

Diurnal desert rodents must devise other ways to cope
with heat stress. Studies of antelope ground squirrels (*Ammospermophilus leucurus*) in the deserts of southern California
have elucidated some unique tactics for surviving heat stress
(Chappell and Bartholomew 1981). Ground squirrels foraging on the surface of the ground during midday in summer
are faced with temperatures approaching 75°C. Daily foraging therefore assumes a bimodal activity pattern, with most
activity taking place during midmorning and late afternoon.
During the day, body temperature is quite labile, varying
from 36.1° to 43.6°C. Squirrels use fluctuating body temperature to store heat during their periods of activity. High temperatures limit the time that squirrels can be active in the
open to no more than 9 to 13 minutes. During this time,
they move rapidly from one patch of shade to the next, paus-
ing only to seize food or monitor predators. Exposure to
midday temperatures is minimized by running rapidly across
open areas and seeking shade in their burrows. In addition,
ground squirrels employ their tail as a sort of parasol, or
"heat shield." An active squirrel holds its wide, flat tail
tightly over its back with the white ventral surface upward.
In this position, the tail shades a large portion of the animal's
back, thus lowering the temperature due to solar radiation.
On a hot day, to maintain their body temperature below
43°C, squirrels must retreat to burrows every few minutes.
Burrows deeper than 60 cm usually have temperatures between 30° to 32°C. The body temperature of the antelope
ground squirrel shows a pattern of rapid oscillations, rising
while the squirrel is in the sun and dropping when it retreats
to its burrow (figure 8.30). Ground squirrels and other rodents do not sweat or pant. Instead, they use this combination of transient heat storage and passive cooling in a deep,
saturated burrow to permit activity during daylight hours
when heat is extreme. As explained in the following section,
antelope ground squirrels employ a strategy for coping with
heat that is very similar to that of camels—saving water by
allowing their body temperature to rise until the heat can be
dissipated passively. The major difference between these
squirrels and camels is a function of body mass. The large
camel can store heat for an entire day and cool off at night,
whereas the antelope ground squirrel goes through the same
cycle many times during the day.

The Dromedary

The biology of camels (see chapter 19) has been reviewed by
Gauthier-Pilters and Dagg (1981), Yagil (1985), and
Kohler-Rollefson (1991). The dromedary (*Camelus dromedarius*), touted as the "ship of the desert," occurs in semiarid
and arid regions of the Old World. Camels take long journeys, some lasting from 2 to 3 weeks, with no opportunity to
drink water. In the Sahara, dromedaries often remain without drinking water from October to May, existing solely on

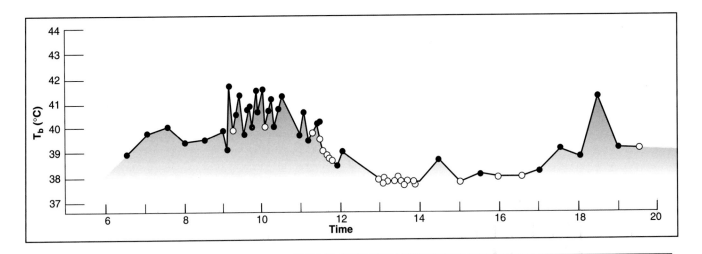

Figure 8.30   **Activity and body temperature of the antelope ground squirrel.**   Short-term cycles of activity and body temperature of *Ammospermophilus leucurus.* Active (●); inactive (○).

*Source: Data from M.A. Chappell and G.A. Bartholomew, "Standard Operative Temperatures and Thermal Energetics of the Antelope Ground Squirrel (*Ammospermophilus leucurus,*)" in* Physiol. Zool., *54:81-93, 1981.*

the water content in the forage. During such arduous trips, they are faced with severe problems of water conservation and temperature regulation. As a result, they have evolved unique mechanisms to cope with desert ecosystems.

One interesting mechanism employed by camels in coping with intense heat and lack of water of arid regions is their ability to vary body temperature as a device for saving water (Schmidt-Nielsen et al. 1957). Researchers have studied the daily cycle of body temperature in camels that received water and those that were dehydrated (figure 8.31). Camels given water show a very small daily fluctuation in body temperature, ranging from 36°C in the early morning to a maximum of 39°C by midafternoon. If a camel is deprived of water, however, the daily variation in body temperature triples—falling to 34.5°C at night and climbing to 40.5°C during the day. This temperature variation is important because when camels are dehydrated, they can lose excess heat at night through radiation, conduction, and convection to a cooler environment, thus saving water! This results in a change in body temperature for dehydrated camels of about 6°C; for camels with water, the change is only half this (see figure 8.31). To determine how much water camels can conserve by storing heat during the day and losing it later that night by nonevaporative means, we use the following calculations. About 0.9 cal is required to increase each gram of tissue 1.0°C. A 500-kg camel that increases its temperature by 6°C can therefore store 2,700,000 cal of heat in its body. If this heat had to be lost entirely by evaporation (each mL of water dissipates about 580 cal), however, a total of 4655 mL of water would be required. Thus, a camel would have to evaporate about 4.5 L of water to maintain a stable body temperature at the nighttime level. By tolerating a high body temperature during the day, camels can therefore conserve considerable amounts of water.

Camels can tolerate a water loss greater than 30% of their body mass, whereas a 15% loss is lethal in most other mammals. Their toleration of dehydration appears to be related to their ability to maintain blood plasma volume near normal levels during dehydration. They can rehydrate quickly, taking in water equal to 30% of their body mass within a matter of minutes. Camels have an impressive ability to concentrate urine and absorb water from fecal material. Their kidneys can produce urine with a high chloride content; dromedary camels can drink a sodium chloride solution even more concentrated than seawater without ill effect.

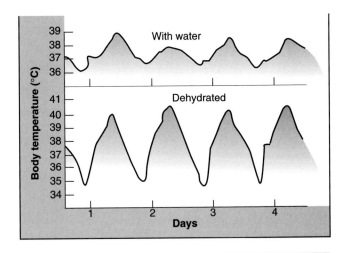

Figure 8.31   **Heat storage and loss in the dromedary.**
Diagram showing the relationship of changes in body temperature compared to heat storage and loss in the dromedary. Camels store more heat when water availability is restricted.

Source: Data from K. Schmidt-Nielsen et al., "Body Temperature of the Camel and its Relation to Water Economy" in American Journal of Physiology, *188:103-112, 1957.*

Many of the camel's morphological, behavioral, and physiological adaptations enhance its survival during periods of heat. As a result of its thick pelage, much of the solar energy that strikes a camel's fur never reaches its skin. During summer, the surface of its fur can reach 70° to 80°C, but skin temperature stays at only 40°C. Abundant sweat glands release secretions onto the skin surface beneath the hair, and evaporative cooling aids temperature regulation. The combination of a thick pelage and high body temperature reduces the rate of heat gain to the core of the body, slows the rate of sweating needed, and thus conserves water. During the day, camels must resort to evaporative cooling through sweating and panting, coupled with deep, slow breathing to eliminate heat and reduce respiratory water loss. A thick fur coat protecting the evaporative surface of the bare skin from unrestrained access by environmental heat could also reduce the vaporization of sweat. So the camel has evolved an optimal thickness to its pelage: it is not so thick or dense as to interfere with dissipation of vaporized water from the skin to the atmosphere nor so thin as to provide inadequate protection. It seems counterintuitive that wearing a fur coat on a hot day may enhance thermoregulatory ability, but it works for camels. The deserts of northern Africa are cold in the winter, and camels have an extremely thick fur. In summer, camels shed their winter coat but retain hair of up to 6 cm long on the back and 11 cm on the hump. On the venter (abdomen) and surface of legs, the hair is only 1.5 to 2.0 cm long. As in guanacos and other large mammals, the lightly furred areas of camels act as thermal windows for heat loss.

The morphological peculiarities of camels couple with behavior to enhance water conservation and temperature regulation. Early in the morning, camels lie down on surfaces that have cooled overnight by radiation. They tuck their legs beneath the body with their ventral surface of short fur in contact with the cool ground. In this position, a camel exposes only its well-furred dorsum and flanks to the hot sun and places its lightly furred legs and venter in contact with cool sand, thus conducting heat away from the body. Camels also form group huddles during the day, in which they lay closely together. The side-by-side contact prevents solar radiation from permeating the body of each animal and raising its body temperature.

The Camel's Hump. The ability of the dromedary to withstand heat and dryness does not depend on water storage in its hump but rather on numerous physiological mechanisms aimed at water conservation. It is a popular misconception that the camel's hump is filled with water. Actually, the hump is filled with fat bound together by fibrous connective tissue. Contrary to myth, however, this fat does not help much as a source of energy for long travels in the desert nor as a lightweight method of carrying water (Hill and Wyse 1989:156). To understand the costs and benefits of fat stored in the camel's hump, the amount of water gained and lost in breaking down (catabolizing) lipids or carbohydrates must be assessed. Consider the production of metabolic water in relation to respiratory water loss. For temperatures and humidities of a desert environment, camels actually loose more water across the lungs in obtaining oxygen to oxidize fats than they gain from the oxidation of those fats. In reality, the breakdown of fat to derive energy actually imposes a water deficit.

Other Desert Ungulates

Desert antelopes (e.g., oryx, gazelles, and elands) possess a number of adaptations for coping with heat and dehydration (Taylor 1972; figure 8.32). The pallid color and glossy fur of the desert antelope reflects direct sunlight, and the fur is excellent insulation against the heat. Heat is lost by convection and conduction from the venter, where the pelage is very thin. Unlike the dromedary, the antelope's coping mechanisms are independent of drinking water. Elands search for shade during the heat of day, whereas dromedaries remain in the full sun all day, with relief from heat not coming until evening. Elands and oryx both pant at high temperatures and can reduce their metabolic rates to conserve water. Furthermore, evaporation through the skin is reduced by 30% in elands and by 60% in oryx. Evaporative water loss through the respiratory tract is also reduced. In comparison to camels, oryx can reduce daytime evaporative water losses by suppressing sweating completely, even when very hot. The fat tissue in elands and dromedaries is localized in the hump rather than dispersed uniformly under the skin, where it would impair heat loss by radiation. The water requirements of antelopes are negligible, plus their ability to conserve water is greatly enhanced by their production of a concentrated urine and dry feces.

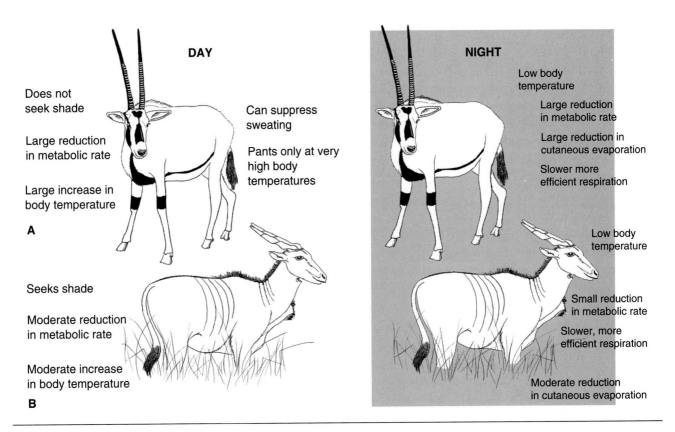

Figure 8.32   **Water conservation in antelopes.**   Adaptations for water conservation in two nondrinking east African antelopes, (A) the oryx and (B) the eland. Figures on the left are for daytime; figures on the right are for night.

Summary

The success of mammals derives from their sophistication in form, function, and, especially, behavior. Mammals are widely distributed throughout the earth—ranging from tundra to deserts, each environment characterized by unique extremes in climate. As a result of such strong selection pressures, mammals have evolved a unique suite of mechanisms to enhance survival. These "adaptive strategies" are diverse and highly sophisticated, with variation exhibited at the levels of ecosystem and species and between individuals.

Basic terminology in the physiology of environmental adaptation includes endothermy, homeothermy, euthermy, heterothermy, poikilothermy, and ambiguous terms such as warm- and cold-blooded. Most mammals maintain their body temperatures between 36° and 38°C. To do this, they must balance heat gain and heat loss against such forces as radiation, conduction, convection, and evaporation. The equilibrium maintained by mammals is achieved through interaction with the environment by the mammalian thermostat—the hypothalamus—with its species-specific set points. Each species has unique coping mechanisms and peculiarities in their thermoneutral zone, thermal conductance, and metabolic rate. These characteristics and the important concept of countercurrent heat exchange help determine the distribution of species as well as such long-standing "biogeographic rules" as Bergmann's, Allen's, and Gloger's.

Mammals employ several mechanisms to defend against cold, including (1) behavioral (communal nesting, nest construction, food hoarding, foraging dynamics, and activity patterns) and (2) anatomical and physiological (body mass reduction during winter, decreasing thermal conductance, nonshivering heat production, and dormancy).

The challenges faced by mammals in desert environments are even more severe than those for species residing in the cold. Water is essential for the survival of all animals, so the role of water economy and the anatomy and physiology of the mammalian kidney are crucial to desert animals. Desert mammals conserve water by eating succulent plants and using metabolic water. Certain small mammals also have ingenious mechanisms for consuming halophytic plants. In addition to conserving water, xeric-adapted mammals regulate body temperature through evaporative cooling by sweating, panting, saliva spreading, and respiratory heat exchange.

As we have seen with mammals faced with cold, mammals residing in hot environments possess certain adaptive strategies that enhance their survival, including unique patterns in insulation, appendages, metabolic profiles, and body size. As animals in the cold use hibernation, desert mammals employ estivation to avoid heat and lack of water. Certain mammals epitomize the ability to survive in heat, namely heteromyid and dipodid rodents, dromedaries, and desert antelopes.

Discussion Questions

1. Distinguish among the following terms: homeothermy, ectothermy, poikiothermy, heterothermy.
2. Small mammals are common inhabitants of deserts and mountains. Describe the different adaptations they employ to maintain homeothermy in each environment.
3. Explain why it is advantageous for certain small mammals to abandon homeothermy during brief or extended periods of their lives.
4. Explain nonshivering thermogenesis. Track this heat-producing mechanism for a small mammal from its initiation in nature to its function in maintaining homeothermy.
5. Compare the advantages and disadvantages of the color white for animals residing in northern environments.
6. What are the benefits of having a larger body size in the Arctic? Compare the advantages and disadvantages of body size in terms of mass-specific metabolism and total energy needs.
7. Define countercurrent heat exchange. What are the advantages of this mechanism for a caribou of the Arctic and a sloth of Panama?
8. Define thermal conductance. Compare the benefits of a thick fur coat for a lemming in the Arctic compared with a dromedary in the Sahara Desert.
9. Compare the different types of dormancy in mammals: hibernation, torpor, and estivation. Following each definition give an example of different mammals that fit into each category. What are the advantages and disadvantages of dormancy?
10. Describe the cycle of hibernation of a ground squirrel residing in northern Montana. Compare hibernation in this ground squirrel with estivation of a kangaroo mouse residing in the desert of California.
11. List the different water conservaton strategies employed by desert mammals.
12. Discuss four major mechanisms of evaporative cooling (insensible water loss).

Suggested Readings

Boyce, M. S. 1988. Evolution of life histories of mammals: theory and pattern. Yale Univ. Press, New Haven, CT.

Dew, E. M., K. A. Carson, and R. K. Rose. 1998. Seasonal changes in brown fat and pelage in southern short-tailed shrews. J. Mammal, 79:271–278.

Folk, G. E., Jr., M. L. Riedesel, and D. L. Thrift. 1998. Principles of integrative environmental physiology. Austin & Winfield Publishers, San Francisco, CA.

Gordon, M. S., G. A. Bartholomew, A. D. Grinnell, C. B. Jørgensen, and F. N. White (eds.) 1982. Animal physiology: principles and adaptations, 4th ed., Macmillan, New York.

Hickman, C. P., Jr. Integrated principles of zoology, 9th ed. Wm. C. Brown, Dubuque, IA.

Hinds, D. S. 1973. Acclimatization of thermoregulation in the desert cottontail, *Sylvilagus audubonii*. J. Mammal, 54:708–728.

Irving, L. 1972. Arctic life of birds and mammals, including man. Springer-Verlag, Berlin.

Koopman, H. N. 1998. Topographical distribution of the blubber of harbor porpoises (*Phocoena phocoena*). J. Mammal, 79:260–270.

Lyman, C. P., J. S. Willis, A. Malan, and L. C. H. Wang. 1982. Hibernation and torpor in mammals and birds. Academic Press, New York.

Michener, G. R. 1992. Sexual differences in over-winter torpor patterns of Richardson's ground squirrels in natural hibernacula. Oecologia, 89:397–406.

Nagy, K. A., and M. J. Gruchacz. 1994. Seasonal water and energy metabolism of the desert-dwelling kangaroo rat (*Dipodomys merriami*). Physiol. Zool., 67:1461–1478.

Schmidt-Nielsen, K. 1990. Animal physiology: adaptation and environment, 4th ed. Cambridge Univ. Press, New York.

Thomas, D. W. 1990. Winter energy budgets and costs of arousals for hibernating little brown bats, *Myotis lucifugus*. J. Mammal, 71:475–479.

Tomasi, T. E. and T. H. Horton (eds.) 1992. Mammalian energetics. Comstock Publ. Assoc., Ithaca, NY.

Whittow, G. C. (ed.) 1971. Comparative physiology of thermoregulation, vol. II, Mammals. Academic Press, New York.

Whittow, G. C. (ed.) 1973. Comparative physiology of thermoregulation, vol. III, Special aspects of thermoregulation, Academic Press, New York.

Yousef, M. K., S. M. Horvath, and R. W. Bullard. (eds.) 1972. Physiological adaptations, desert and mountain. Academic Press, New York.

We refer you to the series of symposia on *Mammalian Hibernation* begun in 1959 by Charles P. Lyman and Albert R. Dawe. Symposia are held approximately every 3 years, each resulting in a volume of proceedings. The proceedings of the latest symposium, held July 1996 in Tasmania, are entitled, *Adaptations to the cold: tenth international hibernation symposium.* These volumes contain a wealth of information on hibernation biology spanning an organismic to molecular scale and will help you keep pace with the changes in this interesting and dynamic field of biology.

CHAPTER

9

Reproduction

Reproduction is the process by which organisms produce new individuals, passing on genetic material and thus maintaining the continuity of the species and of life. All mammals reproduce sexually by way of internal fertilization. Furthermore, all mammals are **dioecious,** that is, the sexes are separate, and a given individual normally has the sex organs and secondary sexual characteristics of only one gender.

As you will see in this chapter, mammals vary greatly in the structure and function of their reproductive system. The monotremes lay eggs and incubate their young, and as is the case for most reptiles and birds, monotremes have a **cloaca**—a common opening for the urinary and reproductive tracts. In addition, as with birds, only the left ovary is functional in monotremes; they have, however, true mammary glands but lack nipples.

The marsupials or pouched mammals, like the monotremes, possess a cloaca. The placenta in these animals is not very efficient due to the limited contact between the maternal and fetal blood supply. As a consequence, marsupials undergo a very brief gestation period and a prolonged period of nursing. The young are born in an altricial (poorly developed) state and require a long period of development in the pouch (**marsupium**) of the mother. The eutherian mammals, commonly called "placental mammals," have made a major evolutionary advance in the evolution of their reproductive patterns. The key innovation was the appearance of a highly sophisticated placenta that facilitates efficient respiratory and excretory exchange between the maternal and fetal circulations. Unlike the marsupials, the eutherian mammals exhibit a longer gestation period and a shorter period of nursing or lactation. Furthermore, at birth, the young are more highly developed (precocial) than are young marsupials.

Mammals vary widely in the complexity of their reproductive patterns, but it is incorrect to interpret the seemingly more primitive reproductive patterns of the monotremes and marsupials as being less successful than those of eutherians. The success of a species can be evaluated only case by case and in the context of its specific ecological relationships. Many examples exist of successful groups of noneutherian mammals, one of which, with which most of us are well acquainted, is the opossum. This marsupial originated in South America and has colonized environments as far north as Ontario, Canada, within only a short time. The great success of this species is demonstrated by its broad geographic distribution and high reproductive potential.

The variations in reproductive patterns of mammals seem endless. For example, gestation ranges from 10 to 14 days in some dasyurid marsupials to over 650 days in elephants. Lactation varies from only 4 days in pinnipeds such as the hooded seal (*Cystophora cristata*) to over 900 days in chimpanzees and orangutans (*Pan troglodytes* and *Pongo pygmaeus;* Hayssen et al. 1993). Most mammals give birth to between 1 and 15 young per litter, and intervals between births may range from 3 to 4 weeks in rodents to as long as 3 to 4 years or more in sirenians, elephants, and rhinos (Hayssen et al. 1993). Voles (Rodentia: Muridae) may hold the record for

biotic potential: the meadow vole (*Microtus pennsylvanicus*) of the North American grasslands may produce up to eight or nine litters of five to eight young in a single year. As a result, a single female may potentially produce up to 72 offspring during one breeding season! Bailey (1924) reported that a captive female meadow vole produced 17 families in a single year. Variation in reproductive patterns may be explained by assessing environmental factors and the risk of raising young in highly seasonal environments. A thorough review of the seemingly endless reproductive strategies employed by mammals is presented by Hayssen and colleagues (1993).

To examine the variation in reproductive patterns in mammals, we divide this chapter into six parts. First, we concentrate on the role of the male and the female in the production of **gametes** (sperm produced by the testes of the male and ova produced by the ovaries of the female). The next five phases take place within the body of the female and include fertilization, implantation, gestation, parturition, and lactation. As we discuss each process, remember that each species of mammal has evolved a unique reproductive capacity for coping with the selective forces of the environment in which it resides.

THE REPRODUCTIVE SYSTEMS

The Male Reproductive System

The reproductive system of males comprises paired testes, paired accessory glands, a duct system, and a copulatory organ.

Testes

The **testes** (sing., *testis*) are the site of production of sperm (the male gametes) and the synthesis of male sex hormones, chiefly testosterone (figure 9.1). The paired testes of mammals are oval-shaped and may be suspended in a pouchlike, skin-covered structure called the **scrotum.** The position of the testes in mammals varies considerably. In many species (e.g., some bats and many rodents), the testes migrate from the body cavity into the scrotum during the breeding season and afterward are withdrawn into the inguinal canal. In some mammals (e.g., monotremes, some insectivores, anteaters, tree sloths, armadillos, manatees and dugongs, all the seals except the walruses, and whales, hyraxes, and elephants), the testes are always in the abdominal cavity. The testes remain in the scrotum permanently in most primates, artiodactyls, perissodactyls, and carnivores. It is possible that the temperature of the abdominal cavity is too high to permit development of sperm (**spermatogenesis**), and thus the scrotum allows for cooling of the testes. Temperature may not be the only critical factor for spermatogenesis, however, and a number of environmental factors interact to initiate and end spermatogenesis (van Tienhoven 1983; Hadley 1988).

Spermatogenesis, the production of spermatozoa by the process of meiosis, occurs within the **seminiferous tubules** of the testes. Within these tubules, spermatozoa

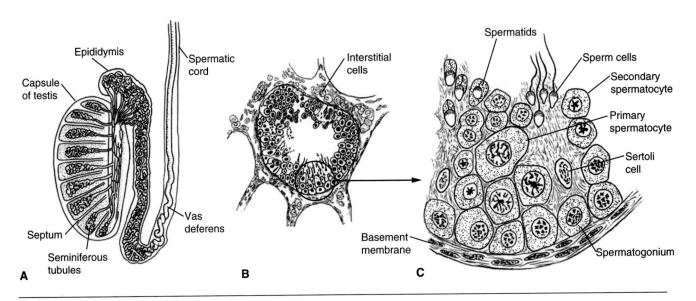

Figure 9.1   **Mammalian testis.**   Section of the mammalian testis showing coils of (A) seminiferous tubules containing spermatozoans (B) in various stages of development. (C) The composite cross section of a seminiferous tubule shows spermatogonia near the basement membrane, progressing from spermatocytes to spermatids with mature spermatozoa (sperm cells) located near the lumen of the tubule. See Curtis (1975) and Hickman et al. (1995) for a description of the maturational process within the seminiferous tubule.

mature by passing through several stages of development and are nurtured by **Sertoli cells** (see figure 9.1). Surrounding the tiny, coiled tubules within the testes are microscopic interstitial cells (**Leydig cells**), which are the site of secretion of the male hormone, **testosterone.** The secretions of Sertoli and Leydig cells provide an optimal microenvironment for spermatogenesis (Gordon 1982; Hadley 1988).

Ducts and Glands

After the sperm are mature, they must be transported to the exterior by a series of ducts (figure 9.2). On leaving the seminiferous tubules, sperm are collected in the **epididymis,** a highly coiled tube located on the surface of the each testis. This tube serves both as a duct for passage of sperm and as a brief storage site where sperm and all glandular secretions are nourished prior to ejaculation. After the epididymis leaves the testis, its shape changes, becoming enlarged and straight to form the **ductus deferens** (vas deferens), which passes to the urethra. Near the junction of the ductus deferens and the urethra, three different glands add secretions to the seminal fluid: the paired **seminal vesicles** (vesicular glands), a single **prostate gland,** and paired **bulbourethral glands** (Cowper's glands). Additional accessory glands, such as the paired **preputial glands** and **ampullary glands,** contribute their products to the seminal fluid (**semen**). Considerable variation in reproductive glands occurs among species of mammals, and not all species possess all the glands listed. In some animals, coagulating glands produce a substance that coagulates the secretions of the seminal vesicles. Following copulation, fluid from this gland contributes to formation of a **copulation plug,** which blocks the entrance to the the female's vagina. Presence of this plug often indicates that

a female has mated; it may persist for up to 2 days following copulation. The copulation plug is thought to be important in retaining the spermatozoa and preventing their loss from the female tract. Copulation plugs have been noted in many species of rodents and in some bats, insectivores, and marsupials (Austin and Short 1972a; DeBlase and Martin 1981).

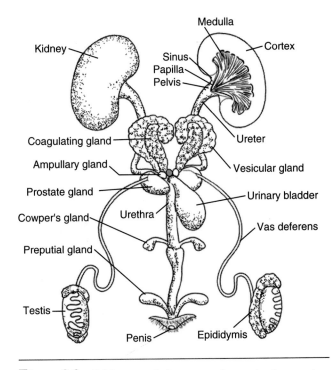

Figure 9.2   **Male urogenital system.**   Composite diagram of the male urogenital system of the rat (*Rattus norvegicus*).

In addition to nourishing sperm, acting as a lubricant, and, for some species, helping to form a copulation plug, glandular secretions enhance the alkalinity of the semen so as to protect the sperm from the acidic environment present in the male urethra and female vagina.

The Penis

Transfer of male sperm to the body of the female is facilitated by the highly vascular, erectile **penis.** The penis is composed of cylindrical bodies, the **corpora cavernosa,** which become filled with blood during sexual excitation, thus causing the organ to become erect. The sperm pass during coitus through the urethra, which is also used for urination. In some mammals (e.g., all carnivores, most primates, rodents, bats and some insectivores), the tip of the penis (**glans penis**) may include a complex bony structure, the **os penis,** or **baculum** (figure 9.3). Mammalogists identify different species by this structure, as well as use it as an index of approximate age of an individual (Burt 1960). The glans penis shows many different configurations, such as the bifurcated penis of some monotremes and marsupials, and the corkscrew shape in the domestic pig. These configurations are compatible with the vagina of females of the same species. Furthermore, the penis of different mammals may be situated either anterior (most eutherian mammals) or posterior (marsupials, rabbits, hares, and pikas) to the scrotum. In mammals such as monotremes and marsupials, the penis is situated in a sheath within the cloaca, and in marsupials, it may be directed posteriorly (Biggers 1966).

The Female Reproductive System

The reproductive organs of females consist of a pair of ovaries, which produce the eggs and certain hormones; a pair of **oviducts** (Fallopian tubes), which act as the channel for travel of the ova from the ovary to one or two enlarged **uteri** (sing., uterus), the site of embryonic development; and a **vagina,** the opening to the outside of the body. A **cervix** (pl., *cervices*) connects the uterus and vagina (figure 9.4).

The Ovary

The primary reproductive organ of the female is the **ovary.** In mammals, the ovaries are a pair of small, oval bodies that lie slightly posterior to the kidneys. They produce the female gametes, called **ova** (sing., *ovum*), and certain hormones. As in the testes of the male, chromosome reduction occurs in female gametes within the ovary. Immediately under the surface of the ovary is a thick layer of spherically grouped cells called **follicles** (figure 9.5), each of which encloses a single egg. At birth, large numbers of follicles are present in the female mammal (about 2 million in the ovaries of humans),

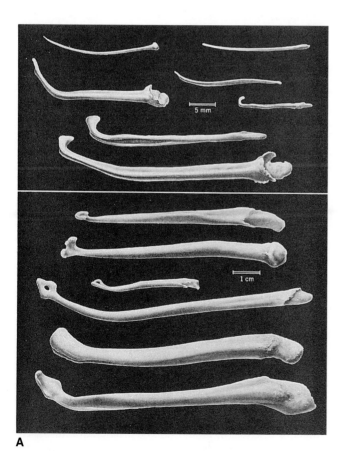

A

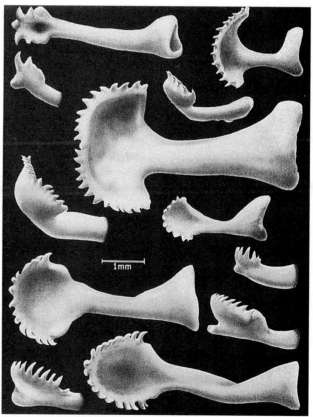

B

Figure 9.3   **The baculum.**   Representative bacula of the Families (A) Mustelidae and (B) Sciuridae.

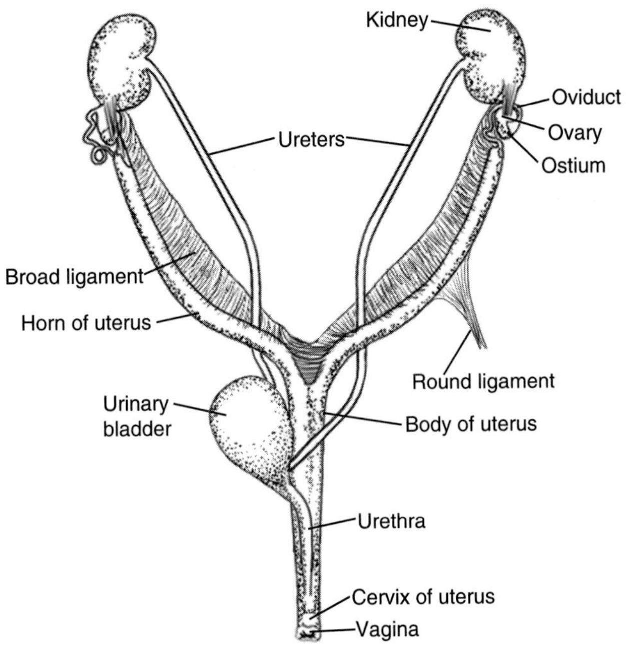

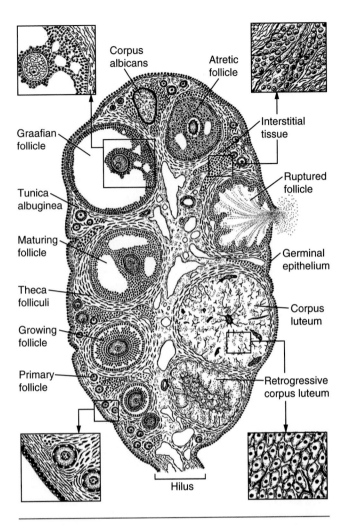

Figure 9.5 **Mammalian ovary.** Composite diagram of the mammalian ovary showing the progressive stages in the differentiation of a Graafian follicle shown on the left. The mature follicle may become atretic (*top*) or ovulate and undergo luteinization, as on the right. Insets show magnification of cellular components in each region of the ovary.

Figure 9.4 **Female urogenital system.** Composite diagram of the female urogenital system of the cat (*Felis catus*).

but the number decreases steadily with age. The ovarian follicle is the source of three types of gonadal steroids: androgens (masculinizing), estrogens (feminizing), and progestins. The amounts of each class of steroid vary throughout the reproductive cycle of the female (Albrecht and Pepe 1990; Strauss et al. 1996).

Ovulation. As the egg matures and the surrounding follicle enlarges, they move closer to the surface of the ovary. **Ovulation** occurs when the follicle bursts, releasing the egg, which penetrates the surface of the ovary and passes into the oviduct, where it may be fertilized if sperm are present. Some sperm may reach the site of fertilization (typically, the oviduct) within a few minutes, but for the most part, it takes several hours for **capacitation** (the physiological changes in the sperm that facilitate penetration of the covering of the egg) to take place (Austin and Short 1972c). For some species of mammals, however, the period between insemination and actual fertilization may be delayed for up to 12 months—a phenomenon called delayed fertilization (discussed in detail in "Reproductive Variations" in this chapter). If fertilization does not occur, the egg enters the uterus and degenerates.

The development of follicular cells is controlled by the action of **follicle-stimulating hormone (FSH)** and **luteinizing hormone (LH)** produced by the anterior pituitary gland (figure 9.6). Ovulation is induced by high levels of LH. Under the influence of these hormones, the ruptured follicle fills with yellow follicular cells and is called the **corpus luteum** (pl., *corpora lutea*), or "yellow body." The corpus luteum continues functioning during early pregnancy and produces the female hormone **progesterone.** This hormone

promotes growth of the uterine lining (**endometrium**) and makes possible the implantation of the fertilized egg (Stouffer 1987; Keys and Wiltbank 1988; Nelson, 1995). Progesterone also stimulates development of the mammary glands. Inactive corpora lutea eventually degenerate into **corpora albicans,** or "white bodies," visible on the surface of the ovary. If we examine the number of corpora lutea and corpora albicans in the ovaries, we can learn a great deal about the reproductive history of a particular female (Perry 1972). The number of corpora lutea is related to the number of ova that have been ovulated and can be used to estimate the number of embryos that were produced. The major female hormone secreted by the developing follicles is **estradiol,** (a form of **estrogen**), under the control of the FSH (see figure 9.6). Estrogen is essential in promoting the proliferation of the endometrium. As the corpus luteum disintegrates, no more progesterone is produced, causing the "sloughing off" of the endometrium. Although this sloughing off and regrowth of

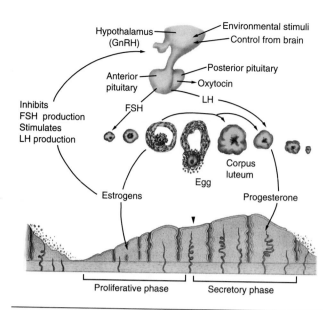

Figure 9.6   **Estrous cycle.**   Hormones influence the estrous cycle of a mammal. FSH secreted by the anterior pituitary gland initiates the onset of follicular development. During the proliferative phase of the cycle, FSH and LH further the maturation the follicle. Estrogens secreted by the follicular cells surrounding the egg cause thickening of the endometrium, further follicular growth, and inhibit the production of FSH. When the egg is mature, the enlarged follicle bulges from the surface of the ovary and bursts, releasing the egg (ovulation; see also figure 9.5). Ovulation, induced by high levels of LH, initiates the secretory phase of the cycle. The egg passes into the oviduct, where, if it is fertilized, it implants into the endometrium. Unfertilized eggs enter the uterus and degenerate. Under the influence of LH, the ruptured follicle fills with yellowish cells and is called the corpus luteum. The corpus luteum produces high levels of progesterone and estrogens that strongly inhibit GnRH secretion by the hypothalamus, thereby inhibiting secretion of FSH and LH necessary to permit maturation of new follicles. High levels of estrogens and progesterone stimulate the uterus to thicken, making final preparations for gestation. Progesterone also stimulates growth of the mammary glands. *Abbreviations:* FSH = follicle-stimulating hormone; LH = luteinizing hormone; GnRH = gonadotropin releasing hormone.

Source: Data from H. L. Gunderson, Mammalogy, *1976, McGraw-Hill Company, New York.*

the endometrium occurs cyclically in all mammals, it is most pronounced in primates. In some mammals, for example in dogs and cats, a false pregnancy, or **pseudopregnancy,** may occur. This results from the maintenance of the corpora luteum, without fertilization, beyond the time of normal regression. Mammals exhibiting a false pregnancy behave as if they are pregnant, even though they are not.

The Estrous Cycle.   All female mammals, except higher primates, restrict copulation to specific periods of the sexual cycle. In nonprimates, a period of brief receptivity shortly before and after ovulation is called estrus, or **heat** (adj., *estrous*). In the majority of mammals, ovulation is **spontaneous,** that is, it occurs without copulation (figure 9.7). In contrast, if the egg is shed within a few hours following copulation, we refer to it as **induced** ovulation.

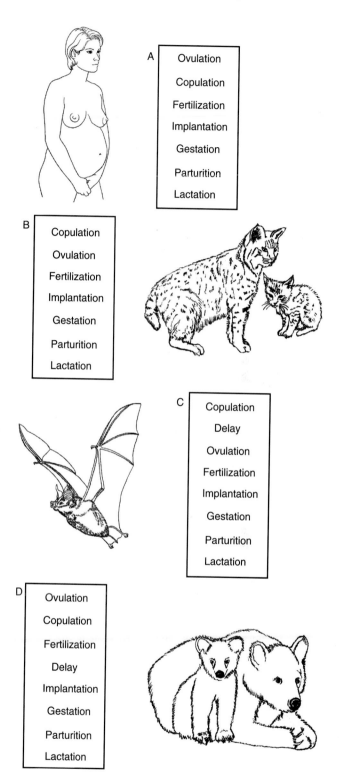

Figure 9.7   **Reproductive events.**   (A) Spontaneous ovulation is the usual sequence of reproductive events occurring in eutherian mammals. Several different groups show variation on this general sequence, including (B) induced ovulation exhibited by cats and some rodents, (C) delayed fertilization displayed by certain insect-eating bats, and (D) delayed implantation shown in a diverse array of groups. Delayed development exhibited by the Neotropical fruit bat and embryonic diapause, exhibited by marsupials such as kangaroos and wallabies, are not shown in this figure.

Many investigators suggest, however, that these two strategies for ovulation cannot be separated and instead represent a continuum (Weir and Rowlands 1973). Some mammals known to be induced ovulators are rabbits, many carnivores, and some ground squirrels. Some desert rodents and montane voles (*Microtus montanus*) may also exhibit induced ovulation in response to the occurrence of green vegetation and associated nutritional factors in their diet (Beatley 1969; Reichman and Van De Graaff 1975; Negus and Berger 1977; Kenagy and Bartholomew 1985).

During the breeding season, the time span from one period of estrus to the next is called the **estrous cycle.** A species that has one such cycle per year, as is true for some carnivores, is termed **monestrous.** Other species, including rodents, rabbits, hares and pikas, are **polyestrous,** meaning they have several cycles a year. In general, several stages may be seen during a given estrous cycle. As an individual moves from **anestrus** (the nonbreeding, quiescent condition) to **proestrus** (the beginning stage), the uterus begins to swell, the vagina enlarges, and ovarian follicles grow in response to several hormones—the animal is said to "come into heat." The ova are usually discharged from the ovary late in this estrous stage, which may be followed by **fertilization,** the successful union of sperm and egg to form a zygote, and implantation, in which the embryo adheres to the uterine wall. Gestation represents the period of development within the uterus. The period of lactation (production of milk by the mammary glands) is reported to "block" the start of another estrous cycle; however, in certain mammals, parturition (birth) is followed immediately by another estrous cycle (postpartum estrus). Many members of the rodent Family Muridae are known to ovulate within an hour after parturition. If, however, mating does not occur, the individual enters the normal condition of **metestrus,** followed by anestrus. Periods between mating in polyestrous species are termed **diestrus,** when the reproductive system recycles to the proestrous condition. Mammalogists can determine the different estrous periods by using the **vaginal smear technique,** which identifies specific cell types associated with a certain phase of the cycle (Kirkpatrick 1980).

The Female Duct System and External Genitalia

The female duct system consists of paired oviducts (or Fallopian tubes), one or two uteri, one or two cervices, and the vagina (see figure 9.4) The external genitalia include the **clitoris,** the **labia majora,** and **labia minora.** The clitoris is the embryonic homologue of the penis, and in those species where males possess a baculum, females possess a small bone, the **os clitoris,** in the clitoris.

The architecture of the female reproductive tract also varies considerably in mammals. The monotremes and marsupials do not possess true vaginas (figure 9.8). In the monotremes, the paired uteri and oviducts empty into a **urogenital sinus,** as do liquid waste products (figure 9.8A). This sinus passes into the cloaca. The female reproductive tract of marsupials is referred to as **didelphous,** meaning that the uteri,

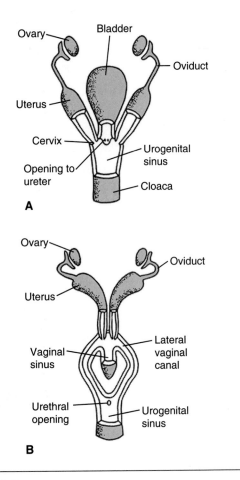

Figure 9.8 **Reproductive tracts.** Diagram showing two types of reproductive tracts of mammals. (A) In female monotremes, the urogenital sinus represents a common chamber for transport of both urine and reproductive cells and opens into the cloaca. (B) The reproductive tract of a female marsupial showing the lateral vaginal canals that receive the penis of the male. The vaginal sinus connects with the urogenital sinus to serve as a passageway for delivery of the young.

oviducts, and vaginal canals are paired (figure 9.8B). During copulation, the forked penis of the male delivers sperm to the paired uteri of the female. In more advanced marsupials, such as the kangaroos, two lateral vaginas are used to conduct sperm, and a medial vagina functions as the birth canal.

In eutherian mammals, four principal uterine types occur that are based on the relationship of the uterine horns (figure 9.9). A **duplex** uterus is found in lagomorphs, rodents, aardvarks, and hyraxes, and is characterized by two uteri, each with a cervix opening into the vagina. A **bipartite** uterus is typical of whales and most carnivores. Here, the horns of the uterus are separate but enter the vagina by a single cervix. The most widespread condition is called the **bicornuate** uterus, which is found in the insectivores, most bats, primitive primates, pangolins, some carnivores, elephants, manatees and dugongs, and most ungulates. The uterine horns are Y-shaped, being separated medially but fused distally, where they form a common chamber, the

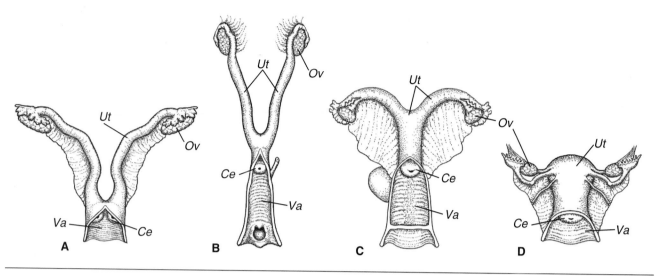

Figure 9.9 Types of uteri. The degree of fusion of the uterine horns varies among different groups of eutherian mammals. (A) A duplex uterus occurs in rodents, rabbits and hares, aardvarks, and hyraxes. Note the separate horns and distinct cervix of each. (B) Most carnivores and whales exhibit a bipartite uterus. (C) Insectivores, most bats, primitive primates, pangolins, some carnivores, elephants, manatees and dugongs, and most ungulates possess a bicornuate uterus. (D) A simplex uterus is typified by higher primates, some bats, anteaters, tree sloths, and armadillos. *Abbreviations:* Ut = uterus, Ce = cervix, Ov = ovary, Va = vagina.

body, which opens into the vagina though a single cervix. The last type of uterus is called **simplex,** in which all separation between the uterine horns is lacking, and the single uterus opens into the vagina through one cervix. This condition is seen in some of the bats, higher primates, and xenarthrans.

Implantation

Fertilization usually occurs shortly after ovulation in the oviduct. **Implantation** is the attachment of the embryo to the uterine wall. Cleavage of the **zygote,** or fertilized egg, occurs as it passes down the oviduct toward the uterus, propelled by ciliary action and rhythmic muscular contractions (peristalsis). Next, the blastocyst (the embryo at the 32 to 64 cell stage) implants on the wall of the uterus. The uterus of mammals is characterized by a thick muscular wall (the **myometrium**) surrounding the highly vascular **endometrium**—the specific site of implantation (figure 9.10). The anterior pituitary gland (which produces FSH and LH) and the corpus luteum (which produces progesterone) collaborate to prepare the endometrium for implantation. The blastocyst differentiates into the embryo and the trophoblast (see figure. 9.10). On contact with the lining of the uterus, the cells of the trophoblast region proliferate rapidly, producing enzymes that break down the **decidua**—the uterine mucosa in contact with the trophoblast. This permits the blastocyst to sink into the vascularized endometrium, and subsequently, it is surrounded by a pool of maternal blood. At this point, the trophoblast thickens, extending thousands of fingerlike roots (villi) that penetrate the lining of the endometrium. The trophoblast further invades the decidua, eventually contributing to the formation of the placenta. The more invasive the trophoblast, the larger the

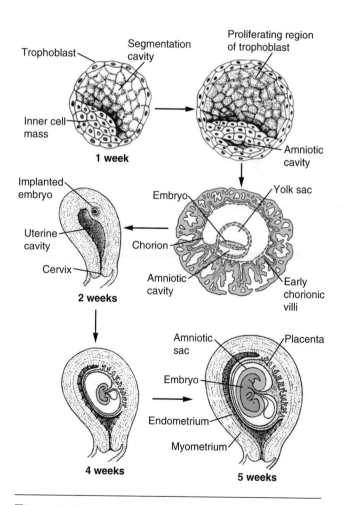

Figure 9.10 Human embryo. Diagram showing the early development of the human embryo and its extraembryonic membranes.

area of decidua through which both the embryo and placenta obtain nutrients, oxygen, water, ions, and hormones from the mother and discharge waste products. The degree of intimacy between the maternal and embryonic parts of the placenta varies in different mammals, and mammalogists employ these differences to help define various groups of mammals.

The Placenta. The placenta, a complex of embryonic and maternal tissues, performs several essential functions during pregnancy. It (1) physically anchors the fetus to the uterus, (2) transports nutrients from the circulation of the mother to the developing fetus, (3) excretes metabolites of the fetus into the maternal compartment, and (4) produces hormones that regulate the organs of the mother and fetus. Because both the placenta and embryo are genetically alien to the mother, uterine tissue should reject the embryo. This does not happen because the placenta has evolved the ability to suppress the normal immune response that the mother's body would ordinarily mount against it and the embryo. This ability to become an **allograft** (successful foreign transplant) is attributed to the production by the chorion of essential lymphocytes and proteins that block the mother's immune response by suppressing the formation of specific antibodies (Adcock et al. 1973; Strauss et al. 1996).

Although eutherians are often referred to as "placental" mammals, marsupials also have a placenta, albeit not one as efficient as that found in the eutherians. Most marsupials have a **choriovitelline placenta** (figure 9.11A). Here, the yolk sac is greatly enlarged, serving the nutritional needs of the developing embryo. The blastocyst of marsupials does not sink deeply into the endometrium but forms a shallow depression. The surface area for adhesion and absorption is increased by a slight wrinkling of the surface of the blastocyst. Besides the yolk sac, the developing embryo gains a limited amount of nutrition from the uterine lining. The connection between the blastocyst and uterine wall is relatively weak, however, and the system of nourishment to the embryo is inefficient. For these reasons, marsupials have short gestation periods and prolonged lactation periods, relative to eutherians (Hickman et al. 1997).

In contrast to marsupials, eutherian mammals (and the marsupial bandicoots) have a **chorioallantoic placenta** (figure 9.11B). In this case, the blastocyst adheres to the endometrium and then sinks very deeply into it, forming a strong adhesion. This adhesion is enhanced by the rapid growth of **chorionic villi** (fingerlike projections of capillaries from the outermost embryonic membrane) that penetrate the endometrium. The uterus becomes highly vascularized at the point of attachment of the developing embryo. It is important to note that although the maternal and fetal circulations are in close contact, these two blood systems are not fused. Fetal blood does not circulate in the mother, and maternal blood does not circulate in the fetus. The complex, extensive network of villi not only provides mechanical strength but also an increased surface area for rapid and efficient exchange of nutrients, gases, and waste products be-

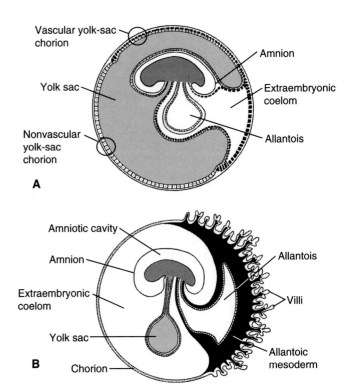

Figure 9.11 **Embryo and extraembryonic membranes.** Diagram showing the embryo and extraembryonic membranes of a (A) marsupial and a (B) placental mammal. Note the greatly enlarged yolk sac of the marsupial that serves the nutritional needs of the developing embryo. In the placental mammal, extensive chorionic villi penetrate the endometrium and provide mechanical strength and increased surface area for rapid and efficient exchange of nutrients, gases, and waste products between the maternal and fetal blood supplies.

tween the maternal and fetal blood supplies. A human placenta is reported to possess about 48 km (30 miles) of villi! Even though the bandicoots (Family Peramelidae) are the only marsupials possessing a chorioallantoic placenta (Dawson 1995), it is not very efficient compared with the placental version due its lack of villi.

The four types of placentas are classified based on the distribution of villi permeating the endometrial lining of the uterus (Ramsey 1982; figures 9.12 and 9.13). Lemurs, non-ruminating artiodactyls, pangolins, and perissodactyls have a placenta that is **diffuse,** characterized by the villi scattered evenly throughout the uterus. A **cotyledonary** placenta, found in ruminating artiodactyls, is characterized by more or less evenly spaced patches of villi scattered within the uterus. Carnivores have a **zonary** placenta, characterized by a continuous band of villi within the uterus. Many groups, including insectivores, bats, most primates (including humans), some rodents, rabbits and hares, and pikas, have a placenta referred to as **discoidal,** in which villi occupy one or two disc-shaped areas within the uterus. A variation on the discoidal arrangement is the **cup-shaped discoid** found in most rodents.

We also categorize mammalian placentas by assessing the speed and efficiency with which nutrients, oxygen, and

generally recognized (figure 9.14):

1. An **epitheliochorial** system is typified by having six tissue layers, with the villi resting in pockets in the endometrium. This system occurs in pigs, lemurs, horses, and whales.

2. The **syndesmochorial** system has one less layer because the epithelium of the uterus erodes at the site of attachment, thus reducing the separation of the maternal and fetal bloodstreams. This system is found in ruminating artiodactyls, such as cows, sheep, and goats.

3. Even further erosion of the maternal tissue is exemplified by the **endotheliochorial** system of the carnivores. Here, the chorion of the fetus is in direct contact with the maternal capillaries.

4. In the **hemochorial** system seen in advanced primates (including humans), insectivores, bats, and some rodents, no maternal epithelium is present and the villi are in direct contact with the maternal blood supply.

5. In an **endothelioendothelial** system, maternal and fetal capillaries are next to each other, with no connective tissue between them.

6. The greatest destruction of placental tissues and least separation of fetal and maternal bloodstreams occurs in the **hemoendothelial** system in which the fetal capillaries are literally bathed in the maternal blood supply. This system occurs in some rodents, rabbits, and hares (Austin and Short 1972b; Gunderson 1976; Ramsey 1982; Martensson 1984; Pough et al. 1996).

Many of us have observed the birth of pets or livestock (some of us may even have witnessed the birth of children) and noticed the appearance of the "afterbirth," which is the expelled placenta and extraembryonic membranes. Sometimes the afterbirth is expelled either with the young or shortly thereafter. Mammals such as pigs, lemurs, horses, and whales possess the epitheliochorial type of placenta, which provides the least intimacy between maternal and fetal membranes. In such cases, the placentas are termed **nondeciduous,** and due to the minimal attachment of villi in the uterine wall, they can

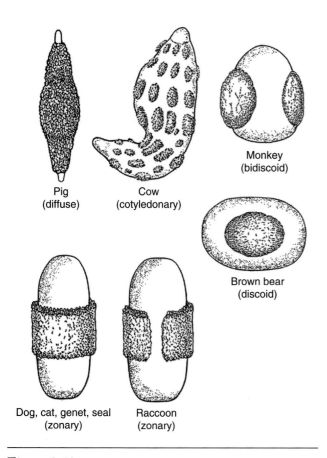

Figure 9.12   Placental villi. The shape and distribution of placental villi vary among different groups of mammals.

waste material are exchanged between maternal and fetal blood supplies (see figure 9.13). Recall that as the blastocyst sinks into the endometrium and implantation proceeds, the chorionic villi grow farther into the endometrium. The uterus becomes greatly vascularized in the area of the implantation. Such an arrangement increases the surface area, allowing a more rapid interchange between the maternal and fetal circulation. Exchange rates depend on the number and thickness of capillary walls, connective tissue, and uterine and chorionic (fetal) epithelial tissues. Six arrangements are

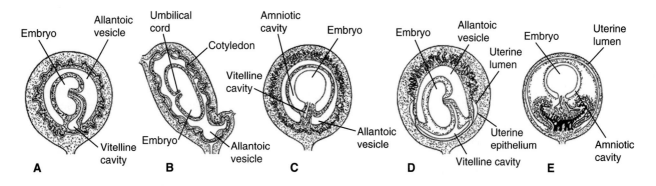

Figure 9.13   Types of placentae. Sectional views of different types of mammalian placentae: (A) diffuse, (B) cotyledonary, (C) zonary, (D) discoid, and (E) cup-shaped discoid. Views (A),(B), and (D) are sagittal and views (C) and (E) are transverse sections.

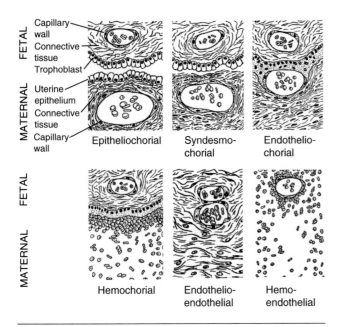

Figure 9.14   **Internal structure of the placenta.**   Diagram showing variation in the number of tissue layers separating maternal and fetal blood supplies in different systems of placental mammals. The number varies from one layer in a hemoendothelial system to six layers in an epitheliochorial system.

pull away from the wall of the endometrium at birth without causing any bleeding. Mammals such as xenarthrans, rodents, carnivores, and primates have more intimate associations between the endometrial and fetal bloodstreams. Such placentas are termed **deciduous** and are typified by the hemoendothelial system. As a result of this close attachment, extensive erosion of the uterine tissue takes place at parturition. A portion of the uterine wall is actually torn away when the placenta separates at delivery. Bleeding occurs because some blood vessels and connective tissue of the uterine wall are shed with the placenta. Hemorrhaging following birth soon stops due to the collapse of the uterus by contractions of the myometrium (the smooth muscle layer of the uterus), which tends to constrict the blood vessels. The previous site of attachment of the fetus on the uterine wall is evidenced by a pigmented area called a **placental scar.** Placental scars form in rodents, shrews, carnivores, bats, rabbits, hares, and pikas, for example, and are useful in studying an individual's reproductive history (Lidicker 1973). Females that are **nulliparous** (never having been pregnant) show no evidence of placental scars, whereas a female that is termed **parous** has scars indicating prior parturition. A **multiparous** individual has placental scars of different ages, thus indicating that several fetuses or litters were produced.

The Trophoblast.   The reproductive patterns of marsupials and placentals are very different. Marsupials bear embryonic young after a brief gestation, whereas placentals produce precocial young following a comparatively long period of gestation. One explanation for this dichotomy is the evolution of the trophoblast and its importance in preventing immuno-

rejection between the fetus and mother (Lillegraven 1975, 1987; Luckett 1975; Eisenberg 1981; Vaughan 1986). Recall that the **trophoblast** (see figures 9.10 and 9.14), the embryonic contribution to the placenta, greatly enhances fetal development by efficiently transferring nutrients and dissolved gasses between the fetal and maternal circulations. The trophoblast also prevents the immunorejection response from developing between the fetal and maternal circulatory systems. During early gestation in marsupials, membranes of the eggshell act as a barrier between antigen-bearing parts of the embryo and the uterine fluid, thus preventing immunorejection. Late in gestation, however, shell membranes are shed, leaving the embryo and maternal systems vulnerable to immunological attack. To circumvent rejection, marsupials have evolved abbreviated gestation periods. In contrast, although placentals lack shell membranes, early stages of the zygote are protected from mixing with maternal blood by the **zona pellucida**—a noncellular layer surrounding the zygote. Later in development, immunorejection is prevented by the trophoblast and the decidua. Throughout gestation, a close association of uterine and fetal tissues is maintained, yet the trophoblast provides a constant "line of defense" against the mixing of fetal and maternal tissues. As a result, the trophoblastic tissues have permitted placentals to evolve longer periods of gestation, thus giving rise to the production of more precocial young and a decrease in the length of the energetically expensive period of lactation. The "appearance" of the trophoblast is of central importance in favoring the evolution of structural diversity and adaptive radiation of the placental mammals.

GESTATION AND PARTURITION

The period of time from fertilization until the birth of the young is referred to as **gestation.** This period of intrauterine growth also varies greatly within mammals. The prototherian mammals (monotremes) lay eggs (oviparous) that hatch outside the body, so, by definition, gestation does not even occur in monotremes. The therian mammals (Infraclasses Metatheria and Eutheria), however, bear live young (viviparous) and retain the embryo within the body. They therefore have a gestation period.

Marsupials bear altricial young following a brief gestation period, and lactation is protracted compared with similarly sized placental mammals (figure 9.15). Virginia opossums (*Didelphis virginiana*) may be familiar to many North Americans. This marsupial has a very short gestation—only about 12.5 days. Following this brief period of intrauterine development, from 4 to 25 embryolike young (the size of honey bees) emerge from the birth canal. The young are altricial, and each weighs only about 0.13 g (0.005 oz)—an entire litter can fit into a teaspoon! Although altricial, the young have well-developed, muscular forelegs with sharp claws and can move hand over hand, up the hair of the mother's belly and into her pouch—traveling up to 50 mm (2 in.) in about 16.5 seconds. After reaching the

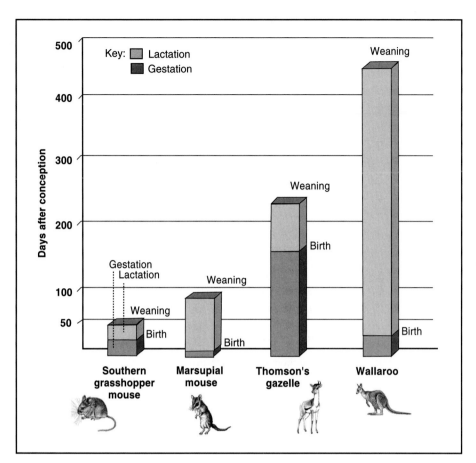

Figure 9.15 **Gestation and lactation.** Diagram comparing the periods of gestation and lactation of marsupial and placental mammals. Note that marsupials have a shorter interval of gestation and longer period of lactation than similarly sized species of placental mammals.

Source: Data from Hickman et al., Integrated Principles of Zoology, *10th ed., 1997, McGraw-Hill Company, Dubuque, IA.*

Body mass is not always the best predictor of gestation period. The blue whale (*Balaenoptera musculus*) is the largest animal that has ever lived. So it might be expected to have the longest gestation period. This huge mammal (about 130,000 kg; 143 tons) has a gestation period of only 10 to 12 months, however, similar to that of a horse. The xenarthrans (anteaters, sloths, and armadillos) are known to have longer gestation periods than predicted by body mass. The two-toed sloth (*Choloepus didactylus*) of the Neotropical region weighs about 9 kg (20 lb) but has a gestation period of almost 11 months. This is about 4 months longer than the wapiti or elk (*Cervus elaphus*) that weighs 22 times more (Vaughan 1986; Hayssen et al. 1993). Other groups of mammals that do not exhibit longer gestation periods as body mass increases are the marsupials, whales and dolphins, higher primates, elephant shrews, and hystricomorph rodents (porcupines, guinea pigs, capybara, agoutis, pacas, cane rats and allies; Sadleir 1973).

Primates tend to exhibit long gestation periods compared with other mammals of similar body size. Some

mother's pouch, each pup takes hold of one of the 13 available nipples (figure. 9.16), which then enlarges, forming a bulb within the mouth of the suckling young. They remain attached to the nipple until about 2 months later, when they are weaned. Although as many as 25 young are born, the entire litter cannot survive. Pouched young average about eight in number and, following weaning, ride on the mother's back for another month (McManus 1974).

The length of gestation, which varies according to taxonomic group, is correlated with body mass, the number of young per litter, and the degree of development of the young. A prime example of a mammal with a large body mass and the record holder for the longest (about 22 months) gestation period is the African elephant (*Loxodonta africana*). Although many rodents have very short gestation periods, the short-nosed bandicoot (Genus *Isoodon*) probably holds the record with a gestation of about 12.5 days. Bats are rather unusual among small mammals in displaying a comparatively long gestation period. Pregnancy generally lasts from 3 to 6 months, and most bats have only one young per litter. Some species, however, such as the red bat (*Lasiurus borealis*), produce litters of up to five young (Hayssen et al. 1993).

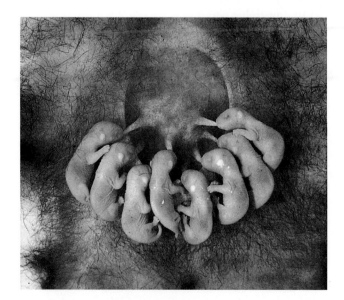

Figure 9.16 **Young Virginia opossums.** Photograph showing eight,week-old Virginia opossums attached to the teats within the pouch of the mother.

of the smaller primates, such as the lemurs, lorises, and marmosets, have gestation periods ranging from 60 to 200 days. Larger primates, such as gorillas, have gestation periods lasting about as long as that of humans (280 days). Newborn primates are precocial but require long periods of maternal care and lactation. These extended periods of caregiving appear to be important in the evolution of socialization in higher primates. In addition to body mass, environmental factors such as latitude, ecosystem, and length of growing season influence the length of gestation period and litter size in mammals.

REPRODUCTIVE VARIATIONS

For most mammals, fertilization occurs within several hours of insemination. The fertilized egg becomes implanted in the uterus, and development continues until birth. Thus, we define the gestation period as the interval between fertilization and parturition. Mammals may exhibit modifications that lengthen the fertilization or gestation period, including delayed fertilization, delayed development, delayed implantation, and embryonic diapause (Weir and Rowlands 1973; Gaisler 1979; Nowak 1994).

Delayed Fertilization

Many animals residing in seasonal environments employ special mechanisms to optimize survival. Bats of the Families Vespertilionidae and Rhinolophidae occupying temperate zones of the New and Old Worlds exhibit a reproductive tactic called **delayed fertilization** or **delayed ovulation** (Wimsatt 1945; Austin and Short 1972a; Racey 1982; Nowak 1994; figure 9.7). Here, copulation takes place in September or October before hibernation commences. Follicular growth has occurred in the ovary, but ovulation does not happen at this time. The sperm are immotile and stored in either the uterus or the upper vagina, and then both sexes enter hibernation. When the female emerges from hibernation in the spring, the eggs are ovulated, spermatozoa become motile, and fertilization takes place. It is noteworthy that some bats do copulate during the period of hibernation, especially species in the Genera *Myotis* and *Plecotus*. Implantation of the blastocyst occurs about a week after fertilization (Kunz 1982). The gestation period may last for 50 to 60 days, as in the silver-haired bat (*Lasionycteris noctivagans*) and little brown bat (*Myotis lucifugus;* Wimsatt 1945; Kunz 1982; Hayssen et al. 1993). Young are born in early summer, when insects are abundant. Delayed fertilization is advantageous for northern species of bats because it provides young with more time to build critical body mass in order to sustain their long period of hibernation.

Delayed fertilization is most common in temperate species of bats that undergo hibernation; however, mammalogists have found sperm storage in bats residing in tropical regions that apparently do not undergo true hibernation. Tropical species may store sperm as an adaptive strategy synchronized with the availability of food (Racey 1982).

Delayed Development

Delayed development differs from delayed fertilization in that the blastocyst implants shortly after fertilization, but development is very slow. A fertilized blastocyst implants in the uterus in summer but may have a 7-month period of development or gestation. Delayed development is reported for both micro- and megachiropterans. *Artibeus jamaicensis* employs delayed development as a unique mechanism to synchronize the birth of young with the end of the dry season, when the availablity of large fruits is at its peak (Fleming 1971; Racey 1982). The Old World insectivorous bat *Miniopterus australis* also exhibits delayed development (Richardson 1977). The reproductive delay in this species is reportedly in response to unpredictable availability of insects. Other species of bats reported to undergo delayed development are *Macrotus californicus* (Bradshaw 1961, 1962); *Haplonycteris fischeri* (Heideman 1988); and *Rhinolophus rouxii, Cynopterus sphinx, Eptesicus furinalis,* and *Myotis albescens* (Racey 1982).

Delayed Implantation

In **delayed implantation,** ovulation, copulation, fertilization, and early cleavage of the zygote up to the blastocyst stage occur normally (see figure 9.7). Development of the blastocyst is arrested, however, and each blastocyst floats freely in suspended animation in the reproductive tract until environmental conditions become favorable for implantation. Unlike delayed development, the blastocyst does not implant into the uterine wall, but rather floats in the reproductive tract. During this free-floating stage, the blastocyst is encased in a protective coat (called the zona pellucida) until the optimal time for its development. Eventually implantation occurs, and development proceeds normally (Enders 1963). We can think of delayed implantation as either **obligate,** as in armadillos, in which the delay occurs as a normal, consistent part of the reproductive cycle, or **facultative,** as when rodents or insectivores are nursing a large litter or are faced with extreme environmental conditions.

Although delayed fertilization and delayed development occur only in bats, delayed implantation is seen in many diverse groups, including insectivores, rodents, bears, mustelids (weasels and allies), seals, armadillos, certain bats, and two species of roe deer (Genus *Capreolus*). An excellent review of delayed implantation in carnivores is provided by Mead (1989). Delayed implantation occurs in the Old World vespertilionids, *Miniopterus schreibersi* and *M. fraterculus* (Kimura and Uchida 1983). The length of gestation in *M. schreibersi* varies geographically, and differences may in part reflect local environmental conditions (Racey 1982). For many species of bats, timing of implantation may be contingent on endocrine controls for different populations (see Nowak 1994).

Delayed implantation probably was first reported in the mid-1600s, as described in the "field notes" of William Harvey who joined King Charles I of England on hunting trips for European roe deer (see Gunderson [1976], for a discussion of the historical account). Many different mammals employ delayed implantation, and this reproductive strategy varies widely even within the same genus (Sandell 1984; King 1990). Siberian and European roe deer (*Capreolus capreolus* and *C. pygargus*) are the only ungulates that exhibit delayed implantation. Following the rut in July and August, implantation is delayed until December or January, and the young are born in the spring (April to June). The gestation period is between 264 and 318 days (Hayssen et al. 1993; Danilkin 1995; Sempéré et al. 1996). In North America, the black bear (*Ursus americanus*) exhibits delayed implantation. In Pennsylvania, for example, black bears mate during the summer, usually sometime from early June to mid-July. After the ova are fertilized, implantation of the blastocyst in the uterine wall is delayed for up to 5 months. Young are born in winter dens in mid-January, following a gestation period of about 7 months (Pelton 1982; Hayssen et al. 1993). Mustelids also represent excellent examples of delayed implantation (King 1983; Mead 1989). Although great variation occurs among different species of mustelids, the western spotted skunk (*Spilogale gracilis*) seems to represent a fairly typical example of delayed implantation (Mead 1968). Female spotted skunks enter estrus in September and mate in September or October. The zygotes undergo normal cleavage but stop at the blastocyst stage, at which time they float freely in the uterus for about 6.5 months. After implantation, gestation lasts only about a month, and young are usually born between April and June. The total period of gestation takes 210 to 260 days. In the same family with the spotted skunk is the fisher (*Martes pennanti*). Of all mammals, this mustelid has the most protracted delay to implantation. Fishers breed between March and April. Following a 9-month delay, blastocysts implant in January or February, and most births occur in March and early April (Wright and Coulter 1967; Powell 1993; Frost et al. 1997). The total pregnancy lasts between 327 and 358 days, about the same as that of the blue whale (Hayssen et al. 1993).

The adaptive advantage of delayed implantation is poorly understood (King 1984). Two closely related species of weasels residing in the same habitat may exhibit very different reproductive patterns. For example, in eastern North America, the long-tailed weasel (*Mustela frenata*) and the least weasel (*Mustela nivalis*) may occupy the same habitat, but the former demonstrates delayed implantation and the latter does not. Many mammalogists believe that by delaying the arrival of the young to spring, when hunting is easiest because small mammals are plentiful, the long-tailed weasel optimizes the likelihood of survival of its young. But why does the smaller least weasel not show such a delay?

The nine-banded armadillo (*Dasypus novemcinctus*) of the New World is another interesting example of variation in reproductive biology. Armadillos in the southern United States breed in July or August. Their copulatory position is rather unusual for a quadruped mammal—the female assumes a mating position on her back. Implantation is delayed until November, followed by a gestation period of about 4 months. In late February, the litter is born and comprises identical quadruplets, all of one sex—they all come from a single fertilized ovum—a phenomenon called **monozygotic polyembryony.** Because only one ovum was fertilized by a single sperm, all the young have exactly the same genetic makeup. We believe that delayed implantation serves to time the birth of the litter for the spring flush of invertebrate food—the staple in the diet of armadillos (Lowery 1974).

Embryonic Diapause

Some kangaroos and other macropodids exhibit another variation on the typical pattern of spontaneous ovulation—**embryonic diapause** (Dawson 1983, 1995). Red kangaroos (*Macropus rufus*) are one of the best examples of embryonic diapause (figure 9.17). The first pregnancy of the season is followed by a gestation period of about 33 days. The single young, called a joey, is born in a very altricial state yet makes its way to the pouch and attaches to a nipple. Under favorable conditions, the female mates again when the pouched young is about 6 months of age. The presence of the suckling young causes the development of the new embryo to be arrested between the 70- and 100-cell stage. This period is referred to as embryonic diapause and may last about 235 days during which time the first joey is growing in the pouch (Dawson 1995). At the end of the 235-day, diapause, the arrested embryo resumes development. The second young is born immediately after the previous young finally exits the pouch, at about 10 months of age, or if the pouched young is prematurely lost. At this point, the mother may again mate and become pregnant for the third time, but because of the suckling joey (young number 2), the development of the new embryo is arrested. While this is happening, the first joey is making periodic trips to and from the pouch yet still suckling. At this point, *three* young are dependent on the mother: (1) one "weaned" joey on foot that still occasionally needs milk to supplement its new diet

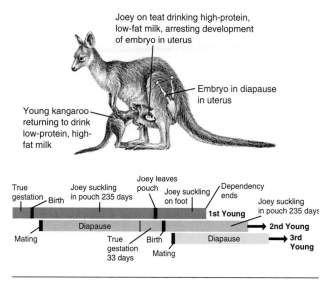

Figure 9.17   Embryonic diapause.   Illustration showing embryonic diapause in the red kangaroo. The complex reproductive pattern of many kangaroos may result in having three young in different stages of development dependent on the mother at one time.

of vegetation; (2) a nursing, pouched joey; and (3) an embryo "on hold." If the first or second joey dies or is removed from the pouch, normal embryonic development (of the third young) resumes almost immediately so that a new individual is born about a month later (Enders 1963; Dawson 1995). The suite of reproductive interactions outlined here is induced by the joey, whose suckling produces nervous stimulation of the hypothalamus. As a result, release of FSH and LH by the pituitary gland is inhibited, thereby reducing follicular development and subsequent ovulation. The production of estrogen and progesterone by the corpus luteum is suppressed by the increase in prolactin.

For macropods (*M. rufus*) residing in harsh environments such as the Australian deserts, embryonic diapause makes a great deal of sense; such opportunistic reproductive behavior has great selective advantage in variable environments. For example, if conditions are favorable, litters can develop in the pouch for long periods before they are weaned. If conditions are poor, as typified by periods of extreme drought and dwindling supplies of green herbage, a female red kangaroo can abandon her young with no delay in producing her next litter because the diapausing blastocysts immediately begin to grow and are born in about a month (Sharman 1963, 1965; Hayssen et al. 1993). Red kangaroos represent a good example of biological control of reproductive output—the suckling joey represents the proximate cause of embryonic diapause, yet the ultimate control is the selective pressures exerted by the environment.

PARTURITION

Before we begin our discussion of parturition; reviewing some terms concerning the developmental stages of young will be helpful. Newborn mammals can be categorized into two groups based on their degree of development. **Altricial** young are born hairless, blind, and essentially helpless. Rabbits, carnivores, and many rodents bear altricial young. **Precocial** young are born fully haired, with eyes open, and are able to get up and walk around shortly after birth. Good examples of precocial young are hares, many large grazing mammals (the ungulates), cetaceans, hyraxes, some rodents, and some primates.

Parturition, or birth, results when the fetus has completed its period of intrauterine growth. The mechanisms that act to terminate gestation and initiate parturition vary among species, are complex, and are not well understood (Gordon 1982; Pough et al. 1996). During the last part of gestation, the adrenal gland of the fetus begins to secrete **adrenocortical hormones,** such as **cortisol,** which initiates parturition. At this time, the placental secretions of estrogen increase and progesterone decreases. As a result of these hormonal changes, the placenta produces hormones called **prostaglandins,** which increase contractions of the uterus. The pressure of the fetus on the cervix then stimulates production of the hormone **oxytocin,** which is produced by the **posterior pituitary gland.** Oxytocin also increases uterine contractions and stimulates milk "letdown." **Relaxin,** a hormone produced by the corpora lutea, softens the ligaments of the pelvis, so it can spread to allow the fetus to pass through the birth canal. After dilation of the vagina, rhythmic contractions of the uterus gradually force the fetus through the vagina to the outside. If the fetal membranes are not ruptured in the birth process, the mother tears them from the young, thus permitting the newborn to breathe. The mother usually severs the umbilical cord and consumes the placenta.

LACTATION

Definition and Physiology

Lactation the production of milk by the mammary glands, is the quintessential feature of Class Mammalia. Probably the most well known characteristic of a mammal is its **mammae** or breasts—the word *mammal* is derived from the Latin word *mammalis,* meaning **breasts.** In females, these specialized skin glands produce milk to nourish the young; in males, they are usually rudimentary and nonfunctional. It is noteworthy, however, that lactation in males has been reported recently in a population of Dayak fruit bats (*Dyacopterus spadiceus*) from Malaysia (Francis et al. 1994).

Although the process is more complicated than we can cover here, lactation, like the estrous cycle, pregnancy, and parturition, is regulated by hormones, as well as the presence of the suckling young. During pregnancy, high levels of the ovarian hormones estradiol and progesterone circulate in the blood and cause enlargement of the mammary glands, making them structurally ready to secrete milk. The actual production of milk does not take place until after parturition. Following parturition and the expulsion of the placenta, the estradiol concentration in the blood decreases. This decrease, in turn, signals the **anterior pituitary gland** to secrete the **lactogenic hormone, prolactin.** Prolactin stimulates milk production but does not by itself cause the delivery, or "letdown," of milk. Suckling by the newborn stimulates nerve receptors in the nipples. This information is transmitted to the hypothalamus and then to the posterior pituitary gland. This gland then releases oxytocin, the same hormone that is associated with uterine contractions during birth. Oxytocin stimulates the alveoli of the breasts to eject milk into the ducts, thus enabling the newborn to remove the milk by suckling. Prolactin is also important in inducing maternal behavior in females and, interestingly, sometimes in males.

Composition of Milk

Milk contains fats, proteins (especially casein), and **lactose** (or milk sugar), as well as vitamins and salts. It provides nutrition for the growth of the newborn, transmits passive immunity, and may support the growth of symbiotic intestinal flora. The first product released by the mammary gland following birth is called **colostrum,** a protein-rich fluid containing antibodies that confer the mother's immunity to various diseases to the young. Other important milk proteins, such as **lysozyme** and **interleukin,** are present in milk throughout lactation. Lysozyme is reported to kill bacteria and fungi, protecting young animals from infection. The composition of milk varies greatly among groups of mammals (Jenness and Studier 1976; Oftedal 1984; Oftendal et al. 1987; Hayssen 1993; Kunz and Sterns 1995). Mammals residing in northern environments have extremely high-fat, high-protein milk. The young of these species must increase weight rapidly to combat the cold after they are weaned and disperse. The milk of whales and seals, for example, may contain 40% to 61% fat and 11% to 12% protein—some milks are 12 times higher in fats and four times richer in protein than that of the domestic cow. For example, the milk of the hooded seal (*Cystophora cristata*) contains about 61% fat and 11% protein, and as a result, pups may gain 20.5 kg in their short 4-day nursing period (Bowen et al. 1985). The extent to which young depend on milk for nutrition also varies greatly. Rodents, such as voles and mice, rely entirely on milk until they are weaned. In contrast, many young ungulates, such as deer (Genus *Odocoileus*), consume grass only a few days after birth, well before weaning occurs. Marsupials, bats, and primates have periods of lactation that are 50% longer than those of other mammals of similar body size (Hayssen et al. 1993). In many species of mammals, the composition of milk changes during the course of lactation. Milk produced during early stages of lactation and as weaning approaches often possesses more protein and less sugar than that delivered to young during the interim stages.

In addition to producing milk with different nutritive values, red kangaroos and wallaroos (Genus *Macropus*) may produce two milks of completely different characteristics at the same time (Dawson 1995). During the earlier discussion of embryonic diapause in red kangaroos, we noted that the mother could simultaneously have two young, widely separated in age, both feeding on her milk. Because each joey has different requirements for nourishment, the mother must produce two different kinds of milk—two active teats are required for the nursing young. Oxytocin, the milk-letdown hormone, is elicited by different suckling stimuli. The first kind of milk is very concentrated and is secreted for the newborn joey that exerts a slight sucking pressure. This youngster requires a great amount of nourishment, but because of its diminutive size, it can only consume a small volume at a time. The second joey is making periodic foraging bouts and requires a more dilute form of milk because it is close to being weaned. This larger joey requires a greater volume of milk and exerts a greater pressure on the teat, thus eliciting a more dilute, less nutritious milk.

All female mammals except the monotremes (platypuses and echidnas) have teats, or nipples, to facilitate the transfer of milk to the young, but their arrangement and number vary greatly. Female echidnas (Genera *Tachyglossus* and *Zaglossus*) typically lay a single egg that is transferred from the cloaca directly into the abdominal pouch, where it is incubated for about 10 days (figure 9.18). Because platypuses (*Orithorhynchus anatinus*) do not have pouches, they keep their eggs (usually

A

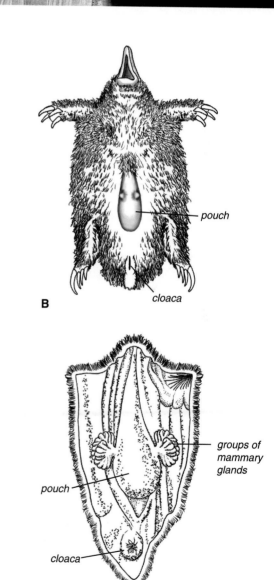

B

C

Figure 9.18   An egg-laying mammal.   Photograph of (A) a young echidna between 2 and 4 months of age. (B) Lower surface of brooding female. (C) Dissection showing a dorsal view of the pouch, mammary glands, and two tufts of hair in the lateral folds of the mammary pouch from which the secretion flows.

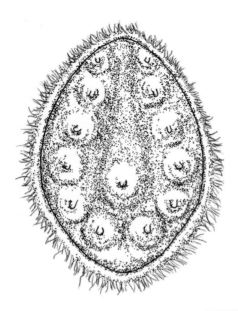

Figure 9.19   **The marsupium.**   The opened pouch of the Virginia opossum, has a horseshoe-shaped arrangement of mammae.

two, each about the size of a robin's egg) in a nest within a burrow, where they are incubated for about 2 weeks. Meanwhile, the embryo is nourished from the yolk. Monotremes lack true nipples, and after "hatching," the young (about the size of a raisin) suck milk that drains from the mother's mammary glands onto tufts of hair located on her abdomen, or in the pouch in the case of the echidna. In monotremes, the milk flows from pores in the skin rather than from nipples.

Marsupials generally have a circular arrangement of nipples in a pouch. The opossum has 13 nipples: 12 arranged in a U-shape, and the thirteenth located in the center (figure 9.19). Newborn marsupials are unable to suckle, so milk is "pumped" to the young by the mother. The numbers of mammary glands or teats also vary among groups of mammals and are usually correlated with the size of the litter. The number of mammary glands ranges from 1 in Australian pygmy possums (*Distoechurus;* Ward 1990) to 29 in the tenrec (*Tenrec ecaudatus*) of Madagascar. This latter species may be the most prolific mammal, with litters of as many as 32 young, but the average litter is between 15 and 20 (Louwman 1973; Eisenberg 1975; Nicoll 1983). Eutherian mammals typically have mammae arranged in two longitudinal ventral rows. Domestic pigs have large litters and, hence, have two rows of teats spanning all the way from near the forelimbs to between the hind limbs. Primates possess a pair of pectoral mammae, and horses have a pair of abdominal teats. The length of lactation varies greatly among mammals, ranging from only 4 days in hooded seals (*C. cristata*) to more than 900 days in orangutans (*Pongo pygmaeus*). The record for "fastest developer" goes to the streaked tenrec (*Hemicentetes semispinosus*), whose young are weaned in about 5 days and begin breeding as early as 3 to 5 weeks of age (Hayssen et al. 1993).

Summary

Mammals reproduce sexually, and fertilization is internal. In the male reproductive system, the testes function to produce sperm (the male gametes) and to synthesize the male sex hormone (testosterone). Testes in mammals are oval-shaped and may or may not be suspended in a pouchlike sac called a scrotum. Sperm are produced in the seminiferous tubules of the testes and travel by a series of ducts. The ducts receive glandular secretions to the spermatozoa to form semen, which passes from the body via a canal (the urethra) into the highly vascular, erectile penis. The female reproductive system consists of a pair of ovaries that produce the ova and the female hormones, estrogen and progesterone. Ovulation occurs when the follicle housing the ovum within the ovary bursts, releasing the egg into the oviduct . The development of follicular cells is stimulated by follicle-stimulating hormone and luteinizing hormone produced by the anterior pituitary gland. Follicular cells secrete estradiol (a form of estrogen) that causes a thickening of the endometrium in preparation for implantation of the egg. Ovulation is induced by high levels of luteinizing hormone. Stimulation of corpora luteum development is accomplished by luteinizing hormone. The corpus luteum, from the follicular cells, produces the hormone progesterone, which combines with estrogens to promote the growth of the uterine lining and makes possible the implantation of the fertilized egg. Progesterone also stimulates mammary gland development and maintenance of pregnancy.

Estrus, or heat, represents the period of receptivity to copulation shortly before and after ovulation. Ovulation may be spontaneous, as in most mammals, or induced, as in rabbits, many carnivores, and some ground squirrels. The female duct system consists of paired oviducts, one or two uteri, one or two cervices, and the vagina. Much variation occurs in this system among mammals. Four principal types of uteri based on the relationship of the uterine horns exist: duplex, bipartite, bicornuate, and simplex. Implantation occurs when the embryo attaches to the endometrium, or the lining of the uterine wall. The highly vascular endometrium passes nutrients and gases to and receives wastes from the extraembryonic tissues called the placenta. The relationship of the yolk sac to the developing embryo is the means of catagorizing all mammalian placentas as either choriovitelline (most marsupials) or chorioallantoic (the bandicoots and all eutherian mammals). Placentas can also be classified by distribution of villi; speed and efficiency of exchange of nutrients, oxygen, and waste materials; and the intimacy of the placenta with the uterine wall.

The period from fertilization until birth of the young is called gestation. The gestation period for marsupials is short and lactation is protracted, whereas the reverse is true for eutherian mammals. Typically, variation in the length of gestation is correlated with body mass, the number of young per litter, and the degree of development of the young. Larger animals tend to have longer periods of gestation. Modifications in the length of fertilization or gestation, however, are exhibited by many mammals. These reproductive variations include delayed fertilization (in bats of temperate zones), delayed development (in some bats in Neotropical areas), delayed implantation (in some insectivores, rodents, bears, weasels and allies, seals, armadillos, and roe deer), and embryonic diapause (in kangaroos and some wallabies). Parturition, or birth, occurs when the fetus has completed its period of intrauterine growth. The degree of development of the newborn may be altricial or precocial. Processes associated with

parturition are influenced by a variety of hormones, including oxytocin and relaxin.

The production of milk to nurse the newborn is called lactation and is aided by the hormones oxytocin and prolactin. Milk is composed of fats, proteins, and milk sugar. The composition of milk varies greatly among mammals, ranging from only 0.2% fat in black rhinos (*Diceros bicornis*) to as high as 61% fat in hooded seals (*C. cristata*). Kangaroo mothers may produce two kinds of milk to nurse two joeys of differing ages. The number of mammae also varies greatly in mammals, ranging from 2 in primates to 29 in the tenrec (*H. semispinosus*). Monotremes do not have true nipples or mammae. Their young suck milk that drains from mammary glands onto tufts of hair on the mother's abdomen.

Discussion Questions

1. Trace the movement of spermatozoa from their origin and development in the seminiferous tubules of the testes to release at the penis.
2. Describe and give the function(s) of the following structures of the female reproductive system: ovaries, uterus, vagina, and mammary glands. How do these structures vary among monotremes, marsupials, and eutherian mammals?
3. Describe and distinguish the patterns in reproduction in monotremes, marsupials, and eutherian mammals. What aspects of reproduction are universal in Class Mammalia but not found in other vertebrates?
4. What reproductive attributes enhance survival of young residing in northern environments. List some such

species, and state how the female makes physiological adjustments to enhance the survival of her young.
5. Delayed implantation is a common reproductive phenomenon for members of the Family Mustelidae (weasels and allies). The occurrence of this reproductive strategy varies widely among closely related species residing in similar environments. How can you explain this disparity, and how does it fit with the theory of natural selection?
6. Many mammals such as the opossum are known to produce more young that can be supported by the number of the mother's mammae. This attribute seems maladaptive. Explain this apparent waste of energy.

Suggested Readings

Altringham, J. D. 1996. Bats: biology and behaviour. Oxford Univ. Press, Oxford.

Bentley, P. J. 1998. Comparative vertebrate endocrinology. Third edition. Cambridge Univ. Press, New York.

Bronson, F. H. 1989. Mammalian reproductive biology. Univ. of Chicago Press, Chicago.

Heideman, P. D., and K. S. Powell. 1998. Age-specific reproductive strategies and delayed embryonic development in an Old World fruit bat, *Ptenochirus jagori*. J. Mammal, 79:295–311.

Hildebrand, M. 1995. Analysis of vertebrate structure, 4th ed. John Wiley & Sons, New York.

Macdonald, D. (ed.). 1984. The encyclopedia of mammals. Facts on File Publ., New York.

Oftedal, O. T. 1985. Pregnancy and lactation. Pp. 215–238, *in* Bioenergetics of wild herbivores (J. R. Hudson and R. G. White, eds.). CRC Press, Boca Raton, FL.

Oftedal, O. T. 1992. Nutrition in relation to season, lactation, and growth of north temperate deer. Pp. 407–417, *in* The biology of deer (R. D. Brown, ed.). Springer-Verlag, New York.

Patterson, B. D., and C. S. J. Thaeler. 1982. The mammalian baculum: hypotheses on the nature of bacular variability. J. Mammal, 63:1–15.

Racey, P. A. and S. M. Swift (eds.). 1995. Ecology, evolution and behaviour of bats. Proceedings of the 67th Symposium of the Zoological Society of London, November 26–27, 1993, Clarendon Press, Oxford.

Romer, A. S., and T. S. Parsons. 1986. The vertebrate body, 6th ed. Saunders College Publ., Philadelphia.

Voelker, W. 1986. The natural history of living mammals. Plexus Publ. Medford, NJ.

Wilson, D. E. (ed). In press. The complete book of North American mammals. Smithsonian Institution Press, Washington, D.C.

PART THREE

Adaptive Radiation and Diversity

Part II detailed the great diversity in the structure and function of mammals. This part, including chapters 10 through 19, explores the life history characteristics, morphology, fossil history, and conservation of living orders and families of mammals. There are only about 4600 extant mammalian species throughout the world—relatively few compared with over 9600 species of birds or 40,000 species of fishes. The number of vertebrate species pales in comparison with the 104,000 species of mollusks or the estimated millions of species of insects.

Structurally and functionally, however, mammals are quite diverse. Consider the difference in size between the smallest and largest living species of birds. The largest, heaviest bird is the ostrich (*Struthio camelus*), which weighs about 135 kg as an adult. The smallest hummingbirds (Family Trochilidae) weigh about 3 g (or 0.003 kg). Although this is a difference of 4.5 orders of magnitude, consider the mass difference between the largest mammal, a blue whale (*Balaenoptera musculus*), at about 160,000 kg (160,000,000 g), and one of the smallest mammals, a pygmy shrew (*Sorex hoyi*), at about 3 g—a range of over seven orders of magnitude! Size, however, is just one aspect of a species that results from numerous selective factors operating through evolutionary time. As you read the chapters in this section, keep in mind not only the obvious differences in structural and functional adaptations within the various mammalian orders but also the many examples of parallel or convergent evolutionary adaptations evident among phylogenetically distant groups. Environmental constraints force many otherwise unrelated taxa to adapt to similar problems in similar ways and, through time, to develop similar morphological features.

River otters (*Lontra* [*Lutra*] *canadensis*) are superbly adapted for semiaquatic habitats.

CLASSIFICATION

The Class Mammalia is divided into two Subclasses of extant taxa, based primarily on reproductive characteristics: the Prototheria and the Theria. Prototherians are a small group of only three living species in the Order Monotremata. Therians make up the vast majority of mammals. They are divided into two Infraclasses: the metatherians and the eutherians. Metatherians are more commonly referred to as marsupials, whereas eutherians are typically called "placental" mammals. The term *placental* should be discouraged for this group as it implies that marsupials have no placenta. Marsupials do have a placenta, although it is not used as an endocrine organ during prolonged gestation, as it is in eutherians. Even the term *eutherian* is somewhat unfortunate because of its subtle implication of "progression" through marsupials to "good" (*eu-*), or more advanced, mammals. For lack of any reasonable alternative, however, we will use the terms *placental* and *eutherian* interchangably. The approximately 270 extant species of marsupials, along with the 3 species of monotremes, make up about 6% of the world's mammalian fauna. The remainder consists of eutherians.

Several questions concerning mammalian classification at the ordinal level and below remain unresolved, but in this text and the following outline, we follow the classification of Wilson and Reeder (1993). Suborders we include that are not used in Wilson and Reeder are given in brackets in the outline. For each family, the number of extant genera and species is given in parentheses. In the text, alternative taxonomic arrangements proposed by other authorities are noted. The conservation status of species considered **endangered** (defined as in danger of extinction if causal factors continue) or **threatened** (i.e., *vulnerable*—likely to move into the endangered category in the near future if causal factors continue), is derived from various regional sources, as well as the *IUCN Red Data Book* (1985).

CLASS MAMMALIA

Subclass Prototheria

Order Monotremata
| Family | Ornithorhynchidae (1, 1) | Duck-billed platypus |
| | Tachyglossidae (2, 2) | Echidnas |

Subclass Theria (Infraclass Metatheria)

Order Didelphimorphia
| Family | Didelphidae (15, 63) | New World opossums |

Order Paucituberculata
| Family | Caenolestidae (3, 5) | Shrew or rat opossums |

Order Microbiotheria
| Family | Microbiotheriidae (1, 1) | Monito del monte |

Order Dasyuromorphia
| Family | Myrmecobiidae (1, 1) | Numbat |
| | Dasyuridae (15, 61) | Marsupial mice, Tasmanian devil |

Order Peramelemorphia
| Family | Peramelidae (3, 8) | Bandicoots and bilbies |
| | Peroryctidae (4, 11) | New Guinea bandicoots |

Order Diprotodontia
Family	Phascolarctidae (1, 1)	Koala
	Vombatidae (2, 3)	Wombats
	Phalangeridae (6, 18)	Cuscuses, brushtail possums
	Potoroidae (4, 7)	Potoroos, bettongs
	Macropodidae (11, 50)	Kangaroos, wallabies, wallaroos, pademelons
	Burramyidae (2, 5)	Pygmy possums
	Acrobatidae (2, 2)	Feathertail glider, possum
	Pseudocheiridae (5, 14)	Ringtailed possums
	Petauridae (3, 15)	Striped possums, wrist-winged gliders
	Tarsipedidae (1, 1)	Honey possum

Order Notoryctemorphia
| Family | Notoryctidae (1, 2) | Marsupial moles |

(Infraclass Eutheria)

Order Insectivora
Family	Solenodontidae (1, 2)	Solenodons
	Tenrecidae (10, 24)	Tenrecs
	Chrysochloridae (7, 18)	Golden moles
	Erinaceidae (7, 21)	Hedgehogs, gymnures
	Soricidae (23, 312)	Shrews
	Talpidae (17, 42)	Moles and desmans

Order Scandentia
| Family | Tupaiidae (5, 19) | Tree shrews |

Order Macroscelidea
| Family | Macroscelididae (4, 15) | Elephant shrews |

Order Dermoptera
| Family | Cynocephalidae (1, 2) | Colugos |

Order Chiroptera

Suborder Megachiroptera
| Family | Pteropodidae (42, 166) | Old World fruit bats |

Suborder Microchiroptera
Family	Rhinopomatidae (1, 3)	Mouse-tailed bats
	Craseonycteridae (1, 1)	Kitti's hog-nosed bat
	Emballonuridae (13, 47)	Sac-winged or sheath-tailed bats
	Nycteridae (1, 12)	Slit-faced bats
	Megadermatidae (4, 5)	False vampire bats
	Rhinolophidae (10, 130)	Horseshoe bats
	Noctilionidae (1, 2)	Fishing bats
	Mormoopidae (2, 8)	Leaf-chinned bats
	Phyllostomidae (49, 141)	New World leaf-nosed bats
	Natalidae (1, 5)	Funnel-eared bats
	Furipteridae (2, 2)	Smoky bats
	Thyropteridae (1, 2)	Disk-winged bats
	Myzopodidae (1, 1)	Sucker-footed bat
	Vespertilionidae (35, 318)	Common bats

Mystacinidae (1, 2)	Short-tailed bats
Molossidae (12, 80)	Free-tailed bats

Order Primates

[Suborder Strepsirhini]

Family Cheirogaleidae (4, 7)	Dwarf and mouse lemurs
Lemuridae (4, 10)	Lemurs
Megaladapidae (1, 7)	Sportive lemurs
Indridae (3, 5)	Indrid lemurs, sifakas
Daubentoniidae (1, 1)	Aye-aye
Loridae (4, 6)	Lorises, potto
Galagonidae (4, 11)	Bushbabies, galagos

[Suborder Haplorhini]

Family Tarsiidae (1, 5)	Tarsiers
Callitrichidae (4, 26)	Marmosets, tamarins
Cebidae (11, 58)	New World monkeys
Cercopithecidae (18, 81)	Old World monkeys
Hylobatidae (1, 11)	Gibbons, siamang
Hominidae (4, 5)	Gorilla, chimpanzees, orangutan, humans

Order Xenarthra

Family Bradypodidae (1, 3)	Three-toed sloths
Megalonychidae (1, 2)	Two-toed sloths
Dasypodidae (8, 20)	Armadillos
Myrmecophagidae (3, 4)	True anteaters

Order Pholidota

Family Manidae (1, 7)	Pangolins

Order Tubulidentata

Family Orycteropodidae (1, 1)	Aardvark

Order Carnivora

Suborder Feliformia

Family Felidae (18, 36)	Lion, tiger, bobcat, cougar
Herpestidae (18, 37)	Mongooses
Hyaenidae (4, 4)	Hyenas
Viverridae (20, 34)	Civets, genets

Suborder Caniformia

Family Canidae (13, 33)	Wolves, coyote, foxes, jackels
Ursidae (6, 19)	Bears
Mustelidae (25,65)	Weasels, skunks, otters
Odobenidae (1, 1)	walrus
Otariidae (7, 14)	Eared seals
Phocidae (10, 19)	Earless seals
Procyonidae (6, 18)	Raccoon, red panda, coati

Order Cetacea

Suborder Mysticeti

Family Balaenidae (2, 3)	Bowhead and right whales
Balaenopteridae (2, 6)	Rorquals
Eschrichtiidae (1, 1)	Gray whale
Neobalaenidae (1, 1)	Pygmy right whale

Suborder Odontoceti

Family Delphinidae (17, 32)	Dolphins
Monodontidae (2, 2)	Narwhal, beluga
Phocoenidae (4, 6)	Porpoises
Physeteridae (2, 3)	Sperm whales
Platanistidae (4, 5)	River dolphins
Ziphiidae (6, 19)	Beaked whales

Order Rodentia

Suborder Sciurognathi

Family Aplodontidae (1, 1)	Mountain beaver
Sciuridae (50, 273)	Squirrels
Castoridae (1, 2)	Beavers
Geomyidae (5, 35)	Pocket gophers
Heteromyidae (6, 59)	Kangaroo rats and mice
Dipodidae (15, 51)	Jerboas, birch mice, jumping mice
Muridae (281, 1326)	Rats and mice
Anomaluridae (3, 7)	Scaly-tailed squirrels
Pedetidae (1, 1)	Springhaas
Ctenodactylidae (4, 5)	Gundis
Myoxidae (8, 26)	Dormice

Suborder Hystricognathi

Family Bathyergidae (5, 12)	Mole rats
Hystricidae (3, 11)	Old World porcupines
Petromuridae (1, 1)	Dassie rat
Thryonomyidae (1, 2)	Cane rats
Erethizontidae (4, 12)	New World porcupines
Chinchillidae (3, 6)	Viscachas, chinchillas
Dinomyidae (1, 1)	Pacarana
Caviidae (5, 14)	Cavies, Patagonian "hare," guinea pigs
Hydrochaeridae (1, 1)	Capybara
Dasyproctidae (2, 13)	Agoutis, acouchis
Agoutidae (1, 2)	Pacas
Ctenomyidae (1, 38)	Tuco-tucos
Octodontidae (6, 9)	Viscacha rats, coruro
Abrocomidae (1, 3)	Chinchilla rats
Echimyidae (16, 71)	Spiny rats
Capromyidae (4, 12)	Hutias
Myocastoridae (1, 1)	Nutria

Order Lagomorpha

Family Ochotonidae (2, 26)	Pikas
Leporidae (11, 54)	Rabbits and hares

Order Proboscidea

Family Elephantidae (2, 2)	Elephants

Order Hyracoidea

Family Procaviidae (3, 11)	Hyraxes

Order Sirenia

Family Dugongidae (1, 1)	Dugong
Trichechidae (1, 3)	Manatees

Order Perissodactyla

Family Equidae (1, 9)	Horses, asses, zebras
Tapiridae (1, 4)	Tapirs
Rhinocerotidae (4, 5)	Rhinoceroses

Order Artiodactyla

[Suborder Suiformes]

Family	Suidae (5, 16)	Pigs, warthogs
	Tayassuidae (3, 3)	Peccaries
	Hippopotamidae (2, 2)	Hippopotamuses

[Suborder Tylopoda]

Family	Camelidae (3, 6)	Camels, llama, guanaco, alpaca

[Suborder Ruminantia]

Family	Tragulidae (3, 4)	Chevrotains
	Giraffidae (2, 2)	Giraffe, okapi
	Moschidae (1, 4)	Musk deer

Cervidae (16, 42)	Deer
Antilocapridae (1, 1)	Pronghorn antelope
Bovidae (45, 137)	Antelope, sheep, goats, bison, cattle

Total: Orders = 26; Families = 133; Genera = 1117; Species = 4604.

Totals in Wilson and Reeder (1993) include some recently extinct taxa not included in this listing.

Eleven orders have a single family, and 20 families have a single species. In contrast, there are 1326 species in the rodent Family Muridae and 318 species of bats in the Family Vespertilionidae.

CHAPTER 10

Monotremes and Marsupials

Both monotremes and marsupials may be distinguished from eutherian mammals (or "placentals") on the basis of reproductive characteristics. Monotremes are the only mammals that are **oviparous,** meaning that like birds and some other vertebrates, they lay eggs. Marsupials give birth to live young (they are **viviparous**) but are characterized by a very brief intrauterine gestation period. This results in reduced parental "investment" in terms of energy expenditure in **neonates** (newborn) at birth. In this chapter, we explore various aspects of form and function in these two lineages of noneutherian mammals.

MONOTREMATA

As noted in chapter 4, mammals evolved from reptilian ancestors over a long period. Monotremes (Subclass Prototheria) differ significantly from marsupials and eutherians (Subclass Theria) in their retention of various reptilian features. Extant monotremes are represented by two families. The Ornithorhynchidae, which includes the duck-billed platypus (*Ornithorhynchus anatinus*), is **monotypic** (a group that includes a single taxon). There are two species within the Tachyglossidae: the short-billed echidna (*Tachyglossus aculeatus*) and long-billed echidna (*Zaglossus bruijni*). On an evolutionary time scale, the monotremes probably owe their continued survival to relative geographic isolation from eutherians. They may have survived because they occupy ecological niches with little or no competition.

The ordinal name Monotremata (meaning "one opening") refers to the cloaca, a common opening for the fecal, urinary, and reproductive tracts. The most notable reptilian features of monotremes are the structure of their reproductive tract and the fact that young hatch from small, somewhat rubbery-shelled eggs. The female reproductive tract includes separate uteri with large ovaries that produce large follicles during the breeding season (figure 10.1). The uteri have separate openings into the urogenital sinus that lead to the cloaca. After an egg is shed into the infundibulum, it passes to the Fallopian tube, where fertilization occurs. The shell is deposited in the oviduct over a period of about 2 weeks and is composed of three identifiable layers (see Griffiths 1978, 1989 for details). Nutrient material is absorbed through the shell; thus, the eggs are permeable (not **cleidoic,** or impermeable, as is the case for birds). The early cleavage stages of the egg are **meroblastic,** that is, restricted to the anterior end (as opposed to holoblastic, in which cleavage occurs throughout). The first cleavage furrow divides the germinal disk into two areas of unequal size. The second cleavage is perpendicular, so that at the four-cell stage, two large cells and two small cells **(blastomeres)** occur on the top of the yolk. Meroblastic eggs also occur in reptiles and birds.

Monotreme eggs are small, about 16 mm long and 14 mm wide. They are incubated for 10 to 11 days. During the final stages of incubation, a sharp egg tooth forms at the end of the snout of a developing young, which it uses to free itself from the egg, again as is true in birds and reptiles. Neonatal monotremes are similar structurally to neonatal marsupials (Griffiths 1989). Development in both is rudimentary and the forelimbs and shoulder muscles are well developed. After the eggs hatch, as in all other mammals, the young are nursed. Monotremes have mammary glands, but unlike other mammals, they have no teats. Milk is secreted from pores on the belly of the platypus and from

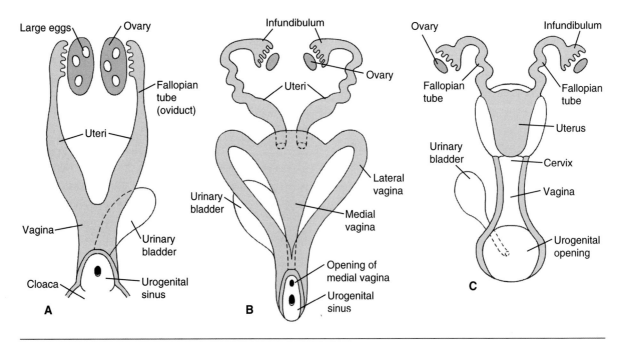

Figure 10.1 **Female reproductive tracts.** The structure of female reproductive tracts varies in the three groups of mammals: (A) prototherians (monotremes); (B) metatherians (marsupials); and (C) eutherians ("placentals").

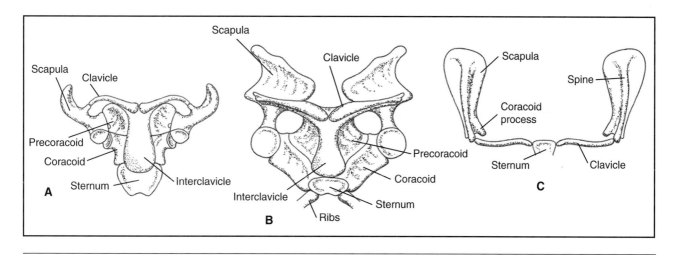

Figure 10.2   **Reptilelike pectoral girdle in monotremes.**   Comparison of the pectoral girdles in a (A) therapsid reptile, (B) short-beaked echidna, and (C) muskrat, a eutherian mammal. Monotremes have pectoral girdles very similar to ancient reptiles.

paired glandular lobes in the pouch of echidnas. The structure of the mammary glands is identical in monotremes and marsupials (Griffiths et al. 1973), and the process of lactation in monotremes is as "sophisticated and highly evolved" (Griffiths 1989) as in any mammal.

Besides laying eggs, monotremes exhibit several other reptilian features. The pectoral girdle (figure 10.2) has a coracoid, precoracoid, and interclavicle bone, as in primitive therapsid reptiles. Although monotremes are homeotherms, their body temperature is lower than that of therians—about 32°C. The chromosomes are unique among mammals in their mix of normal macrochromosomes and reptilelike microchromosomes (Bick and Jackson 1967) and in their meiotic division (Murtagh 1977). Sperm are **filiform** (threadlike) and reptilian in stucture, as is the anatomy of the testes (Carrick and Hughes 1978).

Morphology

Although they are diverse in their outward appearance, the two monotreme families share several characteristics. The cranial features of monotremes are unique, with adults having indistinct sutures. Unlike therian mammals, the jugal bone is reduced or absent. The zygomatic arch is made up of the maxilla and squamosal bones. The dentary bone is greatly reduced, and adults are edentate (without teeth). With their elongated rostrum, lack of teeth, and high-domed cranium, monotreme skulls appear birdlike (figure 10.3). The **cochlea** (semicircular canals of the inner ear) of monotremes are also unique among mammals because they are not coiled.

Monotremes have several other distinctive features. As in marsupials, **epipubic bones** (attached to the pelvic girdle and projecting forward) occur in both sexes. Adult males have a large, hornlike medial spur on the ankle. In the platypus, this spur connects to a poison gland (see Family Ornithorhynchidae). Details of the skull and postcranial

skeleton of the platypus and echidnas can be found in Grant (1989) and Griffiths (1989), respectively. Monotremes, like marsupials, have no corpus callosum, a bundle of nerve fibers that integrate the two hemispheres of the brain. Males have a baculum, permanently abdominal testes, and no scrotum.

Fossil History

The ancestry and relationship of prototherians with therians remain unresolved. Prototherians may have diverged from therians before the end of the Triassic period, over 200 million years ago (Carroll 1988; O'Brien and Graves 1990), but this is still debated. Monotremes are of Mesozoic origin (Woodburne and Case 1996). Archer and colleagues (1985) described a monotreme from the Cretaceous period of Australia. Fossil evidence of Australian ornithorhynchids extends from the Cretaceous and early Tertiary periods (Woodburne and Tedford 1975; Archer et al. 1978, 1992). The earliest tachyglossid, *Megalibgwilia ramsayi*, is from the early Miocene epoch of Australia (Griffiths et al. 1991).

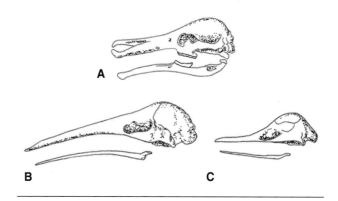

Figure 10.3   **Monotreme skull morphology.**
Skulls of the (A) duck-billed platypus; (B) long-beaked echidna, and (C) short-beaked echidna. All monotremes are edentate, although note the keratinized gum pads in the platypus. (Not to same scale.)

The first evidence of a non-Australian fossil monotreme was a platypus from the early Paleocene epoch of southern Argentina (Pascual et al. 1992).

Economics and Conservation

Although certainly of intrinsic interest, monotremes have no economic importance. In the early days of Australian colonization, the platypus was hunted for its fur, but this was never a major industry. Most populations today are considered secure. The long-beaked echidna once enjoyed a much more widespread range than it does today and is considered threatened.

Families

Ornithorhynchidae

The semiaquatic, semifossorial platypus is the sole extant ornithorhynchid. It occurs near freshwater lakes and rivers at both high and low elevations along the east coast of Australia and throughout Tasmania, where it feeds on a variety of invertebrates, small fish, and amphibians. The platypus's physical appearance is so remarkable among mammals (figure 10.4) that the first specimen brought to London in 1798 was believed to be a hoax. Adult males average 50 cm in total length and 1700 g in body mass; females are smaller. Short, dense fur covers all but the bill, feet, and underside of the tail. The bill is distinctive and quite unlike that of a true duck. It is soft, pliable, very sensitive, and has nostrils at the tip. The bill is highly innervated both for tactile reception and to sense electric fields generated by the muscle contractions of prey (Scheich et al. 1986; Proske et al. 1993; Manger and Pettigrew 1995). The small eyes and ears are situated in a groove extending from the bill. During a dive, this groove closes and the platypus relies on the sensitivity of the bill to locate prey.

Feet of the platypus are **pentadactyl** (five-toed), and the **manus** (forefoot) is webbed. Webbing is folded back when the platypus is on land. The long, sharp claws are used for burrowing. The spur on the hind limb connects to a venom gland (figure 10.5). The function of the spur is not entirely clear, but it may involve competition between

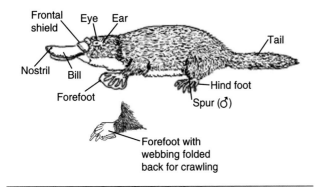

Figure 10.4 **External features of the duck-billed platypus.** The external anatomy of the platypus is unique.

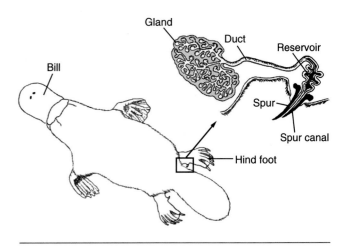

Figure 10.5 **Platypus venom gland.** Anatomy of the venom gland associated with the spur on the hind limb of the male platypus.

males during the breeding season. Unlike echidnas, platypus neonates have three molariform teeth in each quadrant. These are shed before the young emerge from the burrow. Continuously growing, keratinized pads in the gums take the place of teeth in adults. These pads grind food before it is swallowed.

Burrows are built in stream banks and are generally simple, although those constructed by females for nursing are 20 to 30 m long, with a nest chamber at the end. Unlike echidnas, the platypus has no pouch. Females curl their bodies around the eggs to incubate them; average litter size is two. As with birds, only the left ovary is functional. The young are not carried after hatching and remain in the burrow during the 4- to 5-month nursing period. Milk is secreted from the mammary glands onto distinct, protruding tufts of fur.

Tachyglossidae

The short-beaked echidna occurs throughout Australia, New Guinea, and Tasmania in forest, scrub, and desert habitats. The long-beaked echidna is restricted to forested highland areas of New Guinea. Maximum body mass is about 6 kg in *Tachyglossus* and 10 kg in *Zaglossus*. Both species are characterized by a very long rostrum (figure 10.6) and long, extensible tongue. Enlarged submaxillary salivary glands produce a mucus that coats the tongue and makes it sticky. Guard hairs on the back and sides of the body are modified to form barbless spines up to 6 cm long. *Zaglossus bruijni* has much thicker hair and fewer spines than *T. aculeatus*. Large, scooplike claws on the feet enable echidnas to break into anthills and to burrow rapidly and powerfully. They are active anytime during the day or night but avoid extremes of temperature. Using its sticky tongue, *Tachyglossus* (meaning "rapid tongue") consumes ants, termites, and other insects, which are ground to a paste between the tongue and spiny palatal ridges. *Zaglossus* (meaning "long tongue") feeds primarily on earthworms.

A

B

Figure 10.6   **Echidnas, or spiny anteaters.**   The beak is up to two-thirds the length of the head in the (A) long-beaked echidna (*Zaglossus bruijni*) and much shorter in the (B) short-beaked echidna (*Tachyglossus aculeatus*). Also note the greater amount of spines in *Tachyglossus*.

Unlike the platypus, the ankle spurs on male echidnas do not function, and echidnas have no teeth at any stage of development. Also, they have a pouch in which the eggs are incubated and hatched, again unlike the platypus. The mammary glands converge to two small areas in the pouch (see figure 9.18), the milk areolae, where neonates cling to surrounding fur to nurse. Young remain in the pouch about 2 months until their spines begin to develop.

MARSUPIALS

Marsupials (or metatherians) are often characterized by the female's abdominal pouch, or marsupium, which gives rise to the common name of this group. This is a poor diagnostic feature, however, because not all marsupials have a marsupium, and as we have seen, a pouch occurs in echidnas. Marsupials are best distinguished from eutherians on the basis of their reproductive mode, specifically, the relatively small maternal energy investment in young prior to birth. In fact, no marsupials have litters that weigh more than 1% of the mother's body mass (Russell 1982). In contrast, small eutherians, such as rodents or insectivores, may have litters that

weigh 50% of the mother's body mass. Maternal investment in lactation is much greater in marsupials (figure 10.7), however, so by the time young are weaned, total investment in a litter may be similar between marsupials and eutherians. In addition to reproductive characteristics, marsupials differ from eutherians in many skeletal and anatomical features (table 10.1). The two groups also have different dental characteristics. Unlike eutherians, marsupials characteristically have a well-developed "stylar shelf" on the upper molars, as well as a "twinned" hypoconulid and entoconid in the lower molars (figure 10.8).

As noted by Lee and Cockburn (1985), several life history characteristics of marsupials differ from those of eutherians, at least in matter of degree. Different characteristics should not be viewed as "shortcomings," however, but simply as different adaptive strategies. In each case, marsupials are consistently more conservative (less diverse) in their adaptive radiation than are eutherians. For example, marsupials generally have lower basal metabolic rates—about 70% of comparably sized eutherians. Except for bandicoots, they also have slower postnatal growth. Relative brain size also is smaller in marsupials, especially when considering large species, as is the range of body size. For example, the difference in body mass between the smallest living marsupial species, the Pilbara ningaui (*Ningaui timealeyi*—adult body mass, 2–9 g), and the largest, the red kangaroo (*Macropus rufus*—average mass of males, 66 kg), is about four orders of magnitude. This is much less pronounced than the extremes noted in eutherians; that is, seven orders of magnitude difference between pygmy shrews and blue whales. There are no marine marsupials, however, and therefore no parallel adaptation to that of cetaceans, so this type of comparison is questionable. Nonetheless, it illustrates the point of reduced morphological diversity in marsupials. Neither have marsupials developed true flight, like the bats (see the following

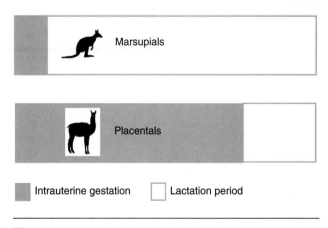

Figure 10.7   **Lactation and gestation in therians.**   Relative lengths of gestation and lactation differ significantly in marsupials and eutherians. The majority of energy expenditure in marsupials occurs during lactation.

Source: Data from M. Archer and G. Clayton, Vertebrate Zoogeography and Evolution in Australia, *1984, Hesperian Press, Carlisle, Western Australia.*

Table 10.1.   General skeletal and anatomical differences between metatherian (marsupial) and eutherian ("placental") mammals

Metatherians	Eutherians
Braincase small relative to body size; minimal development of neocortex; no corpus callosum	Braincase relatively large; greater complexity of neocortex; corpus callosum present
Auditory bullae, if present, formed primarily from alisphenoid bone	Auditory bullae present, formed from tympanic bone
Large vacuities often present in posterior part of palate	Palatal vacuities absent or small
Jugal bone large; jugal and squamosal bones articulate with dentary bone in mandibular fossa	Jugal bone does not articulate with dentary in mandibular fossa
Angular process of dentary inflected 90° (i.e., is perpendicular to the axis of the dentary), except in koala and honey possum	Angular process of dentary not inflected
Primitive dental formula 5/4, 1/1, 3/3, 4/4 = 50. Last premolar is the only deciduous tooth.	Primitive dental formula 3/3, 1/1, 4/4, 3/3 = 44. Incisors, canines, and premolars are deciduous.
Epipubic bones occur in both sexes.	Epipubic bones do not occur.
Female reproductive tract bifurcated (see text); tip of penis (glans) bifurcated	Reproductive tract and glans penis not bifurcated
Marsupium often present enclosing teats; opens either anteriorly or posteriorly	Marsupium not present
Scrotum anterior to penis, except in the mole *Notoryctes;* baculum never present	Scrotum posterior to penis; baculum sometimes present

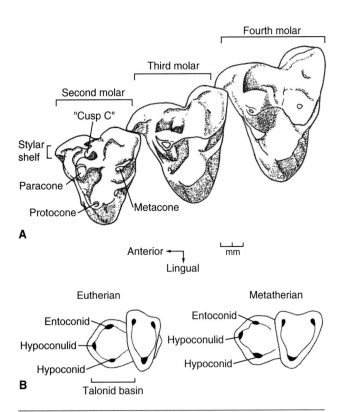

Figure 10.8   **Marsupial molariform dentition.**
(A) Left maxillary of a marsupial containing molars 2-4, showing the characteristic stylar shelf on the labial (cheek) side, large size of the metacone relative to the paracone, and well-developed "stylar cusp C" (B) Lower molars, showing the hypoconulid closer to the entoconid ("twinned") in a marsupial compared with a eutherian, where the hypoconulid is equidistant between the entoconid and hypoconid.

section). Consequently, marsupials have been unable to take advantage of the abundant feeding niches afforded by either plankton or night-flying insects. Likewise, there are no fossorial herbivores among the marsupials, and all the large marsupial carnivores are extinct. Marsupials are therefore more restricted than eutherians in their adaptive radiation and associated structural diversity. Nonetheless, a fascinating array of behavioral and morphological adaptations is evident among the approximately 270 extant species of marsupials.

Reproductive Structure

Females have a **bifurcated** (paired) reproductive tract. Two lateral vaginae are on either side of a medial vaginal canal, or sinus (see figure 9.8; figure 10.1), and the uterus is duplex. The two prongs of a male's bifid penis probably are compatible with the corresponding lateral vaginae during copulation. During copulation, sperm travel up the lateral vaginae. If fertilization occurs, the zygote(s) implant in the uterus (uteri). Following a short gestation period, parturition (birth) occurs through extension of the medial vaginal canal, not through the lateral vaginae.

The typical placenta of marsupials differs from that of eutherian mammals. Marsupials have a **choriovitilline** (yolk-sac) placenta, in which the membranes are less developed and less efficient at facilitating nutrient exchange from the mother to the fetus (see figure 9.11 and associated text). Unlike the **chorioallantoic** placentae of eutherians, the choriovitilline placentae of marsupials have no **villi** (fingerlike projections of capillaries from the embryonic membrane). Because of the tremendous surface area they provide, villi enhance both nutrient exchange and the strength of fetal attachment. Among marsupials, only the bandicoots (Order Peramelemorphia) and koala (Family Phascolarctidae) have chorioallantoic placentae, but they still lack villi. Thus, gestation in marsupials is necessarily short because of inefficient nutrient exchange and a weak structural connection between the fetus and the endometrium (wall of the uterus). The limited adaptive radiation of marsupials compared with eutherians, with a lack of hooves, flippers, or wings in marsupial forms, may be related to their limited intrauterine development time. Also, accelerated development of muscular, clawed forelimbs is necessary so that newborn marsupials can crawl from the vaginal opening to the nipples to nurse and grow. This necessarily precludes the forelimbs from becoming hooves, flippers, or wings.

An interesting feature of two New World families—didelphids and caenolestids—and one not found in Australasian marsupials (or other vertebrates) is paired sperm (figure 10.9). Although not paired in the testes, sperm become coupled at their heads in the epididymis (Tyndale-Biscoe and Renfree 1987). Paired sperm again separate once in the female's oviducts. Although this phenomenon has been known for over 100 years, its significance is unclear. Bedford and colleagues (1984) suggested that such pairing may increase sperm survival and allow individuals to produce fewer sperm without lowering reproductive potential.

The scrotum is anterior to the penis in almost all marsupials. In the marsupial mole (*Notoryctes typhlops*) the testes are abdominal; testes are only scrotal in wombats (Family Vombatidae) during the breeding season.

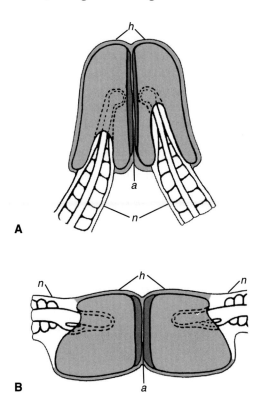

A

B

Figure 10.9   **Paired sperm in marsupials.**   New World marsupials have paired sperm, a feature unknown in any other mammals. Lateral view of the anterior ends of sperm pair typical of (A) a didelphid and (B) a caenolestid. *Abbreviations:* h = head; n = anterior portion of neck; a = acrosomes.

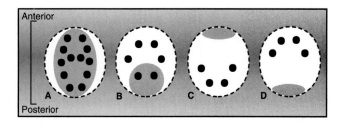

Anterior

Posterior

A   B   C   D

Figure 10.10   **Types of pouches found in marsupials.** Some species have no pouch. Others, such as the mouse opossums, antechinuses, and quolls, have (A) a fold of skin on either side of the teats. (B) Pouches found in the Virginia opossum, Tasmanian devil, and dunnarts enclose most but not all the teats. The deepest pouches are found in arboreal or burrowing species. These may open either (C) anteriorly, as in possums and kangaroos, or (D) posteriorly, as in wombats and bandicoots. Open area of each pouch is shaded.

As noted earlier, not all marsupials have a marsupium. The numbat (*Myrmecobius fasciatus*), some New World possums (Family Didelphidae), rat-possums (Family Caenolestidae), and many small, marsupial mice (Family Dasyuridae) have no pouch. In other didelphids and dasyurids, there is simply a fold of skin on either side of the teats. Pouches often are best developed in arboreal species and those that either burrow or jump. Pouches may open anteriorly or posteriorly (figure 10.10). Also, the condition and appearance of the pouch differ depending on the stage of the reproductive cycle. Woolley (1974) found no relationship between the structure of the marsupium and several life history characteristics. A marsupium therefore probably arose independently in various marsupial lineages (Kirsch 1977). The duration that young remain within the pouch, however, and the time of weaning depend on several factors, including maternal body mass, litter size, and number of litters per year.

Marsupial gestation periods generally are very short (12–13 days in some didelphids and bandicoots) and usually shorter than the interval between maternal estrous periods. At parturition, neonates are tiny, often only a few milligrams, and never more than 1 g (figure 10.11); development of neonatal organ systems is just beginning. Although neonatal marsupials have highly altricial (immature) features, their forelimbs and shoulder muscles are well developed, and the forefeet have deciduous claws. These features allow neonates to climb to a teat, whether in a pouch or not. Once attached, the teat swells, keeping the developing young in place during the early stages of the prolonged lactation period. Thus, "Marsupials have evolved a different but highly successful reproductive strategy when compared with eutherian mammals, in that their major reproductive investment is placed in lactation rather than gestation and placentation . . ." (Tyndale-Biscoe and Renfree 1987:371).

Zoogeography and Early Radiations

Living marsupials occur in North America (only one species); Central and South America; and Australasia, which includes Australia, Tasmania, New Guinea, and surrounding

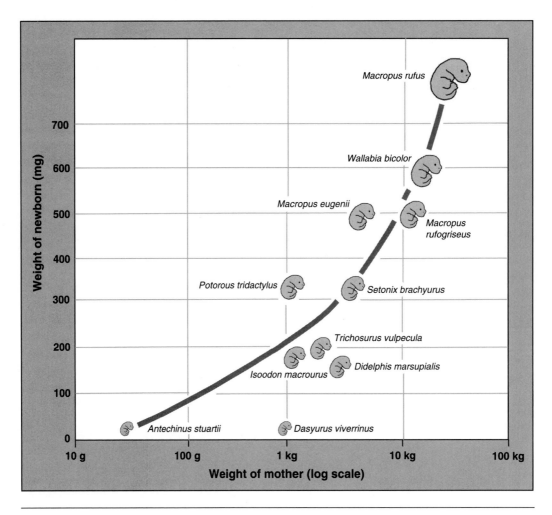

Figure 10.11   **Small size of marsupial newborns.**   Neonatal marsupials are extremely small, never more than 1 g. An entire litter never weighs more than 1% of the mother's body mass.

Source: Data from H. Tyndale-Biscoe, Life of Marsupials, *1973, Elsevier Publishing Company.*

islands. How marsupials came to be where they are today (i.e., their historical zoogeography) remains controversial. The oldest fossil evidence of marsupials is from North America. A recently described marsupial-like mammal, *Kokopellia juddi,* dates from the early Cretaceous period of North America (Cifelli 1993), approximately 100 mya. Although they may have arisen in North America, marsupials eventually declined there as eutherian mammals increased in diversity. Marsupials were extirpated in North America by the mid-Miocene epoch, about 15 to 20 mya. Prior to their extinction in North America, several genera of didelphid marsupials dispersed to Europe about 50 mya in the early Eocene epoch. Like their North American counterparts, these became extinct about 20 to 25 mya.

A diversity of marsupial fossil forms is known from South America by the late Cretaceous period–early Paleocene epoch (Bonaparte 1990; Muizon 1994), and marsupials still persist in South America. During most of this time, South America was isolated from North America. About 9 mya in the late Miocene epoch, representatives of a few families moved along island arcs between the two continents

(Marshall et al. 1982). About 2 to 5 mya, the Panamanian land bridge developed, and a major interchange of North and South American mammalian fauna occurred. Representatives of 17 North American families moved south during the late Pliocene and early Pleistocene epochs. During the same time, representatives of 13 families moved from South America to the north (Webb 1985). These transtropical migrations and the resulting "intercontinental competition" caused many marsupials to go extinct, including large carnivores such as *Thylacosmilus* (figure 10.12). Only a few lineages of small insectivores and omnivores have remained. Of the marsupials that moved from South to North America during this faunal interchange, only the Virginia opossum (*Didelphis virginiana*) persists today.

Where exactly marsupials arose, or their pattern of dispersal, is unknown. Numerous alternative hypotheses have been proposed (see Marshall 1980 for an extensive review). Based on recent fossil discoveries, however, Marshall and colleagues (1990:479) suggest that marsupials originated in North America, but ". . . share a single marsupial fauna with South America in the late Cretaceous." As

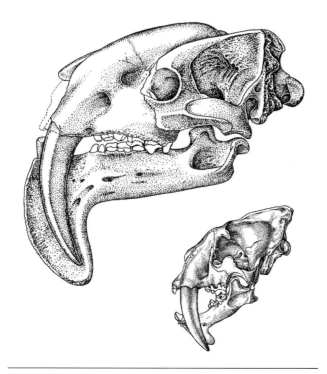

Figure 10.12   **Extinct carnivorous marsupial.**   The large, extinct South American marsupial *Thylacosmilus atrox* (Thylacosmilidae) was very similar to the North American saber-toothed tiger (inset not to same scale).

already noted, they dispersed to Europe from North America. From South America, they dispersed to the Antarctic/Australian continent (figure 10.13). This probably occurred about 65 mya (Woodbourne and Case 1996) but certainly well before the opening of the Drake Passage between South

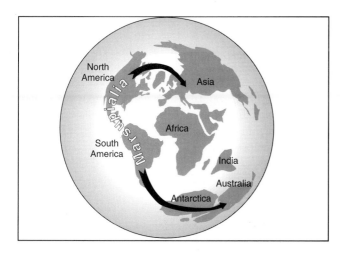

Figure 10.13   **Early marsupial dispersal.**   Hypothesized origin and dispersal routes of ancestral marsupials during the early Cretaceous about 80 mya. Note juxtaposition of Antarctica and Australia.

Source: Data from G. Marshal, "Marsupial Paleobiogeography" in Aspects of Vertebrate History, *L.L. Jacobs, ed., 1980, Museum of Northern Arizona Press, Flagstaff, Arizona.*

America and Antarctica during the late Eocene epoch. Antarctica and Australia then separated, isolating the ancestors of the extant Australian marsupial species. Most marsupial fossils from Australia are from the late Oligocene epoch, about 26 mya, or later (Woodbourne et al. 1993). Compared with South American forms, Australian marsupials evolved in relative isolation from eutherians.

Orders and Families

Until recently, all marsupials were placed in a single order, the Marsupialia. Ride (1970), however, recognized four orders based on characteristics of the dentition (**polyprotodont**—unshortened mandible, lower incisors small and unspecialized, or **diprotodont**—shortened mandible with first pair of lower incisors enlarged to meet upper incisors) and digits (**didactylous**—unfused toes, each in their own skin sheath; or **syndactylous**—skeletal elements of the second and third toes fused, with both digits in a common skin sheath; figure 10.14). These characteristics remain useful in broad descriptions of marsupial orders (table 10.2). Corbet and Hill (1991) and Szalay (1994) proposed alternative taxonomic arrangements of marsupials, based on a variety of research techniques, as have others (see Woodburne and Case 1996). Taxonomic relationships among marsupial families proposed by Luckett (1994) are given in figure 10.15. We follow Wilson and Reeder (1993), who recognized 7 orders (see table 10.2) and 18 extant families. The first three orders we discuss (Didelphimorphia, Paucituberculata, and Microbiotheria) all have a single family and occur in the New World. The remaining four orders are Australasian in distribution.

DIDELPHIMORPHIA

This order has been referred to as Ameridelphia (Marshall et al. 1990), reflecting its New World distribution. The 15 genera and 63 species are in a single family—the Didelphidae. The only extant marsupial north of Mexico, the Virginia opossum, is a didelphid. It ranges from British Columbia south through much of the United States, Mexico, and Central America. The other didelphids occur from Mexico south to southern South America and on islands in the Lesser Antilles.

Life history strategies of didelphids vary; they occur in almost all habitats from deserts to tropical forests and at elevations up to 3400 m. They are primarily terrestrial burrowers, although many are semiarboreal and inhabit tree dens. Generally solitary, most didelphids are opportunistic feeders, their diet being dependent on seasonal forage availability. Gestation periods usually are less than 2 weeks; neonates weigh about 0.1 g (see figure 10.11). Reproductive seasons also vary but, as with all marsupials, are timed so that the young leave the pouch when resources are optimal. The most specialized didelphid is the yapok, or water opossum (*Chironectes minimus*), the only marsupial adapted for an aquatic habitat and a diet of aquatic organisms. The hind feet are webbed, and the female's marsupium becomes watertight during dives.

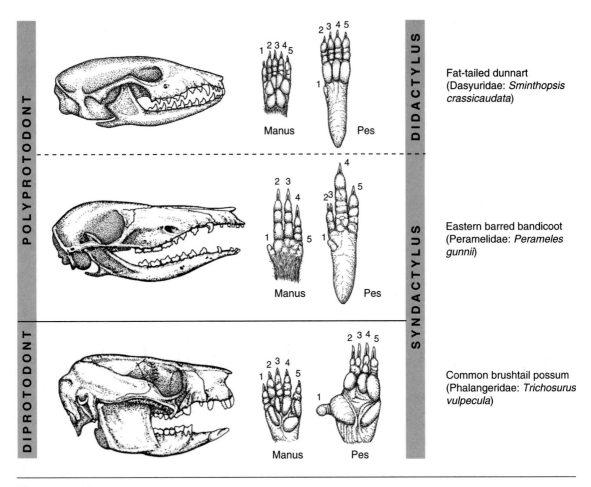

Figure 10.14   Marsupial hind feet.   Within each order of marsupials, digits of the hind feet are either didactylous or syndactylous. This may enhance the grasping ability of arboreal species. Dentition is either polyprotodont or diprotodont.

Morphology.   Didelphids are the most generalized marsupials (Marshall 1984); that is, they retain many plesiomorphic (ancestral) characteristics, believed to approximate the structure of ancestral metatherians. These characteristics include pentadactyly and the primitive metatherian dental formula (5/4, 1/1, 3/3, 4/4 = 50). Didelphids are polyprotodont

Table 10.2.   Structure of the digits and dentition of the seven orders of marsupials discussed in this chapter*

Digits	Dentition	
	Polyprotodont	Diprotodont
Didactylous	Didelphimorphia Microbiotheria Dasyuromorphia	Paucituberculata
Syndactylous	Peramelemorphia Notoryctemorphia†	Diprotodontia

* See figure 10.14.
† Debate continues over the characterization of both the dentition and digits in marsupial moles. Although clearly not diprotodont, neither is the dentition typically polyprotodont. Likewise, whether syndactyly occurs or not is debatable.

and didactylous (see table 10.2), with a long rostrum and well-developed sagittal crest. Body mass ranges from 10 g in mouse opossums (Genus *Marmosa*) to 2 kg in *Didelphis*. The marsupium is well developed in some genera, absent or poorly developed in others. Most (such as the Virginia opossum) have long, sparsely haired, prehensile tails to accommodate their semiarboreal habits and an opposable **pollex** (thumb or first digit on the forefoot). Some species, including the Patagonian opossum (*Lestodelphys halli*), which has the southernmost distribution of any didelphid, and several species of mouse opossums, have **incrassated** tails. That is, they store fat in the base of their tail to sustain them during periods of torpor.

Fossil History.   This order is known from the late Cretaceous period in both North and South America. The group had radiated into 12 recognized genera by the Paleocene epoch (Carroll 1988). Didelphimorphs occurred in Europe by the Eocene epoch but died out there in the Miocene, about 20 mya. They also were extirpated in North America in the Miocene epoch, but recolonized after the Panamanian land bridge formed during the Pleistocene.

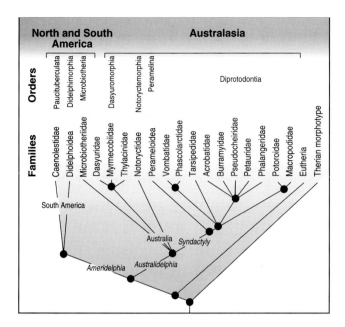

Figure 10.15   **Taxonomic relationships of marsupials.**
The Microbiotheriidae occur in South America but are shown as
Australian based on their phylogenetic relationships.

*Source: Data from W.P. Luckett, "Suprafamilial Relationships Within Marsupialia:
Resolution and Discordance From Multidisciplinary Data" in* J. Mammal Evol.,
2:225-283, 1994.

Economics and Conservation. *Didelphis virginiana* is
one of the few didelphids eaten by humans. It is also taken as
a furbearer and occasionally by sport hunters in North
America (Gardner 1982). In general, however, it is of minor
economic importance for fur, sport, or as a poultry depreda-
tor. Opossums are sometimes used in laboratory research.
Loss of tropical habitat may adversely affect certain Central
and South American didelphids, although none is considered endangered.

PAUCITUBERCULATA

The single family, Caenolestidae, within this order contains
three genera and five species of "shrew," or "rat," opossums.
The three species of *Caenolestes* occur in dense vegetation in
cold, wet, high-elevation forests and meadows of northwestern South America. The Peruvian shrew opossum (*Lestoros
inca*) is found in the Andes Mountains of southern Peru,
whereas the Chilean shrew opossum (*Rhyncholestes raphanurus*) is restricted to Chiloé Island and south-central Chile.
Caenolestids are primarily nocturnal, insectivorous or omnivorous, and terrestrial. With the other two orders of New
World marsupials, caenolestids share the characteristic of
paired spermatozoa within the epididymis—unknown in
Australasian marsupials or eutherians.

Morphology. Caenolestids are small and shrewlike in appearance (figure 10.16). They have a long rostrum, small
eyes, and hind limbs that are longer than the forelimbs.

Adults weigh about 40 g. Total length is about 30 cm, half of
which is the long, fully haired tail. They have no marsupium.
Members of this order are didactylous and are the only New
World marsupials that are diprotodont (see table 10.2). The
dental formula is 4/3–4, 1/1, 3/3, 4/4 = 46–48. The procumbent first lower incisors have enamel only on the anterior
surface, and the lower canines are vestigial. These are the
only marsupials without a deciduous last premolar.

Fossil History. The earliest known fossil caenolestids are
from the early Eocene epoch. This family was extremely diverse during the Miocene epoch, when they were the most
abundant marsupials (Carroll 1988).

Economics and Conservation. Little is known about
most of the species of caenolestids. None are considered endangered, although only a few *Rhyncholestes raphanurus* have
been collected.

MICROBIOTHERIA

The single family in this order, Microbiotheriidae, contains
only one extant species. The monito del monte (*Dromiciops
gliroides*) is nocturnal and arboreal, and inhabits dense, humid forests of south-central Chile. Their distinctive, round
nests are found in fallen logs, tree cavities, and thickets.
Nests are often lined with leaves of water-resistant Chilean
bamboo (*Chusquea* spp.). Microbiotheriids forage primarily

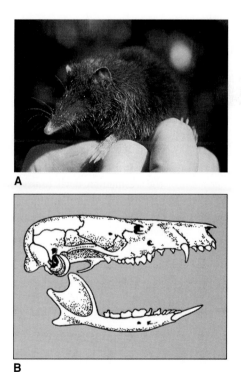

Figure 10.16   **South American caenolestids.**   (A) The rat
opossum. (B) Note the diprotodont lower incisors, small posterior
incisors, reduced size of the posterior molars, and characteristic
preorbital vacuity.

for invertebrates but occasionally may consume herbaceous material. Prior to hibernation, fat accumulates in the base of the tail. Enough fat may be stored in a week to double an individual's body mass.

Morphology. These small (16–30 g), mouselike animals have a maximum head and body length of 13 cm. The well-furred, prehensile tail is also about 13 cm long. The fur is short and thick, and a well-formed pouch is evident in females. The soles of the hind feet have five distinct, transverse pads, or ridges. The dental formula is the same as in didelphids, with a total of 50 teeth. The skull of *Dromiciops gliroides* is noteworthy for its greatly inflated auditory bullae, unlike any other marsupial (figure 10.17). Previously considered a didelphid, this species was placed in a separate order (Aplin and Archer 1987) based on tarsal morphology, serology, and karyotype.

Fossil History. Marshall and colleagues (1997) suggest that *Khasia cordillerensis,* from the early Paleocene epoch of Bolivia, was a microbiotheriid. The family is known in South America from the late Oligocene fossil Genus *Microbiotherium.*

Economics and Conservation. These harmless marsupials, called "colocolos" by natives, are of no economic value but are thought to bring bad luck if observed in or around homes. Although rare throughout their range, they are not considered threatened or endangered.

DASYUROMORPHIA

These small to medium-sized Australasian marsupials include three families of carnivorous species; however, practically all extant species are in the Family Dasyuridae. These include what are usually referred to as marsupial "mice." Given their characteristic foraging behavior, they might more properly be considered marsupial "shrews." With the recent extinction of the thylacine, or Tasmanian wolf (*Thylacinus cynocephalus*), the Tasmanian devil (*Sarcophilus harrisii*) is the largest living carnivorous marsupial.

Morphology. Dasyuromorphians are polyprotodont and didactylous (see table 10.2). Their canines are well developed, and several species, including the Tasmanian devil and the quolls (Genus *Dasyurus*), have specialized carnassial (bladelike, shearing) molariform dentition. Their tails are usually long, often held erect, and never prehensile.

Fossil History. Dasyuromorphian families were differentiated in Australia by the early Miocene epoch. They probably were there much earlier, however, because the oldest fossils date from the Oligocene epoch (Tedford et al. 1975).

Economics and Conservation. Aspects of conservation and economics are noted under the accounts of each family.

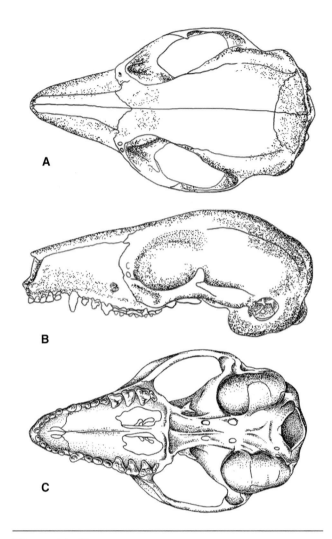

Figure 10.17 Skull of *Dromiciops.* (A) Dorsal, (B) lateral, and (C) ventral views of the skull of a monito del monte. Note the five upper incisors in an arc-shaped pattern, large palatal vacuities, and inflated auditory bullae.

Thylacinidae. This monotypic family included the recently extinct Tasmanian wolf, or thylacine (figure 10.18). Recent fossil thylacines occurred throughout Australia and New Guinea, and the species was common in Tasmania prior to European colonization. The thylacine soon came into conflict with humans, however, because it preyed on domestic livestock. Thylacine populations rapidly declined throughout the 1800s and early 1900s because of predator control, habitat loss, and competition with domestic dogs. The last thylacine died in the Hobart Zoo in 1934. Ironically, the species was given complete legal protection by the Tasmanian government 4 years later. The phylogenetic relationships of thylacinids and proposed affinities with extinct borhyaenids of South America have received much attention (Sarich et al. 1982, Krajewski et al. 1992).

Myrmecobiidae. This monotypic family includes only the numbat (*Myrmecobius fasciatus*), which is highly specialized

Figure 10.18   Extinct Tasmanian wolf.   Very similar to eutherian canids in appearence and habits, the tylacine, or Tasmanian wolf was the largest recent marsupial carnivore, with a body mass of 15-35 kg.

for **myrmecophagy** (a diet of ants and termites). Once ranging throughout southern Australia, the species is now reduced to isolated populations in arid, scrub woodlands in southwest Western Australia. Numbats forage diurnally, which is unusual among marsupials, and search for prey in hollow logs. They are solitary except during the breeding season. The gestation period is 2 weeks. The usual litter size is four. They have no marsupium, and neonates cling to curly hair on the mother's abdomen as they hang from the teats.

Numbats reach about 44 cm in total length, with the fully furred tail equal to the head and body length. Adults weigh up to 700 g. The reddish or grayish brown pelage has seven transverse white-and-black stripes on the dorsum, and there is a black eye stripe. Like other myrmecophageous mammals, they have an elongated, tapered rostrum and tongue to take ants and termites. Their tongue can be extended 10 cm—close to half their head and body length (figure 10.19). Numbats have small, degenerate, peglike teeth. They also have supernumerary (additional) cheekteeth, for a total number of teeth that may reach 52—more than any other terrestrial mammal. Like many other marsupials, both sexes have a large presternal gland on the upper chest, probably used to mark territories (Friend 1989).

Numbats are endangered because of habitat loss and predation by the introduced red fox (*Vulpes vulpes*). They were extirpated from New South Wales by 1857 and South Australia by 1924 (Archer 1978). A recovery program involves numbat reintroductions and fox control (Friend 1995). The fossil record of this family dates only from the late Pleistocene epoch.

Dasyuridae.   Like the New World didelphids, this large, diverse family of 15 genera and 61 species is the most generalized (ancestral) structurally and functionally of the Australasian marsupials. They range in size from the smallest marsupials, the tiny Pilbara ningaui, and other marsupial "mice" to the largest extant marsupial carnivore, the Tasmanian devil. Dasyurids occur throughout Australasia, where they occupy all terrestrial and semiarboreal habitats from deserts to high-elevation rainforests.

Dentition totals 42 to 46 teeth and is specialized for a carnivorous or insectivorous diet. A marsupium usually is absent or poorly developed. Incrassated tails occur in several genera that inhabit deserts, including the dunnarts (Genus *Sminthopsis*), the mulgara (*Dasycercus cristicauda*), and the pseudantechinuses (Genus *Pseudantechinus*). Dasyurids primarily are nocturnal (although the mulgara also may be active during the day) and generally solitary. The fossil history extends to the early Miocene epoch.

Several dasyurids, including the southern dibbler (*Parantechinus apicalis*), the red-tailed phascogale (*Phascogale calura*), and several species of dunnarts, are of concern to conservationists because of reduced distributions and population sizes.

PERAMELEMORPHIA

Bandicoots and bilbies (or rabbit-eared bandicoots) occur throughout Australasia. All are terrestrial omnivores, feeding on invertebrates, small vertebrates, and plant material. They occupy a wide variety of habitat types from arid deserts to tropical rainforests and jungle, often at high elevations.

Figure 10.19   External features of a numbat.   The numbat, or banded anteater, is distinctively striped. Like all species of ant and termite eaters, it has a long, extensible tongue.

Peramelemorphians are unusual among marsupials in having chorioallantoic placentae. Unlike eutherians, however, the placentae have no villi; thus, gestation periods in these species are no longer than in other metatherians. In fact, both the northern brown bandicoot (*Isoodon macrourus*) and the long-nosed bandicoot (*Perameles nasuta*) have gestation periods of 12.5 days—the shortest known among mammals.

Until recently, bilbies were placed in a separate family, the Thylacomyidae (Archer and Kirsch 1977). They are now included with bandicoots in the Peramelidae. The New Guinea bandicoots, formerly placed with the peramelids, are now considered a distinct family, the Family Peroryctidae (see Groves and Flannery 1990).

Morphology.   Peramelemorphians have short, compact bodies, with a long, pointed rostrum. Extremes of body size occur in the peroryctids (see later section). All are adapted for digging, with strong claws on the second, third, and fourth digits of the forefeet. Hind limbs are larger than forelimbs, and the hind feet are long with a well-developed claw on an enlarged fourth digit (figure 10.20). Bandicoots are unique among marsupials in having a well-developed **patella** (kneecap) and no **clavicle** (collar bone; Jones 1968).

Dentition is polyprotodont, with 5 pairs of upper incisors and 48 total teeth (46 teeth in some peroryctids). The canines are well-developed, and the molars are adapted for an omnivorous diet. The marsupium also is well developed and opens posteriorly. There are eight teats, even though mean litter size usually is four. This allows for consecutive litters to be produced, as teats remain swollen after nursing the previous litter, and incoming neonates cannot attach to them.

Fossil History.   Fossil history dates from the early Miocene epoch of Queensland. There is debate, however, as to the affinity of peramelemorphians with other marsupial groups (Gordon and Hulbert 1989).

Economics and Conservation.   Many species of Australian peramelids have suffered significant population declines because of introduced predators and habitat loss to domestic livestock. The western barred bandicoot (*Perameles bougainville*), once common on the mainland, now exists only on Bernier and Dorre islands off western Australia. The pig-footed bandicoot (*Chaeropus ecaudatus*) and the lesser bilby (*Macrotis leucura*) both became extinct within this century, the first in the 1930s and the latter possibly as late as the 1960s. The greater bilby (*M. lagotis*) is endangered. The range of several other species is greatly reduced from presettlement times.

Peramelidae.   There are three genera and eight extant species of bandicoots and bilbies. Peramelids exhibit rather broad ecological flexibility (Gordon and Hulbert 1989), occurring in a wide variety of habitats. Bandicoots are smaller than bilbies, with maximum head and body length of 50 cm and body mass from 0.5 to 2.0 kg. Body mass of bilbies reaches 2.5 kg. Both groups exhibit sexual dimorphism, with males being larger than females. Bandicoots have short, course pelage, often with stiff, quill-like guard hairs, and relatively small ears and tail. Bilbies have a long tail, 50% to 60% of the head and body length (figure 10.21), and longer, silkier pelage than bandicoots. Their long, rabbitlike ears reach beyond the tip of the snout and are folded over the eyes when they sleep. Bilbies are powerful burrowers; unlike bandicoots, they construct their own burrows. In the form of a deeply angled spiral 2 m deep, burrows offer peramelids a refuge from desert heat during the day.

Peroryctidae.   The 4 genera and 11 species of peroryctids are restricted to New Guinea and surrounding islands, with one species extending to the Cape York Peninsula of northern Australia. They range in size from the mouse bandicoot (*Microperoryctes murina*), with a maximum head and body length of 17 cm and tail of 11 cm, to the giant bandicoot (*Peroryctes broadbenti*), which reaches a total length of 90 cm and body mass of 5 kg. The three species of New Guinean spiny bandicoots (Genus *Echymipera*) and the Ceram Island long-nosed bandicoot (*Rhynchomeles prattorum*) lack the fifth upper incisor found in other peramelemorphs and have a total of 46 teeth. All species are nocturnal, terrestrial, and solitary. They are insectivorous or omnivorous and occur in grassland, shrub, and rainforest habitats. Like the closely

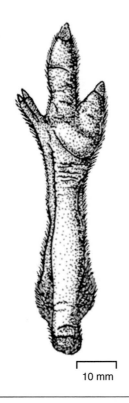

10 mm

Figure 10.20   Enlarged hind foot of a bilby.   The left hind foot of the greater bilby, showing the enlarged fourth digit and syndactylous second and third digits.

A   **B**

Figure 10.21   **Characteristics of bandicoots.**   (A) A greater bilby (or rabbit-eared bandicoot) has a longer tail and longer, silkier pelage than (B) bandicoots.

related peramelids, they have a chorioallantoic placenta and short gestation period (Flannery 1995).

DIPROTODONTIA

This diverse order of eight families includes the familiar kangaroos, koala (*Phascolarctos cinereus*), wombats, and numerous other primarily herbivorous marsupials. As might be expected, given the 116 species in this order, adaptive radiation has been extensive. Many of the pygmy-possums (Genus *Cercartetus*) are insectivorous, whereas the noolbenger (*Tarsipes rostratus*) and eastern pygmy-possum (*C. nanus*) are nectivores. Many species are terrestrial, but some are arboreal.

Morphology.   All species are diprotodont, and the second and third digits are syndactylous. In many arboreal diprotodonts, including the koala, ringtail possums (Family Pseudocheiridae), and cuscuses (Family Phalangeridae), the first two digits of the forefeet oppose the other three digits (figure 10.22), that is, they are **schizodactylous.** On the hind feet, the **hallux** (big toe) is also opposable. This is not the case in the terrestrial species, however. Dental adaptations include a large pair of lower incisors and three pairs of smaller upper incisors (although wombats have a single pair of upper and lower incisors). Upper canines are variable in size and shape; there are no lower canines.

Fossil History.   The earliest fossil diprotodonts date from late Oligocene deposits of Australia.

Economics and Conservation.   Many diprotodonts compete with domestic livestock on grazing lands. Other species have been hunted for meat or hides. Some species of wallabies and kangaroos have been seriously reduced in density and distribution since European settlement, with several recent extinctions being attributed to habitat loss or introduced predators.

Phascolarctidae.   The familiar koala, with a superficial resemblance to a small bear, is the only extant species in this family. Koalas occur in *Eucalyptus* woodlands throughout eastern and southeastern Australia. The koala is most closely related to wombats, and these families share a number of morphological characteristics. Among several common features, both have a marsupium that opens posteriorly and vestigial tails (unusual in an arboreal species), and both lack the first two premolars. Unlike wombats, the koala has three upper incisors, and the dentition is closed-rooted (i.e., not ever growing). The angle of the dentary bone in koalas is not inflected, unlike any other marsupial except the honey possum (see table 10.1).

Koalas are sexually dimorphic, with males being 50% larger than females. With body mass from 6.5 to 12.5 kg, koalas are among the largest arboreal browsers and are on the

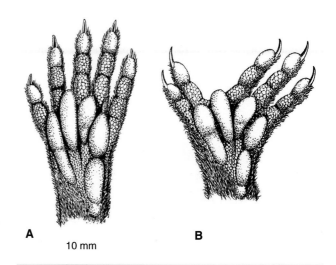

A   10 mm   **B**

Figure 10.22   **Schizodactylous forefeet.**   In many arboreal diprotodonts, including the koala, ring-tailed possums, and cuscuses, the forefoot is schizodactylous. (A) digits closed; (B) digits open for grasping.

ground only while moving between trees. They do not build a nest but simply rest in the forks of trees. Koalas are unusual among herbivores in their highly selective, specialized diet: leaves, stems, flowers, and even bark of numerous species of *Eucalyptus*. This is a very poor-quality forage, but koalas are slow-moving and inactive up to 20 hours a day, and so have reduced energy requirements. Also, their alimentary tract has the largest cecum relative to body size of any mammal (see figure 6.14). Like the bandicoots and bilbies, the koala has a chorioallantoic placenta. But again, because there are no chorionic villi, gestation is only about 35 days. As one of the most recognizable marsupials, the koala is an important tourist attraction. Although not considered either threatened or endangered, populations often are small, and the koala is not common.

Vombatidae.   The two genera and three species of wombats are short-limbed, plantigrade, powerful burrowers. Adult body mass is about 30 kg. The common wombat (*Vombatus ursinus*) is found in forested areas of southeastern Australia and Tasmania. The southern hairy-nosed wombat (*Lasiorhinus latifrons*) inhabits semiarid regions of south Australia, and the endangered northern hairy-nosed wombat (*L. krefftii*) is now restricted to a small area within the Epping Forest National Park in central Queensland.

Although they share several characteristics with the koala, wombats have several distinct features that reflect their terrestrial, grazing, semifossorial existence. Like rodents, they have a single pair of upper and lower incisors, no canines, and a reduced number of premolars (figure 10.23). Dentition is open-rooted and continuously growing, which is unique among marsupials. Wombats are nocturnal and consume primarily grasses and forbs (broad-leafed herbs). Unlike the koala, they have poorly developed ceca. During hot, dry days, *Lasiorhinus* minimizes time above ground by remaining in extensive, interconnected burrow systems. Clusters of numerous burrows, called warrens, may be shared by up to 10 individuals (Wells 1978). The well-developed marsupium, with two teats, opens posteriorly. A single young is born after a gestation period of 21 days and leaves the pouch permanently by about 9 months of age.

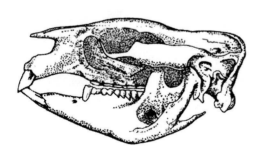

Figure 10.23   Skull of a common wombat.   The single pair of upper and lower incisors—as in rodents and the aye-aye, a primate—are unique among marsupials.

Phalangeridae.   This family includes 3 species of brushtail possums (Genus *Trichosurus*), the scaly-tailed possum (*Wyulda squamicaudata*), and 4 genera and 14 species of cuscuses (Groves 1993). Brushtail possums and the scaly-tailed possum occur only in Australia, whereas the cuscuses are widespread throughout Australia, New Guinea, and surrounding islands. The bear cuscus (*Ailurops ursinus*) is restricted to Sulawesi. All phalangerids occur in scrub or heavily forested areas.

Phalangerids are medium-sized with large eyes, short snout, and soft, dense pelage. Adults range from 0.5 to 1.2 m total length and from 1.1 to 4.5 kg body mass. All are nocturnal; have long, prehensile tails; and are excellent climbers (figure 10.24). *Trichosurus* feed on leaves (folivorous), as well as flowers and nectar; *Phalanger* and *Wyulda* are omnivorous. The marsupium opens anteriorly; litter size is one or two. The gestation period is about 17 days, and young remain in the pouch for 4 to 5 months. Fossil remains date from the late Oligocene epoch.

Potoroidae.   There are four genera and seven extant species of potoroids, including potoroos (Genus *Potorous*), bettongs (Genera *Bettongia* and *Aepyprymnus*), and the musky rat-kangaroo *Hypsiprymnodon moschatus*). These genera were

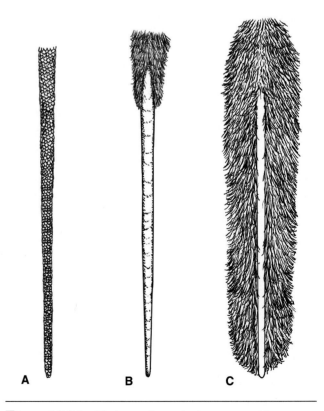

Figure 10.24   Phalangerids vary in the amount of fur on the tail.   (A) Small, nonoverlapping conical scales cover the distal part of the tail in *Wyulda*; (B) strongly prehensile tail in *Phalanger* has fur at the base only; (C) well-furred tail in *Trichosurus* with ventral friction pad.

formerly included as a subfamily within the Macropodidae (see Archer and Bartholamai 1978). Potoroids are small, secretive, densely furred animals. Like the macropodids, the hind limbs are larger than forelimbs, and the hind feet are large. The tail is weakly prehensile (bettongs carry nesting material with their tail); the upper canines are well developed; and there is a large, carnassial premolar. The **sacculated** (several chambered) stomach is not as well defined as it is in macropodids. Mean body mass of the musky rat-kangaroo is only 530 g, whereas the rufous bettong (*A. rufescens*) is about 3 kg. Potoroids are opportunistic omnivores or herbivores. Most include underground fungi as a large part of their diet (Lee and Cockburn 1985). They are essentially solitary and, with the exception of the musky rat-kangaroo, nocturnal. The conservation status of several species of potoroids is of concern. The burrowing bettong (*B. lesueur*) and brush-tailed bettong (*B. penicillata*) are endangered, as is the long-footed potoroo (*P. longipes*)—the first specimen of which was collected in 1968 (Seebeck and Johnston 1980). Two species—the desert rat-kangaroo (*Caloprymnus campestris*) and broad-faced potoroo (*P. platyops*)—are recently extinct.

Macropodidae. This is the largest marsupial family, with 11 genera and about 50 extant species. Body mass ranges from the 1-kg hare-wallabies (*Lagorchestes*) to 80-kg red kangaroos (*Macropus rufus*). Macropodids are browsing or grazing herbivores that occupy practically all terrestrial habitats (tree kangaroos, Genus *Dendrolagus*, are semiarboreal) from deserts to rainforests throughout Australasia.

Their ecology and morphology parallel those of the eutherian artiodactyls. Macropodids have a large, sacculated stomach in which microorganism-aided digestion occurs. As in some artiodactyls, food is regurgitated for additional chewing and swallowed again. Macropodids are diprotodont, but, with the exception of the banded hare-wallaby (*Lagostrophus fasciatus*), the upper and lower incisors do not occlude. Canines are small or absent, and there is a diastema (figure 10.25). The molars are hypsodont (high-crowned), and, as in elephants and manatees, **mesial drift** (forward movement of

the cheekteeth in the jaw, as worn anterior teeth drop out) occurs. Among marsupials, this serial replacement of cheekteeth, most pronounced in *Macropus*, occurs only in macropodids. Only the nabarlek (*Petrogale* [*Peradorcas*] *concinna*), however, has supernumerary molars (more than the usual four) that are shed throughout life.

Macropodids (meaning "big-footed") are characterized by their strong, well-developed hind limbs and large hind feet. Most species have a long, broad tail that acts as a balance during rapid (up to 50 km/h), bipedal, hopping locomotion. The tail acts as an added limb at slow speeds or as a "tripod" while sitting or foraging (figure 10.26). The enlarged hindquarters give macropodids a low center of gravity in the pelvic area, which helps individuals maintain balance while hopping.

The marsupium is large and opens anteriorly. Mammary glands are complex, and the physiology of lactation very sophisticated, with pronounced differences in milk composition during lactation. Thus, a mother nursing a developing pouch young, and a young that has left the pouch but returns to nurse, produces milk of different nutrient composition from different teats (Green 1984). Gestation periods of macropodids are relatively long compared with those of other marsupials—close to the length of the estrous cycle. A single young is produced at a time, and embryonic diapause (see chapter 9) occurs in all species except the western gray kangaroo (*Macropus fuliginosus*).

European settlement has had little adverse effect on some of the larger species and, in fact, has increased numbers and distribution as a result of livestock raising. Conversely, some large and many smaller macropodids, including the bridled nailtail wallaby (*Onychogalea fraenata*), quokka (*Setonix brachyurus*), and rufous hare-wallaby (*Lagorchestes hirsutus*), are considered endangered. Species that have gone extinct within the last 30 to 100 years include the toolache wallaby (*Macropus greyi*), central hare-wallaby (*L. asomatus*),

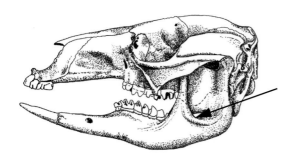

Figure 10.25 **Wallaby skull.** Skull of a wallaby shows the diprotodont dentition and the diastema typical of herbivores. Note also the pronounced masseteric fossa of the mandible (arrow).

Figure 10.26 **Macropodid tail function.** Tails in macropodids may serve a variety of functions in locomotion and foraging. Tail acts (A) as a "tripod" while foraging; (B) as a fifth limb in slow, "pentapedal" locomotion; and (C) as a counterbalance in rapid, bipedal hopping.

eastern hare-wallaby (*L. leporides*), and crescent nailtail wallaby (*O. lunata*).

Burramyidae.   There are two genera and five species of pygmy-possums—the smallest of the possums. Mean body mass of adults ranges from 7 to 50 g, and head and body length is only 5 to 12 cm. The mountain pygmy-possum (*Burramys parvus*) is limited to terrestrial alpine areas above 1300 m elevation in southeastern Australia. Only four populations are known to exist within an area of 10 km². This species was known only from fossil remains until 1966, when a live animal was taken at Mount Hotham, Victoria (Mansergh and Broome 1994). The remaining burramyids, in the Genus *Cercartetus*, are all arboreal and occur in a variety of habitats in Australia and Tasmania. One species, the long-tailed pygmy-possum (*C. caudatus*), also occurs in New Guinea. Tails are long and prehensile, the pouch opens anteriorly, and, like macropodids, females exhibit embryonic diapause.

Pygmy-possums are nocturnal and omnivorous, consuming invertebrates, fruits, seeds, nectar, and pollen. They have long, extensible, "brushed" tongues with an extensive system of papillae (figure 10.27) that are especially well developed in *Cercartetus*. Papillae may serve to increase the surface area for uptake of nectar and pollen. All burramyids enter torpor, but among them, *Burramys parvus* is noteworthy because it is the only marsupial known to undergo prolonged periods of hibernation. Species of *Cercartetus* store fat at the base of the tail. Both of these adaptations probably contribute to the relatively long life spans of these small marsupials.

Acrobatidae.   This family, formerly included in the Burramyidae (Aplin and Archer 1987), includes only the feather-tail glider (*Acrobates pygmaeus*), found in wooded habitats of eastern Australia, and the feather-tailed possum (*Distoechurus pennatus*), which occurs in disturbed forests, gardens, and rainforests of New Guinea (Flannery 1995). They are named for the long, stiff, featherlike hairs on the side of the tail. *Acrobates pygmaeus* is nocturnal and highly arboreal and, at 10 to 14 g body mass, probably the world's smallest gliding mammal. A furred patagium (gliding membrane) extends between the elbows and knees. The weakly prehensile tail aids both in climbing and as a rudder for gliding. *Distoechurus pennatus*, however, is terrestrial and lacks a gliding membrane. Both species are primarily nectivorous and have long, brush-tipped tongues, similar to that of burramyids and tarsipedids, which they use for taking nectar and pollen. The papillae on the tongue of *Acrobates* are longer and finer than those of *Distoechurus*. There is a general reduction in the size and number of teeth. Interestingly, although *Acrobates* is nectivorous, the molars are bunodont for secondary feeding on insects. *Acrobates pygmaeus*, which may nest in groups of up to 20 individuals, has several litters per breeding season and exhibits embryonic diapause.

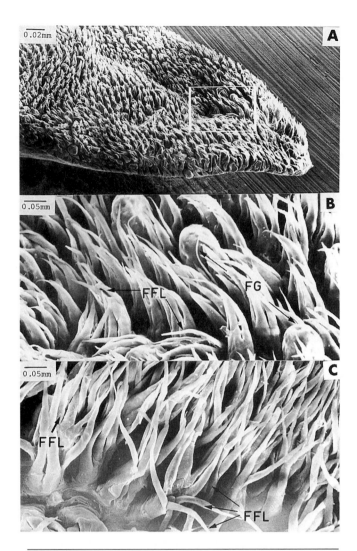

Figure 10.27   **Burramyid tongue papillae.**   Scanning electron micrographs of the papillae found on the tongue of the nectar-feeding eastern pygmy-possum (*Cerartetus nanus*). (A) Dorsal surface of the tip of the tongue. (B) Finer detail of the tip. (C) Finer detail of the back of the tongue. *Abbreviations: FFL*=fine filiform papllae; *FG*=fungiform papillae. Papillae in burramyids probably increases the surface area for absorbing nutrients and moves food more efficiently toward the esophagus. A similar adaptation occurs in the honey possum (see figure 10.29) and nectar-feeding bats.

Pseudocheiridae.   The 5 genera and 14 species in this family are closely related to the Petauridae, in which they were formerly included (Baverstock et al. 1990). Pseudocheirids occur in Australia, New Guinea, and a few surrounding islands (Flannery 1995). Most are slow-moving and inhabit forested areas; the rock ringtail possum (*Petropseudes dahli*) occurs on rocky slopes and outcrops. These arboreal, nocturnal species feed primarily on leaves, and many of their morphological features reflect this adaptation. Unlike petaurids, the molars of pseudocheirids are selenodont (cusps form crescent-shaped ridges) to finely chew leaves. The alimentary tract and very large cecum also are specializations for folivory. The arboreal

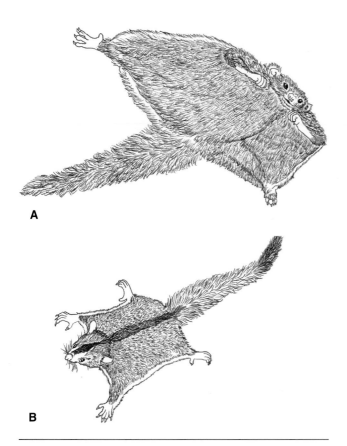

Figure 10.28 **Marsupial gliders.** The gliding membranes extend from the elbow to the ankle in (A) the pseudocheirid and from the wrist to the ankle in (B) the petaurid.

habits of pseudocheirids are aided by the schizodactylous digits of the forefeet. The long, furred, prehensile tail is usually the same length as the head and body. A gliding membrane extends from the elbow to the ankle in *Petauroides*, not from the wrist as in *Petaurus* (figure 10.28; see the following section). The marsupium opens anteriorly and encloses either two or four teats.

Petauridae. This family includes four species of striped possums (Genus *Dactylopsila*), Leadbeater's possum (*Gymnobelideus leadbeateri*), and five species of wrist-winged gliders (Genus *Petaurus*). These arboreal gliders are named for the fact that the patagium extends from the wrist to the ankle (see figure 10.28). *Petaurus* is highly **convergent** (similar traits found in lineages that are not closely related) with the North American gliding squirrels (Genus *Glaucomys*). All petaurids are medium-sized (0.1–2.0 kg), with some type of dark dorsal stripe; long, well-furred, prehensile tails; and a well-developed marsupium that opens anteriorly. In some species, the pouch is partitioned into left and right compartments by a septum. Females usually give birth to a single young that remains in the pouch for 4 months. There are a total of 40 teeth; the diprotodont lower incisors are long and sharp, and the molars are bunodont.

Petaurids occur in forested areas of Australia, New Guinea, and surrounding islands. They are nocturnal and herbivorous or insectivorous. Arboreal locomotion is aided by the prehensile tail and opposable hallux. The endangered Leadbeater's possum was previously believed to be extinct, but populations were rediscovered in 1961 in Victoria, Australia. The mahogany glider (*Petaurus gracilis*) is also endangered.

Tarsipedidae. The unusual honey possum, or noolbenger (*Tarsipes rostratus*), is the sole member of this family. Average body mass of females is 12 g and of males, about 9 g. Many of the adaptations of this tiny species are for its exclusively nectivorous diet. It has a long, pointed rostrum and tubular mouth with extensible, brush-tipped tongue. The 22 (at most) small, peglike teeth are fewer in number than in any other marsupial (figure 10.29). The honey possum is nocturnal, highly arboreal, and occurs in shrubs and woodlands of southwestern Australia. The long tail is prehensile and the hallux opposable. Claws occur only on the two syndactylous digits of each hind foot. Instead, pads on the digits are used to grip branches.

The marsupium is well developed. Honey possums exhibit delayed implantation, and pouch young occur throughout the year (Renfree et al. 1984). They are unique in giving birth to the smallest mammalian young; neonates weigh no more than 5 mg. Their sperm, however, is about 0.3 mm long, the largest known among mammals. Although it shares certain characteristics with other marsupial families, the honey possum is unique in many respects and probably represents the sole surviving member of an otherwise long-extinct lineage. There is no fossil history.

NOTORYCTEMORPHIA

This order encompasses a single family, the Notoryctidae, which includes the marsupial mole, *Notoryctes typhlops*. A second species, *N. caurinus,* has been described (see Johnson

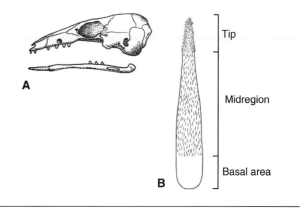

Figure 10.29 **Features of the honey possum.**
(A) Except for the procumbent lower incisors, the honey possum has small, degenerate teeth, reflecting a soft diet. The dental formula is I = (1 – 2)/1, C = 1/0, PM – M = (1 – 4)/(1 – 3) = 11 – 22, with postcanine teeth not clearly differentiated. The palate is ridged to remove pollen and nectar from (B) the long, extensible, brush-tipped tongue.

1995) but may not warrant species status. They represent the only completely fossorial marsupials and are widely distributed over much of western Australia in shrub-desert areas and sandy bottomland soils. Their diet consists of moth and beetle larvae and eggs, which are taken as individuals burrow through the soil. Although the morphological characteristics noted in the following section are remarkably similar to those of eutherian talpids and chrysochlorids (see chapter 11), marsupial moles burrow differently. They essentially "swim" through the ground. Substrate collapses behind them, and they leave no permanent tunnels. They also spend much more time foraging on the surface than eutherian moles and are active both day and night. Little is known of the reproductive biology of *Notoryctes*.

Morphology. Like eutherian moles, the body of *Notoryctes* is **fusiform** (torpedo-shaped, compact, and tapered) and adapted for a fossorial existence. Maximum head and body length of adults is about 14 cm, with a stubby tail 2 to 3 cm long. Adults weigh about 60 g. The iridescent pelage is long and silky. Other fossorial adaptations include strong forelimbs with greatly enlarged, scooplike claws on the third and fourth digits (figure 10.30). The thick, keratinized nasal shield is used to push dirt, and the cervical vertebrae are fused for added rigidity. There are no pinnae. Located under the skin, the vestigial eyes are 1 mm in diameter and have no lens; the optic nerve also is greatly reduced.

Epipubic bones occur in both sexes but are reduced in size. Dentition is variable, with 40 to 44 teeth. The occlusal

Figure 10.30 **External features of the marsupal mole.**
The marsupial mole has large forefeet, reduced eye, no pinnae, and thick, leathery nasal shield.

surface of the molars is **zalambdodont** (V-shaped; see figure 11.1), which is unusual among marsupials. During the breeding season, females have a well-developed marsupium that opens posteriorly. The testes are never scrotal but, according to Johnson (1995), lie between the skin and the abdominal wall.

Fossil History. Notoryctids are known from the Miocene epoch of Queensland. Their affinities with other marsupial lineages remain unresolved.

Economics and Conservation. Marsupial moles are eaten by Australian aborigines, who capture them by following the distinctive trails made by *Notoryctes* when traveling on the surface. Neither species affects grazing lands, and neither is considered endangered.

Summary

Reproductive characteristics, as well as distinctive morphological and anatomical features, serve to differentiate monotremes and marsupials from each other and from eutherian mammals. Monotremes (prototherians), which include the duck-billed platypus, and short- and long-beaked echidnas, have a cloaca—a single opening for the fecal, urinary, and reproductive tracts. Reproduction also is distinctive in these, the only oviparous (egg-laying) mammals. Eggs are small, with a semipermeable shell that allows for nutrient exchange from mother to the developing young prior to the eggs being laid. As with all mammals, neonatal monotremes are nursed. Monotremes have mammary glands but no teats. Like many marsupials, echidnas have a pouch (marsupium) in which eggs are incubated and hatchlings are nursed. The platypus has no pouch. Monotremes exhibit several features characteristic of reptiles, including a lower body temperature than most other mammals, microchromosomes, threadlike sperm, and uncoiled cochlea. The pectoral girdle retains coracoid, precoracoid, and interclavicle bones, as did primitive therapsid reptiles, the early ancestors of mammals.

The semiaquatic platypus is noteworthy in other features, including an ankle spur connected to a poison gland, and a unique bill. The bill is highly innervated and responds to both tactile and electrical stimuli. During dives, the platypus detects aquatic prey by sensing the weak electric fields generated by muscle contractions. Unlike the platypus, echidnas are strictly terrestrial. The short-beaked echidna feeds on ants and termites, and the long-billed echidna on earthworms; both species have large, scooplike forefeet for foraging and burrowing.

Prototherians may have diverged from therians 200 mya, but the relationship of these groups remains unresolved. Most fossil evidence of monotremes is from Australia, with one early Paleocene specimen from southern Argentina. Extant monotremes probably owe their continued survival through evolutionary time to relative geographic isolation from eutherians. Today, monotremes occupy ecological niches with little competition.

The 7 orders and approximately 270 species of marsupials (metatherians) also differ in many respects from eutherians. Marsupials are named for the marsupium, or pouch, which varies in complexity among species. Not all marsupials, however, have a marsupium. The uterus is duplex, with lateral vaginae on either side of a medial vaginal canal. Like eutherians, marsupials have a placenta, but without villi. Thus, nutrient exchange from mother to fetus is poor, and the structural connection between the fetus and the uterine wall is weak. Both of these factors contribute to a very short intrauterine gestation period. Neonates are tiny and undeveloped; total mass of litters is generally less than 1% of maternal body mass. Most marsupial maternal reproductive investment, in time and energy expenditure, occurs during a prolonged lactation period.

Marsupials exhibit many skeletal and anatomical differences from eutherians, including a relatively small braincase and no corpus callosum, reduced or absent auditory bullae, palatal vacuities, articulation of the jugal bone with the dentary bone, an inflected angular process, epipubic bones, and different dental characteristics. Generally, marsupials also have more conservative life history characteristics. They have a lower basal metabolic rate and a smaller range in body sizes than eutherians, and they occupy a narrower range of ecological niches. For example, no marsupials occupy niches comparable to eutherian bats, whales, or fossorial herbivores. All large marsupial carnivores are extinct. The oldest fossil marsupials are from North America, which is probably where they arose. With the exception of a single species, however, marsupials occur today only in South America and Australasia.

Discussion Questions

1. Contrast the various reptilian features of monotremes with those of marsupials. What are the differences and similarities?
2. The platypus uses its bill for locating aquatic prey through electroreception. Why do you think this complex adaptation developed in place of simple visual searching for prey?
3. In species with an enclosed marsupium, pouch young spend prolonged periods breathing air with a very high concentration of carbon dioxide. How do the developing neonates accommodate this "poisonous" environment during the lactation period?
4. What are the advantages in the marsupial mode of producing "expendable neonates" with a prolonged lactation period, compared with the eutherian mode of prolonged gestation with placental involvement? What are the disadvantages?
5. What significance do you attach to paired spermatozoa occurring only in New World marsupials?
6. The suggestion is often made that marsupials are "inferior" to eutherian mammals. What evidence would you present to either support or refute this argument?

Suggested Readings

Archer, M. (ed.). 1982. Carnivorous marsupials, vol. 1. Royal Zoological Society, New South Wales, Sydney.

Archer, M. and G. Clayton. 1984. Vertebrate zoogeography and evolution in Australia. Hesperian Press, Carlisle, Western Australia.

Augee, M. L. (ed.). 1992. Platypus and echidnas. Royal Zoological Society, New South Wales, Sydney.

Meserve, P. L., B. K. Lang, and B. D. Patterson. 1988. Trophic relationships of small mammals in a Chilean rainforest. J. Mammal. 69:721–730.

Seebeck, J. H., P. R. Brown, R. W. Wallis, and C. M. Kemper (eds.). 1990. Bandicoots and bilbies. Surrey Beatty and Sons, Sydney.

Strahan, R. (ed.). 1995. The mammals of Australia. Reed Books, Chatsworth, Australia.

CHAPTER

11

Insectivora, Macroscelidea, Scandentia, and Dermoptera

The four orders discussed in this chapter have had a confused and somewhat chaotic taxonomic history. Historically, the Order Insectivora has been a "wastebasket" taxon for families of uncertain affinities. We can make few generalizations concerning structure or function in this order. In fact, placement of living as well as extinct families within the order often has been made primarily on the basis of convenience. All four orders in this chapter once were included within a single Order Insectivora. The elephant shrews, colugos, and tree shrews have a cecum and were placed together in the Suborder Menotyphla, and the remaining families, which lacked a cecum, were placed in the Suborder Lipotyphla. Because of their structural diversity (along with many other criteria noted in the discussion), the concensus is to consider them as separate orders (Butler 1972). The orders discussed in this chapter, and the families included in them, represent a noteworthy array of structural and functional adaptations and life history patterns. These diverse adaptations are especially evident in the Insectivora.

INSECTIVORA

Six diverse families make up this order: hedgehogs and gymnures (Erinaceidae), tenrecs and otter shrews (Tenrecidae), shrews (Soricidae), moles and desmans (Talpidae), golden moles (Chrysochloridae), and solenodons (Solenodontidae). Many of their morphological characteristics are considered primitive and probably represent characteristics common to the earliest mammals. As such, insectivores probably are near the ancestral stocks of many other orders of eutherians that have advanced, or more recently derived, characteristics. Nonetheless, despite their primitive characteristics, many insectivores have developed highly specialized adaptations and fairly complex behavior. Finally, some obvious examples of convergent evolution occur in this group, especially between the moles and golden moles.

Morphology

Unlike most mammalian orders, no key character or set of characters serves to identify insectivores. This is one of the most anatomically diverse mammalian orders; each family has interesting adaptations for survival. Insectivores are generally small to medium-sized, pentadactyl, with generalized plantigrade locomotion, and long, somewhat pointed snouts. Pelage of adults often is made up only of guard hairs, sometimes modified as spines, as in hedgehogs and tenrecs. The **pinnae** (external ear) and eyes usually are small or absent; the eyes of golden moles are nonfunctional and without external openings. Primitive characteristics include a small braincase and a brain with smooth cerebral hemispheres. A ring-shaped tympanic bone is present instead of auditory bullae, and the anterior vena cavae are paired. In males, the testes are usually abdominal or within the inguinal canal; if external, the scrotum is anterior to the penis. A cloaca is present in some genera. The jugal bone is reduced or absent, and the pubic symphysis is reduced.

Another primitive insectivore characteristic is their dentition. As noted in chapter 4, some insectivores retain tribosphenic molars, including tenrecids, chrysochlorids, solenodontids, and some soricids. Teeth are rooted, so they do not grow throughout life. The deciduous teeth are shed early and are seldom functional. The molars have four or five cusps and usually form a V-shaped (zalambdodont) or W-shaped (**dilambdodont**) occlusal pattern (figure 11.1). In many species, the total number of teeth is often the same as the primitive eutherian pattern: 3/3, 1/1, 4/4, 3/3 = 44.

Fossil History

Fossil insectivores include a diversified assemblage with approximately 150 described genera. As noted by Butler (1972:254), this is because "...any fossil eutherian not clearly related to one of the other orders is classifiable in the order Insectivora." The earliest fossil insectivore, the tiny, poorly known remains of *Batodon*, dates from the mid-Cretaceous period in North America, about 100 mya. The oldest members of clearly recognizable families—soricids and talpids—date from the Eocene epoch, about 50 mya.

Economics and Conservation

Insectivores are of little economic importance today. Until the late 1800s, moles were trapped for their pelts, which were used for hats, apparel trim, and other purposes. Today,

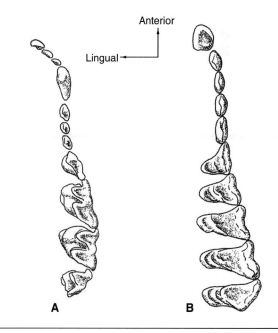

Figure 11.1   Representative occlusal surfaces in insectivores. (A) Occlusal surfaces of the left upper tooth row from a European mole (*Talpa europaea*) with a dilambdodont (W-shaped) cusp pattern. (B) left upper tooth row from a giant otter shrew (*Potamogale velox*) with a zalambdodont (V-shaped) cusp pattern.

although they may damage lawns, fields, and gardens, moles have a relatively minor economic effect compared with rodents, for example. Hedgehogs, moles, and shrews take harmful insects as prey, but their effect is fairly negligible compared with insectivorous bats. Because many insectivores are found in tropical areas where habitats are often rapidly lost to logging and agriculture, several species are considered threatened or endangered (see the discussion of each family).

Families

Erinaceidae

This family encompasses 7 genera and 21 species, and includes hedgehogs, which have barbless spines on the back and sides, and gymnures, which lack spines. Erinaceids are an Old World family found in Africa, Europe, and Asia, including Sumatra, Borneo, and the Philippines. They inhabit many different habitats, including forests, grasslands, fields, and farmland. As their name suggests, desert hedgehogs (Genus *Paraechinus*) inhabit arid areas in North Africa, India, and Pakistan. Long-eared desert hedgehogs (Genus *Hemiechinus*) are found from northern Africa east to the Gobi Desert of Mongolia. Hedgehogs are mainly nocturnal and terrestrial, but some are semiarboreal. Adult *Hemiechinus auritus* weigh only 40 to 50 g, whereas the common European hedgehog (*Erinaceus europaeus*) reaches 1100 g. All are omnivorous, feeding on small vertebrates, eggs, fruits, and carrion, in addition to invertebrates.

A European hedgehog has approximately 5000 spines. Their defensive posture is typical for a mammal with spines or scales. When threatened, they roll into a tight ball with the spines directed outward (figure 11.2). This posture is aided by paired longitudinal "drawstring" muscles, the panniculus carnosus, on either side of the body. Hedgehogs have an interesting "self-anointing" behavior in which they spread large amounts of foamy saliva on the spines. This may be done as a sexual attractant during the breeding season, to reduce parasites, to clean the spines, or as additional protection from predators (Wroot 1984). Several species, including the common European hedgehog, undergo true hibernation throughout the winter—the only insectivores that do. Desert hedgehogs commonly **estivate** (enter a dormant condition in the summer). Litter size is generally four to six, and two litters a year may be produced. The young are altricial (immature). At birth, their short, soft spines have not yet broken through the skin, but the spines quickly grow in length after birth (figure 11.3) and harden within a few weeks.

Gymnures have no spines. They have omnivorous diets similar to those of hedgehogs and often are closely associated with **hydric** (wetlands) habitats. The pelage is long and soft in the Philippines gymnure (*Podogymnura truei*) but very coarse and rough in the moon rat (*Echinosorex gymnurus*), which may reach 460 mm in head and body length and 2 kg in body mass. Moon rats are noted for their strong ammonialike odor, which emanates from a pair of anal scent

A

B

Figure 11.2   Hedgehog spines.   (A) Lateral view of a long-eared hedgehog (*Hemiechinus auritus*). An adult hedgehog has about 5000 spines, each 2-3 cm long. The medulla of each spine is filled with air pockets to reduce weight. (B) The defensive posture of an African hedgehog (*Erinaceus frontalis*) is similar to that of armadillos and scaly anteaters.

glands. The Philippines gymnure is considered vulnerable because of logging and habitat loss.

Talpidae

Forty-two species of moles, desmans, and shrew-moles constitute this family. They are distributed throughout Europe, the Palaearctic region, and Asia, including Japan. In North

Figure 11.3 **Spines in a newborn hedgehog.** Spines on these 2-day-old European hedgehogs (*Erinaceus europaeus*) erupt and harden after birth.

America, they are found in southern Canada and most of the United States. The 17 genera of talpids are fairly diverse, ranging from moles, which are fossorial, to desmans, which are semiaquatic. Shrew moles, the smallest talpids, are terrestrial. They are much more agile than moles, forage on the surface, and occasionally climb small shrubs.

The morphology of talpids reflects their fossorial life. The body shape is fusiform, with short, powerful limbs and short, smooth pelage. The pinnae are reduced or absent, and the eyes are minute. The sternum is keeled for enhanced pectoral muscle attachment, which is useful in digging. The forefeet are large and paddlelike with large claws. Because the radius articulates with the humerus, the forefeet are permanently rotated outward (figure 11.4) and produce the unique lateral digging motion of moles. The humerus is short and broad, again for enhanced muscle attachment, and articulates with both the clavicle and scapula to generate a great deal of force for digging. The olecranon process of the ulna also is very long for enhanced attachment of the triceps muscle. As a result, the digging muscles of the eastern mole (*Scalopus aquaticus*) can generate a force 32 times its own body mass (Arlton 1936).

The only evidence of moles that most people see is their tunneling activity and resulting molehills of excess excavated dirt. Burrowing is generally done in moist soils. Shallow tunnels 4 to 5 cm in diameter and a few centimeters

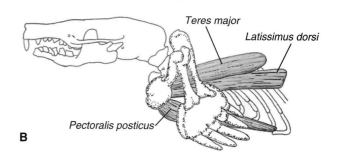

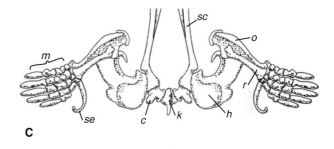

Figure 11.4 **Morphological features of moles.** (A) The short hair, pointed snout, fusiform body shape, and large forepaws are evident in this eastern mole. (B) The massive forearm is supported and rotated by the teres major, latissimus dorsi, and pectoralis posticus muscles. (C) Forelimbs and highly modified pectoral girdle of the European mole. Note the large manus (*m*) with modified sesamoid bone (*se*), the small clavicle (*c*), keeled sternum (*k*), elongated scapula (*sc*), and elongated olecranon process (*o*) of the ulna. The humerus (*h*) is massive, rectangular, and articulates with the radius (*r*).

below the surface are made when moles search for invertebrate prey. These are not as permanent as the complex, branching network of defended tunnels that extend to 150 cm deep and include the nest chamber (figure 11.5).

Restricted to the Old World, desmans have a diet consisting of aquatic invertebrates and fish. The Russian desman (*Desmana moschata*) prefers ponds and marshes with thick vegetation. Conversely, the Pyrenean desman (*Galemys pyrenaicus*) inhabits streams and rivers with clear, fast-moving water. Head and body length is about 200 mm in the Russian

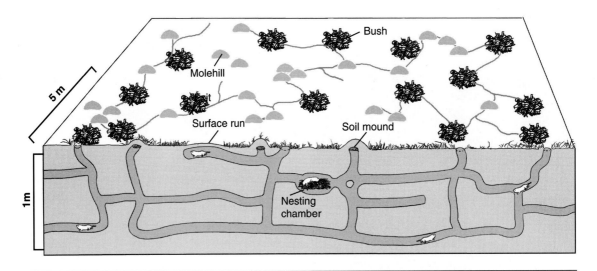

Figure 11.5 **Mole burrow system.** Example of depth and extent of burrow construction of a mole.
Source: Data from A. V. Arlton, "An Ecological Study of the Mole" in J. Mammal., *17:349-371, 1936.*

desman and about 140 mm in the Pyrenean desman. The tail is laterally compressed in both species and equals the head and body length. The hind feet are webbed and **fimbriated** (having a fringe of hairs) for additional surface area. Both adaptations help desmans to swim rapidly. Their long, flexible snouts are very sensitive and are used to locate prey underwater. Desmans echolocate to maneuver and locate prey (Richard 1973), like shrews and some tenrecs. Both species are considered vulnerable because of habitat loss, construction of dams and roads, and water pollution. Additionally, Russian desmans are harvested for their pelts. Pyrenean desmans are further reduced in number because of predation by introduced mink (*Mustela vison*).

Tenrecidae

The diversity of form and function in the Order Insectivora as a whole is reflected in the 24 species of this family. The 21 species of tenrecs are restricted to Madagascar, whereas the 3 species of otter shrews are found in west-central Africa. Some authorities place otter shrews in a separate family— the Potamogalidae. As suggested by their name, they are semiaquatic and closely resemble river otters (Carnivora: Mustelidae). The giant African water shrew (*Potamogale velox*) is the largest living insectivore, with a total length up to 640 mm.

The morphology of tenrecids defies a general description, as do their behavior and habitats. Rice tenrecs (Genus *Oryzorictes*) look like moles, are fossorial, and occur in marshy areas. Long-tailed tenrecs (Genus *Microgale*) resemble shrews and occupy thick vegetation and ground litter in a variety of habitat types. The web-footed tenrec (*Limnogale mergulus*) has a long, laterally flattened tail and webbed hind feet, and looks like a small muskrat. It preys on fish, amphibians, and aquatic invertebrates in rivers, lakes, and marshes. The large (*Setifer setosus*) and small Madagascar "hedgehog"

(*Echinops telfairi*) have sharp, barbed spines on the head, back, and sides like erinaceids (figure 11.6). Like hedgehogs, they have well-developed panniculus carnosus muscles and roll into a ball when threatened. Tenrecs have no auditory bullae and no jugal bone and, thus, an incomplete zygomatic arch. Incisors and canines are usually small and unspecialized, and the upper molars are often zalambdodont.

Several species are heterothermic, and enter torpor during the day or hibernate seasonally. Body temperatures generally are low, ranging from 30° to 35°C while individuals are active. The streaked tenrec (*Hemicentetes semispinosus*) maintains a body temperature 1°C above ambient temperature while it hibernates during much of the winter. Additionally, several species, such as the streaked tenrec and the long-tailed tenrecs are believed to echolocate as part of their foraging activities. The common tenrec (*Tenrec ecaudatus*) has one of the largest litter sizes of any mammal, with as many as 32 young per litter (Eisenberg 1975). As with many mammalian groups in Madagascar, several species of tenrecs are reduced in density and distribution as a result of habitat loss or other factors. The web-footed tenrec is considered vulnerable, and 15 other species probably would be considered threatened or endangered if adequate information were available.

Chrysochloridae

The golden moles encompass 7 genera and 18 species. They are distributed throughout central and southern Africa in forests, fields, and plains with soils suitable for burrowing. Many of the same adaptations to enhance underground movement found in marsupial moles (Family Notoryctidae) and true moles (Family Talpidae) are seen in chrysochlorids—a good example of convergent evolution (figure 11.7). Golden moles have no pinnae, and the vestigial eyes usually are covered with skin. The pelage moves equally well in any

A

B

C

Figure 11.6   **Morphological diversity within the tenrecs.**
A great deal of morphological diversity is evident within the tenrecs, as seen in (A) a long-tailed tenrec (*Microgale dobsoni*); (B) a small Madagascar "hedgehog"; and (C) a giant otter shrew (*Potamogale velox*).

direction. It is a "metallic" or iridescent red, yellow, green, or bronze, depending on the species. A smooth, leatherlike pad covers the nose, which golden moles use for pushing soil. Unlike talpids, the forelimbs of golden moles are under the body and do not rotate outward. They dig by forward extension of the limbs and scratching at the soil with large claws, especially those on the powerful third digit. As in talpids,

burrow depth and construction depends on local soil characteristics. The large golden moles (Genus *Chrysospalax*) also forage on the surface of the ground. Despite their lack of externally visible eyes, they have an excellent sense of direction and are able to return quickly to their burrows if threatened. *Chrysospalax trevelyani* is endangered, and several other species of golden moles are either rare or of undetermined status.

Solenodontidae

Both extant species in the family are narrowly distributed: the Cuban solenodon (*Solenodon cubanus*) is restricted to Cuba, and the Haitian solenodon (*Solenodon paradoxus*) to the Dominican Republic and Haiti. Solenodons are among the largest insectivores, with total lengths approaching 600 mm. The pelage, naked tail, claws, and pinnae are all rather long, and the eyes are small. They have a pointed, highly flexible, and sensitive shrewlike snout used in capturing prey (figure 11.8). They inhabit rocky, brushy, or forested areas, move slowly in a zigzag path with a waddling gait, and are omnivorous. Solenodons emit high-frequency clicking sounds that probably function in echolocation. Along with certain shrews and the platypus, they are one of the few mammals that uses a toxin. It is produced in the submaxillary glands, located at the base of the second lower incisor, which is large and deeply grooved to accommodate the toxic secretions (figure 11.9). Both species are considered endangered because of habitat loss and introduced predators, such as mongoose and feral cats, against which they are defenseless. Recovery of solenodon populations is also hampered by their low reproductive rate.

Although sometimes placed with the solenodontids, the West Indian shrews (Genus *Nesophontes*) currently are placed in a separate family—the Nesophontidae. The eight described species in this family became extinct within the last several hundred years, and some species probably within

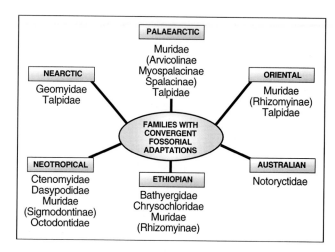

Figure 11.7   **Fossorial mammals.**   Fossorial species from among several families occur in all faunal regions.

Figure 11.8 **Unusual insectivores—the solenodons.**
The small eyes and shrewlike snout are characteristic of the
endangered Cuban solenodon. Prior to the arrival of Europeans,
solenodons were among the largest predators within their restricted
geographic area. They are quite defenseless against introduced
mongoose and house cats, however.

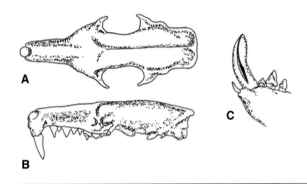

Figure 11.9 **Use of toxin for prey capture by solenodons.**
(A) Dorsal and (B) lateral views of the skull of a Cuban solenodon.
Note the incomplete zygomatic arch and enlarged first upper incisor.
(C) The large second lower incisor is deeply grooved to accommodate
secretions of toxin from the submaxillary gland below it.

the last 50 years. Unfortunately, extinction is a distinct possibility for surviving solenodons.

Soricidae

Shrews constitute the largest and most widely distributed
family of insectivores. There are 23 genera and about 312
species, although many are of uncertain taxonomic status.
This large family is placed in two subfamilies. The Subfamily Soricinae (the red-toothed shrews) include three tribes
found throughout much of the Nearctic, Palaearctic, and
Oriental faunal regions (figure 11.10). The Subfamily Crocidurinae (the white-toothed shrews) are restricted to Old
World faunal regions (see figure 11.10). Shrews are generally
small. Body mass ranges from 3 g and 35 mm head and body
length for adult pygmy white-toothed shrews (*Suncus etrus-
cus*), and pygmy shrews (*Sorex hoyi*), two of the smallest
mammals in the world, to 100 g and 150 mm head and body
length for the musk shrew (*Suncus murinus*). Most species of
shrews weigh 10 to 15 g and have a head and body length of
about 50 mm. Legs are short and the feet are unspecialized,
except in the web-footed water shrew (*Nectogale elegans*) and
the North American water shrew (*Sorex palustris*), which are
semiaquatic. Species associated with wet habitats have fimbriated hind feet. Shrews have small eyes and a long, pointed
rostrum. Their pelage is short, dense, and usually dark-
colored. In many species, lateral glands produce a musky
odor that is most noticable during the breeding season.

Shrew skulls have no zygomatic arch and no auditory
bullae. The teeth are noteworthy in several ways. The first
upper incisor is large, hooked, and has a posterior cusp that
appears to be a distinct tooth (figure 11.11). The first lower
incisor is long and procumbent (projecting horizontally forward). The upper molars are dilambdodont. The deciduous
teeth are shed prior to parturition. In members of the Sub-

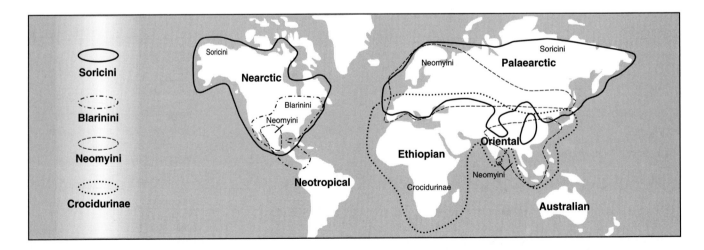

Figure 11.10 **Worldwide distribution of shrews.**   Geographic distribution of red-toothed shrews (the Subfamily Soricinae, which includes
three tribes: Neomyini, Soricini, and Blarinini) and the white-toothed shrews (Subfamily Crocidurinae).
Source: Data from S. Churchfield, The Natural History of Shrews, *1990, Cornell University Press, Ithaca, NY.*

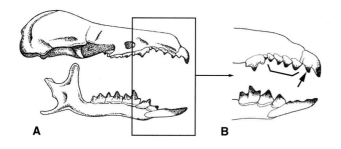

Figure 11.11   Characteristic features of shrew skulls.
(A) A typical shrew skull. Note the lack of the auditory bullae and zygomatic arch. (B) The enlarged anterior portion of a red-toothed shrew showing the distinctively pigmented enamel, procumbent first lower incisor, secondary cusp on the first upper incisor (arrow), and unicuspids (bracket). The incisors function as tweezers picking up insect prey that are then passed to the sharp, multicusped posterior teeth for crushing and chewing.

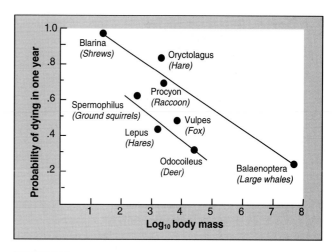

Figure 11.12   Mortality and body mass relationship.
Inverse relationship between body size in a number of different mammals and the probability of dying within the first year of life. Note that the chance of a shrew such as *Blarina* living much beyond one year is very low.

Source: Data from J. Eisenberg, The Mammalan Radiations, *1981, University of Chicago Press.*

family Soricinae, the tips of the teeth are a deep red (see figure 11.11). This color is caused by iron deposits (Dötsch and Koenigswald 1978) and is worn down and reduced as individuals age. Shrews with pigmented teeth are thus distinguished from the "white-toothed" shrews (Subfamily Crocidurinae), which do not have this characteristic. The groups also differ in biogeography, behavior, and physiology (Churchfield 1990; Merritt et al. 1994).

Shrews are mainly insectivorous, but many species are functional omnivores, consuming subterranean fungi. Shrews are found in many terrestrial habitats, usually with a heavy vegetative ground cover and abundance of invertebrates. This family is a good example of the ways body size constrains aspects of life history. Shrews are often closely associated with water or moist habitats because they have a high respiratory water loss—a characteristic directly related to small size. The average life span of most shrews is only about a year (figure 11.12) because they have such high mass-specific metabolic rates, again associated with small size. Shrews are too small to hibernate or migrate, although some may enter torpor. As a result, most forage throughout the year, which exposes them to extremes of weather and predators, and affects life span.

Many other aspects of these small insectivores are fascinating. Like solenodons, several species, including the short-tailed shrews and European water shrews (Genus *Neomys*), have salivary glands that secrete a toxin used to immobilize prey (Churchfield 1990). Many species of soricine shrews use high-frequency sounds for interspecific communication, orientation, and prey detection (Tomasi 1979; Forsman and Malmquist 1988; Churchfield 1990). Echolocation in shrews is not as well-studied as in bats. Crocidurine shrews, and presumably soricine shrews as well, exhibit a behavior called "caravanning." Churchfield (1990:33) described this behavior, which involves a mother and her young, as follows: "Each youngster grasps the base of the tail of the preceding shrew so that the mother runs along with the young trailing in a line behind her."

MACROSCELIDEA

The elephant shrews are in a single family (Macroscelididae), with 4 genera and 15 species. As noted previously, they are sometimes included in the Order Insectivora together with the tree shrews (Family Tupaiidae discussed in the next order) in the Suborder Menotyphla. Recent evidence of skeletal, cranial, and dental features, as well as molecular analyses, supports the current ordinal placement (Dene et al. 1980; Yates 1984).

Elephant shrews are restricted to central and eastern Africa from about 15°N latitude southward. An exception is the North African elephant shrew (*Elephantulus rozeti*), which occurs from Morocco to western Libya. Macrosceledids inhabit a diverse number of habitats, including desert, brushland, plains, forests, and rocky areas. They are strictly terrestrial and generally diurnal, except during hot weather, when they become increasingly nocturnal. Elephant shrews feed on insects and other animal and plant material. Although quadrupedal, they move bipedally in an erratic fashion when alarmed. Elephant shrews are easily alarmed: they are nervous animals and constantly twitch their ears and nose while making squeaking, chirping sounds.

Morphology

The common name of elephant shrews relates to their long, flexible, highly sensitive snout and large eyes and ears (figure 11.13). The pelage is long and soft. The hind legs are longer

Figure 11.13 Body morphology of elephant shrews.
Note the long proboscis, enlarged hind limbs, and kangaroolike
appearence of this short-nosed elephant shrew (*Elephantulus fuscipes*).
Convergence is evident in body form of elephant shrews and several
families of rodents, such as the kangaroo rat (Rodentia: Heteromyidae),
shown in figure 17.10, or the jerboa (Rodentia: Dipodidae),
in figure 17.15.

than the forelegs, which gives rise to the ordinal and family
name ("long limbs"). The kangaroolike hind legs allow for
hopping—a bipedal form of locomotion when moving
rapidly. Forelimbs are pentadactyl, whereas hind limbs have
either four or five toes. Size varies: head and body length and
mass range from 95 mm and a maximum 50 g, respectively, in
the short-eared elephant shrew (*Macroscelides proboscideus*), to
315 mm and about 400 g, respectively, in golden-rumped elephant shrews (Genus *Rhynchocyon*). Tail length is slightly
shorter than head and body length. The skull has a complete
zygomatic arch and auditory bullae, and the palate is noteworthy for its series of large openings (figure 11.14). All genera have functional incisors except *Rhynchocyon*. The molars
in elephant shrews are **quadrituberculate** (four-cusped), and
the occlusal surfaces are dilambdodont. As in tree shrews, a
cecum is present.

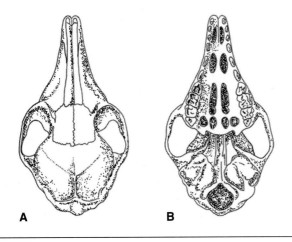

Figure 11.14 Elephant shrew skull characteristics.
(A) Dorsal and (B) ventral views of the skull of the North African elephant
shrew, showing the complete zygomatic arch, large auditory bullae, and
numerous perforations in the palate.

Fossil History

The earliest fossil elephant shrews, the Genus *Metoldobotes,*
date from the Oligocene of Africa. Later fossil macrosceledids are quite variable structurally: *Myohyrax* from the
Miocene is similar to a hyrax (Hyracoidea: Procaviidae), and
a Pliocene form, *Mylomygale,* is similar to a rodent (Patterson 1965).

Economics and Conservation

The golden-rumped elephant shrew (*Rhynchocyon chrysopygus*)
and the four-toed elephant shrew (*Petrodromus tetradactylus*)
are harvested for meat in Kenya. The rufous elephant shrew
(*Elephantulus rufescens*) is known to contract naturally occurring malaria and, with other species in the genus, is used in
medical research. Although an interesting group, elephant
shrews are of little economic importance.

SCANDENTIA

Because they have long been considered as the most primitive living primates, tree shrews have generated debate and
controversy out of proportion to their size as a group. As
such, they probably have a richer and more confused taxonomic history than any other mammalian family. An excellent review summarizing dental, skeletal, and anatomical
evidence against considering tree shrews as primates is given
by Campbell (1974). Anatomy and related systematics of the
group are discussed in detail by Luckett (1980). Most authorities consider tree shrews as a distinct lineage separate
from either the primates or insectivores. The order encompasses a single family, the Tupaiidae, and 5 recent genera
containing 19 species.

Tree shrews are restricted to the Oriental faunal region, ranging from India, southern China, and the Philippines southward through Borneo and the Indonesian
islands. Throughout their range, tree shrews occur in
forested habitats up to an elevation of 2400 m. They are diurnal, except for the feather-tailed tree shrew (*Ptilocercus
lowii*), and omnivorous, feeding on different plants and animals. Vocalizations are limited, but all species scent-mark
extensively. The common name is unfortunate because tupaiids are certainly not shrews. Also, the larger species such
as the Philippines tree shrew (*Urogale everetti*), confined to
the island of Mindanao, and the terrestrial tree shrew (*Tupaia* [*Lyonogale*] *tana*), of Borneo and Sumatra, spend most
of their time on the ground.

Morphology

Tree shrews superficially resemble squirrels, although they
have a more slender snout without vibrissae (figure 11.15).
The limbs are equal in length, pentadactyl, and have long
claws. Maximum total length is about 450 mm, half of which
is the tail. Several characteristics of tree shrews resemble

Figure 11.15   A typical tupaiid.   This common tree shrew (*Tupaia glis*) superficially resembles a squirrel. Tupaiids have had a confused taxonomic history—at various times being placed in the orders Primates and Insectivora, and currently within their own order, Scandentia.

those of primates, including their large braincase, postorbital bar, scrotal testes, and structure of the carotid and subclavian arteries. Unlike elephant shrews, with which they are sometimes grouped, tree shrews have tribosphenic molars and an unperforated palate. Like elephant shrews and several families of insectivores, the occlusal surface of the upper molars is dilambdodont. The dental formula is 2/3, 1/1, 3/3, 3/3 = 38. The lower incisors are procumbent and used for grooming.

Fossil History

The only definite fossil tree shrew, *Palaeotupaia sivalicus,* was described from mid-Miocene deposits of India (Chopra and Vasishat 1979). None of the previous reports of fossils referable to tree shrews, summarized by Jacobs (1980), provides unequivocal association with the family, however.

Economics and Conservation

Despite the continued loss of forest habitat throughout their range, none of the species of tree shrews is considered to be endangered. They do little damage to crops or plantations and are of no economic significance.

DERMOPTERA

Dermopterans (literally, "skin-winged") commonly are called "flying lemurs" or colugos. The latter name certainly is preferable, however, because dermopterans do not fly and are not lemurs. Instead, they glide (**glissant**). Like the tree shrews, this order also has a confusing taxonomic history. Historically, colugos have been grouped taxonomically with the bats, insectivores, and primates. This small order now contains a

single family, Cynocephalidae, which has one genus and only two species.

The Philippines colugo (*Cynocephalus volans*) is found in the southern Philippines, whereas the Malayan colugo (*C. variegatus*) occurs in Indochina, Malaya, Sumatra, Java, Borneo, and small islands nearby. Both species are completely arboreal and inhabit lowland and upland forests and plantations. Colugos are primarily nocturnal, foraging on flowers, leaves, and fruits. They den in tree cavities or hang upside down from branches during the day (with their head remaining upright, unlike bats). They move among trees very efficiently; colugos are the premier gliding mammals, and their morphological adaptations for this means of locomotion are pronounced.

Morphology

The patagium of colugos is more extensive than in any of the gliding marsupials or rodents. It extends from the neck to the digits of the forelimbs, along the sides of the body and hind limbs, and encloses the tail (figure 11.16). Colugos weigh between 1 and 2 kg, with head and body length 340 to 400 mm and tail length 170 to 270 mm. *Cynocephalus variegatus* is the larger species. Despite the colugos' size, glide distances of over 100 m are common, with very shallow glide angles. Thus, while gliding 100 m, a colugo loses less than 10 m in elevation. Colugos are helpless on the ground but adept at climbing high into trees, aided by long, curved claws. Their brownish gray-and-white, mottled pelage camouflages them against tree trunks. Colugos have a keeled sternum, as do bats and other gliding species.

The dental formula is 2/3, 1/1, 2/2, 3/3 = 34. The first two lower incisors are procumbent and **pectinate** ("comblike" with 5–20 distinct prongs from a single root), a feature unique to dermopterans. These lower incisors are used to grate food and groom the fur. The upper incisors are small with distinct spaces between them, and the postorbital processes and temporal ridges are well developed (figure 11.17).

Figure 11.16   Characteristic external features of colugos.
The extent of the patagium is evident in this colugo.

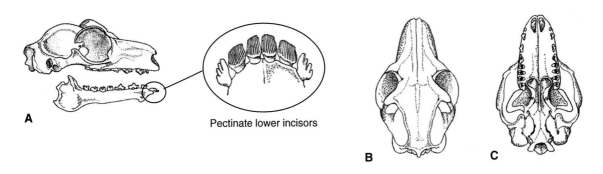

Figure 11.17 **Colugo skull features.** (A) Lateral, (B) dorsal, and (C) ventral views of the skull of a colugo with evident postorbital processes and temporal ridges. (Inset) Pectinate lower incisors, a characteristic feature of the two species in this unusual order.

Fossil History

The extinct Family Plagiomenidae is placed in the Order Dermoptera and dates from the late Paleocene epoch in North America and early Eocene epoch in Europe. There are no fossil remains of colugos from Asia, however.

Economics and Conservation

Colugos are hunted for their fur and for food, but habitat loss to logging and farming is a greater threat to populations. Neither species is presently considered threatened or endangered, however.

Summary

The Insectivora, Macroscelidea, Scandentia, and Dermoptera represent groups rich in diversity of form and function. The insectivores include families that demonstrate convergent evolution—the talpids and chrysochlorids. The insectivores retain many primitive eutherian characteristics, including their brain anatomy, dentition, cranial morphology, postcranial structures, and cloaca. As such, they are thought to be ancestral to other mammalian orders. In contrast to their primitive characteristics, many insectivores have developed highly specialized features, such as echolocation (tenrecs and shrews) and toxin in the saliva (solenodons and some shrews). Unlike in many other orders, no key character serves to identify insectivores.

Elephant shrews, tree shrews, and colugos were formerly included within the Insectivora. Each group is now placed in its own order, each with a single family. The taxonomic history of these orders has been somewhat of a mystery since the early 1800s, one that is still being resolved today.

Discussion Questions

1. What do you think might be the functional significance of the pigmented teeth found in the soricid Subfamily Soricinae—the red-toothed shrews? Likewise, what is the adaptive significance of caravanning behavior in shrews?
2. The text discusses how body size influences the life history characteristics of shrews. What other factors would you expect also are affected by small body size? How does latitude enter the picture?
3. How does body size influence the available range of prey?
4. Why have so few species of mammals developed the use of toxins as part of their life history strategies?
5. The common tenrec may have the largest litters of any mammalian species. Given its extremely large numbers of young, what might you expect are its neonatal survival rate, dispersal of young from the natal area, and associated factors?

Suggested Readings

Crowcroft, W. P. 1957. The life of the shrew. Max Reinhardt, London.

George, S. B. 1986. Evolution and historical biogeography of soricine shrews. Syst. Zool. 35:153–162.

Gorman, M. L. and R. D. Stone. 1990. The natural history of moles. Christopher Helm, London.

Nicoll, M. E. and G. B. Rathbun. 1990. African Insectivora and elephant-shrews: An action plan for their conservation. IUCN/SSC Specialist Group, IUCN, Gland, Switzerland.

CHAPTER 12

Chiroptera

Bats are second only to rodents in the number of recognized species, which now total about 920 in the order. They are one of the more fascinating groups of mammals for both the general public and professional mammalogists. In terms of feeding, reproduction, behavior, and morphology, including structural adaptations for true flight, bats show a greater degree of specialization than any other mammalian order. Bats also exhibit a great range in body size, from the tiny Kitti's hog-nosed bat (*Craseonycteris thonglongyai*) of Thailand, which weighs about 2 g, to fruit bats that weigh up to 1600 g. The vast majority of bats, however, are relatively small, weighing from 10 to 100 g. This chapter examines the structural and functional adaptations of bats.

Flight has allowed bats to be widely distributed and fill many feeding niches. Most species are insectivorous, often taking insects in flight or gleaning them from foliage. Others are carnivorous, taking small vertebrates, such as frogs, mice, and occasionally other bats. A few species (especially the Family Noctilionidae) feed on fish (they are piscivorous). Many species are nectivorous, consuming pollen and nectar from flowers, and the fruit bats are frugivorous. Finally, the well-known vampire bats, which feed only on blood (and have contributed to many popular misconceptions), are sanguinivorous. Different feeding adaptations may occur within a given family. For example, all these feeding modes are seen among species of New World leaf-nosed bats (Family Phyllostomidae). These different feeding habits and foraging styles among bat species result in a greater variety of head shapes, dentition, and facial features than occurs in other mammalian orders.

Variation is also evident in the reproductive patterns of bats. Many species breed in the spring and exhibit the typical pattern of spontaneous ovulation (see chapter 9). Others, however, such as some of the evening bats (Family Vespertilionidae) and horseshoe bats (Family Rhinolophidae) in northern areas, exhibit delayed fertilization. Mating occurs in the fall prior to migration or hibernation. Sperm is stored in the uterine tract, and ovulation and subsequent fertilization occur in spring. Parturition coincides with the emergence of insect prey. Other reproductive variations discussed in chapter 9 occur in different species of bats, including delayed implantation and delayed development (Racey 1982; Ransome 1990; Kunz and Pierson 1994).

From the family descriptions later in the chapter, you will see that bats roost in many different places: caves, hollow trees, under loose bark, buildings, understory vegetation, rock fissures, and many other protected places. Some species roost alone, whereas others form aggregations that can vary from small to huge. The roosting habits of bats are adaptations that reflect the interrelationships of social structure, diet, flight behavior, predation risks, and reproduction of each species (Kunz 1982). The majority of species are strictly nocturnal (Speakman 1995), although diurnal activity is common in species of island-dwelling Old World fruit bats (Family Pteropodidae) and a few species of microchiropterans (Kunz and Pierson 1994).

Small mammals usually have several large litters per breeding season. Bats are unusual, however, in their limited reproductive potential. Although there are exceptions (usually in tropical species), bats most often produce a single litter per year, with only one or two young per litter. Thus, declining bat populations may be slow to rebound. Neonates typically weigh 20% to 30% of the maternal body mass (Kurta and Kunz 1987). Hayssen and Kunz (1996) examined the relationship between litter mass and several maternal morphological characteristics in over 400 species of bats from 16 families. They found different relationships in pteropodids from those occurring in other (microchiropteran) bats, suggesting different selection pressures on reproductive traits in the two suborders (see the next section). Bats also have much longer life spans than is typical for small mammals, sometimes exceeding 30 years. In some species, prolonged life spans may be a function of reduced metabolic rate on both daily (torpor during the day) and seasonal (hibernation in winter) bases.

Because of the role bats fill as the major consumers of night-flying insects and pollinators of plants, chiropterans are important in community structure and in economics. Despite this, and their ability to adapt to various environments, bat populations often are negatively affected by environmental perturbations, both natural and human-induced. As a result, many species throughout the world are in danger of extinction through loss of cave and riparian (riverbank) habitats, exposure to pesticides, and human exploitation.

MORPHOLOGY

The 17 families of bats described in this chapter are divided into two distinct suborders: the Megachiroptera and the Microchiroptera. Megachiropterans include a single family (Pteropodidae), the Old World fruit bats. The other 16 families of bats make up the Suborder Microchiroptera. As you might expect, given the names of these suborders, size is one of the obvious differences between the two groups. There is considerable overlap in size, however, because the largest microchiropterans are much larger than the smallest megachiropterans. Several other morphological differences between these suborders are given in table 12.1.

Bats are the only mammals that fly; consequently, many of their unique morphological features relate to flight. The ordinal name Chiroptera, derived from the Greek words *cheir* (hand) and *pteron* (wing), refers to the modification of the bones of the hand into a wing, the primary adaptation for flight in bats (figure 12.1). The wing is formed from skin stretched between the arm, wrist, and finger bones. Several different muscle groups are used to keep the skin taut over the wing (Vaughan 1970a). Although the skin on the wings is very thin and appears delicate, it is fairly resistant to tears or punctures. The primary modification of the forelimb (wing) is its elongation, especially the forearm (radius and ulna), metacarpals, and fingers. The radius is

Table 12.1.   Major differences between megachiropterans, the Old World fruit bats (Family Pteropodidae), and microchiropterans, the remaining 16 families of bats

Additional differences are detailed by Koopman (1984) and Simmons (1995).

Microchiropterans	Megachiropterans
Echolocation for foraging and maneuvering; primarily insectivores	No echolocation, except in the Genus *Rousettus;* primarily frugivores or nectivores
Tragus often well-developed	No tragus; continuous inner ear margin
Nose or facial ornamentation often evident	No nose or facial ornamentation
No claw on second digit; second finger closely associated with third	Claw evident on second digit except in Genera *Dobsonia, Eonycteris, Notopteris, and Neopteryx;* second finger independent
Cervical vertebrae modified; head flexed dorsally from main axis of body when roosting	Cervical vertebrae not modified; head held ventrally when roosting
Tail and tail membrane (uropatagium) often evident	Tail and uropatagium usually absent
Generally small body size; eyes generally small	Generally large body size; eyes generally large
Mandible with well-developed, long, narrow, angular process	Angular process of mandible absent or broad and low
Postorbital process usually absent	Postorbital process well-developed
Palate usually does not extend beyond last upper molar	Palate extends beyond last upper molar

greatly enlarged in bats—sometimes being twice as long as the humerus—and the ulna is very much reduced. Because of the demands of flight, movement is restricted to a single plane in the wrist, elbow, and knee joints of bats. The radius cannot rotate (as it does in humans), and the wrist (**carpal** bones) moves only forward and backward (flexion and extension). Thus, the wing gains strength and rigidity to withstand air pressures associated with flight. Wing membranes usually attach along the sides of the body in bats, although

there are exceptions (see the section on the Family Mormoopidae in this chapter, for example).

The pectoral area is also highly developed and modified and contributes to a bat's ability to fly. The last cervical and first two thoracic vertebrae are often fused. Along with a T-shaped **manubrium** (the anterior portion of the sternum, or breast bone), the first two ribs form a strong, rigid pectoral "ring" to anchor the wings. In conjunction with the pectoral ring, the articulating **scapula** (shoulder blade) and proximal

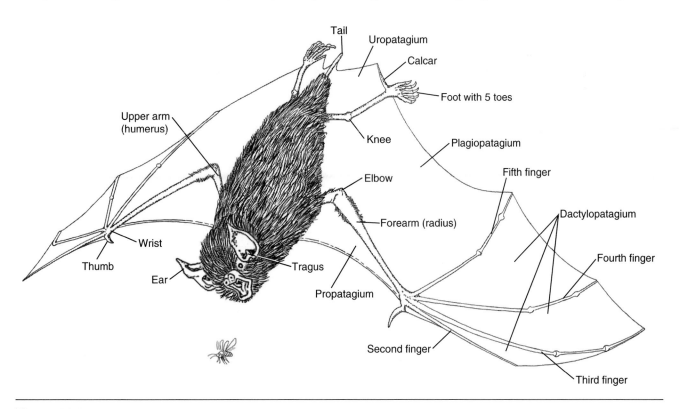

Figure 12.1   The major external features of bats.   Note the humerus, elongated forearm (radius), and fingers that form the wing. The claw on the thumb (digit one) occurs in all families, although it is rudimentary in smoky bats (Furipteridae) and short-tailed bats (Myzopodidae). Not all species have an external tail and uropatagium (see figure 12.2A).

end of the humerus are highly modified for flight (Vaughan 1970b). Besides their primary function in flight, wings also aid bats in thermoregulation. The thin, vascularized wing membranes dissipate excess body heat generated during flight. Additional aspects of the wings will be considered when we examine the mechanics of how bats fly.

In many species, the **uropatagium** (the membrane between the hind limbs that encloses the tail; see figure 12.1) also aids bats in flying. Although not necessary for flight (some species have no uropatagium), this membrane may contribute to lift and help stabilize the body during turns and other maneuvers. Aerodynamic stability is also enhanced in bats by their body mass being concentrated close to their center of gravity. The size and shape of the uropatagium, and whether it completely encloses the tail, vary considerably among and within families (figure 12.2A). The

calcar, a cartilagenous process that extends from the ankle, helps support the uropatagium.

The hind limbs of bats are small relative to the wings and are unique among mammals in being rotated 180 degrees, so that the knees point backward. This aids in various flight maneuvers, as well as in the characteristic upside-down roosting posture of bats. Bats roost head-down, hanging by the claws of the toes. A special locking tendon (Quinn and Baumel 1993; Simmons and Quinn 1994) allows them to cling to surfaces without expending energy.

As noted previously, the head and facial features of bats exhibit a great deal of diversity. Facial ornamentation, in the form of fleshy nose leafs, are especially elaborate in New World leaf-nosed bats, horseshoe bats (see figure 12.16), and the false vampire bats (Family Megadermatidae). Such ornamentation functions in the transmission of echolocation

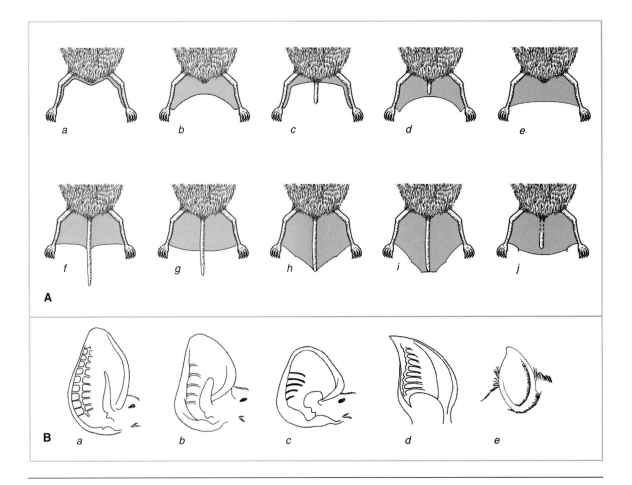

Figure 12.2   **Variation in the uropatagium, tail, external ear, and tragus of representative families of bats.**
(A) Differences in uropatagium and tail in fruit bats (Pteropodidae) (a) in the Genus *Pteropus,* as well as some New World leaf-nosed bats (Phyllostomidae). New World leaf-nosed bats (b); tube-nosed fruit bats in the Genera *Nyctimene* and *Paranyctimene* (Pteropodidae) (c); New World leaf-nosed bats (d); Kitti's hog-nosed bat (Craseonycteridae), with extensive uropatagium but no external tail (e); mouse-tailed bats (Rhinopomatidae) with long, thin tail free of the uropatagium (f); free-tailed bats (Molossidae) with about one-half the tail free of the uropatagium (g); evening bats (Vespertilionidae), similar to tails in horseshoe bats (Rhinolophidae) and false vampire bats (Megadermatidae) (h); slit-faced bats (Nycteridae) with T-shaped tip of the tail (i); sac-winged bats (Emballonuridae), fishing bats (Noctilionidae), and moustached bats (Mormoopidae) have a tail that protrudes through the uropatagium (j). (B) Representative sizes and shapes of the external ears and tragus in bats. Variation in size and shape of the ear and tragus in vespertilionids (a–c); antitragus in a rhinolophid (d); continuous inner ear margin without tragus or antitragus in a pteropodid (e).

pulses, because species in these families, as well as the slit-faced bats (Family Nycteridae), emit ultrasonic pulses through their nostrils.

"Blind as a bat" is a totally misleading phrase. All bats have perfectly functional vision. Eyes are large in megachiropterans but relatively small in microchiropterans. In microchiropterans, however, most of which are nocturnal, visual acuity is secondary. Primary perception of the environment is through acoustic orientation, both through echolocation (see the section "Echolocation" in this chapter) and audible vocalizations important in various social interactions. The importance of hearing to microchiropterans is reflected in the tremendous variety in the size and shape of the pinnae among species (Obrist et al. 1993; see figure 12.2B). Most microchiropterans have a **tragus,** a projection from the lower margin of the pinnae (see figure 12.2B), which is important in echolocation. Some species also may have an **antitragus,** a small, fleshy process at the base of the pinnae (see figure 12.2B). An antitragus is especially evident in groups in which the tragus is absent or reduced, as in the horseshoe bats and free-tailed bats (Family Molossidae).

How Bats Fly

All aspects of the biology and natural history of bats are associated with their ability to fly. Compared with birds, flight in bats is slow but highly maneuverable. To comprehend how a bat flies, it is necessary to understand a few basic aerodynamic terms (see Norberg 1990 for detailed analyses of flight in vertebrates). A bat (like a bird or an airplane) is able to fly because it generates enough **lift** to overcome gravity and sufficient forward propulsive thrust to overcome **drag** (figure 12.3A). Lift and thrust are achieved in bats through the structure of the wings, in conjunction with the pectoral bones and muscles, the uropatagium, and hind limbs.

As in airplanes, the dorsal surface of the wing in bats is convex and the ventral surface is concave, which causes air to move more rapidly over the wing than underneath it. This reduces the relative air pressure above the wing and results in lift (see figure 12.3B). In addition, to facilitate airflow, the surface of the wing is kept smooth and taut by layered elastic tissue in the wing. Generally, the greater the **camber** (the extent of the front-to-back curvature of the wing), the more

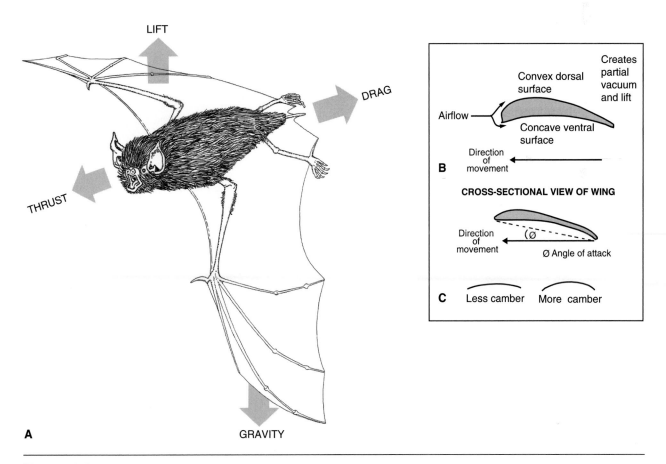

Figure 12.3   Aerodynamics of flight in bats.   (A) In order to fly, the power producing forward thrust must overcome "backward" drag, while upward lift must be greater than the effects of "downward" gravity. Drag increases in proportion to the surface area of the wings and air speed, as well as increased camber and angle of attack. (B) Lift is created as air moves over the convex top of the wing as a bat moves forward. It has a greater distance to travel relative to the concave bottom of the wing. This produces less pressure above the wing and results in upward lift. A decrease in pressure with increased velocity of the air (or fluid) is known as Bernoulli's principle. (C) The angle of attack is the angle of the wings relative to the horizontal plane in the direction of movement; camber is the degree of curvature of the wing.

lift that can be produced. Likewise, the greater the **angle of attack** (see figure 12.3C), the more lift that can be generated. If the camber and angle of attack are *too* great, however, the smooth flow of air over the surface of the wing is disrupted and becomes turbulent; lift is greatly reduced or lost completely. Bats can quickly adjust both camber and angle of attack throughout the wing-beat cycle by using several muscles in the wing, in conjunction with the movement of the wrist, thumb, fifth finger, and the hind limbs (Vaughan 1970c; Strickler 1978; Norberg 1990; Swartz et al. 1992).

As noted by Fenton (1985:16) "In flight the wing area proximal to the fifth digit provides lift, while the portion distal to the fifth digit acts as a variable pitch propeller providing thrust. Power comes from the contraction of nine pairs of flight muscles located on the chest and back. . . ." As a bat flies, thrust and lift are provided during the downstroke, in which the wings are fully extended and move in both a downward and forward motion. Little if any thrust or lift is developed during recovery, when the wings, partially folded to reduce drag, move upward and backward in preparation for the next downstroke. The chest muscles of bats power the downstroke and provide for lift and thrust. They are much larger than the back muscles, which function during the up-

stroke. In many species, the specialized scapula "locks" the movement of the humerus and stops the upstroke phase. See Vaughan (1970c) and Strickler (1978) for detailed analyses of the wing-beat cycle of bats and associated biomechanical factors.

The shape of a bat's wing is another factor that affects aerodynamic properties. When viewed from above, wing shape varies from short and broad to long and thin in different families and species (figure 12.4). The proportion of wing length to width is called the aspect ratio and results from the relative lengths of the metacarpals and phalanges of the third, fourth, and fifth fingers. A high aspect ratio—long, narrow wings—adapts a species for sustained, relatively fast flight, as is seen in many free-tailed bats. These species would be expected to forage high above the ground, predominately in open habitats free of obstructing vegetation. A low aspect ratio—short, wide wings—is seen in bats with slower, more maneuverable flight, as in the false vampire bats and the slit-faced bats (Family Nycteridae). A low aspect ratio is associated with bats that forage more often in habitats with dense, obstructing understory vegetation. Besides aspect ratio, low **wing loading**—the ratio obtained by the body mass of the bat divided by the total surface area of the wing—can

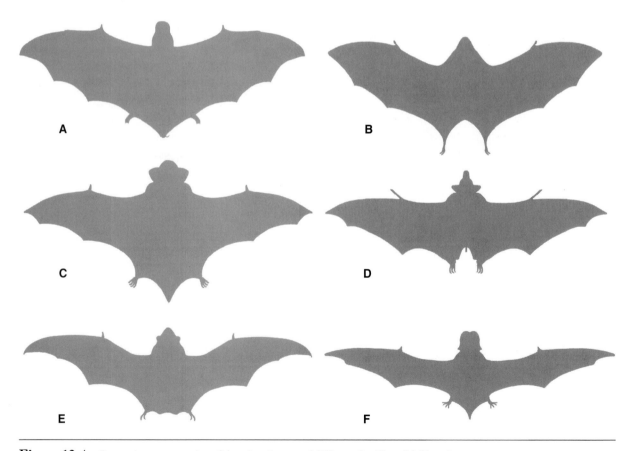

Figure 12.4 **Increasing aspect ratios of the wings in several different families of African bats.** The aspect ratio—calculated as the wingspan squared divided by the area of the wing—increases as wings become longer and thinner in different species of bats. (A) *Nycteris thebaica* (Nycteridae); (B) *Cardioderma cor* (Megadermatidae); (C) *Mimetillus moloneyi* (Vespertilionidae); (D) *Eidolon helvum* (Pteropodidae); (E) *Saccolaimus peli* (Emballonuridae); (F) *Mops midas* (Molossidae). Broader wings generally translate to slower, more maneuverable flight patterns.

enhance slow, maneuverable flight. The lower the wing load-ing, the greater the potential lift and capacity for slow flight. In general, a direct relationship exists between the wing morphology and flight patterns of bats and their foraging and life history characteristics (Vaughan 1959, 1966; Find-ley et al. 1972; Lawlor 1973; Altringham 1996).

Echolocation

To find prey items and maneuver within their environment, most bats **echolocate**—that is, they emit high-frequency sound pulses and discern information about objects in their path from the returning echoes. Many of the structural char-acteristics of bats, including their seemingly bizarre facial fea-tures, relate to echolocation. As with the relationship between wing morphology, flight patterns, and life history features, many aspects of echolocation also are species-specific and re-late to foraging. Most megachiropterans rely on vision and ol-faction to locate food, avoid obstacles, and maneuver in their environment (although Genus *Rousettus* has a limited form of echolocation using low-frequency tongue clicks). Vision in-volves interpreting information about objects in the immedi-ate environment from energy received as light waves produced by or reflected from those objects. Microchiropterans substan-tially augment visual senses with hearing, in the form of a highly sophisticated system of auditory echolocation. Hearing also involves interpreting information about objects in the en-vironment from energy (received as sound waves) produced by or reflected from objects. Hearing can give the same amount of information in terms of size, shape, texture, distance, and movement as vision.

In echolocation, a bat generates and emits high-frequency, high-intensity sounds through the mouth or nose. These sound pulses are then transmitted through, scattered by, or reflected back by objects in their path. When reflected, the returning sound (echo), with altered characteristics, may be picked up by the bat. The characteristics of the returning echo are altered from the original sound pulse depending on the physical characteristics of the object, including its dis-tance, texture, movement, shape, and size. These features are all interpreted by the bat from the echo. Although simple in concept, the actual process is extremely complex, and the sensory apparatus of bats is very sophisticated. For example, the echo from a target 1 m away returns to the bat in 6/1000 of a second. Although 6/1000 sec is very fast, bats are able to discriminate echo delays as short as 70 millionths of a sec-ond! (Suga 1990). Because bats cannot transmit and receive at the same time, outgoing pulses must be of short duration so as not to obscure the returning echo. Many species also can determine the presence of objects as thin as 0.06 mm (0.002 inch) in diameter, the width of a human hair.

High-frequency sound pulses are particularly suitable for bat echolocation, because objects about one wavelength in size reflect sound particularly well. For example, at a fre-quency of 30 kiloHertz (kHz), the **wavelength** (distance from peak to peak in a sound wave; figure 12.5) is about

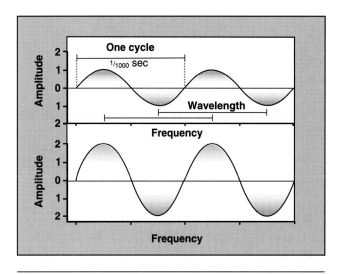

Figure 12.5   **Characteristics of a sound wave.**   The upper sound wave represents a 1 kHz pure tone wave and is the same frequency as the lower one, but the amplitude is twice as great in the lower soundwave. Amplitude, or the intensity of sound, is measured in newtons per square meter. A newton equals 10^5 dynes (a dyne is the force necessary to move 1 g of mass 1 cm/sec).

Source: Data from J.E. Hill and S.D. Smith, Bats: A Natural History, *1984, British Museum, London.*

11 mm—roughly equivalent to the length of a small moth. Also, low-frequency sounds (long wavelengths) can "wrap around" small targets such as insects and eliminate any echo. The negative aspect of high-frequency sound is that it is quickly absorbed by the atmosphere and thus has limited range relative to lower frequency sound. If they were within the range of human hearing (and many species use signals that are), the high-intensity echolocation signals of some species would be very loud, including the signals from rhi-nolophids, molossids, emballonurids, rhinopomatids, and many vespertilionids. Bats avoid self-deafening from loud outgoing pulses by disarticulating the bones of the middle ear and dampening the sound. Other bats, including nyc-terids, phyllostomids, and megadermatids, are sometimes called "whispering bats." Their much lower intensity signals have a less effective range than the higher intensity signals of other species. Interestingly, insects from at least six orders can hear ultrasound (Altringham 1996), and many night-flying insects have developed countermeasures to evade echolocating bats (Yager et al. 1990; Yager and May 1990; May 1991; Fullard et al. 1994; Fenton 1995).

The remarkable aspects of echolocation have been studied by numerous investigators (Simmons et al. 1975; Fenton 1985; Neuweiler 1990; Popper and Fay 1995) fol-lowing Donald Griffin's pioneering work in the 1940s (Grif-fin 1958). The sound pulses, produced by contraction of the cricothyroid muscles of the larynx (Novick and Griffin 1961; Novick 1977), are at frequencies above the range of human hearing, usually greater than 20 kHz. When a bat is taking insects in flight, the frequency and characteristics of echolo-cation pulses vary depending on whether the bat is in the

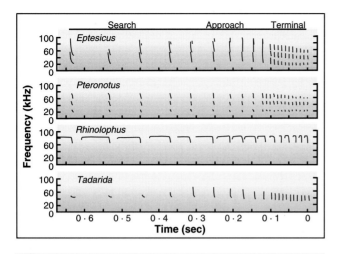

Figure 12.6   **Increased pulse repetition at terminal phase.**
The increased pulse rate during the terminal phase of closing in on a prey item, referred to as the "feeding buzz," is evident in these reproductions of sonograms from four species of bats: *Eptesicus fuscus* (Vespertilionidae), *Pteronotus personatus* (Mormoopidae), *Rhinolophus ferrumequinum* (Rhinolophidae), and *Tadarida brasiliensis* (Molossidae).
Source: Data from J.E. Hill and S.D. Smith, Bats: A Natural History, *1984, British Museum, London.*

search, approach, or terminal phase (figure 12.6). This variation involves changes in signal duration, loudness, and the nature and extent of frequency modulation and **harmonics** (integral multiples of fundamental frequencies that provide a broader scanning ability). All these factors change in different phases of approach to a target and under different foraging situations, as well as among different species of bats.

Bats perceive a great amount of detailed information from returning echoes, including the size, shape, texture, and relative motion of a target, in addition to the distance to it (figures 12.7 and 12.8). The constant-frequency (CF) component of the search phase, a pulse of constant pitch, is particularly useful for discriminating between moving objects, such as prey, and stationary obstacles. Moving objects can be determined by their **Doppler shift**—a change in sound frequency of an echo relative to the original signal caused by the movement of one or both objects: the source (bat) or the target (insect). We observe a Doppler shift when a loud car races by us as we are standing still. The pitch of the engine noise sounds higher as it comes toward us because the sound waves are being "pushed" closer together. They pile up, the wavelength becomes shorter, and the sound frequency (pitch) is higher. As the car moves away from us, the sound waves are now being "stretched out" and become longer. The sound is now at a lower frequency (also, low-frequency sound travels farther than high-frequency sound). Because a bat and its insect prey are moving relative to each other, the returning echoes may be shifted in frequency. Many species of bats are able to compensate for this phenomenon.

Broader band, frequency-modulated (FM) pulses, which go up or down, are better than CF pulses for determining the finer details of a target. Thus, the bat uses a series of FM pulses as it approaches closer to the target. Through-

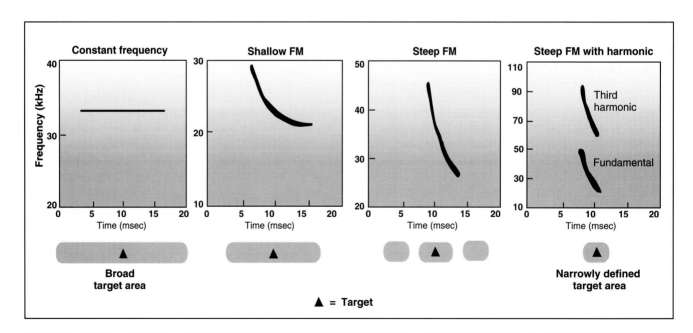

Figure 12.7   Typical pattern of frequency changes in echolocation pulses that continuously narrows the target area perceived by a bat.
Bats may use a narrow-band, long-duration, constant frequency (CF) pulse until a potential prey item (target indicated by ▲) is located. The target area (*shaded*) perceived by the bat becomes increasingly more narrowly defined as the bat shifts to broad-band, shorter duration, echolocation pulses including a shallow-frequency modulated (FM) sweep, a steep FM sweep, and finally a steep FM sweep with harmonics (which are integral multiples of fundamental frequencies) that pinpoint the target and provide fine details.
Source: Data from M.B. Fenton, American Scientist, *Vol. 69, 1981.*

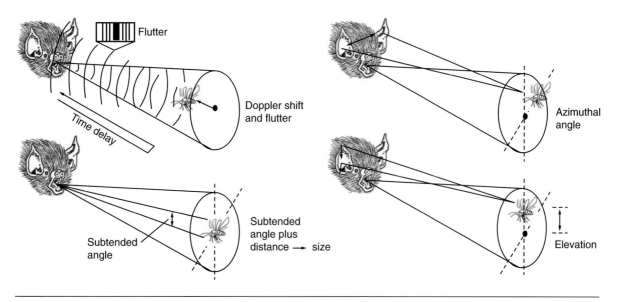

Figure 12.8   Various information gained from returning echolocation signals.   Distance and relative speed of the target are indicated by the time delay and strength (in decibels; dB) of the returning pulse and by the Doppler shift. Flutters indicate the presence of rapidly beating insect wings. The size of the target is reflected in the amplitude (subtended angle) of the returning signal in conjunction with the distance. Time delay and amplitude differences between the ears convey information on azimuth (angular distance from center), and the interference patterns of the sound waves within the inner surface of the ear indicate elevation.

Source:Data from N. Suga, "Biosonar and Neural Computation in Bats" in Scientific American, *262:34-41, 1990.*

out the sequence, the rate of sound pulses increases (see figure 12.6) until it is extremely rapid during the terminal phase. This is because even small movements of the prey quickly create large angular differences in its relative position the closer the bat comes. The search-approach-termination sequence is repeated very rapidly. The highly developed sensory apparatus, specialized neural anatomy, and sophisticated neural pathways of bats (Suga 1990) provide an amazing amount of detailed information about their surroundings.

Echolocation is not unique to bats; it occurs in other mammals, including some rodents, insectivores such as shrews and tenrecs, and seals (Fenton 1984). Only in the toothed whales, however, might echolocation approach the level of sophistication found in bats.

FOSSIL HISTORY

Considering the number of families and species of living bats, their fossil history is meager with fossil evidence existing for relatively few modern families (figure 12.9). Based on occlusal surface patterns of the molars, bats are believed to have evolved from arboreal, shrewlike insectivores (Carroll 1988), although the time of divergence is unknown and no intermediate forms have been discovered. The earliest fossil evidence of a bat is that of *Icaronycteris index,* from the early Eocene epoch of North America (Jepson 1966). Many of the characteristics of *I. index* are considered primitive or unspecialized relative to modern bats. Primitive features include the total number of teeth (38, which is close to the

primitive eutherian number of 44), lack of fusion in ribs and vertebrae, lack of a keeled sternum, and several other features noted by Jepson (1970). Nonetheless, other specialized morphological features of modern microchiropterans did occur in *Icaronycteris,* which was fully capable of flight (figure 12.10). In their essential features then, including flight and echolocation (Novacek 1985; Habersetzer et al. 1994), bats have existed for over 50 million years.

Megachiropterans are known from only two fossil genera: *Archaeopteropus* from the early Oligocene epoch of

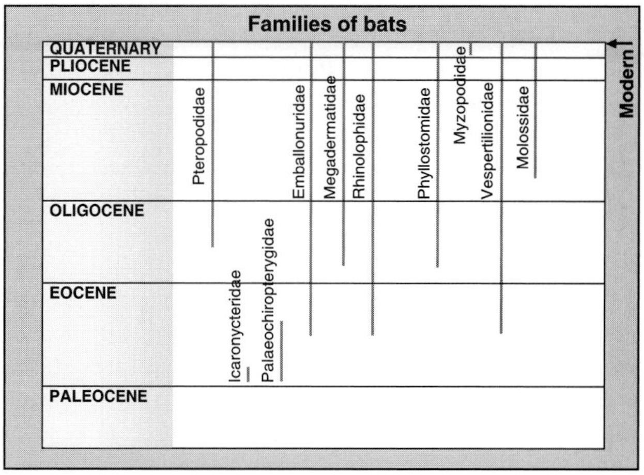

Figure 12.9   Fossil history of living chiropteran families.
The earliest fossil evidence for extant families is from the mid-Eocene epoch; only 8 of the 17 modern families of bats have fossil evidence.

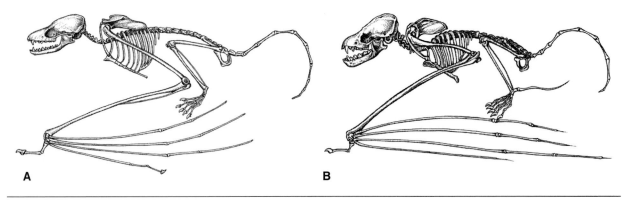

A   **B**

Figure 12.10   **Similarities between the oldest fossil bat and modern bats.**   (A) Fossil skeleton of the bat *Icaronycteris* from the early Eocene epoch. Note the development of the forelimb into a wing, with elongation of the phalanges, partial fusion of the radius and ulna, and dorsal position of the scapula. (B) Skeleton of a modern *Myotis* is similar in many respects to that of *Icaronycteris.*

Europe and *Propotto* from the early Miocene epoch of Africa (Butler 1978). Given the differences noted in table 12.1, it has been suggested that bats are **diphyletic;** that is, megachiropterans and microchiropterans may have evolved independently, with megachiropterans being more closely related to the Order Primates (Pettigrew 1986, 1995; Pettigrew et al. 1989). Based on numerous molecular and morphological studies, however, most investigators reject this hypothesis and consider bats to be **monophyletic,** that is, derived from a common ancestral lineage (Bennett et al. 1988, Baker et al. 1991, Bailey et al. 1992). Speakman (1993) and Simmons (1995) provide excellent summaries of the debate over monophyly verses diphyly in bats.

ECONOMICS AND CONSERVATION

Bats are associated with both positive and negative economic effects. On the negative side, bats are associated with many bacterial, rickettsial, viral, and fungal diseases (Constantine 1970, Sulkin and Allen 1974). One of the fears people have of bats stems from their association with rabies. Rabies has been reported in numerous Old and New World species, but the percentage of infected individuals in any population generally is no more than 1% to 4%, depending on the season and whether species are solitary or colonial (Constantine 1967; Brass 1994). Carnivores such as dogs, foxes, skunks, and raccoons are much more likely to be significant vectors of rabies than are bats (see chapter 27).

Vampire bats are directly responsible for losses of hundreds of millions of dollars to the livestock industry through disease transmission, including rabies. The distribution of vampire bats is restricted to New World tropical areas, however, as is their effect on livestock (see the section on Family Phyllostomidae in this chapter). Histoplasmosis is a fungal disease of the lungs sometimes associated with bat **guano** (fecal droppings) in caves and mines, but this health hazard is limited because few people frequent such places.

The benefits of bats far outweigh their negative attributes, however, and bats are one of the most economically beneficial mammalian groups. They are the only significant predator of nocturnal insects, consuming tremendous numbers of insect pests that would otherwise damage agricultural crops. For example, big brown bats (*Eptesicus fuscus*) feed on a variety of insects that are often vectors of plant diseases (Whitaker 1995). Bats also fill key ecological roles in many forest communities, especially as seed dispersers and plant pollinators (Fujita and Tuttle 1991; Fleming and Estrada 1993). Nearly 200 plant species of economic importance for food, timber, medicines, or fiber are pollinated by bats, including bananas, peaches, dates, and figs. Large accumulations of guano in caves have been mined in the past and used for fertilizer, as well as an ingredient in gunpowder. Bats also are used in a wide variety of medical research programs, including studies of drugs, disease resistance, and navigational aids for the blind.

Population declines and range reductions of many species of bats throughout the world are due to cave closures and loss of foraging and roosting habitats, siltation or draining of riparian areas, insecticide accumulations, and, in the case of large Old World fruit bats, consumption by humans. Many species have gone extinct in recent decades, especially endemic species of pteropodids (Kunz and Pierson 1994). In the United States, the gray bat and Indiana bat (*Myotis sodalis*) are endangered, as are some subspecies of big-eared bats (Genus *Plecotus* [formerly *Corynorhinus*]). A primary goal of conservation programs is to change people's perceptions of bats through education, so that they might learn to appreciate these fascinating mammals rather than fear and persecute them.

SUBORDERS

Megachiropterans

Pteropodidae

Pteropodidae is the only family in the Suborder Megachiroptera. The 42 genera and 166 species of fruit bats recognized by Koopman (1993) are divided into two subfamilies:

the fruit-eating Pteropodinae (36 genera and 154 species) and the pollen- and nectar-feeding Macroglossinae (6 genera and 12 species). Other authorities (see Hill and Smith 1984; Corbet and Hill 1991) consider four subfamilies of pteropodids based on feeding and morphological characteristics. Fruit bats occur throughout tropical and subtropical areas of the Old World, usually in forested or shrubby areas with a predictable supply of fruit. They are distributed south and east of the Sahara Desert in Africa, east through India, southeast Asia and Australia, to the Caroline and Cook Islands in the Pacific.

Major features that distinguish the Old World fruit bats from the other families of bats (microchiropterans) were given in table 12.1. Pteropodids do not echolocate, the exception being the low-intensity tongue clicking of members of the Genus *Rousettus*. Pteropodids have none of the distinctive facial features related to echolocation, such as a nose leaf or enlarged tragus, that are found in microchiropterans (figure 12.11). They navigate primarily using vision, and their large eyes are unusual anatomically. The choroid, which surrounds the retina, has numerous **papillae,** or projections. These extend into the retina and form undulating, uneven contours of unknown function (Suthers 1970). Most of the photoreceptors are rod cells, for black-and-white vision, although a few cone cells for color vision occur in some species, such as *Pteropus giganteus* (Pedler and Tilley 1969).

Body size varies considerably among members of this family. The smallest species, such as the long-tongued fruit bat (*Macroglossus minimus*), pygmy fruit bat (*Aethalops alecto*), or spotted-winged fruit bat (*Balionycteris maculata*), weigh only about 15 to 20 g; the largest species in the Genus *Pteropus* weigh up to 1600 g, a difference of two orders of magnitude. Wingspans may reach 2 m in some species of *Pteropus* and *Acerodon,* the largest of any bat. Sexual dimor-

phism is evident in greater body size in males in the Genera *Eonycteris, Epomops,* and *Epomophorus.* The hammer-headed fruit bat (*Hypsignathus monstrosus*) exhibits a greater degree of sexual dimorphism than any other species of bat, with males being nearly twice the body mass of females (Langevin and Barclay 1991). Male epauleted fruit bats (Genus *Epomophorus*) have light-colored tufts of fur on the shoulders ("epaulets"), associated with glandular patches used to attract females during breeding.

Dental formulas also vary, with the teeth being specialized for a diet of fruit. Although some genera have 34 teeth, the more specialized macroglossine nectar-feeders have a reduced number. Canines are always present, however. The molars generally have smooth surfaces, usually without cusps but with shallow, longitudinal grooves. As noted in table 12.1, the palate extends beyond the last molar. In frugivores, the palate has ridges (figure 12.12) against which the tongue mashes ingested food. The juice is consumed and pulp and seeds discarded.

Fruit bats are primarily nocturnal, although some species may be diurnal. Large species may fly up to 100 km between roosting and foraging areas, depending on the local availability of fruit. Food resources are located primarily through olfaction. Pollen and nectar make up a significant proportion of the diet in species in the Subfamily Macroglossinae. Many of these species have bristlelike papillae on their long, extensible tongues to help collect pollen from flowers. Pteropodids thus serve valuable ecological functions in pollination of many species of tropical plants and in seed dispersal (Rainey et al. 1995; Utzurrum 1995). Seeds usually are spit out or passed through the gut away from the parent tree, thus promoting dispersal (Kingdon 1972).

Bats usually roost in trees or shrubs (figure 12.13), although a few species use caves or buildings. Roosts often are communal, especially in the larger pteropodids. The short-nosed fruit bat (*Cynopterus sphinx*) roosts in "tents" it constructs in the stems or fruit clusters of various tree species

Figure 12.11   Facial features of a Gambian epauleted fruit bat (*Epomophorus gambianus*), Family Pteropodidae.   The large eyes, lack of facial ornamentation, and simple ears with no tragus or antitragus can be contrasted with most echolocating microchiropterans.

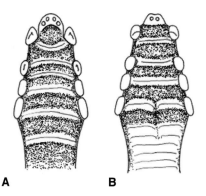

A   **B**

Figure 12.12   Ridges on palatal surface of fruit bats.
(A) Six ridges occur on the palate of epauleted bats (Genus *Epomophorus*). (B) Additional flat, posterior serrated ridges are found in fruit bats of the Genus *Epomops.*

Figure 12.13   **Roosting fruit bats in a tree.**   Communal roosts are particularly common in the larger pteropodids (Genus *Pteropus*).

(Balasingh et al. 1995; Bhat and Kunz 1995). Concentrations of up to a million straw-colored fruit bats (*Eidolon helvum*) may occur at roosting sites (Kulzer 1969). Conversely, roosting dawn bats (Genus *Eonycteris*) are solitary. Whether species are colonial or solitary is a function of several factors, including predation pressure, food availability, roosting site availability, and breeding pattern. Gestation periods range from 100 to 125 days, although some pteropodids such as *E. helvum* (Mutere 1965) exhibit delayed implantation.

Where large congregations of fruit bats occur, they may become serious crop depredators. In other areas, loss of roosting and foraging habitats and use of bats as food by people in Africa and on South Pacific islands (see Wilson and Graham 1992) have severely reduced bat populations. The functional role of pteropodids in community ecology was discussed by Rainey and colleagues (1995) and Richards (1995). Several species recently have become extinct, including the Palau Island flying fox (*Pteropus pilosus*), Guam flying fox (*P. tokudae*), and Panay giant fruit bat (*Acerodon*

lucifer). Others are considered endangered, including the tube-nosed fruit bat (*Nyctimene rabori*), Comoro black flying fox (*P. livingstonii*), Rodrigues flying fox (*P. rodricensis*), and Samoan flying fox (*P. samoensis*).

Microchiropterans

Rhinopomatidae

The mouse-tailed bats are named for their long tails, which are nearly equal to the length of the head and body (see figure 12.2A [f]). They are the only living microchiropterans with a tail longer than their forearm. Adults weigh from 6 to 14 g and have dark dorsal and paler ventral pelage, with naked areas on the face, rump, and abdomen. The ears are large and connected across the forehead by a flap of skin, and the slitlike nostrils are valvular. The single genus, *Rhinopoma*, contains three species found throughout North Africa and the Middle East to India and Sumatra. These insectivorous bats occur in arid regions where they roost in caves, cliffs, houses, and even in Egyptian pyramids. Although mouse-tailed bats do not hibernate during the winter, they enter torpor and do not forage, living off stored body fat. Females are monestrus and give birth to a single young in July or August after a gestation of about 123 days. Qumsiyeh and Jones (1986) summarized life history characteristics of two species of mouse-tailed bats. As noted from figure 12.9, there is no fossil history for this family. They are considered the most primitive microchiropterans, however, based on several morphological features, including two phalanges on the second digit of the manus, unfused premaxillary bones, and an unmodified anterior thoracic region (Koopman 1984), as well as primitive chromosomal characteristics (Qumsiyeh and Baker 1985).

Emballonuridae

This family comprises 13 genera and 47 species of sac-winged or sheath-tailed bats. They enjoy a widespread distribution in tropical and subtropical habitats from Mexico to Brazil, sub-Saharan Africa, the Middle East, India, southeast Asia, Australia, and Pacific islands eastward to Samoa. Emballonurids range in body mass from 5 to 105 g. The total number of teeth varies from 30 to 34, and dentition is adapted for an insectivorous diet; molars have a distinct dilambdodont occlusal pattern.

Emballonurids roost in caves, logs, houses, crevices, or the leaves of shrubby vegetation. Some species, such as ghost bats (Genus *Diclidurus*) are solitary, whereas others form groups up to 50 individuals, including the sharp-nosed bat (*Rhynchonycteris naso*) and the white-lined bats (Genus *Saccopteryx*). The common names of the family refer to their distinct morphological features. They are known as sac-winged bats because of glandular wing sacs on the ventral surface of the wings near the elbow (within the **propatagium;** see figure 12.1). The sacs are most prominent in males and exude a red, odiferous substance that is probably important in pheromone production and attraction of females. The

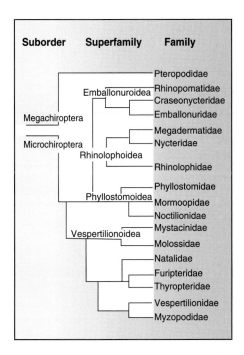

Figure 12.14 **Phylogenetic arrangement of bats.**
This dendrogram suggests possible relationships between the
superfamilies and families of living chiropterans.

family also is referred to as sheath-tailed bats because of the
way the tail protrudes through the uropatagium (see figure
12.2A [j]) and slides up and down with movement of the
hind limbs. Among emballonurids, the Seychelles sheath-
tailed bat (*Coleura seychellensis*) is considered endangered.
This family also is considered primitive and, with the
mouse-tailed bats and Kitti's hog-nosed bat (see next section), is grouped in the Superfamily Emballonuroidea (figure 12.14).

Craseonycteridae

This recently described monotypic family (Hill 1974) includes only Kitti's hog-nosed bat (*Craseonycteris thonglongyai*), known from a few locations in Kanchanaburi
Province, western Thailand. It is the smallest species of bat
and, based on body mass, the smallest mammalian species
in the world. Head and body length is about 30 mm, and
body mass is about 1.5 to 2.0 g. They have large ears and
tragus, and a distinctive plate on the nose. There is no external tail or calcar. Also known as the bumblebee bat, this
species has a large uropatagium and broad wings adapted
for slow, hovering flight. The bats roost in small caves,
where they are solitary. They feed on insects taken in flight
or gleaned from foliage, as well as small spiders (Hill and
Smith 1981). The echolocation characteristics of the
species and their relation to foraging were described by
Surlykke and colleagues (1993). This species is believed
to be most closely related to the sac-winged and mouse-
tailed bats.

Nycteridae

There is a single genus (*Nycteris*) and 12 species of slit-faced
bats in this family. Ten species are distributed throughout
sub-Saharan Africa and Madagascar; *N. tragata* is found in
Malaysia, Sumatra, and Borneo; and *N. javanica* is on the
East Indies islands of Java, Bali, and Kangean. They occur in
habitats ranging from semiarid areas to savannas and tropical forests, where solitary individuals or small colonies roost
in caves, trees, houses, or animal burrows. Nycterids are
small to medium-sized bats weighing 10 to 30 g. Head and
body length is 40 to 90 mm, with tail length up to 75 mm.
They have large ears and a small tragus. The common name
derives from the unusual longitudinal groove throughout the
facial region (figure 12.15), which along with the external
nose leaf, functions in their low-intensity echolocation calls,
emitted through the nostrils. The uropatagium completely
encloses the tail and is partially supported by the unique, T-
shaped tip of the tail (see figure 12.2A [i]). Most species are
insectivorous, but *N. grandis* also feeds on vertebrates, including other bats (Fenton et al. 1981).

Megadermatidae

The false vampire bats (the name reflects the historical, false
belief that they feed on blood) encompass four genera and
five species. They are found in tropical and savanna habitats
throughout central Africa, India, southeast Asia, the East
Indies, and Australia. These are moderate to large bats; the
Australian giant false vampire or ghost bat (*Macroderma gigas*) has a wingspan of 60 cm and body mass up to 170 g
(Breeden and Breeden 1967; Taylor 1984). Megadermatids
have large ears that are united across the forehead, a **bifid**
(divided) tragus, and a large, erect nose leaf. All species lack
upper incisors, which is unusual for bats.

They roost in caves, trees, buildings, and bushes. The
four genera are variable in their feeding and roosting habits.
The African yellow-winged bat (*Lavia frons*) consumes insects and may be diurnal as well as nocturnal. Roosts may
contain single individuals or groups of up to 100. The
African false vampire bat (*Cardioderma cor*) roosts in groups

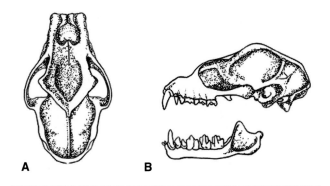

Figure 12.15 **Skull of a slit-faced bat.** (A) The dorsal aspect
shows the deep concavity between the eye orbits. (B) Lateral view of
the skull and mandible.

Table 12.2.   Morphological and behavioral differences separating the two subfamilies of horseshoe bats

Rhinolophinae	Hipposiderinae
Three phalanges in all toes of the foot except the first	Two phalanges in all toes of the foot
A median projection (sella) extends from the nose leaf (see text)	No median projection
No sac behind nose leaf	Males often have a sac behind nose leaf that can be everted and secretes a waxy material
Dentition: 32 teeth, with small lower premolar	Dentition: 28–30 teeth, with no small lower premolar
Generally forage individually	Generally forage in small groups

up to 80 individuals and feeds primarily on ground-dwelling beetles and scorpions (Csada 1996). The two species of Asian false vampire bats (Genus *Megaderma*) and *Macroderma gigas* feed on small vertebrates, including lizards, rodents, small birds, and other bats (Marimuthu et al. 1995). *Megaderma* may form very large colonies, whereas *Macroderma gigas* roosts alone or in small groups. Worthington and colleagues (1994) found extreme genetic subdivision between disjunct maternity sites of *M. gigas*.

Rhinolophidae

Koopman (1993) considered the horseshoe bats to include 130 species in 10 genera. He divided them into two subfamilies, the Rhinolophinae and the Hipposiderinae, based on several distinctive morphological and behavioral characteristics (table 12.2). The latter subfamily includes the Old World leaf-nosed bats, which are considered by some authorities (see Hill and Smith 1984) to merit their own family (Hipposideridae). The family is widely distributed throughout Europe, Africa, the Middle East, Asia, Japan,

the East Indies, and northern Australia. Rhinolophids occur in a wide variety of habitats from deserts to tropical forests.

As might be expected given the large number of species, horseshoe bats exhibit considerable morphological and behavioral variation. They are named for their distinctive facial ornamentation (figure 12.16), which ranges from structurally simple to very ornate. Generally, this includes a nose leaf above the nostrils and a flap of skin shaped like a horseshoe below and on the sides of the nostrils. Between the horseshoe and nose leaf is a median projection called the **sella**. As noted in table 12.2, similar nose ornamentation occurs in the Subfamily Hipposiderinae, but without the sella. Nose ornamentation functions in directing the echolocation pulses emitted through the nostrils, not from the mouth (Robinson 1996). Rhinolophids generally have large ears without a tragus. The tail usually is short and the uropatagium small.

Rhinolophids are insectivores, taking prey in flight or gleaning insects from vegetation. They roost singly or in

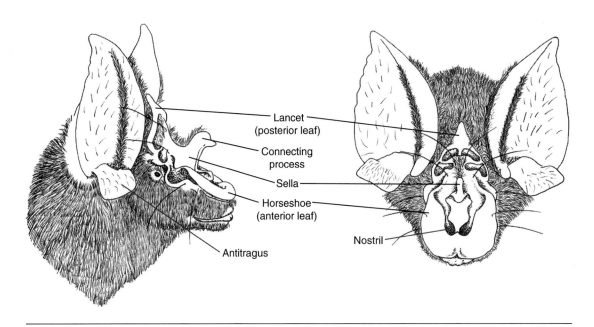

Labels: Lancet (posterior leaf), Connecting process, Sella, Horseshoe (anterior leaf), Antitragus, Nostril

Figure 12.16   Distinctive nose leaf of horseshoe bat.   Note the anterior horseshoe shape, connecting stella, and posterior leaf in *Rhinolophus euryotis*. Horseshoe bats emit echolocation pulses through their nostrils, and this complex structure helps to beam the sound waves.

small groups in caves, crevices, hollow trees, or houses. While roosting, they wrap their wings around themselves and fold up the uropatagium, thus appearing to be large cocoons. Species that hibernate exhibit delayed fertilization, mating in the fall with ovulation and fertilization in the spring. Species from warmer regions ovulate and mate in the early spring. Nevertheless, rhinolophids commonly have a single young, with birth and lactation coinciding with the peak occurrence of insects in the spring.

Mormoopidae

This New World family formerly was considered a subfamily of the Phyllostomidae but gained family status based on work by Smith (1972). The two genera and eight species are known as the mustached bats or naked-backed bats. They occur in semiarid to tropical forest habitats from the southwestern United States through Mexico, Central America, and the Caribbean to northern South America. Mormoopids are small to medium-sized bats. All have a tail, and a tragus with a secondary fold of skin, best developed in the Genus *Mormoops*. This group differs from phyllostomids in lacking a nose leaf and having very small eyes. The lips are enlarged and surrounded by numerous stiff hairs and, with a platelike growth on the lower lip, form a funnel into the mouth. The naked-backed bats (*Pteronotus davyi* and *P. gymnonotus*) actually have thick pelage on their back but appear naked because their wings meet middorsally, obscuring the fur. All mormoopids are insectivorous, have long, narrow wings for rapid flight, and usually forage near water. They roost in warm, humid caves and are gregarious, sometimes forming large colonies.

Noctilionidae

Known as bulldog (figure 12.17A) or fishing bats, this family encompasses a single genus (*Noctilio*) and two species distributed in tropical lowlands from northern Mexico to northern Argentina and on several Caribbean islands. The greater bulldog bat (*N. leporinus*) has a head and body length up to 13 cm. Males may weigh 70 g (Davis 1973) and have a 50-cm wingspan. These bats hunt at night and use the very long hind legs and large, rakelike feet with well-developed claws to scoop up small fish (figure 12.17B), crustaceans, and aquatic insects (Wenstrup and Suthers 1984; Brooke 1994) detected through echolocation near the surface of fresh or saltwater (Schnitzler et al. 1994). They also take terrestrial invertebrates. Small groups usually forage together. The lesser bulldog bat (*N. albiventris*) is smaller, 8 to 9 cm in head and body length, with maximum body mass about 40 g (Davis 1976). The hind limbs are not as well developed as in *N. leporinus,* and *N. albrentris* apparently feeds primarily on insects, again taken by echolocation. Both species roost in cliffs, caves, buildings, and hollow trees (figure 12.18) located near water. Noctilionids, mormoopids, and phyllostomids (see the next section), form the Superfamily Phyllostomoidea (see figure 12.14).

A

B

Figure 12.17   A fishing bat.   (A) From this close-up of the head, it is easy to see why they are also called bulldog bats. Note the long ears and well-developed tragus, with a comblike lateral edge, and the lack of a nose leaf. (B) A fishing bat taking a small fish from the water.

Phyllostomidae

This large New World family of leaf-nosed bats consists of 49 genera and about 141 species. Based on dentition, nose leaf structure, and skull features, Koopman (1993) recognized eight subfamilies (table 12.3), including the vampire bats (Desmodontinae). Authorities disagree, however, about the number of subfamilies (see Baker et al. 1989; Jones and Carter 1979).

Phyllostomids are widely distributed from the southwestern United States through Central America, the Caribbean, and northern South America. They are found from sea level to high elevations, in habitats ranging from

Figure 12.18   Roosts of fishing bats.   These lesser fishing bats are roosting in a hollow tree stump on the bank of the New River in northern Belize.

Table 12.3.   Subfamilies, number of genera, and number of species within the Phyllostomidae

Subfamily	Number of Genera	Number of Species
Phyllostominae	11	33
Lonchophyllinae	3	9
Brachyphyllinae	1	2
Phyllonycterinae	2	3
Glossophaginae	10	22
Carolliinae	2	7
Stenodermatinae	17	62
Desmodontinae	3	3

Source: Data from K.F. Koopman, "Order Chiroptera" in Mammal Species of the World: A Taxonomic annd Geographic Reference, *2nd ed., eds. D.E. Wilson and D.M. Reeder, 1993, Smithsonian Institution Press, Washington, D.C.*

Glossophaginae, Lonchophyllinae, Phyllonycterinae, Brachyphyllinae, Stenodermatinae, and Carolliinae primarily feed on nectar, pollen, or fruit.

The three species of true vampire bats (sometimes included in their own family, the Desmodontidae) are sanguinivorous, that is, they consume only blood. Their dentition is specialized (figure 12.19), with large, sharp incisors for making an incision, and large canines. The cheek-teeth are reduced in size, number, and complexity, because blood does not require chewing. The common vampire bat

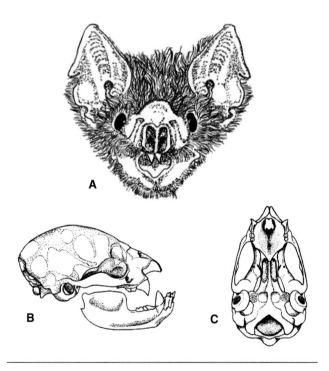

Figure 12.19   The common vampire bat.   (A) Note the flat nose and relatively large eyes. (B) Lateral and (C) ventral views of the skull show the specialized teeth. The incisors and canines are enlarged, with great reduction in the size and number of molars. *Desmodius rotundus* has a total of 20 teeth; other phyllostomids have 32 to 34 teeth.

desert to tropical forest. Nose-leaf ornamentation is present in most genera, although it is not as pronounced as in the Old World leaf-nosed bats (Rhinolophidae: Hipposiderinae). For those genera of phyllostomids without the nose leaf, such as vampire bats and many fruit-eating species, the upper lip usually has some type of plate or other prominent outgrowth. The ears are sometimes connected at the base, and a tragus is present. A tail and uropatagium may or may not be present. Diversity in body size is great, ranging from small species such as *Ametrida centurio*, with a head and body length of 4 cm, to the largest New World species, Linnaeus' false vampire bat (*Vampyrum spectrum*, not to be confused with megadermatids), with a head and body length of 13 cm and a wingspan of about 0.6 m.

Phyllostomids also exhibit a variety of feeding habits. Many of the smaller Phyllostominae are insectivorous. Larger species in this subfamily, such as *V. spectrum*, the spear-nosed bat (*Phyllostomus hastatus*), and Peter's woolly false vampire bat (*Chrotopterus auritus*), are carnivorous, feeding on small vertebrates. Species in the Subfamilies

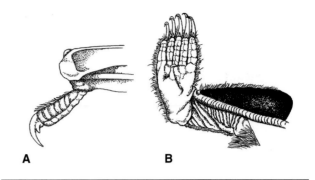

Figure 12.20   **Claws of the New Zealand short-tailed bat.**
(A) The basal talon on the thumb is particularly evident. (B) Small talons also occur on the claws of the feet. Like flightless birds, these bats are highly adapted for terrestrial locomotion.

(*Desmodus rotundus*) often preys on mammals, especially large **ungulates** (hoofed mammals), both domestic and wild. In contrast, the white-winged vampire bat (*Diaemus youngi*) and hairy-legged vampire bat (*Diphylla ecaudata*) probably prey more on birds.

Aspects of the biology of the common vampire bat are detailed in Brass (1994). Vampire bats make an incision about 1 cm long and 5 mm deep and 4 mm wide (Greenhall 1972) in the tail, snout, ears, feet, anus, or other area with sparse amounts of hair. Grooves on the sides and bottom of the tongue produce suction through capillary action while the bat laps up blood. Anticoagulant in the bat's saliva may keep the blood from coagulating for several hours (Hawkey 1966; Cartwright 1974). Nonetheless, blood loss is minimal, and prey individuals usually are unaware that they have been attacked. An individual vampire bat probably consumes less than 8 L of blood a year, based on an average consumption of about 25 mL/day (Wimsatt and Guerriere 1962). The open wound is subject to bacterial infections, parasites, and rabies, however. This can make livestock ranching impractical in areas of Central and South America within the range of vampire bats (Turner 1975; World Health Organization 1984; Delpietro et al. 1993).

Food sharing in vampire bats (Wilkinson 1984, 1990) is a fascinating social behavior that is a rare example of reciprocal altruism. Vampire bats will starve to death if they do not eat for about 3 days. Within roosting colonies of females (males roost individually), a bat that has eaten recently regurgitates blood to a roost-mate that is close to starvation. Food sharing even occurs between unrelated individuals but only in bats that have a close roosting association, that is, between those who can return the favor when necessary.

Phyllostomids roost in caves, trees, buildings, or the burrows of other animals, often with other species of bats. Roosts of single individuals, small groups, or large clusters may occur. The tent-building bats (Genus *Uroderma*), white bat (*Ectophylla alba*), and several species of fruit-eating bats (Genus *Artibeus*) construct a roost shelter by biting through the midrib of palm fronds, which causes the leaf to fold over and form a protective "tent." Kunz and McCracken (1996) suggested that the tent-making behavior may provide resources that aid in polygyny (males mating with more than one female). Eighteen species of tropical bats are known to make tents: 15 in the New World and 3 in the Old World (Kunz et al. 1994). The Puerto Rican flower bat (*Phyllonycteris major*) is recently extinct.

Mystacinidae

The two species of New Zealand short-tailed bats are unusual in several of their structural adaptations. Adapted for terrestrial locomotion to a greater degree than other bats, mystacinids have very sharp claws on the hind feet and thumb, each with a small, basal talon (figure 12.20). They also have a thick membrane along the side of the body. When an individual is not flying, the wings can be folded up, much like furling a sail. This makes the bats much more agile

on the ground. The uropatagium can be "furled" in a similar manner. Short-tailed bats roost in caves, trees, and burrows in forested areas. The only species of bat known to burrow, they use their upper incisors to excavate. The dentition is typical for insectivorous bats, although the diet is diverse; in addition to insects, they eat fruit, nectar, and arthropods. Although both species were once widely distributed throughout New Zealand forests, *Mystacina tuberculata*, the smaller of the two species with a body mass of 12 to 15 g, is now reduced to a small number of populations (Flannery 1987). The larger *M. robusta* (body mass of 25–35 g) has not been found since 1965 and is probably extinct.

Natalidae

The single genus (*Natalus*) and five species of funnel-eared bats occur from northern Mexico to northern South America and on many Caribbean islands. Natalids are small bats. Head and body length is 35 to 55 mm, and adults weigh from 4 to 10 g. The long, soft, dorsal pelage is yellow to reddish brown. The tragus is short, and the large ears are funnel-like, giving the family its common name. The forehead region is domed, and there is no nose leaf. Adult males have an unusual mass of glandular sensory cells, the "natalid organ," below the skin of the forehead (Dalquest 1950; Hoyt and Baker 1980). The wings are long and slim (a high aspect ratio), and natalids have an erratic, butterflylike flight as they consume insects in flight. These bats roost singly or in small clusters far back in caves and tunnels in lowland areas. Natalids are most closely related to two other small New World families, the Furipteridae (the thumbless bats) and the Thyropteridae (the disc-winged or sucker-footed bats), within the Superfamily Vespertilionoidea (see figure 12.14).

Furipteridae

The smoky bats include two genera, each with a single species. *Furipterus horrens* is found from Costa Rica to Peru and Brazil, whereas *Amorphochilus schnablii* occurs along coastal areas and interior valleys from western Ecuador to northwestern Chile. These bats are small; the body mass of

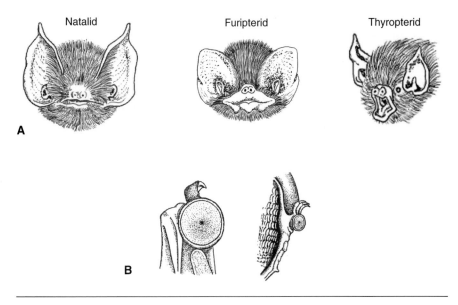

Figure 12.21   Facial features and discs in bat families.   (A) A funnel-eared bat (*Natalus stramineus; left*), a smoky bat (*Amorphochilus schnablii; center*), and a disc-winged bat (*Thyroptera tricolor; right*). Note the characteristic three fleshy projections on the lower lip of the smoky bat. (B) Adhesive discs of a disc-winged bat are larger on the wrist (*left*) than on the foot (*right*).

F. horrens is only 3 g. They resemble funnel-eared bats in their general appearance (figure 12.21A), wing shape, inconspicuous nose leaf, ear shape, small tragus, and dense fur. Furipterids also are known as thumbless bats because this digit is so small and almost completely enclosed in the wing membrane that it appears to be absent. Smoky bats are insectivorous and inhabit moist, low-elevation forested areas, where they roost in caves.

Thyropteridae

Of the two species of sucker-footed, or disc-winged bats, *Thyroptera discifera* occurs from Nicaragua southward to northern South America, and *T. tricolor* is distributed from southern Mexico to Bolivia and southern Brazil. These bats are small; the body mass of *T. tricolor* is about 4 g. They are similar to the funnel-eared bats and smoky bats in features such as wing shape, high domed forehead, ear shape, and tragus (see figure 12.21A, *right*). The common names of these bats refer to their most striking anatomical feature: round, concave, suction discs at the base of the thumbs and on the soles of the feet (Wimsatt and Villa-R. 1970). The larger thumb discs are attached by a short stalk, or **pedicle** (see figure 12.21B). Discs are used to cling to stems, leaves, and other smooth, hard surfaces. A single disc is capable of supporting an individual's body mass. These species occupy humid forests, usually near water, where they feed on insects. They roost during the day, adhering by their discs inside curled leaves of plants, often the fronds of banana trees (Genus *Heliconia*). Several individuals, possibly a family group, sometimes share a single leaf (Findley and Wilson 1974), which usually is used as a roosting site for only a day.

Myzopodidae

This monotypic family includes the Old World sucker-footed bat (*Myzopoda aurita*), the only bat species **endemic** (naturally occurring in a limited geographic area) to Madagascar. Like the New World disc-winged bats (Family Thyropteridae), *M. aurita* has suction discs at the base of the thumbs and soles of the feet. There is no stalk associated with the thumb disc in *M. aurita*, however. The discs in myzopodids and thyropterids also differ anatomically (Schliemann and Maas 1978). Discs probably evolved independently in each family, another example of convergent evolution. A mushroom-shaped process at the base of the ear (figure 12.22) in this species is unique. *M. aurita* is insectivorous and probably restricted to rain forest habitats on the east coast of the island. The species is extremely rare; very few specimens have ever been collected, and little is known of its biology (Goepfert and Wasserthal 1995). Like many other endemic species on Madagascar, it may be endangered, but no definitive information exists on populations.

Vespertilionidae

There are more species of evening bats than any other family of chiropterans (or any mammalian family other than murid rodents). They enjoy a worldwide distribution in tropical, temperate, and desert habitats and are absent only

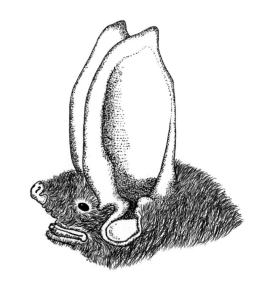

Figure 12.22   Old World sucker-footed bat.   Notice the fleshy, mushroom-shaped structure at the base of the ear. Its function is unknown.

Table 12.4.   Subfamilies, number of genera, and number of species within the Vespertilionidae*

Subfamily	Number of Genera	Number of Species
Kerivoulinae	1 (*Kerivoula*)	22
Vespertilioninae	30	269
Murininae	2	16
Miniopterinae	1 (*Miniopterus*)	10
Tomopeatinae	1 (*Tomopeas*)	1

Source: Data from K.F. Koopman, "Order Chiroptera" in Mammal Species of the World: A Taxonomic and Geographic Reference, *2nd ed., eds. D.E. Wilson and D.M. Reeder, 1993, Smithsonian Institution Press, Washington, D.C.*
* Compare with Hill and Smith (1984)

from polar, high-elevation areas, and some oceanic islands. Koopman (1993) placed the 35 genera and 318 species of vespertilionids in five subfamilies (table 12.4). Within the Subfamily Vespertilioninae, the Genus *Myotis* (with 84 species) is the most widely distributed genus of bat, occurring worldwide.

As might be expected of so large a family, there is a great deal of diversity in size, pelage coloration, and morphological features. Head and body length ranges from 3 to 10 cm and adult body mass from 4 to 50 g. The eyes are small and a nose leaf is absent (except for a rudimentary one in *Pharotis imogene* and the long-eared bats, Genus *Nyctophilus*). The ears of some species are exceptionally large, up to 4 cm in *Plecotus* (figure 12.23). A well-developed tragus usually is present. The total number of teeth varies from 28 to 38, usually through loss of premolars. The number of molars is always 3/3, and the cusp pattern of upper molars is dilambdodont. Species generally are insectivorous, with a few exceptions. For example, the fish-eating bat (*Myotis*

Figure 12.23   Large ears on some vespertilionid bats.
Several subspecies of *Plecotus townsendii* are threatened or endangered, in part because people collect them for the novelty of their large ears.

vivesi) has adaptations similar to those of noctilionids, whereas the pallid bat (*Antrozous pallidus*) feeds on scorpions, beetles, other ground-dwelling prey, and sometimes fruit.

Caves are often favored roosting sites, although vespertilionids use many different types of shelter. Depending on the species, roosting may be solitary, in small clusters, or in large groups. Roosts may contain more than one species, although species do not form mixed clusters. Because insects are unavailable during winter, species in temperate regions either migrate or hibernate. Like other hibernators, bats arouse periodically throughout the winter. Although they may fly, they do not feed (Whitaker and Rissler 1993). Hibernating species may exhibit delayed fertilization, with females segregating into maternity colonies. In tropical species, breeding (without delayed fertilization) may take place throughout the year. Nevertheless, a single neonate is typical, although there are exceptions. Some of the smaller species have two young; litters of up to four young may occur in the Genus *Lasiurus*.

The Tanzanian woolly bat (*Kerivoula africana*) recently became extinct. In North America, both the gray bat and the Indiana bat are endangered.

Molossidae

The 12 genera and 80 species of free-tailed bats are distributed in the Old World from southern Europe, Africa, Asia, and Australia to the Fiji Islands. In the New World, they occur from southwestern Canada through the United States and Caribbean to Central America and most of South America. Molossids are found in habitats from forests to deserts. They are medium-sized to large bats, with head and body length from 4 to 13 cm. The large ears meet on the forehead and point forward; there is a tragus but no nose leaf. The common name results from the tail extending well beyond the outer edge of the uropatagium (see figure 12.2A [g]). The family name derives from the Greek word *mollossus*, meaning "mastiff," a reference to the general doglike snout of these bats.

The hair in molossids is short and fine. The two species of *Cheiromeles*, have hair so short and sparse that they are called "naked bats." Other free-tailed bats often have sensitive, erectile tufts, or crests, of hair between the ears, associated with glands. Crests are especially evident in male lesser mastiff bats (Genus *Chaerephon*), and in the Genera *Tadarida* and *Molossus*, males have throat glands. Molossids have high aspect ratio wings (see figure 12.4F) and are generally swift aerial insectivores.

The dentition is typical of insectivorous bats; total number of teeth ranges from 26 to 32, depending on the species. Molossids do not hibernate, although species in north temperate regions, such as the mastiff bat (*Eumops perotis*) and Mexican free-tailed bat (*Tadarida brasiliensis*), may migrate and go into torpor during the winter. They roost in caves, buildings, trees, crevices, and on foliage. They may be solitary, roost in small groups, or form extremely large aggregations. Mexican free-tailed bats are especially gregarious; large numbers congregate in the southwestern United States. For example, an estimated 20 million *T. brasiliensis* have been reported from a single cave in central Texas; an estimated 50 million individuals once occupied one Arizona cave (Cockrum 1969). An interesting historical note is the experimentation done with *T. brasiliensis* during World War II in an attempt to use them to carry incendiary devices into enemy territory as "bat bombs" (Couffer 1992). This elaborate plan (called Project X-ray), which included dropping caged bats by parachute, was never implemented.

Summary

Second only to rodents in the number of species in the order, bats represent unparalleled variety among the mammals. No order is more diverse than chiropterans in the number of feeding niches they fill. Most bats are insectivores; numerous other species are specialized to feed on fruit, nectar, or pollen. Other bats are carnivores, taking small terrestrial vertebrate prey, including other bats, and a few species feed on small fish or have a diet of blood. Bats also exhibit diversity in their reproductive patterns and roosting ecology. Although many species exhibit spontaneous ovulation, vespertilionids in northern areas employ delayed fertilization in conjunction with hibernation or migration. Other species have delayed implantation or delayed development. Roosting ecology likewise varies among species. Bats may roost in large colonies, in smaller aggregations, in clusters of a few individuals, or as solitary individuals. Roosting habits depend on interrelated characteristics of social organization, foraging ecology, predator avoidance, and reproduction.

Morphological diversity also characterizes the two suborders: the megachiropterans and the microchiropterans. Most bats are small—many less than 10 g in body mass. The largest bats are the megachiropterans (Family Pteropodidae), which may reach 1600 g and have wingspans close to 2 m. Size is only one of the differences between the suborders. Unlike in microchiropterans, the facial characteristics of megachiropterans are simple. They have no nose leaf or other ornamentation, the eyes are relatively large, and the ears are without a tragus. Several other skeletal differences exist between the two suborders as well. Microchiropterans also have a sophisticated system of acoustic orientation called echolocation. They sense their immediate environment and "locate" food or objects from the "echo" returned from high-frequency sound pulses they generate and emit through either the mouth or nostrils. Although many similarities occur in the general aspects of echolocation in bats, species differ in the details of sound pulse duration, timing, frequency modulation, intensity, and length of signals. Echolocation does not occur in megachiropterans. All these differences have led some authorities to suggest that the pteropodids were derived independently through evolutionary time, that is, bats are diphyletic. Recent genetic evidence suggests, however, that bats are monophyletic, that is, they evolved from a common ancestor.

Despite the diversity found throughout the order, there is an obvious unifying characteristic: all bats have highly modified forearms and hands that form wings, and they are the only mammals that fly. Wings are the primary structures used to create the upward lift and forward thrust necessary for flight. Although there are common elements in wing structure for flight, the shape of wings varies. Wing shape and size, quantified in measures of aspect ratio and wing loading capacity, reflect the habitat, foraging characteristics, degree of maneuverability, and general life history characteristics of bat species.

Several species of bats worldwide are endangered or have recently gone extinct because of loss of roosting and foraging habitat, cave closure, insecticide accumulation, and other adverse influences. Despite the many positive economic benefits provided by bats, this fascinating and highly sophisticated group of mammals continues to suffer because of human ignorance and traditional misconceptions.

Discussion Questions

1. From the family descriptions given in this chapter, describe as many examples of convergent features as you can in various families of bats; for example, suction discs found in both myzopodids and thyropterids. Likewise, what convergent characteristics occur among bats, birds, and butterflies?

2. Given the active dispersal abilities of bats, speculate as to why the large fruit bats (Family Pteropodidae) never dispersed to tropical areas of the New World. Likewise, why might the monotypic myzopodid have remained endemic to Madagascar and the short-tailed bats (Mystacinidae) never dispersed beyond New Zealand?

3. What kinds of technological difficulties do you suspect may have limited early investigators in their studies of bat echolocation? What kinds of technological

advances have allowed researchers to study bat echolocation in the field as well as the laboratory?

4. Does the fact that megachiropterans and micro-chiropterans differ so markedly in their use of echolocation and associated morphological adaptations argue against the idea that these two groups are mono-phyletic? What arguments can you make for and against echolocation as a key feature in this debate?

5. Vampire bats have highly specialized diets consisting only of blood. Speculate as to how this specialization might have arisen during the evolution of these bats. Why do you think vampire bats are restricted in their distribution to tropical areas? Given convergent evolution in other features of bats, why has sanguinivory never arisen in any Old World bats?

Suggested Readings

Altenbach, J. S. 1979. Locomotor morphology of the vampire bat, *Desmotus rotundus.* Spec. Publ. No. 6, Am. Soc. of Mammalogists.

Fenton, M. B. 1992. Bats. Facts on File Publ., New York.

Findley, J. S. 1993. Bats: a community perspective. Cambridge Univ. Press, Cambridge, England.

Humphrey, S. R. and J. B. Cope. 1976. Population ecology of the little brown bat, *Myotis lucifugus,* in Indiana and North Central Kentucky. Spec. Publ. No. 4, Amer. Soc. Mammalogists.

Jones, G. 1994. Scaling of wingbeat and echolocation pulse emission rates in bats: why are serial insectivores so small? Functional Ecol. 8:450–457.

Kunz, T. H. (ed.). 1988. Ecological and behavioral methods for the study of bats. Smithsonian Institution Press, Washington, D.C.

CHAPTER

13

Primates

rimates, one of the more ancient mammalian orders, had its origins among the insectivores present in the Cretaceous period in North America. Most of these early forms belong to the extinct Suborder Plesiadapiformes. Following radiations during the late Cretaceous period and the Eocene epoch in both North America and Europe, primates were present on all of the continents except Australia. These radiations resulted in the two recognized suborders of living primates (Kay et al. 1997). The Suborder Strepsirhini consists of seven living families—Lemuridae, Megaladapidae, Indridae, Galagonidae, Daubentoniidae, Cheirogaleidae, and Loridae (table 13.1)—and three extinct ones—Adapidae, Archaeolemuridae, and Palaeopropithecidae (Szalay and Delson, 1979; Fleagle 1988; Wilson and Reeder, 1993). The Suborder Haplorhini consists of six living families—Tarsiidae, Cebidae, Callitrichidae, Cercopithecidae, Hylobatidae, and Hominidae—and four extinct ones—Omomyidae, Parapithecidae, Oreopithecidae, and Pliopithecidae. The fossil evidence concerning the evolution of these primate groups has recently been reviewed by Kay and colleagues (1997). In this chapter, we first review the ordinal traits of primates, examine some general primate morphology and the fossil history for this order, and then note the economics and conservation of primates. Our major focus is on accounts of the families of living primates. We have more interest in the various primates than in any other group of mammals because we, as *Homo sapiens*, are part of this order.

Table 13.1. Living primates are divided into 2 suborders and 13 families comprising a total of 232 species

	Number of Living Species	Distribution
Suborder Strepsirhini		
Family Lemuridae	10	Madagascar, Comoros Islands
Family Cheirogaleidae	7	Madagascar
Family Megaladapidae	7	Madagascar
Family Indridae	5	Madagascar
Family Daubentoniidae	1	Madagascar
Family Loridae	6	Central Africa, Southeast Asia, Sri Lanka
Family Galagonidae	11	Africa
Suborder Haplorhini		
Family Tarsiidae	5	Indonesia, Philippines
Family Callithrichidae	26	Central and South America
Family Cebidae	58	Central and South America
Family Cercopithecidae	81	Africa, Asia, Indonesia
Family Hylobatidae	11	Southeast Asia, China, Indonesia
Family Hominidae	5	Worldwide

Source: Data from D.E. Wilson and D.M. Reeder, Mammal Species of the World, *2nd ed., 1993, Smithsonian Institution Press, Washington, D.C.*

ORDINAL CHARACTERISTICS

The name *primates* means "the first animals," reflecting an early (incorrect) and anthropocentric bias giving the group that contains humans special importance. Primates are characterized by several ordinal traits, although many primates are quite generalized and thus, in a sense, defy the sort of clear listing of specialized traits that can be made for many mammalian orders. In 1873, Mivart characterized the primates as ". . .an **unguiculate, claviculate,** placental mammal with orbits encircled by bone; three kinds of teeth at least at one time of life; brain always with a posterior lobe and a **calcarine fissure;** the innermost digits of at least one pair of extremities opposable; hallux with a flat nail or none; a well-marked caecum; penis pendulous; testes scrotal; always two pectoral mammae" [boldface added]. Although this list characterizes primates, none of these traits is unique to primates. What defines a primate has been the subject of considerable debate in recent decades (Schwartz et al. 1978; Fleagle 1988; Luckett 1980). A major characteristic of primates is the cheekteeth. They are generally bunodont, having four sides like a square with four rounded cusps, and brachyodont, with a low crown.

Various investigators have extended this characterization by discussing the evolutionary trends that help to define the primates (Clark 1959; Macdonald 1984).

1. The hands and digits have become refined, with increased mobility of the digits, nails replacing claws, and sensitive pads on the digits with friction ridges—which are important for grasping.

2. Both the absolute and relative brain sizes have increased, with elaboration of more **cerebral cortex.** A tradeoff has occurred between increased dependence on sight correlated with enlarged brain areas associated with vision and decreases in brain areas associated with olfaction.

3. The muzzle region is shortened in primates, associated with a decline in the use of smell and a concomitant shift to binocular, stereoscopic, color vision.

4. Reproduction occurs at a slower rate, sexual maturity is delayed, and life spans are longer.

5. The diet has progressively shifted to greater reliance on fruits, seeds, and foliage, with a decline in the amount of animal matter consumed.

6. Social and mating systems have changed from ones based on overlapping male and female home ranges or territories to a diverse array of complex sociospatial and breeding patterns.

Primates occupy a wide variety of habitats. Their geographical distribution is primarily tropical and subtropical, although there are some exceptions, such as the Japanese macaque (*Macaca fuscata*), which lives in areas that have considerable snow.

PRIMATE EVOLUTION

At one time, both the tree shrews (Order Scandentia) and the flying lemurs (Order Dermoptera) were included with the primates. The rationale for these relationships has some phylogenetic basis in primate evolution from earlier stocks, but tree shrews and flying lemurs are now placed in separate orders. In some classifications, both past and present (Luckett and Szalay 1975; Fleagle 1988; Martin 1990), the primates are divided into prosimians and anthropoids. Here, we use the scheme proposed by Wilson and Reeder (1993) (figure 13.1). Although no investigators have used the emerging molecular technologies to attempt a phylogeny of all primates, some species within the order have been so tested. We use several results from the application of molecular techniques as we examine the families within this order.

MORPHOLOGY

The primates can be distinguished by certain aspects of their skeletal morphology as well as characteristics of soft anatomy. All modern primates have a bony postorbital bar (figure 13.2) with the eyes generally directed forward. The snout, or muzzle, is reduced in most primates, as are the olfactory lobes in the brain; both traits reflect the relative diminution of the sense of smell in primates relative to their mammalian ancestral stock. The braincase surrounding the relatively large brain is enlarged. All primates share a petrosal-covered **auditory bulla,** the tympanic floor of which is derived from only the petrosal plate and ectotympanic bone. Primate jaws move mostly in the vertical plane, in contrast to many other mammals in whom considerable horizontal jaw movement takes place. The jaw symphysis became progressively more ossified in advanced primates.

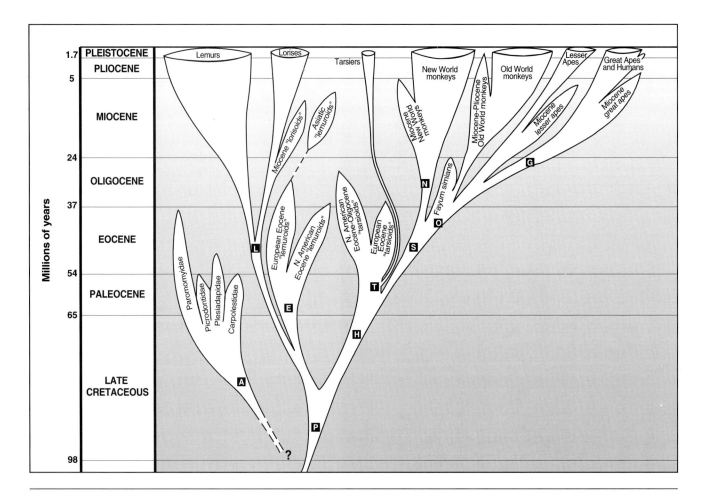

Figure 13.1   **Phylogenetic tree for primates.**   Both living families and fossil forms are shown. In this depiction, the branching of the primates begins about 90 mya; this differs from the time scale noted in the text and illustrates the different approaches of those studying primate phylogeny. *Abbreviations:* A = archaic primate stock; P = primate of modern aspect stock; E = lemuroid stock; L = lemur/loris stock; H = tarsioid/simian stock; T = tarsioid stock; S = simian stock; N = New World simian stock; O = Old World simian stock; G = great ape stock.

Source: Data from R.D. Martin, Primate Origins and Evolution, *1990, Princeton University Press; after data from Haq and van Eysinga, 1987.*

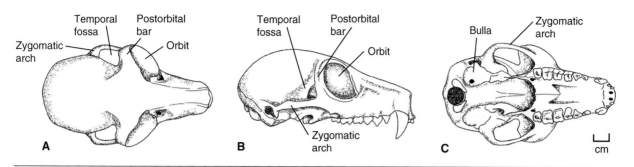

Figure 13.2   Prosimian skull.   Three views (A) dorsal, (B) lateral, and (C) ventral of the skull of a prosimian (*Varecia variegata*). Key characteristics include the forward-facing eyes and the lateral postorbital bar. Strepsirhine primates have an orbital cavity confluent with the temporal fossa, in contrast to haplorhines, whose postorbital plate marks off the rear of the orbit.

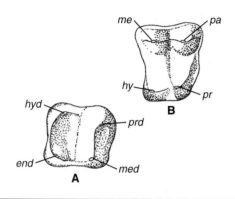

Figure 13.3   Primate teeth.   The basic four-cusped crown pattern of primates: (A) lower left molar, and (B) the upper right molar. *Abbreviations:* end = entoconid; by = hypocone; hyd = hypoconid; me = metacone; med = metaconid; pa = paracone; pr = protocone; prd = protoconid.

The cheekteeth of primates (figure 13.3) have been described as bunodont and brachydont with the sides of the upper molar relatively filled out; the teeth are relatively complex and adapted primarily for grinding and secondarily for shearing (Schwartz 1986).

Locomotion is plantigrade and, for most primates, pentadactyl. In some species, the hallux or pollex or both are reduced or absent. For most primates, the digits terminate in nails rather than claws.

FOSSIL HISTORY

The first primates appeared about 70 million years ago (Martin 1990; see figure 13.1). The early radiations of primate types in Europe and North America during the Paleocene and Eocene epochs produced a number of forms grouped together in the Genus *Plesiadapis* (figure 13.4; Kay et al. 1997). These small, squirrel-like mammals possessed an elongated skull with the eye orbits coming together with the temporal region of the skull, not separated from the back of the skull by a bony plate as in most other primates (Fleagle 1988; Martin 1990). The radius and ulna in the forelimb and tibia and fibula of the hind limb were entirely separate, permitting rotation of the feet. The digits of the hands and feet were long enough to enable these primates to grasp the limbs and branches of trees. The terminal digits still had claws.

The adapids, ancestors of today's lemurs, evolved during the early Eocene epoch, most likely as descendants of the plesiadapids. They ranged through the tropical and subtropical forests that then covered North America and Eurasia. Two examples of this group, the Genera *Smilodectes* and *Notharctus*, had the bony postorbital bar completely enclosing the orbit. No diastema occurred between the incisors and canines, resulting in a continuous tooth row. *Smilodectes* had eyes that were directed forward, making binocular, stereoscopic vision possible. The first digit of each foot was set apart from the others, suggesting a degree of opposability. These small primates were capable climbers, grasping branches as they moved through the trees. Other features associated with their mode of existence were a flexible back and long tail, useful for climbing and balance.

Another group that likely evolved from the plesiadapids was the Omomyidae, arising during the Oligocene and Eocene epochs. *Tetonius* from North America and *Necrolemur* from Europe exemplify this group. These animals had large orbits located in a short, flat face; the large eyes suggest a nocturnal activity rhythm. Short jaws contained large canines and rather primitive molars. The living descendants of the omomyids are the tarsiers (Genus *Tarsius*).

Fossil evidence for New World monkeys (Families: Callithrichidae and Cebidae) is minimal, making an assessment of their origin difficult. Perhaps the best hypothesis about their origin is that these families of primates evolved from adapid ancestors that migrated into South America. These stocks then underwent adaptive radiation in South America.

The chain of fossil evidence for the cercopithecoid primates of Asia and Africa (Old World primates) is more

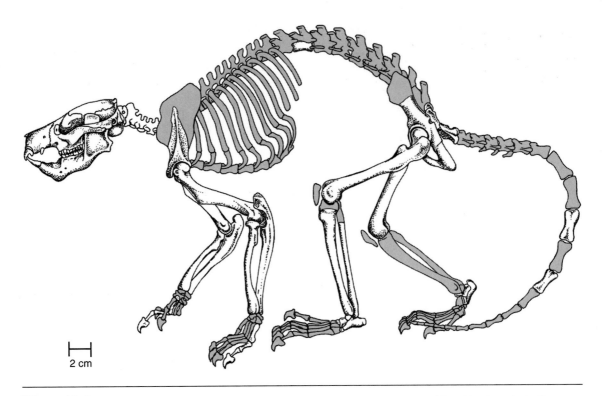

2 cm

Figure 13.4   **Fossil primate.**   One of the earliest ancestral primates, *Plesiadapsis tricuspidens*. Note in particular the flexible spine and long tail. Skeletal portions shown with dark lines are known from fossils; portions shown in gray are reconstructed.

complete, extending from the Oligocene to the Recent epochs (Fleagle 1988; Martin 1990). The fossil evidence from North Africa for one genus, *Parapithecus*, is relatively complete. This primate was characterized by a jaw that was only 5 cm in length but was deep, with a **condyle** located high on an **ascending ramus** to articulate with the skull. This latter trait is characteristic of modern cercopithecoid primates. The Hylobatidae and Hominidae evolved from cercopithecoid ancestors. Another fossil genus, *Miopithecus*, was ancestral to modern langurs (figure 13.5).

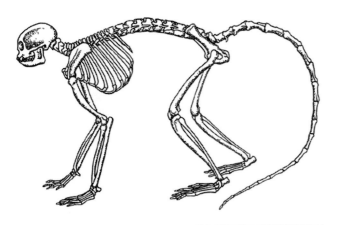

Figure 13.5   **Early primate.**   *Miopithecus*, an early Old World primate. Note the approximately equal length of the forelimbs and hind limbs and the enlarged braincase.

The most widely accepted proposal for the development of traits that characterize primates is the arboreal theory (Martin 1979; but see Cartmill 1972). An arboreal ancestry explains the adaptations of the visual system, including the forward-facing eyes and enlarged visual cortex of the brain. It also helps explain many of the skeletal adaptations of the limbs and digits, which permit easy and safe movement in trees. The theory also accounts for hands gradually replacing the mouth for gathering and handling food.

ECONOMICS AND CONSERVATION

Humans are interested in primates for many reasons. In the Hindu religion, primates are considered sacred and so are protected (figure 13.6). Several species of macaques and langurs enjoy this special relationship with humans. In other cultures, primates are taken for food, as in some areas of West Africa and South America. In these instances, overhunting can reduce or eliminate particular species. Indeed, this has happened to the red colobus monkey (*Colobus badius*) in Liberia, Cameroon, Ghana, and the Republic of the Congo (Gartlan and Struhsaker 1972; Leutenegger 1976). Still other primates have been and continue to be important for human medicine. Species such as *Macaca mulatta*, *Pan troglodytes*, and *Aotus trivirgatus*, among others, are used as animal models for various diseases. Probably the most widespread use of a nonhuman primate has been the testing of the Salk polio vaccine on large numbers of rhesus macaques.

Figure 13.6   Urban monkeys.   A group of rhesus macaques (*Macaca mulatta*) forage on a street in a village in northern India. The monkeys are free to come and go because Hindu beliefs protect them.

The capture and export of these monkeys from India reached alarming proportions by the 1960s and 1970s, eventually leading to a ban on their export. This resulted in the establishment of several colonies of rhesus monkey as breeding "farms" in places such as the Florida Keys and Morgan's Island, South Carolina, for the production of the stocks of monkeys needed for vaccine testing and other research uses.

The most significant threat to the continued existence of many primate populations comes from the destruction of their habitat by humans. In Indonesia, Brazil, and elsewhere, humans destroy vast areas of forest for lumber, other wood products, and fuel. In addition, major forested areas have been cleared in regions of Madagascar, Africa, and South America for agriculture to feed the ever-enlarging human population. These threats, and others such as the recent warfare in areas of Rwanda inhabited by the mountain gorilla (*Gorilla gorilla*), may result in the demise of many species. The future holds little promise for a number of primate species.

Conservation efforts, specifically with respect to primates, include establishing reserves where the habitat should be left undisturbed. This has been done with some positive effect, for example, in Sulawesi, where reserves are designed to protect primate diversity that includes *Tarsius spectrum, Macaca nigra, M. tonkeana,* and several other species. Another type of conservation effort involving a limited number of species has been to develop captive breeding stocks of an endangered species for reintroduction into native habitat, as has been done with the golden-lion tamarin (*Leontopithecus rosalia*) in Brazil (Kleiman 1976; Magnanini et al. 1975;

Mittermeir and Cheney 1986; see chapter 29). This program has progressed to the point where tamarins have been placed into the wild in several locales with suitable habitats of sufficient size that could be preserved. The outcome of both types of conservation measures will be known only several decades from now as we continue to monitor the diminishing population levels of a number of primate species.

STREPSIRHINE PRIMATES

The living Strepsirhini (lemurs and lorises) were formerly called Prosimians (figure 13.7). The seven families described here are concentrated on Madagascar but include species that live in Africa, Southeast Asia, and the Malay Archipelago. Several traits distinguish strepsirhine from haplorhine primates. One obvious external trait involves the **rhinarium,** an area of moist, hairless skin surrounding the nostrils (figure 13.8). The maximum dental formula for lemurs and lorises is 2/2, 1/1, 3/3, 3/3. Strepsirhines have a **bicornate uterus** (uterus with two horns), which contrasts with the fused simplex uterus of the haplorhines. The noninvasive, **epitheliochorial placenta** in lemurs and lorises also distinguishes them from the haplorhines. In general, strepsirhines produce neonates that are smaller relative to the mother's body weight than those of haplorhines (Leutenegger 1973). Based on the foregoing traits, many of which are primitive in a phylogenetic sense, Martin (1990) concludes that both the Strepsirhini (lemurs and lorises) and Haplorhini (tarsiers and simian primates) are of monophyletic origin. This means that each group originated only once from some ancestral primate (see figure 13.7). Various molecular techniques, including immunodiffusion comparisons of albumins (Dene et al. 1976; Sarich and Cronin 1976), amino acid sequences (de Jong and Goodman 1988), and DNA hybridization data (Bonner et al. 1980) support this conclusion.

Families

Daubentoniidae

The sole species in this family is the aye-aye (*Daubentonia madagascariensis;* figure 13.9). Aye-ayes have a body length of about 400 mm and bushy tails that are longer than their bodies (ca. 600 mm). They have coarse dark brown or black fur. The face is short and broad with a tapered muzzle. The large eyes face forward. Studies of chromosomes reveal that the aye-aye likely evolved from the same ancestral stock as lemurs and lorises, diverging quite early from the main lemuriform stock but at some time after the branching that resulted in today's lorises (see figure 13.7; Rumpler and Dutrillaux 1986; Rumpler et al. 1988).

Aye-ayes inhabit lowland rainforest. They are generally solitary, except for mothers with young, and have home ranges of approximately 5 ha (Petter and Petter 1967; Petter 1972). Females nurse their young with a single pair of inguinal mammae. Members of this species are nocturnal

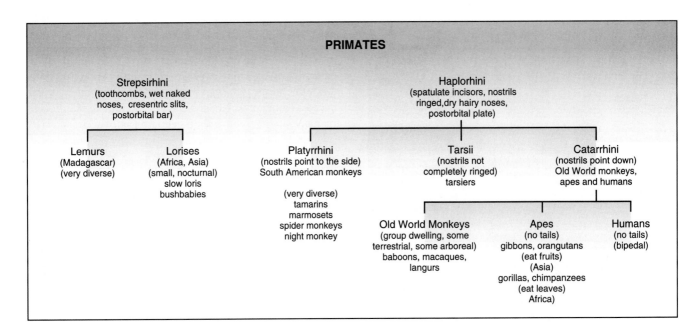

Figure 13.7   **Primate phylogeny and characteristics.**   The tree diagram illustrates the relationships among primate groups and some key traits for each group.

insectivores. The hands have particularly long fingers, and the middle finger bears a long, wirelike claw that is used for extracting insects from wood. They also use this elongated digit and claw to remove the pulp from fruits such as mangos and coconuts. The aye-aye possesses chisel-like incisors used for gnawing and chewing, much in the fashion of rodents.

The species originally inhabited regions of eastern and northwestern Madagascar (Petter 1962a, 1962b, 1977); it is extinct or nearly extinct from all habitats except possibly Nosy Mangabe Island, where it was introduced in 1966 and 1967 (Petter and Petter-Rousseaux 1979). At one time, this species was protected by local custom because the people believed that anyone who harmed one would die, but now they are killed if found because they are thought to be evil and bring bad luck (Tattersall 1972). The primary reason for the demise of the aye-aye, however, has been destruction of its forest habitat through cutting and burning, fostered by the need to use the land for agriculture to feed a growing human population and the planting of imported tree species for managed forests. A relative, the large aye-aye (*Daubentonia robustus*), was driven to extinction within the last 1000 years.

Loridae

Lorises and pottos consist of four genera and six species (Wilson and Reeder 1993; figure 13.10). They are distributed from sub-Saharan Africa to India, Southeast Asia, Indonesia, and the Philippines. The angwantibos are two species of the Genus *Arctocebus* in West Africa. The slender loris (*Loris tardigradus*) is found in southern India and Sri Lanka. The two species of slow lorises (Genus *Nycticebus*) live in much of Southeast Asia and on neighboring islands.

The single species of potto (*Perodicticus potto*) is found in portions of West and Central Africa. The Loridae formerly contained the galagos, which recently have been given separate familial status (Jenkins 1987).

The smallest members of this family are the angwantibos and slender loris, being only 180 to 250 mm in head and body length and weighing 85 to 500 g. The slow lorises and pottos are larger, ranging from 200 to 400 mm in head and body length and weighing 1.0 to 1.4 kg. Tails are very short or absent in the Loridae, except for the potto, which has a tail of about 65 mm in length. All lorids have thick, woolly fur of darker colors, ranging from brown to gray and black with light underparts (Napier and Napier 1967). They have relatively large forward-facing eyes, with generally flattened faces, although some species have distinct muzzles that may be either pointed or short and rounded.

All lorids are nocturnal and arboreal. Their habitat varies, depending on the species, from bamboo and evergreen forests to tropical rain forest and includes, for some species, logged areas and shrublands. They are slow but sure climbers, using a strong grip made possible by prehensile hands and feet that have the thumb (pollex) set at nearly a 180-degree angle from the remaining digits. It is thus termed pseudo-opposable. Many members of the family can climb well in a suspended position, below a limb, as well as locomoting on top of limbs. The index finger (second digit) is reduced in most species to a tubercle (knob), and the second toe is modified as a toilet claw in *Loris*, *Nycticebus*, and *Perodicticus*. Lorids are insectivorous, frugivorous, or both.

Lorids live as single individuals or in pairs. Their spatial relations in the wild are poorly known. They generally have single births, though twinning occurs occasionally for

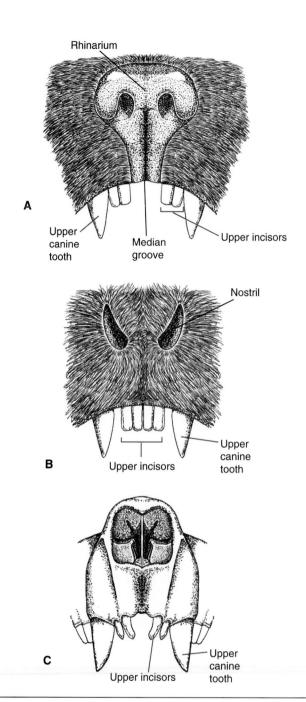

Figure 13.8   **Snout variation.**   The snouts of (A and B) strepsirhines are compared with that of a (C) haplorhine. Strepsirhines retain the primitive rhinarium surrounding the nostrils. The hairy maxillary processes have obliterated the rhinarium in haplorhines.

all but the potto. Species within this family engage in marking behavior, using urine or anal glands. They also communicate using vocalizations and limited facial expressions.

Galagonidae

In addition to this group being set aside as a separate family, its taxonomy has undergone considerable revision in recent years (Jenkins 1987; Groves 1989; Nash et al. 1989): what was one genus has been split into four genera, and a number

Figure 13.9   **Daubentoniidae.**   The aye-aye uses its specially adapted digit to probe for insects.

of groups formerly given only subspecific status have been elevated to species-level taxa. The four genera are *Euoticus* (two species), *Galago* (five species), *Galagoides* (two species), and *Otolemur* (two species). As a group, these primates are known as bushbabies (figure 13.11). All members of this family live in Africa, inhabiting rain forest in West Africa and woodland savanna from Senegal to East Africa and down to southern Africa.

Members of the Galagonidae are arboreal. A key feature of this group is their mode of locomotion, which involves leaping and bounding from branch to branch and between tree trunks. Two adaptations, well-developed hind limbs and a long tail used for balance, aid in this form of locomotion. Bushbabies are nocturnal and have large eyes that allow them to see well at night. They range in size from the diminutive dwarf bushbaby (*Galagoides demidoff*), which has a head and body length of 120 mm, tail of 170 mm, and mass of only 60 g, to the much larger thick-tailed bushbaby (*Otolemur crassicaudatus*), which has a head and body length of 320 mm, tail length of 470 mm, and mass of 1.2 kg. Bushbabies are all pentadactyl, with the second toe modified as a toilet claw. Their diet varies from primarily

Figure 13.10   **Loridae.**   The slender loris is native to India and Sri Lanka. Dark fur accentuates the large eyes.

only on Madagascar (Tatersall 1982). Members of the group are diurnal or crepuscular and arboreal, but some species, *Lemur catta* for example, spend considerable time on the ground (figure 13.12). They are medium-sized, ranging from 260 to 450 mm in head and body length, with tails of 250 to 560 mm, and body mass of 2 to 4 kg. Phylogenetic relationships among three of the genera in this family have been studied using restriction fragment DNA polymorphisms (Jung et al. 1992). The results support the conclusion that *Lemur* and *Hapalemur* diverged as a common stock from the ancestral group that lead to our current Eulemur group. *Lemur* and *Hapalemur* later diverged, resulting in two genera. Additional analyses, attempting to sort out phylogenetic relationships within *Eulemur*, resulted in several possible trees and some conflicts with existing phylogenies based on morphology. These sorts of problems are probably typical when we use an emerging technology. Additional studies using other techniques and samples, including other species to compare DNA-based phylogenies with those derived from morphological traits, should resolve such issues.

insectivorous to omnivorous, including fruits, grains, and even small mammals. Bushbabies have a soft, woolly fur, ranging in color from gray to brown and russet brown, with lighter undersides.

Field study, limited primarily to *Otolemur crassicaudatus* and *Galagoides senegalensis,* suggests that galagonids apparently live in groups of up to seven to nine individuals. For some species, mothers and infants may nest separately. Bushbabies build well-concealed nests in trees or use cavities in hollow trees. Single births or twins occur for all species studied. Mothers nurse their young with two to three pairs of mammae. Early in their infants' development, mothers carry their young with their canines, gripping them by the scruff of the neck. Bushbabies communicate via urine marking and vocalizations, notably the crylike sound made by young bushbabies, from which they get their common name. They also communicate visually with facial expressions and body postures.

Lemuridae

The true lemurs, Family Lemuridae, contains four genera: *Eulemur* (five species), *Hapalemur* (three species), *Lemur* (one species), and *Varecia* (one species). Lemurids are found

Figure 13.11   **Galagonidae.**   Lesser bushbabies inhabit dry forests, savanna habitat with trees, and gallery forests from Senegal to Ethiopia and from Somalia to South Africa.

Figure 13.12   Lemuridae.   The ring-tailed lemur lives in relatively large social groups and is the only prosimian that spends a large proportion of its time in the terrestrial habitat.

Given their daytime activity phase, unique among strepsirhines, the eyes of lemurs are smaller than those of their close relatives. They also have a more prominent muzzle than most strepsirhine primates. Members of *Eulemur*, *Lemur*, and *Varecia* have a diet consisting of fruits, flowers, and some vegetation, primarily leaves. That of *Hapalemur* consists of bamboo shoots and reeds. The lower incisors and lower canines form a forward-projecting dental comb used in both autogrooming and allogrooming.

This is the most colorful of the strepsirhine families, with pelage colors ranging from gray and greenish gray to brown and reds of various hues. Almost all species have a distinct ruff of fur around the neck differing in color from that of the main body pelage. All species possess ear tufts. Some species, such as *Eulemur macaco*, exhibit **sexual dichromatism,** that is, males and females have distinctly different pelage colors. In other species, distinct subspecific differences occur in pelage color, such as in *Varecia v. variegata* (the black-and-white ruffed lemur) and *V. v. rubra* (the red ruffed lemur). All members of the family, except *Lemur catta* (the ring-tailed lemur), locomote by clinging and leaping (chapter 5). Ring-tailed lemurs use considerable quadrupedal walking and climbing. Lemurids have hind limbs considerably longer and more developed than the forelimbs (figure 13.13). They use their long tails for balance during their bounding movements and as they climb among tree limbs.

Socially, members of Lemuridae are more gregarious than other strepsirhines, living in groups of from 3 to 6 for *Hapalemur griseus* to 20 or more, including several adults of both sexes, in several species of *Eulemur* (Napier and Napier 1967). Single births are most common, but twins occur occasionally. Females nurse their young with a single pair of pec-

toral mammae. Young of *Eulemur*, *Lemur*, and *Varecia* carry their infants on their abdomen at right angles to the main body axis. When they are somewhat older, the young may ride on their mother's back. The communication of Lemuridae includes considerable scent marking of branches and tree trunks, using urine, anal glands, and specialized sternal glands. *Lemur catta* possesses cutaneous arm glands, which they use to mark objects. Lemurids also have a modest vocal repertoire and use both facial expressions and body postures for visual communication. Human population and activities threaten several lemurids. *Hapalemur simus*, the broad-nosed gentle lemur, which inhabits a limited range of humid coastal forest in the east central region of Madagascar, is in immediate danger of extinction.

Megaladapidae

A recent taxonomic revision has removed the Genus *Lepilemur* from the Family Lemuridae and placed it in a separate family of its own, Megaladapidae, elevating what were seven subspecies of this genus to species level (Groves 1989; Wilson and Reeder 1993). Members of the family are known by the common name sportive lemurs. They are restricted to the island of Madagascar, where they inhabit both dry deciduous and tropical rain forests. All are nocturnal and arboreal. They have dense, woolly fur that is usually shades of red, mixed with brown or gray. The undersides are pale gray or yellowish

Figure 13.13   Megaladapidae.   Sportive lemurs (*Lepilemur mustelinus*) live in the deciduous and humid forests of Madagascar. They sleep during the day in tree hollows and feed at night on leaves supplemented with fruit, flowers, and bark.

white. Some species have a lengthy spinal stripe from the head to the base of the tail. *Lepilemur* is medium-sized, with head and body length of 280 to 350 mm, tail ranging from 250 to 280 mm, and body mass of 0.5 to 1.0 kg (see figure 13.13). The diet of sportive lemurs is quite different from that of most other strepsirhines (except possibly the lemurids); they primarily consume leaves but also eat bark, fruits, and flowers.

Megaladapids locomote by vertical clinging and leaping (chapter 5), moving from one tree trunk to the next with occasional hops on the ground. As an adaptation for this mode of locomotion, the hind limbs are considerably longer than the forelimbs. *Lepilemur* also has a prehensile thumb, which is pseudo-opposable and capable of strong grips on vertical branches.

Socially, the sportive lemurs live solitary lives except for mothers with their infants. Some marking occurs with urine and the glands in the circumanal region. The primary means of communication is a relatively extensive vocal repertoire. Mothers give birth to single young and have been observed carrying their young in their mouths as they leap around.

Cheirogaleidae

The Cheirogaleidae comprise four genera and seven species, all inhabiting Madagascar. Three of the genera, *Cheirogaleus* (two species), *Allocebus* (one species), and *Phaner* (one species) are all given the same common name, dwarf lemurs, whereas members of the fourth genus, *Microcebus* (three species), are called mouse lemurs (figure 13.14). As the common names imply, all of these primates are small, the dwarf lemurs being 190 to 300 mm in head and body length, having tails the same length or slightly longer than the head and

body length, and weighing from 300 g to slightly less than 1 kg. The mouse lemurs are even smaller, with head and body length of 130 to 170 mm, tail length of 170 to 280 mm, and body mass of about 60 g. Members of this family inhabit both wet and dry tropical forests. Because of the forest destruction throughout Madagascar, several of these species, particularly *Allocebus trichotis*, are thought to be nearly extinct in their natural habitat. The fur of Cheirogaleidae is dense and woolly, varying from gray to reddish brown to cinnamon, with lighter underparts. *Phaner* is characterized by a well-defined spinal stripe that bifurcates on the crown, where it joins dark eye rings. The eyes are large and forward-facing. Muzzles are clearly present in all members of the family, though how pronounced varies from one taxon to another.

All members of this family are arboreal and nocturnal. Locomotion of these small, agile primates is described as scurrying in short runs and darting squirrel-like from place to place. Mouse lemurs do much more leaping than dwarf lemurs. *Cheirogaleus* and *Allocebus* are primarily frugivorous, *Microcebus* is primarily insectivorous, and *Phaner* consumes both insects and fruits.

Cheirogelids are solitary or found in pairs. Females give birth to one to three young per pregnancy and nurse them with two pairs of mammae, one inguinal and one pectoral. Communication apparently involves less scent marking than in most other strepsirhines, although more research is needed. Cheirogelids also communicate through vocalizations, facial expressions, and some body postures.

Indridae

This family consists of three genera—*Avahi* (one species), *Indri* (one species), and *Propithecus* (three species)—which occur only on Madagascar (figure 13.15). All indrids are arboreal and nocturnal. *Avahi* is the smallest, with head and body length of 260 to 300 mm, tail length of 550 to 700 mm, and a mass of about 1 kg. *Propithecus* (sifakas) are intermediate in size, with head and body length of 400 to 600 mm, tail length of 450 to 600 mm, and body mass of 3.5 to 8.0 kg. *Indri* is the largest strepsirhine primate, measuring 530 to 700 mm in head and body length, with a very short tail, and weighing 7 to 10 kg. The diet of all Indridae consists of leaves, fruits, flowers, and bark. The families differ in their general habitat preferences; members of *Avahi* and *Indri* frequent tropical rain forests, whereas *Propithecus* live in evergreen and deciduous forests (Macdonald 1984).

Locomotion of all members of this family involves vertical clinging and leaping between tree trunks and shrubs. In conjunction with this pattern of movement, the hind limbs are considerably more developed than the forelimbs. The hand is prehensile with the thumb pseudo-opposable. The lower incisors are arranged in a comblike configuration for grooming. The eyes are of moderate size and are directed forward, with a distinct but modest muzzle. The thick fur varies in color, depending on the species, from white and black to brown, maroon, reddish, and orange.

Figure 13.14   Cheirogaleidae.   Gray lesser mouse lemurs (*Microcebus murinus*) are among the smallest of the primates. They build nests and are primarily arboreal.

Figure 13.15   Indridae.   Indri (*Indri indri*) is the largest of the prosimians. They live in the coastal rain forest in Madagascar and are endangered due to habitat destruction.

There are no distinct sexual differences in either body size or color patterns.

Indris live in small social groups of three to six individuals, containing adults of both sexes and young. Most births are of single young, and mothers nurse their infants with a single pair of pectoral mammae. Young are generally carried crosswise to the body axis and under the abdomen as small infants, switching to riding on the backs of their mothers after a few weeks or months. Indrids, particularly *Propithecus*, have distinct home ranges, which they defend against intruders. Communication is via scentmarking, vocalizations, and visual expressions and postures. The indri is considered endangered.

HAPLORHINE PRIMATES

The living haplorhine primates are divided into six orders (Wilson and Reeder 1993), which live in Africa, Asia, and Central and South America (see figure 13.7). In addition to the previously mentioned characteristics that distinguish haplorhines from strepsirhines, haplorhines have an invasive, **hemochorial** form of **placenta;** a postorbital plate, and spatulate incisors. The maximum dental formula is 2/2, 1/1, 2/2, 3/3 in simian primates from Asia and Africa (Old World primates) and 2/2, 1/1, 3/3, 3/3 in primates from Central and South America (New World primates). Distinctive differences in the visual and olfactory systems of Haplorhini also separate them from Strepsirhini (Fleagle 1988; Martin 1990).

Families

Tarsiidae

The tarsiers constitute a single genus (*Tarsius*) with five species all inhabiting the islands of Indonesia, the Malay Archipelago, and the Philippines (Musser and Dagosto 1987). Fossil evidence indicates that tarsiids once inhabited Europe, North America, and mainland Asia. Head and body length ranges from 95 to 140 mm, with tail from 200 to 260 mm, and body mass of 100 to 130 g (figure 13.16). A distinguishing trait of tarsiers is their ability to rotate their heads

Figure 13.16   Tarsiidae.   Western tarsiers (*Tarsius bancanus*) live in pairs on Borneo, Bangka, and Sumatra. They inhabit rain forests, shrub areas, and plantations, where they consume insects and some vertebrates.

almost 180 degrees, which is a function of the flexible neck vertebrae. The pelage is gray to gray-brown, with the face more ochre-colored in some species. A distinct tuft of fur occurs on the distal third of the tail.

All species are crepuscular and nocturnal. Their large eyes face forward, and they possess a reduced snout. The hind limbs, associated with the vertical clinging and leaping mode of locomotion, are considerably more developed than the forelimbs. Tarsiers bound rapidly from one tree trunk to another, sometimes covering distances of 2 m in one leap. They can also hop on the ground. Tarsiers inhabit primary and secondary rainforest, with some species also found in shrubland habitats. They are primarily carnivores, consuming insects, lizards, and spiders.

From studies of captive tarsiers, it appears that they are territorial, defending core areas (territories) within their overlapping home ranges. They spend considerable time patrolling and marking their territory boundaries. They live in pairs, although sometimes females and young are found alone. They give birth to single, precocial young, which are fully furred and capable of movement through the trees. Youngsters ride on the mother's back and also can be carried in her mouth.

Cebidae

There are two groups of New World primates: the Cebidae (capuchinlike monkeys) and Callitrichidae (marmosets and tamarins). The Cebidae are distributed from southern Mexico southward to Paraguay, northern Argentina, and southern Brazil. The family comprises 11 genera with 58 species: *Alouatta* (howler monkeys, 8 species), *Aotus* (night monkeys, 10 species), *Ateles* (spider monkeys, 6 species), *Brachyteles* (muriqui, 1 species), *Lagothrix* (woolly monkeys, 2 species), *Callicebus* (titi monkeys, 13 species), *Cebus* (capuchins, 4 species), *Saimiri* (squirrel monkeys, 5 species), *Cacajao* (uakaris, 2 species), *Chiropotes* (bearded sakis, 2 species), and *Pithecia* (saki monkeys, 5 species). Schneider and colleagues (1993) have proposed a phylogeny of the New World primates (Cebidae and Callitrichidae) using DNA sequence data for the ε-globin gene (figure 13.17). The maximum parsimony tree, derived using this technique, is quite similar to that obtained from a morphological analysis.

The thick, woolly fur of cebids is of various colors, ranging from green, gray, and brown to black, white, and various shades of gold and red. The red uakari (*Cacajao rubicundus*) has a hairless head ranging in color from pink to scarlet. The congeneric *C. melanocephalus* has black facial and head skin. The diet of most cebids includes leaves, fruits, flowers, and buds. The notable exceptions to this regimen are *Cebus*, which do not consume leaves, and *Chiropotes*, which consume only fruit. In addition, *Cebus* and *Saimiri* eat insects, and *Aotus* eats insects and some small mammals.

Size varies considerably among the Cebidae. Species of *Saimiri*, with head and body length of 230 to 370 mm, tail of 370 to 460 mm, and body mass of 0.5 to 1.1 kg, are the

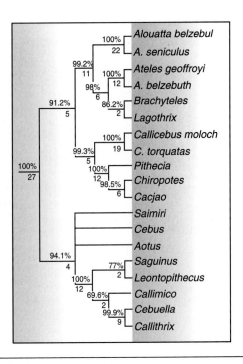

Figure 13.17   New World monkeys.   Cladogram of relationships of New World primates. A consensus maximum parsimony tree for ε-globin gene sequences. Numbers above each node represent values obtained from bootstrap simulations. Numbers below each node represent the minimum number of substitutions necessary to break up each grouping.

Source: Data from H. Schneider, et al., "Molecular Phylogeny of the New World Monkeys (Platyrrhini, Primates)" in Mol. Phylogen. Evol., *2:225–242, 1993.*

smallest, although *Aotus* and *Pithecia* are of about the same size. Species of *Brachyteles* are the largest members of the family, with head and body length of 460 to 630 mm, tail length of 650 to 800 mm, and body mass of 8 to 10 kg. Several genera exhibit varying degrees of **sexual dimorphism,** with males larger than females, in *Alouatta, Brachyteles, Saimiri,* and *Cacajao; Ateles* females are larger than males. For the remaining five genera, no obvious sexual dimorphism occurs.

Most species prefer rain, montane, and deciduous forests, and many are also found in disturbed forest. Some species live from sea level to 2500 m elevation, whereas others are more restricted in elevational range. The variants are *Cacajo,* which live primarily in lowland swamp forest, and *Pithecia,* which live in forests at higher elevations and in disturbed forests. Most cebids forage and live in the midcanopy and understory layers of the forest. *Aotus* is the only truly nocturnal primate; all others are diurnal.

Four of the genera—*Alouatta, Ateles, Brachyteles,* and *Lagothrix*—have prehensile tails. Members of these genera are classified as **semibrachiators** for some of their locomotion. The other genera are characterized by quadrupedal locomotion with varying use of leaping and hopping. Most cebids have prehensile hands, though only in some groups is the thumb pseudo-opposable, and *Ateles* has only a rudimentary thumb.

Social systems among cebids are of two general types. Members of genera other than *Aotus* and *Callicebus* form groups of varying sizes (from 5 to more than 100 individuals) that include multiple numbers of both sexes. These species are all **polygamous**. Members of *Aotus* and *Callicebus* form **monogamous** family groups. All female Cebidae give birth to a single young from each pregnancy, though occasional twinning has been reported in a few species. Young are nursed via a single pair of pectoral mammae. Most species carry the infant clinging to the belly for the first weeks or months after birth, after which the young ride on their mother's back, often using her tail to help them hang on. Many species engage in different forms of marking behavior, including the use of urine.

Callitrichidae

The Callitrichidae consists of 5 genera with 26 species: *Cebuella* (pygmy marmoset, 1 species), *Callithrix* (marmosets, 9 species), *Saguinus* (tamarins, 12 species), *Leontopithecus* (lion tamarins, 4 species), and *Callimico* (Goeldi's monkey, 1 species) (figure 13.18; Wilson and Reeder 1993). Callitrichids range from southern Central America southward to central South America. They primarily in-

Figure 13.18   Callitrichidae.   Lion tamarins (*Leontopithecus rosalia*) live in remnants of primary forest in Brazil. They have a varied diet that includes animal material, fruits, flowers, and gum from trees.

habit tropical rain forest but can also be found in deciduous forest and patches of forest in the savanna. They are smaller than cebids, ranging from one of the smallest primates, *Callithrix pygmaea*, which has a head and body length of 175 to 190 mm, a tail of 180 to 200 mm, and mass of only 120 to 190 g, to *Leontopithecus rosalia*, measuring 340 to 400 mm in head and body length, with a tail of 260 to 380 mm, and mass of 260 to 380 g.

The pelage of callitrichids is possibly the most colorful of any family of primates. In addition to having fur colors ranging from black and white to red, russet, and gold, many marmosets and tamarins have distinct color patches on their faces, heads, and ear tufts. All members of the family use quadrupedal locomotion, though many are also capable of hopping or leaping. They all have prehensile hands with nonopposable thumbs.

The diet of all callitrichids includes insects and fruit. Some genera, notably *Callithrix*, consume gums and saps from trees. Callitrichids live in extended family groups ranging from female-male pairs to about 15 individuals. Spatially, the marmosets and tamarins live in and defend areas of 10 to 40 ha, depending on the availability of resources. Within a family group, only one female breeds annually, producing nonidentical twins twice each year. The exception is *Callimico:* two females may breed in the group, and single births occur twice each year. Carrying the young involves all members of the group. Considerable olfactory communication occurs through urinary chemosignals and marking behavior. Marmosets and tamarins also use visual communication based on their varied color patterns.

Several species in this family are endangered, including the cotton-top tamarin, (*Saguinus oedipus*) and the golden-lion tamarin (*Leontopithecus rosalia*). The combined pressures of habitat destruction and growing human populations threaten these and several other species of callithrichids.

Cercopithecidae

Also called "typical," or Old World, monkeys, the Cercopithecidae contains roughly a third of all primate genera (18) and about 40% of all known primate species (81). They are distributed throughout much of Africa and southern Asia, including the Malay Archipelago. Cercopithecids live as far north as northern Japan and as far south as southern Africa. Some species have cheek pouches for food storage, and males of all genera and female guenons have large canines. All have powerful muscles to provide good grinding action for the teeth, and **perineal swelling** in sexually receptive females.

The family can be divided into two groups (table 13.2). One group, the cercopithecine monkeys, consisting of 11 genera (48 species), predominates in Africa, although some species occur in Asia. No molecular analyses have been completed to establish the phylogenetic relationships among all cercopithecids. Mitochondrial DNA, however, from eight species of *Macaca* has been tested and compared using

Table 13.2.　Taxonomy and characteristics of the Cercopithecidae

Genus (no. species)	Distribution	Body Mass (kg)	Habitat	Mating/Social System(s)
Subfamily Cercopithecinae				
Allenopithecus (2)	Africa	4–8	Swamp forest	Groups of 10–30 of mixed sexes
Cerocebus (3)	Africa	6–10	Evergreen and rain forest	Groups of 20–40 of mixed sexes, polygynous
Cercopithecus (19)	Africa	2–9	Rain forest, montane forest, swamp forest	Groups of 10–50 of mixed sexes, mostly polygynous
Chlorocebus (1)	Africa	3–5	Wooded savanna	Groups of 10–50 of mixed sexes, polygynous
Erythrocebus (1)	Africa	4–13	Wooded savanna, open savanna	One-male groups
Lophocebus (1)	Africa	6–9	Evergreen forest	Groups of 20–40 of mixed sexes, polygynous
Macaca (16)	Africa, Asia	4–18	Montane forest, riverine forest, forest edge	Groups of 10–100+ individuals of mixed sexes, mostly polygynous
Mandrillus (2)	Africa	12–50	Rain forest	One-male units of 12–20, groups of 20–40 of mixed sexes
Miopithecus (1)	Africa	1–2	Swamp forest	Groups of 70–100 of mixed sexes, polygynous
Papio (1)	Africa	12–25	Savanna, rocky scrubland, wooded savanna	Groups of 10–80 of mixed sexes, one-male groups
Theropithecus (1)	Africa	14–21	Grassland	One-male units within large herds of 100+
Subfamily Colobinae				
Colobus (4)	Africa, Asia	5–15	Forest, wooded grassland	Groups of 4–15 of mixed sexes polygynous, some territorial
Nasalis (2)	Asia	8–24	Mangrove, lowland rain forest	Groups of 12–20 polygynous, territorial
Presbytis (8)	Asia	5–8	Rain forest, swamp forest	Groups of 12–30 of mixed sexes, polygynous, some territorial
Procolobus (5)	Africa	3–11	Forest, savanna woodland	Groups of 5–20 of mixed sexes, polygynous, some territorial
Pygathrix (5)	Asia	5–10	Rain forest, conifer forest	Groups of 15–60 of mixed sexes, polygynous
Semnopithecus (1)	Asia	5–24	Forest, scrub, cultivated areas	Groups of 15–50 of mixed sexes, polygynous, some territorial
Trachypithecus (9)	Asia	4–14	Plantations, forest	Groups of 12–30 of mixed sexes polygynous, some territorial

restriction endonucleases (figure 13.19). There appear to be at least four major groups within *Macaca* (Ya-Ping and Li-Ming 1993). We should also note that this genus is the most widespread of any nonhuman primate, extending from North Africa (*M. sylvanus*) to Japan (*M. fuscata*).

Cercopithecines live in a wide range of both forested and terrestrial habitats. The family ranges in size from the small talapoin monkey (Genus *Miopithecus*), with a head and body length of 340 to 370 mm, tail length of 360 to 380 mm, and body mass of 1.1 to 1.4 kg, to the largest members of the group, *Mandrillus* (drills and mandrills), with a head and body length of 800 mm, tail length of 70 mm, and body mass of 12.0 to 25.0 kg. Sexual dimorphism is pronounced in some genera (e.g., *Mandrillus*), intermediate in other genera, (e.g., *Macaca*), and essentially absent in other genera (e.g., *Miopithecus*).

Coat colors include shades of brown, gray, green, and red, as well as black and white. Some genera such as *Mandrillus* and *Cercopithecus* (guenons, cercopithecus monkeys) and some macaques (*Macaca*) have brightly colored patches of skin on the nose and face, the scrotum, or the rump. They also may have brow ridges or other patterns of hair in the head region

that accentuate facial expressions or eye movements. Cercopithecines are all quadrupedal, although some (e.g., Genus *Papio*, baboons) are better adapted for a terrestrial existence, with relatively longer front limbs, whereas others (e.g., Genus *Cerocebus*, mangabeys) that are primarily arboreal have larger, more developed hind limbs. The guenons and mangabeys, as well as some less well-known genera, such as *Allenopithecus* (Allen's swamp monkey) and *Lophocebus* (gray-cheeked mangabey), have a diet consisting primarily of leaves, whereas other genera, such as *Macaca* and *Papio*, eat more fruits and seeds and may, on occasion, catch and consume other animals.

Social organization, as well as other traits of cercopithecine monkeys, varies greatly (see table 13.2). As discussed in greater detail later (chapter 22), the varying forms of social organization in this taxon reflect a combination of evolutionary selection pressures related to finding resources, such as food and shelter, and to the threat of predation. Vocal and visual communication of cercopithecines have been the subject of considerable research, including studies of the differentiation of calls by vervet monkeys (*Chlorocebus aethiops*) in response to aerial versus ground predators (Seyfarth et al. 1980a, 1980b).

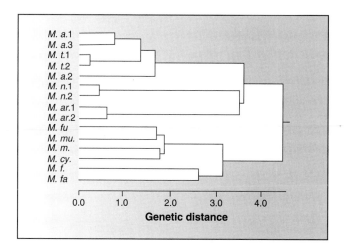

Figure 13.19   **Macaque phylogenetic tree.**   Cladogram for eight species of *Macaca.* Multiple numbers by a species abbreviation indicate multiple samples for that species. *Abbreviations: M. mu = M. mulatta; M. cy. = M. cyclopis; M. fa. = Macaca fascicularis; M. n. = M. nemestrina; M. a. = M. assamensis; M. fu. = M. fuscata; M. t. = M. thibetana; M. a. = M. arctoides.*

Source: Data from Z. Ya-Ping and S. Li-Ming, "Phylogenetic Relationships of Macaques Inferred from Restriction Endonuclease Analysis of Mitochondrial DNA" in Folia Primetologica, 60:7–17, 1993.

The second group of Cercopithecidae, the colobines, consists of 7 genera and 34 species (figure 13.20). Although found predominately in Asia, some colobines, notably in the Genera *Colobus* (black colobus monkeys) and *Procolobus* (red and olive colobus monkeys), occur in Africa. Colobines are characterized by the absence of cheek pouches and the presence of a sacculated stomach and large salivary glands. The upper portion of the stomach has a nearly neutral pH; bacteria found here aid in fermentation and breakdown of leafy vegetation. Colobines are generally more slenderly built than cercopithecines. The molar teeth possess high pointed cusps, and the outside of the lower molars and insides of the upper molars are less convexly buttressed than in the cercopithecines. These differences in dentition and the more specialized stomach can be related to the diet of colobines, which for many genera consists of leafy vegetation. Not all colobines, however, are folivores. Many include fruits, seeds, soil, flowers, and other nonleafy materials in their diets, although few records exist of consumption of other vertebrates and only a few species have been observed eating insects.

Colobines live in a variety of forest types, but they also occur in other habitats, including cultivated areas, rural and urban areas, and dry scrublands. Their coat colors are as varied as those of cercopithecines. Some of the common names for particular species exemplify the special pelage patterns that characterize many of them, including the golden snub-nosed monkey (*Pygathrix roxellana*), white-rumped black leaf monkey (*Trachypithecus francoisi*), and white-fronted sureli (*Presbytis frontata*). Most colobines are arboreal, although some, like the Hanuman, or common langur (*Semnopithecus entellus*), spend more than half their time on the ground. All colobines move by quadrupedal locomotion.

Figure 13.20   **Cercopithecidae.**   Proboscis monkeys (*Nasalis larvatus*) live in the mangrove swamps and lowland rain forests of Borneo. The function of the prominent nose is not fully understood.

They can walk bipedally along branches, supporting themselves by grasping other branches with their forelimbs, and they often leap between trees.

Colobines live in social groups ranging in size from occasional solitary individuals to aggregations of more than 100. Average size may be smallest in the Mentawi Islands sureli (*Presbytis potenziani*), which form groups of only three to four individuals. These monkeys are uniquely monogamous among colobines. All other colobines live in groups that average 10 to 20 individuals with two or more adult males and adult females with a polygynous mating system.

Hylobatidae

The apes consist of the Families Hylobatidae and Hominidae. Goodman and colleagues (1990) have used genetic data (figure 13.21) to examine phylogenetic relationships among the apes. Their findings are similar to results obtained with DNA hybridization (Sibley and Ahlquist 1984). Note that Goodman and colleagues (1990) place the gibbons (Hylobatidae) in the Family Hominidae, whereas Wilson and Reeder (1993) place them in their own family, with 1 genus, *Hylobates,* and 11 species.

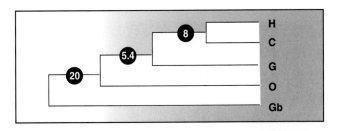

Figure 13.21   **Phylogenetics of Hylobatidae and Hominidae.**
Maximum parsimony tree diagram for the ψη region nucleotide
sequence orthologues for the Families Hylobatidae and Hominidae.
The number above each node is the difference in tree lengths between
the tree shown and the nonparsimonious tree that adds the least length
in breaking up the clade at that node.

Source: Data from M. Goodman, et. al., "Primate Evolution at the DNA Level and a Classification of Hominids" in J. Mol. Evol., *30:260–266, 1990.*

Hylobatids inhabit evergreen rain forests and monsoon deciduous forests of southeast Asia and portions of the Malay Archipelago. The 10 species known as gibbons range from 450 to 650 mm in head and body length and have a body mass of 5 to 7 kg (figure 13.22). The siamang (*Hylobates syndactylus*) is larger, with head and body length of 750 to 900 mm and a body mass of 9 to 12 kg. Hylobatids do not have tails. Their fur is dense and shaggy, and pelage color ranges from white and gray to black and brown. Distinct facial rings often differentiate the eyes, nose, or mouth, and some have tufts on the crown of the head.

One of the most distinguishing traits of the Hylobatidae is their true brachiation mode of locomotion. They also climb quadrupedally and walk, bipedally, on branches. Their arms are considerably longer than their legs. Their hands are prehensile with a fully opposable thumb used to grasp branches during their travels through the forest. About two-thirds of the diet of gibbons is fruit, and much of the remainder is leaves. The diet of siamangs is about half leaves and half fruit.

Gibbons and siamangs live in family groups with a monogamous adult pair and several juveniles. A single young is produced every 2 to 3 years. Male siamangs provide considerable parental care. All species are highly territorial. Visual and vocal communication are infrequent within the family group. Hylobatids engage in daily bouts of singing, which includes duets of great calls involving the adult male and female. These appear to serve at least two functions: (1) advertising their presence and territorial defense, and (2) establishing and maintaining pair bonds.

Hominidae

The phylogeny of this last family has been debated and revised several times in recent decades (figure 13.23; Fleagle 1988; Groves 1989; Wilson and Reeder 1993). The Hominidae consists of four genera, *Gorilla* (one species), *Pan* (two species), *Pongo* (one species), and *Homo* (one species). These are the largest primates, and all lack tails. Most births are of single young, and young have the longest developmental periods (usually 2–3 years until weaning and several more

years to attain sexual maturity) of any primate. All species of Hominidae exhibit sexual dimorphism, with males larger than females. All hominids, other than *Homo sapiens,* are threatened or endangered.

The orangutan (*Pongo pygmaeus*) is found in lowland and hilly tropical rain forests of Borneo and northern Sumatra (see figure 13.23A). Males have head and body lengths of about 970 mm and weigh 60 to 90 kg. Females have a head and body length of 780 mm and a body mass of 40 to 50 kg. They are covered by coarse, sparse, long hair that varies in color from reddish to orange, chocolate, or maroon. Orangs spend most of their lives in trees and locomote by brachiation. Their diet consists of about 60% fruit augmented by large quantities of leaves and shoots. They also consume insects, tree bark, eggs from bird nests, and small mammals. Orangs live solitary lives, except for sexual consortships and mothers with infants. Males have large home ranges that overlap the ranges of as many females as possible. The most common vocalization is the long call of the male, usually given early in the morning. The call may function to attract mates, signal ownership of a particular area, or simply alert other orangutans of the male's location.

The two species of chimpanzee (*Pan troglodytes,* common chimpanzee, and *Pan paniscus,* pygmy chimpanzee or bonobo) live in central Africa in a variety of habitats, ranging from woodland savanna to deciduous and humid forests (see

Figure 13.22   **Hylobatidae.**   Pileated gibbons (*Hylobates pileatus*) inhabit tropical rain forest and semideciduous forest in southeast Thailand and western Kampuchea. Note especially the long arms used for brachiation through the arboreal habitat.

Figure 13.23   Hominidae.   The great apes include
(A) orangutans, (B) common chimpanzees, and (C) gorillas.

figure 13.23B). Common chimpanzees have a head and body
length of 770 to 920 mm for males and 700 to 850 mm for females; males weigh about 40 kg and females about 30 kg.
Pygmy chimpanzees, in spite of their name, are only slightly
smaller. They have long, coarse, sparse hair that is generally
black but can turn gray in older animals. Their diet consists
primarily of fruits supplemented with leaves, seeds, flowers,
and several other plant foods. In addition, common chimpanzees, but not bonobos, consume meat, sometimes caught
through cooperative hunting (Goodall 1986). Chimpanzees
use tools in several contexts, such as to obtain food and water.
Chimpanzees live in a complex community with approximately 12 to 100 individuals, splitting into smaller parties of
3 to 6 (common chimpanzees) or 6 to 15 (bonobos) that frequently change composition. Chimpanzees communicate
through a full range of vocalizations and facial expressions, as
well as demonstrative behavior, including throwing branches
and rocks. Morin and colleagues (1994) suggest that two
species of common chimpanzees may exist. Using hair collected from nests vacated by known individuals, Morin and
colleagues extracted nuclear DNA to score banding patterns
for the polymorphism lengths of simple sequence repeat loci.
The results indicate a wide divergence between *Pan troglodytes
verus,* the West African subspecies, and *P. t. troglodytes* and *P. t.
schweinfurthii,* the two East African subspecies.

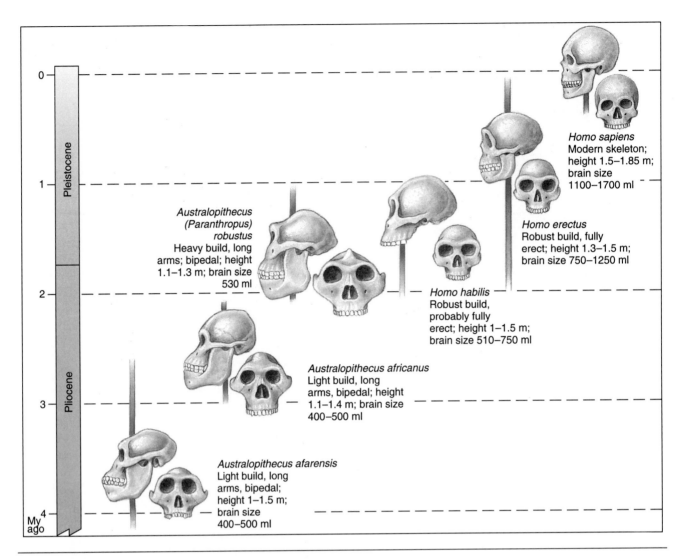

Figure 13.24   Human evolution. The pattern of evolution of the various hominids during the course of human evolution illustrates that (1) at times several forms coexisted and (2) a number of hominid forms have gone extinct during the evolution of this lineage.

The gorilla (*Gorilla gorilla*) is the largest primate, with males having a head and body length of 1600 to 1800 mm and a body mass of 140 to 180 kg, and females having a head and body length of 1400 to 1550 mm and body mass of 80 to 95 kg (see figure 13.23C). They live in tropical rain forests in widely separated populations in East and West Africa. The relatively dense fur is black to gray-brown. Older adult males have silver hairs on their backs. The diet of the gorillas in East Africa is almost entirely folivorous, consisting of leaves and stems, whereas West African gorillas eat proportionally more fruit. They live in relatively stable social groups of 5 to 30 individuals in East Africa, usually with one silver-backed male, several black-backed males, females, adolescents, and younger gorillas; groups in West Africa are smaller, averaging about 5 individuals. Gorillas are primarily terrestrial, moving quadrupedally, often using the knuckles of the hand. They can climb trees, although this activity is more prevalent in younger animals. Some gorillas nest in trees at night. Males stand bipedally to perform a chest-beating display. They communicate by facial expressions and a series of barks,

grunts, and other sounds. Their social behavior has been studied extensively by Schaller (1963) and Fossey (1983).

Humans (*Homo sapiens*) evolved from an ancestral African ape stock some time between 5 and 10 mya (Sarich and Wilson 1967a, 1967b; Uzzell and Pilbeam 1971; Fleagle 1988; Martin 1990). The ancestral forms of hominids in the evolutionary sequence have been debated, but consensus has not been reached. New archeological data contribute to new interpretations several times each year (figure 13.24). The evolutionary sequence includes several fossil forms of *Australopithecus*, several fossil forms of *Homo*, and a parallel line involving *Paranthropus*. Humans are characterized by erect bipedalism; reduced sexual dimorphism relative to other hominids; a large brain, averaging about 1300 cm^3, contained in a large vaulted cranium; and a skeleton that is less robust and more gracile than those found in our recent hominid ancestors. Human social organization is varied, including monogamy, polygyny, and, in a few instances, polyandry. Young are born in a relatively altricial state, and there is an extended period of physical and behavioral development.

Summary

Primates, the order of mammals that includes humans, comprises 13 families, 60 genera, and 227 species. Primates are largely tropical and subtropical in distribution, ranging through Africa, Asia, South and Central America, the Malay Archipelago, Japan, and Madagascar. Major characteristics of the group include refined hands and digits with nails replacing claws, binocular stereoscopic vision, a complete postorbital bar, a reduced muzzle, and slower rates of reproduction with increased developmental time. They exhibit a progression of sociospatial systems ranging from overlapping home ranges to a diverse array of social and mating systems. Cheekteeth are bunodont and brachydont. Primates, as a group, are generalists compared with most other mammal groups.

The first primates appeared about 70 mya and are grouped together as plesiadapids. The adapids probably evolved from the plesiadapids and are ancestral to modern lemurs. Another fossil group, *Omomyidae,* is ancestral to the tarsiers. *Parapithecus,* a fossil genus from Africa and Asia, is ancestral to today's cercopithecoid primates. The arboreal theory is the most accepted concerning the evolution of traits that characterize primates. Molecular techniques have been used to examine some phylogenetic relationships; ongoing investigations should help clarify questions concerning primate phylogeny.

Living primates are divided into the Strepsirhini and Haplorhini. Strephsirhines are characterized by a bicornate uterus, epitheliochorial placentation, relatively small neonates, and a maximum dental formula of 2/2, 1/1, 3/3, 3/3. The group includes Daubentoniidae (aye-aye), Loridae (lorises), Galagonidae (galagos), Lemuridae (lemurs), Megaladapidae (sportive lemurs), Cheirogaleidae (dwarf and mouse lemurs), and Indridae (avahi, indri, and sifakas). Although the majority of strepsirhines are concentrated on Madagascar, some groups live in Africa and Asia, including the Malay Archipelago. Many of the Madagascan species are endangered due to habitat destruction.

Haplorhines are characterized by a fused simplex uterus, hemochorial placentation, neonates that are larger relative to the mother's size, differences in the rhinarium compared with sterpsirhines, and maximum dental formulae of 2/2, 1/1, 3/3, 3/3 (New World species) or 2/2, 1/1, 2/2, 3/3 (Old World species). The group includes Tarsiidae (tarsiers), Cebidae (capuchinlike monkeys), Callitrichidae (tamarins and marmosets), Cercopithecidae (cercopithecine and colobine monkeys), Hylobatidae (gibbons and siamang), and Hominidae (apes). Haplorhines, other than humans, are widely distributed throughout Africa, Asia, the Malay Archipelago, and Latin America.

Discussion Questions

1. Primates are often characterized by a mixture of generalized and specialized traits. In terms of the morphology of six genera of living primates of your choice, write down those traits that you consider to be generalized and those that you think are specialized. Discuss your reasons for these classifications.
2. What relationships can you discern and describe between the physical size of different primates and their (a) diet, (b) habitat, and (c) social system?
3. Why do you think that virtually all primate distributions are limited to the tropics and subtropics?
4. Using your knowledge of molecular techniques to assess phylogenetic relationships, describe the studies you would carry out to investigate the relationships of the New World primates (Cebidae and Callitrichidae). How would you integrate the information you obtain with what is already known about phylogeny in these groups from morphological evidence on living primates and fossil history?

Suggested Readings

Eisenberg, J. F. 1981. The mammalian radiations. Univ. of Chicago Press, Chicago.

Fleagle, J. G. 1988. Primate adaptation and evolution. Academic Press, New York.

Macdonald, D. (ed.) 1984. The encyclopedia of mammals. Facts on File Publ., New York.

Martin, R. D. 1990. Primate origins and evolution. Princeton Univ. Press, Princeton, NJ.

Richard, A. F. 1985. Primates in nature. W. H. Freeman, New York.

Szalay, F. S. and E. Delson. 1979. Evolutionary history of the primates. Academic Press, New York.

Wolfheim, J. H. 1983. Primates of the world, distribution, abundance, and conservation. Univ. of Washington Press, Seattle.

CHAPTER

14

Xenarthra, Pholidota, and Tubulidentata

The Orders Xenarthra, Pholidota, and Tubulidentata are not closely related phylogenetically, but they share several characteristics related to a common feeding mode. Specifically, all are, or tend toward being, myrmecophagous, that is, they have diets composed predominantly of ants and termites. Because these insects form large colonies and are common in tropical and semitropical areas throughout the world, they offer an excellent potential energy source to various mammalian groups. Not only do the three orders discussed in this chapter eat ants and termites, so do mammals within an array of other orders: monotremes (the echidnas), marsupials (the numbat [*Myrmecobius fasciatus*], and rabbit-eared bandicoots), and even some carnivores (the aardwolf [*Proteles cristatus*] and the sloth bear [*Melursus ursinus*]). Seven mammalian orders have ant-eating representatives distributed throughout all tropical or semitropical land masses. Many other species also incorporate ants and termites as part of their diets.

Certain morphological features of each of the three groups in this chapter are quite distinctive, including their pelages. The xenarthrans have either an armored carapace or are heavily furred. The pholidotes are covered with reptilelike scales, whereas the aardvark (Order Tubulidentata) has a tough, sparsely haired hide. Several morphological similarities also are apparent in many (not all) mammalian species adapted for a myrmecophagous diet. Most have long snouts and long, powerful, sticky tongues used to collect insects. Dentition is reduced or absent. Strong, heavily clawed forepaws are used to dig into anthills and termite mounds. Their stomachs are generally simple with thickened, muscular, often **keratinized** (made of a tough, fibrous protein) pyloric regions to aid in digestion and protect against the formic acid contained in many ant species. Also, the groups discussed in this chapter usually have low reproductive capacities—generally one young per litter. The reproductive pattern in the xenarthrans, pholidotes, and tubulidentates is an example of a "*K*-selected" strategy of small litter size and extended parental care (see chapter 24). These similar characteristics are an excellent example of convergent evolution, because, as noted, these diverse mammalian orders are not closely related.

XENARTHRA

This order, formerly referred to as Edentata (meaning "without teeth"), encompasses four extant families found only in the Western Hemisphere: armadillos, anteaters (sometimes referred to as vermilinguas), and two families of tree sloths. Xenarthrans are a morphologically diverse group primarily restricted to tropical and semitropical habitats from Mexico southward throughout South America. Only the long-nosed or nine-banded armadillo (*Dasypus novemcinctus*) occurs as far north as the southern United States. All xenarthrans have low metabolic rates (figure 14.1) and low body temperatures. Body temperature averages about 34°C, compared with 36° to 38°C in other mammals. Members of this order are lim-

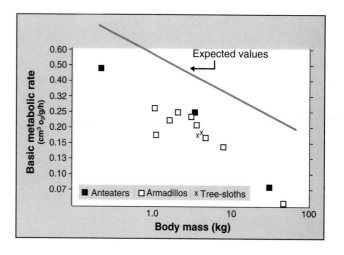

Figure 14.1   Metabolism in xenarthrans.   The expected relationship of mass-specific basal metabolic rate of other mammals (black line), calculated as BMR = constant (body mass $^{-0.25}$), and the much lower metabolic rates found in xenarthrans.

Source: Data from B.K. McNab, The Evolution and Ecology of Armadillos, Sloths, and Vermilinguas, *(G.G. Montgomery, ed.), 1985, Smithsonian Institution Press, Washington, D.C.*

ited geographically to warm regions due to their low basal metabolic rates and poor thermoregulatory abilities.

Morphology

The name *Edentata* actually is inappropriate because only the anteaters (Family Myrmecophagidae) are "without teeth." The name *Xenarthra* is more appropriate because it refers to one of the distinguishing characteristics of the order—the presence of at least two accessory, or supplemental, "xenarthrous" intervertebral articulations, located primarily on the lumbar vertebrae (figure 14.2). These give added rigidity to the axial skeleton (Gaudin and Biewener 1992). Additional characteristics shared by this otherwise diverse group include loss of incisors and canines, and cheekteeth (if present) that are single-rooted and without enamel. In terrestrial species, the acromion and coracoid processes of the scapula, which are rudimentary, fused processes in most therian mammals, are separate and well-developed to enhance muscle attachment for digging. Also, the transverse processes of the anterior caudal vertebrae and the ischia are fused in all xenarthrans, except the silky anteater (*Cyclopes didactylus*). Several other features of the skeleton and musculature define this order, including dermal ossicles in the skin, position of the infraorbital canal, and a secondary scapular spine, among others (Engelmann 1985).

Fossil History

Using evidence from mitochondrial DNA, Hoss and colleagues (1996) concluded that xenarthran lineages diverged prior to the Paleocene epoch. The earliest known fossil xenarthrans are armadillos from the late Paleocene epoch of

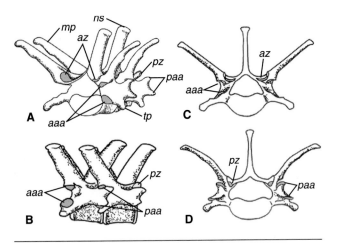

Figure 14.2   Vertebrae of xenarthrans. Third and fourth lumbar vertebrae in the nine-banded armadillo, showing the extra articular surfaces (shown in black throughout). (A) Anterior three-quarters view. (B) Lateral view (anterior to the left). (C) Anterior view. (D) Posterior view. *Abbreviations: az* = anterior zygapophysis; *mp* = metapophysis; *ns* = neural spine; *pz* = posterior zygapophysis; *tp* = transverse process; *aaa* = anterior accessory intervertebral facets; *paa* = posterior accessory intervertebral facets.

South America, although the order probably dates from even earlier times (Engelmann 1985). The fossil history of this group is rich and diverse. Extinct xenarthrans include giant armadillos (Family Glyptodontidae) that were over 3 m long with heavily armored head, back, and tail, and weighed over 1800 kg (figure 14.3). The giant ground sloth (*Megatherium americanum*), another extinct xenarthran, weighed over 2700 kg.

The 13 genera of xenarthrans living today are equal to only a tenth of the known number of extinct genera. Several of these forms became extinct recently, killed off by humans

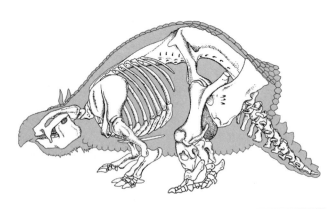

Figure 14.3   The extinct glyptodont. The extent of the armored carapace of this early Miocene xenarthran (*Propalaeohoplophorus*) is shown in outline, with a portion of the scute (scalelike) pattern shown in the center of the figure. Note the massive nature of the limbs and heavy, inflexible vertebrae, necessary to support the carapace, which made up 20% of the body mass. Total length approached 2 m.

only a few hundred years ago. These include the ground sloth (*Mylodon listai*), whose bones, hide, and reddish hair have been found with human artifacts in a cave in southern Argentina. The Puerto Rican ground sloth (*Acratocnus odontrigonus*) and the lesser Haitian ground sloth (*Synocnus comes*) also survived until 400 to 500 years ago.

Economics and Conservation

Arboreal sloths are greatly affected by loss of tropical rain forests, which reduces their habitat. All three-toed sloth species (Genus *Bradypus*) are declining, and the maned sloth (*B. torquatus*) of Brazil is endangered. The giant armadillo (*Priodontes maximus*) also is endangered due to habitat loss and hunting pressure. The pink fairy armadillo (*Chlamyphorus truncatus*) and several other species of armadillos are threatened, as is the giant anteater (*Myrmecophaga tridactyla*).

Families

Megalonychidae

This family includes two species of arboreal two-toed sloths and the recently extinct ground sloths. The two-toed sloths traditionally were included with the three-toed sloths in the Family Bradypodidae and then in their own family, the Choloepidae. Based on a number of cranial characteristics (figure 14.4), authorities now include them as the only extant members of the Family Megalonychidae (Webb 1985; Wetzel 1985). Hoffmann's two-toed sloth (*Choloepus hoffmanni*) occurs from Nicaragua southward through Peru and central Brazil. Linne's two-toed sloth (*C. didactylus*) is found east of the Andes Mountains in northern South America.

As with all xenarthrans, members of this family have no incisors or canines, and the cheekteeth are usually 5/4. The anterior upper premolar is **caniniform** (shaped like a canine tooth; see figure 14.4) and separated from the rest of the molariform dentition by a diastema. These teeth are sharp because of unique occlusion of the anterior surface of the lower caniniform premolar with the posterior surface of the upper one.

Two-toed sloths have two toes on the forefeet, each with a long (80–100 mm), sharp claw, and three toes on the hind feet. *Choloepus* differs from three-toed sloths in the number of digits on the forefeet (figure 14.5) and in being somewhat larger and heavier. Their total length ranges up to 740 mm, and body mass reaches 8.5 kg. Unlike most mammals, which have seven cervical vertebrae, individual two-toed sloths may have five, six, or occasionally eight.

Two-toed sloths also are more active than three-toed sloths. They move to a different tree each day, and they have a broader range of feeding habits. They are almost entirely arboreal and folivorous and spend most of their adult life hanging upside down from tree branches. They do, however, descend to the ground to defecate (at intervals of about 4 days in captivity) and apparently are capable swimmers. Their long, brownish gray pelage often has a greenish color because of green algae growing in it.

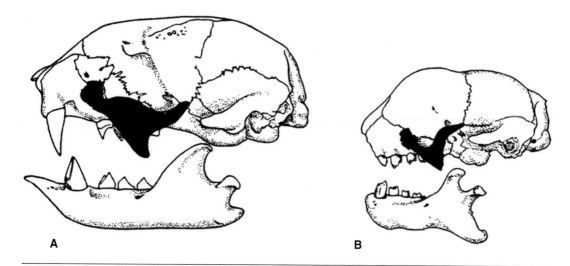

Figure 14.4   Features of sloth skulls.   Skulls of (A) a two-toed sloth and (B) a three-toed sloth. Note the distinctive anterior caniniform premolar (a premolar that looks like a canine tooth) in the two-toed sloth. Both skulls have an incomplete zygomatic arch with the jugal bone (*shaded*) a flattened plate with upper and lower processes on the posterior edge. Also note the difference in the shape of the mandibles. (Both skulls are about half scale.)

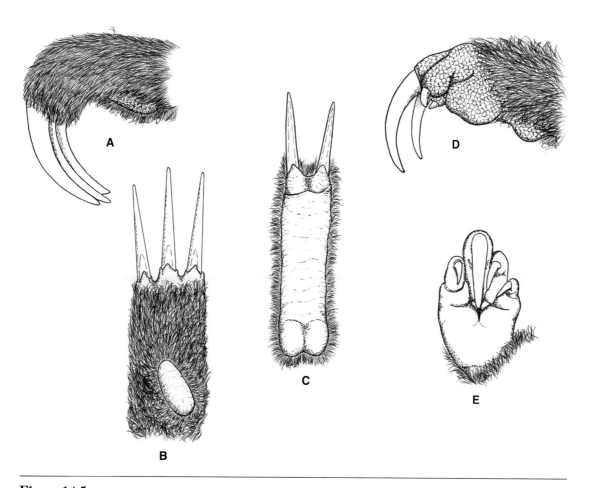

Figure 14.5   Large recurved claws in xenarthrans.   Right front foot of several species showing similarity of the claws in each family. (A) Lateral view and (B) ventral view with toes spread of a three-toed sloth; (C) ventral view of a two-toed sloth; (D) lateral view of a giant anteater; and (E) ventral view of a southern tamandua, showing the large central digit. (Not to relative scale.)

Two-toed sloths give birth to a single young; females do not reach sexual maturity until 3 years of age, males not until 4 or 5 years old. They are very long-lived; captive individuals reach over 31 years of age. Both *Choloepus* and *Bradypus*, discussed in the following section, are poor thermoregulators. Their body temperature tracks the ambient temperature to a much greater extent than is the case in other mammals. This is especially evident in two-toed sloths, whose body temperature varies between 24° and 33°C, a factor that limits their geographic distribution to warm, tropical regions.

Bradypodidae

Three-toed sloths in the Genus *Bradypus* are distributed from Honduras south through northern Brazil. As with *Choloepus*, the three species of *Bradypus* are arboreal folivores. They are active both day and night and have more narrowly restricted movement and feeding habits than two-toed sloths. Individuals may spend prolonged periods in the same tree. The teeth of three-toed sloths are cylindrical, with a central core of soft dentine, surrounded by harder dentine and then cementum. There is no enamel. The stomach has several compartments, and, as in other sloths, cellulose digestion is aided by microfauna. Foley and colleagues (1995) found that *B. tridactylus* retained a large mass of digesta in the gut. Passage is slow because of low metabolic rate and slow fermentation. Three-toed sloths are smaller (mean about 60 cm total length and 4.5 kg body mass in the brown-throated three-toed sloth, *B. variegatus*) and generally more common than *Choloepus* in areas of sympatry. They have three well-clawed toes on the forefeet (see figure 14.5). Litter size is one, and gestation is 5 to 6 months (compared with about 11 months in *Choloepus*). Three-toed sloths have eight or nine cervical vertebrae, which allows for greater flexibility in the neck; they can rotate their heads in a 270-degree arc.

Myrmecophagidae

The four species in this family are truly edentate and highly specialized for myrmecophagy, as suggested by the family name. Anteaters are found in forested or savanna habitats from southern Mexico south into South America east of the Andes Mountains as far south as Paraguay. They have long, tapered skulls, with a long tongue and very small mouth. The giant anteater's tongue has a maximum width of only 13 mm but can extend up to 600 mm. It is anchored on the sternum (see figure 6.3) and covered with a viscous secretion produced in the submaxillary glands. The tongue also has tiny, barblike spines directed posteriorly. Both the spines and the secretion aid in trapping ants. The giant anteater has coarse, shaggy gray hair with a dark diagonal stripe on the shoulders and a bushy tail (figure 14.6). Total length averages about 2 m, and they may weigh as much as 40 kg. They are entirely terrestrial and active throughout the day or night. Silky anteaters are much smaller—about 430 mm in total length and 230 g in body mass. They have woolier, grayish to yellow fur, with a darker dorsal band. They are nocturnal and almost entirely arboreal, even seeking shelter in trees. The two species of lesser anteaters, or tamanduas, are intermediate in size, active day or night, and forage both on the ground and in trees. They have coarse tan or brown pelage, and in the northern tamandua (*Tamandua mexicana*) and in southern specimens of the southern tamandua (*T. tetradactyla*), black fur forms a "vest" (see figure 14.6). Both the silky anteater and tamanduas have a prehensile tail that aids in climbing. All species have long, sharp, powerful claws for foraging, with the middle claw often enlarged (see figure 14.5). Stomachs are simple, with the pyloric portion strengthened for digesting insects. As in sloths, litter size in anteaters is generally only one.

Dasypodidae

The 8 genera and 20 species of armadillos occur in different habitats from the southeastern United States through Central America to the tip of South America. The long-nosed armadillo is the only xenarthran in North America (figure 14.7). It has extended its distribution significantly since the late 1800s through natural dispersal (Galbreath 1982), as well as by introduction to Florida. Farther distribution northward probably is limited by cold ambient temperatures, the high thermal conductance of armadillos, their inability to enter torpor, and the lack of food in winter (McNab 1985). Armadillos burrow extensively. They feed opportunistically on a variety of invertebrates (Bolkovic et al. 1995) and consume various amounts of vegetation and carrion.

Armadillos have several unique reproductive and morphological characteristics, the best known of which is the hard, armorlike carapace that gives the group its common name. Plates of ossified dermal "scutes" cover the head, back and sides, and tail in most species (figure 14.8). Scutes are covered by nonoverlapping, keratinized epidermal scales. The bands of armor are connected by flexible skin. The outside of the legs also have some armored protection, but not the inside of the legs or the ventral surface. The tough skin in these areas is covered by coarse hair. In the United States, *Dasypus novemcinctus* is commonly called the nine-banded armadillo because its carapace usually has nine movable bands, although band number varies throughout its geographic range.

The vertebrae in armadillos are modified for attachment of the carapace, which actually articulates with the metapophyses of the lumbar vertebrae (figure 14.9). Species vary in size from the tiny (100 g or less) pink fairy armadillo to the giant armadillo, which weighs up to 60 kg. Armadillos share the usual xenarthran reduction in dentition, having only peglike molariform teeth, which are small, open-rooted, homodont, and without enamel. The giant armadillo is exceptional among terrestrial mammals in having up to 100 small, somewhat vestigial cheekteeth. Although incisors form in embryonic *Dasypus novemcinctus*, they degenerate and rarely persist by the time of birth.

Reproduction in *Dasypus* is also noteworthy. Delayed implantation occurs, as does monozygotic polyembryony in

Figure 14.6   **Features of representative anteaters.**   (A) Coarse hair, diagonal shoulder stripe, and long, tapered skull of the giant anteater. (B) Distinctive "vest" of the northern tamandua, a smaller, semiarboreal anteater.

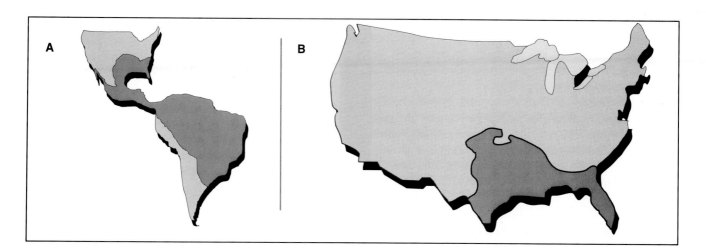

Figure 14.7   **Distribution of the nine-banded armadillo.**   (A) Range within the Western Hemisphere and (B) the northern-most extent of the species range in the United States.

Source: (A) Data from B.K. McNab, The Evolution and Ecology of Armadillos, Sloths, and Vermilinguas, *(G.G. Montgomery, ed.), 1985, Smithsonian Institution Press, Washington, D.C. (B) Data from Chapman and Feldhamer,* Wild Mammals of North America, *1982, Johns Hopkins University Press.*

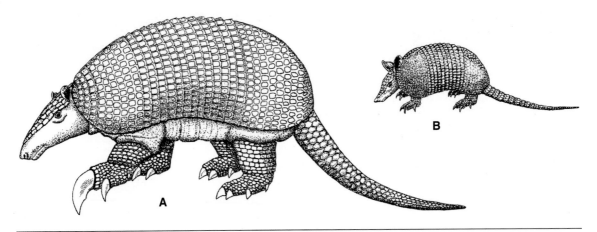

Figure 14.8   **Size comparison.**   Approximate relative sizes of (A) the giant armadillo and (B) the nine-banded armadillo.

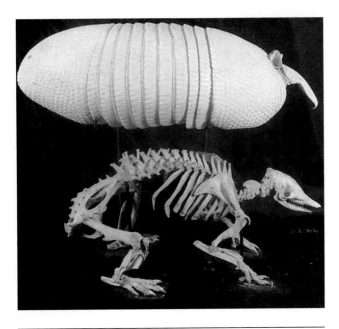

Figure 14.9   **Carapace and articulated skeleton.**   Note the distinctive metapophyses of the lumbar vertebrae for support of the carapace in the nine-banded armadillo.

several species. In *D. novemcinctus,* four young of the same sex are produced after division of a single fertilized ovum; in *D. hybridus,* eight young are normally produced. From a common placenta, each embryo develops its own placenta, with no mixing of blood or nutrient material between embryos (Talmadge and Buchanan 1954). Eisenberg (1981:55) suggested that this is a "novel way of increasing the reproductive capacity" of *Dasypus* toward a more "*r*-selected" mode (see chapter 24; also Gleeson et al. 1994). Finally, because *D. novemcinctus* is subject to naturally occurring leprosy, it is a valuable model for a wide range of biomedical research projects (Storrs 1971; Pan American Health Organization 1978).

PHOLIDOTA

Pangolins, or scaly anteaters, are in a single family, Manidae, with a single genus (*Manis*) and seven species. As was true for the aardvark, pangolins were once placed in the Edentata based on morphological features associated with diet. Four species of pangolins are found in Africa south of the Sahara Desert (these sometimes are placed in a different genus—*Phataginus*), and three other species occur in Pakistan, India, Sri Lanka, southeast Asia, southern China, and Indonesia. Their habitats include forests, savannas, and sandy areas, with distribution directly related to the occurrence of ants and termites, their primary prey (Heath 1992, 1995). Some species are terrestrial and are strong diggers, living in large, deep burrows, whereas others are arboreal, living primarily in trees and having semiprehensile tails. Even the terrestrial species are good climbers and occasionally may forage in trees. Pangolins may be active during the day but are primarily nocturnal. Occurring as solitary individuals or in pairs, they have limited vocal, visual, and auditory acuity. Olfactory communication plays a significant role in their behavior, however. Strong scent is produced from paired anal glands, and feces and urine are deposited along trails and trees. Like skunks (Carnivora: Mustelidae), they may eject an unpleasant-smelling secretion from the anal glands. Gestation is about 140 days. Litter size is usually one, but occasionally twins are produced. The young are born with soft scales that do not harden until 2 days after birth. Newborn young cling to the female's back or tail, and, if threatened, the female curls up around the neonate.

Morphology

The ordinal name means "scaly ones" and refers to the major diagnostic characteristic of this group. With the exception of the sides of the face, inner surface of the limbs, and the ventral surface, pangolins are covered with **imbricate** (overlapping) scales, somewhat like those of a pine cone

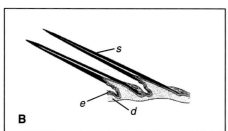

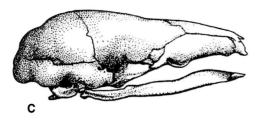

Figure 14.10   Pangolin characteristics.   (A) Dorsal view of a pangolin, showing its distinctive scales. (B) Diagram of a section through two scales, showing the scales (*s*), epidermis (*e*), and dermis (*d*). (C) Skull of a pangolin, with incomplete zygomatic arch and very simple mandible.

(figure 14.10). Scales, which are composed of keratinized epidermis and are dark brown to yellow in color, serve the same function (protection) as the armor of armadillos or the spines of echidnas. A pangolin has the same number of scales throughout life; they grow larger as the individual grows. Skin and scales make up 20% or more of the body mass of most species, although, as might be expected, scales on arboreal species such as the tree pangolin (*Manis tricuspis*) are lighter and thinner than those on terrestrial species. When alarmed, pangolins curl up in a ball, with the sharp-edged, movable scales directed outward. Pangolins may reach 1.6 m in length, with the tail comprising half the total. External ears do not occur in the four African species, but they are found in Asian pangolins.

Pangolins are truly edentate. Instead of teeth, they use a long, powerful tongue for foraging. A sticky, viscous saliva is secreted onto the tongue by a large salivary gland in the chest cavity. Tongue muscles are enclosed in a sheath and pass over the sternum and anchor on the pelvis (Kingdon 1971). A pangolin's tongue is longer than its head and body length and is structurally similar to that of the giant anteater (Chan 1995). The skull is long and tapered, with a straight slender mandible and incomplete zygomatic arch (see figure 14.10). The morphology of the skull reflects the pangolin's diet, lack of teeth, and associated lack of any strong facial muscles for chewing. Pangolins are plantigrade and pentadactyl, with large, sharp, recurved claws to break into ant and termite mounds, which they locate primarily through scent. As in other ant-eating species, pangolins have a stomach with a muscular, gizzardlike pyloric region for grinding ants and termites.

Fossil History

Fossil material is meager and all outside the current distribution of pangolins. A mid-Eocene pangolin (Genus *Eomanis*) is known from Europe, and remains characteristic of pangolins (Genus *Patriomanis*) from Oligocene sites were found in North America and Europe.

Economics and Conservation

African species are eaten by native people, and their scales are used for adornment. Scales also are used as good-luck charms, with entire skins being extremely valuable (Kingdon 1971). The African Cape pangolin (*Manis temminckii*) is endangered. In Asia, because powdered scales are believed to be of medicinal value, populations of the three species in this region are greatly reduced.

TUBULIDENTATA

This is the smallest eutherian order, with a single living member. The Family Orycteropodidae contains only the aardvark (*Orycteropus afer*), which occurs in Africa south of the Sahara Desert in habitats ranging from dry savanna to rain forests. As with pangolins, the aardvark is closely associated with the distribution of ant and termite mounds. The same mound may be visited repeatedly, with the aardvark foraging at previously excavated sites. Larvae, locusts, and even wild cucumbers (*Cucumis humifructus*) are also eaten. Aardvarks are solitary. They are predominately nocturnal (rarely crepuscular) and travel as much as 30 km a night as they forage. Prey is probably located with the aid of highly acute olfactory and auditory senses. Their burrow system is extensive; several openings occur throughout a small area. Females give birth to a single young after 7 months of gestation. As with other myrmecophagous species, aardvarks have several distinct morphological features that adapt them to their prey base.

Morphology

The word *aardvark* is Afrikaans for "earth pig," and the species superficially resembles a pig (figure 14.11). It weighs up to 60 kg and is about 1.5 m in head and body length. The long, square snout is somewhat flexible, with tufts of hair that protect the nostrils during digging. The nasal septum has short, somewhat fleshy tentacles that probably serve an olfactory function. This idea is further supported by the fact that the aardvark has more **turbinate** bones (scroll-like bones in the nasal passages) than any other mammal. Aardvarks have small eyes and large, erect ears; the latter is unusual in myrmecophagous species. The yellowish brown hide is very tough, sparsely haired, and insensitive to ant and other insect bites. The forefeet have four toes, and the hind feet five, each with a heavy, strong clawlike nail. Aardvarks can burrow rapidly as well as break into ant and termite mounds. As in anteaters and pangolins, the aardvark's skull is elongated (see figure 14.11), and it has a small tubular mouth with a long, sticky tongue. Aardvarks chew their food, however, and the dentition is fairly unusual. Incisors and canines are found only during the fetal stage, and adults typically have 20 to 22 cheekteeth. These teeth are composed of numerous pulp tubules surrounded by hexagonal prisms of dentine (see figure 14.11C) and account for the ordinal name ("tubule-toothed"). Teeth are open-rooted, without enamel, and covered with cementum. Despite these chewing teeth, aardvarks also grind ingested material in a muscular pyloric region of the stomach, similar to that of pangolins.

Fossil History

As in pangolins, the fossil record of the aardvark is fragmentary. The earliest known tubulidentate (Genus *Myorycteropus*) is from early Miocene deposits of East Africa. *Orycteropus gaudryi* dates from the late Miocene epoch and, except for more cheekteeth, is similar to the extant species. A relatively unspecialized form, *Leptorycteropus*, dates from the mid-Pliocene epoch. Pleistocene remains are known from France, Greece, Turkey, India, and Madagascar (Patterson 1975).

Economics and Conservation

Aardvarks are eaten by natives, and the teeth are used as jewelry and good-luck charms. Populations have been reduced throughout the range.

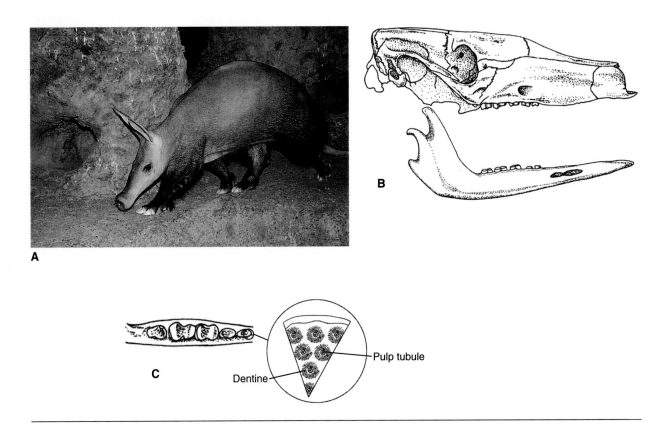

Figure 14.11   Features of the aardvark.   (A) Piglike snout, large ears, and heavy claws of an aardvark. (B) Aardvark skull. (C) Diagrammatic representation of the occlusal surface of the toothrow of an aardvark with a section of the pulp tubules and surrounding dentine.

Summary

The three orders in this chapter offer an excellent example of phylogenetic convergence because many of the species "make a living" in similar ways. Although quite distinct morphologically, the aardvark, pangolins, and anteaters share a number of characteristics related to their myrmecophagous diets. The armadillos and sloths, along with the true anteaters in the Order Xenarthra, exhibit interesting structural diversity within a relatively small order. Xenarthrans occur only in the New World and are restricted to warm regions because of their poor thermoregulatory abilities. Formerly known as the Edentata, the order is named for the accessory vertebral articulations that give added rigidity to the axial skeleton.

The pholidotes occur in both Africa and Asia. They are also called pangolins, or scaly anteaters, and are covered with keratinized, overlapping scales. These scales serve the same protective function as the carapace does in armadillos. Like the true anteaters, pangolins are edentate.

The aardvark is not closely related to either xenarthrans or pholidotes, but like these groups, the aardvark's occurrence within its range is closely associated with ants and termites. The aardvark exhibits morphological and behavioral characteristics similar to other myrmecophagous species. Unlike true anteaters and pangolins, however, aardvarks have cheekteeth.

Discussion Questions

1. Consider the symbiotic relationship between sloths and the algae that grow on their pelage. What are the benefits to both groups?
2. Discuss the relationship between low metabolic rates, poor thermoregulatory abilities, geographic distributions, and small litter sizes in xenarthrans.
3. Why might you expect pangolins that are primarily arboreal to have lighter, thinner scales than terrestrial species?
4. Why might you expect to find "altruistic" behavior among populations of the nine-banded armadillo rather than in some of the other species discussed in this chapter?
5. What is the significance of small eyes and ears in most myrmecophagous mammalian species?

Suggested Readings

Goffart, M. 1971. Function and form in the sloth. Pergamon Press, New York.

Loughry, W. J., P. A. Prodöhl, C. M. McDonough, and J. C. Avise. 1998. Polyembryony in armadillos. Am. Sci. 86:274–279.

Montgomery, G. G. and M. E. Sunquist. 1978. Habitat selection and use by two-toed and three-toed sloths. Pp. 329–359 *in* The ecology of arboreal folivores (G. G. Montgomery, ed.). Smithsonian Institution Press, Washington, D.C.

Rose, K. D. and R. J. Emry. 1993. Relationships of Xenarthra, Pholidota, and fossil "edentates": the morphological evidence. Pp. 81–102 *in* Mammal phylogeny: placentals (F. S. Szalay, M. J. Novacek, and M. C. McKenna, eds.). Springer-Verlag, New York.

CHAPTER

15

Carnivora

MORPHOLOGY

FOSSIL HISTORY

ECONOMICS AND CONSERVATION

FAMILIES
Feliformia
Caniformia
Pinnipedia

A great deal of diversity exists among the 11 families and approximately 271 species in this order. Most of these species eat meat. Although it is easier to digest than vegetation, meat is much more difficult to locate, capture, kill, and consume. Thus, carnivores generally are thought of as the major group of mammalian predators because they feed primarily on animal flesh (including other mammals). Whereas numerous orders of mammals are herbivores, whales, insectivores, many bats, and a few rodent and marsupial species also consume animals. Carnivores occur naturally on all continents except Australia and have adapted to diverse niches in a variety of terrestrial and aquatic habitats.

Although most carnivores eat meat, this is not a defining characteristic of the order. Living carnivores share several morphological characteristics, including specialization of the teeth (see the "Morphology" section of this chapter). Carnivores are arranged into two suborders (Wozencraft 1993), based on the structure of their auditory bullae and carotid circulation (Flynn et al. 1988; Wozencraft 1989a). The Suborder **Feliformia** (meaning "catlike") includes four families: the Felidae (cats), Herpestidae (mongooses), Hyaenidae (hyenas), and Viverridae (civets). The other Suborder, **Caniformia** (meaning "doglike") includes the Families Canidae (dogs), Ursidae (bears), Mustelidae (weasels), Procyonidae (raccoons), and three families of aquatic carnivores: Odobenidae (walrus), Otariidae (eared seals, including fur seals and sea lions), and Phocidae (earless, or true, seals). The aquatic carnivores are referred to as **pinnipeds** (meaning "feather-footed"), based on the modification of their limbs into flippers. In the past, pinnipeds were considered either as a separate order of their own or as a suborder within the Order Carnivora, with the remaining members grouped in the Suborder Fissipedia (meaning "split-footed") because of the individual toes on each foot. Wozencraft (1989a, b; 1993) suggests a single order (figure 15.1) is most appropriate. Unlike whales, pinnipeds leave the water to rest, breed, and give birth. Also, the sea otter (*Enhydra lutris*) is essentially aquatic. Several species of otters, and the polar bear (*Ursus maritimus*), spend much of their time in the water.

Smaller species of carnivores generally have larger litters and breed more frequently. Larger species, such as bears and the pinnipeds, have small litters (often a single young), and females may breed at intervals of several years. Both induced ovulation or delayed implantation (see chapter 9) occur in certain species and are generally related to environmental or life history factors. Few carnivores have precocial young. Juveniles must learn to hunt successfully under a variety of circumstances, and an extended period of maturation and learning is often required before these skills are acquired and individuals disperse (Holekamp et al. 1997). This need to learn and adapt, as well as to develop a high degree of coordination and dexterity, is reflected in the high brain-to-body mass ratio of carnivores (Eisenberg 1981).

When hunting, individuals may be solitary, paired, or in small groups. Smaller carnivores, such as weasels, are generally solitary and restricted to taking smaller prey.

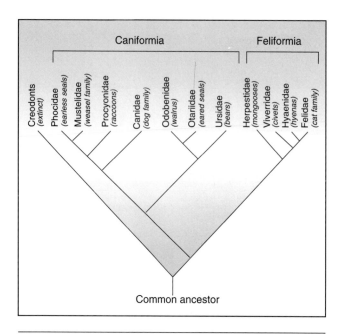

Figure 15.1   **Phylogeny of modern carnivore families.**
This phylogeny is based on 100 dental, cranial, skeletal, and anatomical characteristics. The herpestids (mongooses) are recognized as a distinct family, whereas the greater and lesser pandas are included in the ursids (bears). Wozencraft (1989b) grouped the walrus with the otariids but later (Wozencraft 1993) placed the species in a separate family (the Odobenidae).

Larger species, such as wolves, lions, hyenas, and African hunting dogs (*Lycaon pictus*), may hunt in packs. In addition to smaller prey, packs are able to prey on larger, more dangerous species (figure 15.2) that may be several times the size of individual members of the pack (Cooper 1991; Scheel and Packer 1991; Holekamp et al. 1997). It has been suggested (Mech 1970; Kruuk 1972; Schaller 1972) that social behavior evolved primarily to improve hunting success and, through communication of an individual's dominant or submissive status, to reduce intraspecific competitive pressure within groups. Groups also may provide communal infant care, reduce predation, and defend territories from rival packs. More recent investigators (Mills 1985; Packer et al. 1990) have found cooperative hunting is just one of many factors that may favor group living. Caro (1994) provides a concise summary of factors affecting group living in carnivores.

Methods of hunting vary from concealment and a surprise pounce (stealth and "ambush" seen in many felids) to a stalk followed by a short, swift run (weasels) to a prolonged chase (wolves or hyenas). One of the functions of pelage in carnivores is concealment (as it is in potential prey species), and coloration often is related to hunting behavior. Generally, in cats and other groups in which concealment is critical, the pelage often has spots or stripes. Conversely, in predators such as canids, pelage is typically plain because concealment is less critical (figure 15.3). There are exceptions to both cases, however.

Figure 15.2 **Wolves often attack moose.**   The most common outcome of such encounters is that the moose escapes unharmed.

A

B

Figure 15.3 **Pelage coloration of carnivores.**   Pelage often is adapted to hunting technique. Species that hunt from concealment, such as (A) the leopard (*Panthera pardus*), often have spots or stripes to help camouflage them. Species that do not hunt from concealment, such as (B) the red fox (*Vulpes vulpes*), usually have pelage without stripes or spots.

Species in the Order Carnivora exhibit several basic morphological and behavioral adaptations for searching out, capturing, and handling their prey so as to minimize the possibility of injury during the process. This chapter explores general characteristics of carnivores, as well as life history strategies within lineages that have evolved to deal with the diverse evolutionary developments between predators and their prey.

MORPHOLOGY

The defining morphological characteristic of carnivores is the specialization of their fourth upper premolar (P^4) and first lower molar (m_1) as carnassial (shearing, or cutting) teeth. Carnassials are especially well developed in the predaceous felids, hyaenids, and canids but much reduced in the omnivorous ursids and procyonids (figure 15.4). Regardless of the relative development of carnassials, all carnivores have well-developed, elongated canine teeth.

Skulls are usually heavy, with strong facial musculature for crushing, cutting, and chewing flesh, ligaments, and bone. The relative development of facial muscle groups (see figures 4.19 and 6.6) and associated skull shape reflect different life history patterns and relative use of canines and cheekteeth. Carnivores often have a deep, sharply defined, C-shaped **mandibular fossa** (the portion of the cranium that articulates with the mandible). This strong hinge joint, particularly evident in mustelids (figure 15.5), minimizes lateral movement of the mandible as captured prey struggles

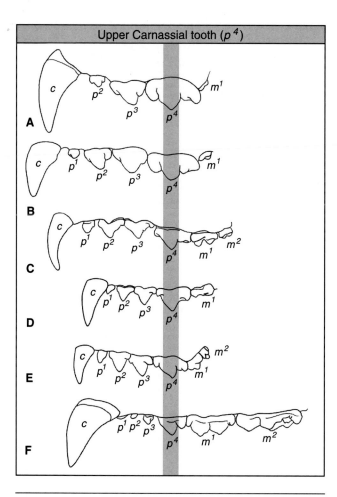

Figure 15.4   Carnassial dentition.   The relative development of the carnassial dentition in various families of carnivores is represented by the vertical line through the maxillary carnassial tooth (P^4) in a: (A) lion (Felidae); (B) hyena (Hyaenidae); (C) dog (Canidae); (D) marten (Mustelidae); (E) mongoose (Herpestidae); and (F) bear (Ursidae). Note the large canine tooth at the left (anterior) and the fact that the carnassial teeth are posterior (toward the back of the jaw). Mammalian dentition functions like a nutcracker, with the greatest force generated closest to the articulation of jaw and skull (the fulcrum point in a lever system). Thus, carnassial teeth are posterior in the dental arcade, rather than anterior, to better facilitate their crushing and shearing function.

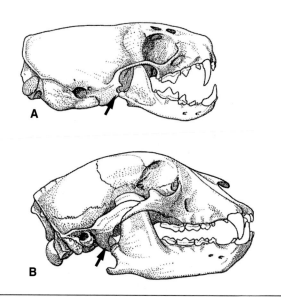

Figure 15.5   Carnivore jaw articulation.   (A) The strongly C-shaped mandibular fossa of a mustelid restricts lateral movement of the lower jaw. (B) The fossa of a bear is flatter.

and permits only a vertical, or up-and-down, motion. Omnivorous carnivores, such as bears and raccoons, have a relatively flatter mandibular fossa that allows more lateral motion of the jaw as the animal chews.

The auditory bullae in carnivores (as in other eutherians) house the tympanic membrane and inner ear. Bullae are formed either entirely from the tympanic bone (derived from the reptilian angular bone; see figure 4.6) or from the tympanic and endotympanic bones. The structure of the bullae is a criterion used to differentiate the two suborders of carnivores. In feliforms, both the tympanic and endotympanic bones form the bullae, with a septum occurring where the two meet. In caniforms, the bullae are formed almost entirely from the tympanic bone, and there is no septum.

All but a few carnivore species have a well-developed **os baculum** (penis bone; figure 15.6). Although the function of the baculum is open to question, it may serve to prolong copulation in species with induced ovulation. Most carnivores have distinctive **anal sacs** associated with secretory anal scent glands. These occur on both sides of the anus and produce substances that function in defense and intraspecific communication. They are especially well developed in mustelids, herpestids, and hyaenids. Skunks are well-known for ejecting the anal gland secretion as a defensive mechanism. Weasels also have strong-smelling scent glands, but they are not used defensively. Anal sacs are relatively small in canids and felids and are absent in ursids and some procyonids.

Carnivores usually have well-developed claws on all digits. Even the unusual clawless otters (Genus *Aonyx*) have vestigial claws. In most felids and some viverrids, claws are retractile. This helps keep them sharp because they have less contact with the ground. Neither the pollex nor the hallux is opposable. The centrale, scaphoid, and lunar bones of the wrist are fused (figure 15.7) to form a scapholunar bone, which may add support for cursorial locomotion in some terrestrial species. As in ungulates, the clavicle is reduced or lost, which serves to increase the length of the stride and allow for faster running in cursorial species. The postcranial skeleton of terrestrial carnivores is generalized, with the differences that occur among families often being only a matter of proportion (Ewer 1973). Differences in limb structure reflect locomotor adaptations: cursorial canids and felids are digitigrade, whereas ursids and procyonids are plantigrade. Specialized morphological adaptations of pinnipeds are noted in the "Pinnipedia" section of this chapter.

Carnivores range in size from maximum body mass of 70 g in the least weasel (*Mustela nivalis*) to 800 kg in grizzly

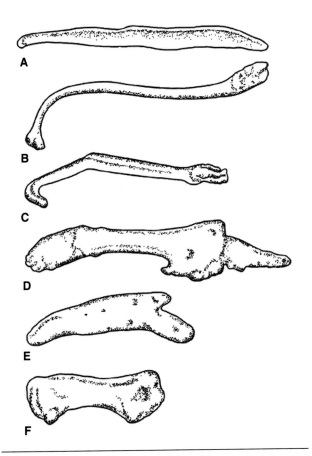

Figure 15.6   **Shapes and sizes of bacula.**   The size and shape of the baculum varies considerably among carnivores. Lateral view of the baculum in a (A) canid (red fox [64 mm]); (B) procyonid (raccoon, [90 mm]); (C) mustelid (least weasel, 19 mm]); (D) herpestid (Egyptian mongoose, *Herpestes ichneumon* [18 mm]); (E) viverrid (common genet, *Genetta genetta* [6 mm]); and (F) felid (lion, [7 mm])

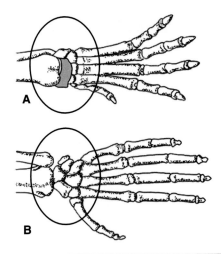

Figure 15.7   **Carnivore wrist structure.**   (A) In carnivores, the centrale, scaphoid, and lunar bones of the wrist are fused (*black area*), whereas in other mammals, such as a (B) primate, they are not.

bears (*Ursus arctos*) and polar bears—over 11,000 times heavier. Secondary sexual dimorphism is often evident, with males being larger than females. This dimorphism is particularly pronounced in some pinnipeds; male elephant seals (*Mirounga angustirostris*) may be six times the size of females. Evolution in size variation from small to large species within and among carnivore families reflects the size range of potential prey species available to them.

Because meat is easy to digest, carnivores have simple stomachs with an undeveloped cecum. Nonetheless, some specialized feeding habits have evolved. Besides those that eat only flesh, species are insectivorous, piscivorous, frugivorous, omnivorous, or almost completely herbivorous (Eisenberg 1981). The structure of the dentition, including the carnassials, is correspondingly modified (see figure 15.4). Canids, felids, and mustelids subsist mainly on freshly killed prey. These families show correspondingly greater development in "tooth and claw"; they also have greater carnassial development and cursorial locomotion. In addition to live prey, canids, ursids, and hyaenids take a large amount of **carrion** (dead, often decaying animal matter). Because not all

carnivores are strictly carnivorous (most ursids and procyonids are omnivorous), their diet varies depending on season and local availability of food. The giant panda bear (*Ailuropoda melanoleuca*) eats primarily bamboo shoots and roots and only occasionally eats animal matter (Schaller 1993).

FOSSIL HISTORY

The earliest known mammalian genus generally adapted for carnivory was *Cimolestes* (figure 15.8), from the late Cretaceous period, over 65 mya. *Cimolestes* was small, about the size of a weasel, and is considered the basal group (Martin 1989) for both modern carnivores and an archaic group of terrestrial carnivores, the Creodonta. The creodonts (figure 15.9) extended from the late Cretaceous period to the Miocene epoch, when they became extinct, possibly through competition with modern carnivore lineages. Modern carnivores may have originated with two early "miacoid" families of small carnivores (Wyss and Flynn 1993): the Viverravidae

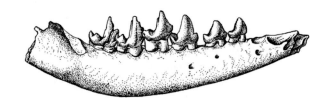

Figure 15.8   **Ancestral carnivore.**   Lateral view of the lower jaw of the late Cretaceous *Cimolestes*. These early eutherian mammals may have been ancestral to both the creodonts (extinct by the Miocene epoch), and modern carnivores. Actual length of the jaw fragment is about 3 cm.

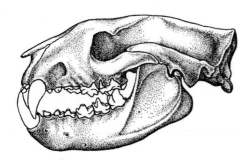

Figure 15.9   Creodont carnassial dentition.   Unlike the modern lineage of carnivores, the carnassial teeth in creodonts, such as *Oxyaena* shown here, involved the first or second upper molars and the second or third lower molars. Skull length is 21 cm.

and Miacidae. These first recognizable carnivores, with P^4/m_1 carnassial dentition, appeared in the early Eocene epoch, about 50 mya. Unlike modern carnivores, the centrale, scaphoid, and lunar bones of the wrist in these early specimens were not yet fused. Although modern families of carnivores probably began to diverge by the late Eocene epoch, most are recognizable only by the Oligocene epoch, including canids, felids, viverrids, mustelids, and ursids (Carroll 1988). This adaptive radiation reflected corresponding diversification of prey groups, which was in turn related to the development of more diverse vegetative biomes in the early Cretaceous period (Eisenberg 1981).

The earliest ancestral pinnipeds, the enaliarctids, date from the late Oligocene and early Miocene epochs (Ray 1976; Berta 1991). For example, the earliest walrus, *Neotherium*, dates from the mid-Miocene epoch, about 14 to 15 mya (Repenning and Tedford 1977; Carroll 1988). Odobenids and otariids are closely related (see figure 15.1) and are often grouped in the Superfamily Otarioidea. Both represent a distinct lineage from the phocids, however.

ECONOMICS AND CONSERVATION

Many species of carnivores have close associations with humans. The dog (*Canis [familiaris] lupus*) was the first domesticated mammal species (see chapter 28), and dogs and cats (*Felis [catus] silvestris*) have been popular pets throughout the world for thousands of years. Likewise, numerous species of carnivores have been important as furbearers throughout human history, including most species of felids and pinnipeds. Today, raccoons (*Procyon lotor*) are trapped for their pelts, as are foxes and many species of mustelids, including mink (*Mustela vison*), river otter (*Lontra [Lutra] canadensis*), marten (*Martes americana*), fisher (*M. pennanti*), and sable (*M. zibellina*). Many other species are useful to the agricultural industry as predators on rodents and other agricultural pests.

Conversely, carnivores often are in conflict with human economic interests. Vilified as "red in tooth and claw,"

they are viewed by some people as viciously preying on innocent grass-eating livestock. As a result, the density and distribution of many species have been greatly reduced. The wolf (*Canis lupus*) is a prime example. It has been extirpated throughout much of its former range in the Old World. In the contiguous 48 United States, it occurs in only about 5% of its historic range (Paradiso and Nowak 1982). Similarly, in the last 200 years, the grizzly bear was practically extirpated in the United States. In response to changing sentiments toward conservation, however, many species, including the wolf, are beginning to recover (Mech 1995). Historically, large carnivores throughout the world also have been important as big-game trophies. This has resulted in serious overexploitation and range reduction, as in the case of the tiger (*Panthera tigris*).

FAMILIES

Feliformia

Felidae

The cat family is distributed worldwide except for Australia, New Zealand, and surrounding islands; polar areas; Madagascar; Japan; and most oceanic islands. The classification of this family, especially regarding the Genus *Felis,* is controversial. Some authorities recognize only a few genera; however, Wozencraft (1993) considered 18 genera and 36 species. Nevertheless, all cats are characterized by a shortened rostrum (figure 15.10), well-developed carnassials, and large canine teeth that are "highly specialized for delivering an aimed lethal bite" (Ewer 1973:5). Felids kill their prey by suffocation or by biting the prey's neck so the canines enter between the vertebrae and separate the spinal cord. Loss or reduction in size of the other teeth is more evident in felids

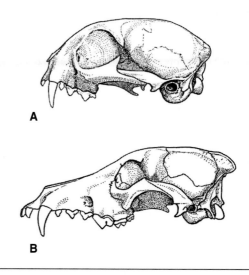

Figure 15.10   Shape of felid and canid skulls.   (A) A felid skull typically has a short, rounded rostrum compared with (B) the long rostrum typical of canids.

than in any other carnivores; total number of teeth is reduced to 28 or 30. The shortened toothrow and rostrum add to the force generated through the enlarged canines. The cheekteeth are bladelike and adapted for seizing and slicing meat from prey, rather than crushing bones, as occurs in hyaenids or canids.

Body mass ranges from about 2 kg in the black-footed cat (*Felis nigripes*) of Africa to 300 kg in tigers. Large felids are noted for their ability to roar (because flexible cartilage replaces the hyoid bone at the base of the tongue), whereas smaller species purr. Most cats are at least semiarboreal. They are digitigrade and have strongly curved, sharp claws to hold prey. Claws are retractile, except in the cheetah (*Acinonyx jubatus*). The dorsal surface of the tongue is covered by posterior-directed papillae that give the tongue a "sandpaper" feeling and may help to retain food in the mouth.

Felids prey almost exclusively on mammals and birds (Kruuk 1986), although the Asian flat-headed cat (*Prionailurus planiceps*) and fishing cat (*P. viverrinus*) consume fish, frogs, and even mollusks. Most felids are nocturnal. They are very agile and either stalk their prey or pounce from ambush. Their pelage is often spotted or striped, an adaptation to cryptic hunting behavior and habitat. The diurnal cheetah, the fastest mammal in the world, relies on its ability to outrun prey over short distances. Species generally are solitary or form pairs; however, lions (*Panthera leo*) associate in "prides" that include up to 18 related females, their offspring, and several unrelated males (Packer 1986). The larger species are intolerant of other species; lions are known to kill leopards, and leopards kill cheetahs (Kingdon 1972). Seventeen species of felids are endangered either because of overhunting, habitat loss, or the fur trade, and most other species are threatened.

Hyaenidae

Hyaenids are native to the Old World in the Middle East, India, and Africa, where they are the most abundant large carnivore. They inhabit grassy plains and brushy habitats. The four genera and four species include the aardwolf (*Proteles cristatus*), which is often placed in a separate family, the Protelidae, because of morphological and behavioral differences from hyenas. Hyenas are large carnivores; the spotted hyena (*Crocuta crocuta*) reaches 80 kg in body mass. Well-developed canines and cheekteeth make spotted hyenas highly competent hunters capable of killing large prey (Holekamp et al. 1997). They also are highly adapted to scavenging carcasses and feeding on carrion, using their large cheekteeth to crush bones (figure 15.11A). Their scavenging specialization may allow hyenas to minimize competition with sympatric canids. Hyenas regurgitate pellets of undigested material, including bone fragments, ligaments, hair, and horns. Unlike other hyaenids, the small, nocturnal aardwolf is myrmecophagous (feeds on ants and termites), which is unusual among carnivores. It primarily takes snouted harvester termites (Genus *Trinervitermes*), which it locates through both sound and scent. Reflecting

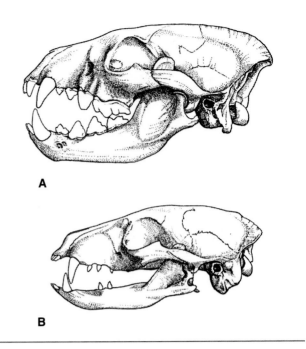

Figure 15.11   **Hyaenid dentition.**   (A) The dentition of the striped hyena (*Hyaena hyaena*) is adapted for crushing bones, unlike (B) the reduced dentition of the myrmecophagous aardwolf.

these feeding habits, the dentition in *P. cristatus* is weakly developed, with smaller, more widely spaced cheekteeth than other hyaenids (figure 15.11B).

Hyaenids are digitigrade and have nonretractile claws. The long neck and forelimbs, and shorter hind limbs produce a characteristic sloping profile. Pelage is coarse, with a prominent mane in all but the spotted hyena, in which the mane is less apparent. This species is also noteworthy for its "laughing" vocalizations, one of several types of calls used in communication. Unlike most carnivores, females are slightly larger than males. Their external genitalia resembles that of males, such that it is very difficult to determine the sex of spotted hyenas in the field (figure 15.12). All hyenas have **protrusable** (can be turned inside out) anal scent glands, which are dragged along the ground, depositing a pastelike substance to scent-mark territory (see figure 20.3). Hyaenids are unusual among carnivores in lacking a baculum. The endangered brown hyena (*Parahyaena brunnea*) has been reduced in numbers because of perceived livestock depredation. The aardwolf is also reduced in numbers.

Herpestidae

This Old World family is native to Africa, the Middle East, and Asia. The 18 genera and 37 species of mongooses were included within the Viverridae (civets and genets) in the past, but they differ from viverrids based on the structure of the auditory bullae and anal sacs, which are similar to those of hyaenids and mustelids. Anal sacs can be everted, exposing cutaneous glands (Kingdon 1972). Several other features distinguish herpestids from viverrids (table 15.1). Mongooses generally are small (figure 15.13); body mass ranges

Figure 15.12   Body morphology.   Hyenas have a characteristic sloping profile. The high, strong shoulders provide leverage for individuals as they pull meat and hide from carcasses. This female has an erect pseudopenis.

Table 15.1.   Morphological characteristics differentiating herpestids (mongooses) from closely related viverrids (civets and genets)

	Herpestidae (mongooses)	Viverridae (civets and genets)
Tail length	Less than head and body length	Equal to or greater than head and body length
Digits	Four or five; webbing reduced or absent	Five; webbing between toes
Claws	Nonretractile	Retractile
Ears	Short and round; no bursae (pockets) on lateral margins	Long and pointed; bursae on lateral margins
Pelage	Usually uniform coloration	Usually spotted or striped
Behavior	Social, often forms groups; diurnal or nocturnal; terrestrial	Generally solitary; nocturnal; arboreal

from the dwarf mongoose (*Helogale parvula*) at 0.3 kg to the white-tailed mongoose (*Ichneumia albicauda*) at about 5.0 kg.

Herpestids occur in a variety of terrestrial and semiarboreal habitats. They are feeding generalists, with associated generalized dentition (Ewer 1973). Some species are solitary, whereas others are highly social and form groups (Rood 1986). As in many other carnivore families, several diseases occur in herpestids, including rabies. The small Indian mongoose (*Herpestes javanicus*) was introduced to the West Indies and Hawaii, where it is considered a pest. It preys on poultry and native fauna, especially nesting birds, and carries rabies. The Liberian mongoose (*Liberiictis kuhni*) is endangered.

Viverridae

Like the closely related herpestids, the civets and genets occur only in the Old World. The 20 genera and 34 very diverse species occupy tropical and subtropical habitats in Europe, Africa, the Middle East, and Asia. Viverrids and mongooses are the only mammalian carnivores on Madagascar. Dental adaptations are diverse and reflect the variety of morphological adaptations in this family. In physical appearance and life history, viverrids resemble members of other carnivore families and fill many equivalent niches. For example, some viverrid genera are carnivorous and parallel mustelids or felids. Others are omnivorous like procyonids, some are frugivorous, and others are scavengers like hyenas. Species may be diurnal or nocturnal, occur singly or in groups, and may be terrestrial, semiaquatic, or arboreal. Civets retain several primitive features, including limbs that are generally short and unspecialized. The pelage usually is spotted or striped (see table 15.1).

Most civets have perianal glands, unique among carnivores. These are "... composed of a compact mass of glandular tissue, typically lying between the anus and vulva or penis and opening into a naked or sparsely haired area, which may be infolded to form a storage pouch" (Ewer 1973:92). These glands produce a fluid called "civet," which is used for scent

Figure 15.13   The banded, or striped, mongoose.
This species (*Mungos mungo*) has a body mass of 1.6–2.5 kg and occurs in packs throughout central and southern Africa.

marking and functions in intraspecific communication among individuals. Civet has been economically important in making perfume for thousands of years. Both the spotted linsang (*Prionodon pardicolor*) and the Malabar large spotted civet (*Viverra civettina*) are endangered.

Caniformia

Canidae

Canids include the wolves, coyotes, foxes, dingo, dholes, jackals, and the dog. The 13 genera and 33 species occur naturally throughout North and South America, Africa, Asia, and Europe. The dingo (*Canis lupus dingo*) was introduced to Australia, New Guinea, and parts of Asia 3500 to 4000 years ago (Corbett 1995). Habitats of canids range from hot, dry deserts to tropical rain forests to arctic ice. Canids generally take animal prey throughout the year; however, plant material may be taken seasonally by some species. Jackals often eat carrion, and the bat-eared fox (*Otocyon megalotis*) consumes a large amount of insects. Many canids are solitary. Four species—the wolf, African hunting dog, Asian dhole (*Cuon alpinus*), and bush dog (*Speothos venaticus*)—form packs and engage in cooperative hunting (Moehlman 1986), although solitary wolves are also successful hunters (Messier 1985; Thurber and Peterson 1993).

Generally, canids have long limbs relative to head and body length and are adapted to pursue prey in open habitats. They range in body mass from 1 kg in the fennec (*Fennecus zerda*) to about 80 kg in the gray wolf. They are generally digitigrade and have nonretractile claws. The pollex and hallux are reduced. Skulls characteristically have an elongated rostrum (see figure 15.10), with well-developed canines and carnassial teeth. Although there are a few exceptions, the typical dental formula of 3/3, 1/1, 4/4, 2/3 = 42 is close to the primitive eutherian number of 44. The earliest fossil remains of identifiable canids, Genus *Hesperocyon,* date from the early Oligocene epoch of North America.

Several species of canids (e.g., foxes) are hunted for sport or trapped for their fur. In the United States, the coyote has been the target of state and federal predator control programs because of livestock depredations. The wolf is listed as endangered in the lower 48 United States and in India, Nepal, Pakistan, and Bhutan. The red wolf (*Canis rufus*), Simien jackal (*C. simensis*), and African hunting dog also are endangered species. The red wolf recently has been reintroduced in the United States (Phillips 1990). The Falkland Island wolf (*Dusicyon australis*) was driven to extinction in the 1870s.

Mustelidae

This large, diverse family of 25 genera and 65 species includes weasels, skunks, badgers, otters, and the wolverine (*Gulo gulo*). They are most commonly found throughout the Northern Hemisphere and are absent from Australia, Madagascar, the Celebes, and some other oceanic islands.

Mustelids are highly specialized predators. Kingdon (1972:2) suggested their strict carnivory and distribution may be related because ". . . winter diets do not allow for any buffering from insects, fruit, or other foods. . . ." They inhabit both terrestrial and arboreal habitats, as well as fresh and saltwater.

Mustelids have long bodies with relatively short legs. They are digitigrade, pentadactyl, and have nonretractile claws. They range in size from the 30- to 70-g least weasel, the smallest carnivore in the world, to the 55-kg wolverine. Males generally are about 25% larger than females. The carnassials are well developed, and, as noted by Stains (1984), no extant mustelid has more than one molar after the carnassial teeth. Nonetheless, the diverse dental adaptations within the mustelids reflect the varied life histories and diets in this family. The mandibular fossa is strongly C-shaped, which restricts the lateral movement of the mandible and allows little "give" for struggling prey (see figure 15.5).

Mustelids are noteworthy for their enlarged anal scent glands. The thick, powerful-smelling secretion (musk) is used for communication and defense. Some species, such as skunks, marbled polecat (*Vormela peregusna*), and zorilla (*Ictonyx striatus*), also have a striking contrast in their black and white pelage (figure 15.14). This pattern probably serves as "warning coloration" to potential predators that their anal scent glands make them hazardous to capture.

Mustelids are monestrous, and many genera display both induced ovulation and delayed implantation. Although the actual gestation period is usually 1 or 2 months, because of delayed implantation, in some species the total period of pregnancy may be about 1 year. A single litter per year is typical.

The black-footed ferret (*Mustela nigripes*) is endangered. Once believed to be extinct, it has recently been reintroduced (Williams et al. 1992; Miller et al. 1994). Several species of otters are threatened.

Figure 15.14   Convergence in pelage characteristics.
The zorilla, an African mustelid, has the same warning coloration of the more familiar striped skunk of the United States.

Procyonidae

All of the 6 genera and as many as 18 species in the raccoon family are restricted to the New World, where they typically inhabit forested temperate and tropical areas, usually near water. Body mass ranges from about 1 kg in the ringtail (*Bassariscus astutus*) and olingo (*Bassaricyon gabbii*) to 18 kg in raccoons; males usually weigh about 20% more than females. Procyonids typically have long, bushy tails (prehensile in the kinkajou [*Potos flavus*]), with alternating light and dark rings, and obvious facial markings. They are plantigrade; some have semiretractile claws, and all are adept at climbing trees. Procyonids have 40 teeth, except the kinkajou, which has 38. Dentition is generalized and adapted for an omnivorous diet, with fruit predominating in the kinkajou and olingo. The carnassials are fairly well developed only in the ringtails and cacomistle (*Bassariscus sumichrasti*). The raccoon, a popular game animal with both hunters and trappers, is one of the most commonly harvested furbearers in North America. The species also commonly contracts rabies (see chapter 27).

Ursidae

Bears historically occurred throughout North America, the Andes Mountains of South America, Eurasia, and the Atlas Mountains of North Africa. Their habitats vary from tropical forests to polar ice floes. Wozencraft (1993) recognized 6 genera and 19 extant species of ursids, including both the red, or lesser, panda (*Ailurus fulgens*) and the giant panda. Both these latter species have a controversial taxonomic history; the lesser panda has been included in the Family Procyonidae, and the giant panda is sometimes considered in a separate family, the Ailuropodidae.

Body mass in bears ranges from 5 kg in the red panda to 800 kg in grizzly and polar bears, the largest terrestrial carnivores. Sexual dimorphism is evident. Males are about 20% heavier than females in monogamous species such as the sun bear (*Helarctos malayanus*) and sloth bear (*Melursus ursinus*) and up to twice as large in polygamous species. All bears are plantigrade and pentadactyl and have nonretractile claws. There usually are 42 teeth. Canines are large, but the last upper molar is reduced, and carnassials are not well developed. Molars are broad, flat, and relatively unspecialized, reflecting an omnivorous diet. Only the polar bear is strictly carnivorous, feeding on fish and seals. The sloth bear, like the aardwolf, is highly myrmecophagous.

In northern areas, black bears (*Ursus americanus*), grizzlies, polar bears, and Asiatic black bears (*U. thibetanus*) den in hollow trees, caves, or burrows. They sleep through the winter, especially pregnant females, and live off their stored body fat. This process is called "winter lethargy" rather than hibernation (see chapter 8) because body temperature, heart rate, and other physiological processes are not reduced to the same extent as in true hibernation (Hellgren et al. 1990; Brown 1993). Bears are monestrous and exhibit delayed implantation. Young, often twins, are usually born between November and February while the female dens.

Historically, bears have been hunted for their hides, meat, and fat. Most species have been eradicated through much of their ranges because of predation on domestic livestock. Recently, large numbers of black bears have been illegally killed for their gallbladders, which are valuable in traditional Asian medicine. The giant panda and Asiatic black bear are endangered; the Mexican subspecies of grizzly bear (*U. a. nelsoni*) is recently extinct.

Pinnipedia

The last three families—the pinnipeds—exhibit specialized adaptations for an aquatic existence. Pinnipeds are not as totally adapted for life in the water as are whales, manatees, and the dugong, however. Whales and sirenians (chapter 18) spend their entire lives in the water, but pinnipeds must "haul out" onto land or ice floes (where they are slow and vulnerable) to breed, give birth, or rest.

Being adapted to both terrestrial and aquatic conditions has caused all pinnipeds to be morphologically similar, although as seen from figure 15.1, phocids are not closely related to the otariids and odobenids. Pinnipeds have a coarse pelage of guard hairs that helps protect them when they are out of the water. As in whales, body shape is adapted to reduce turbulence and resistance (drag) as they swim through the water (see chapter 16). Thus, bodies are fusiform, with no constriction in the neck region (figure 15.15). External genitalia are concealed in sheaths within the body contour, as are the teats, and external ears are reduced or lost. A subcutaneous layer of fat (blubber) provides energy, insulation, and bouyancy (see figure 8.8). It also serves to maintain a streamlined body shape, which enhances hydrodynamic properties and further reduces drag.

Figure 15.15   **Pinniped body shape.**   All pinnipeds have a streamlined, torpedo-shaped body for minimal resistance as they swim, as in this harbor seal.

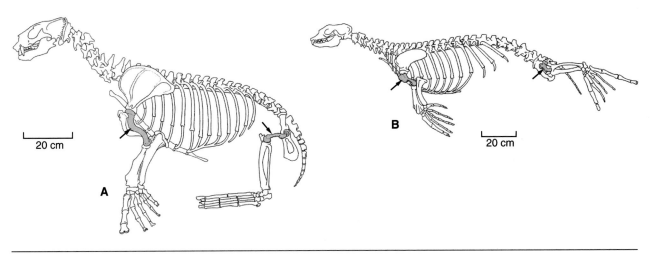

Figure 15.16   **Pinniped skeletal characteristics.**   Lateral view of the skeleton of (A) a New Zealand fur seal (*Arctocephalus forsteri*), an otariid, and (B) a West Indian monk seal (*Monachus tropicalis*), a phocid. Note the very short, broad humerus and femur (*arrows*) and elongated foot bones in both. In otariids and the walrus, the cervical and thoracic vertebrae and scapula are enlarged to support the primary muscle groups that power the forelimbs for propulsion. The hind limbs propel phocids through the water, and the lumbar vertebrae are enhanced for muscle attachment.

The limbs are relatively short, stout, and modified to form paddlelike flippers (figure 15.16). The forelimbs provide the propulsive force in otariids and odobenids; the hind limbs serve this function in phocids. These differences are reflected in skeletal anatomy. Otariids and the walrus have enlarged cervical and thoracic (neck and upper chest) vertebrae that support the large muscle groups associated with the forelimbs. Because the hind limbs provide propulsion in phocids, the lumbar (lower back) vertebrae are relatively large (see figure 15.16).

Most pinnipeds have generalized feeding habits, and the primary function of the teeth is to grasp and hold prey rather than to chew. Teeth in most species approach homodonty, with the premolars and molars being similar and somewhat conical (figure 15.17). There are exceptions, however, in species that are more specialized feeders. For example, the crabeater seal (*Lobodon carcinophagus*) has distinctive cheekteeth that form a sieve to filter krill from the water. Unlike terrestrial carnivores, the interorbital area of the pinniped skull is long and narrow, and the braincase is longer in proportion to the facial area (see figure 15.17).

The eyes of seals are relatively large and modified to focus underwater by means of a greater corneal curvature than occurs in the eyes of terrestrial mammals. On land, however, seals are quite nearsighted. Pinnipeds see effectively under conditions of reduced light. Like felids and

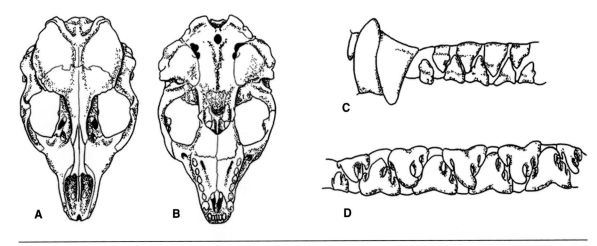

Figure 15.17   **Pinniped skulls and cheekteeth.**   Dorsal view (A) and ventral view (B) of a phocid, the harbor seal (*Phoca vitulina*). Note the narrow interorbital regions, shortening of the nasal bones on the rostrum, and lack of a postorbital process in the phocid. The cheekteeth of otariids usually are single-cusped and peglike, as in (C) Stellar's sea lion (*Eumetopias jubatus*), whereas those of phocids are multicusped and particularly well-developed in (D) the crabeater seal. Also see figure 6.8.

some other terrestrial groups, they have a well-developed tapetum lucidum, a specialized membrane behind the retina. The tapetum lucidum increases light-gathering efficiency by reflecting back to the retina light that has passed through the retina but not been absorbed (Lavigne et al. 1977). All pinnipeds hear well underwater and are able to determine the direction from which sounds come. This is critical for locating prey, especially under reduced light conditions or in total darkness. King (1983) details several unique pinniped adaptations for directional hearing underwater.

Many of the same physiological processes for diving seen in whales also occur in pinnipeds, including **bradycardia** (reduced heart rate) and shunting blood from peripheral areas to the brain and heart. Like whales, seals also have more oxygen-binding hemoglobin and myoglobin than do terrestrial mammals. Pinnipeds do not dive as deep or remain submerged as long as whales, however (Kooyman 1981; DeLong et al. 1984; Lavigne and Kovacs 1988). Among pinnipeds, phocids dive deeper and remain submerged longer than do otariids. Methods used to investigate the diving ecology of seals, including time-depth recording equipment and analyses of data, are discussed by Gentry and Kooyman (1986), DeLong and Stewart (1991), and DeLong and colleagues (1992).

Until the nineteenth century, pinnipeds had been harvested for thousands of years on a subsistence basis for fur, food, and oil, with little effect on populations. Large-scale commercial hunting and resultant overharvest of many species began in the early 1800s, and many species were "commercially extinct" by the end of the century. Today, several species are still harvested, but most populations are secure. The most serious threats to pinnipeds include incidental drowning in fishing nets, habitat degradation, and environmental contamination of marine habitats (Reijnders et al. 1993).

Odobenidae

This monotypic family, sometimes considered a subfamily of otariids, includes only the walrus (*Odobenus rosmarus*). Walruses have a circumpolar distribution in shallow arctic waters, where they remain near ice floes and rocky shorelines (King 1983). Body mass of adult bulls is about 1000 kg but can reach 1600 kg. Typical of pinnipeds and most other carnivores, females are significantly smaller than males. The skin is thick and wrinkled, and the layer of blubber, which is usually 6 to 7 cm thick, can reach 15 cm. Certain morphological features of walruses are shared with otariids, whereas others are similar to phocids (table 15.2). In addition, walruses, like otariids, have naked ventral surfaces on all flippers, and the nails on the first and fifth digits of the hind flippers are rudimentary.

The significant feature in both male and female walruses is their tusks, which are enlarged upper canines (figure 15.18). These grow throughout life and can reach as long as 100 cm in males and 60 cm in females. The crown portion of the tusk is composed internally of dentine and externally of cementum. Other than a small cap at the end of the tooth that wears away during the first couple of years, there is no enamel. Tusks are used for defense, raking the bottom for mollusks, breaking through ice, hanging from ice floes while remaining in the water, and, among males, as weapons for establishing dominance hierarchies.

Walruses remain in shallow water, usually at depths of between 80 and 100 m (King 1983); they feed primarily on clams, other mollusks, and a variety of invertebrates taken from the muddy bottom. They are gregarious, often huddle together, and form groups of 100 to 1000 individuals. Walruses are polygamous, mating during February and March on near-shore ice floes. Following a 3-month delayed implantation period and 12-month gestation period, females haul out to give birth on land, usually to a single calf, in May

Table 15.2.   Morphological characteristics differentiating the three families of pinniped carnivores*

	Phocidae (earless seals)	Odobenidae (walrus)	Otariidae (fur seals/sea lions)
External pinnae	No	No	Yes
Testes	Abdominal	Abdominal	Scrotal
Tip of tongue notched	Yes	No	Yes
Hind limbs rotate forward	No	Yes	Yes
Guard hairs with medulla	No	No	Yes
Underfur	Essentially absent	Essentially absent	Present in sea lions
Alisphenoid canal	Absent	Present	Present
Auditory bullae	Inflated	Small and flattened	Small and flattened
Transverse groove on upper incisors	No	No	Yes
Lower incisors present	Yes	No	Yes
Total number of teeth	26–36	18–24	34–38
Fused symphysis of lower jaw	No	Yes	No
Postorbital process	Absent	Absent	Present
Chromosome number	32–34	32	36

* For additional differences of the skull, middle ear, and inner ear, see Repenning, C. A. 1972. Underwater hearing in seals: functional morphology. Pp. 307–331 *in* Functional anatomy of marine mammals (R. J. Harrison, ed.). Academic Press, London.

Figure 15.18   Walrus tusks.   The large tusks of the walrus are canines. Also note the vibrissae.

or June. Thus, given the lactation period, females normally mate every 2 to 3 years.

Walruses are hunted for meat, oil, and hides, and the tusks (ivory) have been highly prized for artwork carvings (called scrimshaw) for hundreds of years. Although from 10,000 to 15,000 individuals are harvested annually from the Pacific population (Reijnders et al. 1993), walrus populations appear to be stable.

Otariidae

Fur seals and sea lions occur in subpolar, temperate, or coastal waters of western North America, South America, Asia, southern Australia and New Zealand, and oceanic islands. They only inhabit marine communities, unlike phocids, which also occur in freshwater and estuarine communities. Several general characteristics of the 7 genera and 14 species of otariids have been noted previously and in table 15.2. Two subfamilies are recognized: five genera and five species of sea lions (Subfamily Otariinae), with blunt noses and little underfur, and two genera and nine species of fur seals (Subfamily Arctocephalinae), with pointed noses and abundant underfur. All otariids are highly dimorphic, with males being larger than females. For example, maxi-

mum body mass of male northern fur seals (*Callorhinus ursinus*) is five times greater than that of females; male southern sea lions (*Otaria byronia*) are twice the size of females (figure 15.19).

Otariids feed on fish, cephalopods, and crustaceans. Generally, they are much more gregarious than terrestrial carnivores, and breeding colonies of up to a million individuals may occur within limited areas. All otariids breed on land in rocky, isolated areas that are inaccessible to potential predators. Breeding males are polygynous and defend a territory with a group of 3 to 40 females. Most species are known to exhibit delayed implantation.

The most familiar species of otariid is the California sea lion (*Zalophus californianus*), commonly displayed in circuses and zoos. Historically, all species were harvested for meat, hides, and oil from blubber. Today, relatively few individuals of most species are harvested for subsistence purposes; about 6500 northern fur seals are taken annually (Reijnders et al. 1993). A remnant population of the Juan Fernandez fur seal (*Arctocephalus philippii*), once believed extinct due to overharvest, is considered threatened. The Japanese sea lion (*Z. c. japonica*) is probably extinct.

Phocidae

The true seals include 10 genera and about 19 recent species. They are found primarily in polar, subpolar, and temperate waters around the world; the monk seals (Genus *Monachus*) are the only pinnipeds that inhabit tropical areas. Besides in oceans, phocids occur in inland freshwater lakes and estuaries. Phocids in the Northern Hemisphere constitute the Subfamily Phocinae and have well-developed claws on all flippers. The monk seals, elephant seals, and antarctic seals make up the Subfamily Monachinae and have reduced claws on the hind flippers. The flippers of phocids are furred

Figure 15.19   Pinniped size dimorphism.   Male pinnipeds are often much larger than females, as is evident in these southern sea lions.

on all surfaces, and the nails are all the same size on the hind flippers.

Among pinnipeds, phocids are the most diverse in size. The smallest is the Baikal seal (*Phoca sibirica*), restricted to freshwater Lake Baikal, with a body mass of only 35 kg. The largest phocid (and the largest carnivore) is the northern elephant seal; male body mass is about 3700 kg. Unlike otariids, phocids generally lack underfur (see table 15.2) and the cheekteeth are multicusped (see figure 15.17D). Phocids are not gregarious and do not form large breeding colonies. Almost all phocids breed on ice. Because they are clumsy and vulnerable on land, ice floes offer several benefits to breeding individuals. It is easier for seals to move on ice than on rocks, and they have quick access to the relative safety of deep water. Like other pinnipeds, they exhibit delayed implantation.

Fish and cephalopods form the diet of most species. The leopard seal (*Hydrurga leptonyx*), the only pinniped that regularly feeds on warm-blooded prey, takes penguins and the young of other seals. As noted earlier, phocids dive deeper for prey and remain submerged longer than otariids and thus are the most aquatically adapted of the carnivores. On land, however, they are less agile and mobile. They are unable to raise themselves on their front flippers and simply hunch their bodies, moving forward like inchworms.

Commercial sealing of many species is controversial, including the harvesting of young harp seals (*Phoca groenlandica*). The Mediterranean monk seal (*Monachus monachus*) and Hawaiian monk seal (*M. schauinslandi*) are endangered, and the Caribbean monk seal (*M. tropicalis*), last seen in the early 1950s, is most likely extinct.

Summary

Carnivores are terrestrial or aquatic predators that usually consume other animals as a major part of their diet. Most morphological and behavioral characteristics of carnivores involve adaptations to enhance locating, capturing, killing, and consuming their prey without being injured in the process. Nonetheless, some specialized feeding habits have evolved. These include insectivory, as in the aardwolf and the sloth bear, scavenging on carrion by hyenas, omnivory in a number of species, and almost complete herbivory in the greater and lesser pandas.

All carnivores have digits with well-developed claws and dentition with enlarged canine teeth. The defining ordinal characteristic, however, is carnassial teeth—specialization of the fourth upper premolar (P^4) and first lower molar (m_1) for cutting and shearing. Carnassial dentition is especially well developed in highly predaceous families, such as felids, canids, and hyaenids, and less developed in more omnivorous groups, such as ursids and procyonids. Facial musculature is well developed. The articulation of the jaw with the cranium typically is hinged so as to preclude any lateral motion as captured prey struggle to escape.

Size diversity is pronounced both within the order and within various families. The largest carnivores are the elephant seals. The largest terrestrial carnivores are the polar bear and grizzly bear, which are 11,000 times heavier than the smallest carnivore, the least weasel. With few exceptions, secondary sexual dimorphism is evident; males are often many times larger and heavier than females. Size of individual carnivores and method of hunting (solitary, paired, or in groups) generally relate to the size of the prey species the carnivore is able to capture. Although most species are solitary hunters, wolves, spotted hyenas, lions, and some others generally hunt in packs. Thus, they are able to prey on species that are several times larger than themselves. In addition to improved foraging efficiency, group membership may provide several additional benefits to individuals: communal infant care, reduced predation, and defense of feeding areas from rival packs.

The earliest carnivorous mammal, *Cimolestes*, from the late Cretaceous period, is considered the basal group for modern carnivores. An early group of carnivores, the creodonts, extended from the Cretaceous period to the Miocene epoch. Creodonts are considered a sister group to the modern order. Most families of modern carnivores had developed by the Oligocene epoch, concurrent with the adaptive radiation of herbivorous groups on which they preyed. Two suborders of modern carnivores are generally recognized. Four families occur within the Suborder Feliformia: felids, herpestids, hyaenids, and viverrids. The Suborder Caniformia includes canids; ursids; mustelids; procyonids; and three families of aquatic carnivores (the pinnipeds): otariids, phocids, and an odobenid. Pinnipeds have distinctive morphological adaptations consistent with their aquatic life histories and in the past have been placed in their own order.

Carnivores have had a long association with humans. Dogs and cats are the most popular domestic pets throughout the world. Numerous other species of carnivores, mink and sable for example, are valuable in the fur industry. Conversely, most large predators, such as wolves, cougars, and bears, have been eliminated throughout much of their range because of conflict with human economic interests, often livestock depredation. Other species have been overexploited through trophy hunting or commercial harvest and are endangered as a result.

Discussion Questions

1. What are the adaptive benefits of delayed implantation to such carnivores as ursids and pinnipeds?
2. What adaptive benefits accrue to bears in northern areas that enter "winter lethargy"?
3. As occurs in several other mammalian orders, sexual dimorphism often is evident among carnivores, with males being larger than females. Consider such factors as feeding and reproduction to hypothesize why size dimorphism might evolve.
4. Some carnivores are diurnal, whereas others are nocturnal. Based on what you have learned from earlier chapters, what types of hunting tactics would you expect from diurnal and nocturnal predators? Which families of carnivores fit into each category?
5. How might the striking black-and-white "warning coloration" of skunks and other species have evolved? How do potential predators learn to avoid these species?

Suggested Readings

Bonner, N. 1994. Seals and sea lions of the world. Facts on File Publ., New York.

Dunstone, N. and M. L. Gorman (eds.). 1993. Mammals as predators. Symp. Zool. Soc. Lond. No. 65. Clarendon Press, Oxford.

Gittleman, J. L. (ed.). 1989. Carnivore behavior, ecology, and evolution, vol. I. Cornell Univ. Press, Ithaca, NY.

Gittleman, J. L. (ed.). 1996. Carnivore behavior, ecology, and evolution, vol. II. Cornell Univ. Press, Ithaca, NY.

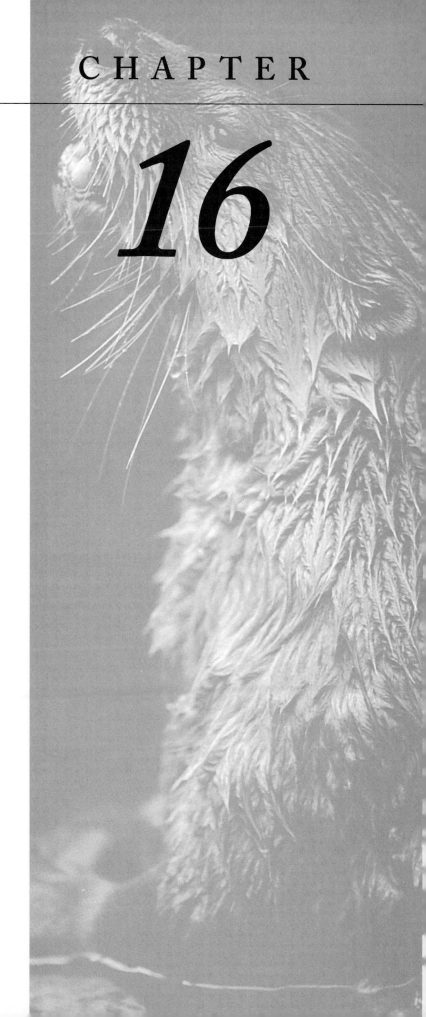

CHAPTER 16

Cetacea

Whales are among the most fascinating and, in some ways, least understood of all mammals. The Order Cetacea (from the Greek word for "whale") is characterized by extremes. It includes the largest animals that have ever lived. Adult blue whales (*Balaenoptera musculus*) are heavier than the biggest dinosaurs (although probably not longer; see Gillette 1991). Whales have the loudest voices, they endure tremendous water pressure as they dive deeper than any other mammal, and certain species make some of the longest migrations. But much of the life history of whales remains unknown. Studying whales has often been difficult because of their relative inaccessibility and the wide-ranging movements of individuals. Only recently have advances in technology allowed researchers to more thoroughly investigate the physiology, behavior, sound communication, social interactions, reproduction, and other aspects of whale biology. Conversely, because the whaling industry has provided carcasses for hundreds of years, probably more is known about the anatomy of whales than of many other large mammals.

In addition to the manatees and dugong (Order Sirenia; see chapter 18), whales are the only group of mammals that are entirely aquatic—they never leave the water. This chapter examines many of the interrelated morphological, physiological, locomotor, and behavioral adaptations of whales that have evolved in response to the demands of life in an aquatic environment.

MORPHOLOGY

The 10 families of whales are divided into two well-defined extant suborders: the **Mysticeti,** or baleen whales, and the **Odontoceti,** or toothed whales. The primary feature of mysticetes is their baleen plates (see the section "Mysticetes: Baleen and Filter Feeding" in this chapter), which take the place of teeth. Baleen is used to strain small marine organisms from the water for food. The teeth in odontocetes generally are homodont, simple, and peglike. Unlike most mammals, which are diphyodont (having two sets of teeth during a lifetime: deciduous and permanent), odontocetes are monophyodont. That is, they have a single set of teeth (Matthews 1978; Slijper 1979). Odontocetes also echolocate to orient within their environment and find prey; mysticetes do not. Several other differences exist between these two extant suborders (table 16.1). A third suborder of whales, the extinct **Archaeoceti,** is also recognized. These ancient whales had many features that were intermediate during the transition from primitive terrestrial mammals to fully aquatic whales.

Unlike fish, whose entire evolutionary history was in water, the adaptations of whales for life in the water are all secondary. That is, the morphological characteristics seen in whales today are all derived from those of ancestral land mammals that made a gradual transition from land to sea about 50 mya (see the section "Fossil History" in this chapter). Water, especially saltwater, provides a particularly buoyant environment. Thus, whales can be larger than any terrestrial mammal and still remain mobile because they are supported by a buoyant environment they never leave. Nonetheless, body size varies considerably among the species of cetaceans; more than half of all whale species are the smaller dolphins and porpoises. Keep in mind, however, that even "small" whales are quite large relative to the average size of most terrestrial mammals. Whales range in size from Heaviside's dolphin (*Cephalorhynchus heavisidii*), which weighs about 40 kg with a total length of 1.7 m, to the blue whale with a mass of 200,000 kg and a total length of 30 m (figure 16.1). No terrestrial mammals are the size of the large baleen whales because their skeletal support system would have to be so massive that they would be rendered immobile; there probably are metabolic and reproductive constraints against such size as well. Several interrelated features of whales, including body and skull shape, thermoregulation, and the physiology of diving, are associated with life in the water.

Table 16.1.   Primary differences between the two suborders of cetaceans: the mysticetes (baleen whales) and odontocetes (toothed whales)

Mysticeti	Odontoceti
Baleen present instead of teeth; teeth present in fetus lost before birth	Teeth present, usually homodont; exceed the primitive eutherian number in some species; monophyodont
Paired external nares ("blowholes"); located anterior to eye	A single external nare; located posterior to eye in all species except sperm whales
Facial profile convex, with no fatty "melon" present	Facial profile concave, with depression occupied by a "melon," or fatty organ
Skull symmetrical	Skull generally asymmetrical
Do not echolocate; auditory bullae (tympanoperiotic bones) attached to skull	Echolocate; tympanoperiotic bones not attached to skull
Nasal passages simple	Nasal passages with a complex system of diverticula
Mandibular condyle directed upward	Mandibular condyle directed posteriorly
Sternum consists of a single bone	Sternum consists of three or more bones

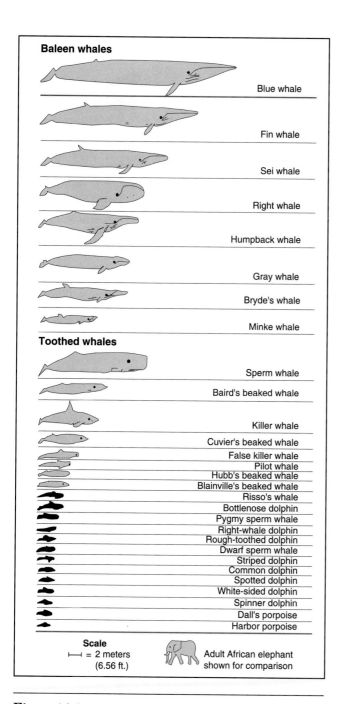

Baleen whales

Blue whale

Fin whale

Sei whale

Right whale

Humpback whale

Gray whale

Bryde's whale

Minke whale

Toothed whales

Sperm whale

Baird's beaked whale

Killer whale

Cuvier's beaked whale
False killer whale
Pilot whale
Hubb's beaked whale
Blainville's beaked whale
Risso's whale
Bottlenose dolphin
Pygmy sperm whale
Right-whale dolphin
Rough-toothed dolphin
Dwarf sperm whale
Striped dolphin
Common dolphin
Spotted dolphin
White-sided dolphin
Spinner dolphin
Dall's porpoise
Harbor porpoise

Scale
⊢—⊣ = 2 meters
(6.56 ft.)

Adult African elephant
shown for comparison

Figure 16.1   Relative sizes of baleen and toothed whales.
The scale represents approximately 2 m. Adult African elephants are
shown for comparison.

Source: Data from R. T. Orr and R. C. Helm, Marine Mammals of California,
Revised edition, 1989, University of California Press.

Streamlined Body Shape

The general body shape of whales is the same whether they
are large mysticetes or small odontocetes (figure 16.2A, B).
Whales are fusiform (streamlined, or torpedo-shaped),
which allows them to move forward through the water with
less drag. Drag is affected by several factors, including the
density and viscosity (resistance to flow) of the fluid (water

in this case) and the size, shape (cross-sectional area), and
speed of the object moving through it. How these factors interact in fluid dynamics is suggested by the Reynolds number (Re), a dimensionless value. Of primary importance in
determining Re is the product of the size and speed of an object moving through a fluid. Thus, a 0.3-mm-long zooplankton moving at 1 mm/s results in Re = 0.3. By comparison, a
large whale moving at 10 m/s results in Re = 300,000,000.
An excellent account of the mechanics of organisms in fluids, both water and air, is provided by Vogel (1994).

As a whale moves forward, water flows smoothly over
it with minimal turbulence (referred to as **laminar flow**) because of several structural features. Nothing that would increase drag and impede the smooth, continuous flow of
water protrudes from a whale's body. It has no external ears
or hind limbs, for example. Hind limbs have been reduced to
small, vestigial innominate bones. These are not attached to
the vertebral column (figure 16.2C) but are imbedded in
ventral muscle anterior to the anus (Slijper 1979). Likewise,
the male has no scrotum, and testes are permanently abdominal. The penis is retractile and lies in a fold of skin (figure
16.3A). Streamlining is further achieved by almost complete
absence of body hair. Insulation is provided by the subcutaneous blubber (the layer of fat under the skin), and thermoregulation is maintained by several physiological features
noted in the section in this chapter on that topic.

The forelimbs of whales are modified as flexible flippers. The humerus, radius, and ulna are shortened, whereas
the digits are greatly elongated by having additional **phalanges**, or finger bones (figure 16.3C-E). The flippers do not
provide power for forward thrust, however, but only act as
stabilizers. Power for locomotion is generated by the elongated tail that ends in a dorsoventrally flattened, horizontal
fluke (see figure 16.2). The **fluke** is supported only by connective tissue, rather than by any skeletal element. Because
water is denser than air and has much greater viscosity, the
large surface area of the tail fluke moving up and down,
which is termed **caudal undulation**, "pushes" against the
water and propels a whale forward (Fish and Hui 1991; Fish
1992). The dorsal spinous vertebral processes provide sites
for muscle attachment to raise the tail, and the **chevron
bones** on the ventral portion of the anterior caudal vertebrae
(see figure 16.2) provide attachment for the muscles that depress the tail.

The simple postcranial skeleton of cetaceans is composed primarily of the backbone, which provides attachment
sites for the large tail muscles. Limbs are not needed for support, because the aquatic environment provides the needed
buoyancy. Cervical vertebrae are reduced in size and are often fused; most whales have little mobility of the head.

Skull Structure

Both suborders of whales have highly modified, **telescoped
skulls** in which posterior bones of the cranium are compressed
and overlap each other. The anterior, or rostral portion, of the

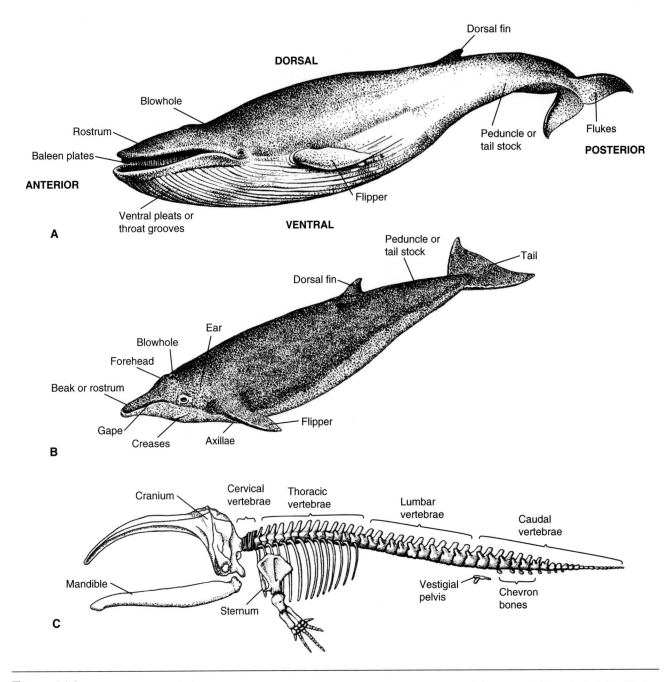

Figure 16.2   Whale body morphology.   Lateral view of the general body plan and terminology for (A) baleen and (B) toothed whales. Whales have a streamlined body shape with few external appendages. (C) Skeleton of a mysticete whale. Note the vestigial hind limbs and chevron bones on the caudal vertebrae.

skull is greatly elongated through extension of the nasal, pre-maxillary, maxillary, and frontal bones. These bones extend posteriorly such that they overlap the parietal bones (figure 16.4). Telescoping has accommodated the posterior displacement of the external **nares** (nostrils, or "blowholes") to the top of the skull, so that only this portion needs to be above water for a whale to breath. In addition, odontocetes have an asymmetrical skull structure, especially around the internal nares (figure 16.5). Asymmetry probably is related to echolocation and movement of the nares.

Thermoregulation

Whales face certain challenges living in an aquatic environment because of the physical properties of water. One of these is water's very high thermal conductivity: water absorbs heat from a warm body about 27 times faster than air does. Thus, one of the challenges faced by whales is to maintain their body temperature in frigid polar waters or the deep, cold waters of temperate and tropical areas. Loss of body heat represents loss of energy, which must be made up by increased

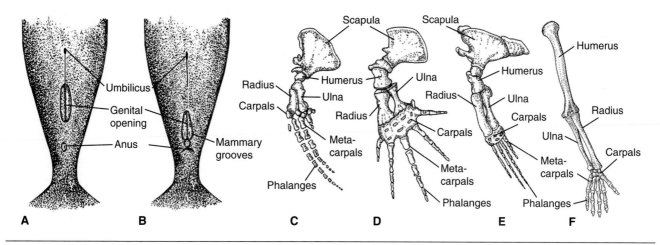

Figure 16.3   Genital grooves and forelimbs.   Ventral view of the genital grooves in (A) a male whale and (B) a female whale. Forelimbs of whales have been modified to form paddlelike flippers: (C) a pilot whale (Genus *Globicephala*), (D) a right whale, and (E) a blue whale. Compared to a human arm (F), whales have a relatively short humerus, radius, and ulna. The phalanges have been lengthened and increased in number. Drawings are not to the same scale.

food intake. Thus, it benefits individuals in cold water to minimize the amount of heat they radiate to the environment. Unlike terrestrial mammals, whales have no insulating fur, nor can they burrow or construct nests to help maintain their body temperature against cold ambient conditions.

Whales maintain thermal equilibrium in several ways. By virtue of their large size, they have a favorable surface-to-volume ratio (the "scale effect"). Even though they have a large surface area from which they lose heat, this area is small relative to their heat-producing body mass. Also, the subcutaneous layer of blubber acts to insulate whales. The thickness of blubber varies both seasonally and among species, from 5 cm in small dolphins to as much as 50 cm in the bowhead whale (*Balaena mysticetus*). When large whales are feeding at high latitudes, blubber may represent up to 70% of their body mass (Bonner 1989).

Key:
■ nasal
▨ pre-maxilla
□ maxilla
▨ frontal
□ parietal
▦ occipital

Figure 16.4   Telescoping of the skull in modern whales.
Skulls of (A) a terrestrial mammal (the horse, *Equus caballus*); (B) an archaeocete whale (*Basilosaurus*) with heterodont dentition; (C) a modern odontocete (common dolphin, *Delphinus delphis*) with homodont dentition; and (D) a modern mysticete (fin whale, *Balaenoptera physalus*). In modern whales, the nasal, premaxilla, maxilla, and frontal bones have extended posteriorly to overlap the parietal bones. Arrows indicate the resultant posterior movement of the nares or nostrils.

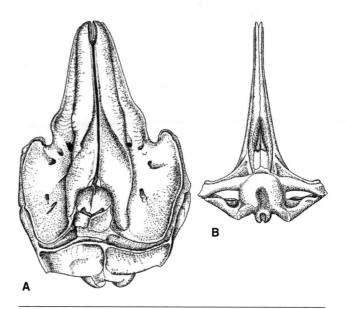

Figure 16.5   Odontocete verses mysticete skulls.
Dorsal views of the skulls of (A) an odontocete, Risso's dolphin, and (B) a mysticete, the northern right whale. Note the asymmetrical skull typical of odontocetes. Skulls are not to the same scale.

Although blubber forms a passive means of heat retention for thermoregulation, heat also may be actively retained in whales through their well-developed countercurrent heat exchange system. This system is associated with areas of the body with little blubber or muscle, such as the flippers, flukes, and head. In these areas, major arteries are surrounded by a closely associated network of veins. This rete mirabile ("wonderful net") acts as a heat exchanger (Bonner 1989). As the warm blood in the arteries moves from the body core outward toward the body surface, heat transfers to the adjacent colder blood in the returning venous system. Also, whales have somewhat of an advantage in thermoregulation compared with terrestrial species because large bodies of water generally maintain fairly constant temperatures.

Physiology of Diving

Whales must also face the challenges associated with diving. Because pressure increases by 1 atmosphere (atm; 14.7 lb/in.^{2}, or approximately 1 kg/cm^2) for every 10 m increase in depth, deep-diving species face tremendous physical as well as physiological demands. At a depth of 30 m, the pressure is 4 atm, or about 60 lb/in.^{2}; at 1000 m, the water pressure is over 1500 lb/in^2. Whales have adapted in several ways to meet these challenges. They are able to survive the extreme pressures encountered during deep dives because their bones are relatively noncompressible. Their upper airways are fairly rigid as well and are supported by bundles of cartilage, although their lungs collapse at depth.

As is true for any mammal, whales must breath air or they will die. Large species, however, are able to remain submerged without breathing for prolonged periods, up to 2 hours in sperm whales (see the section "Physeteridae" in this chapter) and bottlenose whales (*Hyperoodon ampullatus;* Benjaminsen and Christensen 1979). Whales can do so because they use oxygen more efficiently than terrestrial mammals. Land mammals use 4% of the oxygen they inhale with each breath, but whales use 12%. Oxygen is carried to the cells of the body more efficiently as well, because the average **hematocrit** (number of erythrocytes per volume of blood) of whales is twice that of terrestrial mammals. Additionally, whales have two to nine times the amount of **myoglobin** (oxygen-binding protein in muscles) found in terrestrial mammals. Rapid gas exchange is facilitated by extra capillaries in the alveoli of the lungs. Whales also exhibit a characteristic "diving response," which includes bradycardia while submerged. They maintain normal arterial blood flow to the brain and heart but with vasoconstriction and reduced peripheral circulation. Nonetheless, constant blood pressure is maintained to vital organs throughout the dive—the combined result of increased blood pressure through vasoconstriction and decreased blood pressure because of bradycardia.

When humans dive while breathing compressed air, the increased water pressure causes nitrogen, which makes up about 79% of air, to dissolve in tissues and body fluids. If a diver ascends too quickly from a deep dive, the decreasing water pressure causes the nitrogen to come out of solution faster than it can be taken to the lungs and exhaled. The nitrogen forms bubbles in joints or other areas of the body, resulting in the "bends," or "decompression sickness," a condition that is not only painful but can be fatal. Whales dive deep and return to the surface rapidly but are able to avoid this problem. Cetacean lungs are relatively small compared with their body size. With increased depth and pressure, their lungs begin to collapse; at a depth of about 100 m, they are completely collapsed and contain no air. Any residual air is pushed to more rigid, cartilaginous portions of the trachea and bronchioles with little or no gas exchange. Thus, an increased invasion rate of nitrogen is no longer a problem with increased depth.

Although metabolic wastes accumulate during a dive, the respiratory center of the brain has a high tolerance to carbon dioxide buildup. Whales also tolerate high levels of lactic acid, produced by anaerobic respiration, and repay their "oxygen debt" when they surface and breath again, rapidly ventilating their lungs. This results in the characteristic "blow." This is a vapor cloud produced by the condensation of exhaled warm air contacting cooler outside air, as well as spray from a small amount of water on the blowhole. Species sometimes can be identified by the shape, height, and direction of their blow.

It is interesting that the stomach structure of most whales is similar to that of ruminating artiodactyls in having three chambers (figure 16.6), although all whales are carnivores and do not ruminate. In whales, food is physically broken down by the highly muscular walls of the first chamber, or forestomach. No digestive enzymes are secreted in the first chamber, but they are in the second chamber, which is the main stomach. Numerous reticular folds allow the main stomach to greatly expand. The third chamber, or pyloric

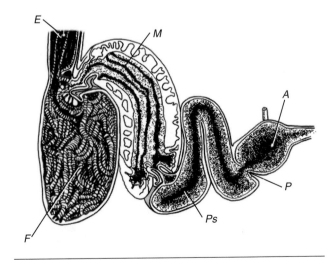

Figure 16.6   Whale stomachs. Several compartments are evident in the stomach of a bottlenose dolphin. *Abbreviations:* E = esophagus; F = forestomach; M = main stomach; Ps = pyloric stomach; P = pylorus; A = ampulla of duodenum.

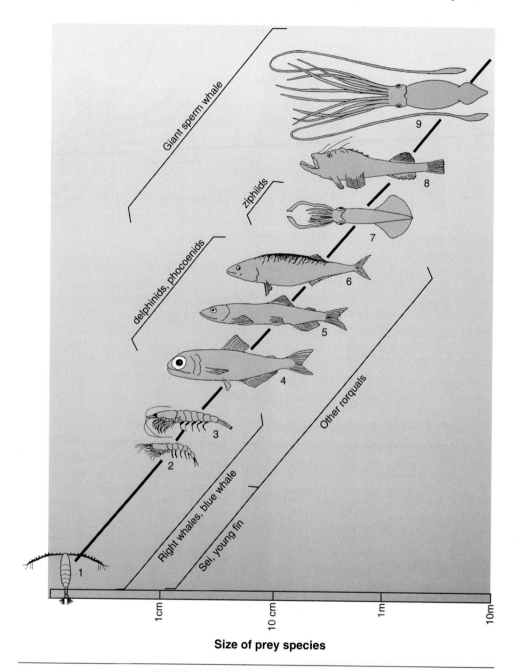

Figure 16.7   Size ranges of food items of baleen and toothed whales.   Shown on a logarithmic scale are (1) calanoid copepods; (2–3) zooplankton; (4) lantern fish; (5) capelin; (6) mackerel; (7) small squid; (8) deep-water angler fish; and (9) large squid.

Source: From D. E. Gaskin, Ecology of Whales & Dolphins, 1982, Heinemann Publishing.

stomach, is named for its numerous pyloric glands (Slijper 1979; Langer 1996). Mysticetes and odontocetes have different morphological adaptations and feed in distinctly different ways on different types of prey (figure 16.7).

Mysticetes: Baleen and Filter-Feeding

The baleen whales are named for their most characteristic feature. Baleen is composed of **keratin,** the same cornified protein material that makes up the horns of rhinoceroses and the fingernails of humans. Plates of baleen hang in comblike fashion from the upper jaw only (figure 16.8). Depending on the species, from 200 to almost 500 plates occur on each side. Baleen plates in the front of the mouth are shorter; posterior baleen plates become progressively longer. The inner edge of each plate has a fringe of long, frayed filaments. These strainerlike filaments overlap each other from plate to plate and form a continuous filter. Baleen continues to grow throughout the life of an individual as the inner portion is continuously worn away. Again, depending on the species'

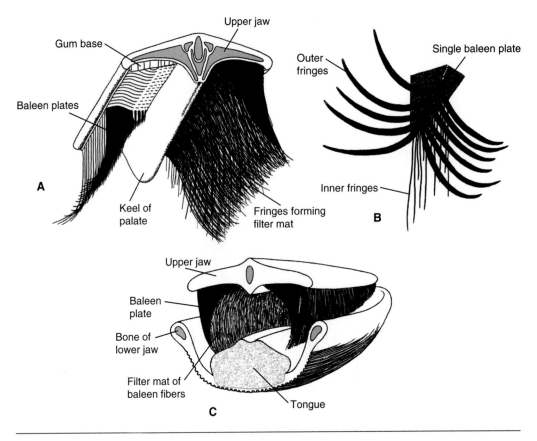

Figure 16.8   Baleen.   (A) Arrangement of baleen in the upper jaw of mysticetes. Hundreds of plates may occur on each side of the mouth. (B) Example of a single plate. The size and number of plates vary among species. Growth is continuous as baleen is worn down from scraping by the tongue. (C) Transverse section through the head of a baleen whale showing the plates in place in relation to the large tongue.

feeding habits, baleen can be fairly short or up to 5.2 m long, as in the bowhead whale (figure 16.9). Fetal mysticetes actually have vestigial teeth that are lost before birth. Details of the formation and fine structure of baleen are given by Slijper (1979), Pivorunas (1979), and Bonner (1989).

Mysticetes feed on a variety of small marine organisms. **Plankton** (often referred to as krill, although not all plankton are krill) have limited locomotion and form large, floating aggregations. **Necton** is another category of slightly larger and more mobile marine organisms, including fish. Baleen whales feed on both plankton and necton, collectively referred to as **zooplankton** (see figure 16.7), as well as phytoplankton. Mysticetes generally do so by either "skimming" or "gulping." Skimming occurs in right whales (Family Balaenidae) and the gray whale (Family Eschrichtiidae). Individuals remain at or just below the surface. With their rostrum out of the water and mouth open, they move through large swarms of zooplankton. Water passes through the mouth and baleen, leaving a mass of zooplankton behind on the baleen. This material likely is scraped off by the tongue and swallowed.

The larger mysticetes (Family Balaenopteridae) are known as **rorquals,** which means "tube-throated." This refers to the longitudinal grooves or pleats on the throat

and chest that allow for great expansion of the oral cavity as the throat fills with water during "gulping." In contrast to skimming, individuals remain submerged and open their mouth. A huge oral cavity is created (figure 16.10A, B) as the lower jaw is distended (aided by a loosely joined, ligamentous mandibular symphysis; also see Lambertsen et al. 1995), the pleated throat is expanded, and the tongue is withdrawn into the large ventral pouch under the chest. Enormous amounts of water, filled with plankton and necton, flow into this oral cavity. Blue whales can gulp 16,000 gallons (about 64 tons!) of water at a time. This water is not swallowed but is expelled through the baleen and out the sides of the mouth as the throat and thoracic cavity contract (figure 16.10C, D) and the tongue protrudes forward. The mass of zooplankton trapped on the baleen is scraped off with the tongue and swallowed.

Unlike most other predators, baleen whales do not actively pursue their prey. Krill and other zooplankton passively float on ocean currents—they cannot attempt to outrun or outmaneuver whales. Thus, baleen whales do not need great speed or agility to capture them; locating concentrations of prey is all that is necessary. Baleen whales can become large at the expense of speed because increased size affords them thermoregulatory and other benefits.

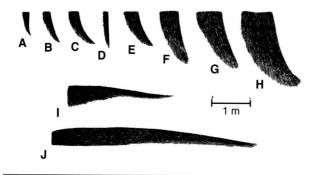

Figure 16.9   **Size and shape of baleen plates.**   Size and shape of plates vary greatly among different species. Representative plates of baleen from (A) minke; (B) sei; (C) Bryde's; (D) pygmy right; (E) gray; (F) humpback; (G) fin; (H) blue; (I) right; and (J) bowhead whales. Scale = 1 m.

Odontocetes: Dentition and Echolocation

All odontocetes have teeth. The number varies among species, from a single pair of teeth in *Mesoplodon* (see the section "Ziphiidae" in this chapter) to well over 100 pairs in some dolphins. Teeth are used to hold fish and squid, prey items that are found through echolocation. Opinion differs about where echolocation sound pulses are produced (see Morris 1986 for review). Purves (1967), Purves and Pilleri (1983), and Pilleri (1990) argue that echolocation sounds are produced in the larynx, as is true in bats and other mammals. In contrast, Norris (1968) suggests that the larynx is not involved but that echolocation pulses originate in the complex system of nasal sacs that occur in the forehead region of odontocetes. This idea is supported by most experimental evidence (see Au 1993 for review). However the sound is generated, it is reflected by the parabolic (dish-shaped) skull and focused through the oil-filled "melon" in the forehead (figure 16.11). The low-frequency echolocation sound pulses that whales emit are not as variable as those of bats, but the effective range of echolocation signals is much greater. The returning echoes are received via the relatively small, thin mandible. The mandible has an oil-filled sinus that channels

In contrast, toothed whales are similar to more typical active predators: they must pursue and catch their prey. This necessitates a greater degree of agility and speed with associated restrictions on body size. Not surprisingly, of the approximately 67 species of odontocetes, two-thirds are relatively small dolphins, river dolphins, and porpoises.

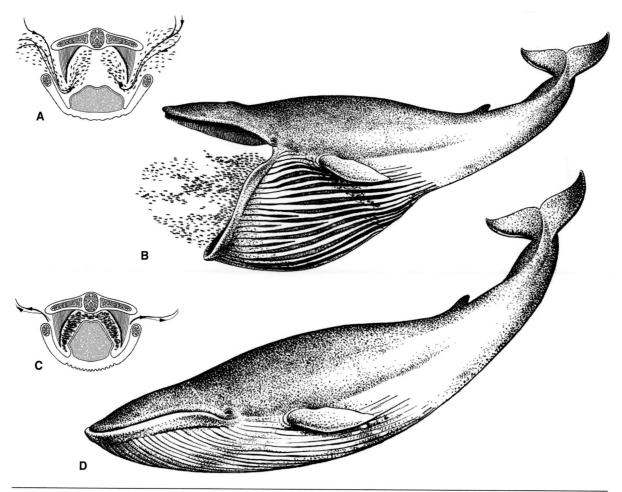

Figure 16.10   **Gulping in baleen whales.**   (A) As the mouth opens, huge amounts of water pour in along with vast quantities of plankton and necton, with the (B) throat grooves allowing for expansion of the oral cavity. (C) This water is then expelled through the filterlike baleen mat as (D) the throat contracts, trapping the food, which is scrapped off by the tongue and swallowed.

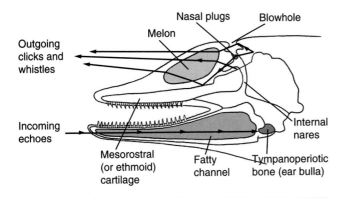

Figure 16.11 **Echolocation in toothed whales.** As in the dolphin shown here, sound in the form of clicks and whistles probably is produced in the complex nasal area. The sound is reflected from the skull and beamed out (in ways not fully understood) through the "melon." The returning echo moves through the oil-filled channel of the lower jaw to the ear bullae.

the sound to the auditory bullae. The oil in the mandible and the melon is the same.

Because water conducts sound much better than air, whales receiving sound echoes could not achieve directionality or sensitivity unless they had special adaptations (not found in other mammals) that allow them to localize and discern the direction of incoming sound waves. They are able to do so because their auditory bullae, or **tympanoperiotic** bones (containing the middle ear), are not fused to the skull. Instead, the bullae are isolated by connective tissue and a system of sinuses filled with a unique foamed, mucous emulsion (Norris and Harvey 1974; Popper 1980). As noted by Au (1993:30), "Having the bullae physically isolated from the skull and therefore isolated from each other allows [them] to localize sounds received by bone conduction."

FOSSIL HISTORY

The morphological adaptations for aquatic life we described for whales are all secondarily derived. Whales evolved from terrestrial land mammals, probably from among the early Condylarthra Family Mesonychidae (figure 16.12). The condylarths were the early ancestral ungulates (see chapter 19), and the mesonychid assemblage extended from the mid-Paleocene to the Oligocene epochs. Mesonychids may have moved into the shallow-water feeding niche left empty by the demise of the large aquatic reptiles, such as plesiosaurs and ichthyosaurs.

Certain early groups of whales showed intermediate morphological stages during transition from terrestrial to fully aquatic mammals. Within the primitive Archaeoceti, the earliest known whale is *Pakicetus inachus* in the early Eocene Family Protocetidae (Gingerich et al. 1983). It was small and structurally intermediate between mesonychids and later Eocene whales. The ear structure of *P. inachus* was primitive, with none of the adaptations for deep diving or di-

rectional hearing underwater. In addition, the dentition was heterodont. Other archaeocete specimens also point to their intermediate structure between terrestrial Paleocene species and whales that were fully aquatic. Such a transitional specimen from the middle Eocene epoch was *Basilosaurus isis,* which retained functional pelvic limbs and foot bones (Gingerich et al. 1990). Archaeocetes may have been amphibious during their transition from land to sea, returning to land after feeding in the water (Fordyce 1980). By the middle Eocene epoch, however, whales were totally aquatic and highly adapted for life in the water.

The relationship of the Suborder Archaeoceti to the two modern extant suborders is unresolved, as is the relationship of mysticetes and odontocetes with each other (see figure 16.12), although the two modern groups likely had a common ancestor (Carroll 1988). Both suborders were clearly distinct by the early Oligocene epoch. The earliest definite mysticete is *Aetiocetus* (Family Aetiocetidae) from the late Oligocene epoch. Although identifiable as a primitive mysticete, it had teeth rather than baleen. The Family Aetiocetidae, which appears to be the stem group from which mysticetes arose, became extinct by the late Oligocene epoch. Another family from the Oligocene epoch, the Agorophiidae, appears intermediate between the earlier archaeocetes and primitive true odontocetes. Based on their skull characteristics, even these early Oligocene epoch odontocetes probably echolocated (Carroll 1988). Most of the extant families of whales were recognizable by the Miocene epoch.

ECONOMICS AND CONSERVATION

Aboriginal whaling dates back thousands of years with Alaskan, Scandinavian, and other people using the meat and oil for subsistence. These early hunting efforts were land-based and typically restricted to slower species inhabiting coastal waters. Using small boats, handheld harpoons, and nets, hunting often involved trying to drive whales ashore to be killed. These efforts probably had a minimal effect on populations. The extent of whaling operations soon increased, however, as did their effect on populations. Basques were taking significant numbers of right whales during the first millenium, and by the 1500s and 1600s, most European countries had entered the whale trade using large ships that operated throughout the world. Whaling in New England began during the 1600s as well. By this time, exploitation of whales had moved from a subsistence level to a commercial enterprise (figure 16.13) because of the high economic value of the meat, oil, and "whalebone" (baleen). Commercial exploitation had a pronounced effect on whale populations, and significant declines began.

Technological advances in the whaling industry during the 1800s increased the efficiency of whaling even more, allowing additional species to be taken (figure 16.14). A modern, explosive harpoon gun was in use by 1864, concurrent

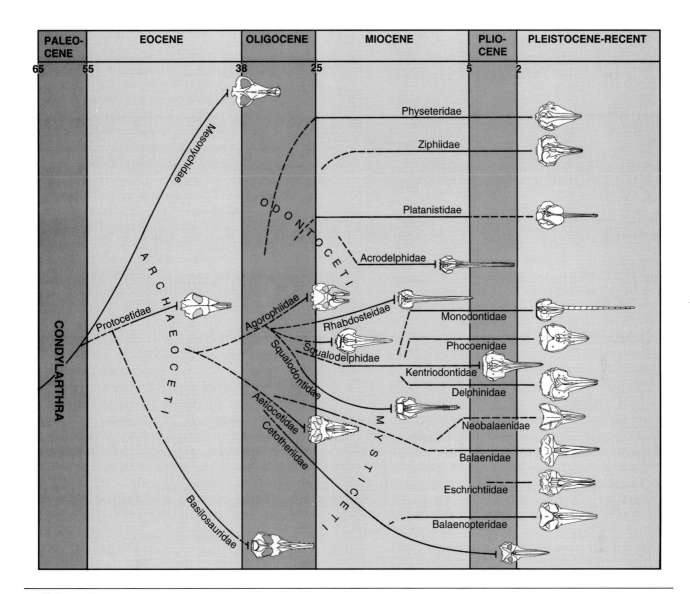

Figure 16.12   **Proposed phylogeny of the Cetacea.**   Current and extinct families are shown. Note the unresolved relationship of the two living suborders with the Archaeoceti, as well as the uncertain relationship of odontocetes and mysticetes with each other.

Source: Data from R. Carroll, Vertebrate Paleontology and Evolution, *1988, W.H. Freeman and Company, New York.*

with the replacement of sailing ships by steam-driven vessels. The early 1900s saw the development of floating factory ships. The stern slipway, in which whales were hauled directly aboard factory ships, was in use by 1925. This allowed whalers to operate more easily on the high seas. Many species were so severely depleted that they became "commercially extinct" and began to be protected from harvest.

Created in 1946, the International Whaling Commission (IWC) established harvest quotas on certain species and protected other species from any harvest, but it had no enforcement powers. Whale stocks, especially the larger species, continued to be overharvested due to the short-term economic interests of some countries. Today, the IWC, public opinion, political pressure, various governmental agencies, and nongovernmental conservation groups work toward

conservation of whales and protection of ocean habitat (see chapter 29). In addition, the availability of alternative products to whale meat and oil has reduced the demand for whales. Nonetheless, some commercial whaling continues, generally for smaller species, by Japan, Russia, Norway, and occasionally other countries. Unfortunately, these commercial ventures are often conducted under the guise of "scientific collections." Also, continued subsistence hunting of bowhead whales by Inuit peoples remains highly controversial. Nonetheless, some large species, including the gray whale and even the blue whale, may be increasing in number. Other species, however, including bowhead and northern right whales, have not recovered. Slijper (1979), Tonnessen and Johnsen (1982), and Evans (1987) provide excellent summaries of the history of whaling.

Figure 16.13   Early whaling.   Using handheld harpoons and open boats, early whaling was a hazardous profession.

SUBORDERS

Mysticetes

Baleen whales are large, with the females generally longer and heavier than the males. The smallest mysticetes, male pygmy right whales (*Caperea marginata*), are about 6 m in length. During the summer, baleen whales generally feed in northern or southern polar latitudes, where they accumulate vast stores of subcutaneous fat (blubber). During the winter, they often migrate long distances to warmer, more equatorial areas. Feeding is greatly reduced during the winter because

tropical waters contain less food, and whales subsist off their stored blubber. Baleen whales do not echolocate, although they produce a variety of sounds. These include "moans" and "grunts" in the low-frequency range, higher-frequency "chirps" and "whistles," and very high-frequency (as much as 30 kHz, or 30,000 cycles per second) pulsating clicks. Because these vocalizations are so loud and water is such an excellent conductor of sound, they may transmit for hundreds, and possibly thousands, of kilometers. Different sounds serve a variety of communication functions (Herman and Tavolga 1980), some of the most important of which may be identification of sex, social status, and location. See Evans (1987; his table 1.1) for a summary of the vocal characteristics of various cetacean species.

Although there are 4 extant families in this suborder, they include only 6 genera and 11 species. Thus, in terms of the number of species, baleen whales comprise only about 14% of living cetaceans.

Balaenidae

We follow Mead and Brownell (1993), who include two genera and three species in this family: the bowhead whale, occasionally called the Greenland right whale, which inhabits northern polar waters throughout the year; the black, or northern right whale (*Eubalaena glacialis*), which is found in subpolar, temperate, and subtropical waters of the Northern Hemisphere; and the southern right whale (*E. australis*), which is found in temperate and antarctic waters. The latter two species are considered conspecific by many authorities, although populations are genetically distinct.

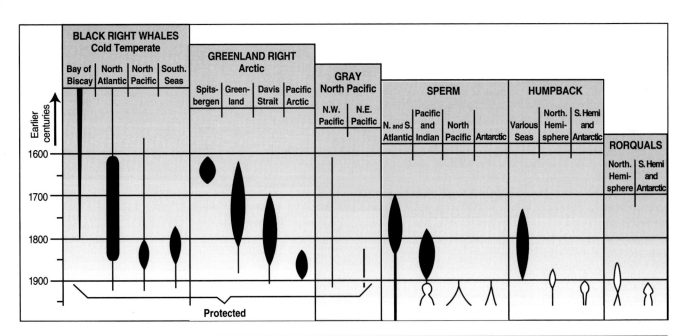

Figure 16.14   Schematic history of whaling.   Solid bars indicate relatively primitive whaling operations—those from open boats with handheld harpoons; open bars indicate modern whaling operations. The width of the bars indicates only the development and decline of regional phases, not the relative numbers of each species taken.

Source: Data from P.G.H. Evans, The Natural History of Whales and Dolphins, *1987, Facts on File, New York, NY.*

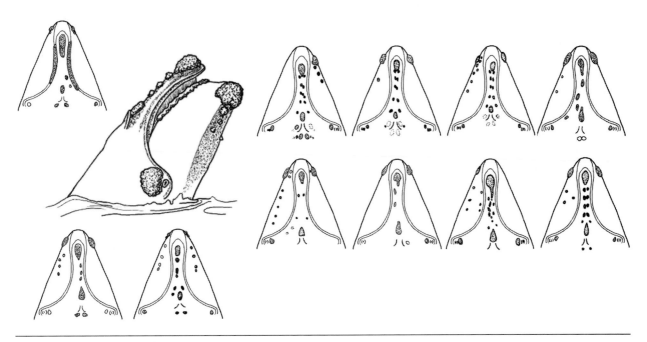

Figure 16.15   **Excrescences on the head of right whales.**   Variation in the pattern of excrescences can be used to identify individuals, as in the dorsal view of the head of 11 right whales.

Balaenids are characterized by their stocky body shape (see figure 16.1); large head; a narrow, highly arched rostrum that accommodates very long, narrow baleen plates; a massive lower lip; lack of throat grooves; and no dorsal fin. Abundant **excrescences,** or **callosities,** occur on the head of *Eubalaena* (figure 16.15). These are rough, horny, white or yellow patches that are covered with barnacles and whale lice (small amphipod crustaceans). The largest excrescence, termed the "bonnet," is situated just anterior to the blowholes. The pattern of excrescences is specific to each whale and can be used to identify individuals. The origin and function, if any, of excrescences is unknown (Cummings 1985).

Mysticetes feed by either skimming or gulping. Balaenids are highly specialized for skimming plankton at or just below the surface. Skimming is facilitated by the highly arched rostrum and 250 to 350 plates of long baleen, up to 5.2 m in bowhead whales, the longest of any mysticete (see figure 16.9). Skimming also is aided by the anterior separation of the baleen plates on each side of the jaw, which allows water to flow more easily into the open mouth (figure 16.16). This feeding behavior also explains why throat grooves found in the rorquals (see the following section), which feed by gulping, do not occur in balaenids.

These were the "right" whales to hunt because they were slow moving and found close to land. They had a great amount of blubber (and thus oil), meat, and baleen, and therefore were economically valuable. Because of their high oil content, they have a tendency to float longer than other species after being killed. Bowhead and right whale populations were the first to be overexploited and dramatically reduced (see figure 16.14), and populations do not appear to be recovering. Today, bowhead and northern

right whales remain endangered; the southern right whale is threatened.

Balaenopteridae

This family encompasses six species of rorquals, or tube-throated whales. They all have parallel grooves, or pleats, along the throat and chest that allow expansion of the oral cavity during feeding. Unlike balaenids, the rostrum is not highly arched, the baleen plates are short, and they meet anteriorly. There are five species in the Genus *Balaenoptera.* All are long and slender, with short pectoral fins and dorsal fins that are posterior to the midpoint of the body. These include

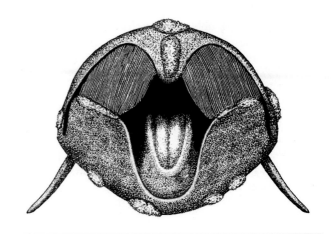

Figure 16.16   **Feeding by skimming.**   Anterior view of a right whale feeding by skimming. The separation of the baleen plates in the front of the mouth is evident, as is the large tongue and lower lips. Note the excrescences.

the blue whale, distributed throughout all oceans; the fin whale (*B. physalus*) and sei whale (*B. borealis*), also cosmopolitan in generally temperate waters; Bryde's whale (*B. edeni*), which occurs in tropical and warmer temperate areas; and the small (7 m total length) minke whale (*B. acutorostrata*), which is found in all oceans. The other species in the family is the humpback whale (*Megaptera novaeangliae*), which enjoys a cosmopolitan distribution but is often more sedentary and found closer to shore than *Balaenoptera*. Humpbacks have a more robust body shape than other rorquals, and the flippers are very long, up to one-third the body length (see figure 16.1). Numerous bumps cover the head. Each bump has a single, short, sensory hair projecting from it.

Balaenopterids feed on zooplankton by gulping; sei whales use both gulping and skimming. The larger species feed almost excusively on krill, but the smaller species also may consume small fishes (see figure 16.7). Humpback whales are known to use a "bubble-netting" technique as well (Hain et al. 1982; Würsig 1988). In bubble-netting, the whale exhales a stream of air bubbles under water as it swims in a circle. This practice tends to congregate the plankton inside a curtainlike column of rising bubbles. The whale, with mouth open, then moves up through the column and consumes the concentrated plankton. Sometimes, coordinated groups of up to 20 humpbacks work together to concentrate prey (Jefferson et al. 1993). Rorquals feed during the summer, and they live off stored body fat during winter. As in the balaenids, seasonal feeding and breeding activities of northern and southern populations of rorquals are out of phase because they are in different hemispheres. As a result, they are genetically isolated, although they are considered to be conspecific.

All species vocalize, but humpback whales are well known for their complex, repetitive, long-lasting songs. There are six basic songs with variations, and all the whales in a given region sing the same song. The song changes throughout the course of a season (Payne et al. 1983; Guinee et al. 1983); all individuals in a region adapt to the changes. It is probable that songs function to help males compete for breeding females, as only mature males sing and only during the breeding season. Roger Payne and Howard Winn pioneered the study of songs in humpback whales (Payne and McVay 1971; Winn et al. 1981; see chapter 20).

Humpbacks, blue whales, and the other large rorquals were drastically overharvested throughout the early and mid-1900s. Blue and humpback whales are endangered, and fin whales are threatened. Along with the sei whale, they are all protected from whaling. Humpbacks have been protected since 1944, and populations are increasing; blue whales have been protected since 1965. Recovery has been slow, however, and most species almost certainly will never regain their former numbers.

Eschrichtiidae

Gray whales (*Eschrichtius robustus*) constitute a monotypic family. Males may reach a total length of about 13 m, females about 14 m. They have fairly slender bodies, broad flippers, and no dorsal fin, although the posterior third of the body has a series of low humps or **crenulations**. There are two to five short throat grooves. *E. robustus* is gray, with lighter mottled areas, and covered with an unusual number of barnacles and associated whale lice.

Gray whales feed during the summer in shallow waters of the North Pacific, specifically in parts of the Arctic Ocean, Bering Sea, and Okhotsk Sea. They roll on their sides and suck muddy sediment from the bottom to strain out worms, invertebrates, and amphipods, predominately *Ampelisca macrocephala*. The 140 to 180 pairs of short baleen plates are narrow, stiff, and coarse. There is a gap in the baleen in the anterior part of the mouth. Small groups of gray whales are most common, although up to 150 individuals may congregate in good feeding areas.

Because their feeding areas freeze during the winter, in autumn gray whales migrate up to 18,000 km, one of the longest migrations of any mammalian species. They follow the Pacific coastline to calving grounds in Baja California or the Sea of Japan. They conceive during migration and give birth in January or February after a 13.5-month gestation period. A single calf is born in alternate years. Calves are born in shallow lagoons where the water is calmer and warmer than in the open ocean.

Gray whales swim slowly in shallow waters near shore, so they have always been easy prey for whalers. Populations probably were eliminated from the Atlantic Ocean by the late 1700s (Bryant 1995). The eastern Pacific population has been protected since 1946, and gray whales appear to be increasing in numbers. Rice and Wolman (1971) summarized aspects of the life history of this species.

Neobalaenidae

This family includes only the pygmy right whale, which is included in the Family Balaenidae by some authorities (see Fraser and Purves 1960). One of the least known cetaceans, it occurs only in temperate waters of the Southern Hemisphere. The pygmy right whale is rarely observed and easily confused with the minke whale. As suggested by the name, it is the smallest of the mysticetes—females are only about 6.5 m in total length, males about 6 m—about one-third the size of the northern right whale. The flippers are small; there is a small **falcate** (curved toward the tail) dorsal fin and about 213 to 230 baleen plates per side. As in the closely related right whales, the rostrum is highly arched, and the baleen is long and narrow (Baker 1985). Unlike right whales, however, the pygmy right whale has two shallow throat grooves.

Odontocetes

Toothed whales generally are smaller than baleen whales. Unlike mysticetes, male odontocetes usually are larger than females. The forehead houses a complex system of nasal sacs and a fatty melon, both of which function in echolocation. All toothed whales have an asymmetrical skull, and their single external nare is often left of the center line. In all other respects, odontocetes exhibit bilateral symmetry. Skull asymmetry also functions in echolocation. Odontocetes echolocate to find food and orient within their environment.

Delphinidae

The 17 genera and 32 species of dolphins are distributed practically worldwide in all but polar waters, including some tropical river systems in Asia and South America. Delphinids are the most varied family of cetaceans; they range in size from the smallest cetacean, Heaviside's dolphin (1.7 m in length and 40 kg), to the killer whale (*Orcinus orca*), at 9 m in length and 7000 kg. Most delphinids have a rostrum that forms a beak and a falcate dorsal fin in the middle of the back. Delphinids vary from uniform gray or black to stripes, spots, or bands to the contrasting black-and-white patterns of the killer whale. Many morphological differences in this family are related to feeding. The total number of teeth varies among species (figure 16.17) from 2 to 7 pairs in Risso's dolphin (*Grampus griseus*) to over 120 pairs in the spinner dolphin (*Stenella longirostris*). Feeding habits of delphinids vary in terms of the size of prey species taken, distance from shore, and diving depth. Most dolphins, however, dive for short periods to depths of less than 200 m. Delphinids generally take small fishes and squid, but killer whales feed on fishes, seals, marine mammals, other cetaceans, birds, and even sharks.

Several **pelagic** (open ocean) species are gregarious and form **pods** (groups or schools) of up to 1000 individuals, although typical group sizes are much smaller (see table 7.2 in Evans 1987). Species found closer to shore generally form smaller groups. Spinner dolphins and bridled dolphins (*Stenella attenuata*) are the species most often captured and drowned in purse-seine nets set for tuna. Recently, special fishing techniques have been developed in attempts to minimize losses of dolphins (Macdonald 1984; Evans 1987). Other species of dolphins are taken for food in various parts of the world. Most species are insufficiently known to determine their conservation status.

Monodontidae

The two species in this family—the white whale or beluga (*Delphinapterus leucas*) and the narwhal (*Monodon monoceros*)—are circumarctic in distribution. Both species occur in polar waters along the coast or pack ice. Both have somewhat robust bodies, rounded heads, and no dorsal fin (figure 16.18A, B). Belugas generally are found in shallow waters, are 3 to 5 m in length, and weigh up to 1500 kg. Calves are gray at birth and become progressively paler as they mature; adults are white. The total number of teeth range from 14 to 44. Narwhals are similar in size to belugas, mottled grayish black dorsally, and paler on the ventral surface. Narwhals generally inhabit deeper waters than belugas. Dentition consists of one pair of upper incisors. In male narwhals, the left upper incisor develops into a long tusk (up to 3 m) with a counterclockwise spiral (figure 16.18C). The right tooth usually does not erupt. In females, neither of the incisors usually erupts. It has been suggested that the tusk is used to spear prey items or to focus echolocation signals. Most authorities agree, however, the tusks are just what they appear to be—weapons for establishing dominance breeding hierarchies among males (Hay and

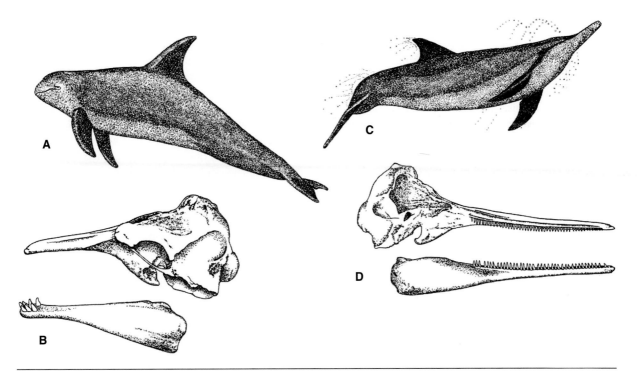

Figure 16.17   Variation in dolphin tooth number.   (A) Risso's dolphin, or grampus, has (B) from 2 to 7 pairs of teeth, whereas (C) the spinner dolphin may have (D) a total of over 120 pairs of teeth. Note the lack of an external beak in Risso's dolphin.

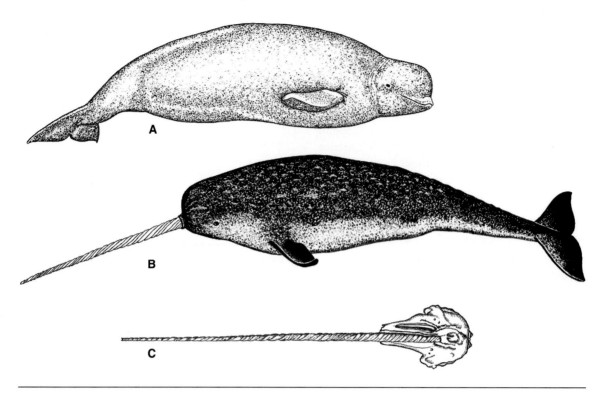

Figure 16.18   Monodontids.   (A) Narwhal and (B) beluga whales, and (C) the spiral tusk of the narwhal. The dorsal portion of the cranium has been removed to show the root of the large, spiral left tusk of males and the typically unerupted right tusk.

Mansfield 1989). Both species are gregarious, and aggregations over 1000 individuals may form, although groups of 10 to 15 are more common (Jefferson et al. 1993). They feed at or near the bottom on a variety of fishes, small squid, and mollusks. Both species are taken by Inuits for food.

Phocoenidae

Porpoises sometimes are considered a subfamily within the Delphinidae; here, we consider them a separate family. Unlike dolphins, which generally have a beak, the four genera and six species of porpoises have no distinct beak. The total number of teeth range from 60 to 120 and are blunt with flattened crowns, in contrast to the sharply pointed, conical teeth in dolphins. Phocoenids generally are small and stocky, ranging in total length from 1.5 m in the California harbor porpoise, or vaquita (*Phocoena sinus*), to about 2.2 m in Dall's porpoise (*Phocoenoides dalli*). Most species have a short, triangular dorsal fin. They are distributed in Northern Hemisphere oceans and seas, including the Black Sea, and in Asian rivers. In the Southern Hemisphere, they occur in coastal waters of South America and around several island groups. The finless porpoise (*Neophocaena phocaenoides*) and *Phocoena* are found close to shore in bays and estuaries; Dall's porpoise occurs in deeper offshore waters. The spectacled porpoise (*Australophocaena dioptrica*) is found in both inland and offshore waters of the Southern Hemisphere.

Dall's porpoises, like certain dolphins, are killed accidentally in fishing nets. Other species, including the harbor porpoise (*Phocoena phocoena*), are taken for food. The Cali-

fornia harbor porpoise is limited to the northern portion of the Gulf of California, the most restricted range of any cetacean. It is considered endangered (Jefferson et al. 1993), as is the finless porpoise.

Physeteridae

This family consists of the giant sperm whale (*Physeter catodon*) and two smaller species: the dwarf sperm whale (*Kogia simus*) and the pygmy sperm whale (*K. breviceps*). The latter two species are sometimes placed in their own family (Kogiidae). The giant sperm whale occurs in all but polar oceans. Males may reach 18 m in total length, the largest of any odontocete (see figure 16.1). The large, blunt head constitutes one-third of its total length. Their flippers are small, and they have a small dorsal hump and series of posterior crenulations similar to gray whales. The thin mandible, with 18 to 25 pairs of teeth, is much shorter than the upper jaw, which contains sockets but no teeth.

Giant sperm whales have a huge, highly distinctive **spermaceti** organ (figure 16.19) that may account for close to 10% of the total mass of an individual (Bonner 1989). Composed mainly of two structures, the "case" and the "junk," this organ contains a waxy liquid, the spermaceti. In conjunction with the elaborate and convoluted nasal passages, the spermaceti organ may function to control and regulate bouyancy (Clarke 1978a, b) or as a lens to focus echolocation signals (Norris and Harvey 1972). Both functions are necessary to accommodate the very deep dives of sperm whales. They may reach depths of 3200 m or more,

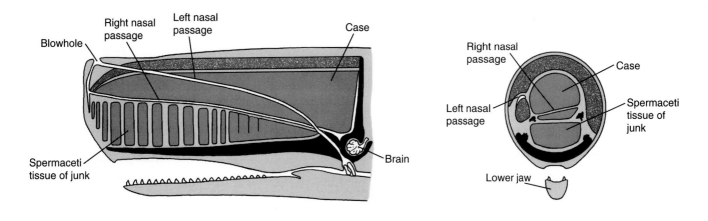

Figure 16.19 **Spermaceti organ of a sperm whale.** (A) Lateral and (B) transverse views. The dark gray portion represents the two structures that make up this unusual organ: the "case" and the "junk." The single blowhole is in the extreme anterior part of the skull and to the left of the midline. See Bonner (1989) and Clarke (1978a) for detailed descriptions. Clarke suggests that the density of the spermaceti changes depending on its temperature and affects buoyancy for rapid diving and surfacing. Cooling and warming is accomplished by water flow through the nasal passages.

apparently diving straight up and down, and remain submerged for up to 2 hours (Leatherwood et al. 1983; Jefferson et al. 1993). Sperm whales feed mainly on large squid but also take a variety of fishes. Historically and recently, they have been one of the most important species in the whaling industry because of the vast amounts of spermaceti oil they contain. Sperm whales also are known for ambergris, a brownish, pliable, organic substance produced in the intestinal tract of some individuals. The function of ambergris remains unknown, although it may aid in digestion. Ambergris is extremely valuable as a perfume fixative ("worth its weight in gold"); as noted in chapter 6, pieces up to 421 kg have been found (Slijper 1979). Sperm whale hunting has been banned by the IWC since 1981.

Both pygmy and dwarf sperm whales are significantly smaller than *Physeter catodon*. They reach a maximum length of 3.4 m and 2.7 m, respectively (see figure 16.1). Both species may be fairly common in tropical and temperate waters. Only the bottlenose dolphin (*Tursiops truncatus*) is found stranded on beaches more often than pygmy sperm whales. Pygmy and dwarf sperm whales are rarely seen at sea, however, and are poorly known. They feed on small squid, fishes, and crustaceans (Caldwell and Caldwell 1989).

Platanistidae

We follow Mead and Brownell (1993), who consider this family to include four genera and five species of river dolphins. The taxonomy of river dolphins is uncertain, however,

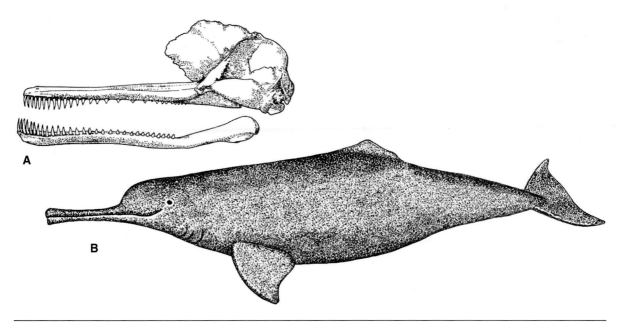

Figure 16.20 **The Ganges river dolphin.** (A) Lateral view of the skull shows the numerous homodont teeth and the distinctive maxillary crests overhanging the rostrum. (B) The extremely long rostrum and robust body are evident.

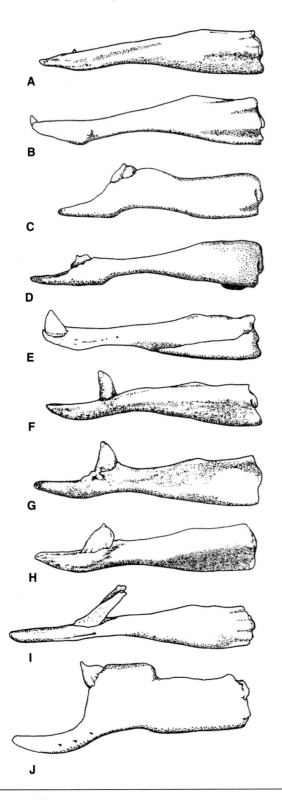

Figure 16.21   Specialized ziphiid dentition.   Highly specialized and variable dentition occurs in the lower jaws of the ziphiid Genus *Mesoplodon.* Lateral view of the right mandible of (A) Gervais' beaked whale, *M. europaeus;* (B) True's beaked whale, *M. mirus;* (C) ginkgo-toothed beaked whale, *M. ginkgodens;* (D) Gray's beaked whale, *M. grayi;* (E) Hector's beaked whale, *M. hectori;* (F) Stejneger's beaked whale, *M. stejnegeri;* (G) Andrew's beaked whale, *M. bowdoini;* (H) Hubbs' beaked whale, *M. carlhubbsi;* (I) strap-toothed whale, *M. layardii;* and (J) Blainville's beaked whale, *M. densirostris.*

and many of these species have been placed in their own families by other authorities (see Rice 1984; Heyning 1989). Platanistids are small cetaceans 2 to 3 m long and are found in estuaries, turbid rivers, and inland lakes. The South American river dolphin, or boto (*Inia geoffrensis*), occurs in the Orinoco and Amazon River basins throughout northern South America. The Chinese river dolphin, or baiji (*Lipotes vexillifer*), is restricted to the Yangtze River system; the La Plata river dolphin, or franciscana (*Pontoporia blainvillei*), occurs in coastal waters of Argentina and southern Brazil. Two other species occur in India, Pakistan, and Bangladesh: the Ganges river dolphin, or susu (*Platanista gangetica*), which often swims on its side with a foreflipper touching the bottom, and the Indus river dolphin, or bhulan (*P. minor*). Despite their widespread distribution, all river dolphins are convergent in appearance, with long beaks and small eyes (figure 16.20). Their eyes are without lenses, so they are virtually blind. They locate prey, generally small fishes and crustaceans, through echolocation. All platanistids are negatively affected by dam construction and development; the Indus river dolphin is endangered.

Ziphiidae

The beaked whales occur in all oceans. This diverse family of 6 genera and 19 species of deep divers feeds on smaller species of squid and fishes (see figure 16.7), apparently capturing them through suction (Heyning and Mead 1996). Ziphiids are slender, with a pronounced beak and small dorsal fin. Total length ranges from 4 to 13 m. Unlike other toothed whales, female ziphiids are larger than males. Although Shepherd's beaked whale (*Tasmacetus shepherdi*) may have up to 98 teeth, most ziphids have only 1 or 2 teeth in each side of the lower jaw. These teeth usually occur only in males and are used primarily for intraspecific fighting. In the Genus *Mesoplodon*, the single pair of teeth is quite specialized among species (figure 16.21). Like river dolphins, the taxonomy of ziphiids is unresolved; they are poorly known because they are uncommon and difficult to identify at sea.

Summary

All whales exhibit similar morphological features because they are all adapted for life in the water. Compared with terrestrial mammals, whales are very large. Their large size can be accommodated because water provides a buoyant, supportive environment. Whatever their size, all whales have a streamlined, fusiform body. There are no external ears, hind limbs, or fur to disrupt the smooth flow of water over the body as a whale moves. Forelimbs have been modified into paddlelike flippers, and a horizontally flattened tail fluke provides propulsion. The skull has also been modified by "telescoping," with resultant movement of the external nares to the top of the head. In place of fur, insulation is provided by a layer of subcutaneous fat—the blubber. Whales also have several physiological adaptations for diving. Compared with terrestrial mammals, they have a high hematocrit and more myoglobin, use oxygen more efficiently, and while submerged undergo bradycardia and reduced blood flow to peripheral areas of the body.

Despite these similarities, there are several differences in the two cetacean suborders. Mysticetes, or baleen whales, are named for the baleen plates that hang from the upper jaw in place of teeth. Baleen is used as a filter to strain small marine organisms (krill) from the water. Mysticetes, which account for only 14% of extant cetacean species, are the largest whales; they have sacrificed speed and agility for the benefits of increased body size. This was adaptive because their primary prey, krill, essentially is immobile. Species within the six families of toothed whales generally are smaller and more agile than baleen whales. They feed on fishes, squid, and larger prey items rather than krill. Unlike mysticetes, odontocetes have an asymmetrical skull with a single external nare, a complex system of nasal sacs, and a large, bulbous forehead with an oil-filled melon. These structural features all relate to the ability of toothed whales to echolocate.

The specialized morphological characteristics of all whales are secondarily derived. That is, whales were not always in the water but evolved from terrestrial land mammals, specifically mesonychid condylarths, which made a gradual transition from land to sea during the Eocene epoch. Intermediate stages during this transition are evident in several lineages of the Archaeoceti, an extinct suborder of whales. The exact relationship of the two living suborders is uncertain, even though both groups were distinct by the early Oligocene epoch.

Whales have been hunted for thousands of years for their meat, oil, and other valuable products. Subsistence hunting, which has a relatively minimal effect on populations, eventually gave way to large-scale commercial exploitation. Right whales and bowhead whales probably were the first species to experience significant population declines because of overharvesting. The International Whaling Commission was formed in 1946 to help conserve stocks of many species. Through the efforts of the IWC, governments, and many conservation agencies, most nations no longer harvest whales. Although populations of some species have begun to recover, others have not.

Discussion Questions

1. In contrast to whales, the seals and walrus, which spend the majority of their lives in water, are heavily furred. What life history aspects can you think of that make retention of fur adaptive in seals and the walrus but not in whales?

2. Can you offer any suggestions as to why humpback whales sing but apparently no other species of baleen whales do?

3. Sexual dimorphism is evident in whales. Why might females be larger than males in baleen whales, while the opposite situation occurs in toothed whales?

4. We discussed some of the ways that whales maintain their body heat in cold polar waters. Consider the other extreme: how do they dissipate heat while in tropical waters?

5. Despite conservation programs, many species of large whales have not recovered their former numbers. What negative factors affect oceans today that might mitigate against future recovery of whales?

Suggested Readings

Boyd, I. L. (ed.). 1993. Marine mammals: advances in behavioural and population biology. Oxford Univ. Press, New York.

Burns, J. J., J. J. Montague, and C. J. Cowles (eds.). 1993. The bowhead whale. Spec. Publ. No. 2, Society of Marine Mammals, Lawrence, KS.

Harrison, R. and M. M. Bryden (eds.). 1988. Whales, dolphins, and porpoises. Facts on File Publ., New York.

Hoelzel, A. R. (ed.). 1991. Genetic ecology of whales and dolphins. International Whaling Commission, Cambridge, England.

Jones, M. L., S. L. Swartz, and S. Leatherwood (eds.). 1984. The gray whale. Academic Press, New York.

Scammon, C. M. 1874. The marine mammals of the northwestern coast of North America. John H. Carmany, San Francisco.

Thomas, J. A. and R. A. Kastelein (eds.). 1990. Sensory abilities of cetaceans: laboratory and field evidence. Plenum Press, New York.

Thomas, J. A., R. A. Kastelein, and A. Y. Supin (eds.). 1992. Marine mammal sensory systems. Plenum Press, New York.

CHAPTER
17

Rodentia and Lagomorpha

RODENTIA

Morphology
Fossil History
Economics and Conservation
Sciurognath Rodents
Hystricognath Rodents

LAGOMORPHA

Morphology
Fossil History
Economics and Conservation

This chapter includes two structurally and functionally similar orders. The Order Rodentia is characterized by a single pair of upper and lower incisors and contains more species than any other mammalian order. The Order Lagomorpha is characterized by a second pair of small, round "peg teeth" posterior to the upper incisors and is a much smaller group. These orders are similar, however, in that individuals have specially adapted, enlarged incisors for gnawing and a diastema without canines between the incisors and the cheekteeth. On this basis, early taxonomists combined them in the formerly recognized Order Glires. The origins and relationship of the two orders are unresolved today. We have included rodents and lagomorphs in a single chapter because of their similarities in feeding and associated structural characteristics. Given the size of the Order Rodentia, we will examine an array of structural and functional adaptations that have enabled them to flourish throughout most of the world. At the same time, we will explore the considerable amount of convergence among unrelated rodent families that "make a living" in similar ways.

RODENTIA

Rodents constitute the largest mammalian order, with 28 extant families and approximately 2016 currently recognized species. Over 43% of living mammalian species are rodents. Rodents enjoy a **cosmopolitan** (worldwide) distribution and are native everywhere except Antarctica, New Zealand, and a few oceanic islands. They have very successfully adapted to a wide range of different habitats, including terrestrial, arboreal, scansorial, fossorial, and semiaquatic. Rodents are found in all biomes, often as **commensals** (in close association with humans). Also, they exhibit a diverse array of locomotor adaptations, including plantigrade, cursorial, swimming, fossorial, jumping, and gliding. The vast majority of rodents are small (20–100 g), although the largest, the capybara (*Hydrochoerus hydrochaeris*), may reach 50 kg.

Given the large number of rodent species, their degree of diversity and adaptability, and convergent evolutionary trends, it is not surprising that the systematic relationships of many families and subfamilies are complex and unresolved. Despite the number of species and their widespread distribution and diversity, rodents are surprisingly uniform in several general morphological characteristics.

Morphology

The diagnostic characteristic that defines all rodents is a single pair of upper and lower incisors. Their large incisors are open-rooted and ever-growing and are used for gnawing (the name *Rodentia* is derived from the Latin *rodere*, "to gnaw"). As gnawing quickly wears down the tips of the incisors, a chisel-like edge forms because the anterior side of each incisor is covered with enamel and wears more slowly than the

posterior side, which lacks enamel (figure 17.1). In many species, the mouth can close behind the incisors, and the animal may have either internal or external cheek pouches for transporting food. Rodents also have a diastema, which is a gap between the incisors and cheekteeth that allows for maximum use of the incisors in manipulating food. Canine teeth are absent, and the number of molariform teeth is reduced. Molariform dentition may or may not be ever-growing, and a variety of occlusal cusp patterns occur (figure 17.2).

Rodents have dental formulae greatly reduced from the primitive eutherian number. A typical dental formula is 1/1, 0/0, 0/0, 3/3 = 16. Rodents never have more than two pairs of premolars, and no species has more than 22 teeth, except the sand rat (*Heliophobius argenteocinereus*), with 1/1, 0/0, 3/3, 3/3 = 28. Even then, not all the premolars are in place at the same time. Rodents generally are herbivorous or omnivorous, depending on the season and availability of food items. Females have a duplex uterus, males have a baculum, and the testes may be scrotal only during the breeding season. The jaw musculature and skull structure associated with the dentition serve as important criteria for grouping rodents.

Despite confusion and lack of consensus in rodent classification, three groupings, first suggested by Simpson (1945), generally are accepted: **sciuromorph** (squirrel-shaped), **myomorph** (mouse-shaped), and **hystricomorph** (porcupine-shaped) rodents. Each group is distinguished by skull structure and jaw musculature, specifically the origin of the masseter muscles (figure 17.3). These three groups are often given subordinal status, although not all families readily fit into one of the groups. The terms *sciuromorph, myomorph,* and *hystricomorph* have only anatomical implications and may not necessarily suggest systematic relationships. Often, however, they are used for both.

Besides these cranial characteristics, all rodents, whether living or extinct, have one of only two types of lower jaw, depending on the insertion of the masseter muscle. They are either **sciurognathous,** with a relatively simple mandible

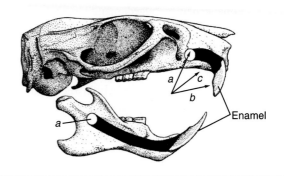

Figure 17.1 Skull of a rodent. Note the characteristic features of rodents, including a single pair of upper and lower incisors, which are open-rooted (*a*) and grow throughout an individual's life; the incisors' beveled edges (*b*); and the diastema (*c*) between incisors and cheekteeth.

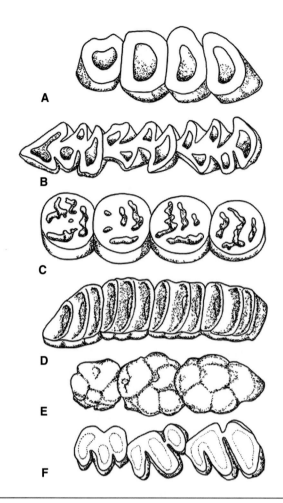

Figure 17.2   Rodent molariform occlusal surfaces.   A variety
of patterns occur in the occlusal surfaces of molariform teeth of rodents.
(A) Ring of enamel in a mole-rat; (B) prismatic in murids, voles, and
lemmings; (C) flat with lakes of dentine surrounded by enamel in Old
World porcupines; (D) transverse ridges in a chinchilla; (E) cuspidate
pattern in a murid; and (F) folded enamel in a murid.

with insertion directly ventral to the molariform dentition,
or **hystricognathous,** with a strongly deflected angular
process and flangelike, or ridged, mandible for masseter insertion ventral and posterior to the teeth (figure 17.4). On
this basis, it is probably most correct in discussing family
characteristics to consider two suborders (Sciurognathi and
Hystricognathi), with the very large group of myomorph
rodents considered an infraorder within the Sciurognathi.
Carleton (1984) provides an excellent review of the higher
level classification of rodents.

Fossil History

The reduced dentition, enlarged incisors, and diastema in all
modern rodents appeared early in the fossil record. Rodentlike reptiles with these characteristics, the **Tritylodonts,** had
developed by the late Triassic period, concurrent with the
appearance of seed-bearing vegetation. Tritylodonts were
succeeded by the multituberculates about 50 million years

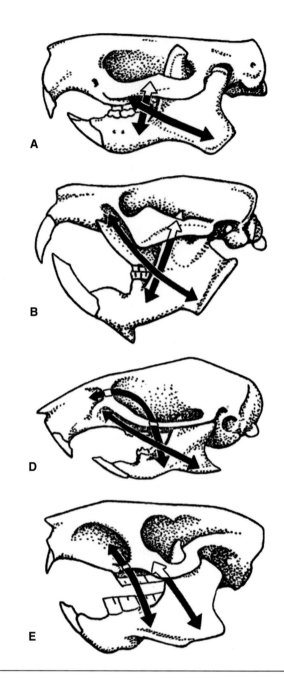

Figure 17.3   Masseter muscle complex in rodents.
Diagrammatic representation of the origin and insertion of the masseter
muscles in (A) an aplodontidlike, primitive protrogomorph form with the
origin of the masseter muscle entirely on the zygomatic arch and not
anterior on the rostrum. (B) Sciuromorphs have a small infraorbital
foramen through which no masseter muscles pass. Origin of the middle
muscle is anterior to the eye, and the deep muscle is beneath the
zygomatic arch. (C) A slightly larger infraorbital foramen is found in
myomorphs, with the deep portion of the masseter muscle passing
through it and originating on the rostrum, whereas the middle masseter
is anterior to the eye, as in sciuromorphs. (D) In hystricomorphs, the
infraorbital foramen is very large; the deep masseter muscle passes
through it and attaches anterior to the eye, and the middle masseter
originates on the zygomatic arch.

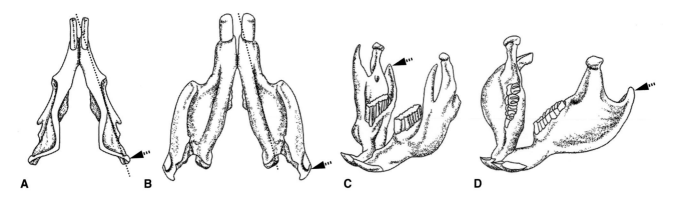

Figure 17.4   Types of lower jaws.   Ventral view of the mandible of (A) a sciurognathus rodent and (B) a hystricognathus rodent showing the relationship of the very long incisors (*dotted lines*) to the angular process (*arrows*). (C) In the dorsolateral view of the sciurognath, note the large coronoid process and surface directly below teeth for insertion of the masseter muscles. (D) In the hystricognath, the coronoid process is greatly reduced, and insertion of the masseter is on the large, bony flange of the angular process.

later, in the mid-Jurassic period. These early rodentlike, "prototherian" mammals also had enlarged incisors, no canines, and cheekteeth with multiple cusps (figure 17.5; also see figure 4.9). Neither of these groups evolved into rodents, however. The oldest known members of the Order Rodentia, the sciuromorph Paramyidae (figure 17.5C) represented by teeth from *Paramys atavus*, date from the late Paleocene epoch in North America and Eurasia. The paramyids were ancestral to several rodent families, and their closest living descendant, based on jaw and muscle structure and dentition, is the mountain beaver (Aplodontidae: *Aplodontia rufa*).

Economics and Conservation

Rodents had and continue to have both positive and negative effects on humans (figure 17.6). Several species are of economic importance as food for humans, and the fur of other rodents is valuable in the garment industry. North American rodents important as furbearers include the muskrat (*Ondatra zibethicus*) and beaver (*Castor canadensis*), as well as the introduced nutria (*Myocastor coypus*). The South American chinchilla (*Chinchilla lanigera*) is another important furbearer, with countless individuals bred in captivity throughout the world. On the negative side, numerous rodent species worldwide severely damage crops and grain stores. Rodents consume an average $30 billion worth of cash crops and cereal grains each year!

Historically, rodents, specifically European rats (*Rattus* spp.), were vectors for epidemics of bubonic plague that ravaged human populations periodically throughout the late Middle Ages, killing millions of people. Bubonic plague remains a problem throughout many less developed areas of the world. Rodents are vectors or reservoir hosts for a wide variety of other viral, bacterial, fungal, and protozoan infectious diseases, including murine typhus, leptospirosis, listerosis, rickettsial diseases, Lassa fever, Q fever, histoplasmosis, Lyme disease, and hantaviruses (Cox 1979; see chapter 27).

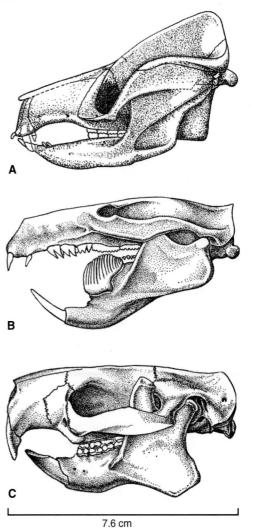

Figure 17.5   Rodentlike fossils.   A mammal-like reptile (A) *Oligokyphus* (actual length about 7.6 cm), from the Jurassic period, with a diastema and enlarged incisors. These characteristics are also evident in a multituberculate mammal (B) *Ptilodus* (7.6 cm in length) from the Paleocene epoch. Compare with (C) *Paramys* (7.6 cm in length), from the oldest known rodent family, the Paramyidae.

7.6 cm

A

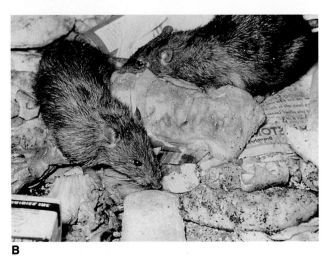

B

Figure 17.6   Economic effects of rodents.   Rodents may have both significant positive and negative economic effects on humans. (A) Muskrats and several other species are important sources of fur in the United States, and millions are harvested yearly. (B) Norway rats (*Rattus norvegicus*) and numerous other rodent species cause billions of dollars in crop and property damage and are reservoirs for numerous diseases throughout the world.

Today, rodent populations are important for scientific purposes as well. Much of the work in population dynamics, animal behavior, physiology, and psychology has relied heavily on rodents as experimental animals. Researchers often use them in field or laboratory studies because of their rapid turnover rates, short gestation periods, large litter sizes, rapid sexual maturity, and easy handling. Many species of voles (Genus *Microtus*) or lemmings (Genus *Lemmus*) show predictable, 3- to 4-year population cycles throughout large geographic areas. Factors that cause these cycles, and how they operate, are still being studied to determine how external and internal influences affect population dynamics. Rodents such as guinea pigs (Genus *Cavia*), the house mouse

(*Mus musculus*), and Norway rat (*Rattus norvegicus*) are important research animals in biological laboratories throughout the world.

Sciurognath Rodents

Families in this suborder generally have a small to medium-sized infraorbital foramen with either no passage of any part of the masseter muscle through it (or limited passage, as in the myomorphs). The springhare (Pedetidae), gundis (Ctenodactylidae), and scaly-tailed squirrels (Anomaluridae) are exceptions, however, because they have very large infraorbital foramina—more characteristic of hystricomorphs—for passage of much of the masseter muscle. All have typical sciurognath structure of the lower jaw, however. Thus, these three families may be considered as "hystricomorphous sciurognaths" and typify the difficulties involved in classifying rodents.

Aplodontidae

This is a monotypic family of one extant genus and species. The mountain beaver (also called a boomer, or sewellel) is endemic to the Pacific northwest from British Columbia to northern California. The species is of special interest because it is the most primitive living rodent, with a fossil history dating to the Paleocene epoch. Mountain beavers (which are not limited to mountains and definitely are not beavers) occur in humid forested areas with a dense understory associated with heavy rainfall. They are restricted to wet areas, in part because their primitive kidneys cannot produce concentrated urine. Although *Aplodontia* is considered here as a sciuromorph, the masseter muscles originate entirely on the zygomatic arch (see figure 17.3); as such, mountain beavers can be considered the only living "protrogomorph" rodent (Wood 1965). They are stocky, have no external tail, and weigh up to 1.5 kg. The flattened skull is distinctive, appearing triangular from above, with flask-shaped auditory bullae and teeth that have unique projections (figure 17.7). Mountain beavers are found in small colonies and burrow extensively, often near streams or in moist ground. Burrows may be fairly complex and can extend for 300 m. Mountain beavers also climb as part of their foraging and are capable swimmers. They are considered a pest and cause severe problems by damaging or consuming seedlings planted as part of forest regeneration programs (Feldhamer and Rochelle 1982).

Sciuridae

The squirrels are a large and diverse group of about 50 genera and 273 living species distributed worldwide except for Australia, Madagascar, southern South America, and some desert regions. Sciurids are conveniently divided into two subfamilies that reflect the diversity of habitats and behavioral characteristics of this large family: the tree squirrels and ground squirrels (Sciurinae), and the "flying" squirrels (Petauristinae). Tree squirrels, including the gray squirrel

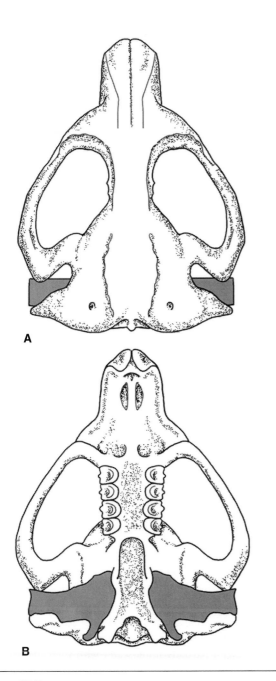

A

B

Figure 17.7 **Mountain beaver skull.** (A) Dorsal and
(B) ventral views of the mountain beaver showing the triangular skull
shape, flask-shaped auditory bullae (*shaded*), and labial projections on
upper cheekteeth. Lingual projections occur on the lower cheekteeth.
Actual size.

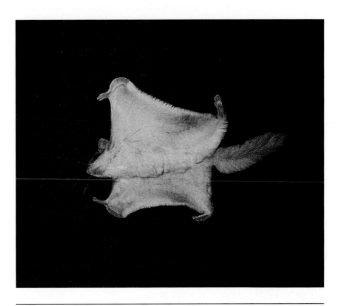

Figure 17.8 **Gliding adaptations.** The patagium and
flattened tail, essential for maneuverability and extending time in the air,
are clearly seen in this southern flying squirrel (*Glaucomys volans*).

(*Sciurus carolinensis*) and the fox squirrel (*S. niger*), den in tree
cavities or build nests among tree limbs. They generally are
diurnal and arboreal. In boreal and cool temperate areas,
most species hibernate. Ground squirrels and marmots are
terrestrial and burrow, and many species both hibernate and
estivate. Sciurids can be serious crop pests in many agricultural areas throughout the world.

The Petauristinae occur in both the Eastern and
Western hemispheres. They are distinguished by a furred
patagium between the front and hind limbs, which in some

species also includes the neck and tail. The patagium increases the surface area for extended gliding locomotion (figure 17.8) and enhances maneuverability. Members of the
Genus *Petaurista*, for example, glide for distances up to 450 m
and can turn 90 degrees in midair. In North America, the best
known members of the subfamily are the "flying" squirrels
(Genus *Glaucomys*). In Africa, the gliding niche has been
filled by the scaly-tailed flying squirrels (see the section
in this chapter on the Anomaluridae). In Australia, several
species of marsupial phalangers (Phalangeridae) are also gliders. These cases of similar morphological adaptations and behaviors in similar habitats are good examples of evolutionary
convergence.

Geomyidae

The 5 genera and approximately 35 species of pocket gophers are restricted to North and Central America from
southern Canada through Mexico to extreme northern
Colombia. They occur in a variety of habitats with soil conducive to burrowing. Geomyids are generally small, 120 to
440 mm in total length, including a tail 40 to 140 mm long.
Morphology reflects their fossorial mode of existence. Visual
and auditory acuity is reduced in favor of enhanced tactile
and olfactory sensitivity. They have thickset, chunky bodies
with small eyes and pinnae (figure 17.9). There are claws on
the forefeet and hind feet for digging; those on the forefeet
are larger. The incisors extend anteriorly so the lips can be
closed behind them to keep dirt out of the mouth when the
animal is digging. The pectoral girdle is highly developed for
digging, with a short, powerful humerus and a keeled sternum for enhanced muscle attachment, similar to that in the
three families of moles. The pelvic girdle is small and relatively undeveloped.

Figure 17.9   Fossorial adaptations.   This lateral view of a plain's pocket gopher (*Geomys bursarius*) shows the fusiform body shape of these fossorial species.

Gophers feed on subterranean parts of plants, primarily roots and tubers, although occasionally they leave their tunnels and forage above ground. Food is brought back to storage chambers in large, externally opening, furred cheek pouches that extend from the mouth to the shoulder. Tunnels may be quite extensive and include chambers for shelter, nesting, food storage, and fecal deposits. Tunnels are most easily noted by a series of aboveground mounds. Pocket gopher mounds are differentiated from those of moles by having a fan rather than a conical shape and by the entrance hole not being in the center of the mound. Although gophers may damage croplands, gardens, and seedlings, they also aerate soil, increase water penetration into the soil, and promote early successional plants.

Heteromyidae

The kangaroo rats, kangaroo mice, and pocket mice include 59 species in 6 genera. They are distributed from southwestern Canada through the western United States and Central America to northwestern South America. Kangaroo rats (Genus *Dipodomys*) and kangaroo mice (Genus *Microdipodops*) take their name from their superficial similarity to marsupial kangaroos (figure 17.10). These external characteristics are adaptations for their distinctive, bipedal jumping locomotion. The spiny pocket mice (Genera *Liomys* and *Heteromys*) and pocket mice (Genera *Perognathus* and *Chaetodipus*) are quadrupedal. Like the geomyids, all heteromyids have externally opening, furred cheek pouches, which they use to carry seeds back to their burrows for long-term storage, although they may eat more perishable food items while foraging.

Heteromyid skulls are often quite distinctive. A characteristic feature of the family is that the infraorbital foramen pierces the **rostrum** (the anterior region of the cranium), and several genera have greatly inflated auditory bullae (see figure 17.10). Large auditory bullae increase auditory acuity, which is of value in detecting potential nocturnal predators. They also aid in maintaining balance while heteromyids hop, often erratically, to elude those predators. Heteromyids are often found in arid regions; in the western United States, they are the rodents best adapted to dry, desert conditions. Kangaroo rats and pocket mice can exist for prolonged periods without free water. They survive on the moisture in food (mostly seeds) as well as metabolic water production. They also conserve water by producing a highly concentrated urine, being nocturnal, and remaining inactive in relatively humid burrows throughout the day. Spiny pocket mice occur in wetter, tropical habitats of Central and South America.

Castoridae

This family contains only two extant species, both restricted to the Northern Hemisphere. The Canadian beaver (*Castor canadensis*) occurs in Alaska, Canada, and throughout much of the United States, and the European beaver (*C. fiber*) occurs throughout northern Eurasia. The family name comes from the two pairs of castor glands and a pair of anal glands located near the cloaca in both males and females. These glands produce material with strong pheromones that is deposited on logs and mud piles to demarcate territories.

Beavers are well adapted to aquatic habitats and prolonged periods in icy water. They are large (up to 35 kg) with long, heavy guard hairs overlaying fine, insulating underfur. Both body size and pelage enhance their heat retention capabilities. The familiar spatulate tail is used in swimming, construction of dams and lodges, and certain behaviors. Additional aquatic adaptations include a **nictating membrane** (a thin membrane that can be drawn over the eyeball), **valvular** nostrils and ears that close while under water, and fully webbed hind feet (figure 17.11). Beavers cut and feed on a variety of trees, but especially aspen. They eat the cambium, bark, leaves, and roots, and store sections of trees and twigs underwater for food during the winter. How do beavers swim partially underwater with their mouths open, carrying cut branches, without water entering the lungs? They breathe through the nose, not the mouth. The epiglottis is above the soft palate, so as they breathe, air moves through the nostrils to the trachea and lungs. Also, the posterior part of the tongue is convex and blocks the pharynx, except when the animal swallows (Cole 1970).

Beavers cut trees and alter environments to a greater extent than any other mammal except humans. Dams are often very large and complex, and may reach 600 m in length, although 25 m is about average. Beavers impound large areas, creating ponds of still, often deep water. In the early 1800s, the quest for beaver pelts for the European market was a prime motivating factor in the early exploration of western North America by trappers. Beaver populations were overexploited and extirpated throughout most of their range. Many current populations are the result of

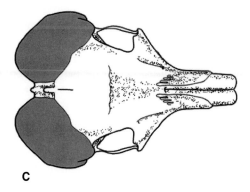

Figure 17.10 **Kangaroo rat.** (A) Long, tufted tail, elongated hind limbs, and kangaroolike bipedal body shape of a desert kangaroo rat (*Dipodomys deserti*). (B) Lateral view of the cranium of *Dipodomys* showing the threadlike zygomatic arch and infraorbital foramen that pierces the rostrum (arrow), and (C) dorsal view showing very large, inflated auditory bullae (shaded).

Figure 17.11 **Beaver body morphology.** Note the animal's fully webbed hind feet, thick pelage, and dorsoventrally flattened tail.

reintroductions to the former range. Because of low fur prices, populations are abundant in some areas, and beavers have become pests.

Pedetidae

The single species in this family, the springhare (*Pedetes capensis*), occurs in sandy soils or cultivated areas throughout semiarid regions of southern Africa. Total length reaches 900 mm, including a tufted tail up to 500 mm, and body mass ranges up to 4 kg. The skull is heavy with long nasal bones that extend beyond the premaxillary bones, and there is a large infraorbital foramen. As already noted, pedetids may be considered "hystricomorphous sciurognaths" based on their infraorbital structure. The scaly-tailed flying squirrels and gundis also fall into this category.

As with North American kangaroo rats, springhares have large hind feet and muscular hind limbs modified for bipedal, jumping locomotion (figure 17.12). They exhibit erratic hopping when frightened or escaping predators. Springhares burrow and tunnel extensively, and if they sense danger as they are about to leave their burrow, they "spring" immediately into the air from the exit, a behavior that gave rise to the common name. They breed throughout the year, but litter size is usually one. Springhares are nocturnal and herbivorous, feeding on a variety of cultivated crops. When foraging, they move slowly in a quadrupedal manner. They are hunted both because of the damage they do to crops and as a source of food by the Xhoisan people of South Africa.

Anomaluridae

The scaly-tailed flying squirrels are restricted to West and Central Africa where they inhabit tropical forests. There are seven species in three genera, and with one exception (the

Figure 17.12   Springhare.   The reduced forelimbs, long, powerful hind legs, and long tail evident in these various postures of the springhare resemble those of the smaller kangaroo rat (heteromyid) and larger kangaroo (macropodid).

scaly-tailed squirrel [*Zenkerella insignis*]), all have a well-formed patagium that encloses the forelimbs, hind limbs, and tail. The fur is brightly colored and silky. The common name is derived from two rows of keeled scales found at the ventral base of the tail (figure 17.13). These scales probably

Figure 17.13   Convergence in gliding.   The hairless, overlapping scales at the base of the tail are apparent in this scaly-tailed flying squirrel. Note the similarity in patagium and body form between this African anomalurid and the North American flying squirrel in figure 17. 8, a good example of phylogenetic convergence.

add traction while climbing or landing on a tree trunk after a glide. Anomalurids are arboreal and den in tree cavities up to 35 m above ground. They are primarily nocturnal and herbivorous, and may associate in colonies of up to 100 individuals. The taxonomic position of this group is uncertain; their resemblance to sciurid flying squirrels is due to convergence, not to phylogenetic relationship.

Ctenodactylidae

The four genera and five species of gundis are found in sparsely vegetated, rocky, semiarid regions of northern Africa from Morocco east through Somalia. They are similar to pikas (Lagomorpha: Ochotonidae), which are discussed later in this chapter. Their skulls have a large infraorbital foramen. They are unusual in that the paroccipital processes are very long and curve under and contact the auditory bullae. Gundis are up to 240 mm in head and body length, usually with very short tails, except in Speke's pectinator (*Pectinator spekei*). They have extremely dense pelage, which is groomed with the comblike hind toes (the family name means "comb-toes"). Females have an unusual lateral pair of cervical mammae at the base of the forelimb, as well as a pair on the anterior part of the thorax. Gundis are herbivorous and are able to exist only on the water obtained in vegetation. They are diurnal and live in small colonies or family groups. Colonies vary in size depending on habitat and associated food availability. Runways and group defecation sites are often seen. They do not burrow but seek shelter among rocks during the hottest part of the day. When threatened or alarmed, gundis "play possum," a fear response in which they may remain motionless for up to 12 hours.

Muridae

Murids include several taxa that formerly were recognized as distinct families. Thus, the "muroid" rodents encompass a variety of structurally diverse groups (figure 17.14) and include the mole rats (former Family Spalacidae); the New World mice, voles, lemmings, and allies (former Family Cricetidae); the bamboo rats (former Family Rhizomyidae); and the spiny dormice (former Family Platacanthomyidae), in addition to the Old World rats and mice (formerly the only members of the Family Muridae). This huge family has a very confusing and unresolved taxonomic history, due in part to parallelism, convergence, and reversal throughout the evolutionary history of the group. Although alternative classifications (Chaline et al. 1977) may better represent this group of rodents in terms of their structural diversity, for consistency we follow Wilson and Reeder (1993), who place the 1326 species in this family into 17 recent subfamilies (table 17.1). Carleton and Musser (1984:300) noted that this type of arrangement "... reflects not our conviction that this is the preferred nomenclature, but rather our uncertainty of the hierarchical pattern and our desire to focus upon the distinctive and richly varied geographic and biological properties of the groups...."

With the exception of Antarctica and some oceanic islands, murids are distributed worldwide, either naturally or through introduction, and have been introduced onto many formerly unoccupied islands. As might be expected of a family that includes over 28% of all living mammalian species, murids are found in a diverse array of habitats and show a wide range of locomotor adaptations. The usual dental formula is 1/1, 0/0, 0/0, 3/3 = 16. Molar structure within subfamilies and genera is quite variable. Molars may be rooted or rootless, cuspidate or prismatic (see figure 17.2). Tooth structure reflects the plant materials, invertebrates, or small vertebrates consumed. Likewise, few generalizations can be made concerning the diverse social and breeding behavior that occurs among and within subfamilies of murids. Murids illustrate the dynamic nature of systematics as new, more

A

B

C

D

Figure 17.14 **Structural diversity of murids.** (A) the Damaraland mole-rat resembles North Amercan pocket gophers; (B) the maned rat (*Lophiomys imhausi*), with a glandular patch under the erect mane; (C) the spiny mouse (*Acomys cahirinus*), with coarse, stiff spines on the back; and (D) the eastern woodrat (*Neotoma floridana*), a typical sigmodontine.

Table 17.1.   Extant subfamilies of murid rodents

Subfamily (number of genera, species)	Common name(s)	Distribution
Sigmodontinae (79, 423)	New World rats and mice	Nearctic, Neotropical
Cricetinae (7, 18)	Hamsters	Eastern Europe, Palaearctic, Oriental
Arvicolinae (26, 143)	Voles, lemmings, muskrats	Nearctic, Palaearctic, Northern Oriental
Gerbillinae (14, 110)	Gerbils, sand rats	Ethiopian, Middle East, central Asia, India
Cricetomyinae (3, 6)	Pouched rats and mice	Ethiopian south of Sahara
Petromyscinae (2, 5)	Rock mice, climbing swamp mouse	East central and Southwestern Africa
Dendromurinae (8, 23)	African climbing mice, gerbil mice, fat mice, forest mice	Ethiopian south of Sahara
Lophiomyinae (1, 1)	Maned rat	East Africa
Nesomyinae (7, 14)	Malagasy rats and mice	Madagascar
Murinae (122, 529)	Old World rats and mice	Ethiopian, Palaearctic, Oriental, Australian, Nearctic
Otomyinae (2, 14)	African swamp rats, karoo rats, whistling rats	South and East Africa
Platacanthomyinae (2, 3)	Malabar spiny tree mouse, blind tree mouse	Southwestern India, Southeast Asia
Myospalacinae (1, 7)	Zokors	Northern China, Siberia
Spalacinae (2, 8)	Blind mole rats	Southeastern Europe, Middle East, Northeastern Africa
Rhizomyinae (3, 15)	Bamboo rats, African mole-rats	East Africa, Eastern India, China, Southeast Asia
Calomyscinae (1, 6)	Mouselike hamsters	Middle East, Pakistan
Mystromyinae (1, 1)	White-tailed rat	South Africa

Source: Data from D. E. Wilson and D. M. Reeder (eds.) Mammal Species of the World: A Taxonomic and Geographic Reference, *2nd edition, 1993. Smithsonian Institution Press, Washington, D. C.*

sophisticated molecular techniques allow for more informed analyses of relationships. The three largest subfamilies (see table 17.1), each of which occurs in North America, illustrate in a little more detail the diversity of this large family.

The Subfamily Murinae is a huge group. Originally found throughout most of the Eastern Hemisphere, the two largest, most diverse genera (*Rattus* and *Mus*) have been inadvertently introduced by humans throughout most of the world. These two genera include the ubiquitous Norway rat and the house mouse, two species commensal with humans almost everywhere. The diversity of habitats, feeding modes, and locomotor adaptations of the Murinae parallels that of the family as a whole, about which few generalizations can be made. Most murines are nocturnal, but some genera, such as the striped mice (Genus *Lemniscomys*), are diurnal. Still others, such as the grass mice (Genus *Arvicanthis*), beaver rats (Genus *Hydromys*), groove-toothed creek rats (Genus *Pelomys*), and brush-furred mice (Genus *Lophuromys*), are active both day and night. Likewise, mean litter size in most species is usually 2 to 4, but it is 10 to 12 in the multimammate soft-furred rat (*Mastomys natalensis*). Young are altricial except in the spiny mice (Genus *Acomys*).

The Sigmodontinae encompass the New World rats and mice that were included, for the most part, in the former Family Cricetidae. Again, the size and heterogeneity of this group make generalizations on form and function difficult. The largest genera among the sigmodontines include the

well-known and ubiquitous deer mice and white-footed mice (Genus *Peromyscus*), the South American field mice (Genus *Akodon*), and the rice rats (Genus *Oryzomys*). Members of this subfamily are found from the subarctic to the tropics. Their habitats, feeding habits, reproduction, and activity periods are as varied and diverse as the murines. Most species are terrestrial, but this group also includes the arboreal golden mouse (*Ochrotomys nuttalli*), the semiaquatic fish-eating rats (Genus *Ichthyomys*), the fossorial mole mice (Genus *Notiomys*), and the woodrats (Genus *Neotoma*), which occur in caves and rocky outcrops, scrublands, and forests (see figure 17.14D).

The Arvicolinae are distributed throughout the Northern Hemisphere in a wide variety of habitat types. The cheekteeth generally have a prismatic occlusal surface (see figure 17.2B) and are open-rooted. Voles and lemmings exemplify the typical body shape with short ears and tails; blunt, rounded snout; and fairly short legs. Again, most species are terrestrial, although the heather voles (Genus *Phenacomys*) are arboreal, muskrats (Genera *Ondatra* and *Neofiber*) are aquatic, and mole voles (Genus *Ellobius*) are fossorial.

Dipodidae

This family includes 11 genera of jerboas and 4 genera of birch mice and jumping mice. Birch mice (Genus *Sicista*) and jumping mice (Genera *Zapus, Eozapus,* and *Napaeozapus*) formerly were included in their own family (Zapodidae). Dipodids exhibit a great deal of structural diversity based on the hind foot and associated life history, although in all species the tail exceeds head and body length. Birch mice have a simple, unmodified hind foot and are terrestrial or semiarboreal. These small mice inhabit forests, meadows, and steppes from Europe east through central Asia. They have prehensile tails and are excellent climbers. Jumping mice have elongated hind feet used in ricochetal locomotion and a very long tail; *Napaeozapus* can jump up to 2 m when alarmed. Pelage is brown on the dorsum, golden on the sides, and white on the venter. They occur in a variety of habitat types from Alaska, throughout Canada and much of the United States. One species (*Eozapus setchuanus*) is found in central China.

Jerboas are strongly bipedal and highly adapted for jumping, with the long, tufted tail used for balance. In most species, the three central metatarsals of the hind foot are fused, forming a cannon bone, and the first and fifth toes are lost. The strong hind legs are four times longer than the front legs. Hind foot length in these species is often half the head and body length. The largest, the five-toed jerboa (*Allactaga major*), has a hind foot up to 98 mm long and is capable of leaping 3 m. The jerboas are distributed from the Sahara Desert east across southwestern and central Asia to the Gobi Desert. They are medium-sized rodents (figure 17.15), with head and body length ranging from about 35 to 260 mm. They have large eyes and ears; light, sandy-colored pelage; and inhabit arid, semidesert, and steppe regions, where they burrow in sandy or loamy soils. A vertical process on the jugal bone protects the eye when the head is used in digging, and a fold of skin closes the nostrils. Many of the same adaptations found in the Heteromyidae to conserve water also occur in jerboas. Dipodids are dormant or hibernate in burrows during the winter for prolonged periods, up

Figure 17.15   Convergence in desert rodents.   Long hind feet and tail and reduced forelegs of the five-toed jerboa (*Allactaga elater*) are similar to the pattern seen in desert rodents such as the kangaroo rats (figure 17.10) and the springhare (figure 17.12).

to 9 months in some species. Likewise, jerboas may enter torpor in the summer during particularly hot, dry periods.

Myoxidae

This group includes 8 genera and 26 species of dormice, and encompasses two former families: the Seleviniidae and Gliridae. The desert dormouse (*Selevinia betpakdalaensis*), first described in 1939, is endemic to the small area north and west of Lake Balkhash in east Kazakhstan (central Asia), where the species is considered to be rare. It burrows in sandy or clay soils and avoids desert heat by being nocturnal. Dormice prey on spiders and insects.

Dormice occur in Europe east to central and southern Asia, Africa, and southern Japan. They are small to medium in size, with head and body length from 60 to 190 mm and tail length 40 to 165 mm. They are nocturnal, semiarboreal, and occupy habitats including forests, shrublands, residential areas, and rocky outcrops, where they forage for nuts, fruits, and animal material. They have a simple stomach without a cecum, suggesting a diet low in cellulose. Species may nest in tree cavities, attics, or burrows. Dormice put on weight in the fall and enter extended periods of hibernation (up to 7 months) or dormancy until spring. Many species are associated with residential areas, where they may become pests in gardens, orchards, or houses. The fat, or edible, dormouse (*Myoxus glis*) in Europe is the largest member of the family and weighs up to 200 g. It is trapped for its fat and for use as food. The species has been considered a delicacy since ancient Roman times.

Hystricognath Rodents

Uncertainty also surrounds the classification of this particular group of rodents. The mole-rats, discussed next, are a good example of the problems of placing families within this suborder. Although this group is clearly hystrico*gnathus* in its jaw structure, it has a small (i.e., nonhystrico*morphous*) infraorbital foramen. As noted previously, the anomalurids, pedetids, and ctenodactylids, which are considered "hystricomorphous sciurognaths," also provide examples of this difficulty in terms of a clear subordinal association.

Bathyergidae

Although the infraorbital foramen is small, with little passage of masseter muscle, jaws of bathyergids are hystricognathus with a strongly deflected angular process. This group also has an unusual crown pattern on the molars (see figure 17.2A). The 5 genera and 12 species of mole-rats and sand rats are strictly fossorial. They occur in sandy soils in the hot, dry regions of Africa south of the Sahara Desert. All species burrow extensively, with tunnel length of the common mole-rat (*Cryptomys hottentotus*) reaching over 300 m, the longest of any mammal. Tunnels are complex, with numerous secondary branches and chambers for nesting, feeding, and defecation. Mole-rats feed primarily on underground bulbs and tubers of perennial plants. Several species of bathyergids

Figure 17.16 **Naked mole-rat.** Many morphological adaptations of this adult naked mole-rat are similar to those found in other fossorial species such as pocket gophers, but the wrinkled and nearly hairless skin is unique among terrestrial species.

are solitary or are found in pairs or small groups. Other species, however, form very large colonies of 80 or more individuals.

The naked mole-rat (*Heterocephalus glaber*) is of special interest. Whereas most bathyergids are fully furred, the naked mole-rat has bare, wrinkled skin with only a few tactile hairs present (figure 17.16). Along with its reduced metabolic rate, its lack of fur allows for easier dissipation of body heat while underground. Naked mole-rats have the most highly developed **eusocial** system known among mammals (see chapter 22), a colonial system similar to certain insects (Jarvis 1981; Honeycutt 1992). They cooperate in tunnel digging, predator defense, and reproduction. Individuals vary greatly in body size depending on their function in the colony—smaller workers verses larger nonworkers. The one reproductively active female (the "queen") is the largest individual and mates with one of very few reproductively active males in the colony. Other adults do not breed but take part in foraging, caring for young, and other cooperative functions. Nonbreeders are not sterile, however, and may disperse to found new colonies or replace breeders that die. Individuals in a colony are very inbred (Reeve et al. 1990), and the genetic evolution of eusociality in mole-rats is a fascinating area of current research.

Hystricidae

The 3 genera and 11 species of Old World porcupines inhabit a variety of habitat types, including deserts, forests, and steppes. They occur throughout Africa, the Middle East, India, and Asia, including Indonesia, Borneo, and the Philippines. They are also found in Italy, where they probably were introduced thousands of years ago. Pelage is variable among the three genera. The long-tailed porcupine (*Trichys fasciculata*) has weak spines, or bristles, without quills and a long tail. Brush-tailed porcupines (Genus *Atherurus*) have short, soft spines on the head, legs, and ventral surface. Very long, flat, grooved spines are on the back. Crested porcupines (Genus *Hystrix*) have a short tail and long, hollow quills. These are grouped in clusters of five to six over the posterior two-thirds of the body and produce a rattling sound as they move. Their head and shoulders are covered with long, stiff bristles. Unlike New World porcupines, the tips of hystricid quills do not have barbs. These species are herbivorous, terrestrial, and generally nocturnal. They are considered pests in portions of their range because of the damage they do to plantations, but they are also hunted or raised for food in many areas.

Erethizontidae

New World porcupines encompass 4 genera and about 12 species. The North American porcupine (*Erethizon dorsatum*) occurs throughout Canada and the northeastern and western United States, and the rest occur from southern Mexico south to northern Argentina. They are found in mixed coniferous forests, tropical forests, grasslands, and deserts. Despite their generally chunky, heavyset bodies, erethizontids are more arboreal than the Old World porcupines. The North American porcupine is both terrestrial and semiarboreal, whereas South American species (Genera *Coendou* and *Sphiggurus*) have prehensile tails and spend most of their time above ground. Heavy spines with a barbed tip (figure 17.17), embedded singly and not in clusters as in Old World porcupines, occur over much of the dorsum and sides of the body. The ventral surface has coarse, long hair and no spines. Contrary to popular belief, porcupines do not intentionally throw their quills at attackers. The quills are loosely embedded, however, especially in the tail, and can easily become detached in predators. Because of the barbed tip, the quill works continuously deeper into the wound and can even cause death. Nonetheless, porcupines have a variety of potential predators, with the fisher (*Martes pennanti*) the most adept at flipping porcupines over and attacking the unprotected ventral area.

Petromuridae

The monotypic dassie rat (*Petromus typicus*) is restricted to southern Angola, Namibia, and northwest South Africa in rocky desert habitats. It is fairly small and squirrel-like, with soft, yellow-orange to brown pelage and no underfur. The skull is rather flat and the ribs flexible, which allows the dassie rat to squeeze through rocky crevices. Dassie rats are diurnal, feed on grasses and plant material, and are coprophagous. The crowns of the cheekteeth are unusual; the labial side of the lower teeth and the lingual side of the upper teeth are raised above the rest of the occlusal surface. The family name Petromyidae also occurs in the literature for this species (Macdonald 1984).

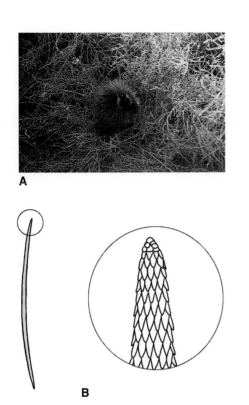

A

B

Figure 17.17 **Porcupine quills.** Familiar pelage of (A) the North American porcupine with guard hairs modified as quills and (B) a diagrammatic enlargement of the overlapping barbs on the tip of a quill.

Thryonomyidae

Cane rats, or "grasscutters," occur throughout Africa south of the Sahara Desert. The greater cane rat (*Thryonomys swinderianus*) is a semiaquatic inhabitant of marshy areas, whereas the lesser cane rat (*T. gregorianus*) is found in drier upland areas. Both species have the typical chunky body shape of hystricomorphs, with the greater cane rat reaching a body mass of 10 kg. Their skulls are heavy, with a very large infraorbital foramen, and there are three grooves on the anterior surface of each upper incisor. Pelage is coarse and bristlelike. Both species may reach very high population densities and cause significant damage to cultivated crops and plantations. Cane rats also are an important food source for native people; they are hunted or farmed, and large quantities of the meat are sold in markets.

Chinchillidae

The three fairly diverse genera in this family are distributed in South America from the Andes Mountains of Peru south to southern Argentina. All have a large head with large eyes and ears, and very dense, soft pelage (figure 17.18). The plains viscacha (*Lagostomus maximus*) lives in large colonies in scrub and grassland areas of Argentina, Paraguay, and Bolivia, where it creates large, extensive burrow systems. Size dimorphism is evident, with males twice as large as females.

The three species of mountain viscachas (Genus *Lagidium*), which look like rabbits with long, strongly curved tails, form colonies in dry, rocky, rather barren areas at high elevation. The two species of chinchillas (Genus *Chinchilla*) also form colonies in these types of habitat. Populations of plains viscacha are greatly reduced because they are pests on crops and compete for rangeland forage with domestic livestock. They also are taken for their fur and meat. Likewise, population densities of mountain viscacha have declined because they

A

B

Figure 17.18 **Representative chinchillids.** (A) Populations of chinchilla are greatly reduced in numbers throughout their native range. (B) A rabbitlike mountain viscacha from Argentina.

are hunted for their meat and fur. Chinchilla fur has been highly prized for centuries and is one of the most valuable of any mammal in the world. Both species are extirpated throughout much of their original ranges and are considered endangered.

Dinomyidae

The monotypic pacarana (*Dinomys branickii*) is a rare inhabitant of forested areas at higher elevations in the slopes and valleys in Colombia and Bolivia. Pacaranas look like huge guinea pigs. The body is heavyset (about 15 kg) with coarse, brownish-black pelage and four lines of white spots on each side from the shoulders to the rump. Pacaranas are slow, lumbering, nocturnal herbivores, and their population levels apparently have never been high. They have been reduced even further because of habitat destruction and consumption by humans, and the species may be near extinction.

Caviidae

The guinea pigs, cavies, and Patagonian "hares" encompass 5 genera and about 14 species. Their distribution in the wild ranges throughout South America, except for the Amazon Basin, in a variety of habitat types. Members of the family share a short, robust body form, with short limbs and ears, and a vestigial tail. The two species of Patagonian hares, or maras (Genus *Dolichotis*), are exceptions (figure 17.19): they have longer ears and are adapted for cursorial locomotion, with long hind legs, a reduced clavicle, and reduced number of digits. The dentary bones of caviids have a prominent lateral groove. The upper cheekteeth converge anteriorly; thus, the toothrows form a V-shape. Caviids are probably the most abundant and widespread rodents in South America. Each genus is found in a different habitat, however, with only yellow-toothed cavies (Genus *Galea*) sympatric with other genera. The origin of the domestic guinea pig (*Cavia porcellus*) is uncertain, although it has been raised for consumption for thousands of years. Ironically, although guinea pigs are common in pet stores and laboratories throughout the world, they are rare in the wild.

Hydrochaeridae

The family name ("water pig") suggests both the habitat affinity and general body form of the monotypic capybara. It is closely associated with water throughout its range, which extends from Panama south to Argentina east of the Andes Mountains. Capybaras are semiaquatic and have partially webbed feet. They swim and dive with agility and feed on aquatic plants in and around lakes, ponds, swamps, and other wet areas (figure 17.20). They are heavyset, with a large head, and small eyes and ears placed high on the head. Their pelage is short, coarse, and fairly sparse. Capybaras are the largest living rodents. Head and body length is about 1.3 m, and mean body mass is 28 kg, although individual body mass may reach 50 kg or more. The skull is distinctive with a very long paroccipital process, grooved dentary bones, and an

A

B

Figure 17.19   Morphological differences in caviids.   These are evident when comparing (A) the rock cavy (*Kerodon rupestris*), and (B) the Patagonian "hare" (*Dolichotis patagonum*).

Figure 17.20 **Capybaras.** This South American species feeds on a variety of plant material, and in large groups can become a serious agricultural pest.

upper third molar that is longer than the combined length of the other three cheekteeth. Capybaras are fairly gregarious, living in groups of 20 or more individuals. As a result, they often become serious agricultural pests. They are hunted either for food or to reduce crop damage. In Venezuela, capybaras are ranched for their meat and leather.

Dasyproctidae

This family is made up of about 11 species of agoutis (Genus *Dasyprocta*) and 2 species of acouchis (Genus *Myoprocta*). They occur from southern Mexico south to northern Argentina east of the Andes Mountains. Agoutis were introduced into Cuba and several other Caribbean islands. Dasyproctids are generally diurnal, solitary, burrowing herbivores. Both genera are similar in appearance with a coarse, glossy orange-brown to black pelage. The pelage is long, thick, and often of a contrasting color in the posterior area (the family name means "hairy-rumped"). Agoutis weigh up to 4 kg and inhabit forests, brushlands, and grasslands, always in association with water. They are hunted by humans for food, and populations have declined in many areas. Acouchis are smaller, have a slightly longer tail, and are restricted to tropical rainforests of the Amazon Basin. Dasyproctids are sometimes considered to be in the same family as the pacas (Genus *Agouti*) discussed in the following section. Unfortunately, species with the common name "agouti" do not have the scientific name *Agouti*.

Agoutidae

The two species of pacas (Genus *Agouti*) occur from central Mexico south to Paraguay and Argentina in forested areas from lowlands to high elevations, usually near rivers or streams. Pacas have the typical robust hystricomorph form and weigh up to about 10 kg. Their coarse pelage is brown-black above with a paler venter and four rows of white spots on each side of the body. The skull is unique in that a portion of the zygomatic arch is enlarged and contains a large sinus

(figure 17.21), possibly for amplification of vocalizations or tooth-grinding sounds. The dental formula is 1/1, 0/0, 1/1, 3/3 = 20. Pacas are nocturnal, terrestrial, burrowing herbivores that eat a variety of plants and fruits. They also may be serious pests on crops, gardens, and plantations and sometimes are killed for this reason. More often, however, they are harvested for their excellent meat. Pacas are considered to be the best tasting, most edible South American rodent. Hunting and habitat loss have greatly reduced population numbers in large parts of their range.

Ctenomyidae

This family is composed of about 38 species of tuco-tucos in the single genus *Ctenomys*. They are distributed from central South America south to Tierra del Fuego and occur in habitat and soil types suitable for burrowing. These fossorial rodents are similar to North American pocket gophers in habits and appearance. Like geomyids, tuco-tucos have broad, thick incisors; small eyes; short, thick pelage; and enlarged claws. Unlike geomyids, however, they lack external cheek pouches. The family and genus name ("comb mouse") is derived from the stiff fringe of hairs around the soles of the hind feet and toes used to groom dirt from the fur. The burrow systems can be very extensive, and population densities may reach 200 per hectare. Tuco-tucos are quite vocal, and the "tloc-tloc" alarm call, which gave rise to the common name, can be heard when the animals are underground. Colonies may become serious pests in plantations because they destroy roots and girdle trees. Because their fossorial habits have resulted in relative isolation and disjunct distributions of species, *Ctenomys* is a diverse, rapidly evolving taxon.

Octodontidae

The six genera and nine species of octodontids occur in the Andean region of Peru, Bolivia, Chile, and Argentina. They are found in several habitat types from sea level to high

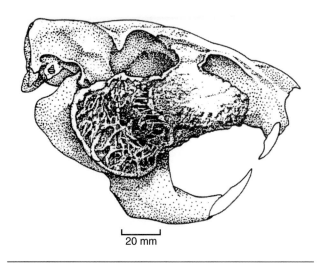

20 mm

Figure 17.21 **Skull of the paca.** The large, distinctive zygomatic arch of *Agouti paca* is evident.

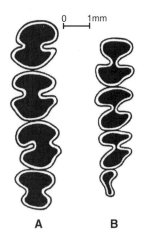

Figure 17.22   Molariform tooth shape.   Left upper cheekteeth in (A) the coruro and (B) the rock rat (*Octomys mimax*) showing characteristic figure-8 shape in octodontids.

elevation. The two species of viscacha rats (Genera *Octomys* and *Tympanoctomys*) are very rare. Octodontids are small to medium-sized, with thick, silky pelage, and often have tufted tail tips. Unlike many other rodent families, the occlusal surfaces of the molars usually are simple, with a single lingual and labial fold forming a figure-8 pattern (figure 17. 22). This pattern gives rise to the family name. Habitats, life history, and feeding habits vary among species, from fairly fossorial communal burrowers to semiarboreal species or rock dwellers. Coruros (*Spalacopus cyanus*) maintain nomadic colonies that move to a new area after depleting food resources at a site.

Abrocomidae

Sometimes considered as a subfamily of Octodontidae, the single genus (*Abrocoma*) and three species of chinchilla rats have the same geographic distribution as the octodontids. They burrow among crevices in remote mountainous thickets and rocky areas. Chinchilla rats are ratlike in appearance, with a pointed snout and long tail. They have long, soft, thick pelage but not of the same quality as true chinchilla fur. With 17 pairs of ribs, *A. bennettii* has more ribs than any other rodent.

Echimyidae

This is a large family of about 16 genera and 71 extant species of spiny rats. They occur from Honduras south to central South America. Almost all species have spiny, sharp, stiff, bristlelike pelage. They appear ratlike with a pointed rostrum. Habitat preferences vary but are closely tied to availability of water. Some genera are semifossorial burrowers, others terrestrial, and some arboreal. Several recently extinct species of spiny rats in the genera *Boromys*, *Brotomys*, and *Heteropsomys* from Cuba, Puerto Rico, and the Dominican Republic have been described from remains found with human artifacts (Guarch-Delmonte 1984).

Capromyidae

About 4 genera and 12 living species of hutias exist in remote forested or rocky valleys and small islands in the West Indies. Hutias are chunky with short legs, a large head, and small eyes and ears. They range in head and body length from 200 to over 650 mm and in tail length from vestigial to long and prehensile. The coarse pelage varies from yellowish gray to black on the dorsum and is paler on the venter. Some hutias (Genus *Capromys*) are arboreal folivores, whereas others (Genus *Geocapromys*) feed on terrestrial plants. Hutias also consume small lizards.

Several genera and species of capromyids have become extinct during the last few hundred years in Cuba, Haiti, Puerto Rico, the Virgin Islands, and the Dominican Republic. Most extant species are considered threatened or endangered, not only because of habitat loss and overharvesting for food, but also from introduction of the mongoose (*Herpestes javanicus*) and feral house cats (*Felis* [*catus*] *sylvestris*). In addition, the closely related giant hutias (Heptaxodontidae), which were also distributed throughout the Caribbean islands, have gone extinct in recent times, many since humans arrived on the islands.

Myocastoridae

The monotypic coypus, or nutria (*Myocastor coypus*), is often considered a member of either the Capromyidae or Echimyidae, but is included in its own family based on morphological and serological differences. Coypus are native to South America from Chile and Argentina north to Bolivia and Brazil. The species is large (up to 10 kg body mass) and heavy-bodied, with rather coarse pelage and a long, round, sparsely furred tail. Their webbed hind feet adapt them to aquatic habitats in and around fresh or brackish marshes or slow-moving streams. They make floating platforms that they use as feeding stations from which they consume aquatic vegetation (figure 17.23).

Figure 17.23   Nutria morphology.   Long, somewhat coarse pelage, webbed hind feet, and round tail are evident on this adult nutria.

Coypus have been widely introduced into Europe and North America. In the United States, most were introduced in the early to mid-1900s for fur farming. They escaped or were released and primarily occur in the Gulf Coast states and along the West Coast to Washington, and the East Coast to Maryland. Populations also are established in several parts of southern Canada. Coypus fur is of major economic importance in South America, and in the United States, millions of pelts were harvested annually (Willner 1982) until the decline of the fur industry.

LAGOMORPHA

Lagomorphs (meaning "hare-shaped") include 11 genera and about 54 species of rabbits and hares in the Family Leporidae and 2 genera and about 26 species of pikas in the Family Ochotonidae. Lagomorphs occur worldwide except for the southern portions of South America; Australia and New Zealand; and islands, such as Madagascar, the Philippines, and those in the Caribbean. A few species of leporids have been introduced into most of these places. The introduction of the European wild rabbit (*Oryctolagus cuniculus*) into Australia and New Zealand is a prime example of the potential pitfalls of introducing an **exotic** (a species outside its native range; see the section in chapter 27 on biological control).

Morphology

All lagomorphs are small to medium-sized. They are terrestrial and herbivorous, and in them, as in many rodents, coprophagy is common (see figure 6.11). The diagnostic feature of the order is the occurrence of "peg teeth," a second pair of small incisors without a cutting edge, immediately behind the larger, rodentlike first incisors (figure 17.24). A third pair of lateral incisors are lost before birth or immediately thereafter. The cutting edge of the primary incisors is notched in pikas but not in rabbits and hares (figure 17.25). The dental formula is 2/1, 0/0, 3/2, 2-3/3 = 26 to 28. The cheekteeth are hypsodont with two transverse enamel ridges, whereas rodents usually have several transverse ridges. The cheekteeth and incisors are open-rooted and ever-growing. In leporids, but not ochotonids, the rostral portion of the maxilla is **fenestrated** (having small, latticelike perforations in the bone), and the frontal bone has a supraorbital process. Rabbits and hares generally have large ears and elongated hind limbs to accommodate their saltatorial locomotion. Rabbits have a well-known "cotton-ball" tail, but the tail in hares is longer. Pikas are rodentlike in appearance and have short limbs, small ears, and no tail. Although agile, they do not have the running ability of leporids. Unlike in rodents, the feet of lagomorphs are fully furred. A cloaca is present, the uterus is duplex, and there is no baculum.

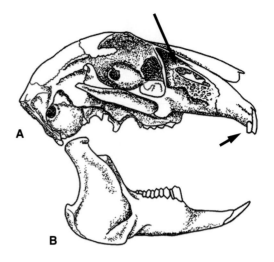
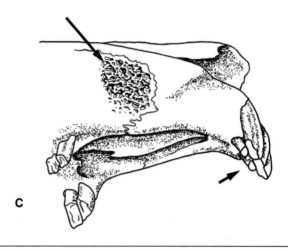

Figure 17.24   Peg teeth.   Lateral view of the (A) cranium and (B) mandible of a leporid, the arctic hare. Although superficially rodentlike, note the peg teeth (*arrows*) and fenestrated rostrum, especially evident in the enlarged view (C).

Fossil History

The earliest known lagomorphs, from the Family Eurymylidae, date from the late Paleocene epoch of Mongolia. Both Old and New World fossil leporids are known from the late Eocene epoch, whereas fossil ochotonids are known from the mid-Oligocene epoch in Asia. Both families were much more diverse in the Tertiary period than they are today, with 21 fossil genera of leporids and 23 fossil genera of ochotonids recognized (Carroll 1988). Chuan-Kuen and colleagues (1987) argue that lagomorphs and rodents are closely related and share a common ancestry, a view held by most nineteenth-century natural historians.

Economics and Conservation

Pikas are of limited economic importance, with a few Asian species considered to be agricultural pests. Leporids are often important as game species. Most species of cottontail

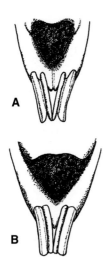

Figure 17.25   Pika upper incisors.   (A) Anterior view of the upper incisors of an ochotonid showing the characteristic notch on the cutting surface. (B) The upper incisors of a leporid do not have a notch.

Figure 17.26   North American pika.   *Ochotona princeps* in typical rocky habitat, carrying cut vegetation.

rabbits (Genus *Sylvilagus*) are hunted; the eastern cottontail rabbit (*S. floridanus*) ranks first among game mammals in terms of the numbers taken and hours spent hunting them in North America. The snowshoe hare (*Lepus americanus*) historically has been an important furbearer, and several other species of hares are hunted for food and sport.

Local populations of leporids may increase to such densities that they become serious crop depredators. Lethal methods such as poisoning and shooting, along with nonlethal methods such as fencing and repellents, are used in attempts to control local populations. In the western United States, ranchers and farmers occasionally have "rabbit drives" in which large numbers of people work together to round up and kill thousands of black-tailed jack rabbits (*Lepus californicus*) in an effort to reduce their density and resultant crop damage.

Ochotonidae

Pikas commonly inhabit either steppe or forest areas, or steep, rocky (talus) slopes in alpine areas, where they live under and among the boulders. Distinct differences mark species within each habitat type. Pikas in talus areas do not burrow, are relatively asocial, and have low population densities and fecundity rates. Conversely, those in steppe or forest habitats dig burrows, are more social, and have higher population densities and fecundity. In both habitats, pikas are extremely vocal. Calls and "songs" are used to maintain distinct territories and social organization. These are enhanced with scent marking from apocrine glands on the cheeks. Pikas are well known for their "haymaking" activity (figure 17.26), in which an individual or breeding pair cut and gather vegetation throughout the summer and fall, letting it cure in piles in the sun and storing it in a traditional place within the territory. This cached hay is then used for winter food. These

caches can weigh up to 5 kg and may directly affect the dynamics of local plant communities and sympatric species of herbivores (Smith et al. 1990).

Leporidae

This family includes the rabbits and hares, which, although morphologically similar, differ in several ways. Rabbits usually build fairly well-constructed, fur-lined nests and give birth to altricial young. Hares do not construct nests but instead make shallow depressions on the ground called "forms" (figure 17.27), and they have precocial young. Neonatal hares are fully furred, have their eyes open, and are able to run a few hours after birth. Rabbits also have an interparietal bone in the skull, but hares do not. The karyotype of most rabbits is $2N = 42$; most hares have $2N = 48$.

Figure 17.27   Black-tailed jack rabbit "form."   This is nothing more than a shallow, cleared depression—in this case, under sagebrush (*Artemisia tridentata*).

A

B

Figure 17.28 **Pelage dimorphism.** This is evident in (A) the brown summer coat of the snowshoe hare and (B) the white winter coat of the same species.

Leporids are found in habitats from the snow and cold of the Arctic to deserts, grasslands, mountain areas, swamps, and tropical forests. One of the adaptations of species in temperate or arctic areas, such as the arctic hare (*Lepus timidus*) or snowshoe hare (*L. americanus*), is pelage dimorphism. Coats are white in the winter and brown in the summer (figure 17.28). "Blending in" with the habitat is beneficial because leporids are potential prey to a wide variety of predators, including lynx (Genus *Lynx*), coyotes (*Canis latrans*), red fox (*Vulpes vulpes*), marten (*Martes americana*), and fisher.

Like rodents, leporids have a high reproductive potential. They may have several litters per breeding season, with several individuals per litter. Postpartum estrus occurs, so that litters are produced in fairly rapid succession. Leporids also exhibit induced ovulation, further enhancing the probability that reproductively mature females will conceive. Sim-

ilar to certain species of rodents that have 3- to 4-year cycles in population density, snowshoe hare populations exhibit 8- to 11-year cycles (see chapter 24). Density may change during this period by two orders of magnitude. Population increases result from increased birth rates and survival of young; the opposite occurs during population declines. The effect of food, predators, and other factors on causation of cycles is an active area of research (Krebs 1996).

In contrast to the high population densities of some leporids, five species of hares currently are considered to be endangered. These include the riverine rabbit (*Bunolagus monticularis*) in South Africa, the Tehuantepec hare (*Lepus flavigularis*) and volcano rabbit (*Romerolagus diazi*) in Mexico, the hispid hare (*Caprolagus hispidus*) in the foothills of the Himalayan Mountains, and the Amami rabbit (*Pentalagus furnessi*) in the Ryukyu Islands of Japan.

Summary

Rodents comprise over 43% of the living species of mammals in the world today. As might be expected of the largest mammalian order, the structural and functional characteristics of their locomotion and morphology vary greatly. The order is also highly adaptable in the variety of habitats occupied in an almost worldwide distribution. Despite their overall diversity, all rodents have a single pair of upper and lower chisel-shaped incisors, a diastema, and reduced numbers of molariform teeth. Convergence and parallelism of behavior and associated morphology is a key theme among rodent taxa. For example, behavior, structural similarity, and kangaroolike bipedal locomotion is evident among some of the heteromyids, pedetids, dipodids, and certain murids in open, arid habitats. Likewise, the anomalurids and certain gliding sciurids show convergent adaptations; as do the fossorial geomyids, bathyergids, ctenomyids, and some murids.

Rodents influence our daily lives as they have throughout history. The classification of rodents offers a continuing challenge and opportunity for worthwhile investigation, and the 28 rodent families discussed in this chapter may well be revised in the near future as additional phylogenetic evidence is gathered.

Although structurally and functionally similar in many respects to rodents, the lagomorphs constitute a much smaller and more restricted order in terms of habitats, morphology, and locomotion. As with rodents, a single key characteristic may be used to define the order. Both families of lagomorphs have peg teeth, a small pair of incisors posterior to the large, upper incisors. Several species of leporids are significant game animals, but the economic importance and effect on humans of the two families of lagomorphs are minor compared with those of rodents.

Discussion Questions

1. Discuss the concept of kin selection and various factors that could contribute to eusocialty and reproductive "altruism" in species such as the naked mole-rat, including fossorial living (with associated relative protection from predators) or patchy distribution of resources.

2. How might breeding be suppressed in the majority of members in a colony of naked mole-rats, such that only the "queen" and one male breed?

3. Discuss the relationship between rabbits and hares in terms of the complexity of nest construction and whether neonates are altricial or precocial.

4. How are the structural differences in the jaw structure of rodents, described in figures 17.3 and 17.4, reflected in the exploitation of food resources by representative members of each group?

5. Convergence in rodents in terms of bipedal locomotion, fossorial adaptations, and gliding was noted in the summary. Think of two or three rodent taxa that show similar convergence in terms of adaptations for an aquatic niche.

6. What is the significance of coprophagy in mammals? Besides lagomorphs, what other mammalian orders exhibit coprophagy?

Suggested Readings

Anderson, P. K. 1989. Dispersal in rodents: a resident fitness hypothesis. Spec. Publ. No. 9, American Society of Mammalogists.

Eisenberg, J. F. 1989. Mammals of the Neotropics, vol. I. Univ. of Chicago Press, Chicago.

Genoways, H. H. and J. H. Brown. 1993. Biology of the Heteromyidae. Spec. Publ. No. 10, American Society of Mammalogists.

King, J. A. (ed.). 1968. Biology of *Peromyscus* (Rodentia). Spec. Publ. No. 2, American Society of Mammalogists.

Rowlands, I. W. and B. J. Weir (eds.). 1974. The biology of hystricomorph rodents. Academic Press, London.

Tamarin, R. H. (ed.). 1985. Biology of the New World *Microtus* Spec. Publ. No. 8, American Society of Mammalogists.

CHAPTER

18

Proboscidea, Hyracoidea, and Sirenia

PROBOSCIDEA

Morphology
Fossil History
Economics and Conservation

HYRACOIDEA

Morphology
Fossil History
Economics and Conservation

SIRENIA

Morphology
Fossil History
Economics and Conservation

The rationale for including in a single chapter elephants, the largest living terrestrial mammals; the rabbit-sized hyraxes; and the dugong and manatees, which never leave the water, may appear obscure. These three orders are usually grouped together as "subungulates" based on their perceived evolutionary relationships. Along with the true ungulate Orders Perissodactyla and Artiodactyla (chapter 19), the orders in this chapter are derived from the primitive **Condylarthra,** a generalized ancestral order of land mammals that arose in the early Paleocene epoch about 65 million years ago. One of the recognizable groups within the Condylarthra was the **Paenungulata** ("near-ungulates"). By the early Eocene epoch of Africa, about 54 mya, the Paenungulata had given rise to the Proboscidea, Sirenia, and Hyracoidea (Carroll 1988). Although the groups began to diverge in the Eocene epoch and appear very different today, they share certain anatomical characteristics. None has a clavicle, the digits have short nails (no nails in the Amazonian manatee, *Trichechus inunguis*), and there are four toes on the forefeet (five in Asian elephants). Females have two pectoral mammae between the forelegs (hyraxes have two inguinal pairs as well) and a bicornuate uterus. Males have abdominal testes with no external scrotum and no baculum. All are nonruminating, herbivorous, hind-gut fermenters. The symbiotic **microfauna** (ciliated protozoans and bacteria) that break down vegetation occur in an enlarged cecum. The ceca in hyraxes are particularly complex. Finally, the dugong, manatees, and elephants are unusual among mammals in their pattern of molariform tooth replacement, which is horizontal, not vertical as in other mammals. Proboscideans and sirenians flourished during the Oligocene and Miocene epochs, but with only three extant families among them, they are today mere remnants of what were once very diverse and abundant groups. Because populations are declining throughout most of their range, the future of elephants, manatees, and the dugong is of concern to conservationists.

PROBOSCIDEA

Elephants today are represented by a single family, the Elephantidae, with two extant species. The African elephant (*Loxodonta africana*) is distributed throughout Africa south of the Sahara Desert. The Asian elephant (*Elephas maximus*) occurs south of the Himalayan Mountains in India, Sri Lanka, Indochina, and Indonesia and has been introduced to Borneo. The two species exhibit several differences, some of which are familiar to many people. African elephants are larger than Asian elephants and have much larger ears; the back is concave with the shoulder higher than the head, and the tip of the trunk has two lips (figure 18.1). The species have different occlusal surfaces on the cheekteeth (see figure 18.1) and different numbers of nails on the hind feet: four in Asian and three in African elephants. There are 19 pairs of ribs in *Elephas*, 21 in *Loxodonta*.

Two subspecies of African elephants are recognized, the savanna elephant (*Loxodonta africana africana*) in eastern, central, and southern Africa, and the forest elephant (*L. a. cyclotis*) in central and western Africa. There are four subspecies of Asian elephants: the Indian (*Elephas maximus bengalensis*), Ceylon (*E. m. maximus*), Sumatran (*E. m. sumatrensis*), and Malaysian (*E. m. hirsutus*). The different subspecies are found in different portions of each species' overall range. As with many other groups, however, some question remains about the validity of subspecies classification in elephants (Eisenberg 1981).

Each species occurs in a variety of habitats, from grasslands and shrublands to forests. Regardless of habitat type, all elephants are closely tied to the availability of water. They are nonruminant herbivores; microbial action takes place in the cecum. During the wet season, herbaceous vegetation and grasses are eaten. Shrubs, leaves, and tree bark are taken in the dry season, with flowers and fruits eaten when available. In agricultural regions, elephants can do extensive damage to cultivated crops. Large adults may consume up to 150 kg of vegetation a day, but because the digestive process is relatively inefficient compared with that of ruminants (see figure 6.9), half of this may pass through the gut undigested.

For the following reasons, large home ranges are needed to sustain groups of elephants: the amount of food necessary to maintain an individual elephant; the often dry, inhospitable habitat conditions; and the fact that aggregations of up to 50 individuals of mixed sex and various ages may form during portions of the year. As a result, home range size may be 1600 km² during seasons when resources are scarce. Elephants may move up to 30 km a day to reach better habitat conditions. Extent and distance moved depend on the quantity and quality of food and water available. Elephant herds are capable of quickly degrading habitats and affecting the availability of resources to other herbivores.

Elephants are gregarious, with herds normally made up of family units of about 10 individuals (Laws 1974). Females and their young, led by a matriarch, remain in these units. Different units made up of related individuals often join temporarily to form "bond groups." Depending on the time of year, males may be either solitary or in temporary all-male groups, with overlapping home ranges. Adult males join females when the latter enter estrus. Juvenile males leave herds when they become sexually mature, between 10 and 17 years of age.

For both species, sexual maturity generally is reached in females between 9 and 12 years old, with peak fecundity from 25 to 45 years of age. Few, if any, females 50 years and older breed. Reproduction is tied to the wet season and the availability of food and water. Females are in estrus during the later part of the rainy season and first part of the dry season. Estrus is extremely brief, lasting only 2 to 4 days. The interval between estrous periods averages 4 years (Moss 1983) because of the long length of gestation and lactation. Following an average 22-month gestation, a single young is born (rarely twins) at the beginning of the wet season, when

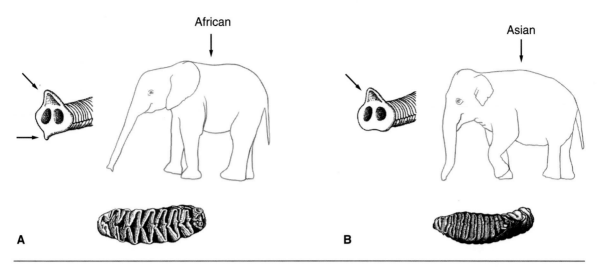

Figure 18.1 Characteristics of elephants. (A) An African elephant and (B) an Asian elephant. Note the different postures, with the shoulder above the head and a concave back in the African elephant. The Asian elephant has a larger, more bulbous skull, which is above the shoulder, and a convex back. The tip of the trunk has two lips (*arrows*) in African elephants and one lip in Asian elephants. The pattern of the laminar ridges on the occlusal surface of cheekteeth differ in African and Asian elephants.

habitat conditions are optimal. Newborn African elephants weigh about 120 kg, Asian elephants 100 kg. They nurse for 3 to 4 years (with the mouth, not the trunk) and weigh 1000 kg by 6 years of age. Although growth rate decreases by age 15, elephants grow throughout their lives. Males exhibit a reproductive period of 2 to 3 months each year called **musth,** during which hormone levels, sexual activity, and aggressive behavior increase. Although males are physiologically capable of breeding by 10 to 15 years of age, most successful matings are by mature males 30 to 50 years old.

Elephants make a variety of vocalizations audible to humans, including trumpeting, growling, roaring, and snorting. They also communicate with one another through extremely low frequency sound, as low as 14 to 35 cycles per second (Payne et al. 1986; Langbauer et al. 1991), well below the range of human hearing. Low-frequency sound allows groups or individuals up to 4 km apart to communicate and coordinate movements and helps males find females during their brief and unpredictable estrous periods. Low-frequency sound is used to communicate because it can travel over long distances and through obscuring vegetation much better than higher frequency sound. Also, a large animal like the elephant physically would have a difficult time producing high-frequency sound.

Morphology

Elephants are the largest living terrestrial mammals. Aspects of anatomy, movement, dentition, and behavior relate to their size and associated long life span. The shoulder height of adult African bull elephants may reach 4 m, with maximum body mass up to 7500 kg. Sexual dimorphism is evident, with body mass smaller in females. Because they grow throughout

life, the oldest elephant in a group often is the largest. Asian elephants are smaller than African elephants; mean body mass of male *Elephas* is about 4500 kg. Size also depends on the subspecies, with African forest elephants being smaller than savanna elephants. The structure necessary to support such large mass gives rise to several modifications resulting in graviportal locomotion and a massive skeleton that makes up 15% of an individual's body mass, about twice that of most terrestrial mammals. The head is very large, in part to help support the trunk and ever-growing tusks. The bones surrounding the brain are thick but made less heavy by a series of air-filled pneumatic cavities, or sinuses (figure 18.2). The feet are broad, with the phalanges embedded in a matrix of elastic tissue to help cushion the weight (figure 18.3).

Throughout their evolutionary history, elephants may have benefited from increased body size in avoiding competition with the large herbivorous African perissodactyls that preceded them, as well as the artiodactyls that arose later. Large body size confers other benefits in addition to reduced competitive pressure. It allows elephants to move greater distances in response to habitat conditions with relatively less energy expenditure. It also reduces predation pressure; only humans threaten adult elephants. Along with benefits, large size also has its drawbacks. Elephants live in warm climates but have very few sweat or sebaceous glands. Heat dissipation without dehydration is a problem, especially in areas with limited water availability. Thus, elephants have very sparse body hair, a characteristic common to other large mammals in warm climates. Also, the wrinkled skin of elephants acts to hold water and facilitate its movement on the body surface, increasing the evaporative cooling effect (Lillywhite and Stein 1987). The large ears, especially of African elephants, are highly vascularized, and when moved back and forth, they act

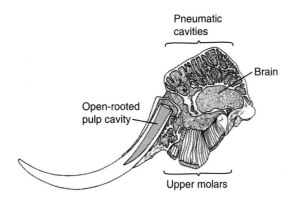

Figure 18.2 **Elephant skull.** Sagittal cross section of the skull of an African elephant, showing the extensive network of pneumatic cavities that help reduce the weight. Note the open-rooted pulp cavity of the tusk. The tusks grow throughout the life of the animal.

as radiators to dissipate heat. Behavioral characteristics such as seeking shade and reducing activity during the middle of the day also help reduce heat load.

The trunk is the most recognizable feature of elephants and gives rise to the ordinal name. It is an elongated, flexible, muscular upper lip and nose, with the nasal canal throughout its length. The trunk has numerous uses. Because an elephant's head is so heavy, its neck is very short; thus, the animal cannot touch the ground with its mouth. The trunk is used to grasp food from the ground as well as from tall trees. It is extremely strong, yet the tip is sensitive enough to pick up small objects (such as peanuts in a zoo). Elephants drink by sucking up to 4 liters of water into their trunk at a time and squirting it into the mouth. Adults may drink 100 liters of water a day. Water is also sprayed from the trunk over the body to keep the animal cool. Mud and dust also can be sucked into the trunk and sprayed on the body for this purpose, as well as to reduce insect infestations. Also,

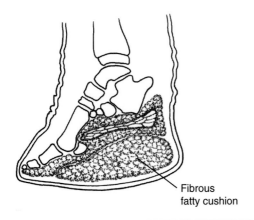

Figure 18.3 **Forefoot of an African elephant.** Although elephants are functionally plantigrade, they actually are digitigrade, with the bones of the feet, and the massive weight of the animal, cushioned by fibrous, fatty connective tissue that uniformly distributes the weight over a broader area.

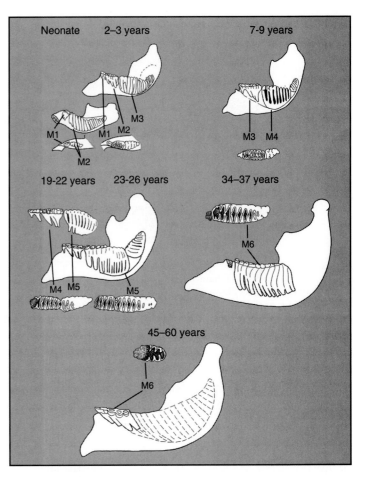

Figure 18.4 **Proboscidean cheekteeth.** Unlike most other species of mammals, cheekteeth enter the jaw horizontally in elephants. Because all the molariform teeth are deciduous in elephants, M1–M3 are equivalent to premolars, while M4–M6 are equivalent to molars in other mammals. Progression of teeth is shown for the African elephant, with ages approximated, especially after 30 years of age. Note the larger size and greater number of laminar ridges in the posterior cheekteeth, M4–M6.

Source: Data from J. Kingdon, East African Mammals, *Vol. IIIB Large Mammals, 1979, University of Chicago Press.*

elephants have an excellent sense of smell, and the trunk may be held upright in the air like a periscope to gain olfactory information from the surroundings.

The dental formula is 1/0, 0/0, 3/3, 3/3 = 28, and the teeth are highly specialized. The remaining upper incisor in extant elephants is the second (I^2) and forms the characteristic tusk. The deciduous I^2 is replaced by the permanent tooth between 6 and 12 months of age, when it is only about 5 cm long. Tusks (often collected by poachers as "ivory") are composed of dentine and calcium salts, with enamel only on the terminal portion. They are open-rooted (see figure 18.2) and grow throughout life. Tusks are largest in African bull elephants, reaching up to 3.5 m in length and 200 kg in weight in very old adults; 100 to 120 kg is more common. In female Asian elephants, the tusks are smaller and may not extend beyond the lower lip. Tusks are used in foraging, defense, and social displays.

The cheekteeth are large and hypsodont, with transverse **laminae** (ridges). These ridges are composed of dentine overlaid with enamel. Cementum occurs between the ridges. Posterior cheekteeth (equivalent to molars) are larger than those anterior (equivalent to premolars) and have a greater number of ridges (figure 18.4). Asian elephants have more laminae than African elephants. Because the mandible is short and the cheekteeth long, only one upper and one lower molar (or parts of two) are active in each jaw at a time. Replacement of cheekteeth is horizontal from the back of the jaw. The new tooth moves forward as the worn anterior tooth is pushed out. The third and final molars begin to come in by about 30 years of age and last for the remainder of an individual's life (see figure 18.4).

Fossil History

This order was widespread throughout most of the Cenozoic era. Elephants occurred not only in Africa and Asia as today, but throughout the Pleistocene epoch in Europe and North America (figure 18.5) and even reached South America. The earliest fossil evidence is from the early Eocene epoch of Africa (Mahboubi et al. 1984), and extensive fossil remains exist of proboscideans throughout the Eocene epoch, with several families recognized. Two of these extinct families, the Moeritheriidae and the Deinotheriidae, diverged early and may have "tenuous connections" with proboscideans (Sikes 1971). The moeritheriids (figure 18.6A), known from the Eocene and Oligocene epochs of northern Africa, were only about a meter in height and probably amphibious.

Deinotheriids occurred in Asia and Europe from the late Miocene to Pliocene epochs. Referred to as "hoe-tuskers," they were elephant-size with large, downward-curving tusks in the lower jaw (figure 18.6B). Three other extinct families have much closer affinities to today's elephants (Carroll 1988). The Gomphotheriidae, which were contemporary with moeritheriids and deinotheriids; the Mammutidae (mastodons) from the early Miocene epoch; and the Stegodontidae from the mid-Miocene epoch all had the large body size and many of the characteristic specializations of elephants today. Gomphotheriids had a pair of tusks in both the upper and lower jaws. Mastodons probably survived until about 8000 years ago. The North American mastodon (*Mammut americanum*) was contemporary with the arrival of humans on the continent. Skeletal material of mastodons was first collected in North America from the Hudson River in 1705 (Sikes 1971). Although members of the Stegodontidae survived until the late Pleistocene epoch, only one proboscidean family remains extant.

The family that survives today, the Elephantidae, is recognizable from the late Miocene Genus *Stegotetrabelodon*. The extinct Genus *Primelephas*, from the late Miocene-early Pliocene epochs, gave rise to the two genera still extant today, as well as the extinct Genus *Mammuthus*, the mammoths that were contemporary with early humans. The woolly mammoth (*M. primigenius;* figure 18.6D) was among the subjects of cave paintings by Paleolithic humans (figure 18.7), who were no doubt a factor in its extinction. Complete specimens of woolly mammoths have been found frozen in Siberian ice.

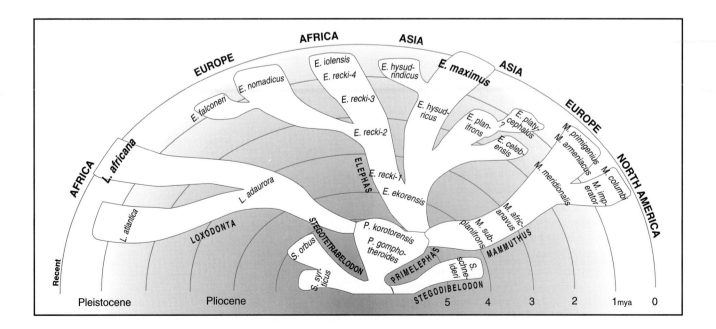

Figure 18.5   Phylogeny of the Family Elephantidae.   Note the occurrence of the Genus *Elephas,* restricted to Asia today, in Europe and Africa during the late Pleistocene epoch.

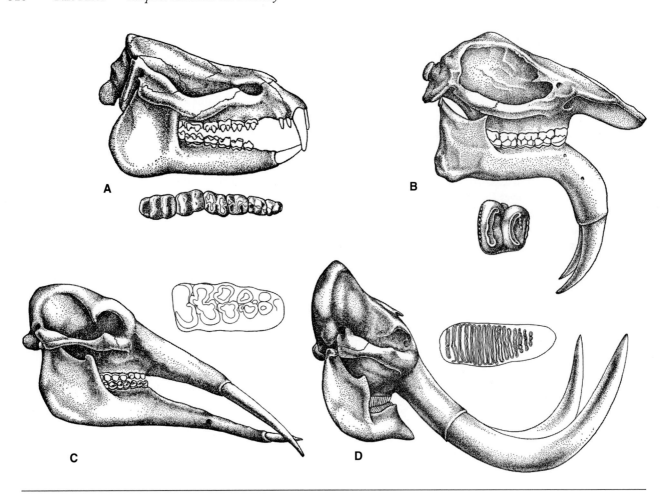

Figure 18.6   **Fossil skulls of early proboscideans.**   (A) The late Eocene *Moeritherium,* actual length about 33 cm, and (B) the Miocene *Deinotherium,* actual length about 1.2 m. (C) *Gomphotherium,* actual length about 1 m, occurred in the Miocene epoch. (D) The actual length of the woolly mammoth skull is about 2.7 m. Note the changes in the molar cusp patterns of the early proboscideans, with the occlusal surface of mammoths being very similar to modern elephants.

Economics and Conservation

The involvement of elephants in human culture, religious tradition, and history is extensive. Humans hunted mammoths as long as 70,000 years ago (Owen-Smith 1988). We are all aware of the central importance of elephants in circuses and as zoo animals since ancient times and of their role in the famous march of Hannibal over the Alps in 218 BC. Asian elephants have been important as draft animals throughout many parts of Asia for over 5000 years, although they can be considered "exploited captives," as opposed to domesticated species (see chapter 28). Currently, the range of both species is much reduced (Cumming et al. 1991). The Asian elephant is considered endangered, and the African elephant is listed as vulnerable.

Both African and Asian elephants remain important in terms of conservation efforts throughout their range. Population declines have been attributed to drought, loss of habitat associated with increased human population growth and desertification, and poaching for meat and ivory. When prices for ivory increased significantly during the 1970s and 1980s to over $100 per kg (Douglas-Hamilton 1987),

poaching escalated. At the same time, several African countries must cull elephants from areas where their population densities are too high relative to limited forage resources. The animals may destroy crops, increase soil erosion, and depress the resource base for other species (Owen-Smith 1988). Countries that cull elephants may depend on the legal harvest and sale of ivory to fund their wildlife management programs. Problems arise because legally harvested ivory cannot be distinguished from poached ivory. The argument has been made that legal trade promotes more illegal poaching by providing a market. Recent developments in molecular genetics, such as DNA "fingerprinting" (see chapter 3), may allow investigators to pinpoint the source population of ivory and help alleviate this problem in the future.

HYRACOIDEA

Referred to in the Bible as "rock badgers," the hyraxes are composed of a single family, Procaviidae, with 3 genera and 11 species. The family name means "before the caviids (guinea pigs)" and points up the taxonomic confusion that

Figure 18.7 **Paleolithic sketch.** This drawinig of a woolly mammoth is from the cave wall of Les Combarelles aux Eyzies, France. Not actual size.

has surrounded hyraxes. Because of their superficial resemblance to rodents (figure 18.8), hyraxes were initially grouped with guinea pigs by taxonomists. Even the common name "hyrax" is unfortunate because it means "shrew mouse." Neither of these associations is accurate, because based on fossil evidence, hyracoids historically have been considered to be most closely related to the other "subungulates," elephants and manatees (although see the section in this chapter on the fossil history of hyraxes).

Hyraxes are distributed in central and southern Africa, Algeria, Libya, Egypt, and parts of the Middle East, including Israel, Syria, and southern Saudia Arabia. The five species of rock hyraxes in the Genus *Procavia* are the most widely distributed geographically and elevationally. They are found in rocky outcrops from sea level to 4200 m elevation in Africa and the Middle East. The three species of bush hyraxes (Genus *Heterohyrax*) are found in similar rocky habi-

tats in East Africa. The three species of arboreal tree hyraxes (Genus *Dendrohyrax*) inhabit forested areas of Africa up to 3600 m elevation (Kingdon 1974). The terrestrial species are diurnal or crepuscular and form large colonies. Conversely, tree hyraxes are nocturnal and solitary.

All hyraxes are herbivorous and feed on a variety of vegetation. Grasses make up a large part of the diet of rock hyraxes; their hypsodont dentition grinds this abrasive material. Tree and bush hyraxes have more brachyodont dentition because they consume less abrasive vegetation. Although they do not ruminate, hyraxes have a unique digestive system involving one large cecum and a pair of ceca on the ascending colon (figure 18.9).

Colony size varies according to species, with rock hyraxes maintaining group sizes up to about 25 and bush hyraxes up to 35. Hyraxes are poor thermoregulators. Individuals in a colony may huddle together to help conserve heat and maintain body temperature. As might be expected in colonial species, hyraxes are very vocal and make a variety of different types of sounds, including whistles, screams, croaks, and chatter. The bush hyrax (*Heterohyrax brucei*) and Johnston's hyrax (*Procavia johnstoni*) are rather unusual among mammals in that they are very closely associated species. They share common burrows at night, huddle together in the morning (see figure 18.8), urinate and defecate at the same sites, and juveniles play together. They have different foraging strategies, however, and do not interbreed.

Reproductive maturity in both sexes occurs at about 16 months of age. Females come into estrus once a year. Gestation is a relatively long 8 months, with litter size ranging from one to four. Neonates are precocial, with most

Figure 18.8 **Hyrax characteristics.** Hyraxes superficially resemble rodents but are not closely related to them. Two different species of hyraxes are often found in close association with one another. Coexistence is facilitated by different feeding strategies.

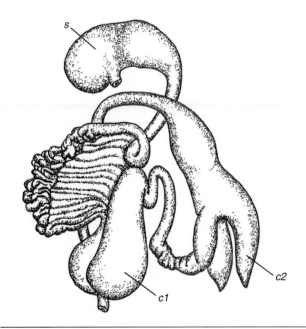

Figure 18.9 **Unusual alimentary tract.** The alimentary tract of a hyrax includes the stomach (*s*), a large cecum at the beginning of the large intestine (*c1*), and a second, paired cecum (*c2*) at the end of the large intestine.

births occurring during the wet season. Young males disperse from their natal area between the ages of 16 to 30 months and try to establish their own breeding territories.

Morphology

Hyraxes are short, with compact bodies and very short tails. Pelage color is brown, gray, or brownish yellow. Total length varies from 32 to 60 cm and body mass from 1 to 5 kg, with rock hyraxes being larger than bush hyraxes. There is no sexual dimorphism. Hyraxes from warm, arid regions have shorter, less dense pelage than tree hyraxes and those from high-elevation alpine areas. Hyraxes have a prominent middorsal gland surrounded by lighter colored hair (figure 18.10). The gland varies in size among species, being most noticeable in the western tree hyrax (*Dendrohyrax dorsalis*) and least so in the Cape rock hyrax (*Procavia capensis*).

Because hyraxes inhabit rocky cliffs or move through trees, good traction for climbing and jumping has obvious adaptive value. This is achieved through specialized pads on the soles of the feet. These pads are kept moist by secretory glands that make the feet similar to suction cups. The toes have short, hooflike nails except for the second digit on the hind feet, which has a claw used for grooming (figure 18.11).

Unlike in elephants and manatees, the molariform dentition is not replaced horizontally. The dental formula for the permanent dentition is 1/2, 0/0, 4/4, 3/3 = 34. Deciduous canines may be retained in rare cases, but there is usually a diastema between the incisors and cheekteeth (figure 18.12A). The upper incisors are long, pointed, triangular in cross section, and have a gap between them. They are ever-growing and stay sharp because the posterior sides do not have enamel. Some species have hypsodont cheekteeth, but in others, they are brachyodont.

Fossil History

Hyracoids first appear in the upper Eocene epoch of North Africa. Several genera, some as large as modern tapirs, are placed in the extinct Family Pliohyracidae (figure 18.12B). The first specimens attributable to the modern Procaviidae are from the Miocene epoch (Carroll 1988). By this time, only smaller forms of hyraxes survived. Larger species may have died out because they could not compete with ungulates. Although hyraxes generally are considered to be descended from the early Condylarthra, Sudre (1979) has questioned their relationship with proboscideans and sirenians.

Economics and Conservation

Tree hyraxes are hunted for both their meat and fur (from which blankets may be made), and all tree hyraxes are affected by the loss of forest habitat. None currently is considered to be endangered, however, and in South Africa, they may be an agricultural pest. Because hyraxes in a colony

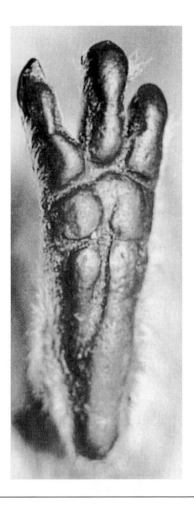

Figure 18.11   Hyracoid digits.   The first and fifth toes are absent on the hind feet of a hyrax. Their toes have hooflike nails, except for the second digit on the hind feet (pes). These have a claw that is used for grooming the fur. The soles have glandular pads that make them moist and increase adhesion to steep, rocky inclines.

Figure 18.10   Middorsal gland.   A Cape rock hyrax in typical habitat showing the prominent white fur surrounding the middorsal gland.

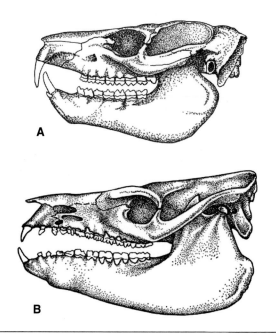

Figure 18.12   Hyracoid skulls.   (A) The skull of a modern tree hyrax, with a well-defined diastema. Actual length about 12 cm. This can be contrasted with (B) an early Oligocene hyracoid *Megalohyrax*. Actual length about 30 cm.

defecate and urinate in a traditional place, massive caked deposits form on the rocks. This material is used by natives as a decorative dye and by Europeans as a fixative in perfumes (Kowalski 1976).

SIRENIA

This order is represented by two families: the Dugongidae, with one extant species, the dugong (*Dugong dugon*), and the Trichechidae, with three species of manatees in the Genus *Trichechus*. Like whales, sirenians never leave the water. Unlike whales, however, they are strictly herbivorous and represent the only mammalian marine herbivores. They inhabit coastal areas, estuaries, bays, and inland river systems in tropical regions, feeding on submerged and emergent vegetation. Dugongs feed on softer, less abrasive vegetation than manatees, however. Sirenian distribution is restricted to (1) relatively shallow coastal areas because the plants they depend on require sunlight and (2) tropical regions with water temperatures around 20°C because of their low metabolic rates and poor thermoregulatory abilities. In contrast to terrestrial herbivores, sirenians have limited competition from other mammals within their feeding niche of shallow-water vegetation. The four extant species represent vestiges of an order that was abundant and diverse during the Tertiary period, with fossil remains of close to 20 genera known. The ordinal name is derived from the sirens, or sea nymphs, of mythology, and dugongs and manatees may have been the basis for the myth of mermaids.

Morphology

Sirenians exhibit many of the same adaptations for life in the water as do cetaceans. They are large with a fusiform body shape (figure 18.13) and are devoid of fur except for very short, stiff bristles around the snout. There is no external ear, the nostrils are valvular and located on the top of the rostrum, the lips and snout are very flexible, and the tail is horizontally flattened. The forelimbs are paddlelike, and no external hind limbs are present. Only small, paired vestigial bones, suspended in muscle, represent the remains of the pelvis. The rostrum and lower jaw are deflected downward, especially in the dugong, to facilitate bottom-feeding. The skeletal bones are very dense and massive (**pachyostotic**), an adaptation to increase body mass and overcome the buoyant effects of living in shallow water habitats. The lungs are long and thin, extending for much of the length of the body cavity (figure 18.14). This helps to evenly distribute the bouyant effects of the air the animals breath. As in elephants, the cheekteeth are replaced horizontally from the back of the jaw.

Fossil History

The first fossils recognizable as sirenians, the Genus *Prorastomus*, are from the early Eocene epoch, by which time aquatic adaptations were well underway. Middle and late Eocene remains of the Genus *Protosiren* occur in India, Europe, and North America. Eocene fossil remains of sirenians are noteworthy in the retention of a fifth premolar. This is unique among Tertiary eutherian mammals (Carroll 1988), as sirenians were the last eutherians to retain five premolars (Domning et al. 1982). Better fossil evidence exists for the dugongids than for the trichechids. Although dugongids are represented as early as the mid-Eocene epoch by *Eotheroides*, there is no direct fossil evidence for the Genus *Dugong*. The early Miocene *Dusisiren* (figure 18.15) is the earliest known ancestor of the recently extinct sea cow, Genus *Hydrodamalis* (see the section in this chapter on Dugongidae). Fossil remains of trichechids date from the middle to late Miocene Genus *Potamosiren*. As noted, Tertiary sirenians were abundant, diverse, and widespread and reached their peak diversity in the Miocene epoch. However, as noted by Domning (1978), this diversity was somewhat confined by the lack of diversity in available marine feeding niches and of effective geographic barriers.

Economics and Conservation

There is reason for concern today about the immediate future of dugongs and manatees. Historically, hunting for meat, bones, hide, and fat has caused severe population reductions. Such reductions are not easily overcome by the very slow reproductive rate of this group. All four species are considered vulnerable, and in the United States, all are protected by the Marine Mammal and the Endangered Species

Figure 18.13 **Sirenian sizes.** The relative sizes of a (A) recently extinct Steller's sea cow, (B) manatee, and (C) dugong are shown in relation to a 6-foot-tall person. Sirenians have a fusiform body shape, forelimbs modified into flippers, and no hind limbs. The most noticeable external differences between manatees and the dugong are in the shape of the tail and the extreme downward deflection of the rostrum in dugongs.

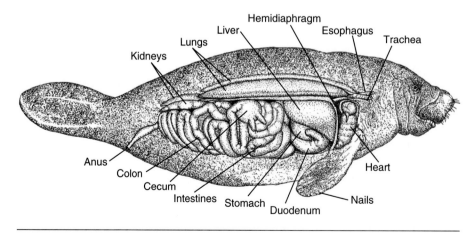

Figure 18.14 **Internal anatomy of a manatee.** Note the lungs lying in a horizontal position along the back.

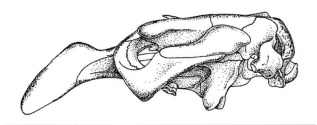

Figure 18.15 **Sea cow ancestor.** Lateral view of the cranium of *Dusisiren,* an early Miocene ancestor of the recently extinct sea cow. Note the strongly deflected rostrum, characteristic of dugongs today.

acts. In other portions of their range, however, subsistence hunting continues. In Florida, the West Indian manatee (*Trichechus manatus*) has been protected by the state since 1893. But animals are still lost because of poaching, being pinned by flood control gates and drowning, and being struck accidentally by boat propellers (figure 18.16). Loss of habitat to development presents a continuing threat to pop-

Figure 18.16 **Boat propeller wounds.** Contact with boat propellers is one of the primary mortality factors affecting manatee populations in Florida. Most manatees show evidence of scarring from propellers, with wounds such as these most often being fatal.

ulations. Because of their feeding habits, Amazonian manatees in South America have been used successfully to clear aquatic vegetation from inland canals (Allsopp 1960; although see Etheridge et al. 1985).

Dugongidae

Dugongs are found in coastal areas of the Pacific Ocean throughout Micronesia, New Guinea and northern Australia, the Philippines, and Indonesia northward to Vietnam. In the Indian Ocean, they occur around Sri Lanka and India, and from the Red Sea south along the east coast of Africa to Mozambique. They attain a maximum body length of 4 m. Average mass is 420 kg, with a maximum of about 900 kg. Unlike in manatees, the tail fluke is notched (see figure 18.13), and the rostrum is much more strongly deflected downward (figure 18.17).

Although technically the adult dental formula is 2/3, 0/1, 3/3, 3/3, the anterior pair of upper incisors, all the lower incisors, and the canine are represented only by **vestigial** (remnant) alveoli. Thus, the dentition actually seen is 1/0, 0/0, 3/3, 3/3. The upper incisor in males forms a short, thick tusk, but in females it does not erupt. The molars are cylindrical and cement-covered and move in horizontally from the back of the jaw as anterior teeth are worn away. In old adults, only one or two remain. These are open-rooted and grow throughout the life of the individual.

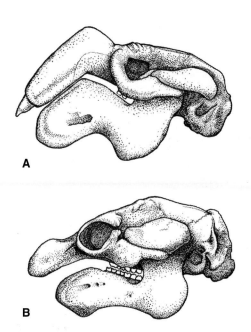

Figure 18.17 **Sirenian skulls.** Skull of a (A) dugong, showing the reduced dentition and strongly deflected rostrum, actually a greatly enlarged premaxilla bone. Note also the anterior portion of the lower jaw (mandibular symphysis) is also strongly deflected downward. Both are adaptations for bottom-feeding. (B) A West African manatee skull shows the greater number of cheekteeth than in dugongs. Teeth are replaced from the back of the jaws as they slowly move forward and anterior teeth are worn out and lost.

Dugongs avoid freshwater much more than manatees. They generally are solitary or live in small groups. They forage by using their forelimbs to walk along the bottom, feeding in the substrate and pulling up plants with the deflected rostrum. Dugongs can remain submerged up to 10 minutes while feeding (Kingdon 1979). Unlike manatees, they are restricted to the less abrasive seagrasses in the Families Hydrocharitaceae and Potamogetonaceae that have less silica content than true grasses (Gramineae). Sexual maturity is attained by both sexes between 7 and 14 years of age. A single young is born following a 12-month gestation.

The recently extinct Steller's sea cow (*Hydrodamalis gigas*) was much larger than the dugong or manatees (see figure 18.13). Total length was about 7.5 m, greatest circumference was over 6 m, and body mass was about 5000 kg, which was five to six times greater than sirenians today. This great size was an adaptation for the cold North Pacific waters they inhabited around the Commander Islands and Bering Island, where the species was first discovered in 1741. By then, its range had been greatly reduced, with a remnant population of 1000 to 2000. Sea cows had no teeth, but instead used rough plates in the mouth to forage on kelp. Slow moving in shallow water with few natural predators, they were easy prey for sailors seeking fresh meat and hides. Relentless slaughter followed their discovery, and Steller's sea cow was extinct by about 1768—only 27 years after its discovery.

Trichechidae

The West Indian manatee occurs from Florida south through the Caribbean Sea to northeast Brazil. Individuals have been reported as far north as New Jersey (Odell 1982). The Amazonian manatee is found throughout the Amazon river basin of South America, but it does not tolerate saltwater. The West African manatee (*Trichechus senegalensis*) is distributed in fresh- or saltwater rivers and estuaries, and along the west coast of Africa from Senegal to Angola. Average body length is from 2.5 to 4 m, with body mass typically between 150 and 360 kg. Maximum body mass may approach 1600 kg.

Unlike dugongs, manatees have a rounded, spatulate tail (see figure 18.13). Manatees have small nasal bones, unlike dugongs. They are also unusual among mammals in having six cervical vertebrae instead of seven. Adult dentition includes only cheekteeth; as in elephants, these are replaced consecutively from the rear of the jaw as anterior molars are worn down and lost. Molars are brachydont, close-rooted, and have enamel. Unlike elephants, manatees have four or five teeth in place at a time on each side of the jaws. Again, in contrast to elephants, an indefinite number of molars, between 10 and 30, resulting from indefinite tooth germ formation, may move through each jaw quadrant during an individual manatee's lifetime. Teeth move forward as the bony interalveolar septa between them are constantly resorbed and redeposited. This "functional polyphyodonty" is an adaptation to the aquatic grasses and other abrasive, coarse, submerged vascular plants consumed by manatees, often in sand and mud, that quickly wear down teeth.

Manatees are usually seen as solitary individuals, paired, or in small groups. Their social ecology and behavior are discussed by Hartman (1979). In Florida, larger groups often form in the winter as the animals congregate around warm-water discharge sites, but the primary social unit is a cow and her calf. Manatees are sexually mature between 8 and 10 years of age. A single calf is born after a gestation of about 13 months. Calves are weaned at 12 to 18 months of age. Thus, the reproductive potential of populations is limited, and losses to boats and other causes may not be easily overcome.

Summary

These three seemingly disparate orders arose from a common terrestrial ancestor and began to diverge by the early Eocene epoch. Yet they continue to share several skeletal and anatomical features. They have a large cecum where vegetation is broken down by symbiotic microorganisms. Elephants, dugongs and manatees, and hyraxes all are hind-gut fermenters. In this regard, their reduced digestive efficiency is similar to that of the perissodactyls. Similarities of anatomy, locomotion, dentition, and behavior in elephants and sirenians relate to size and their long life span. For example, replacement patterns of the cheekteeth in the elephants and the manatees are similar. Teeth are replaced horizontally from the back of the jaw, moving forward slowly as older, anterior teeth are worn out and lost. This "functional polyphyodonty" is an adaptation for herbivory in these large, long-lived animals. The two groups differ in that elephants have only one or two molariform teeth that are functional at any one time. Manatees have several functional cheekteeth active at a time and produce an indefinite number of cheekteeth throughout their life. Elephants are terrestrial and have a graviportal structure to accommodate their bulk. Sirenians have the bouyancy of their water environment to help accommodate their large size. Proboscideans and sirenians are also similar in that they were much more diverse and widespread throughout the Tertiary period than they are today, with numerous fossil genera recognized. Today, the survival of elephants and sirenians is a concern as populations continue to decline in most areas. This is especially noteworthy for sirenians, as there are few other competitors for their feeding niche of shallow-water vegetation. Hyracoids are much smaller than either elephants or sirenians, and population densities appear to be secure.

Discussion Questions

1. Why do elephants use very low frequency sound, whereas bats use very high frequency sound? How might body size be a factor?

2. If today's "environmental ethic" and concern for endangered species were in vogue 250 years ago, do you think the Steller's sea cow would have gone extinct? Should we be concerned today for species with only remnant populations that may be "on their way out" anyway from an evolutionary standpoint?

3. What factors contribute to the low reproductive rate of both proboscideans and sirenians? How does this affect conservation efforts?

4. How does body size influence composition and group size in elephants? How about for hyraxes? What differences are evident?

5. What are the possible benefits to each species of the unusual association between the bush hyrax and Johnston's hyrax? Can you think of potential drawbacks?

6. Contrast the differences in the dentition of dugongs and manatees. What do these differences in families within the same order suggest about the evolutionary process in general?

7. What are the ultimate constraints on the upper limit of body size in terrestrial mammals? Think in terms of surface area-to-mass ratio and skeletal support.

Suggested Readings

Eltringham, S. K. 1991. The illustrated encyclopedia of elephants: from their origins and evolution to their ceremonial and working relationship with man. Salamandar Books, London.

Haynes, G. 1993. Mammoths, mastodonts, and elephants: biology, behavior, and the fossil record. Cambridge Univ. Press, New York.

Maglio, V. J. 1973. Origin and evolution of the Elephantidae. Transactions of the American Philosophical Society. 63:1–149.

Maglio, V. J. and H. B. S. Cooke. (eds.). 1978. Evolution of African mammals. Harvard Univ. Press, Cambridge, MA.

Osborn, H. F. 1942. Proboscidea (2 vols). American Museum of Natural History, American Museum Press, New York.

CHAPTER 19

Perissodactyla and Artiodactyla

The perissodactyls and artiodactyls are the modern ungulates, which are generally large, hoofed, terrestrial herbivores. The unifying characteristic of both orders is the structure of the limbs (figure 19.1). The term *ungulate* refers to mammals that walk on the tips of their toes, which end in thick, hard, keratinized hoofs. Ungulates often have a reduced number of toes and a lengthened foot such that the **calcaneum** (heel bone) does not articulate with the fibula. The limbs are restricted to movement in a single plane. Thus, ungulates are adapted for cursorial (running) locomotion. Although dentition varies among families, cheekteeth often are hypsodont, with complex surfaces. These and other morphological and life history characteristics of ungulates adapt many of them for existence on large, open expanses of land where they must be able to feed efficiently and outrun potential predators. The distribution and adaptations of large herbivores are dictated to a large extent by their forage resources. Conversely, many of the structural characteristics of forage plants are the result of coevolutionary pressures from large herbivores.

The two orders in this chapter include a great deal of structural diversity among taxa, the result of a rich, well-documented evolutionary history for both orders. The perissodactyls ("odd-toed" ungulates) today are a small order in terms of the number of extant species—a remnant of a group that flourished during the early- to mid-Tertiary period. In contrast, modern artiodactyls ("even-toed" ungulates) encompass a diverse array of species. Today, ungulates are the most important group of mammals in terms of human commerce and economics. Many ungulate species have been translocated from their native areas and introduced throughout the world, where they are important as domesticated animals (chapter 28), for sport hunting, and for ecotourism. Many other species, however, are on the verge of extinction because of illegal poaching and habitat destruction.

PERISSODACTYLA

The three families in this order are diverse in terms of their method of locomotion, life history, and morphology and initially may appear to have little in common. The Family Equidae includes the horses, zebras, and asses. The other two families are the more closely related Tapiridae (tapirs) and Rhinocerotidae (rhinoceroses). All perissodactyls are large, terrestrial herbivores. These hind-gut fermenters feed on fibrous vegetation that is often of poor quality. Like the artiodactyls, they share a common morphological feature—foot structure—that defines the order.

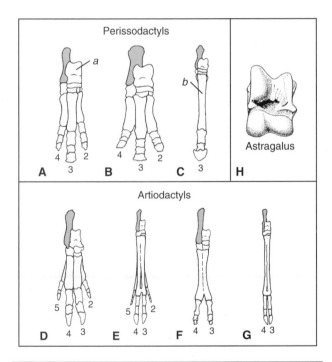

Figure 19.1 **Hind foot structure of perissodactyls and artiodactyls.** The hind feet of perissodactyls, including the (A) tapir, (B) rhinoceros, and (C) horse. In these "odd-toed" ungulates, the main axis of the limb passes through the enlarged third digit (i.e., they are mesaxonic). In equids, the third digit is the only one remaining (also see figure 19.4). Representative artiodactyls include the (D) pig, (E) deer, (F) camel, and (G) pronghorn. In these "even-toed" ungulates, the axis of the limb passes between the third and fourth digits (i.e., they are paraxonic). Note the vestigial second and fifth digits remain as "dew claws" in the pig and deer. For all examples, the heel bone (calcaneum) is shaded and articulates with the astragalus (*a*). (H) The grooved, pulleylike anterior surface of the astragalus in the caribou limits motion of the foot to a single plane. Note the fused metapodials (cannon bone, *b*) in C, E, F, and G.

Morphology

The ordinal name means "odd-toed" and refers to the main weight-bearing axis of each limb passing through the enlarged third digit, a condition called **mesaxonic.** Tapirs have four digits on the forefeet and three on the hind feet, whereas rhinos have three digits on all feet. The third digit is the only one remaining in equids (see figure 19.1). Perissodactyls have a deep groove in the proximal surface of the **astragalus** (ankle bone), which creates a pulleylike surface that limits the limbs to forward-backward movement. The skull in all species of perissodactyls is elongated by the lengthening of the rostrum. The cheekteeth are hypsodont and usually lophodont, adaptations that occur in large grazers to enable the grinding of vegetation. Three upper incisors are retained in the equids and tapirids. Upper incisors are reduced in number or absent in the rhinos. Perissodactyls have a simple stomach, but there is an enlarged cecum at the junction of the small and large intestines, where the majority of the microorganism-aided breakdown of cellulose occurs. Food passes through the digestive system of a perissodactyl about twice as fast as through that of a ruminating artiodactyl. Because food is retained for less time, digestion is less efficient. For example, the digestive efficiency of a horse is only about 70% that of a cow. Perissodactyls compensate for reduced efficiency by consuming more food per unit of body mass. The enlarged cecum and colon provide storage and surface area for absorption of nutrients (see figure 6.9). Perissodactyls have a bicornuate uterus, diffuse placentation, and no baculum.

Fossil History

The perissodactyls (and artiodactyls) originated from the Condylarthra, the dominant mammalian herbivores of the early Paleocene (about 65 million years ago). Condylarths are considered to be the ancestors of many of the other lineages of large mammals, including proboscideans, cetaceans, and sirenians. The oldest identifiable perissodactyl fossils are from the early Eocene epoch (50 mya). By this time several lines of radiation are evident, and about 14 families are recognized (figure 19.2). During this period, perissodactyls far outnumbered the smaller, less diverse artiodactyls. By the end of the Oligocene epoch (25 mya), however, 8 of the 14 families of perissodactyls were extinct. By the early Miocene epoch, only the equids, tapirids, rhinocerotids, and the Chalicotheriidae remained. This last family included unusual ungulates with large forelimbs and short hind limbs adapted for standing semierect to feed on tall trees (Coombs 1983). Several genera had large, retractable claws instead of hoofs (figure 19.3).

Tapirs are among the more primitive extant large mammals, based on their four toes on the forefeet and brachyodont cheekteeth (Dawson and Krishtalka 1984). Originating in the early Eocene epoch of North America, tapirs migrated both north into Asia and south into Central and South America. Tapirs were extirpated throughout most of North America by the late Pleistocene epoch. As was the case for camels, the combination of migration and extirpa-

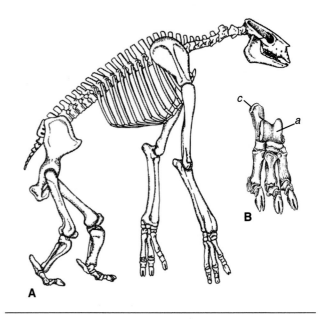

Figure 19.3 **Example of a recently extinct perissodactyl.** Chalicotheriids remained extant in Asia and Africa until the late Pleistocene epoch. (A) Skeleton of the large *Chalicotherium* from the Miocene epoch of Europe and (B) the clawed hind foot of *Moropus*. Note the calcaneum (*c*) and astragalus (*a*).

tion resulted in a discontinuous distribution today. The current Genus *Tapirus* dates from 20 mya in the Miocene epoch and has changed little since then.

Fossil evidence of the rhinocerotids dates from the late Eocene epoch in Asia and North America. Most of the genera extant today date from the Miocene epoch (10–25 mya). They were extinct in North America by the end of the Pliocene epoch (2 mya), however, and never dispersed to South America. Rhinocerotids were abundant and widespread in the Old World until the late Pleistocene epoch (about 60,000 years ago). The largest land mammal that ever lived was a rhinocerotid. *Indricotherium transouralicum* [*Baluchitherium grangeri*] was at least 5 m high at the shoulder. Although many estimates suggest the maximum body mass of *Indricotherium* was 30,000 kg or more, an upper estimate of 15,000 to 20,000 kg probably is more reliable (Economos 1981; Fortelius and Kappelman 1993).

The fossil history of equids is one of the best documented for any mammalian family. This history shows increasing body size and skull proportions, increasing size and complexity of the cheekteeth, and reduction in the number of digits (figure 19.4). Keep in mind, however, that the evolution of the horse was not a ladderlike, directed, progressive process, as suggested by figure 19.4, but a complex radiation of numerous divergent, overlapping lineages (figure 19.5). Equids passed most of their evolutionary history in North America, with migration to the Old World during the

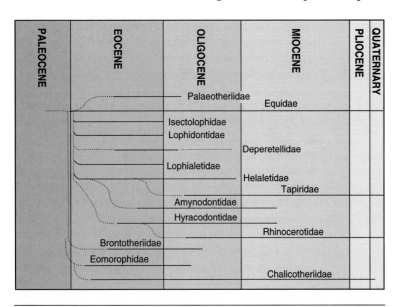

Figure 19.2 **Geologic ranges of the Perissodactyla, including modern and extinct families.** Several families of perissodactyls flourished in the early Eocene epoch, but by the middle Miocene epoch, only four families remained. Today, only three families are extant, which include only six genera.

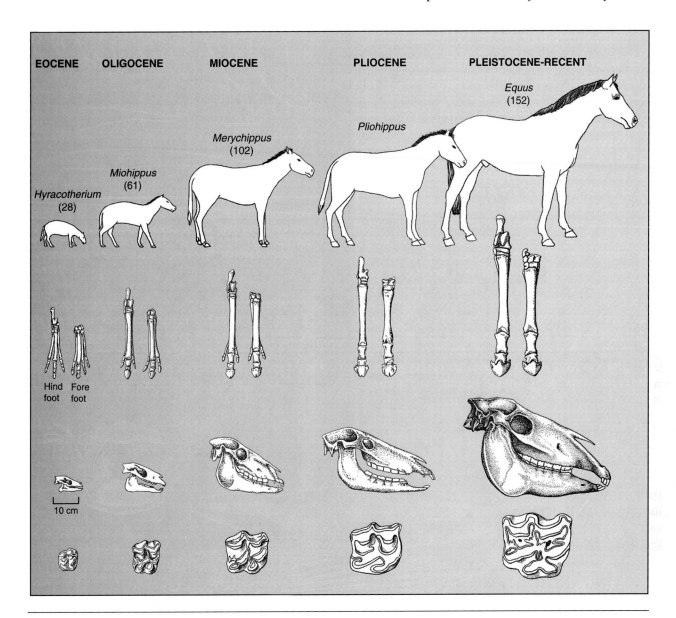

Figure 19.4   Evolution of the horse.   The early Eocene *Hyracotherium* was a small, forest-dwelling browser. The forefeet had four toes, the skull had confluent temporal and orbital openings (scale bar equals about 10 cm), and the brachyodont cheekteeth had simple enamel patterns. Note the progressive increase in size (shoulder heights in centimeters are in parentheses), with reduction in the number of digits, changed proportions of the skull with formation of a postorbital bar, and increasing complexity of the occlusal surface of the cheekteeth in the grazing Genus *Equus* that now inhabits open areas.

Miocene epoch and to Central and South America in the Pliocene and Pleistocene epochs. Equids left no Pleistocene descendants in the New World, however, becoming extinct about 10,000 years ago. They were reintroduced to the New World by the Spanish conquistador Hernando Cortés in 1519.

Economics and Conservation

As a domesticated species, horses probably are second only to cattle in their importance in the development of cultural and economic systems of humans. Domesticated in southern Ukraine about 5000 years ago (see chapter 28), horses have been introduced throughout most of the world and have been pivotal as an aid to travel, exploration, and warfare throughout history. In contrast to the cosmopolitan distribution of domestic horses through introductions, several other species of equids are either recently extirpated in the wild or endangered. Likewise, tapirs are much reduced throughout their ranges because of hunting pressure and destruction of habitat. The plight of rhinoceroses is well known; all species have become prime targets of poachers for their horns and other body parts. Details of endangered species and conservation efforts are discussed in each family account.

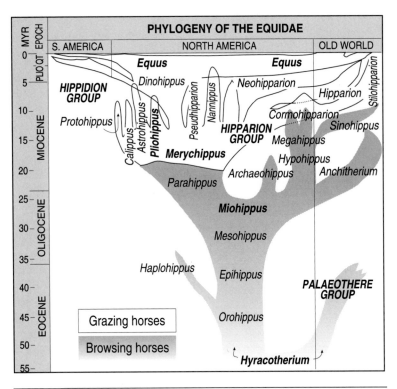

Figure 19.5   **Lineages of the horse.**   Phylogeny of the horse was complex, with overlapping and divergent adaptive radiations.

Families

Equidae

The number of extant equid species is open to question (Corbet 1978; Bennett 1980; Groves and Willoughby 1981), but Grubb (1993a) recognized eight species, all in the Genus *Equus*. All have long, slender limbs, and only the third digit remains functional. Three upper and lower incisors occur in each quadrant, and the cheekteeth are large, hypsodont, and have complex occlusal surfaces. Pelage color is variable in most equids, although the pattern of stripes in zebras depends on the species. Stripes are narrow and close together and extend down to the hooves in Grevy's zebra (*E. grevyi*). The mountain zebra (*E. zebra*) has broad stripes that do not extend to the ventral surface. The plains zebra (*E. burchellii*) has a variable pattern, generally with broad stripes posteriorly that become narrower anteriorly. The neck mane on most equids is erect; it falls to the side only in the domestic horse (*E. caballus*). The body mass of wild equids ranges from about 250 kg in the African ass (*E. asinus*) to 400 kg in Grevy's zebra. The body mass of domestic horses may reach 1000 kg.

The natural distribution of equids includes eastern Africa and central Asia from the Middle East to Mongolia. They inhabit short grasslands and desert scrublands and are never far from water. The basic social unit is the family group, generally 10 to 15 individuals made up of a highly territorial male, several females, and their offspring (Berger 1986). Young females leave the family group when they become sexually mature at about 2 years of age. Young males become sexually mature at the same age. They do not breed until they leave the family group and gain access to other females, which is by about 5 years of age. In the plains zebra, temporary aggregations of 100,000 individuals may form, depending on ecological conditions.

A single offspring is usual after a gestation period of about a year. Birth and subsequent mating 7 to 10 days later occur during the wet season, when vegetation is most abundant. Neonates are precocial. They begin to graze at about 1 month of age and are weaned at 8 to 13 months of age.

Several species of equids are endangered. These include Przewalski's horse (*Equus przewalskii*), although this may be a subspecies of the domestic horse (Bennett 1980), and the African ass, both of which probably are extinct in the wild. The onager, sometimes considered a distinct species (*E. onager*), is endangered, as is Grevy's zebra, whereas the mountain zebra and the Asiatic ass (*E. hemionus*) are threatened. The quagga (*E. quagga*) of South Africa, uniform in color on the posterior and striped on the anterior, became extinct in 1872.

Tapiridae

There are four species in the single Genus *Tapirus*. The family has a discontinuous distribution: Baird's tapir (*T. bairdii*) occurs in Mexico, Central America, and northern South America; and two other species, the Brazilian tapir (*T. terrestris*) and the mountain tapir (*T. pinchaque*), occur in northern South America. The Malayan tapir (*T. indicus*) occurs in Myanmar (Burma), Thailand, Malaya, and Sumatra. Tapirs have a chunky body with short legs and an elongated head with small eyes and ears (figure 19.6A). The nose and upper lip form a pronounced, flexible proboscis. Mean head and body length is 180 to 250 cm, and body mass reaches as much as 300 kg. Pelage color in New World species is a uniform reddish brown to gray or black. The Malayan tapir is white on the trunk of the body and black on the head, shoulders, and limbs. A short, bristly neck mane is characteristic of both Baird's and the Brazilian tapir; the hide is very tough in all species.

Tapirs inhabit heavily forested areas. The mountain tapir lives at elevations of 2000 to 4500 m; the other species range up to 1200 m. All tapirs are nocturnal and feed on understory shoots, twigs, fruit, grass, aquatic vegetation, and occasionally on cultivated crops. All but the mountain tapir are associated with swamps, rivers, or other wet areas. They are good swimmers and feed or seek refuge in water.

A

B

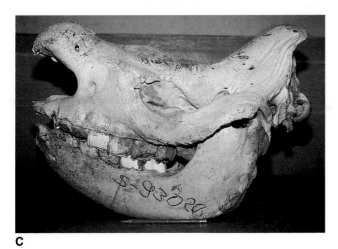

C

Figure 19.6   **Representative perissodactyls.**   (A) Malayan tapir and (B) black rhinoceros. (C) The rhino skull shows the typical elongated nasal bones without any horn core or attachment site for the horns. Scale: pencil = 16 cm.

Tapirs are generally solitary. Sexual maturity is reached at 3 to 4 years of age. Breeding occurs at any time during the year (Padilla and Dowler 1994). Usually, a single young is born after a gestation of about 395 days. Young have a reddish brown coat and are camouflaged with white spots and lines. They stay with their mother for 6 to 8 months, by which time the juvenile pelage is replaced with adult pelage. All species of tapirs suffer from loss of habitat due to logging, agriculture, and forest clearing and are declining in number and distribution. New World species are hunted for meat and hides. The Malayan tapir is considered endangered.

Rhinocerotidae

There are four genera in this family, with five living species. Rhinoceroses are well known for their large, heavyset, graviportal structure (see figure 19.6B). They have small eyes and a prehensile upper lip that extends past the lower lip in black (*Diceros bicornis*) and Asian rhinos. The upper lip is used to gather vegetation. The white rhino (*Ceratotherium simum*) reaches 400 cm at the shoulder, with maximum body mass of 1700 kg. Body mass of adult male Indian or one-horned rhinos (*Rhinoceros unicornis*) may be 2000 kg (Dinerstein 1991).

The family name refers to the rhino's horns, which have no bony core or keratinized sheath (figure 19.7A) but instead are a dermal mass of agglutinated, keratinized fibers (fused hairs). They are conical, often curve posteriorly, and may reach 175 cm in length in the white rhino. Asian rhinos have shorter horns. The anterior horn is positioned medially over the nasal bones. If two horns occur, the shorter, posterior one is over the frontal bones. Neither horn is attached to the bone, however, but to the skin over a roughened section of the skull bones. The nasal bones of the skull are large and project well above and anterior to the maxillae (see figure 19.6C).

Both white and black rhinos are found in sub-Saharan central and east Africa. The Indian rhino occurred in Pakistan and northern India, and the Javan rhino (*Rhinoceros sondaicus*) originally was in southeastern Asia from eastern India to Vietnam, Sumatra, and Java. The Sumatran, or hairy, rhino (*Dicerorhinus sumatrensis*) also originally was distributed throughout southeastern Asia, Sumatra, and Borneo. The geographic range of all species is greatly reduced to tropical and subtropical habitats due to human interference, poaching, and habitat destruction. Depending on the species, they occupy tropical rain forests, floodplains, grasslands, and scrublands. All are dependent on a permanent water supply for frequent drinking and bathing. Wallowing probably is necessary to help control body temperature (Owen-Smith 1975) and to reduce insect harrassment. Rhinos forage on woody or grassy vegetation and occasionally fruits but prefer leafy material when available.

Aside from mother-and-offspring pairs, rhinos generally are solitary. Small groups of immature individuals may form in Indian and white rhinos. Females become sexually

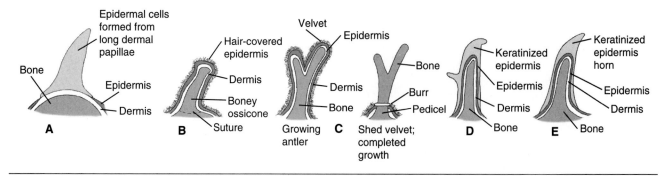

Figure 19.7   **Head ornamentation in five ungulate families.**   (A) Rhinocerotidae, a perissodactyl, and four families of artiodactyls: (B) Giraffidae; (C) Cervidae; (D) Antilocapridae; and (E) Bovidae.

mature at 5 years of age and bear their first calves when 6 to 8 years of age. Gestation is about 8 months in the Sumatran rhino and about 16 months in the other species. Births, usually a single calf, occur at intervals of 2 to 4 years (Dinerstein and Price 1991). Young nurse for 1 to 2 years, although *Ceratotherium simum* begins to eat solid food by 1 week of age. Males generally do not breed before 10 years of age (Ryder 1993).

Populations of all species have declined during the last 150 years. As is the case with many large mammals, the quantity of forage required and low reproductive rates mitigate against recovery if populations are reduced. All species are considered to be endangered, with the Asian species near extinction. Rhinos are illegally harvested for their horns (figure 19.8), which, with other body parts, are valued in traditional Asian medicine for supposed aphrodisiac and medicinal properties. Horns also have been used traditionally for making dagger handles in the Middle East. White rhinos have been successfully translocated to parts of their former range in southern Africa.

ARTIODACTYLA

Artiodactyls are much more selective feeders than perissodactyls, a factor in their greater adaptive radiation. In contrast to the 3 families and 17 extant species of perissodactyls, the artiodactyls include 10 living families, 80 genera, and approximately 220 species. Artiodactyls are distributed almost worldwide, either naturally or through introduction. As might be expected in such a large group, there is tremendous diversity in body size and structure, and three suborders are recognized. The Suborder Suiformes includes three families: the Suidae (pigs and warthogs), the Tayassuidae (peccaries), and the Hippopotamidae (hippopotamuses). The Suborder Tylopoda includes one family: the Camelidae (camels, llamas, and vicuña). Six extant families make up the Suborder Ruminantia: the Tragulidae (chevrotains, or mouse deer), the Giraffidae (giraffe and okapi), the Cervidae (deer), the Moschidae (musk deer), the Antilocapridae (pronghorn), and the Bovidae (antelope, bison, goats, sheep, etc.). All ruminants, with the exception of chevrotains and musk deer, have some type of head ornamentation in the form of horns or antlers. Despite the vast array of species, artiodactyls share a common morphological characteristic that defines the order.

Morphology

Like perissodactyls, the artiodactyls are defined by the structure of the foot. The main weight-bearing axis passes through the third and fouth digits, a condition termed **paraxonic.** The

Figure 19.8   **Poaching for horns.**   In the past, illegal rhino horn sold for tens of thousands of dollars a kilogram.

second and fifth digits are reduced and nonfunctional or absent. There is a definite trend toward cursorial locomotion in the more derived families. The Suiformes exhibit plantigrade locomotion, with unfused **metapodials** (metacarpals and metatarsals), in contrast to members of the Ruminantia, which are cursorial (specifically unguligrade, that is, walking on the tips of the toes), with metapodials fused to form a **cannon bone** (see figure 19.1C and E–G). The astragalus bone has a pulleylike surface above and below (see figure 19.1H). This "double pulley" system is above the distal portions of the limbs and allows great flexion and extension. At the same time, however, the astragalus limits distal limb motion so that it is parallel to the body. The clavicle is reduced or absent. Dentition varies in this order, with the cheekteeth ranging from bunodont and brachyodont to selenodont and hypsodont. The number of teeth varies, although in most families, the upper incisors and canines are reduced or absent. In some species, however, the canines form enlarged tusks. Artiodactyls are diverse in their digestive anatomy, with simple, nonruminating stomachs occurring in suids and tayassuids, grading to much more complex, four-chambered ruminating stomachs in the more derived families. Size ranges from a maximum head and body length of 0.5 m in the mouse deer to almost 6 m in the giraffe and maximum body mass from 2.5 kg in the lesser mouse deer to 4500 kg in the hippopotamus. Characteristics specific to individual families are noted in the following sections.

Fossil History

The earliest ancestors of artiodactyls were the Condylarthra. The oldest recognized genus is the rabbit-sized *Diacodexis* (figure 19.9) in the early Eocene epoch (Carroll 1988). The order then was

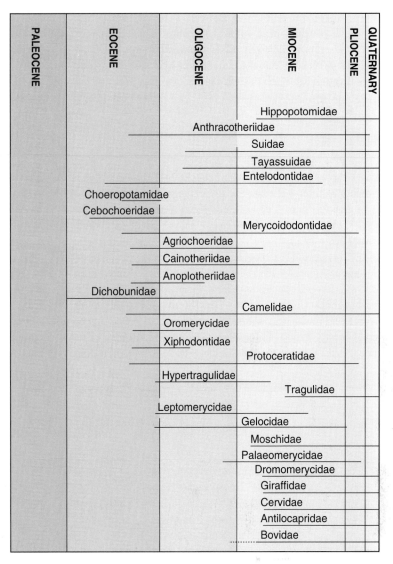

Figure 19.10   Geological ranges of the Artiodactyla, including modern and extinct families.   Relationships among the families are uncertain and difficult to document.

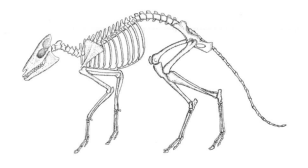

Figure 19.9   **An early artiodactyl.**   The rabbit-sized *Diacodexis* from the early Eocene epoch was already highly adapted for cursorial locomotion. The hind limbs were elongated, as were the metapodials and the third and fourth digits. The astragalus restricted movement of the limbs to the vertical plane.

relatively insignificant compared with the perissodactyls but radiated a great deal by the Oligocene epoch. Of the modern families, fossil evidence for the camels dates from the mid-Eocene epoch, for the suids and tayassuids from the early Oligocene epoch (figure 19.10), and for the remaining families from the Miocene epoch.

Economics and Conservation

Most families of artiodactyls have had some type of economic importance to human civilizations for thousands of years. Domestic artiodactyls include pigs, camels, llamas, and cattle. Some species are no longer found in the wild but survive only in domestication (see chapter 28). Deer, the pronghorn antelope (*Antilocapra americana*), mountain sheep and goats, numerous African antelope, and other bovids all provide meat, hides, sport hunting, and an economic base unparalleled by other mammalian groups. Many of these species are managed intensively on a sustained yield basis throughout much of their ranges.

Families

Suidae

The 5 genera and 16 living species of pigs (Grubb 1993b) have simple stomachs, bunodont cheekteeth, and large, ever-growing canines—the upper pair curving up and outward to form tusks (figure 19.11). Suids have short legs; heavyset bodies; thick skin with short, coarse pelage; small eyes; a relatively large head; and a prominent snout truncated at the end with a round, cartilaginous disk. Several layers of muscles are associated with the snout, which is used in rooting for food. Several species have large facial warts, most prominent in males (figure 19.11A). Maximum body mass ranges from 10 kg in the pygmy hog (*Sus salvanius*) to 200 kg in the wild boar (*S. scrofa*). The native distribution of suids is Europe, Africa (except the Sahara Desert), and Asia, including

Indonesia, Borneo, and the Philippines. They have been introduced to North and South America, Australia, and New Zealand, where both feral and domestic pigs now flourish.

Suids are gregarious and often forage in groups, although solitary individuals may occur, especially of the warthog (*Phacochoerus aethiopicus*). Their habitats include tropical forests, woodlands, scrubby thickets, grasslands, and savannas. Wild boars and the bushpig (*Potamochoerus porcus*) are omnivorous, whereas the babirusa (*Babyrousa babyrussa*) of Sulawesi and nearby islands, the giant forest hog (*Hylochoerus meinertzhageni*), and the warthog are more specialized herbivores. The large head and mobile snout are used to root for food. Suids are sexually mature by 18 months of age, although males may not have access to females until they are 4 years old. Gestation is about 100 days in the pygmy hog, 115 days in domestic pigs, and about 175 days in the warthog. Litter size is 1 to 2 in the babirusa and up to 12 in domestic pigs. Feral hogs are an important hunted species in California and throughout much of the southeastern United States. Unfortunately, they carry a number of diseases, including swine brucellosis, African swine fever, and pseudorabies. These can be transmitted to domestic pigs, which

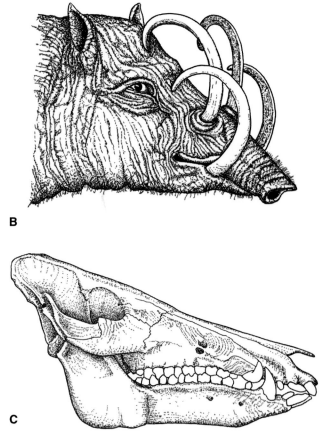

A　**B**　**C**

Figure 19.11　**Representative suids.**　(A) Large facial warts, composed of dense connective tissue, are evident on this warthog. Warts may protect the facial area from an opponent's tusks during aggressive interactions involving ritualized head-to-head pushing contests. (B) Tusks are especially pronounced in the babirusa. The upper canines protrude through the skin of the rostrum and do not occlude with the lower canines. (C) The typical wedge shape of suid skulls is evident in the wild boar. Canine tusks are less pronounced than in the warthog and babirusa.

creates potential problems for livestock operations. Feral hogs also are responsible for habitat damage in many national parks throughout the United States. The pygmy hog, found in the foothills of the Himalayan Mountains, is endangered, as is the babirusa.

Tayassuidae

Peccaries are the least specialized of the suiforms. The three genera, each with a single species, superficially resemble pigs. They have large heads and long, mobile, piglike snouts, but with thin legs and small hooves. Peccaries are smaller than pigs; they range up to 30 kg body mass. They have a total of 38 teeth and fewer tail vertebrae than pigs. Their tusklike upper canines are small, sharp-edged, and point downward, again unlike those of pigs. Also, peccaries are found only in the New World, from the southwestern United States to central Argentina, where they inhabit various areas from desert scrublands to tropical rain forests. They are primarily diurnal herbivores. They root with their snouts as do pigs but occasionally take small vertebrates, invertebrates, eggs, fruit, and carrion.

Peccaries generally form small groups of 5 to 15 individuals, although herds of the white-lipped peccary (*Tayassu pecari*) may number several hundred. A rump gland is used in social communication. Breeding may occur throughout the year; in arid environnments, it is affected by rainfall (Hellgren et al. 1995). The gestation period varies from 115 days in the collared peccary (*Pecari tajacu*) to 162 days in the white-lipped peccary. Litter size averages two. Female collared peccaries attain sexual maturity by about 8 months of age, males by 11 months old. The Chacoan peccary (*Catagonus wagneri*), until recently known only from fossil evidence (Wetzel et al. 1975), has been studied by Taber and colleagues (1993). The species is endangered by loss of habitat and possibly overharvesting. The collared peccary is hunted in Texas, New Mexico, and Arizona, where populations generally are secure.

Hippopotamidae

The two species in this family differ greatly in size. The hippopotamus (*Hippopotamus amphibius*) has a total length up to 4.5 m and maximum body mass of 4500 kg. The more primitive pygmy hippo (*Hexaprotodon [Choeropsis] liberiensis*) has a total length of only 2 m and body mass of about 250 kg. Both species are essentially without hair except for a few bristles around the snout. Without temperature-regulating sweat glands, each species has glandular skin that exudes a pigmented fluid that appears red and gives rise to the misconception that they "sweat blood." The fluid helps protect against sunburn and may help keep wounds from becoming infected in the water. The cheekteeth are bunodont, and *H. amphibius* has ever-growing, tusklike lower canines and incisors, with the alveoli of the canines being anterior to those of the incisors (figure 19.12). Although the canines are not used in foraging, they are important in ritualized fighting

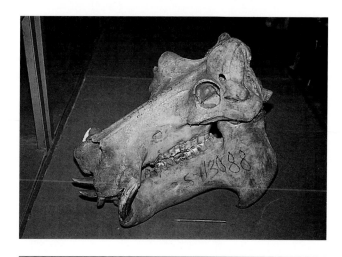

Figure 19.12   **Skull of a hippopotamus.**   Note the eye orbits on the top of the cranium in the hippopotamus, the low-crowned cheekteeth, and the alveoli of the lower canines anterior to those of the incisors. Also note the very large mandible. Scale: pencil = 16 cm.

to establish dominance, functioning as visual signals (Kingdon 1979).

Both species are closely associated with rivers, lakes, and estuaries and are excellent swimmers and divers. Part of their dependence on water is because of rapid evaporative water loss through the epidermis. The hippopotamus remains submerged for up to 30 minutes while walking along the bottom and feeding on plants. In *Hippopotamus amphibius*, the more aquatic of the two species, the eyes and nostrils are high on the head, allowing individuals to remain submerged throughout much of the day with only the eyes and nostrils above water.

The hippopotamus was originally distributed throughout most of Africa south of the Sahara Desert and along the Nile River. The pygmy hippo is restricted to the coastal regions of West Africa from Guinea to Nigeria. The aquatic specialization of *Hippopotamus amphibius* and the geographic restriction of the more terrestrial pygmy hippopotamus may have been in response to pressure from other grazing artiodactyls, specifically bovids. *H. amphibius* grazes on land for 5 to 6 hours each night, with grass as the primary food. Pygmy hippos also consume leaves and fruit. Hippopotamuses are not ruminants, but the stomach has septa and several blind sacs to slow the passage of food for more efficient digestion.

Hippopotamus amphibius is gregarious, forming herds of up to 40 individuals. Pygmy hippos are solitary or form pairs. Although there is a great deal of variability, males generally are sexually mature when 7 years old, females by 9 years of age. A single young is born after a gestation period of 200 days in *Hexaprotodon liberiensis* and 240 days in *Hippopotamus amphibius*. Parturition may occur in the water. Calving intervals are about 2 years but may be affected by drought. Both species have been extensively overhunted with populations further reduced by habitat destruction. The hippopotamus is extirpated from the Nile River valley in Egypt and western and southern Africa. The pygmy hippo probably has never been common and currently is considered threatened. Hippos on Madagascar became extinct within the last 1000 years (Stuenes 1989).

Camelidae

This family has three genera and six species that, like the hippopotamuses, differ greatly in size. Head and body length in both the one-humped, or dromedary, camel (*Camelus dromedarius*) and the two-humped Bactrian camel (*C. bactrianus*) approaches 3.5 m, with a body mass close to 700 kg. The guanaco (*Lama guanicoe*) may reach 2.2 m in head and body length and 140 kg body mass. The vicuña (*Vicugna vicugna*) is much smaller, about 55 kg. The domesticated alpaca (*L. pacos*) and llama (*L. glama*) are considered separate species but may be subspecies or hybrids (Grubb 1993b). All camelids have a small head with long snout and cleft upper lip; a long, thin neck; and long legs with the medapodials fused to form a cannon bone. The toes, with nails on the upper surface, spread out as they contact the ground, with a broad pad acting to support the mass of the animal on soft, loose sand. Upper and lower canines are present, and the cheekteeth are selenodont. Only the outer, spatulate upper incisor is retained in adults. The vicuña is the only artiodactyl with ever-growing lower incisors. Camelids have a three-chambered, ruminating stomach and a short, simple cecum.

The dromedary camel may once have ranged throughout the Middle East but now survives only in domestication. It is the only domesticated mammalian species for which there is no information on wild or fossil forms (Corbet 1978). The Bactrian camel originally ranged throughout much of central Asia but now is restricted in the wild to the western Gobi Desert. Guanacos are found from sea level to 4000 m elevation in various habitats in the Andes Mountains from southern Peru to Tierra del Fuego. Vicuñas occur in the grasslands of Peru, western Bolivia, northeast Chile, and northwest Argentina at elevations from 3700 to 4800 m. All camelids are gregarious, diurnal, and herbivorous and are best suited to dry, arid climates. They can eat plants with a high salt content not tolerated by other grazers. They are well known for their ability to go long distances under difficult conditions, conserving water better than other large mammals (see chapter 8) and may lose up to 40% of their body mass through desiccation without harm (Gauthier-Pilters 1974). When camels are well-fed, their humps are firm and erect. The hump is a fat reservoir and shrinks, leaning to one side, when camels are nutritionally stressed. In camels, both limbs on either side of the body move in unison. This **pacing** locomotion allows for long strides with consequent lower energy expenditure. Wild camelids are social, forming groups of up to 30 individuals with different sex and age compositions depending on the season. A single

young is born after a gestation of 300 to 320 days in *Lama* and 365 to 440 days in *Camelus*.

This family arose in North America in the late Eocene epoch and was restricted to North America throughout most of the Tertiary period. It expanded to both Eurasia and South America during the Pliocene epoch and became extirpated in North America in the late Pleistocene epoch. Camelids have been introduced throughout the world for use as pack animals and for meat, wool, or milk. Both the llama and dromedary camel have been domesticated for up to 5000 years (see chapter 28). In the wild, the Bactrian camel and the vicuña are considered threatened, although populations of the vicuña have recovered from their endangered status.

Tragulidae

This family is the most ancestral of the extant ruminants, that is, they have changed the least since the Oligocene epoch, when they enjoyed a worldwide distribution. Today, they include three genera and four species: the water chevrotain (*Hyemoschus aquaticus*) of west-central Africa and three species distributed from India through Southeast Asia, Indonesia, and Borneo. The Asiatic chevrotain, or lesser mouse deer (*Tragulus javanicus*), is the world's smallest artiodactyl and superficially resembles a tiny, chunky deer. Maximum head and body length is 50 cm, and body mass is 2.5 kg. Other species are slightly larger. Unlike deer, chevrotan males have no antlers and no facial or other body glands. The legs are thin with a cannon bone (except in the water chevrotain, the most ancestral form); dentition is selenodont and brachyodont; and there is a three-chambered, ruminating stomach. As in species of deer without antlers, tragulids have an enlarged, curved upper canine that extends below the upper lip (figure 19.13). Tragulids generally are solitary and

Figure 19.13   A chevrotain, or mouse deer.   In these small artiodactyls, males have enlarged upper canines.

nocturnal and spend most of the day hidden in brushy undergrowth, usually not far from water. Their spotted pelage helps camouflage them. They feed on grass, leaves, and fallen fruit and are in many ways the ecological equivalent of hares (Genus *Lepus*). Gestation in *Tragulus* is 4 to 6 months, and a single, precocial young is born each year. Tragulids are physically and sexually mature by 9 months of age. Hunting and habitat destruction have reduced populations of all species.

Giraffidae

This family includes both the giraffe (*Giraffa camelopardalis*) and the okapi (*Okapia johnstoni*). The giraffe occupies savannas, grasslands, and open woodlands throughout sub-Saharan Africa. The okapi is restricted to dense forested areas of Republic of Congo (formerly, Zaire). Both species are characterized by long legs and neck that allow them access to forage out of reach of other large herbivores. In male giraffes, the top of the head may be 5.5 m off the ground, and the average shoulder height is 3 m (figure 19.14). Okapis are smaller, with a

shoulder height about 1.7 m. Females are smaller than males in both species. These species are found where forage is available year-round at levels between 2 and 5.5 m (Kingdon 1979). The teeth are small and brachyodont, which is unusual for a large herbivore. This suggests that they are selective feeders; forage selection is aided by their very long, muscular, prehensile tongue that, in the giraffe, can be extended almost 0.5 m. The long, muscular neck of both species is also used in intraspecific aggressive interactions of males as they establish dominance hierarchies. Interactions involve standing side by side and "necking"—pushing each other with intertwined necks and occasionally swinging the neck and using the horns to strike an opponent's head and neck. Extensive cranial sinuses help protect individuals from hammering blows. The horns are unique in being short, permanent, unbranched processes (**ossicones**) over the frontal and parietal bones (see figure 19.7B; figure 19.15). They are not extensions of the frontal bone but are distinct, fused to the cranium, and covered with hairy skin.

Both species have unusual pelage patterns. There is a great deal of variation in giraffe patterns and color shades, with white and melanistic individuals known. The okapi is reddish brown dorsally with distinctive striping on the legs and back of the thighs (figure 19.16). Okapis do not form herds but are solitary or live in small family groups (Bodmer and Rabb 1992). Group size in giraffes may be 30 to 50 but is more a reflection of forage availability than gregariousness. Gestation is variable in both species, from 427 to 488 days in the giraffe to a somewhat shorter period in the okapi (Kingdon 1979; Dagg and Foster 1982). Females give birth to a single, precocial calf approximately every 2 years. Calf

Figure 19.14   **The reticulated pattern of giraffes.** This pattern varies considerably both in shape and color. Giraffes feed selectively and sex can be determined at a distance by feeding behavior. Males always stretch with the head extended upward, in contrast to females, which feed with their head down over the vegetation. Sexes therefore forage at different heights, reducing potential competition. Note the long, tufted tail.

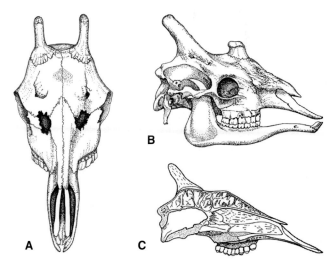

Figure 19.15   **Skull of a giraffe.**   (A) Dorsal and (B) lateral views of the "horns" (ossicones) of a giraffe, showing distinct sutures on the frontal/parietal area. The horns are not extensions of the frontal bone as in bovids. (C) Sagittal section of a giraffe skull showing the extensive sinus cavities in the roof of the cranium. Sinuses allow blood to collect when the head is lowered so blood pressure does not increase to dangerously high levels.

Figure 19.16   The okapi.   This rare member of the giraffe family has a very distinctive coloration pattern.

mortality is very high due to predation. Subsistence hunters have long taken giraffes for meat. Giraffes also are harvested illegally for their meter-long tail tufts, which are used to make bracelets for tourists. The okapi was not discovered by scientists until 1900 and has been protected by the government since 1933 (Hart and Hart 1988).

Moschidae

The classification of musk deer (Genus *Moschus*) is uncertain. The four species now are considered a distinct family (Groves and Grubb 1987; Janis and Scott 1987) but have been included within the Cervidae in the past. Musk deer are small, with head and body length between 80 and 100 cm and maximum body mass no more than 18 kg. The males do not have antlers, but as in the Chinese water deer (*Hydropotes inermis*), the upper canines are long and curved (figure 19.17). Musk deer are distributed from Siberia to the Himalaya Mountains in forested areas with dense understory vegetation. They are named for the musk gland, or "pod,"

that develops slightly anterior to the genital area of males. In adults, this gland contains 18 to 32 g of reddish brown, gelatinous, oily material, probably important during the breeding season for attracting females and marking a territory. Musk is used both in fine perfumes and as an ingredient in Asian traditional medicine. As a result, hunting pressure on musk deer has always been exceptional, and several species are considered endangered in portions of their range. Musk deer have been produced on game farms for about 40 years to help ensure a reliable supply of musk and to reduce harvest pressure on wild populations.

Cervidae

The deer family includes 16 genera and 42 extant species, ranging in size from the southern pudu (*Pudu puda*), with a maximum body mass of 8 kg, to the moose (*Alces alces*) at 800 kg. Deer are widely distributed and are absent only from sub-Saharan Africa and Antarctica. They are not native to Australia or New Zealand but have been introduced there, as well as to many other areas. The habitats occupied by deer are as varied as their size and include deciduous forests, marshes, grasslands, tundras, arid scrublands, mountains, and rain forests.

Deer exhibit sexual dimorphism, with males often being 25% larger in body mass and body dimensions than females. Male deer are also well known for their characteristic antlers, which are made of bone, are usually branched, and are supported on pedicels, which are raised extensions of the frontal bone (see figure 19.7C). They usually occur only in males. Exceptions include Chinese water deer, in which males do not have antlers, and caribou (*Rangifer tarandus*), in which both sexes have antlers. Antlers are deciduous; that is, they are shed each winter following the rut and regrown in the spring. Growing antlers are covered with haired, highly vascularized skin known as **velvet.** Antlers are the fastest growing tissue known other than cancer. Antler size and the number of **tines** (points) are a function of nutritional condition, genetic factors, and age (Brown 1983). There is little correlation between individual age and the number of antler points, however. Different species of deer have a characteristic antler shape and size (figure 19.18), from single spikes in pudu, tufted deer (*Elaphodus cephalophus*), and brocket deer (Genus *Mazama*) to large, **palmate** (flattened) antlers with numerous tines in fallow deer (*Dama dama*), caribou, and moose. Antlers are important in mating behavior and success of males and, therefore, are a highly selected trait (Geist and Bromley 1978; Geist 1991).

Pelage color varies among and within species. Most newborn cervids have white spots and lines on a darker background, providing excellent camouflage. Certain species maintain this pattern as

A   **B**

Figure 19.17   Musk deer skulls.   (A) Male musk deer have no antlers, but enlarged, tusklike upper canines are evident. (B) Upper canines are significantly smaller in female musk deer. Scale: pencil = 16 cm.

Figure 19.18   Relative size and diversity of antler structure in several species of deer.   (A) Palmate antlers of the moose, the largest living cervid. (B) Large antlers of a male caribou. This is the only species of deer in which females also have antlers. (C) White-tailed deer (*Odocoileus virginianus*), a popular game species throughout North America. (D) Pére David's deer, no longer found in the wild. (E) The small, spike antlers of the pudu. (F) Male Chinese water deer lack antlers, but like musk deer (see figure 19.17) they have enlarged upper canines. (G) Skull of a sika deer showing the extension of the frontal bone (*p* = pedicel) on which antlers form and the typical points (*t* = tines) in this commonly introduced species.

adults, including fallow, sika (*Cervus nippon*), and axis deer (*Axis axis*). Deer have no upper incisors; they browse or graze by cutting off forage between the lower incisors and a calloused upper pad. They are herbivores, and many of the temperate species change their diet depending on the season. Deer are gregarious, with species such as caribou forming herds of 100,000 or more. Visual, auditory, and olfactory senses are acute. Several glands may be present from which herd members gain reproductive and other information about each other. Glands may be located on the face (prelacrimal), between the toes (interdigital), or on the lower hind legs (tarsal or metatarsal glands). Males often fight each other, using their antlers, to establish territories on which to attract a group of females for breeding. Maintenance by males of female breeding groups depends on the tendency of females to remain grouped, and therefore defensible, and the extent to which estrus is synchronized (Eisenberg 1981). Gestation is generally 6 to 7 months. Among cervids, only the roe deer (*Capreolus capreolus*) exhibits delayed implantation (see chapter 9). Depending on habitat conditions and body condition

of the female (called a doe, hind, or cow, depending on the species), from 1 to 3 young are typical.

Deer have been hunted by humans for thousands of years for both meat and antler trophies. Today, many species are domesticated to harvest both meat and antler velvet. Others have been introduced as free-ranging exotics to numerous countries outside their native range, often with unanticipated negative effects on native species (Feldhamer and Armstrong 1993). Overharvesting and habitat loss have resulted in several species becoming endangered, including the spotted deer (*Cervus alfredi*), swamp deer (*C. duvaucelii*), several subspecies of red deer (*C. elaphus*), Eld's deer (*C. eldii*), and Pére David's deer (*Elaphurus davidianus*).

Antilocapridae

This monotypic family includes only the pronghorn, which is endemic to western North America. Pronghorns have barrel-shaped bodies and long, thin legs. Shoulder height reaches about 1 m, and maximum body mass is 70 kg, with males being larger than females. The horns consist of a keratinized

Figure 19.19   The horn structure of pronghorn antelope. The black jaw patch, found only in males, is important in male-male and male-female behavioral interactions.

sheath over a permanent, bony core extension of the frontal bone. Unlike in most members of the Family Bovidae, however, a new horn sheath grows each year under the old one, which splits and is shed following the breeding season. Horns are upright, with a posterior hook and a short, anterior branch, or prong (see figure 19.7D; figure 19.19). Horns on males may reach 250 mm in length. Horns may be absent on females; if present, they are much shorter and do not have the prong.

Pronghorns occur in open grasslands and semidesert areas, where they forage on grasses, forbs, and low shrubs, especially sagebrush (*Artemisia tridentata*). The pronghorn is the fastest New World mammal. Maximum speeds of 86 km/h can be maintained for up to 6 km (Kitchen and O'Gara 1982). Males are unusual in that they defend territories from March or April until after the rut, even though breeding occurs only in the fall. However, rutting behavior is variable among popu-

lations (Maher 1991). The gestation period is about 250 days, and twins are common. During winter, large herds may form. Pronghorns migrate up to 160 km between distinct summer and winter ranges. The pronghorn is a popular game animal. In the 1920s, populations were very low but have recovered in the United States because of successful management programs. Populations of the Mexican subspecies (*Antilocapra americana peninsularis* and *A. a. sonoriensis*) are endangered, however.

Bovidae

The largest family of artiodactyls includes 45 genera and about 137 species. Indicative of the number of species and diversity of the group, five subfamilies traditionally have been recognized (Simpson 1945). More recently, Simpson (1984) proposed 10 subfamilies, Gentry (1990) proposed 6, and Grubb (1993b) divided bovids into 9 subfamilies (table 19.1). Designation of subfamilies and tribes within subfamilies is based on considerations of horn structure, cranial and skeletal features, behavior, genetics, feeding strategies, and other factors, all of which are interrelated. Generally, bovids have hypsodont and selenodont cheekteeth, with no upper incisors or canines. All species have four-chambered, ruminating stomachs. Size varies greatly, from the dwarf antelope (*Neotragus pygmeus*) with a shoulder height of 250 to 300 mm and body mass to 2.5 kg, to the bison (*Bison bison*), several species of cattle (Genus *Bos*), and elands (Genus *Taurotragus*) with body mass approaching 1000 kg. All bovids have a pair of horns with the exception of the four-horned antelope (*Tetracerus quadricornis*). Horns are present on males and often on females. They have a bony core, which is an extension of the frontal bone (see figure 19.7E), covered with a keratinized sheath that is unbranched and rarely shed. Horns may be straight, spiraled, or curved (figure 19.20) and grow throughout life. Like antlers in deer, horns may function in defense against predators as well as in intrasexual fighting for access to breeding females. Variation in size and shape of horns reflects the fighting behavior of the species.

Table 19.1.   Subfamily designations of the Bovidae, according to various authors, with the number of genera in parentheses[*]

G. G. Simpson (1945)	C. D. Simpson (1984)	Gentry (1990)	Grubb (1993b)
Bovinae (8)	Bovinae (5)	Bovinae (7)	Bovinae (8)
Cephalophinae (2)	Cephalophinae (2)	Cephalophinae (2)	Cephalophinae (2)
Hippotraginae (9)	Hippotraginae (3)	Hippotraginae (5)	Hippotraginae (3)
Antilopinae (13)	Antilopinae (8)	Antilopinae (14)	Antilopinae (14)
Caprinae (13)	Caprinae (11)	Caprinae (10)	Caprinae (10)
	Alcelaphinae (3)	Alcelaphinae (5)	Alcelaphinae (4)
	Reduncinae (2)		Reduncinae (2)
	Aepycerotinae (1)		Aepycerotinae (1)
	Neotraginae (7)		Peleinae (1)
	Tragelaphinae (3)		

[*] As is often true for other large families, opinions diverge among authorities about the relationships among bovid genera. Future advances in molecular techniques and additional research in other areas will no doubt result in alternative classifications.

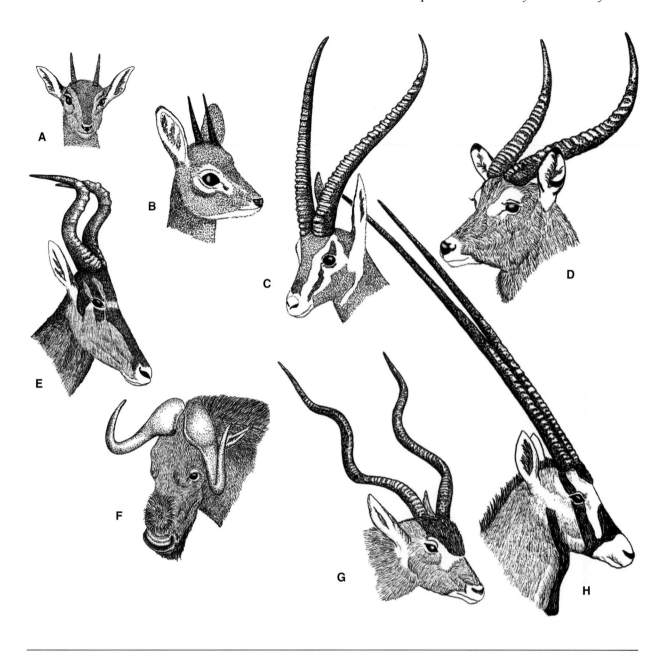

Figure 19.20   Variety of horn shapes and sizes in bovids.   (A) suni (*Neotragus moschatus*); (B) klipspringer (*Oreotragus oreotragus*); (C) Grant's gazelle (*Gazella granti*); (D) waterbuck (*Kobus ellipsiprymnus*); (E) hartebeest (*Alcelaphus buselaphus*); (F) gnou (*Connochaetes gnou*); (G) addax (*Addax nasomaculatus*); and (H) oryx (*Oryx gazella*). Not to the same scale.

Absent naturally only from South America and Australia, domesticated bovids are distributed practically worldwide. Wild species are found primarily in Africa and Eurasia, where they occupy grasslands, savannas, scrublands, and forests. They also occur in harsher environments, including tundras, deserts, and swamplands. All are herbivores, with different subfamilies having different feeding strategies (figure 19.21). Social systems are related to body size, feeding strategy, and predation pressure (Jarman 1974). Generally, small species are specialized feeders in dense, closed habitats and tend to be solitary or paired. Larger species are more gre-

garious and occupy more open areas, where they are generalist feeders on high-fiber (more cellulose) vegetation.

Cattle, sheep, and goats have been domesticated for over 5000 years and are an integral part of agricultural economies throughout the world. Many other species of bovids have been introduced or are hunted for meat, hides, and sport. Numerous others throughout the world are adversely affected by habitat loss and overharvesting and are considered endangered. A previously unknown bovid, the saola (*Pseudoryx nghetinhensis*), was recently discovered in Vietnam (Henshaw 1997; Kemp et al. 1997).

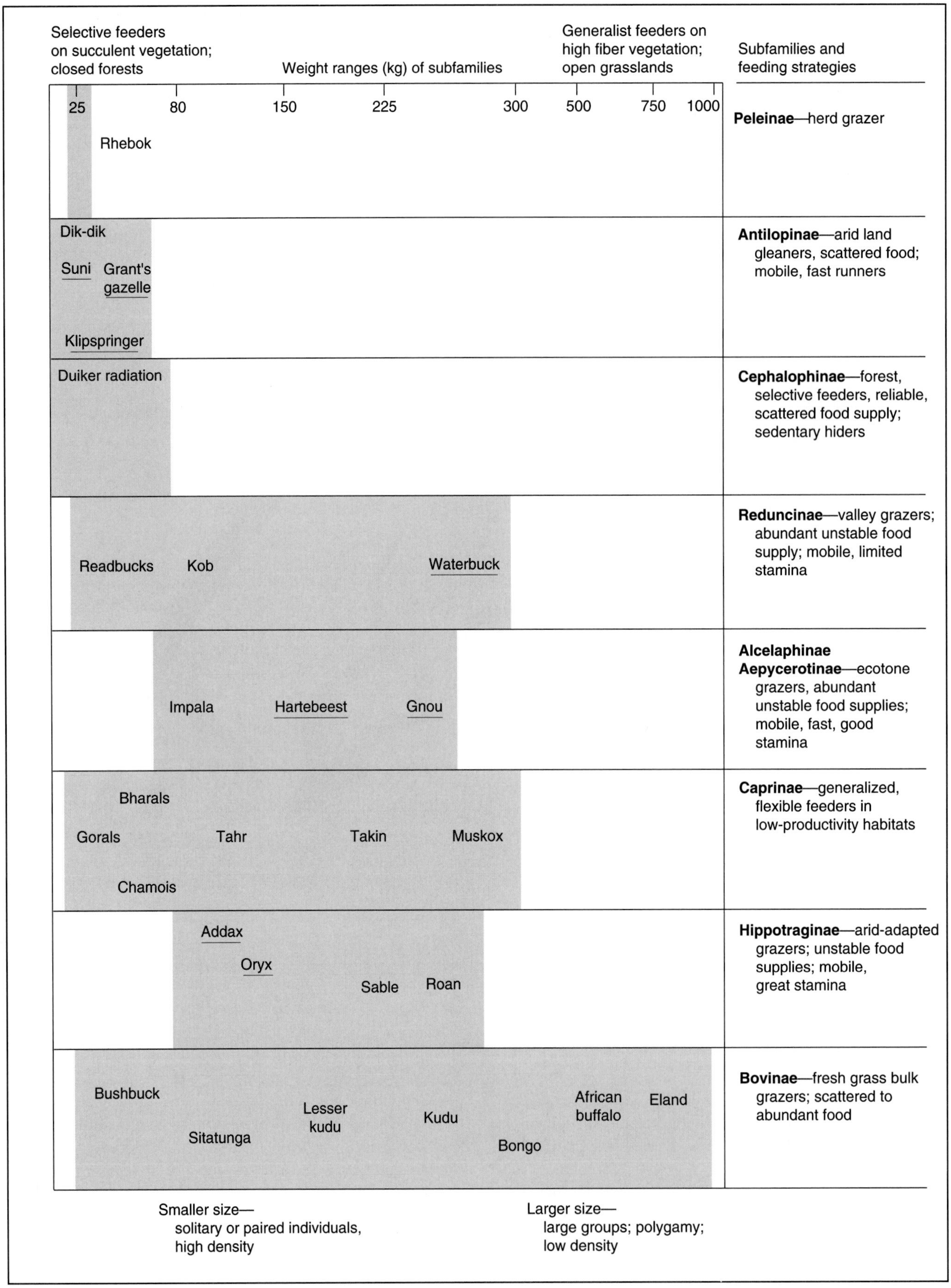

| Selective feeders on succulent vegetation; closed forests | Weight ranges (kg) of subfamilies | Generalist feeders on high fiber vegetation; open grasslands | Subfamilies and feeding strategies |

Weight axis: 25 80 150 225 300 500 750 1000

Peleinae—herd grazer

Rhebok

Antilopinae—arid land gleaners, scattered food; mobile, fast runners

Dik-dik
Suni Grant's gazelle
Klipspringer

Cephalophinae—forest, selective feeders, reliable, scattered food supply; sedentary hiders

Duiker radiation

Reduncinae—valley grazers; abundant unstable food supply; mobile, limited stamina

Readbucks Kob Waterbuck

Alcelaphinae Aepycerotinae—ecotone grazers, abundant unstable food supplies; mobile, fast, good stamina

Impala Hartebeest Gnou

Caprinae—generalized, flexible feeders in low-productivity habitats

Bharals
Gorals Tahr Takin Muskox
Chamois

Hippotraginae—arid-adapted grazers; unstable food supplies; mobile, great stamina

Addax
Oryx
Sable Roan

Bovinae—fresh grass bulk grazers; scattered to abundant food

Bushbuck
Sitatunga Lesser kudu Kudu African buffalo Eland
Bongo

Smaller size— solitary or paired individuals, high density

Larger size— large groups; polygamy; low density

Figure 19.21 **Feeding strategies of bovid subfamilies.** Ecological niches vary, depending on the size of the species. Underlined names are species shown in figure 19.20. Subfamilies are based on Grubb (1993b).

Sources: Data from J. Kingdon, East African Mammals, Vol. IIIB: Large Mammals, 1972, University of Chicago Press. Subfamilies; from P. Grubb, Mammal Species of the World, 2nd edition (D.E. Wilson and D.M. Reeder, eds.), 1993, Smithsonian Institution Press, Washington, D.C.

Summary

Both perissodactyls and artiodactyls are defined on the basis of limb structure. Perissodactyls are mesaxonic, with the main weight-bearing axis of the limbs passing through the third digit, whereas artiodactyls are paraxonic, with the axis of the limbs passing between the third and fourth digits. The superior digestive efficiency of ruminant artiodactyls probably led to the displacement of most lineages of perissodactyls early in the Cenozoic. The digestive efficiency of equids, for example, is only about 70% that of bovids. With the exception of the domestic horse, most species within the three surviving families of perissodactyls are severely depleted in numbers and geographic distribution. The order today represents a remnant of a much more diverse group that was extant in the early Tertiary period. In contrast, artiodactyls, especially the cervids and bovids, are widely distributed. Artiodactyls are very diverse in terms of both size (from the mouse deer to hippopotamus) and life history characteristics. Their success can be traced to superior digestive efficiency, especially the remarkable adaptation of rumination, which allows individuals to select large quantities of high-quality forage that is then processed in relative safety. Predation pressure and foraging behavior have led to major morphological developments in many families of these two orders toward cursorial, unguligrade locomotion.

Horns and antlers occur in five of the families discussed in this chapter for defense against predators and as part of intraspecific social behavior. Rhinoceroses are the only perissodactyls with horns. These horns are unusual in that they are not made of bone and are not permanently attached to the skull bone. They are made of agglutinated keratin fibers and are attached to the skin over the nasal and frontal area. Giraffes also have an unusual horn structure. The horns are not extensions of the frontal bones but form from separate bones (ossicones) that fuse to the cranium. Antilocaprids and bovids have horns formed from keratinized sheaths over bony extensions of the frontal bone. These sheaths are deciduous in pronghorns but not in bovids. Cervids have antlers that, unlike horns, are made of bone, are branched, deciduous, and generally occur only on males. Horses and several species of artiodactyls are found worldwide, in many cases because of introductions. In contrast, other species such as camels and some bovids are no longer found in the wild but only in domestication. Still others are critically endangered because of habitat loss, poaching, or overharvesting.

Discussion Questions

1. Why is it adaptive for deer to shed their antlers every year following the rut, even though it means a considerable energy expenditure to regrow a new set beginning a few months later in the spring? What are the benefits of having deciduous antlers rather than keeping the same set of antlers throughout life?
2. Camels and tapirs both passed most of their evolutionary development in North America. Representatives of both families radiated both north and south during the Pliocene epoch, so they currently are distributed in Central and South America, as well as in Asia, even though they died out in North America during the Pleistocene epoch. What reasons might explain why the highly mobile pronghorn antelope never left North America?
3. Consider figure 19.4 depicting the evolutionary development of the horse. Why will you never see a similar figure for rodents or bats?
4. What is the adaptive significance of antlers in female caribou? Consider the northern latitudes and the extreme climatic conditions where caribou occur. Also, keep in mind that males develop antlers before females each year and subsequently cast them in the winter before females do.
5. What kinds of characteristics have plants evolved in response to pressure from herbivores such as perissodactyls and artiodactyls?
6. Considering the length of their neck, what physiological mechanisms do giraffes use to maintain blood flow to their brain?

Suggested Readings

Berger, J. 1994. Science, conservation, and black rhinos. J. Mammal. 75:298–308.
Bubenik, G. A. and A. B. Bubenik. (eds.). 1990. Horns, pronghorns, and antlers. Springer-Verlag, New York.
Clutton-Brock, T. H., F. E. Guinness, and S. D. Albon. 1982. Red deer: behavior and ecology of two sexes. Univ. of Chicago Press, Chicago.
Evans, J. W., A. Borton, H. F. Hintz, and L. D. Van Vleck. 1990. The horse, 2nd ed. W. H. Freeman, New York.
Geist, V. 1974. On the relationship of ecology and behaviour in the evolution of ungulates: theoretical considerations. Pp. 235–247 in The behaviour of ungulates and its relation to management, 2 vols. (V. Geist and F. Walther, eds.). IUCN Publ. No. 24. Morges, Switzerland.
Groves, C. P. 1974. Horses, asses, and zebras in the wild. David & Charles, London.
Owen-Smith, R. N. 1988. Megaherbivores: the influence of very large body size on ecology. Cambridge Univ. Press, New York.
Putman, R. 1988. The natural history of deer. Cornell Univ. Press, Ithaca, New York.
Wemmer, C. M. (ed.). 1987. Biology and management of the Cervidae. Smithsonian Institution Press, Washington, D.C.

PART FOUR

Behavior, Ecology, and Biogeography

In this part, we tie the anatomical and physiological pieces together and consider what mammals do as they interact with their environment. The approach used in these chapters is to introduce the concepts and illustrate them with mammalian examples. The first four chapters (20–23) deal with animal behavior, a topic that is frequently taught as a separate course. Our emphasis here is on **behavioral ecology,** which deals with the animal's struggle for survival as it exploits resources and avoids predators and with how the animal's behavior contributes to its reproductive success. The first chapter (20) lays the groundwork for how individual mammals interact and communicate with other members of their species. In chapter 21 we deal with reproductive behavior—the problems of getting mates and caring for young—then, in chapter 22, with living in groups. How mammals orient in the environment and find a place to live is the subject of chapter 23.

Chapters 24 and 25 treat population and community ecology. These chapters form a continuum that begins with single-species populations, expands in scope to include interactions among different species, and finally includes the structure and function of mixed-species communities. The last chapter (26) in part IV expands the scale to include species distributions on whole continents and the role of history in explaining the distribution of mammals on earth.

Antlers and vocal communication both function in the behavior of this bull elk (*Cervus elaphus*).

CHAPTER

20

Communication, Aggression, and Spatial Relations

ehavior has been called the evolutionary pacemaker because an animal's behavior determines whether it can survive and reproduce (Wilson 1975). Natural selection acts on individuals interacting with the environment and with other organisms. Studying the behavior of mammals is particularly challenging because of the great diversity of forms and life-styles in the class, ranging from solitary species such as northern short-tailed shrews (*Blarina brevicauda*) to highly social forms such as lions (*Panthera leo*) and chimpanzees (*Pan troglodytes*). A second difficulty in studying mammals is that most species are small, nocturnal, and often secretive (Eisenberg 1981), making them much harder to observe than, say, insects or birds. In this chapter, we explore the ways mammals communicate, then we consider how they interact and distribute themselves in space.

Mammals have a wide range of sensory and neural abilities, centered on the neocortex, a structure that has taken over many functions from other, more primitive parts of the brain of other vertebrates (chapter 7). Evolution of this structure is generally credited with giving mammals an ability greater than their reptilian ancestors had to modify their behavior in response to environmental circumstances (i.e., to learn).

COMMUNICATION

Just as nerves and hormones (chapter 7) convey messages from one part of the body to another, **displays** are behavior patterns that convey messages from one individual to another. Displays carry an encoded **message** that describes the sender's state. The recipient of the message makes **meaning** of the message. Mammals send messages to members of their own species and to other species through a diversity of sounds, colors, odors, and postures. A **signal** is the physical form in which the message is coded for transmission through the environment.

What Is Communication?

When a male lion (or domestic house cat [*Felis catus*], for that matter) sniffs a bush, breathing through open mouth and flared nostrils, then backs up, turns, raises its tail, and sprays urine on the bush, more than simple excretion is going on. Wilson (1975) defined **biological communication** as an action on the part of one organism (the sender) that alters the probability of occurrence of behavior patterns in another organism (the receiver) in a fashion adaptive to either one or both of the participants. The word **adaptive** implies that the signal or response is partly under genetic control and influenced by natural selection.

Who benefits from communication: the sender, the receiver, or both? Behavioral ecologists generally argue that the sender must benefit (Slater 1983), and some have compared animal communication with advertising, a method of sending a message in which the sender manipulates the behavior

of others. The receiver may benefit or may be harmed. Signals become exaggerated, or **ritualized,** so that the sender can control the behavior of the receiver with a minimum of wasted energy (Dawkins and Krebs 1978; Krebs and Dawkins 1984).

Signalers may send misleading information. **Deceit** is relatively rare in most species, probably because natural selection favors those individuals that ignore or devalue fake signals. An example of deceit in mammals is the behavior of young male northern elephant seals (*Mirounga angustirostris*). They may enter the harems of territorial males, look and act like females, and attempt to sneak copulations. Females, however, respond to any mounting attempt by vocalizing loudly, thereby alerting the harem master, which then intervenes (Cox and Le Boeuf 1977).

Properties of Signals

Discrete versus Graded

Animal signals can be divided into two general types: discrete and graded. **Discrete,** or digital, signals are sent in a simple either/or manner. For example, equids (zebras, horses, and asses) communicate hostility by flattening their ears and communicate friendliness by raising their ears (figure 20.1). **Graded,** or analog, signals, on the other hand, are more variable and communicate motivation by their intensity. Thus, equids indicate the intensity of hostile or friendly emotions by the degree to which the mouth opens: the more open the mouth, the more intense the signal. Note, however, that the mouth-opening pattern is the same for both hostility and friendliness.

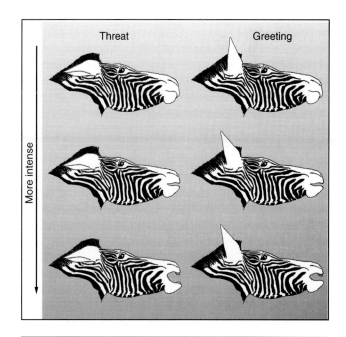

Figure 20.1 Facial signals in zebras. Ears convey a discrete signal. They are either laid back as a threat or pointed upward as a greeting. The mouth conveys a graded signal; degrees of openness indicate the intensity of hostility or friendliness.

Distance and Duration

Although mammalian species typically have only 20 to 40 different displays (Wilson 1975), signals vary in several ways that increase their information content. The distance a signal travels may vary. Low-frequency sounds, such as those made by elephants and some species of whales, travel many kilometers. Visual displays, at the other extreme, usually operate over much shorter distances. The duration of a signal may also vary. Alarm signals, such as those produced by ground squirrels, have a localized, short-term effect and thus a rapid fade-out time. Bright pelage on primates and other male adornments, such as antlers on cervids, may last the entire breeding season.

Composite Signals, Syntax, and Context

Two or more signals can be combined to form a **composite signal.** In the zebra (*Equus* spp.) example, ear position, with the ears forward (friendly) or backward (hostile), is coupled with the degree to which the mouth is open, providing additional information on motivation (see figure 20.1).

Animals possibly convey additional information by changing the **syntax**—the sequence of signals. In human speech, syntax is very important. Contrast the different meanings when the order of words is changed in the following phrases: "The bear eats Steve" versus "Steve eats the bear." Little evidence exists that nonhuman animals use syntax naturally. Laboratory-reared chimpanzees, however, use syntax as they communicate with one another and with their human companions (Savage-Rumbaugh 1986; Savage-Rumbaugh and Brakke 1990).

The same signals can have different meanings depending on the **context**—that is, on what other stimuli are impinging on the receiver. For example, the lion's roar can function as a spacing device for neighboring prides, an aggressive display in fights between males, or a means of maintaining contact among pride members.

Modes of Communication

We know that mammals have several sensory systems, including visual, auditory, and olfactory, by which they get information about their environment. Mammals use different **sensory channels** to send signals; the choice of channel depends on the animal's environment and the type of information being sent. Table 20.1 summarizes some of the properties of signals traveling via different channels.

Odor

From an evolutionary standpoint, one of the earliest channels of communication was chemical; odor is used throughout the animal kingdom, with the exception of most bird species. **Pheromones** are airborne chemical signals that elicit responses in other individuals, usually of the same species. Most pheromones are involved in mate identification and attraction, spacing mechanisms, or alarm. Chemical signals have probably evolved and become widespread for several reasons: such signals can transmit information in the dark, can travel around solid objects, can last for hours or days, and are efficient in terms of the cost of production (Wilson 1975). Because pheromones must diffuse through air or water, however, they are slow to act and have a long fade-out time. Compared with other channels, odors also are somewhat limited in the number of signals that can be sent at one time.

As indicated in many studies on rodents, mammals make extensive use of pheromones. Two general classes of substances, which differ in effect, have been identified: **priming pheromones** that produce a generalized response, such as triggering estrogen and progesterone production that leads to estrus, and **signaling pheromones** that produce an immediate motor response, such as the initiation of a mounting sequence. Bronson (1971) suggested that pheromones in mice have the following functions:

Priming pheromones	Signaling pheromones
Estrus inducer	Fear inducer
Estrus inhibitor	Male sex attractant
Adrenocortical activator	Female sex attractant
	Aggression inducer
	Aggression inhibitor

Table 20.1.   General properties of the major sensory channels of communication used by mammals

	Sensory channels			
Signal property	Olfactory	Auditory	Visual	Tactile
Range	Long	Long	Medium	Short
Transmission rate	Slow	Fast	Fast	Fast
Travel around objects	Yes	Yes	No	No
Night use	Yes	Yes	Little	Yes
Fade-out time	Slow	Fast	Fast	Fast
Locate sender	Difficult	Varies	Easy	Easy
Cost to send signal	Low	High	Medium	Low

Source: Data from John Alcock, Animal Behavior: An Evolutionary Approach, *4th edition, 1989, Sinauer Associates, Inc.*

Examples of priming pheromones are substances in male house mouse (*Mus domesticus*) urine that speed up sexual maturation in young females (Lombardi and Vandenbergh 1977). Other urinary products stimulate the adrenal cortex and block implantation of embryos. Different signaling pheromones either decrease or increase aggression. It is unclear how many different chemicals are involved; different functions may be served by the same pheromone (Drickamer 1989). The sources of these products include urine, feces, the sexual accessory glands, and a number of specialized skin glands.

Mule deer (*Odocoileus hemionus*) and other cervids produce pheromones from the tarsal and metatarsal glands on the hind legs, from the tail, and from urine. The leg is rubbed against the forehead, then the forehead on twigs, thus transmitting odors to the twigs (Muller-Schwarze 1971; figure 20.2). It turns out, however, that the primary source of the tarsal gland scent is urine and that the gland has to be recharged about once a day with urine, a behavior pattern referred to as rub-urination (Sawyer et al. 1994). Apparently fat-soluble pheromones present in the urine dissolve in sebum from the tarsal gland and convey information about individual identity and social status. Olfactory investigation in ungulates is characterized by **flehmen,** a retraction of the upper lip exhibited soon after sniffing the anogenital region of another or while investigating freshly voided urine (Estes 1972). This behavior pattern is especially common in males during the breeding season but also occurs in females.

Many of the pheromones produced by mammals function as a means of establishing territories or home ranges, much as does bird song. The advantage of pheromones is that the odor may last for many days. Because these substances are often associated with the urinary and digestive systems, eliminative behavior is often highly specialized. For example, spotted hyena (*Crocuta crocuta*) clans mark the boundaries of their territories by establishing latrine areas. Clan members defecate in a particular area, then paw the ground to deposit additional scents from digital glands. The feces, which are loaded with digested bone, turn white and become quite conspicuous. Hyenas also engage in "pasting." Both sexes possess two anal glands that open near the rectum. When pasting, the hyena straddles long stalks of grass; as the stems pass underneath, the animal everts its anal gland and deposits a strong-smelling yellowish substance on the grass stems (Kruuk 1972; figure 20.3).

Figure 20.3   **Pasting as a form of territorial marking.**
Spotted hyena pasting by depositing anal gland secretions on a grass stalk.

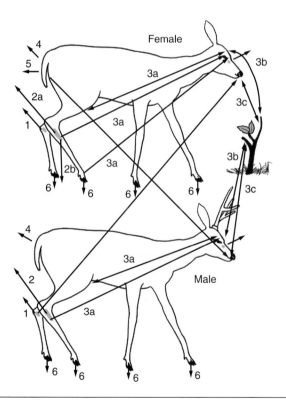

Figure 20.2   **Sources of scents used in intraspecific communication in the mule deer.**   The scents of the following are transmitted through the air: tarsal gland (1), metatarsal gland (2*a*), tail (4), and urine (5). When the animal lies down, the metatarsal gland marks the ground (2*b*). The tarsal gland, scented with urinary pheromones, is rubbed against the forehead (3*a*), and the forehead is rubbed against twigs (3*b*). Marked objects are sniffed and licked (3*c*). The interdigital glands (6) deposit scent on the ground.

Sound

Much more information about immediate conditions can be transmitted faster by sound than by chemicals. Sound can be produced by a single organ, can travel around objects and through dense vegetation, and can be used in the dark. Information can be conveyed by both frequency and amplitude modulation. The best frequency for a particular animal to use seems to depend on the environment. Brown and Waser (1984) demonstrated that there is a **"sound window"** for blue monkeys (*Cercopithecus mitis*) living in the forests of Kenya and Uganda. At around 200 hertz (Hz), sounds are attenuated very little and are relatively unaffected by background noises. This frequency corresponds to that of the "whoop-gobble" call produced by the adult male, which has a special vocal sac. Blue monkeys also seem to be much more sensitive to sounds in this range than the more terrestrial rhesus monkey (*Macaca mulatta*). Howler monkeys (*Alouatta* spp.) in the Neotropical rain forest also signal to groups many kilometers away with low-frequency calls, in the 40-to-100-Hz range. Animals with smaller home ranges, for example, squirrel monkeys (*Saimiri sciureus*), use higher frequency sounds, which dissipate rapidly after striking leaves and branches. Such calls serve to maintain contact among group members or neighboring troops without attracting predators from a distance.

Ultrasonic sounds (frequencies above those audible to humans; above 20 kHz) are used by a variety of animals, particularly mammals. The distress calls of young rodents and some of the vocalizations of dogs and wolves are well above the range of human hearing, as are the echolocation sounds of most bats. Although bat echolocation is used mainly to locate food and other objects, communication also occurs between predator and prey. Some noctuid moths, for example, do not produce sounds themselves, but they possess tympanic membranes on each side of the body that receive sonar pulses from bats (Roeder and Treat 1961). Depending on the location and intensity of sound stimulation, the moth may take evasive action—fly away in the opposite direction, dive, or desynchronize its wing beat to produce erratic flight.

Underwater sound has properties somewhat different from those of sound transmitted though the air, traveling much farther and faster. Marine mammals produce clicks, squeals, and longer, more complex sounds incorporating many frequencies. The short-duration sounds function in echolocation. Baleen whales (Suborder Mysticeti) produce lower and longer sounds than do the toothed whales (Suborder Odontoceti), such as dolphins. Payne and McVay (1971) analyzed the sounds of the humpback whale (*Megaptera novaeangliae*) and determined that their calls are varied and occur in sequences of 7 to 30 minutes duration and are then repeated (figure 20.4). The songs have a great deal of individuality, and each whale adheres to its own for many months before developing a new one. Researchers have not as yet ascribed a clear function to these sounds, but they may serve to maintain group cohesion across great distances because of the low attenuation of low-frequency sounds in water.

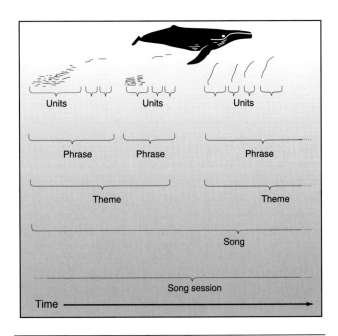

Figure 20.4　Song of the humpback whale. The song of a humpback whale can be broken up into units, phrases, themes, songs, and song sessions. Each whale sings its own variation of the song, which may last up to a half an hour.
Source: Data from R. S. Payne and S. McVay, "Songs of Humpback Whales" in Science, *17:585–597, August 13, 1971.*

Some species of terrestrial mammals produce very low (infrasonic) frequency sounds. Desert rodents such as kangaroo rats have greatly inflated auditory bullae, making them sensitive to low-frequency sounds and vibrations such as those produced by snakes and other predators. Banner-tailed kangaroo rats (*Dipodomys spectabilis*) defend their territories by foot drumming (Randall 1984), producing sounds in the 200-to-2000-Hz range. Both Asian elephants (*Elephas maximus*) and African elephants (*Loxodonta africana*) use very low frequency "rumbles," in the range of 14 to 35 Hz, to communicate over distances of several kilometers (Payne et al. 1986; Poole et al. 1988). Such infrasonic sounds have very high pressure levels (over 100 decibels [db]) and suffer little loss due to the environment. These calls seem to be used mainly for long-distance coordination of group movements and location of mates.

Vision

Visual displays enable the receiver to locate the signaler precisely in space and time and to identify very subtle signals, but they must usually operate in daylight and over short distances. Social and diurnal mammals such as many primates rely heavily on visual cues (chapter 13). Many cursorial, or running, mammals have white rump or tail patches; in the presence of predators, the hairs are erected and the tail waved, making a conspicuous display (figure 20.5). Observers have suggested several functions for this **flagging behavior,** such as:

Figure 20.5 **Rump patches in deer.** These patches may function to signal social status or danger.

1. Distracting the predator from other members of the group
2. Warning other group members
3. Confusing the predator when many group members are displaying
4. Signaling the predator that it has been detected
5. Eliciting premature pursuit

In the case of (4) or (5), the predator may go off in search of less-alert prey or may be lured into an unsuccessful pursuit (Smythe 1977). Others have argued that rump patches have little to do with defense against predators; rather they function in intraspecific social communication (Guthrie 1971). See Hasson (1991) for a review of communication between predators and prey.

Touch

In most primates and in some gregarious carnivores, grooming is an important social activity and functions not only to remove ectoparasites but also as a "social cement" in the reaffirmation of social bonds (figure 20.6). Although most grooming takes place between close relatives, long-term grooming patterns exist between some nonrelatives (Sade 1965; Smuts 1985). In the large, multimale groups characteristic of macaques (*Macaca* spp.) and baboons (*Papio* spp.), grooming between the sexes increases during the mating season. Widespread use of tactile stimuli occurs during copulation. In many rodents, stimulation of the rump region of a female in estrus produces concave arching of the back and immobility (lordosis). In mammals such as cats and rabbits, vaginal stimulation induces ovulation (chapter 9).

Electric Field

At present, no evidence exists that mammals use electric fields for communication as do some species of fish, but several recent studies suggest that some species can detect the

Figure 20.6 **Female rhesus monkey grooming offspring.** In addition to removing ectoparasites and other foreign matter from the skin and hair, grooming acts as "social cement," solidifying social bonds among group members. Although most grooming occurs between close relatives, it increases between unrelated males and females during the mating season.

weak electric fields produced by prey animals. The platypus (*Ornithorhynchus anatinus*) uses sensors on its bill to detect the electric field produced by earthworms (Scheich et al. 1986), and the star-nosed mole (*Condylura cristata*) has detectors in the tentacles of the star that may serve the same function (Gould et al. 1993). In the latter species, however, the main neural connections are to the somatosensory cortex, suggesting that touch is the more important sense (Catania and Kaas 1996).

Functions of Communication

Although we have briefly reviewed some of the properties of signals and the channels typically used by mammals, we still need to know what these signals are used for. All communication ultimately functions to increase fitness, but here we consider its more immediate, or proximate, functions (Wilson 1975; Smith 1984).

Group Spacing and Coordination

Cebus monkeys of the South American rain forest forage in dense vegetation. A group of 15 may spread out over an area 100 m in diameter as they search the treetops for fruits. In addition to the sound of moving branches, an observer hears a continual series of "contact" calls from the different members of the group. An individual that becomes isolated utters a "lost" call, which is much louder than the contact calls. Marler (1968) suggested that primates use the following types of spacing signals:

1. Distance-increasing signals, such as branch shaking, which may result in another group's moving away
2. Distance-maintaining signals, such as the dawn chorus of howler monkeys, which regulates the use of overlapping home ranges
3. Distance-reducing signals, for example, the contact or lost calls of *Cebus* monkeys
4. Proximity-maintaining signals such as occur during social grooming within groups

Recognition

Species recognition is important prior to mating to avoid infertile matings between members of closely related species. Mammals, however, communicate information that is more individualized than that needed for species recognition alone. Recognition of kin has now been demonstrated in a number of mammalian species—in some cases, even in the absence of interactions with kin early in life (Holmes and Sherman 1983). Kin recognition enables social species to behave nepotistically and thus increase their own inclusive fitness (chapter 22). It may also allow individuals to avoid possible costs of inbreeding incurred by mating with close relatives.

How might kin identify each other even if they have never interacted? One possibility is **phenotype matching,** a process in which the individual uses as a referent, or template, kin whose phenotypes are learned by association. The referent is then compared with the stranger; if the two are similar, then the stranger is treated like a relative (Holmes and Sherman 1983). Belding's ground squirrels (*Spermophilus beldingi*) recognized littermates later in life, even when separated from each other as neonates (Holmes 1988). The cues may come from products of major histocompatibility complex (MHC) genes, part of a cell-recognition system that is used by the immune system to distinguish self from nonself. In rodents, and possibly other mammal species, genetic differences in this region produce urinary odor cues that can be used to assess genetic relatedness among individuals (Brown and Eklund 1994).

Laboratory evidence suggests that house mice choose mates on the basis of subtle genetic differences at the level of the MHC (Yamazaki et al. 1983; Potts et al. 1991). In white-footed mice (*Peromyscus leucopus*), females avoided mating with close relatives such as parents and siblings and with individuals from other populations, preferring as mates individuals from their own population (Keane 1990). These examples suggest that mammals communicate information about genetic relatedness at both the kin and population levels.

Reproduction

Sexually receptive females or males may communicate to advertise their identity or condition, court a member of the opposite sex, form a bond, copulate, or perform postcopulatory displays. As described earlier, these behavior patterns are part of species identification, but they may also communicate an individual's reproductive condition. An example of such behavior is the roaring of red deer (*Cervus elaphus*) stags (figure 20.7). Males with high roaring rates obtain more copulations than do males with low roaring rates (Clutton-Brock et al. 1982).

Aggression and Social Status

In social groups in which members of a species are in close proximity, it is sometimes in an individual's best interest to compete with and fight with others for possession of a resource, be it food, space, or access to another individual. Physical combat is expensive in terms of energy and carries the risk of death or injury, even for winners. To avoid unnecessary energy expenditure and risk of injury, social species have evolved displays that communicate information about an individual's mood and about how the animal is likely to behave in the near future (figure 20.8). As a result of previous encounters, the animal may be dominant or submissive to another, and its behavior is thus predictable. Presented with a limiting resource, the submissive individual will yield to the dominant without an overt fight. Many species compete for resources only infrequently; thus, aggressive interactions make up only a small proportion of their behavioral repertoire.

Alarm

Mammals use a variety of signals to alert group members to danger. Although vocalizations used for territory defense (i.e., "songs") are often complex and vary greatly within and among closely related species, alarm calls are likely to be simple, hard to locate, and differ little among species. Members of species that live together and are endangered by the same predators benefit mutually by minimizing divergence among alarm vocalizations (Marler 1973). For example, Marler was unable to tell the difference between the "chirp" alarm calls of African blue monkeys and red-tailed monkeys (*Cercopithecus ascanius*), whereas the male songs differed greatly between the two species. Vervet monkeys (*Chlorocebus aethiops*) communicate **semantically** in that they use different signals to warn about different objects in their environment. Group members climb trees when they hear alarm calls given in response to leopards, look up when they hear eagle alarms, and look down when they hear snake alarms (Struhsaker 1967; Seyfarth et al. 1980). Young vervets give alarm calls in response to a variety of animals and often not to dangerous ones; however, their ability to classify predators and give appropriate alarm calls improves with age (Seyfarth et al. 1980).

Figure 20.7   **Roaring by red deer stag.**   During the rut on the Isle of Rhum, Scotland, the male roars to defend his harem of females against other males.

Mice and rats excrete a substance in their urine when they are electrically shocked, attacked by another mouse, or otherwise stressed. This substance acts as an alarm pheromone and may cause others to avoid the area (Rottman and Snowden 1972).

Figure 20.8   **Yawn display in baboon.**   Threat is communicated by dominant male.

Hunting for Food

A selection pressure in favor of group living is increased efficiency in finding food, which involves both communication about its location and cooperation in securing it. This attribute is exemplified in the African wild dog (*Lycaon pictus*), a canid distantly related to domestic dogs and wolves. Just prior to hunting prey that often consists of large ungulates, members of the pack engage in an intense greeting ceremony, or "rally," consisting of a frenzy of nosing, lip-licking, tail-wagging, and circling (Creel and Creel 1995). The rally ensures that pack members are alert and ready to hunt in a coordinated fashion (figure 20.9). Social carnivores may even communicate information about what type of prey they are about to hunt. For example, Kruuk (1972) noted that hyenas hunting zebra sometimes passed by prey they had hunted on other occasions. The hyenas' behavior made it apparent even

Figure 20.9   **Greeting by wild dogs.**   Wild dogs "rally" before setting out on a hunt.

to the human observer that they were interested only in hunting zebra.

Giving and Soliciting Care

A wide variety of signals is used between parent and offspring or among other relatives in the begging and offering of food, as can be observed with domestic cats and dogs. Baby mice, when they are chilled, emit high-frequency sounds that are inaudible to humans but can be heard by adult mice, who can assist them.

Soliciting Play

Play consists of behavior patterns that may have many different functions in the adult, such as sex, aggression, and exploration. The play bow in canids (figure 20.10) is communication about play and informs others that the motor patterns that follow are not the real thing (Bekoff 1977). Play behavior occurs in a wide variety of mammals and is rare or absent in other taxa. The function of play itself has been debated. Observers usually agree that they can recognize play but have had great difficulty ascribing definitions or functions to it. They most often suggest that the function of play is to practice motor skills and behavior patterns used later in adult life. Play may also serve the immediate function of enhancing muscle development and coordination (Fagen 1981).

Figure 20.10   Soliciting play in domestic dogs.   The play bow performed by the dog on the right communicates that the behavior patterns that follow are play.

AGGRESSION AND COMPETITION

Definitions

Aggression is behavior that appears to be intended to inflict noxious stimulation or destruction on another organism (Moyer 1976). The word *aggression* emphasizes offensive behavior. Behavioral ecologists take a more functional approach and consider aggression as a form of resource **competition,** in which rivals are actively excluded from some limited resource, such as food, shelter, or mates (Archer 1988).

Another term often used in the same way as aggression is **agonistic behavior.** This term includes all aspects of conflict, both attack (offensive) and escape (defensive). It includes threats, submissions, chases, and physical combat, but it specifically excludes predatory aggression, which is included in ingestive behavior (Scott 1972).

The most important forms of agonistic behavior (Wilson 1975; Moyer 1976) include

Territorial: exclusion of others from some physical space

Dominance: control, as a result of a previous encounter, of the behavior of a conspecific

Sexual: use of threats and physical punishment, usually by males, to obtain and retain mates

Parental: attacks on intruders when young are present

Parent-offspring: disciplinary action by parent against offspring, usually associated with weaning

Competition for Resources

Most agonistic behavior involves competition for some limited resource—namely, food; water; access to a member of the opposite sex; or space, such as sites for nesting, wintering, or refuge from predators. Competition can be divided into two forms: **exploitation,** in which individuals use resources and deprive others of them without directly interacting; and **interference,** in which organisms interact so as to reduce one another's access to or use of resources. In the tropics, many species of birds and mammals compete for fruits of various kinds. Fruit-eating bats forage by night, whereas fruit-eating birds, primates, and many rodents forage by day. In this instance, bats reduce the resource for the other taxa by exploitative competition with no direct interaction.

More often, competing individuals are not passive but interfere with others seeking the same resource. Some may establish territories and defend resources against others, or they may establish dominance and control the access of others to the resources, as happens during the breeding season when dominant male deer drive subordinate males away from territories containing females. The importance of competition in structuring populations and communities of mammals will become evident in later chapters in this part of the book.

SPATIAL RELATIONS

Home Range

Most organisms spend their lives in a relatively restricted part of the available habitat and learn the locations of food, water, and shelter in the vicinity. This area, which is used by an animal in its day-to-day activities and in which the animal spends most of its time, is its **home range.** It may be difficult to determine the actual boundaries of the home range because an animal or group may occasionally wander some distance away to a place it never revisits. Such excursions are rare, however, and identifying the boundaries of an individual's home range is generally fairly straightforward.

The size of the home range depends on the size of the animal as well as on the quality of resources the home range contains. Its size is an approximate function of the mass of the animal and its metabolic rate (McNab 1963). For mammals, $A = 6.76W^{0.63}$, where A (the area) equals the expected home range in acres and W equals the mass in kilograms. Thus a 20-g mouse should have a home range of about 0.57 acres, or 0.23 hectares (ha); this figure is within the observed range. The productivity of the habitat is also important: white-footed mice range farther in less-productive habitats; males also range farther than females within the same habitat as they search for mates. If the area is defended against conspecifics, we might expect a smaller value for the equation. This relationship between home range size and body mass suggests that ultimately what determines home range size of most mammals is the food supply.

The area of heaviest use within the home range is the **core area.** This location may contain a nest, sleeping areas, water source, or feeding site. As with home range, the designation of a core area is somewhat arbitrary but useful in understanding the behavior and ecology of different species or the same species in different habitats or at different population densities. Figure 20.11 illustrates home ranges and core areas of baboons in Africa.

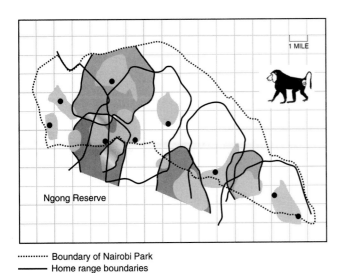

Figure 20.11   **Home ranges and core areas of groups of baboons in Nairobi Park, Kenya.**   Although home ranges of these nine groups overlap extensively, core areas overlap little.

Source: Data from Irven DeVore, Primate Behavior: Field Studies of Monkeys and Apes, *1965, Holt, Rinehart & Winston, Inc.*

Territory

An area occupied more or less exclusively by an individual or group and defended by overt aggression or advertisement is a **territory.** To demonstrate territory, we must show that an individual, mated pair, or group has exclusive use of some space, and we must also observe defense of that area.

Territorial animals spend much time patrolling the boundaries of their space, vocalizing, visiting scent posts, and making other displays. Such behavior would seem to take more time and energy than would simple exploitative competition. In fact, however, these displays often evolved so as to require relatively little energy, for once the territory is established, the neighbors are conditioned and need only occasional reminders to keep out. The cost of defense, then, could be less than the benefit of having exclusive use of a resource.

The key to when an animal should establish a territory seems to be **economic defendability** (Brown 1964), meaning that the costs (energy expenditure, risk of injury, etc.) are outweighed by the benefits (access to the resource). Such things as the distribution of the limited resource in space and whether or not the availability of the resource fluctuates seasonally are important factors. A limited resource—say, food—that is uniformly distributed in time and space is most efficiently used if members of the population spread themselves out through the habitat, possibly defending areas. A resource that is clumped in space and that is unpredictable might favor overlapping home ranges, colonial living, or possibly nomadism. The quality of food may also be important; high-energy food sources are more likely to be defended than low-energy sources. Species that construct elaborate burrow systems (e.g., ground squirrels of the Genus *Spermophilus*) or food caches (e.g., woodrats of the Genus *Neotoma*) should also be more likely to defend territories (Eisenberg 1981).

A mating system involving a peculiar type of territory is the **lek.** In a lek, the only resource that the organism defends is the space where mating takes place. Feeding and nesting occur away from the site. Lek, or arena, systems are characterized by promiscuous, communal mating; the males are likely to have evolved elaborate ornaments such as horns or antlers. The same area typically is used year after year. Males arrive early in the breeding season and, through highly ritualized agonistic behavior, stake out their plots. Certain territories or displaying males appear to be more attractive to females than neighboring ones, in the sense that some males do much more breeding than others. Females move through the areas while the males display, mate with one or more males, then leave. While on the lek, the males do little or no feeding; they spend all their time and energy patrolling the boundaries, displaying to other males, and attempting to attract females into their area. Although lek mating systems are rare among mammals, they do occur in a wide range of species, from some African antelope species (Balmford 1991) to African hammer-headed bats (*Hypsignathus monstrosus;* Bradbury 1977; figure 20.12).

Figure 20.12   Male hammer-headed bat on lek. In this type of territory, no resources are defended; the male displays to attract mates.

Dominance

When individuals live in a social group, as do many species of primates, carnivores, and ungulates, access to resources may be determined through dominance interactions rather than territoriality. One individual is dominant to another if it controls the behavior of the second individual (Scott 1966). In another sense, dominance is a prediction about the outcome of future competitive interactions (Rowell 1974). If four or five mice that are strangers to one another are put together, several outcomes are possible. Most likely a single mouse becomes a despot, being dominant to the rest, while all the subordinates are more or less equal.

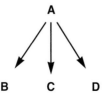

Another possibility is a linear hierarchy, or pecking order, often seen in primates such as baboons (*Papio* spp.), macaques (*Macaca* spp.), and spotted hyenas, in which A dominates B, B dominates C, and so on.

Sometimes triangular relationships form in primates, in which individual A dominates B, who dominates C, who, in turn, dominates A, creating a circular relationship.

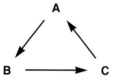

Also in primates, especially in chimpanzees, coalitions may affect dominance, such that A dominates B or C alone, but C and B together dominate A.

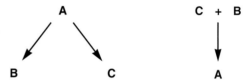

Among nonhuman primates, third parties may intervene, sometimes supporting the aggressor and sometimes the recipient of the aggression. The term **alliance** is used when two individuals repeatedly form coalitions. Alliances have been documented not only in nonhuman primates but also in bottlenose dolphins (*Tursiops truncatus;* Connor et al. 1992).

Much variation occurs in the intensity of dominance and the frequency of reversals, in which the subordinate individual wins an encounter with a dominant individual. Among closely related species and subspecies, one may see a clear-cut hierarchy in one species and not in the other. For example, captive African green monkeys (*Chlorocebus aethiops sabaeus*) from West Africa show little or no dominance hierarchy, even when access to some highly prized food is limited; however, the closely related vervet monkey (*C. aethiops*) from East Africa has a pronounced linear hierarchy. Furthermore, species that are territorial in the wild may, when crowded in captivity, change to a dominance hierarchy. Sometimes the dominance rank order is a function of the resource for which the animals are competing; for example, one individual may have first access to water, another to a favored breeding site.

Costs and Benefits of Dominance

Studies of mammals living in groups indicate that dominant animals are well fed and healthy. Subordinates may be malnourished or diseased and thus suffer higher mortality rates. Christian and Davis (1964) reviewed data for mammals showing that low-ranking individuals, frequent losers in fights, have higher levels of adrenal cortical hormones than do dominants. These hormones elevate blood sugar and prepare the animal for "fight or flight." The cost of this elevation in hormones is a reduction in antigen-antibody and inflammatory responses—the body's defense mechanisms—and a reduction in levels of reproductive hormones.

Field studies of baboons (*Papio anubis*) generally support these findings: in groups and stable dominance hierarchies with few rank reversals, dominant males had lower concentrations of adrenal cortical hormones than did subordinates (Sapolsky 1990). Furthermore, although basal testosterone levels were similar in high- and low-ranking males, those levels plummeted in low-ranking males under the stress of being darted and held captive for several hours. High-ranking males actually showed an increase in testosterone under such conditions (Sapolsky 1991).

Dominance is often directly correlated with reproductive success, as in northern elephant seals, in whom the highest-ranking males do virtually all of the mating (Haley et al. 1994). But in some systems, the highest-ranking male may be so busy displaying that lower-ranking males copulate with the females. In many species, the male dominance hierarchy is age-graded, with younger, lower-ranking males working their way up the hierarchy as older males die off or leave the group. Thus, low rank does not necessarily mean low lifetime reproductive success. Much-needed studies on lifetime reproductive success began to be published only in the last decade (Clutton-Brock 1988).

Popular literature has emphasized the importance of dominance and aggression in "keeping the species fit," because the survivors of fierce battles are the most vigorous and therefore "improve" the species by passing those traits on. The simplest and most likely explanation of aggression and dominance, however, is that dominant individuals benefit. As with other traits, an optimum level of aggression exists, depending on the individual's particular social and physical environment. Individuals that are too aggressive are selected against, as are those that are too passive. For example, female olive baboons (*Papio cynocephalus anubis*) show a linear dominance hierarchy, and dominant females have priority of access to scarce resources. Although high-ranking females have shorter interbirth intervals and higher infant survival rates than low-ranking females, they also suffer more miscarriages and long-term infertility (Packer et al. 1995). These results suggest that qualities essential to achieving high rank may also carry reproductive costs.

Figure 20.13 Infanticide in langur monkeys. The male monkey on the right has stolen the infant of a third female (not visible) and is being attacked by two other females from the same group. Males that have newly taken over a group frequently kill or injure infants sired by the old males.

Extreme Forms of Aggression: Infanticide and Siblicide

The hanuman langur monkey (*Semnopithecus entellus*) lives in a variety of habitats in India. In some areas, social groups of langurs contain only one adult male, 5 to 10 females, and their young. Several observers reported instances in which a new male came into the group, chased out the old male, and killed some or all of the infants (figure 20.13). One interpretation of such behavior is that the group is socially disorganized by the change in males, and the usual restraints on overt aggression are absent. Once the new male has established himself, he ceases his attack on the young. From the standpoint of the species or group, such behavior is maladaptive and should be rare, which, in fact, it is. A second interpretation of these events is that **infanticide** is adaptive, at least for the new male, because he removes the offspring of the presumably unrelated male and causes the females to come into estrus sooner to bear his own offspring (Hrdy 1977a; Hrdy 1977b). This view is consistent with the notion that social behavior results from the action of natural selection on the individual.

When new male lions first enter a pride in an attempted takeover, they kill the smaller cubs and evict the older cubs and subadults (Pusey and Packer 1987). Infanticide by incoming males accounts for about one-fourth of cub mortality (Packer et al. 1990); females therefore group their young into crèches to protect them from nomadic males. Males benefit by killing cubs because they eliminate the offspring of competing males and bring the females into estrus sooner.

Infanticide is also performed by a variety of rodent species. Under crowded conditions rodents often eat their young, usually those already dead. We could interpret such behavior as social pathology, which results from crowded conditions and leads to maladaptive behavior. We should note, however, that under crowded conditions or when food is scarce, the young would probably not survive anyway, so the parents save their reproductive investment for more propitious times by consuming their young. Under adverse conditions, rodents and lagomorphs resorb developing embryos in the uterus, with the same effect as cannibalism. In kangaroos, most of the development of the young takes place in the pouch, so females can terminate "pregnancy" simply by throwing the young out of the pouch.

Such killing of young is not limited to adults. Among spotted hyenas, the young fight vigorously while in the den, forming dominance relationships within hours. One sibling often kills the other, especially when the two are of the same sex (Frank et al. 1991). When food supplies are adequate, however, both cubs typically survive and become close allies (Smale et al. 1995). For additional papers on infanticide, see Hausfater and Hrdy (1984) and Parmagiano and vom Saal (1994).

Summary

Communication may be defined as an action on the part of one organism that alters the behavior of another organism in a fashion adaptive to either the sender alone or both the sender and the receiver. Behavior patterns that are specially adapted to serve as social signals are termed displays. Some definitions of communication imply that both sender and receiver must benefit and that evolution moves toward maximization of information transfer. An alternative view argues that communication is a means by which the sender manipulates the receiver, who may benefit or may be harmed; the purpose of a display is to persuade, not to inform.

Social signals, which vary in fade-out time, effective distance, and duration, convey information by being discrete or graded. They may be combined to form composite signals, and the order in which they appear may affect the information transmitted (syntax).

Channels of communication include odor (mainly via pheromones), sound, touch, and vision. Communication functions in group spacing and coordination; species, population, mate, and kin recognition; reproduction; agonism and social status; alarm; hunting for food; giving and soliciting care; and soliciting play.

Agonistic behavior, defined as social fighting among conspecifics, includes all aspects of conflict, such as threat, submission, chasing, and physical combat, but excludes predation. Aggression emphasizes overt acts intended to inflict damage on another and may include predation, defensive at-

tacks on predators by prey, and attacks on inanimate objects. Agonistic behavior involves competition, one form of which is exploitation, as individuals passively use up limited resources. A second form is interference, in which individuals actively defend resources.

Much conflict behavior involves the social use of space. The home range is the area used habitually by an individual or group; the core area is the zone of heaviest use within the home range. Territory is the space that is used exclusively by an individual or group and is defended. Individual territories occur most often when a needed resource is predictable and uniformly distributed in space. Some territories include food and water supply, nest site, and mates; other territories contain only one defended resource, for example, a place where only mating takes place, such as in the lek.

Species that live in more or less permanent groups usually develop dominance hierarchies in which individuals control the behavior of conspecifics on the basis of the results of previous encounters. Dominance hierarchies are not always determined by fighting ability; age, seniority, maternal lineage, and alliances with friends have been shown to be important factors in some mammals.

Agonistic behavior is usually highly ritualized and communicative in nature; killing or wounding is infrequent. Recent observations interpret overt violence, such as infanticide, as an expression of genetically selfish behavior on the part of an individual rather than as a maladaptive response to abnormal conditions.

Discussion Questions

1. Distinguish among each of the following terms: *communication, message, meaning, signal,* and *display,* giving examples of each.
2. Table 20.1 illustrates the properties of different signal channels in terms of propagation through the environment. Construct a second table that shows the primary channels used by each order of mammal. Do you see any trends in the results based on the evolutionary history (phylogeny) or ecology of the groups?
3. Contrasting views have been put forth about the evolution of communication systems in animals: (1) displays evolve so as to maximize information transfer between sender and receiver (i.e., both sender and receiver benefit), and (2) displays evolve so as to maximize the ability of the sender to manipulate the behavior of the receiver. Give examples of mammalian communication systems illustrating both of these views.
4. In this chapter, we saw examples of how aggressive, dominant individuals have greater reproductive success than low-ranking individuals. Why doesn't this trend lead to the evolution of higher and higher levels of aggression?

Suggested Readings

Archer, J. 1988. The behavioural biology of aggression. Cambridge Univ. Press, New York.

Halliday, T. R. and P. J. B. Slater, eds. 1983. Animal behavior, vol. 2: communication. W. H. Freeman, New York.

Harcourt, A. H. and F. B. M. de Waal, eds. 1992. Coalitions and alliances in humans and other animals. Oxford Univ. Press, New York.

Harper, D. G. C. 1991. Communication. Pp. 374–397 *in* Behavioural ecology: an evolutionary approach (J. R. Krebs and N. B. Davies, eds.), 3d ed. Blackwell, Boston.

Sebeok, T. A., ed. 1984. How animals communicate, 2d ed. Indiana Univ. Press, Bloomington.

CHAPTER

21

Sexual Selection, Parental Care, and Mating Systems

Chapters 7 and 9 dealt with how hormones and environmental factors interact to influence reproduction (proximate causation). This chapter seeks to explain why natural selection has led to the reproductive patterns we observe in nature (ultimate causation). Why do males in some species of mammals fight vigorously for access to females, but those in others fight little and form long-term pair bonds with females? Why is the male northern elephant seal (*Mirounga angustirostris*) more than three times bigger than the female, the male silver-backed jackal (*Canis mesomelas*) about the same size as the female, and the male roof rat (*Rattus rattus*) smaller than the female (Ralls 1976)? In this chapter, we deal with questions that address the ultimate evolutionary forces that have led to sexual dimorphism in some species of mammals. We will attempt to determine why males of some species provide little care for young but those of other species share parental responsibility equally with the female, and why males and females of some species form pair bonds but in others one sex has multiple mates.

ANISOGAMY AND THE BATEMAN GRADIENT

Many organisms reproduce asexually during one part of their life cycle and reproduce sexually at other times, but mammals are less flexible and rely exclusively on sexual reproduction. In mammals, the two sexes are anatomically different and are produced in about equal numbers. Female mammals produce few large, sessile, and energetically expensive eggs, and males produce many small, motile, and energetically cheap sperm. This difference in gamete size, referred to as **anisogamy,** sets the stage for a host of differences in the reproductive behavior patterns of males and females (Trivers 1972). The relatively small number of expensive female gametes are likely to be a limited resource for which males compete. A male's reproductive success is likely to be a function of how many different females he can inseminate, whereas a female's success depends on how many eggs and young she can produce. Several features of mammalian reproduction further promote differences in reproductive behavior between the sexes: fertilization occurs in the oviducts, so it is certain that the female is the biological parent of offspring born to her, but paternity is not as certain. Also, only female mammals gestate and lactate, thus limiting the investment males can make in offspring.

One of the results of these differences is that the sexes are likely to follow different reproductive strategies. Bateman (1948) demonstrated in laboratory populations of fruit flies (*Drosophila melanogaster*) that nearly all the females mated. Some males, however, mated several times, and others failed to mate at all. In other words, the variance in copulatory success was higher for males than for females. Males that copulated most also sired the most offspring, the **Bateman gradient** (Andersson 1994). Females, on the other

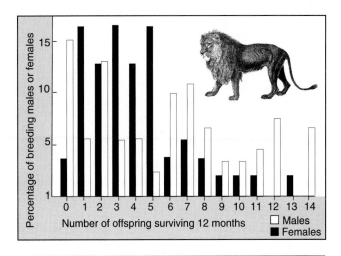

Figure 21.1   **Lifetime reproductive success for African lions.** Female reproductive success varies little; most females produce 1–5 surviving offspring. Male reproductive success varies widely; males sire 0–14 offspring.

Source: Data from C. Packer et al., "Reproductive Success of Lions" in Reproductive Success, *T.H. Clutton-Brock, ed., 1988, University of Chicago Press, Chicago, IL.*

hand, needed only mate once to produce the maximum number of offspring.

When reproductive success is determined over an animal's entire lifetime, the Bateman gradient becomes evident, especially in species where male–male competition for access to grouped females is intense. For example, male lions (*Panthera leo*; Packer et al. 1988; figure 21.1) and northern elephant seals (Haley et al. 1994) have much a higher variance in lifetime reproductive success than females.

SEX RATIO

The sex ratios of populations of most species tend to be about 50:50 at birth or hatching but may deviate significantly among adults. The **operational sex ratio** considers only the reproductively active members of the population. Deviations in this ratio can have a large influence on the mating system, as members of the abundant sex compete for access to the scarcer one.

According to a model based on maternal condition proposed by Trivers and Willard (1973), the Bateman gradient sets the stage for another possible deviation from a 1:1 sex ratio. The model assumes that mothers in the best physical condition produce healthier offspring that are better able to compete for mates or other resources. Because males are more likely than females to compete for additional mates, the reproductive success of males is highly variable, and males should require greater maternal investment. For example, male antarctic fur seal pups (*Arctocephalus gazella*) are heavier at birth, grow faster, and weigh more at 60 days than their sisters (Goldsworthy 1995). Trivers and Willard (1973) argue that reproductive success of sons should be high if their mothers are in good condition, but low, perhaps zero, if their mothers are in poor shape. Female offspring are likely to breed anyway, regardless of their mother's condition. Citing data from mammals, such as mink, deer, seals, sheep, and pigs, these investigators predicted that a female would produce male offspring if she were in good condition and female offspring if she were in poor condition.

More recent studies lend support to Trivers and Willard's model. Dominant red deer (*Cervus elaphus*) hinds (i.e., females) have access to the best feeding sites and are able to invest more in their offspring via lactation (Clutton-Brock et al. 1984). Male offspring in this species grow faster than do females and would seem to benefit more from this greater investment. As adults, males compete intensely to control harems. A female that produces a successful son can achieve more than twice the reproductive success of a female producing a daughter. As predicted by Trivers and Willard (1973), a positive correlation exists between dominance rank of hinds and the percentage of sons produced (figure 21.2). Similarly, in a field study of opossums (*Didelphis marsupialis*) in Venezuela, females receiving supplemental food produced more sons than daughters (Austad and Sunquist 1987). Furthermore, in a sample of old females in poor condition, the sex ratio was skewed toward female offspring, as predicted.

Models different from that of Trivers and Willard have been proposed to explain biased sex ratios in other mammals. In studies of a South African prosimian primate, the bushbaby (*Otolemur crassicaudatus*), Clark (1978) notes that the sex ratio was male-biased. Female offspring tended to remain near the mother's home range, but males dispersed, the usual patterns for mammals (see chapter 23). Daughters therefore compete with their mothers and sisters for food. By producing fewer daughters, Clark (1978) argues, local resource competition is reduced. A troop of baboons (*Papio anubis*) living in Amboseli National Park, Kenya, has been studied for many years by Altmann and colleagues (1988). Dominant females produce more daughters than sons, whereas subordinate females produce more sons than daughters, exactly the opposite result predicted by Trivers and Willard. Altmann and colleagues (1988) point out that daughters, who remain in their natal group for life, share the social rank of their mothers and can benefit from their mother's high rank. Presumably high-ranking families get larger and even more powerful as more and more daughters are born. On the other hand, as with bush babies, sons usually leave the natal troop and thus do not benefit by their mother's rank. In these baboons, the best strategy for dominant females is to produce daughters, whereas for subordinate females, it is to produce sons.

It is not clear how such adjustments of sex ratios occur at the proximate level. Some evidence points to the timing of insemination within the estrous cycle, such that more of one sex is conceived earlier and more of the other sex later in the cycle, as in white-tailed deer (*Odocoileus virginiana;* Verme and Ozoga 1981), Norway rats (*Rattus norvegicus;* Hedricks and McClintock 1990), hamsters (*Mesocricetus auratus;* Huck et al. 1990), and humans. A number of studies, including those in humans, suggest that levels of sex hormones present at the time of conception affect offspring sex ratios (James 1996). In many species of mammals, the sex ratio at conception is male-biased. Intrauterine mortality rates are typically higher for male than for female embryos, and these rates are probably highest among mothers in poor condition. In this way, the maternal condition hypothesis could be explained, as females in poor condition would lose their male embryos and possibly come into estrus again. After birth, other mechanisms may operate to select for one sex or the other.

SEXUAL SELECTION

The success of an individual is measured not only by the number of offspring it leaves, but also by the quality or probable reproductive success of those offspring. Thus, it becomes important who its mate will be. Darwin (1871) introduced the concept of **sexual selection,** a process that produces anatomical and behavioral traits that affect an individual's ability to acquire mates. He believed that sexual selection was a different process than natural selection because the former could produce traits that might reduce an individual's survival and thus oppose natural selection. Sexual selection can be divided into two types: one in which members of one sex choose certain mates of the other sex (**intersexual selection**) and a second in which individuals of one sex compete among themselves for access to the other sex (**intrasex-**

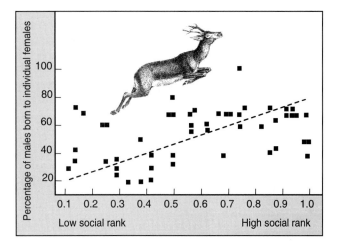

Figure 21.2   Red deer birth sex ratios.   Birth sex ratios produced by individual red deer females differing in social rank over their life spans. Measures of maternal rank were based on the ratio of animals the subject threatened or displaced to animals that threatened or displaced it. High-ranking females tended to have sons.

Source: T. H. Clutton-Brock, S. D. Albon, and F. E. Guinness, "Maternal Dominance, Breeding Success and Birth Sex Ratios in Red Deer" in Nature *308:358–360, 1984.*

ual selection). One result of either mechanism is that the sexes come to look different, that is, they are **dimorphic.**

Intersexual Selection

Often, individuals of one sex (usually the males) "advertise" that they are worthy of an investment; then members of the other sex (usually the females) choose among them. Most naturalists after Darwin discounted the importance of mate choice in evolution, but it has recently become a popular topic of study. Fisher (1958) developed a model for how exaggerated traits could evolve. Suppose some trait (e.g., antlers) that appears in males is preferred by females, for whatever reason. Assuming that the trait is to some extent heritable, these females are likely to produce sons with that trait and daughters that prefer that trait when choosing a mate of their own. Further development of the trait proceeds in males, as does the preference for it in the females, resulting in a **runaway selection** process. Traits produced by runaway sexual selection are arbitrary and are not linked to a male's fitness, except by female choice.

Sexually selected traits may become so exaggerated that survival of the males is reduced. Counterselection in favor of less ornamented males should then occur, and the effects of natural and sexual selection would be brought to a steady state. An often-cited example of runaway sexual selection is the extinct Irish elk (*Megaloceros giganteus*). The enormous antlers of Irish elk—even larger than those of the moose, shown in figure 21.3—apparently functioned for social display, but little evidence supports the notion that heavy antlers directly caused their extinction (Barnosky 1985).

As alternatives to Fisher's runaway selection hypothesis are several "good genes," or **indicator models,** which assume that the trait favored by females in some way indicates male fitness. One is the **handicap** hypothesis proposed by Zahavi (1975). Agreeing that sexual selection can produce traits that are detrimental to survival, he adds that they are both costly to produce and linked to superior qualities in the males. Important to this hypothesis is the notion of "**truth in advertising**": the male's handicap must be honest and linked to overall genetic fitness. Only in this way do females benefit by picking a male with this handicap. Antlers of deer could be viewed as a handicap because any male that can bear the cost of such structures and still survive must be fit indeed. One problem with this model is that when a male with a handicap is picked by a female, not only are his favorable traits passed on to his offspring but so are the genes for the handicap, which should be selected against.

Although Darwin and Fisher thought of sexual selection as a process distinct from natural selection, some biologists argue that the two are inseparable. Kodric-Brown and Brown (1984) claim that most sexually selected traits are aids, rather than handicaps, to survival. Thus, a male deer with a big set of antlers may be dominant over other males and may have better access to a food supply. They further argue that the sexually selected trait must be a reliable indicator of the male's condition. Thus, the trait must not be entirely genetically fixed but rather must be influenced by environmental conditions.

Evidence for an indicator role for a sexually selected trait comes from a hypothesis put forth by Hamilton and Zuk (1984). Working with birds, they argued that sexually selected traits have evolved to reveal an animal's state of health, specifically whether or not it is free from disease or parasitism. They predicted that species in which the males have showy plumage would be more prone to infestation with blood parasites. Comparisons of plumage showiness among museum specimens confirmed their prediction. They also predicted that within such species, brightness of plumage is linked to overall condition, such that brighter males are relatively disease-free and are therefore preferred as mates. Some experimental evidence, mostly from birds, supports their model, but the idea is still controversial, and alternative explanations have been proposed. So far, there are few tests of the Hamilton-Zuk hypothesis in mammals. In humans, Low (1990) found no direct evidence that sexual selection was linked to pathogen stress, although societies with high levels of pathogen stress tended to be polygynous, with some males having more than one mate.

In addition to the size or color of sexually selected traits, the symmetry of paired traits may also indicate fitness.

Figure 21.3   A sexually selected trait. Antlers of a bull moose during the breeding season.

Fluctuating asymmetry refers to random deviations from bilateral symmetry in paired traits (Andersson 1994). This means, for example, that when a paired trait, such as horns or canines, is measured for length, thickness, or some other attribute, the right and left sides may differ. These deviations are thought to reflect the inability of the organism to maintain developmental homeostasis (i.e., symmetry) in the presence of environmental variation and stress. Greater asymmetry is associated with low food quality and quantity, pollution, disease, and genetic factors, such as inbreeding, hybridization, and mutation. Symmetrical individuals are likely to be dominant and preferred as mates. Thus, male fallow deer (*Dama dama*) with symmetrical antlers were dominant over those with asymmetrical antlers (Maylon and Healy 1994), and male oribi (*Ourebia ourebi*) with symmetrical horns had larger harems than did asymmetrical males (Arcese 1994).

Fluctuating asymmetry may indicate fitness in both sexes. Among gemsbok (*Oryx gazella*), both males and females with asymmetrical horns were in poorer condition and lost more aggressive encounters than did those with symmetrical horns (Møller et al. 1996). Furthermore, symmetrical males were more often territorial breeders than were asymmetrical males, and symmetrical females more often had calves than did asymmetrical females.

Fluctuating asymmetry is most pronounced in sexually selected traits. For example, canines are used as weapons in male–male competition in a variety of primates. In general, males (but not females) from species subject to the strongest sexual selection (as indicated by size dimorphism and male–male competition) showed the highest asymmetry in canines (Manning and Chamberlain 1993). No relationship was found between canine asymmetry and body mass or diet type. The connection between fluctuating asymmetry and environmental stress is suggested by the marked increase in asymmetry of canines (but not of premolars) of lowland gorillas (*Gorilla gorilla*) during the twentieth century (Manning and Chamberlain 1994). Such an increase is consistent with environmental degradation associated with increasing rates of deforestation during this century.

Intrasexual Selection

Intrasexual selection involves competition within one sex (usually males), with the winner gaining access to the opposite sex. Competition may take place before mating, as with ungulates such as deer (Cervidae) and African antelope (Bovidae). Typically, males live most of the year in all-male herds; as the breeding season approaches, they engage in highly ritualized battles, using their antlers or horns. The winners of these battles gain dominance and do most of the mating (figure 21.4). Antlers are better developed in those cervid species in which males compete strongly for large groups of females (Clutton-Brock et al. 1982).

It is often difficult to determine which type of sexual selection is operating to produce an observed effect, because

Figure 21.4 **Use of antlers in combat between Père David deer.** Winners of contests become dominant and control access to resources such as mates.

members of both sexes may be present during courtship. As an example, note that the antlers of deer demonstrate the effects of both female choice and male–male competition. Females may incite competition among males and thus maintain some control over the choice of mate. For example, female northern elephant seals vocalize loudly whenever a male attempts to copulate. This behavior attracts other males and tests the dominance of the male attempting to mate (Cox and Le Boeuf 1977). In response to the female's sounds, the dominant harem master drives off low-ranking, potentially inferior mating partners.

Sperm Competition

Competition among males to sire offspring does not necessarily cease with the act of copulation. Females of some species may mate with several males during a single estrous period, creating the possibility of **sperm competition,** that is, a situation in which one male's sperm fertilize a disproportionate share of eggs (Parker 1970). Rather than being thought of as sperm actually fighting it out to gain access to eggs, sperm competition can be considered a selection pressure leading to two opposing types of adaptation in males: those that reduce the chances that a second male's sperm will be used (first-male advantage) and those that reduce the chances that the first male's sperm will be used (second-male advantage; Gromko et al. 1984).

Adaptations of first males include mate-guarding behavior and the deposition of copulatory plugs, both of which reduce the chance of sperm displacement by a second male. In mammals, adaptations of second males are probably restricted to dilution of the first male's sperm by frequent ejaculation of large amounts of sperm from a second male. Female Rocky Mountain bighorn sheep (*Ovis canadensis*) usually mate with more than one male during estrus. Dominant rams seek to guard estrous females from forced copulations by subordinate males but are not always successful. If a

Figure 21.5   Sperm competition in sheep. Dominant rams copulate at high frequency immediately after a "sneak" copulation by a subordinate ram.

subordinate male does achieve a copulation, the dominant male immediately copulates with that female himself (figure 21.5), probably reducing the chances that the subordinate male's sperm will fertilize the egg (Hogg 1988).

Copulatory plugs occur in rodents (Dewsbury 1988), bats (Fenton 1984), and some primates (Strier 1992). Although the plugs in guinea pigs (*Cavia porcellus*) appear to block subsequent inseminations (Martan and Shepherd 1976), those in deer mice (*Peromyscus maniculatus*) are ineffective (Dewsbury 1988). In the latter species, the plugs probably function to retain the sperm within the female's reproductive tract. Dewsbury (1984) found that in the muroid rodents he tested, the last male to mate or the male ejaculating most often sired most of the offspring.

Sperm competition has been suggested as an explanation for the relatively large number of deformed sperm in mammals—up to 40% in humans, for example. Baker and Bellis (1988) argue that these deformed sperm play a "kamikaze" role, staying behind to form a plug to inhibit passage of sperm from a second male. Meanwhile, a small number of "egg-getter" sperm proceed to the oviduct to attempt fertilization. The kamikaze sperm hypothesis has been criticized on several grounds: selection should favor use of seminal fluids rather than sperm to form the plug, and there should be more deformed sperm in ejaculates from species in which the female is likely to mate with several males. No such relationship seems to exist, however (Harcourt 1989). Clearly, more studies are needed to test this hypothesis adequately.

In primates, the amount and quality of sperm that the male produces are related to the type of mating system. In gorillas (*Gorilla gorilla*) and orangutans (*Pongo pygmaeus*), the winners of male–male competition have relatively free access to females. In chimpanzees (*Pan troglodytes*), however,

several males may attempt to mate with a female in estrus. Møller (1988) argues that in this case, competition takes place in the female's fallopian tubes and that the male with the most and best sperm fertilizes the egg. In fact, chimps have larger testes than other apes and produce a high-quality ejaculate—greater numbers of and more motile sperm.

Among woolly spider monkeys (*Brachyteles arachnoides*), males sometimes form aggregations, taking turns copulating with an estrous female at intervals of about 20 minutes (Milton 1985; Strier 1992). Copious amounts of ejaculate are produced, and a plug forms in the female's vagina. This plug is rather easily removed by subsequent males, so it may not function as a deterrent. As might be predicted from the previous discussion, these monkeys have extraordinarily large testes. Rather than competing aggressively for access to females, these males appear to engage in sperm competition (Milton 1985).

Among mammals in general, a similar relationship exists: species in which only one male has access to one or more females have smaller testes than do species in which more than one male has access to females (Kenagy and Trombulak 1986; figure 21.6). Some notable exceptions do occur, however. The supposedly monogamous western grasshopper mouse, *Onychomys torridus*, has extremely large testes (nearly five times the predicted size), in spite of the presumed lack of sperm competition. Lions, on the other hand, have rather small testes, in spite of the observation that one male may copulate more than 50 times in 24 hours with the same female, and a female often mates with more than one pride male (Kenagy and Trombulak 1986).

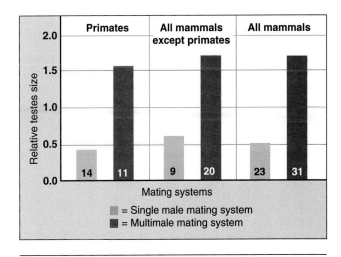

Figure 21.6   Mean relative testes size of mammals in relation to mating system. Sample sizes are at the bottom of each column. Species in which more than one male mates with a female have relatively large testes.

Source: Data from G.J. Kenagy and S.C. Trombulak, "Size and Function of Mammalian Testes in Relation to Body Size" in Journal of Mammalogy, *67:1–22, 1986.*

Brown and colleagues (1995) have suggested that factors other than sperm competition among males may explain the differences in relative testes mass among primates. They argue that the size and length of the female's vagina may determine how much sperm is needed to effect fertilization: a larger or longer vagina requires more ejaculate. They note that the vagina of a female gorilla is approximately 10 cm long, but that of a chimpanzee is about 17 cm. The latter is made even longer by the genital swelling during estrus. The factors that affect vagina size and shape are yet to be explored.

Postcopulation Competition

Following conception, male–male competition may take a different form. In some species of mice, the **Bruce effect** operates early in pregnancy: a strange male (or his odor) causes the female to abort and become receptive (Bruce 1966). Among langur monkeys (*Semnopithecus entellus*), strange males may take over a group, driving out the resident male (see figure 20.13). The new male may then kill the young sired by the previous male (Hrdy 1977a). Females who have lost their young soon become sexually receptive, and the new resident male can inseminate them. Infanticide by adult males thus may be viewed as a second-male adaptation. Similar findings have been made for lions—coalitions of males kill cubs on taking over a pride (Packer 1986).

PARENTAL INVESTMENT

Sexually reproducing species face a number of reproductive decisions, although not necessarily conscious ones: how much of the resources available should be spent on reproduction versus continued growth and survival? Of those resources spent on reproduction, how should they be allocated to individual offspring? Once young are born, what may the parents do to improve their offspring's chances to survive? Although most animals provide no care for their offspring, one parent or both provide at least some care in all species of mammals. We define **parental investment** as any behavior pattern that increases the offspring's chances of survival at the cost of the parent's ability to rear future offspring (Trivers 1972). Because an egg requires a greater investment of energy than does a sperm, male and female parental strategies are expected to differ. Because eggs are likely to be limited in number, we expect that males compete for the opportunity to fertilize them and thus are subject to sexual selection, as we have just seen. Figure 21.7 shows that the cost of parental investment rises more quickly for females than for males. The optimum number of offspring is thus lower for females than for males. A female is likely to mate, but, given her already large investment, her ability to invest further may be limited; thus, she is choosy about which male fertilizes her eggs. Males, on the other hand, try to inseminate as many females as possible. Mammalian mating systems in which the male mates with more than one female should be the most common (Trivers 1972).

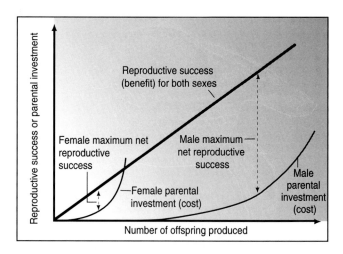

Figure 21.7 **Parental investment and reproductive success as a function of number of offspring produced.** Because the parental investment (cost) for females usually rises more steeply than for males, the optimum number of offspring (highest reproductive success at lowest cost) for females is less than for males. Males, consequently, must seek more than one mate to attain maximum net reproductive success. Broken lines indicate maximum female and male net reproductive success.

Which Sex Invests?

Even though only one sperm fertilizes an egg, millions are generally required in each ejaculation to ensure fertilization of even a single egg (Dewsbury 1982). Also, the number of times most males can ejaculate within a certain period is limited. Remember also that sperm competition in males of some species leads to selection for increased sperm production. Thus, a male's investment in sperm is not necessarily trivial, and he too can be expected to be somewhat choosy about his mate.

Factors other than anisogamy may affect the contribution of each sex to parental care. Certain taxonomic groups of animals are predisposed to a particular pattern: male mammals typically mate with more than one female, and the male contributes little to raising the offspring. Trivers (1972) reasons that confidence of parentage might explain which sex cares for the young. In species in which internal fertilization occurs and in which sperm competition could take place, the male has no way of "knowing," consciously or otherwise, whether or not his sperm fertilized the egg. He might therefore be inclined to desert the females with which he has copulated and seek additional mates. The female, on the other hand, is certain of her genetic relationship to the offspring, so she invests further in current offspring. Another reason why females might stay with the young, suggested by Williams (1975), is that parental care evolved in the sex that is most closely associated physically with the embryos. In mammals, that sex is always female.

Because gestation and milk production are restricted to the female in mammals, the male can do relatively little to provide direct care for the young. In mammals whose young

Figure 21.8   **Extreme sexual dimorphism in elephant seals.**   Among a herd of females, two males fight to establish dominance. Males differ strikingly from females; are about three times larger; possess an enlarged snout, or proboscis; and have cornified skin around the neck. In this highly polygynous species, males invest nothing in their offspring other than DNA from sperm.

are relatively advanced (precocial) at birth, the opportunities for male investment are even lower. In these species males compete for multiple mates more than they do in species whose young are immature (altricial) at birth and in which the male can share more evenly the investment with the female (Zeveloff and Boyce 1980).

Competition among males of some mammalian species is intense. For example, mating of northern elephant seals takes place in colonies; males establish dominance hierarchies, and only the high-ranking males breed (Le Boeuf 1974; Le Boeuf and Reiter 1988; Haley et al. 1994). Typically, less than one-third of the males copulate at all, and the top five males do at least 50 percent of the copulating (figures 21.8 and 21.9). The males make no investment in offspring beyond the sperm, as evidenced by the fact that males may trample pups as they strive to inseminate females. The males probably have no way of knowing which young are their own because pups are born a year after copulation.

At the opposite extreme are species in which the sexes share more or less equally in care of young, as in canids such as silver-backed jackals (Moehlman 1983). Both parents defend the territory and hunt cooperatively. The males play a crucial role, because in cases in which the father disappears,

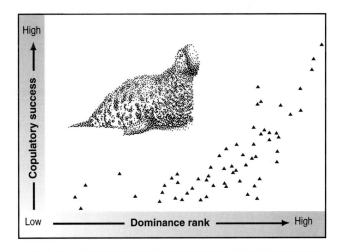

Figure 21.9   **Dominance and sex in elephant seals.**
Copulatory success as a function of dominance in elephant seals for 72 males.

Source: Data from M.P. Haley et al., "Size, Dominance and Copulatory Success in Male Northern Elephant Seals, Mirounga angustirostris,*" in* Animal Behaviour, *48:1249–1260, 1994.*

the female and her offspring die (Moehlman 1986). Among tamarins and marmosets (Family Callitrichidae), females produce twins. Adult males typically show strong interest in the infants soon after birth and carry them as much or more than the female does (Vogt 1984).

Parental Care and Ecological Factors

The extent of parental care varies widely among different mammalian species. Parental care in primates may last for several years, up to 25 percent of the offspring's life span. The kinds of parent–offspring relationships also vary, however, and are not simple to describe.

The term **reproductive effort** denotes both the energy expended and the risk taken for breeding that reduces reproductive success in the future. Finding mates and caring for young take extra energy, which may make parents more vulnerable to predators. Individuals are faced with the decision, conscious or otherwise, of whether to breed now or wait until later. If they choose to breed now, they need to decide if some effort should be spared for another attempt later.

Several environmental factors influence the investment parents make in their young after birth. For example, species adapted to stable environments have a tendency toward larger body size, slower development, longer life span, and having young at repeated intervals (**iteroparity**) rather than all at once (**semelparity**). Semelparity is very rare in mammals (see chapter 24). Typically, iteroparous individuals occupy a home range or territory. These stable conditions favor production of small numbers of young that receive extensive care and thus have a low mortality rate. Such species are said to be *K*-**selected,** in reference to the fact that populations are usually at, or near, *K*, the **carrying capacity** of the environment (Pianka 1970). Intraspecific competition is likely to be intense, and the emphasis is on producing few, high-quality offspring rather than large numbers (e.g., in gorillas).

Species that are adapted to fluctuating environments have high reproductive rates, rapid development, small body size, and provide little parental care. Their populations tend to be controlled by physical factors, and their mortality rate is high. Such species are said to be *r*-**selected,** where *r* refers to the reproductive rate of the population (Pianka 1970). For example, meadow voles (*Microtus pennsylvanicus*) are considered to be *r*-selected: they mature at an early age and have larger, more frequent litters than other, similarly sized rodent species.

In environments in which survival of offspring is low and unpredictable, parents may be expected to "hedge their bets" and put in a small reproductive effort each season. The predictions of **"bet-hedging"** thus seem to contradict those of *r*- and *K*-selection. In unstable, unpredictable environments, *r*-selection should operate, yielding high reproductive rates. Bet-hedging theory, however, predicts low reproductive rates and the spreading of reproductive effort across many breeding seasons. These and other problems have led many researchers to reject the whole concept of *r*- and *K*-

Figure 21.10   Extended family of rhesus monkeys. The 3-year-old son at left will soon leave the group, but will probably not breed for several more years.

selection (Stearns 1992), but most mammalogists still find it a useful way to classify patterns of mammalian life history. For a further discussion of life history traits, see chapter 24.

Prolonged dependency and extensive parental care are also favored when a species, such as many larger felids and canids, depends on food that is scarce and difficult to obtain. Much effort is spent searching for prey, and in some species, cooperation is needed for the kill. During the prolonged developmental period, the young benefit from considerable learning through observation of parents and play.

The Old World monkeys (Cercopithecidae) and the great apes (Pongidae) have the longest period of dependency. Typical of these species is an infancy of 1.5 to 3.3 years and a juvenile phase of 6 to 7 years, making up nearly a third of the total life span (figure 21.10). The prolonged dependency in these large-brained species may be related to their complex social life, in which they must recognize and remember interactions with many individuals across long spans of time.

PARENT–OFFSPRING CONFLICT

Anyone who has raised children or grown up with a sibling has observed frequent disagreement between parent and child. It is not unusual to see a mother rhesus monkey (*Macaca mulatta*) swat her 10-month-old infant or raise her hand over her head, thereby pulling her nipple from its mouth. Frequently, the infant responds by throwing a "temper tantrum." We can view much of the conflict as a disagreement over the amount of time, attention, or energy the mother should give to the offspring (i.e., the infant wants more than the parent wants to give).

We often interpret such conflict in humans as being maladaptive and related to psychological problems of parent or child or to some negative cultural influence. Work with nonhuman primates, however, has led primatologists to the idea that the conflict is a natural part of the weaning process and is necessary for the infant to become an independent and functioning member of the social unit (Hansen 1966). Nevertheless, it is still unclear from an evolutionary standpoint why the infant should resist the weaning process.

Trivers (1974) puts forth the hypothesis that conflict arises because natural selection operates differently on the two generations. To maximize her lifetime reproductive success, the mother should invest a certain amount of time and energy in current offspring and then wean the young and invest in new young. The mother's optimum investment is a tradeoff between investment in current young and effort toward future offspring. The current offspring, however, profits from continued care until the cost to its mother is twice the benefit (since the offspring shares half its mother's genes). When the cost is twice the benefit to the mother, then the offspring's own fitness starts to decline also, because the offspring benefits to some extent by having its mother produce future, related offspring. Trivers' (1974) approach, based on the coefficient of relationship and kin selection (discussed further in chapter 22), makes intuitive sense but has not been rigorously tested.

Conflict is not always limited to parents and offspring; siblings may aggressively compete with one another over the distribution of parental care. Pigs (*Sus scrofa*) are born with razor-sharp teeth that they use in fights with siblings to gain access to favored teats for nursing (Fraser 1990; Fraser and Thompson 1991). Among spotted hyenas (*Crocuta crocuta*), litters are usually twins, and pups are also born with fully erupted canines and incisors (figure 21.11). Fighting begins immediately after birth, and if littermates are of the same sex, one is often killed (Frank et al. 1991). Spotted hyenas are unusual in that young are precocial and both sexes have high levels of circulating androgens at birth.

MATING SYSTEMS

Anisogamy prevails in all mammals, with females investing more in each egg than males invest in each sperm. This difference sets the stage for male–male competition for access to females and for attempts by males to mate with more than one female, a condition referred to as **polygyny.** Polygyny results in greater variation in reproductive success for males than for females; for each male that fertilizes the eggs from a second female, another male is likely to fertilize none, as we saw earlier for lions (see figure 21.1). Sexual selection tends to act more strongly on males than on females among polygynous species; however, not all species are polygynous.

In trying to evaluate the adaptive significance of differences in mating systems, we must look at ecological factors as well as historical ones. For example, group size may be related to predator pressure and food distribution. In the open plains, where large predators are present and food is widely distributed and often clumped, omnivorous primates and grazing mammals such as ungulates live in large groups in which mating with several members of the opposite sex is likely for both males and females. In densely forested areas, where communication over long distances is difficult, small family units and monogamy are more common; likewise where food is uniformly distributed. Thus, the spatial distribution of resources (food, nest sites, or mates) influences the type of mating system (figure 21.12).

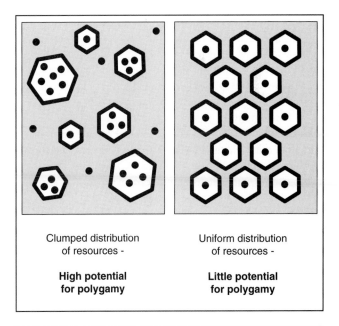

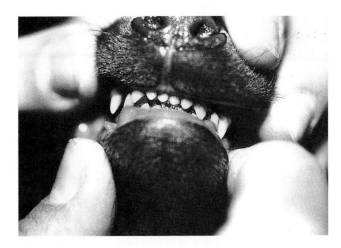

Figure 21.11 **Dentition of spotted hyena on day of birth.** Canines are 6–7 mm long; incisors are 2–4 mm long.

Figure 21.12 **The influence of the spatial distribution of resources on the ability of individuals to monopolize those resources.** Dots are resources and hexagons are defended areas. Uniform distribution of resources on the right offers little opportunity for monopolization; monogamy is the likely mating system here.

368 Part Four   *Behavior, Ecology, and Biogeography*

Classifications of mating systems traditionally have been based on the extent to which males and females associate (bond) during the breeding season. **Monogamy** refers to an association between one male and one female at a time and includes an exclusive mating relationship between pair members. **Polygamy** incorporates all multiple-mating, nonmonogamous mating systems. The following multiple mating systems are subsets of polygamy. Polygyny is association between one male and two or more females at a time. **Polyandry** is association between one female and two or more males at a time. **Promiscuity** refers to the absence of any prolonged association and multiple mating by at least one sex.

Emlen and Oring (1977) developed the following new ecological classification of mating systems based on the ability of one sex to monopolize or accumulate mates. Although developed primarily for birds, the system seems applicable to mammals as well.

Monogamy

In monogamous systems, neither sex is able to monopolize more than one member of the opposite sex. Monogamy is relatively rare in mammals. It is found in less than 5% of mammal species (Kleiman 1977). Facultative monogamy may occur when densities are low and the home range of a male overlaps only that of a single female. Obligate monogamy typically occurs when investment from the male is necessary for the female to rear offspring. When the habitat contains scattered, renewable resources or scarce nest sites, monogamy is the most likely strategy. If there is no opportunity to monopolize mates, an individual benefits from remaining with its initial mate and helping to raise the offspring. The formation of long-term pair bonds also seems advantageous because less time is needed to find a mate during each reproductive cycle. Predation risk is another factor promoting monogamy. Some species live in small social units and behave secretively in order to reduce the chances of being eaten.

Monogamy is reported in most mammalian orders, with the bulk of examples coming from Primates, Carnivora, and Rodentia (Kleiman 1977). Among primates, the forest-dwelling marmosets and tamarins (Callitrichidae) are all monogamous, with extensive male care in the form of carrying of infants. The larger carnivores, especially canids such as jackals, foxes, and wolves, are nearly all monogamous; males carry, feed, defend, and socialize offspring (Moehlman 1986). Among the Rodentia, the prairie vole (*Microtus ochrogaster*) is monogamous (Getz and Carter 1980), as is the California mouse (*Peromyscus californicus*). In the latter species, males and females form long-term pair bonds and occupy overlapping ranges distinct from other pairs, and males spend as much time in the nest with young as do females. Ribble (1991) analyzed the parentage of 82 offspring from 22 families of *P. californicus* using DNA fingerprinting and found no instances of mixed paternity among litters. The male and female associated with each litter were, in all cases, the biological parents.

Polygyny

In polygynous systems, individual males monopolize more than one female. In **resource defense polygyny,** males defend areas containing the feeding or nesting sites critical for reproduction, and a female's choice of a mate is influenced by the quality of the male and of his territory. Territories that vary sufficiently in quality may cross the **polygyny threshold**—the point at which a female may do better to join an already mated male possessing a good territory than an unmated male with a poor territory (Orians 1969).

Many mammalian species are probably facultatively (i.e., optionally) polygynous. In habitats where feeding or nesting resources cannot be monopolized by males, monogamy is likely, whereas habitats with clumped resources that are defensible would favor polygyny (see figure 21.12). Thus, we may expect to find variation in mating systems of a single species.

Female defense polygyny may occur when females are gregarious for reasons unrelated to reproduction, as when females herd for protection against predators or gather around resources such as food or nesting sites. Some males monopolize females and exclude other males from their harems. In many species of seals (Otariidae, Phocidae), the females haul out on land to give birth, and they mate soon after. The females are gregarious because there are limited numbers of suitable sites, and the males monopolize the females for breeding. Intense competition among males results in marked sexual dimorphism and a large variance in male reproductive success, as we have already seen for northern elephant seals (see figure 21.9; Haley et al. 1994).

If males are not involved in parental care and have little opportunity to control resources or mates, **male dominance polygyny** may develop. If female movements or concentration areas are predictable, the males may concentrate in such areas and pool their advertising and courtship signals. Females then select a mate from the group of males. These areas are called leks, and males congregate and defend small territories within them in order to attract and court females. Leks do not contain resources (food or nesting sites) but are purely display sites for mate choice and copulation. Females select a mate, copulate, then leave the area and rear their young on their own. Often older or more dominant males occupy the preferred territories or have the most attractive displays and thus do most of the copulating.

Male hammer-headed bats (*Hypsignathus monstrosus*) from central and western Africa display at traditional sites along riverbanks (Bradbury 1977). Each territory is about 10 m apart, and the males emit a loud clanking noise to attract females. Once chosen, a male copulates with the female and resumes calling immediately. Some males, for whatever reason, have much higher success than others at attracting females, and in one year, 6% of the males achieved 79% of the copulations. As might be expected, this species shows extreme sexual dimorphism, with males nearly twice the size of

Figure 21.13   **Sexually selected facial characteristics in the male hammer-headed bat.** Males establish small territories at leks. Note the highly modified lips and inflated rostrum, products of sexual selection. Males make loud, clanking sounds that attract females.

females. Each male has a huge muzzle that ends in flaring lip flaps and a large larynx associated with the clanking sound they make (figure 21.13).

Leks also are seen in several species of ungulates, such as Uganda kob (*Kobus kob;* Buechner and Roth 1974) and topi (*Damaliscus korrigum;* Gosling and Petrie 1990). In topi, the largest and reproductively most successful males defend single territories, while the smaller males cluster in leks. In Uganda kob, a dual male strategy also exists, except that males on single territories are less successful than those in leks (Balmford 1991). The reproductive success of lekking male fallow deer (*Dama dama*) varies widely, with a few

males having spectacular success while most males do no breeding at all (Appolonio et al. 1992). Competitively inferior males follow a low-risk strategy and defend single territories; in these, they can count on getting a few copulations but spend less time and energy fighting and displaying.

In the absence of territory or dominance, **scramble polygyny** may operate, as males try to obtain copulations. Where females are widely dispersed, as with 13-lined ground squirrels (*Spermophilus tridecemlineatus*), males become highly mobile during the breeding season. Rather than actively competing with other males for territory or dominance, the most successful breeders are those males that cover the most area in search of estrous females (Schwagmeyer 1988).

Polyandry

In polyandrous systems, females monopolize more than one male. Because female investment in gametes and offspring exceeds that of males, polyandry is expected to be rare, especially in mammals, where only females gestate and lactate. We have already seen that in most mammal species, females provide the bulk of parental care, with males seeking new mates and investing little in offspring. Under certain circumstances, polyandry might be expected if food availability at the time of breeding is highly variable or if breeding success is very low due to high predation on the young. True polyandry should result in role reversal, with large females competing for smaller mates, female ornamentation via sexual selection, male parental care, and female dispersal in search of mating opportunities. Such a pattern does not occur in mammals, although there are dozens of species of birds in which, except for egg-laying, the male does all the parenting alone.

In many species of mammals, however, females mate with more than one male, and sometimes littermates have more than one father, for example in deer mice (Birdsall and Nash 1973). Such systems, however, are usually referred to as promiscuous rather than polyandrous, because males also mate with multiple females, males provide little or no care of offspring, and no lasting bond exists between the partners. Several species of larger canids show most of the features of true polyandry. Although mainly monogamous, the African wild dog (*Lycaon pictus*) sometimes exhibits polyandry; females are occasionally mated by several males, males provide extensive care of pups, and females are the dispersing sex (Moehlman 1986).

Summary

In mammals, as with almost all other vertebrates, sexual reproduction is the only option. The gametes of the two sexes differ anatomically, with the female producing few, large, nonmotile eggs and the male producing many small, motile sperm. In most species, males attempt to mate with more than one female and vary greatly in their reproductive success, whereas females usually mate only once per breeding bout and are often reproductively successful. In some species of mammals, females may adaptively adjust the sex ratio of their offspring to produce the sex that will have the highest reproductive success.

Sexual selection affects anatomy and behavior at the time of mating. Intersexual selection involves choices made between males and females, with the females usually making the choice. Sexual selection may lead to the evolution of elaborate secondary sexual characteristics, particularly in males. The evolution of such traits is thought to have come about by either:

1. A process of runaway selection, in which some arbitrary trait evolves because females prefer it and so produce sons with the trait and daughters that prefer the trait. The trait thus continues to evolve until countered by natural selection; or

2. A linkage of the trait with superior genes in the male, so that females selecting males with the trait obtain more fit mates.

In intrasexual selection, competition takes place within one sex (usually the male), with the winner gaining access to the opposite sex. Competition may take place before copulation, or after copulation in the form of sperm competition, blocked implantation, or infanticide.

Parental investment in offspring is generally greater by females than males, as males seek additional mates and leave the female to care for young; however, in some species, males provide extensive care. Species living in harsh, unpredictable environments or suffering high juvenile mortality rates tend to have high reproductive rates and provide little care compared with those adapted to stable environments.

Offspring often seem to want more attention or food than the mother is willing to give and resist attempts by the mother to wean them. Conflicts may reflect the fact that parents only share half their genome with an offspring. What is best for the parent is not necessarily what is best for the offspring.

Most mammals have polygynous mating systems, in which males are able to monopolize more than one female. In such systems, males are usually large, may be more ornamented than the smaller females, and provide little if any care of offspring. In a few species, males cluster and defend a small territory that functions solely to attract females for mating. In some orders, especially Carnivora, food is difficult to obtain, and both parents share in caring for offspring—processes that lead to monogamy. Polyandry, in which females monopolize more than one male, is rare in mammals. It has been reported in some of the larger canids, where females enlist the help of several males to raise offspring. Complete sex role reversal, with males as the sole parent, is unknown in mammals.

Discussion Questions

1. Differences between the sexes are usually assumed to have come about via sexual selection. Can you think of selection pressures other than competing for mates that might produce such differences?
2. As we have seen in this chapter, polygyny is common in mammals. In birds, however, monogamy is the rule (Greenwood 1980). Discuss the possible reasons for this taxonomic difference in mating systems.
3. Discuss the possible relevance to humans of Trivers' concept of parent–offspring conflict.
4. Why is polyandry so uncommon in mammals? What sorts of ecological and phylogenetic circumstances might favor polyandry?
5. Using the figure shown here, describe the general relationship between testes weight and body weight, and then explain why some genera are above the line and some below it. What can you surmise about human mating systems from this graph (Harcourt et al. 1981)?

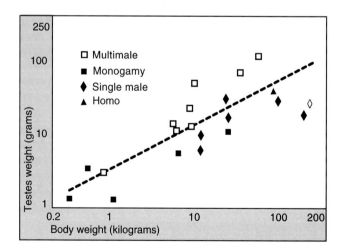

Paired testes weight (g) versus body weight (kg) for different primate genera.

Source: Data from A.H. Harcourt et al., "Testis Weight, Body Weight and Breeding System in Primates" in Nature, 293:55–57, 1981.

Suggested Readings

Andersson, M. 1994. Sexual selection. Princeton Univ. Press, Princeton, NJ.

Clutton-Brock, T. H. 1991. The evolution of parental care. Princeton Univ. Press, Princeton, NJ.

Daly, M. and M. Wilson. 1983. Sex, evolution, and behavior, 2d ed. Willard Grant, Boston.

Krebs, J. R. and R. Dawkins, eds. 1991. Behavioural ecology: an evolutionary approach, 3d ed. Sinauer Assoc., Sunderland, MA.

Trivers, R. L. 1972. Parental investment and sexual selection. Pp. 136–179 *in* Sexual selection and the descent of man, 1871–1971 (B. Campbell, ed.). Aldine, Chicago.

Trivers, R. L. 1974. Parent-offspring conflict. Am. Zool. 14:249–264.

CHAPTER

22

Social Behavior

A **society** is a group of individuals of the same species that is organized in a cooperative manner, extending beyond sexual and parental behavior. In this chapter, we first explore several examples of cooperative social behavior in mammals, then discuss the evolutionary costs and benefits of social behavior in general. Finally, we consider the way natural selection has brought about the evolution of social systems in mammals. For overviews of sociality in different groups of mammals, refer to the chapters in part 3.

Sociality has evolved independently in many groups of animals, ranging from invertebrates to primates. We might expect an increase in the complexity of social behavior as we move from simpler to more sophisticated organisms, yet by some criteria, just the opposite is true: some invertebrates, such as the Portuguese man-of-war (*Physalia*), form colonies of individuals that cooperate much more extensively than mammals (Wilson 1975).

EXAMPLES OF COOPERATIVE SOCIAL BEHAVIOR

As we saw in the previous chapter, male mammals typically provide little care for the young. Instead they seek additional mates in polygynous relationships (Eisenberg 1981). The predominant role of the mother in rearing young is no doubt due, in part, to the fact that only female mammals gestate and lactate. Milk is the "cement" of the young mammal's first social relationship (Wilson 1975). Most mammalian social systems are organized matrilineally; mothers and offspring may stay together, and groups are thus composed of mothers, daughters, sisters, aunts, and nieces. Because of the prevalence of polygyny and the associated tendency of male mammals to disperse as they reach sexual maturity (chapter 23), adult males are usually unrelated to other adults in the group. Complex social organization has evolved in some species in nearly all mammalian orders, but especially among carnivores, cetaceans, and primates. Within each order, the most highly social species tend to be large-sized and have large brains, and the highly social terrestrial species tend to forage above ground in open habitats in the daytime (Wilson 1975). The cooperation that characterizes social living takes one or more of the following forms that extend beyond parental care.

Alarm Calling

When a Belding's ground squirrel (*Spermophilus beldingi*) sees a predator such as a weasel (*Mustela* sp.), it may emit a shrill alarm call, alerting other nearby ground squirrels to run for cover (Sherman 1977). The caller, however, is now more likely to fall prey to the predator. Behavior that appears to be costly to the individual but beneficial to others is said to be **altruistic.** How such behavior evolves has been the focus of much controversy, as we shall see later in this chapter.

Cooperative Rearing of Young

Although cooperative rearing of young—that is, individuals other than the young's mother nursing or providing food for the young—is not common, it does occur in social carnivores and some rodents. Among carnivores, lionesses (*Panthera leo*) share the nursing of cubs in the pride (Packer et al. 1992; figure 22.1), and subordinate wolves (*Canis lupus*) regurgitate food for the alpha female and her litter (Mech 1970). In the banded mongoose (*Mungos mungo*) of Africa, several females breed synchronously, giving birth in a communal den and nursing each other's offspring (Rood 1986). Although the dwarf mongoose (*Helogale parvula*) is monogamous, it lives in packs of about 10 individuals. As occurs in wolves, breeding is suppressed among subordinate females. These nonreproductive members guard the den and bring insects to the young (Rood 1980; figure 22.2). Subordinate

Figure 22.1   Communal nursing in lions.   Females nurse young born to other pride members, a trait that may have evolved via kin selection because pride females are closely related.

Figure 22.2   A family of dwarf mongooses.   All members of a group help to raise the young, even though the offspring are usually produced by a single, dominant pair.

Figure 22.3   Nonoffspring nursing in the white-footed mouse.   The older pups' mother is the sister of the nursing female.

females occasionally do become pregnant, but most of their offspring die, probably due to infanticide by the dominant female. Subordinate females may also become "pseudopregnant," a condition in which the female is in a hormonal state of pregnancy, although no embryo is present. She may then lactate and nurse the dominant female's young (Rood 1980; Creel et al. 1992).

Rodents, such as house mice (*Mus musculus*) and white-footed mice (*Peromyscus leucopus*), may form communal nests in which several litters of different ages are present. In some cases, females nurse their own young and those of other females simultaneously (Jacquot and Vessey 1994; figure 22.3). Lactation doubles a female's energy need, so such costly and seemingly altruistic behavior demands an explanation.

Coalitions and Alliances

Most of the species that live in large groups, from herds of wildebeests to schools of dolphins, have evolved in open habitats, such as savannas or oceans. Such groups may consist of members of both sexes and associated young; they may persist throughout the year or only during the breeding season. The degree of cooperation seen in such groups varies widely from none, except between mothers and offspring, to complex coalitions and alliances among both related and unrelated members.

In the large, multimale groups found in many Old World monkeys and apes, cooperation in caring for young does not extend to nursing or giving food to offspring other than one's own. Group membership is usually highly stable, however, and dominance hierarchies are prominent in some species. Males and females may play different social roles within the group. In rhesus monkeys (*Macaca mulatta*), the highest-ranking males act as "control" animals (Bernstein and Sharpe 1966). These males protect the group against serious extragroup challenges and reduce intragroup conflict by intervening in fights among group members (Vessey 1971).

Learning and early experience play a large part in determining the social structure in primates. Although **kinship,** the sharing of a common ancestor in the recent past, is important in structuring primate social systems, coalitions and "friendships" among nonrelatives are also evident, as in baboons (*Papio cynocephalus;* Smuts 1985; figure 22.4). Chimpanzees (*Pan troglodytes*), among other nonhuman primates, form coalitions in competitive situations, such as when two individuals cooperate to defeat or take some resource away from a third. They also reciprocate, meaning that after individual A intervenes on behalf of individual B, individual B is later likely to intervene on behalf of individual A (de Waal 1992).

Bottlenose dolphins (*Tursiops truncatus*) live in large groups that vary in membership. Males form stable, first-order alliances of two or three in order to herd estrous females and keep them away from other males (figure 22.5). Two first-order alliances may combine to form a second-order alliance of five or six males. Second-order alliances are able to take females from first-order alliances and are also effective at protecting females from other alliances (Connor et al. 1992).

Eusociality

The epitome of social organization, seen most often in insects of the Order Hymenoptera, is referred to as eusociality. Three traits characterize this pattern: (1) cooperation in the care of young; (2) reproductive castes, with nonreproductive members caring for reproductive nestmates; (3) overlap between generations such that offspring assist parents in raising siblings (Wilson 1975). Jarvis (1981) first demonstrated mammalian eusociality in the naked mole-rat (*Heterocephalus glaber;* figure 22.6, see also figure 17.16). She captured 40 members of a colony from their burrow system in Kenya and studied them for 6 years in an artificial burrow system in the laboratory. Only one female in the colony ever had young; mother and young were fed but not nursed by

Figure 22.4   An alliance between two male olive baboons.
The two males on the right are challenging the male on the left in the foreground.

Figure 22.5   **Alliances in bottlenose dolphins.**   Males traveling in formation with a herded female. The males remain on either side and just behind the female. Males form alliances to prevent other males from gaining access to females and to take females from other alliances.

Figure 22.6   **Naked mole-rats.**   The large female in the center is the only reproductive female in the colony. Other adults feed her, care for the young, or maintain the burrow system.

male and female adults of the worker caste; members of this caste were not seen to breed. Another caste of nonworkers assisted in keeping the young warm; males of this caste bred with the reproducing female (or "queen"). This species fits all of the criteria for eusociality mentioned earlier. Another, distantly related, species of mole-rat, *Cryptomys damarensis*, also appears to be eusocial (Jarvis et al. 1994).

WHY MAMMALS LIVE IN GROUPS

It is often assumed that living in complex social groups is somehow superior to living a more solitary life, yet costs and benefits are associated with each. Most of the benefits of sociality listed in the following section can be related to two ecological factors: predation pressure and resource distribu-

tion (Alexander 1974). Keep in mind that these advantages may not, by themselves, have led to the evolution of sociality; rather, they may have become secondarily advantageous once sociality had already evolved via one of the other selective pressures.

Benefits

Protection From Physical Factors

White-footed mice frequently form communal nests in winter (Wolff and Durr 1986); huddling has been shown to conserve significant amounts of energy (Hill 1983). This benefit leads to the formation of aggregations, as with large clusters of bats, but not necessarily to organized social groups.

Protection Against Predators

Detection of and communication about danger are more rapid when individuals are in groups, and predator deterrence may be enhanced by mobbing and group defense. According to the "many eyes" hypothesis, individuals in large groups spend less time watching for predators and so can spend more time in other activities such as feeding. Examples include prairie dogs (*Cynomys* spp.; Hoogland 1979b) and arctic ground squirrels (*Spermophilus undulatus;* Carl 1971). In the latter study, observers could approach within 3 m of isolated individuals but no closer than 300 m to grouped individuals before waves of alarm calls swept through the colony. Musk oxen (*Ovibos moschatus*) and other ungulates form a defensive perimeter in response to wolf attacks (Gunn 1982); adults form a line or circle, keeping themselves between the predator and dependent young.

Finding and Obtaining Food

Living in groups may make it easier to find and obtain food. The best examples of this strategy are used by wolves (Mech 1970) and lions (Schaller 1972). These species are able to capture large species, such as moose (*Alces alces*) in the case of wolves and buffalo (*Syncerus caffer*) in the case of lions, that would be difficult to capture if hunting alone. In bighorn sheep (*Ovis canadensis*), the locations of feeding areas are remembered by older members of the band; this information is transmitted to subsequent generations via tradition (Geist 1971).

Group Defense of Resources

Lion prides are territorial, defending space containing food resources against other prides (Schaller 1972). Similarly, wolf packs defend space against neighboring packs (Mech 1970).

Assembling Members for Location of Mates

As we saw in chapter 21, hammer-headed bats (*Hypsignathus monstrosus;* Bradbury 1977) and fallow deer (*Dama dama;* Appolonio et al. 1992) breed on leks, where males defend small territories and display to attract females for copulation.

Division of Labor Among Specialists

This feature of advanced social behavior is found in eusocial insects but is rare among mammals. Recent studies of naked mole-rats, discussed previously, suggest that only one female is reproductively active in the colony and that the other adults specialize in tasks such as maintaining the tunnel system or nurturing the young (Sherman et al. 1991).

Richer Learning Environment for Young

This advantage is frequently suggested as important for mammals in general and primates in particular. Dependence on learning provides for greater behavioral plasticity, but it requires a long period of physiological and psychological dependence. Large-brained and highly social species, such as dolphins and primates, spend as much as 25% of their lives dependent on parents or other relatives.

Costs

Only a few studies have directly attempted to assess the possible disadvantages of sociality. The following are several obvious costs of living in groups:

Increased Intraspecific Competition for Resources

In prairie dog colonies, the amount of agonistic behavior per individual increases as a function of group size. Also, black-tailed prairie dogs are more highly social and have higher rates of aggression than do the less social white-tailed prairie dogs (Hoogland 1979a).

Increased Chance of Spread of Diseases and Parasites

Ectoparasites, such as fleas and lice, are more numerous in larger and denser prairie dog colonies than in smaller ones (Hoogland 1979a). Fleas transmit bubonic plague, epidemics of which periodically devastate prairie dog colonies. Hence, members of dense colonies are more at risk.

Interference with Reproduction

Parental care misdirected to nonoffspring and killing of young by nonparents exemplify this cost. Female white-footed mice with young are aggressive toward strange adults; in the absence of the mother, pups are usually killed by intruders (Wolff 1985). One of the factors influencing dispersal by male lions is defense of their cubs against infanticide by new coalitions of males (Pusey and Packer 1987); small cubs are almost invariably killed when new males take over a pride. Brazilian free-tailed bats (*Tadarida brasiliensis*) roost in caves in dense colonies containing millions of bats (see chapter 12). Mothers, returning from a night's foraging for insects, have to find their own infant among the thousands present. Most of the time they find their own young by vocalization, but 17% of the time mothers suckle someone else's offspring (McCracken 1984).

HOW SOCIAL BEHAVIOR EVOLVES

Individual Versus Group Selection

A basic element of sociality is cooperation. Individuals work together, often sacrificing personal gain, to achieve a common goal that benefits the social group or species. How such altruistic behavior could evolve was not a topic of great concern for biologists prior to the 1960s. It was generally assumed that groups with cooperating individuals would be more successful than those without cooperators, a type of **group selection.** This thinking culminated in the publication of a book by V. C. Wynne-Edwards (1962). In that book and a sequel (Wynne-Edwards 1986), he argued that the evolutionary significance of social behavior is that organisms can track resources in the environment more efficiently. Intraspecific competition became ritualized into contests whose intensity was proportional to the supply of the limiting resource. Because the result of such competition leads to reduced reproductive success of those participating, Wynne-Edwards believed that natural selection must be acting at the level of the group. Note that individuals must sacrifice personal reproduction for the good of the group.

Not everyone, however, believed that group selection was an important force for evolution. R. A. Fisher (1958), in his book *The Genetical Theory of Natural Selection,* pointed out that his fundamental theorem referred strictly to "the progressive modification of structure or function only in so far as variations in these are of advantage to the *individual* [emphasis added]." His theorem offered no explanation for the existence of traits that would be of use to the species to which an individual belongs. Much earlier, Darwin (1859) had stated that "if it could be proved that any part of the structure of any one species had been formed for the exclusive good of another species, it would annihilate my theory." If Fisher and Darwin were correct, then explanations for the evolution of social behavior had to be sought based on natural selection at the level of the individual and its offspring, rather than at the group or species level.

The publication of *Adaptation and Natural Selection* by Williams (1966) led to a rapid shift in thinking among researchers working on social behavior. Williams argued that when considering any adaptation, we should assume that natural selection operates at that level necessary to explain the facts, and no higher—usually at the level of parents and their young. The argument is based on evidence that natural selection at the group or population level is so weak that it is almost always outpaced by selection of individual phenotypes. In other words, genes promoting altruistic behavior are swamped by genes favoring selfish behavior. Group selection is not impossible: if some genes decrease individual fitness but make it less likely that a group, population, or species will become extinct, then group selection will influence evolution (Williams 1966). Most behavioral ecologists today follow Williams' rule of parsimony (derived from Occam's razor), which adopts the simplest theory explained by the facts and argues that group selection is weak and should be invoked only when lower-level selection has been ruled out. New, more realistic models of group selection continue to be developed (Wilson 1980), however, and conditions favoring evolution by means of group selection may not be as restrictive as was once thought. For further discussion of the controversy, see the books by Brandon and Burian (1984) and Sober (1984).

The Selfish Herd

One way to explain gregarious behavior in mammals at the individual level is to suppose that it is a form of cover-seeking, in which each individual tries to reduce its chances of being caught by a predator. Hamilton (1971) suggested how this might work, using a hypothetical lily pond in which live some frogs and a predatory water snake. The snake stays on the bottom of the pond most of the time and feeds at a certain time of day. Because it usually catches frogs in the water, the frogs climb out on the edge before the snake starts to hunt. They do not move inland from the rim because of even more-threatening terrestrial predators. The snake surfaces at some unpredictable place and catches the nearest frog. What should each frog do to minimize the chances of its being eaten? Hamilton demonstrated mathematically that it should jump around the rim, moving into the nearest gap between two other frogs. The end result would be an aggregation.

Such an example could be also applied to herds of ungulates; when each individual behaves selfishly and moves into the center of the herd to minimize its chances of being picked off, the end result is a tightly packed group. Although the selfish herd might explain some aggregations, it does not explain the defensive groupings of musk oxen mentioned previously. Adult musk oxen put themselves between their young and the wolves, rather than trying to selfishly minimize the chances of being attacked.

Kin Selection

To explain the evolution of cooperative behavior, Hamilton (1963, 1964) presented a theory incorporating both the gene and the individual as units of selection. In its simplest form, this **kin selection** theory suggests that if a gene that causes some kind of altruistic behavior appears in the population, the gene's success depends ultimately not on whether it benefits the individual carrying the gene but on the gene's benefit to itself. If the individual that benefited by the act is a relative of the altruist and therefore more likely than a nonrelative to be carrying that same gene, the frequency of that gene in the gene pool increases. The more distant the relative, the less likely it is to carry that gene, so the greater must be the ratio between the benefit to the recipient and the cost to the altruist, if the gene is to spread. This relationship, called Hamilton's rule, is expressed algebraically as follows:

$$b/c > 1/r$$

where b is the benefit to the recipient, c is the cost to the altruist, and r is the coefficient of relationship, that is, the proportion of genes shared by the two participants by way of descent from a common ancestor. In full siblings, who share half their genes, $r = 1/2$, and therefore b/c must exceed 2 for altruistic genes to spread. In other words, if an individual more than doubles the fitness of a sibling through an altruistic act that causes that individual to leave no offspring, genes promoting that behavior could spread through the population. The more distant the relative, the lower r is, and the higher the benefit-to-cost ratio (b/c) must be. Thus, for first cousins, $r = 1/8$, so b/c must exceed 8. Of course, the control of an altruistic behavior pattern most likely involves more than one gene, but it is assumed that the same principles apply.

It now becomes necessary to consider an individual's fitness as including both a direct component, measured by the reproductive success of one's own offspring, and an indirect component, measured by the reproductive success of one's relatives other than one's own offspring. These two components make up one's **inclusive fitness.**

Hamilton's idea about the evolution of cooperative behavior through kin selection stimulated considerable research because it remained to be seen if cooperative acts are in some way distributed to relatives according to the degree of relatedness. Such studies, however, require information about paternal and maternal relationships that is usually not available to the observer of a natural population. It is particularly difficult to determine paternity in mammals because of internal fertilization. Molecular techniques such as DNA fingerprinting (see chapter 3), however, now make it possible to determine parentage in a variety of species.

Another problem is the recognition factor. How do animals recognize kin in order to "correctly" distribute their altruistic acts? Perhaps the simplest and most common way is by familiarity. When given a choice in a laboratory arena, young spiny mice (*Acomys cahirinus*) prefer to huddle with littermates rather than nonlittermates (Porter et al. 1981). Siblings separated soon after birth and raised by foster mothers treated each other as strangers, but unrelated young raised together responded to each other as siblings. Some species of mammals can recognize relatives with whom they have never associated. A possible mechanism for this is called phenotype matching, in which the individual uses its own kin as a "template" to compare with strangers who might be kin (Holmes and Sherman 1983). Such a mechanism may explain instances when siblings recognize each other even when reared apart. Thus, golden-mantled ground squirrel (*Spermophilus lateralis*) juveniles preferred to play with littermates over nonlittermates, supporting the familiarity mechanism, but siblings reared apart preferred each other as playmates over nonsiblings reared apart (Holmes 1995).

A third mechanism of kin recognition that does not require learning is the presence of "recognition" genes that enable individuals to recognize others with those same genes and to behave cooperatively toward those individuals. This mechanism has been called the "green beard effect" because such a trait could be selected for if genes controlling it confer not only the green beard, but also the preference for green beards on others. Cues used to assess genetic relatedness may come from genes in the major histocompatibility complex (MHC), a cell recognition system used by the immune system to distinguish self from nonself. In rodents, and possibly other mammal species, genetic differences in this region produce urinary odor cues that can be used to recognize genetic similarities or differences in other individuals (Yamazaki et al. 1976; Brown and Eklund 1994). These cues have been shown to affect mate choice in wild house mice (Potts et al. 1991).

One of the most thorough tests of kin selection was Sherman's study of alarm calls in Belding's ground squirrels (Sherman 1977; figure 22.7). He found that whenever a terrestrial predator (such as a weasel or a coyote) was seen, the calling squirrel stared directly at the predator while sounding the alarm. Sherman suggested many hypotheses to explain this behavior, two of which we discuss here. First, the predator may abandon the hunt once it is seen by the potential prey (in this hypothesis, the caller is behaving selfishly). Sec-

ond, other ground squirrels in the area may benefit from the warning, even though the caller may be harmed (in this hypothesis, the caller is behaving altruistically). Sherman demonstrated that callers attract predators and are more likely to be attacked after calling, and thus they are not behaving selfishly.

Because he kept records on mothers and offspring, Sherman knew that the males leave the natal area several months after birth and that the females are philopatric, remaining near their place of birth to breed. He also found that adult and yearling females are much more likely to call than would be expected by chance and that males are less likely to do so. Furthermore, females with female relatives in the area (such as mothers or sisters, but not necessarily with offspring) call more frequently in the presence of a predator than those with no female relatives living nearby (figure 22.8). Sherman concluded that the most likely function of the alarm call is to warn family members.

Working with primates, Massey (1977) tested the kin selection model in rhesus monkeys. She found that within an enclosed group, monkeys aid each other in fights in proportion to their degree of relatedness. In observations of the same species in a free-ranging situation, Meikle and Vessey (1981) found that when males leave the natal group, they usually join groups containing older brothers. They associate with brothers in the new group, aid each other in fights, and avoid disrupting each other's sexual relationships, in contrast to their behavior toward nonbrothers.

Bertram (1976) estimated that males in a pride of lions are related on the average almost as closely as half siblings ($r = 0.22$), and females are related as closely as full cousins

Figure 22.7   **Belding's ground squirrel giving an alarm call.** This conspicuously calling squirrel, a lactating female, is more likely to be attacked by a predator than is a noncaller. Nearby ground squirrels benefit because they can remain hidden or take cover. Females with mothers, sisters, or offspring in the colony are most likely to call.

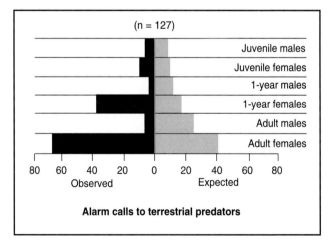

Figure 22.8   Expected and observed frequencies of alarm calls by Belding's ground squirrels to terrestrial predators. Expected frequencies are those that would be predicted if the animals called in proportion to their frequency of occurrence in the population. Calls to terrestrial predators are given disproportionately by females.

Source: Data from P.W. Sherman, "Alarm calls of Belding's Ground Squirrels" in Behav. Ecol. and Sociobiol., *17:313–323, 1985.*

Figure 22.9   Black-backed jackal helper.   The helper, probably from the previous year's litter, is about to regurgitate food to pups.

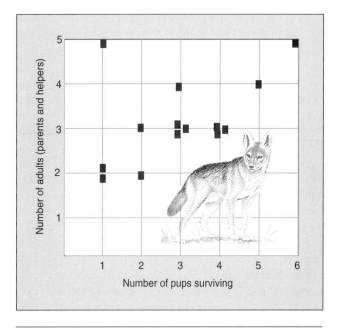

Figure 22.10   Black-backed jackal pup survival as a function of the number of adults in the family.   Young from previous years may assist their parents in rearing the current litter. Each helper increases the number of surviving pups by an average of about 1.5. Helpers may, therefore, increase their inclusive fitness more by rearing siblings than by attempting to breed on their own.

Source: Data from P. D. Moehlman, "Jackal helpers and pup survival" in Nature, *277:382–383, 1984.*

($r = 0.15$). Lions cooperate in several ways, including hunting, driving out intruders, and caring for young. He found that males show tolerance toward cubs at kills and seldom compete for females in estrus and that cubs suckle communally from the pride females. Based on the degree of relatedness among pride members, Bertram concluded that kin selection is in part responsible for these cooperative behaviors. He was not able, however, to evaluate its importance compared with other selective pressures.

Black-backed jackals (*Canis mesomelas*) live in the brushland of Africa, where monogamous pairs defend territories, hunt cooperatively, and share food. Frequently, offspring from the previous year's litter help rear their siblings by regurgitating food for the lactating mother as well as for the pups (Moehlman 1979; figure 22.9). The number of pups surviving is directly related to the number of helpers (figure 22.10). One-and-a-half extra pups survived, on the average, for each additional helper—a yield of one pup per adult involved. Only one-half pup per adult survived when just the parents were involved. Given the fact that helpers are as closely related to their siblings as they would be to their own offspring ($r = 0.5$), yearling jackals gain genetically by aiding their parents. In addition to increasing their indirect fitness by aiding siblings, helpers may increase their direct fitness by gaining experience in rearing young, becoming familiar with the home territory, and possibly gaining a portion of the home territory, all of which would increase their own chances of breeding successfully at a later time (Moehlman 1983, 1986).

Kin selection may also explain the nursing of nonoffspring by female mammals, such as in lions. Recall that r between adult females in the pride is about 0.15. Among rodents, female white-footed mice are occasionally observed nursing young from two different-aged litters (see figure 22.3). In some instances, the mother of the older pups is the sister of the nursing female (Jacquot and Vessey 1994). For a review of nonoffspring nursing, see Packer and colleagues (1992).

Cooperation Among Nonkin

A critical feature of a theory such as Hamilton's is that it is testable and falsifiable. Although the theory cannot be confirmed simply by finding cases of nepotism, it can be falsified in specific cases by observing cooperation among nonrelatives in a natural population. For example, McCracken and Bradbury (1977) demonstrated that the degree of relatedness of colonial bats (*Phyllostomus hastatus*) is too low to explain their coloniality on the basis of kin selection.

Recall that in lions, the males that take over prides are frequently related; their cooperation can be explained by kin selection. Additional studies of these same prides, however, suggest that at least some of the coalitions are among unrelated males, so some additional explanation is needed (Packer and Pusey 1982).

Reciprocal Altruism

Individuals may cooperate and behave altruistically if there is a chance that they will be the recipients of such acts at a later time. Such a situation, called **reciprocal altruism,** is similar to mutualism (see the following section) except that a time delay is involved. According to Trivers (1971), natural selection acting at the level of the individual could produce altruistic behaviors if in the long run these behaviors benefit the organism performing them. He first showed that if altruistic acts are dispensed randomly to individuals throughout a large population, genes promoting such behavior disappear, because there is little likelihood that the recipient will pay back the altruist. If, however, altruistic acts are dispensed nonrandomly among nonrelatives, genes promoting them could increase in the population if some sort of reciprocation occurs. The factors that affect that likelihood are (1) length of life span—long-lived organisms have a greater chance of meeting again to reciprocate; (2) dispersal rate—low dispersal rate increases the chance that repeated interactions will occur; and (3) mutual dependence—clumping of individuals, as occurs when avoiding predation, increases the chances for reciprocation. In any social system, nonreciprocators (cheaters) can be expected, but as cheating increases, altruistic acts become less frequent.

A few field studies have suggested the importance of reciprocal altruism. Working with olive baboons (*Papio anubis*) in Africa, Packer (1977) studied coalitions among males in which two presumably unrelated males joined forces against a third. If that third male was in consort with a female in estrus, one of the attackers might gain access to her. The pair of males tended to maintain the previously established coalition, and the next "stolen" female would be taken over by the other member of the male pair.

A different sort of reciprocity has been demonstrated in vampire bats (*Desmodus rotundus*) in Costa Rica by Wilkinson (1984; figure 22.11). At night, these bats feed on blood, primarily from cattle and horses, and then return to a hollow tree to roost during the day. Wilkinson marked nearly 200 bats that roosted in 14 trees and spent 400 hours observing them in their roosts. He recorded 110 cases of blood-sharing, where one bat ate blood that was regurgitated by another bat. Not surprisingly, most of these exchanges were between mothers and offspring. In most of the other feedings, he was able to determine both the coefficient of relationship between the pair and an index of association based on how often the pair had been together in the past. Wilkinson demonstrated that both relatedness and association con-

tributed significantly to the pattern of exchange (figure 22.12). Close relatives and associates were fed more often than would be expected by chance.

For reciprocity to persist, (1) the pairs must persist long enough to permit reciprocation; (2) the benefit to the receiver

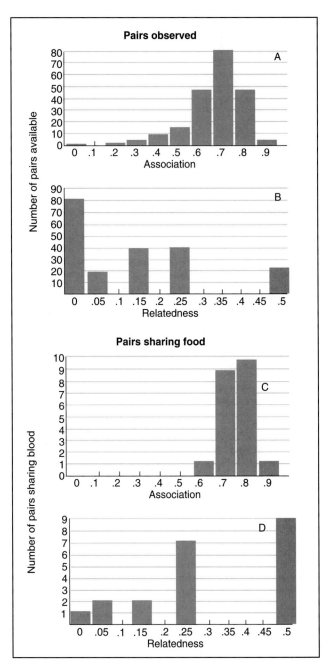

Figure 22.12   Reciprocal blood-sharing in vampire bats. Graphs (A) and (B) show the frequency of pairs observed, based on the degree of association and relatedness. Graphs (C) and (D) show the frequency of blood-sharing, excluding mother–young pairs. Both degree of association and relationship independently predict blood-sharing, implicating both reciprocal altruism and kin selection in the evolution of this behavior.

Source: Data from G. S. Wilkinson, "Reciprocal Food Sharing in the Vampire Bat" in Nature, *308:181-184, 1984, MacMillan Magazine, Ltd.*

Figure 22.11   Vampire bat.   Blood is the sole source of food for this species.

must exceed the cost to the donor; and (3) donors must recognize cheaters (those that do not reciprocate) and not feed them. Through additional studies on captive animals, Wilkinson demonstrated that vampire bats meet these conditions. In this example, both kin selection and reciprocal altruism contribute to the cooperative behavior pattern of blood-sharing.

Behavioral ecologists have used game theory to predict the circumstances in which unrelated individuals might cooperate. In a game called the Prisoner's Dilemma, two players have a choice of cooperating with each other or defecting in a series of moves. Depending on the relative costs and benefits of cooperating and defecting, it sometimes pays to follow a tit-for-tat strategy in which one player cooperates on the first move and then does whatever the other player does on his or her move (Axelrod and Hamilton 1981). For example, lions cooperate in the defense of the pride territory against strangers. When recordings of the roaring from strange females were played within the pride territory, several resident females usually advanced to check it out (Heinsohn and Packer 1995). Some lionesses were leaders and advanced boldly, but others were laggards, hanging back and letting the other member of the pair incur the risk. According to the tit-for-tat strategy, on subsequent pairings of the same leader and laggard, the leader should defect and also become a laggard. In fact, although leaders were more hesitant to advance when paired with known laggards, they did so regardless, contradicting the predictions of the model. Clearly, the relationships among the pride members are more complicated than those assumed by the model. Possibly laggards excel at hunting or in nursing cubs and so are "forgiven" for being bad at pride defense.

Mutualism

In some situations, both individuals benefit from the relationship, and there is no apparent cost to either. An example might be the huddling of mice in cold weather described earlier (Wolff and Durr 1986). Such interactions, referred to as **mutualism,** could explain some cases of cooperation and might be included with the concept of the selfish herd. Another example might be the lion that helps another bring down a wildebeest; both benefit immediately by feeding on the prey. These cases are also similar to reciprocal altruism in the long run, except that there is no delay in returning the favor. Little research has actually been done to test the importance of mutualism in the evolution of social behavior; it is difficult to distinguish mutualism from selfish behavior because there is never a cost to the individuals involved. Usually, however, mammalogists look for examples of apparent altruism, in which the costs to the actor seem to outweigh the benefits, and tend to ignore cases of mutualism.

Parental Manipulation of Offspring

In the discussion of parent–offspring conflict in chapter 21, we pointed out that because offspring share only half of each parent's genes, the interests of the parents may differ from those of their offspring. This difference is manifested by the offspring's demanding more investment from the parents than the parents are willing to give (Trivers 1974). Our first reaction might be that offspring should win such conflicts because they are the ones that must survive and pass the genes on to future generations. But Alexander (1974) argued that parents should win in the long run. Suppose that some offspring have genes that give the offspring a competitive advantage over siblings to the extent that they reduce the parents' lifetime reproductive success. For example, a highly competitive baby mouse that pushes all of its siblings out of the nest would receive more food from its parents and probably increase its direct fitness, although its indirect fitness would suffer. The parents' fitness would also suffer, however, because they would only raise one young that year. When the young mouse grew up and had young of its own, it would pass those competitive traits on, and its fitness would be less than that of a mouse with less competitive young that could coexist in the nest. Genes that will be favored are those that cause offspring to behave so as to maximize the lifetime reproductive success of the parent. Alexander pointed out that the parents thus "manipulate" the offspring to the parents' advantage. He listed the following types of behavior as examples of **parental manipulation:**

- Limiting the amount of parental care given to each offspring so that all have an equal chance to survive and reproduce
- Restricting parental care or withholding it entirely from some offspring when resources become insufficient for an entire brood
- Killing some offspring or feeding some offspring to others
- Causing some offspring to be temporarily or facultatively sterile helpers at the nest
- Causing some offspring to become permanent (obligately sterile) workers or soldiers

The last type of behavior, the extreme form of parental manipulation, is rare or nonexistent in mammals, with the possible exception of the mole-rats discussed previously.

Alexander's reasoning was criticized by Dawkins and Carlisle (1976) because the argument can be turned around and viewed from the perspective of the offspring: Parents that allow themselves to be manipulated to provide more care than they otherwise would will leave more offspring themselves. It becomes a difficult task to sort out whether parents or offspring win in such situations. Although parental manipulation is not inevitable, it needs to be considered as an explanation for some types of cooperation in mammals.

Ecological Factors in Cooperation

Emlen (1982, 1984) developed an "ecological constraints" model to explain the evolution of cooperative breeding in birds and mammals. He considered environmental factors that restrict the chances for individuals to breed independently. One condition that might favor staying at home and helping the parents or others rear offspring is a stable, predictable environment. In these cases, unoccupied territories are absent or rare; young have little chance of dispersing and breeding on their own (table 22.1, #1). An example of this might be the helpers among jackals discussed previously.

Having explained helping on the basis of habitat saturation in stable environments, it is ironic to find that cooperative breeding is most common in arid regions of Africa and Australia, where rainfall is highly variable and unpredictable. In this case, Emlen argues that the same behavior results for different reasons (see table 22.1, #2). The African mole-rat belongs to the Family Bathyergidae, all but two species of which are solitary (see also chapter 17). The two eusocial species are distantly related; both occupy arid regions of Africa where rainfall is unpredictable and prolonged droughts are the rule. Foods are underground tubers, bulbs, and corms that are rich but patchily distributed resources. Most of the time, the soil is too dry to dig burrow systems. Jarvis and colleagues (1994) argue that these conditions favor coloniality because many individuals are needed to dig on those rare occasions when it rains enough to make digging possible. The presence of a protected nest and genetic relatedness among colony members may then promote the evolution of eusociality (Honeycutt 1992; figure 22.13).

Although the ecological constraints model may explain why some animals stay home and refrain from breeding, why should they bother to help? We have already mentioned the kin selection argument, whereby indirect fitness is enhanced. But often nonrelatives or very distant relatives are aided, as with the dwarf mongooses, in which nonrelatives guard and feed young of the dominant pair (see figure 22.2). Reciprocity may be involved: Rood (1983) observed an instance in which a female was later assisted by the young she had helped rear.

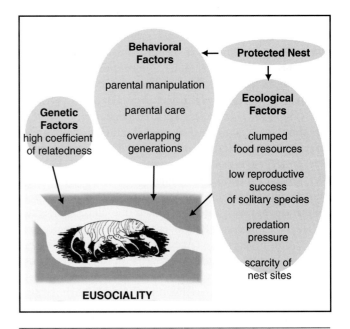

Figure 22.13 **Genetic, ecological, and behavioral factors promoting the evolution of eusociality.** All of the factors shown above may play a role in the evolution of colonial living in the naked mole-rat.

Source: R. L. Honeycutt, "Naked Mole Rats" in American Scientist, *80:43–53, 1992.*

Comparative Studies

In trying to understand the mechanisms involved in the evolution of social behavior, it is necessary to consider the type of environment in which the population lives. Several attempts have been made to relate mating system and social structure to habitat and distribution of resources. The relationship between feeding behavior and group size is evident in the great diversity of African antelope (see figure 19.21). Jarman (1974) classified these herbivores on the basis of five feeding styles, ranging from selective species that feed on only a few highly nutritious parts of localized plants to unselective species that feed primarily on grasses and forage of low nutritive value. The selective species are small, solitary, monogamous, and monomorphic, and they defend small territories. The least selective species are large, gregarious, polygamous, and sexually dimorphic, and they occupy large home ranges. Jarman argued that feeding style is an important determinant of group size; both group size and the pattern of movement over the home range affect reproductive strategies and social behavior.

Several attempts have been made to classify primate societies on the basis of habitat and niche variables. The first was by Crook and Gartlan (1966), with a later version by Eisenberg and colleagues (1972). At one end of a continuum are nocturnal, arboreal, forest-dwellers that feed mostly on insects. They tend to either be solitary, such as some lemurs (Genus *Microcebus*), or to live in small, monogamous groups (e.g., tamarins of the Genus *Saguinus*). At the other end are diurnal, terrestrial, plains-dwellers that feed mostly on seeds,

---

Table 22.1 **Ecological constraints that severely limit personal reproduction**

Type of Constraint	Cause of Constraint
1. Breeding openings are nonexistent	Species has specialized ecological requirements; suitable habitat is saturated, and marginal habitat is rare (stable environments)
2. Cost of rearing young is prohibitive	Unpredictable season of extreme environmental harshness (fluctuating, erratic environments)

Result: Grown offspring postpone dispersal and are retained in the parental unit. The population becomes subdivided into stable, social, kin groups.

From S. T. Emlen, "The Evolution of Helping, I. An Ecological Constraint Model" in American Naturalist, *119:29–39. Copyright©1982 by the University of Chicago Press, Chicago, IL. Reprinted with permission.*

tubers, or other nutritious plant parts. They tend to live in large, multimale groups (e.g., baboons of the Genus *Papio*).

One problem with such studies is that the role of phylogeny, or the evolutionary history of the group, is usually ignored. Some species may have a particular mating and social system not so much because of present circumstances but because it has been retained from their ancestors (Brooks and McLennan 1991). Using both morphological and molecular data and the methods of cladistics, a phylogenetic tree can be prepared. Various traits, such as mating system (monogamy versus polygamy), presence or absence of sexually selected traits, and type of parental care, can be independently mapped on the tree in an effort to understand how often and in what sequence the traits evolved.

One such study addresses the question of the function of **concealed ovulation** in primates (Sillén-Tullberg and Møller 1993). As we saw in chapter 9, most female mammals undergo estrus, becoming sexually attractive and receptive to males around the time of ovulation. In some primates, however, including humans, no obvious behavioral or morphological changes are associated with ovulation. Concealed ovulation has been thought to have two possible functions: one is a female tactic to promote paternal care. The male in a monogamous mating system is "forced" to stay around to impregnate the female and guard her against other males. A second hypothesis is that in a polygynous system with concealed ovulation, no male is certain that he is the father, so males are less likely to harm any infants in the group on the chance that they are their offspring. In other words, concealed ovulation may function to conceal ovulation from either the male in a pair or from all the males in the group. The analysis showed that ovulatory signs tend to disappear under polygyny but rarely under monogamy, suggesting that the second hypothesis is correct and that concealed ovulation functions to confuse paternity and thereby to reduce the chances of infanticide by males.

Summary

A society may be defined as a group of individuals of the same species, organized in a cooperative manner that extends beyond sexual behavior. Types of cooperation include alarm calling, cooperative rearing of young, formation of coalitions and alliances, and eusociality, with the formation of reproductive castes. The potential benefits of social behavior include protection from physical elements, predator detection and defense, group defense of resources, and division of labor. The costs are increased competition, spread of contagious diseases, and interference in reproduction.

Most early researchers of animal social systems assumed that traits favoring cooperation could be selected for even when they were detrimental to individuals. Research on the evolution of social behavior in the last two decades has demonstrated that many types of cooperation can be explained at the level of parents and their young: organisms cooperate for genetically selfish reasons rather than for the good of the group.

Aggregations may occur as individuals attempt to reduce their own chances of being selected by a predator. Kin selection may explain cooperation among individuals that share genes by common descent. Individuals may behave so as to lower their direct fitness but increase their inclusive fitness. Reciprocal altruism may be important in long-lived organisms that do not disperse extensively. Thus, an individual stands a good chance of being "paid back" if he or she cooperates with others. Parents may manipulate their own offspring to behave in ways beneficial to the parents but not to the offspring. Genes giving an offspring a competitive edge against its siblings could be selected against if the reproductive success of the parent is lowered.

Ecology plays a large role in determining the observed types of social behavior, as evidenced by comparative studies of closely related species that occupy different habitats. It is also necessary to consider the evolutionary history of these species, however, because some social traits may have evolved in response to selective pressures in the distant past.

Discussion Questions

1. How might you argue that colonial invertebrates such as the Portuguese man-of-war have a more "perfect" society than does a group of chimpanzees?
2. What are the differences between kin selection and group selection? In what ways can they be considered similar?
3. Woodchucks (*Marmota monax*) live in hay fields and fencerows in the eastern United States. They live a solitary life and are aggressive toward one another. They breed each year, and the young disperse by 1 year of age. Yellow-bellied marmots (*M. flaviventris*) live at intermediate elevations in the Rockies. They are colonial but moderately aggressive, breeding annually but occasionally skipping a year. Juveniles disperse in the second year. Olympic marmots (*M. olympus*) inhabit meadows at high elevations on the West Coast. They are highly colonial, breed in alternate years, are highly tolerant of others, and juveniles do not disperse until their third year. Design a series of observations and experiments to explain the large differences in social behavior of these closely related species. Compare your approach with that of David Barash (1974), who made detailed comparisons of these species.

Suggested Readings

Alexander, R. D. 1974. The evolution of social behavior. Ann. Rev. Ecol. Syst. 5:325–383.

Alexander, R. D., and D. W. Tinkle, eds. 1981. Natural selection and social behavior. Chiron Press, New York.

Harcourt, A. H. and F. B. M. de Waal. 1992. Coalitions and alliances in humans and other animals. Oxford Univ. Press, New York.

Krebs, J. R. and N. B. Davies, eds. 1997. Behavioural ecology: an evolutionary approach, 4th ed. Blackwell Science, Oxford.

Rubenstein, D. I. and R. W. Wrangham, eds. 1986. Ecological aspects of social evolution. Birds and mammals. Princeton Univ. Press, Princeton, NJ.

Wilson, E. O. 1975. Sociobiology: the new synthesis. Harvard Univ. Press, Cambridge, MA.

CHAPTER 23

Dispersal, Habitat Selection, and Migration

Most mammalian movement occurs in a relatively small area, such as when an individual acquires resources within its home range or marks and defends the area against conspecifics. These local movements are sometimes referred to as **station keeping.** They may include relatively long round trips, as, for example, when a mouse harvests seeds and returns them to a cache. We dealt with station keeping in chapter 20. On a larger scale, **ranging** behavior includes forays outside the home range, usually in search of suitable habitat or mating opportunities. According to Dingle (1996), this type of movement includes **natal dispersal,** movement from the natal site to a site where reproduction takes place. Movement on the largest scale is **migration,** persistent movement across different habitats in response to seasonal changes in resource availability and quality. It involves special physiological changes during which the animal does not respond to the presence or absence of local resources. In this chapter, we first treat natal dispersal, discussing both proximate and ultimate reasons for it, and then treat the process of habitat selection, or the finding of a place to live. Finally, we consider migration and homing, the process of getting back to a home range or nest site.

DISPERSAL FROM THE PLACE OF BIRTH

In many species of animals, members of one sex disperse from the place of birth before breeding, whereas members of the other sex are **philopatric,** breeding near the place where they were born. Among mammals, usually the males disperse, but the opposite is true in birds (Greenwood 1980). Natal dispersal means leaving the site of birth or social group (emigration), traversing unfamiliar habitat, and settling into a new area or social group (immigration). Although Dingle (1996) considers dispersal a type of ranging, others have grouped it under migration or treated it separately (McCullough 1985). Moving away from known ground is risky because the individual is unfamiliar with the location of food and shelter and is no longer in the presence of familiar neighbors and relatives. Recall from chapter 22 that cooperative behavior can evolve by both kin selection and reciprocal altruism. Both of these behavior patterns require that individuals remain in the vicinity of relatives or associates, however. Given these costs, there must be considerable benefits for dispersal behavior to be so widespread.

Causes of Dispersal

The causes of dispersal can be understood at several different levels. At the proximate level, we wish to know the immediate reasons why an individual leaves the natal area. For instance, a male might be forced out by its parents or other residents, or it might respond involuntarily to increases in testosterone levels associated with sexual maturation. At the ultimate level, we wish to know the long-term, evolutionary causes of dispersal. For instance, individuals that fail to disperse may have lower reproductive success because their offspring are inbred and therefore less viable. Natural selection would then favor dispersers.

The ultimate cause of dispersal from the natal site has been argued by many to be the avoidance of inbreeding. The costs of inbreeding, referred to as **inbreeding depression,** have been documented in many laboratory and zoo populations (Ralls et al. 1979) but only recently studied in natural populations. Inbreeding depression manifests itself through reduced reproductive success and survival of offspring from closely related parents compared with offspring of unrelated parents. It is caused by increased homozygosity of the inbred offspring and the resulting expression of deleterious recessive alleles. For example, African lion (*Panthera leo*) males typically leave the natal pride at sexual maturity and attempt to breed with females from other prides. Males from a small, inbred population, however, showed lower testosterone levels and more abnormal sperm than did males from a large, outbred population (Wildt et al. 1987; figure 23.1). Presumably, these inbred prides are prone to extinction. When both inbred and outbred white-footed mice (*Peromyscus leucopus*) were released back into the natural habitat, the inbred stock survived less well than the outbred stock, although differences between the two stocks had not been great in the laboratory environment (Jiménez et al. 1994).

If one or the other sex disperses, the chance of matings between related individuals lessens. Among black-tailed prairie dogs (*Cynomys ludovicianus*), young males leave the family group before breeding, whereas females remain. Also, adult males usually leave groups before their daughters mature (Hoogland 1982). Among primates such as vervet monkeys (*Chlorocebus aethiops*) and baboons (*Papio anubis*), males usually leave the natal group at, or shortly after, sexual maturation. They usually transfer to a neighboring group with age peers or brothers (figure 23.2). Several years later, they may again transfer alone to a third group. Cheney and Seyfarth (1983) argued that this pattern of nonrandom followed by random movement minimizes the chances of mating with close kin. Packer (1977) reported that a male baboon that failed to disperse at sexual maturity and then mated with relatives sired offspring with lower survival rates than the offspring of outbred males. Thus, generally, a real cost in reduced fitness seems to be associated with inbreeding.

The effects of inbreeding depend on past population size and mating patterns. Populations with a long history of outbreeding tend to show the most severe effects once inbreeding takes place. This is because recessive mutations accumulate in the population during outbreeding without ill effect but are more likely to be present in both parents and thus passed on to offspring when inbreeding takes place. Populations that have survived episodes of inbreeding in the past, however, such as the cheetah (*Acionyx jubatus*), may tolerate current inbreeding with few ill effects, in part because

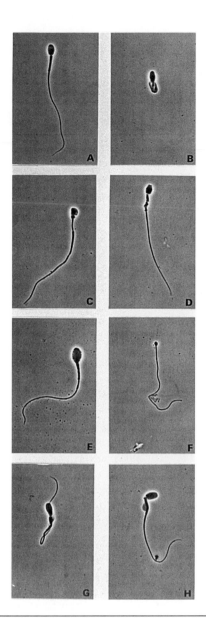

Figure 23.1   Abnormal sperm from an inbred lion population in Africa.   (A) Normal; (B) tightly coiled flagellum; (C) missing mitochondrial sheath; (D) abnormal acrosome and deranged midpiece; (E) macrocephalic with abnormal acrosome; (F) microcephalic with missing mitochondrial sheath; (G) bent flagellum; (H) bent neck with residual cytoplasmic droplet.

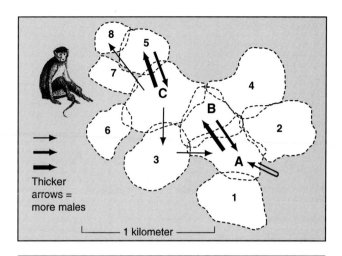

Figure 23.2   Group transfer by natal and young adult male vervet monkeys from three social groups between March 1977 and July 1982.   Arrows indicate direction of movement. Letters indicate ranges of the three main study groups; numbers indicate ranges of regularly censused groups. Only groups with ranges adjacent to the main study groups are shown. Males usually transferred to neighboring groups with brothers or age peers.

Source: Data from D. L. Cheney and R. M. Seyfarth, "Nonrandom Dispersal in Free-ranging Vervet Monkeys: Social and Genetic Consequences," in The American Naturalist, *122:392-412, 1983.*

the deleterious recessive alleles already have been selected out of the population (Shields 1982; see chapter 29).

A second cause of dispersal from the natal site may be the reduction of competition with conspecifics for food, shelter, or mates (Dobson 1982; Moore and Ali 1984). As we saw in chapter 21, gestation and lactation by only females means that males tend to provide little or no direct care for offspring. Most species of mammals are polygynous, with males mating with more than one female. Males may be forced to disperse as they compete for access to females. Although the inbreeding-avoidance hypothesis predicts that

one sex should disperse, it does not predict which sex should disperse; the competition hypothesis predicts that males should be the dispersing sex in polygynous species. According to Greenwood (1980), in such systems, the reproductive success of males is limited by the number of females with which they can mate, and males are likely to range farther than females as they search for mates. Females, on the other hand, are limited by resources (food and nesting sites) that can best be obtained and defended by staying at home. Among group-living mammals, females typically form the stable nucleus, and the males attempt to maximize their access to them, frequently moving from one group to another.

Hamilton and May (1977) proposed a somewhat different competition model in which animals disperse so as to avoid local resource competition with close relatives and thus avoid lowering their indirect fitness. In a new habitat, they are likely to be competing with nonrelatives and therefore would suffer no such cost. These models are not contradictory, but more research is needed on both the proximate and ultimate causes of dispersal. The role of dispersal in population regulation is discussed in chapter 24.

Examples of Dispersal

Lions

Dispersal in lions follows the typical mammalian pattern, in that females usually remain in or near their natal pride, whereas males always leave, usually before 4 years of age, to become nomads or to form coalitions that take over new

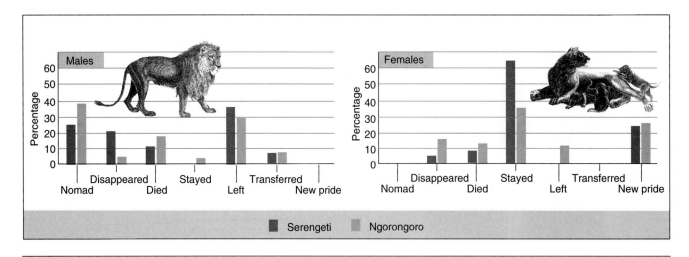

Figure 23.3   **The fate of subadult lions by 4 years of age at two African sites.**   Note the differences between the sexes in dispersal pattern.
Source: Data from Anne E. Pusey and C. Packer, "The Evolution of Sex-biased Dispersal in Lions" in Behaviour, *101:275–310, 1987, E.J. Brill Publishers, Leiden.*

prides (Pusey and Packer 1987; figure 23.3). Competition with other males seems to be an important factor because departures most often occur when a new coalition of males takes over the pride. Some males appear to leave voluntarily, however, either to find mating opportunities or to avoid breeding with kin. Coalitions of males often consist of close relatives. A coalition controls a pride until it is ousted by another coalition a few years later or leaves to take over another pride. In all cases, males leave the pride before their daughters start mating. An additional factor is that new coalitions of males usually kill the young cubs in the new pride (see chapter 20). Thus, breeding males must remain in a new pride long enough to ensure the survival of their cubs. Pusey and Packer (1987) concluded that male–male competition, mate acquisition, protection of young cubs, and inbreeding avoidance all play roles in the evolution of dispersal patterns of lions.

Belding's Ground Squirrels

The question of why individuals disperse was addressed at several different levels of analysis by Holekamp and Sherman (1989). Belding's ground squirrels (*Spermophilus beldingi*) also

follow the typical mammalian pattern, in that females remain in the natal area for life, whereas males disperse as juveniles. The proximate causes of male dispersal seem to be the prenatal effects of testosterone on the male embryo during a critical period in development in the mother's uterus (organizational effects of hormones, see chapter 7) and the attainment of a critical body mass after birth. Testosterone seems less important later in life (activational effects); castration of males just prior to natal dispersal did not prevent dispersal. Holekamp and Sherman were not able to test the inbreeding-avoidance and competition hypotheses directly but concluded that inbreeding avoidance was the more likely means by which dispersal increased fitness (table 23.1).

Inbreeding Versus Outbreeding

If inbreeding depression were the only factor involved in natal dispersal, we might expect individuals to disperse as far as possible from relatives. Such is not usually the case, however. According to Shields (1982), most species that have been adequately studied are relatively philopatric and remain close to the place of birth. He cites apparent cases of **outbreeding**

Table 23.1.   **Why juvenile male Belding's ground squirrels disperse**

Levels of Analysis	Summary of Findings
Physiological mechanisms	Dispersal by juvenile males is apparently caused by organizational effects of male gonadal steroid hormones. As a result, juvenile males are more curious, less fearful, and more active than juvenile females.
Ontogenetic processes	Dispersal is triggered by attainment of a particular body mass (or amount of stored fat). Attainment of this mass or composition apparently also initiates a suite of locomotory and investigative behaviors among males.
Effects on fitness	Juvenile males probably disperse to reduce chances of nuclear family incest.
Evolutionary origins	Strong male biases in natal dispersal characterize all ground squirrel species, other ground-dwelling sciurid rodents, and mammals in general. The consistency and ubiquity of the behavior suggest that it has been selected for directly across mammalian lineages.

From "Why Male Ground Squirrels Disperse," by Kay E. Holekamp and Paul W. Sherman, American Scientist, *77:232—239. Copyright©1989. Reprinted by permission of* American Scientist, *Journal of Sigma Xi, The Scientific Research Society.*

depression, in which matings between members of different populations within a species yield less-fit offspring. Members of a population may possess adaptations to local conditions that are lost through outbreeding. Thus, two areas might differ slightly in temperature, humidity, or the types of food available. If each population is genetically adapted to these conditions, then they would be better off mating with individuals with those same adaptations.

A certain degree of inbreeding may be advantageous. In sexually reproducing organisms, the loss of genes in the offspring can be reduced by half if the parents are related. Furthermore, according to Shields (1982), coadapted gene complexes are less likely to be disrupted in matings between relatives. Finally, kin selection (chapter 22), which results in the evolution of cooperative behaviors, can operate only when relatives are in close proximity. We predict more cooperative behavior within philopatric species and within the philopatric sex. Indeed, among Belding's ground squirrels, females are philopatric and engage in altruistic alarm calling. Males who disperse away from relatives do not make alarm calls (Sherman 1981; see also chapter 22).

From the preceding arguments, we might predict some sort of "optimal" inbreeding strategy in which matings between very close relatives (siblings, or parents and offspring) are avoided but matings with more distant relatives are favored. In laboratory tests, female white-footed mice in estrus preferred males who were first cousins over nonrelatives or siblings (Keane 1990). Heavier pups and larger litters resulted from matings between first cousins than between individuals of other degrees of relationship. Based on electrophoretic analysis of blood enzymes in natural populations, however, little if any mating between relatives seems to occur in this species (Wolff et al. 1988).

Female Dispersal

Although males are the dispersing sex in most species of mammals, exceptions exist in several orders. Among lagomorphs, pikas (*Ochotona princeps*) live in relatively isolated patches of talus (rock debris) on mountain slopes (see chapter 17). Most juvenile pikas remain close to their birth site for life. Individuals occasionally disperse both within and between patches of talus, however. Of those that moved more than 100 m, females moved more than males at sites in Alberta (Millar 1971) but not in Colorado (Smith 1987). Similarly, among banner-tailed kangaroo rats (*Dipodomys spectabilis*) in Arizona, most juveniles stayed home, sharing all or part of the maternal home range. Of those that did disperse, females had a tendency to move farther (Jones 1987). This species is solitary, living in dispersed mounds that are "inherited" from the previous occupant. Other examples of species in which females are the dispersing sex include the chimpanzee (*Pan troglodytes*), African wild dog (*Lycaon pictus*), and white-lined bats (*Saccopteryx* spp.). These groups have widely differing social systems, and it is not yet clear why the typical mammalian pattern is reversed (Greenwood 1983).

HABITAT SELECTION

Generally, plants depend on natural agents such as currents of air or water or other organisms for dispersal. The result is an opportunistic dissemination of plant individuals; few ever reach environments conducive to survival and reproduction. In contrast, the well-developed locomotive abilities of mammals allow them to play more active roles in finding places to live. **Habitat selection** can be defined as choosing a place to live, which does not necessarily imply a conscious choice or that individuals make a critical evaluation of the entire constellation of factors confronting them. More often the choice is an "automatic" reaction to certain key aspects of the environment.

If a species of mammal occupies an area and reproduces there, we know that all its needs are met and that it can compete with other species successfully. Any one of several factors can prevent a species from occupying particular habitats, however. One obvious factor is dispersal ability. Oceanic islands such as New Zealand and Hawaii provide much habitat suitable for mammals; yet, other than marine mammals, only a few species of bats have made it there on their own. Many species such as rats (*Rattus* spp.) and domestic ungulates have been introduced by humans and have thrived.

Sometimes behavior patterns keep species from occupying apparently suitable habitats. The white-footed mouse lives in woodlands, but it is rarely found in adjacent fields, which are inhabited by a closely related species, the deer mouse (*Peromyscus maniculatus bairdii*). In fact, *P. leucopus* seems perfectly capable of living in fields also. Why should a species not take advantage of suitable habitat? One possibility is that the habitat is not actually suitable, perhaps because of competition, predation, or other factors we have failed to detect. It is also possible that such habitats may not have been suitable in the past; if organisms responding to certain environmental cues in previous optimal habitats left more offspring, their genetically influenced behavior patterns would become widespread and persist. New environments, although suitable, might not contain those cues and therefore not be used. We know that white-footed mice prefer the vertical structure that is present in woodlands but generally lacking in field habitats (M'Closkey 1975). This mouse orients toward large trees (Joslin 1977b) and avoids areas with high densities of woody stems (Barry and Francq 1980).

Even if an individual can and "wants" to get to a place, other factors may prevent its becoming established. These factors include predators, parasites, disease agents, **allelopathic** agents (plant toxins or antibiotics), or competitors. Although it is difficult to demonstrate conclusively that one species prevents an area from being colonized by another, experiments and observations lend support to this idea (see chapter 25).

Risk of predation and competition may restrict habitat use. Hairy-footed gerbils (*Gerbillurus tytonis*) live in vegetated islands in a sea of sand in the Namib Desert of southwestern Africa (Hughes et al. 1994). Individuals preferred

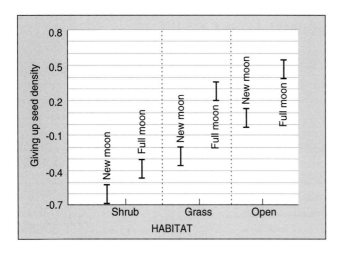

Figure 23.4   **The effect of predation risk on habitat use in hairy-footed gerbils.**   Values along the *y* axis are giving up seed densities (GUDs) ($\log_{10}$ scale) in food trays containing seeds mixed with sand that were placed in different habitats. High values mean that gerbils left the trays when many seeds were still present. Bars denote 95% confidence intervals. Gerbils gave up at higher seed densities in open areas and on full-moon nights.

Source: Data from J.J. Hughes et al., "Predation Risk and Competition Affect Habitat Selection and Activity of Namib Desert Gerbils" in Ecology, *75:1397–1405, 1994, Ecological Society of America.*

that limit the distribution of life on earth, but physical factors, such as light, soil structure, or fire, and chemical factors, such as pH and nutrients, may be important as well. Adaptations of mammals to some of these conditions were discussed in chapter 8.

Determinants of Habitat Preference

Genes and Environment

How can we sort out the roles that genetics and learning play in habitat choice? If two animals reared from birth in identical environments differ in habitat preference when they are tested as adults, we can conclude that those differences result from hereditary factors.

Wecker (1963) conducted one of the classic studies on habitat selection in mammals on deer mice (*Peromyscus maniculatus*). This species, a common North American rodent, is divided into many geographically variable subspecies; two are the larger, long-eared, long-tailed forest form (*P. m. gracilus*) and the smaller, short-eared, short-tailed grassland (prairie) form (*P. m. bairdii*). The grassland subspecies does well in forestlike laboratory conditions where its food preference and temperature tolerance are similar to those of the forest subspecies. Thus, experimenters assumed that the avoidance of forests in the prairie deer mouse is a behavioral response (Harris 1952).

Wecker's objective was to assess the genetic basis of this behavior and to test the importance of "habitat imprinting" (Thorpe 1945). He constructed an enclosure halfway in a forest and halfway in a grassland, released the grassland subspecies in the middle, and recorded their locations. The animals he tested were of three basic types: (1) wild-caught in grassland; (2) offspring of wild-caught parents, reared in a laboratory; and (3) reared in a laboratory for 20 generations. Both wild-caught mice and their offspring selected the grassland half of the enclosure, whatever their previous experience. Laboratory stock and their offspring showed no preference, whether or not they had been raised in forest conditions. Laboratory stock reared in a grassland enclosure until after weaning, however, showed a strong preference for the grassland when tested later (table 23.2).

sites around bushes or grass clumps over open areas and were more active on new-moon nights than on full-moon nights. They also gave up feeding at seed trays sooner in open areas and on full-moon nights (figure 23.4). These differences were likely caused by greater risk of predation in open areas and when the moon was full. When striped mice (*Rhabdomys pumilio*), a close competitor of the gerbil, were removed, gerbils increased foraging activity, especially in the grass clumps.

Finally, a species may be absent from an area because of abiotic (nonbiological) factors. Each organism has a range of tolerances for a variety of physical and chemical factors, and much of its behavior is directed toward staying within these limits. Temperature and moisture are the main factors

Table 23.2.   **Habitat selection by deer mice as a function of hereditary background and early experience***

Number of Mice Tested	Hereditary Background	Early Experience	Habitat Preference
12	Grassland	Grassland	Grassland
13	Laboratory	Grassland	Grassland
12	Grassland	Laboratory	Grassland
7	Grassland	Forest	Grassland
13	Laboratory	Laboratory	None
9	Laboratory	Forest	None

Source: Data from S.C. Wecker, "The Role of Early Experience in Habitat Selection by the Prairie Deer Mouse, Peromyscus maniculatus bairdi*" in* Ecological Monographs, *33:307–325, 1963, Ecological Society of America.*

*The outdoor test enclosure was forest on one side and grassland on the other. Preference was measured by the percentage of time, amount of activity, and depth of penetration by mice in each side of the enclosure.

Wecker reached the following conclusions:

1. The choice of grassland environment by grassland deer mice is predetermined genetically.

2. Early grassland experience can reinforce this innate preference but is not a prerequisite for subsequent habitat selection.

3. Early experience in forest or laboratory is not sufficient to reverse the affinity of this subspecies for the grassland habitat.

4. Confinement of these deer mice in the laboratory for 12 to 20 generations results in a reduction of the hereditary control over the habitat selection response.

5. Laboratory stock retains the capacity to "imprint" on early grassland experience but not on forest.

Wecker also suggested that learned responses, such as habitat imprinting, are the original basis for the restriction of this subspecies to grassland environments; genetic control of this preference is secondary.

In a series of laboratory experiments designed to explore the importance of early experience on bedding preference in inbred house mice (*Mus musculus*), Anderson (1973) raised animals either on cedar shavings or on a commercial cellulose material. When he tested them later, he found that the mice preferred the bedding on which they had been raised, although females raised on cellulose "drifted" toward cedar shavings in subsequent tests. Naive mice preferred cedar shavings. Both this experiment and the one by Wecker illustrate the complex interaction of genes and environment in the development of habitat preferences in mammals.

Tradition

Inherited tendencies and imprinting may be involved in restricting habitat choice to a small part of the potential range, but **tradition**—behavior passed from one generation to the next through the process of learning—may also be important. Such behavior seems to be important in the movements of many species of ungulates.

Mountain sheep (*Ovis canadensis*) live in unisexual groups; females are likely to stay in the natal group but may switch to another female group when they are between 1 and 2 years of age (Geist 1971). Mothers do not tend to chase their young away at weaning, as most other mammals do. Young rams desert the natal group at 2 years of age and join all-ram bands. Females follow an older, lamb-leading female, whereas males follow the largest-horned ram in the band. When rams mature, they are followed by younger rams and pass on to them their habitat preferences and migration routes.

Until the nineteenth century, mountain sheep occupied a much larger range in North America and Asia than they do today. Measures enacted to protect sheep, mainly hunting regulations, have done little to increase their numbers. Formerly inhabited parts of the range that appear intact often have not been recolonized, and transplants to suitable areas often have

been unsuccessful. In contrast, deer (*Odocoileus* spp.) and moose (*Alces alces*) have recolonized areas rapidly and have reached population densities higher than ever (Geist 1971).

Why have sheep failed to extend their range, but moose and deer have done so? Geist pointed out that deer and moose, which are relatively solitary, establish ranges by individual exploration after being driven out of the mother's range; sheep, in contrast, transmit home-range knowledge from generation to generation and often associate with group members for life.

Understanding the niches of these ungulate species can help explain the sheep system, which may seem a rather poor adaptation to the environment. Moose are associated over much of their range with early successional communities that follow forest fires, where moose are a "pioneer species" that disappears when the climax coniferous forest regenerates. Moose therefore must continually colonize these newly formed early successional habitats. Each spring when her new calf is born, the cow drives away her yearling, which may wander some distance before establishing its new range.

Sheep habitats consist of stable, long-lasting, climax grass communities that exist in small patches. Geist argued that, given the distance between patches and the ease with which wolves can prey on sheep, the best strategy for the sheep is to stay on familiar ground. Sheep also have at least two and as many as seven seasonal home ranges that may be separated by 30 km or more. These areas are visited regularly by the same sheep year after year at the same time. Knowledge of the location of these ranges and the best times to visit them is transmitted from one generation to the next. Because new habitats rarely become available, there is no advantage to an individual's dispersing and attempting to colonize other areas.

Tradition may also play a role in habitat use by mantled howling monkeys (*Alouatta palliata*), which are Neotropical folivores. Many trees within their home ranges produce leaves that contain secondary compounds that make them unpalatable or even toxic to the monkeys (Glander 1982). Glander (1977) postulated that older group members provide a reservoir of knowledge about the location and timing of seasonal foods that reduces the chances of the group's encountering toxic leaves.

Theory of Habitat Selection

Theoretical approaches to the problem of habitat selection are in a rather early stage of development, and no single general theory is currently accepted (Rosenzweig 1985). Several different approaches have been used. One is to think of habitats as patches, or areas of suitable habitat interspersed among areas of unsuitable habitat, and to apply optimal foraging theory, first developed by MacArthur and Pianka (1966; see chapter 6). This theory enables us to predict which habitat patches an animal should select and when it should leave one habitat and move to another so as to get the greatest benefit for the least cost. This economic model incorporates such factors as the availability of resources in various patches and the costs of getting from one patch to another. Although the resource is usually assumed to be food (i.e., energy), nest sites or mates are other possibilities. For a review of optimal foraging theory, see Stephens and Krebs (1986).

A second approach is the **ideal free distribution,** which predicts how individuals distribute themselves so as to have the highest possible fitness (Fretwell and Lucas 1970). It assumes that animals have complete and accurate knowledge about the distribution of resources (ideal) and that they are passive toward one another and can go to the best possible site (free). Individuals settle in habitats so that the first arrivals get the best resources. As density increases, less desirable areas are occupied, and animals spread themselves out so that all have the same fitness in the absence of intraspecific competition. One obvious result of such a distribution is that rich habitats will have more individuals than poor ones.

If intraspecific competition occurs via dominance or territory (see chapter 20), a despotic distribution develops, with some individuals monopolizing the best resources (Fretwell 1972). Another variation on the ideal free distribution is the ideal preemptive distribution (Pulliam and Danielson 1991). Potential breeding sites differ in quality— that is, in the expected reproductive success of their occupants—and individuals choose the best unoccupied sites. These best sites are thus preempted and are no longer available to others, but their occupancy does not influence the expected reproductive success of occupants of other sites. Several studies have shown that individuals in the preferred habitat have higher fitness, as measured by reproductive success and survival, than those in less preferred habitat. For instance, Grant (1975) found that meadow voles (*Microtus pennsylvanicus*) in the preferred grassland habitat had higher survival and reproductive success than those in the less preferred woodland.

Habitat selection also affects the growth rates and densities of populations, a topic considered in chapter 24. It may even lead to the formation of new species. Although the usual models of speciation require the formation of geographic barriers to gene flow, reproductive isolation could evolve if members of two parts of a population came to prefer different microhabitats in the same region. If the preferences were heritable and individuals mated assortatively (i.e., with others having similar habitat preferences), a barrier to gene flow would be created that could eventually lead to the formation of new species (Rice 1987). Such a mechanism has yet to be demonstrated in mammals, however.

MIGRATION

Of all the movements that animals make, perhaps none has generated so much interest and controversy as migration. Migratory behavior differs from station-keeping movements associated with resources and maintenance of home ranges or territories. Migration takes an animal out of its home range and habitat type. It is triggered by proximal cues, such as photoperiod, that are linked to ultimate factors, such as a shortage of resources. Migration may also occur in response to endogenous rhythms (Dingle 1996). Earlier definitions of migration require that the individual make a round trip, but one-way movements are included in most recent definitions. In migrating, mammals may use all three means of vertebrate locomotion: flying, swimming, and walking.

Bats

As the only mammals with true flight, we might expect bats to show migratory behavior comparable to the only other flying vertebrates, the birds. Such is not the case for most species. As we saw in chapter 12, bats have evolved relatively slow, maneuverable flight, necessary to catch insects on the wing. The wings are even used to trap insects. Bat wings are thin airfoils of high camber (curvature) that produce high lift at slow speeds but produce excessive drag at the high speeds necessary for long-distance migration. Rather than migrate, many species use hibernation, an energetically less-costly way of dealing with cold temperatures and lack of food.

In spite of these flight constraints, some species of bats do migrate, most often to and from caves and other shelters used as hibernation sites (Griffin 1970). Information about bat migration comes from the seasonal appearances and disappearances at roosting sites coupled with recoveries of banded individuals (Fenton and Thomas 1985). Radiotelemetry has been used in a few instances, mostly to study foraging trips and roosting behavior (Fenton et al. 1993). In a study involving the banding of more than 73,000 little brown bats (*Myotis lucifugus*), individuals migrated more than 200 km from hibernation caves in southwestern Vermont. They generally moved southeast into Massachusetts and neighboring states for the summer (Davis and Hitchcock 1965). The endangered Indiana bat (*M. sodalis*) migrates from hibernation caves in Kentucky and southern Indiana as far north as Michigan (Barbour and Davis 1969).

Swifter flying species such as the hoary bat (*Lasiurus cinereus*) and free-tailed bat (*Tadarida brasiliensis*) move even greater distances. Hoary bats migrate from summer ranges in the Pacific Northwest, as far north as Alaska, south into central California and Mexico for the winter. In winter, they are not found above 37°N latitude, a limit probably set by the distribution of flying insects (Griffin 1970). Free-tailed bats seem to have both migratory and nonmigratory populations. Those in southern Oregon and northern California are year-round residents, but those in the southwestern United States migrate south into Mexico for the winter (Dingle 1980). For instance, a population in the Four Corners area (where Colorado, Utah, New Mexico, and Arizona meet) has a well-established flyway through the Mexican states of Sonora and Sinaloa west of the Sierra Madre Oriental Mountains. The routes of some southwestern U.S. populations have yet to be identified.

Long-distance migration is not restricted to bat species in temperate regions. Seasonal shifts in rain patterns trigger migration in some species of African bats (Fenton and Thomas 1985). For instance, West African fruit bats of the Family Pteropodidae migrate distances of 1500 km each year, following rains into the Niger River basin (Thomas 1983).

Cetaceans

Most species of baleen whales (suborder Mysticeti) spend summer months at high latitudes, feeding on plankton in the highly productive antarctic and arctic waters. As winter approaches in each area, whales migrate to warmer subtropical and tropical waters. Food supply does not drive this migration; tropical waters are relatively unproductive and, in fact, whales do not feed during migration or at their wintering grounds. Instead, they rely on fat deposits. The benefit of moving to warmer water is likely the energy savings from reduced heat loss, especially for calves. Calves are born in the tropical breeding grounds, and lactating females with their newborn calves move to feeding areas at higher latitudes as spring approaches. Breeding cycles of species that breed inshore, such as the humpback whale (*Megaptera novaeangliae*) and the gray whale (*Eschrichtius robustus*), are fairly well known, but little is known about the breeding habits of the offshore species, such as the blue whale (*Balaenoptera musculus*) and fin whale (*B. physalus;* Dingle 1980).

California gray whales spend their summers feeding in the North Pacific and Arctic oceans. In autumn, they migrate south to subtropical breeding grounds off the coast of Baja California (Orr 1970). Humpback calves are born in September, and lactating females begin the trek back north with their calves in the spring, usually after males and newly pregnant females have already left (Dingle 1980).

Pinnipeds

Many species of seals and sea lions (chapter 15) migrate thousands of kilometers from island breeding and molting areas to oceanic feeding areas. Island breeding sites are chosen because they are relatively free of predators. Northern elephant seals (*Mirounga angustirostris*) breed on island rookeries off California, migrate to foraging areas in the North Pacific and Gulf of Alaska, and later return to the islands to molt. Using geographic location-time-depth recorders, Stewart and DeLong (1995) found that seals travel linear distances of up to 21,000 km during the 250 to 300 days they are at sea. Each individual makes two round-trip migrations per year, returning to the same foraging areas during postbreeding and postmolt movements. Males migrate farther north than females, where they feed off the Alaskan coast (figure 23.5).

Ungulates

Large ungulates migrate long distances as well. The best-studied northern species is the barren-ground caribou (*Rangifer tarandus*). Herds migrate north to calving grounds above the timberline in spring and return south in winter, covering distances of more than 500 km (Orr 1970; figure 23.6). More recently, the movements of individuals have been monitored by satellite tracking (Craighead and Craighead 1987). Migrations seem to be made up of series of fairly straight movements that are little influenced by landmark features such as rivers (Bovet 1992; figure 23.7).

The mass migrations of wildebeests (*Connochaetes taurinus*) in East Africa are spectacular (figure 23.8). The Serengeti population spends the wet season, usually December through April, in the southeastern Serengeti plains of Tanzania, where short grasses are lush and calving takes place. Large migratory herds form at the beginning of the

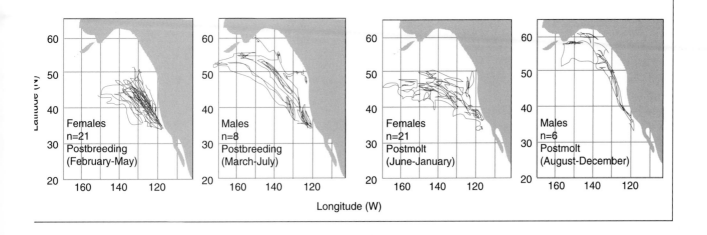

Figure 23.5   Seasonal migratory tracks of northern elephant seals in the eastern North Pacific.   Each seal makes two migrations per year, returning to land for breeding and again for molting. These seals were marked on San Miguel Island, California.

Source: Data from B.S. Stewart and R.L. DeLong, "Double Migrations of the Northern Elephant Seal, Mirounga angustirostris," *in* Journal of Mammalogy, *76:196–205, 1995.*

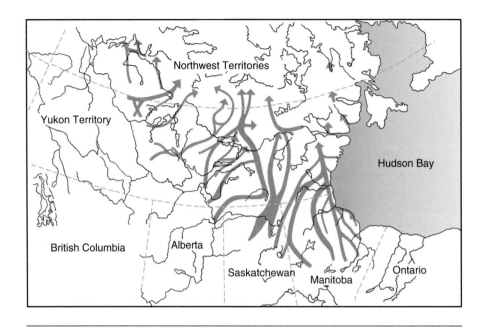

Figure 23.6   **Migration routes of barren-ground caribou.**   In spring, herds move north to calving grounds above the timberline; in autumn, they return south to the shelter of forests.

Source: Data from H.L. Gunderson, Mammalogy, *1976, McGraw-Hill, Inc., New York, NY.*

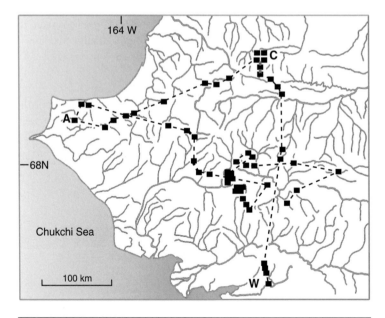

Figure 23.7   **Migration route of an adult, female caribou in northwest Alaska in 1984.**   The dashed line connects successive satellite-fixes (filled squares). The caribou left the winter range (W) on 15 May and arrived on the calving ground (C) on 30 May, where she calved on 5 June, and stayed until 16 June. She then moved to the herd's aggregation area (A), where she stayed on 4 and 5 July. She spent the summer traveling east, with occasional 1- to 2-week stays in localized areas. The last fix was on 7 October, while she was moving toward the winter range. Due to the hydrographic features of the area (thin lines = rivers), the caribou's route was probably as often across as it was parallel to valleys.

Source: Data after Craighead & Craighead, 1987, and Fairbanks of the World Map (1:2500000) of the USSR Main Admin. of Geodesy and Cartography, Moscow, 1973; in J. Bovet, "Mammals" in Animal Homing, *(F. Papi, ed.), 1992, Chapman and Hall, New York.*

Figure 23.8   **Wildebeests migrating in the Serengeti.**   The mass migration of wildebeests is associated with seasonal rainfall and the consequent growth of grasses.

dry season, in May and June, as millions of animals move, sometimes in single file, northwest toward Lake Victoria. In July and August, near the end of the dry season, herds move northeastward into the Masai-Mara of Kenya and return south to the breeding grounds between November and December. These patterns vary considerably, however, depending on the timing of rainfall (Dingle 1980).

Other ungulates engage in elevational migration; elk (*Cervus elaphus*) and mule deer (*Odocoileus hemionus*) move into high-elevation summer ranges that are relatively free of snow and then return to milder winter ranges at lower elevations (McCullough 1985). Mountain sheep follow the same routes each year, climbing up into isolated patches of lush grazing meadow as the snow melts in spring and returning to lower elevations in autumn and winter.

HOMING

Whether migrating thousands of kilometers or simply foraging within the home range, most species of mammals return to a home range, nest site, or den, a process called **homing.** For example, mule deer have shown site fidelity for both summer and winter home ranges, with straight-line distances between seasonal ranges of as much as 115 km (Thomas and Irby 1990). Of 30 deer radiotracked for more than a year, 29 returned to the same home ranges in both summer and winter. Similar results have been obtained for

elk, white-tailed deer (*Odocoileus virginianus*), moose, and mountain sheep (Bovet 1992). Nonmigratory species must also find their way home each time they return to a nest or den after foraging or searching for mates.

Orientation and Navigation

How do mammals find their way home? Most of what we know about the process of homing comes from studies of homing pigeons, but enough work has been done with mammals to suggest that they use many of the same mechanisms. Most research has been done on rodents and bats. One technique for studying homing is to displace the animal from its home range and record such variables as the direction it heads when released (the "vanishing bearing"), the time it takes to get home, whether or not it makes it home at all, and in some cases, the route it takes to get back home. The roles of various senses can be studied via manipulations (e.g., use of blindfolds).

In one of the earliest studies, in which deer mice were released at distances well beyond their home range (Murie and Murie 1931), some individuals made it back home. Three possible mechanisms were suggested: (1) the mice were sufficiently familiar with the terrain; (2) the mice had some sort of homing instinct, or sense of direction; or (3) the results were due to chance. These possibilities form the basis for much of the later work on homing mechanisms (Joslin 1977a).

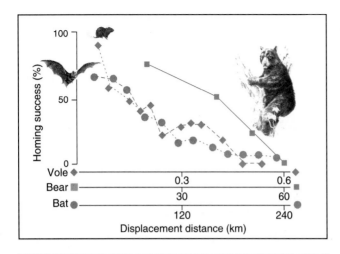

Figure 23.9   **Relationship between homing success and displacement distance.**   Meadow voles (*Microtus pennsylvanicus*), *n* = 460; black bears (*Ursus americanus*), *n* = 112; Indiana bats (*Myotis sodalis*), *n* = 700.

Source: Data after Robinson and Falls, 1965; MacArthur, 1981; Hassell, 1960; in J. Bovet, "Mammals" in Animal Homing, *(F. Papi, ed.), 1992, Chapman and Hall, New York.*

As might be expected, homing success decreases as a function of the distance the individual is displaced (Bovet 1992; figure 23.9). Homing success increases as a function of home-range size, even within the same species. For instance, house mice with large home ranges homed over greater distances than did those with small home ranges (Anderson et al. 1977).

Several studies have added sensory deficits to displacement, with mixed results. Making white-footed mice **anosmic** (eliminating the sense of smell) had little effect on homing ability, whereas blinding them produced negative effects (Cooke and Terman 1977; Parsons and Terman 1978). Blindfolding bats produced little deficit over short distances but reduced success over large distances (>32 km; Mueller 1966).

Path Integration

One way a mammal might home is to somehow keep track of all the turns and accelerations on the outward trip, integrate them, and use that information to get back home. This route-based orientation mechanism does not require actually retracing the outbound path. Hamsters (*Mesocricetus auratus*) placed in the center of a 2-m arena were able to get back to their peripheral nestbox using information collected on the way out to the center (Etienne et al. 1988). It is unlikely, however, that such a mechanism would be used for long-distance homing.

Distant Landmarks

Several laboratory studies have demonstrated that rodents can locate food or shelter based on their relationship to distant visual cues, a mechanism sometimes referred to as **piloting**

(Bovet 1992). Landmarks would seem best for locating objects within the home range, such as when gray squirrels (*Sciurus carolinensis*) locate a food cache (McQuade et al. 1986). Piloting requires the existence of some sort of cognitive, or mental, map of the terrain, but it does not require a compass. Landmarks are assumed to be most useful within or close to the home range. In principle, however, they could be used on longer trips as well. An individual could head toward some distant landmark such as a mountain on the way out and away from it on the way back. On the return trip, however, the animal would have to maintain a constant angle 180 degrees away from the landmark, that is, have a mental compass.

Sun Compass

Another way to maintain direction is to use the sun as a compass and maintain a constant angle to it while traveling. Of course, the sun is not fixed in the sky, so an individual must have some sort of internal clock that enables it to compensate for the movement of the sun across the sky (about 15 degrees/h). One way to test for the existence of a sun compass is to "clock-shift" the animal in the laboratory by delaying the onset of the light-dark cycle and then test it in the field. For instance, if the test animal is shifted 6 hours in the laboratory, it should head 90 degrees off course when tested. Thirteen-lined ground squirrels (*Spermophilus tridecemlineatus*) were tested in an outdoor arena 100 m west of their home cages with only the sky visible (Haigh 1979). When released in the arena, they burrowed in the direction of the home cages. These same individuals were then clock-shifted 6 hours in the laboratory. When retested, most shifted their burrowing direction 90 degrees in a clockwise direction, as predicted (figure 23.10). The ability to use the sun as a compass with time compensation has also been

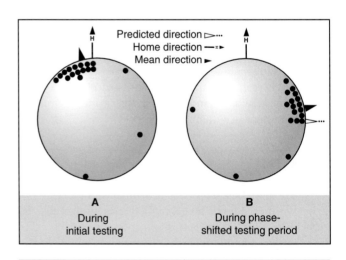

Figure 23.10   **Digging responses of 13-lined ground squirrels in outdoor enclosures.**

Source: Data from G.R. Haigh, "Suncompass Orientation in the Thirteen-lined Ground Squirrel, Spermophilus tridecemlineatus" *in* Journal of Mammalogy, *60:629-632, 1979.*

demonstrated in several other species of small rodents (Bovet 1992). Big brown bats (*Eptesicus fuscus*) use the post-sunset glow to travel from their roost to favorite foraging areas (Buchler and Childs 1982). They depart the roost in the evening at a colony-specific angle to the glow that is independent of landmarks. Other celestial cues such as the stars are used by migrating birds but have not been demonstrated in mammals.

Magnetic Compass

Much less clear is whether mammals can obtain directional information from geomagnetic cues, although many other organisms do. Frequent reports of otherwise healthy whales that strand themselves on beaches suggest that these whales have made navigational errors in areas where magnetic minima intersect the coast (Klinowska 1985). When aerial sightings of fin whales off the northeast U.S. coast were plotted on maps, no association could be drawn between location and measurements such as bottom depth or slope. Sightings of migrating, but not feeding, animals, however, *were* associated with areas of low geomagnetic-field intensity and gradient in autumn and winter (Walker et al. 1992). On the other hand, sightings of common dolphins (*Delphinus delphis*) off

the coast of southern California were related to bottom topography but not to magnetic patterns (Hui 1994).

When white-footed mice were displaced 40 m from their home areas in woods and released in a circular arena in adjacent fields, exploratory and escape behavior was concentrated in the homeward direction. A second group of mice was treated exactly the same, except that a magnetic field opposite of that of Earth's was established in the transport tube. Those mice concentrated their activity in the opposite direction from home, suggesting that they had a magnetic sense and used geomagnetic fields as a compass cue (August et al. 1989). Similar results have been found in several other species of rodents (Bovet 1992) and apparently even in humans (Baker 1987).

If a mammal is going to get home in unfamiliar terrain without the use of landmarks, it needs a map as well as a compass to know its location in relation to home. For instance, it might be able to use the magnetic isoclines to get information about longitude. It would, however, need another gradient along a second axis in order to get information about latitude, to fix itself in two-dimensional space. Such a grid-based bicoordinate navigation system has not yet been demonstrated conclusively in any animal.

Summary

Movements are made by mammals at several different levels of spatial scale. At the smallest scale—station keeping—the animal obtains and defends resources in its home range. Ranging involves forays outside the home range in search of new habitat or mating opportunities and may include natal dispersal. Migration is the persistent movement across different habitats without regard to presence or absence of resources. Among mammals, migration usually consists of a round trip.

Within a species' range, individuals may disperse or remain in the natal area to breed. In mammals, males typically disperse, but females are philopatric. By dispersing, the chances of matings between close relatives are reduced and inbreeding is minimized. Mammals may also disperse to reduce competition for resources among relatives and neighbors. Most mammal species are polygynous; females defend resources needed for reproduction, and males range over a larger area in search of mates. Although there are demonstrated costs to extreme inbreeding, outbreeding may also incur genetic costs; some studies show an optimal degree of inbreeding, in which first cousins are favored as mates.

Habitat selection refers to the choice of a place in which to live. Organisms may fail to colonize an otherwise suitable area because of the inability to get there, as evidenced by the lack of mammals on oceanic islands. Behavior patterns may restrict mammals to a fraction of the habitat that they seem, to us, to be adapted to occupy. Other factors

such as competitors, parasites, predators, diseases, or physical and chemical factors can restrict a species' distribution.

Experiments with mammals have been conducted to determine the roles of genes and experience in habitat selection. Selection of the "correct" habitat is under some degree of genetic control, as has been demonstrated by studies in which animals have been reared in isolation and later tested in various habitats. Early experience can modify later choices, however. Tradition, the transmission of knowledge of habitats from one generation to the next, is thought to be important in some ungulate species.

Theoretical approaches to habitat selection include optimal foraging models, in which individuals choose patches (habitats) and stay in them so as to maximize the gain of some resource. The ideal free distribution assumes that, in the absence of competition, individuals settle in the best habitats first, with later arrivals moving into suboptimal sites.

Migration in mammals has been well documented in bats, cetaceans, pinnipeds, and ungulates. Many bats with temperate distributions migrate from hibernacula, often caves, to summer roosts and then return the following autumn. Baleen whales move from high-latitude feeding grounds to tropical seas to breed. Pinnipeds such as elephant seals spend most of their time at sea as they migrate from pelagic feeding grounds to island sites that are used for both breeding and molting. Ungulates such as barren-ground caribou migrate between northern breeding and southern wintering grounds. Elk and deer make vertical migrations as they spend summers at high elevations and then return to lower elevations in winter. Wildebeests in East Africa engage in mass migrations in association with seasonal rainfall patterns and the growth of grasses.

Migrants often return to the same home ranges they occupied previously, and nonmigrants traversing their home ranges in search of resources return to a nest or den, a process called homing. Individuals that are experimentally displaced from their home ranges often traverse unfamiliar terrain and return home. Short-distance homing can occur by path integration, in which the animal stores and integrates information from the outward trip. Landmarks may be used for trips within familiar ground, a process called piloting. Some species of mammals use the sun as a compass and are able to compensate for the apparent movement of the sun across the sky. The ability to detect and use geomagnetic cues has also been demonstrated in a few species of mammals.

Discussion Questions

1. In contrast to mammals, most species of female birds disperse farther from the natal site to breed than do males. Why might this be so? Consult the references by Greenwood (1980, 1983).
2. Review the life history of a mammalian species of your choice. Try to explain the dispersal patterns of each sex in light of our discussion of both proximate and ultimate causation. Include ecological as well as genetic factors in your answer.
3. Although tradition is claimed to play a role in the habitat preference of some animals, such as mountain sheep (Geist 1971), firm data are lacking. Design an experiment to demonstrate the role of tradition in habitat choice.
4. African mole-rats (*Cryptomys hottentotus*) are colonial, subterranean rodents that dig the longest underground burrow systems of any mammal. Burrow systems are linearly arranged, with the main tunnel often more than 200 m long. The main tunnel is usually oriented in a north-south direction. Burda and colleagues (1990) explored the possible role of the geomagnetic field as a cue for underground orientation, considering that these animals are virtually blind. Family groups were provided with nesting material in a light-proof arena. They were then exposed to the local geomagnetic field (control) or to experimental fields produced by Helmholtz coils surrounding the arena.

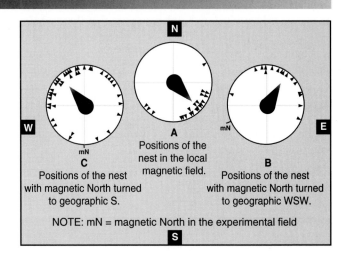

Arrows represent the mean vectors for nest location.

Source: Data from H. Burda, et al., "Magnetic Compass Orientation in the Subterranean Rodent Cryptomys hottentotus *(Bathyergidae)" in* Experientia, *46:528–530, 1990.*

The symbols in the figure denote the location of nests along the walls of the arena. What conclusions can you draw from these results? What additional experiments might be necessary?

Suggested Readings

Bovet, J. 1992. Mammals. Pp. 321–361, *in* Animal homing (F. Papi, ed.). Chapman and Hall, New York.

Chepko-Sade, B. D. and Z. T. Halpin, eds. 1987. Mammalian dispersal patterns. The effects of social structure on population genetics. Univ. of Chicago Press, Chicago.

Dingle, H. 1996. Migration: the biology of life on the move. Oxford Univ. Press, New York.

Krebs, C. J. 1994. Ecology: The experimental analysis of distribution and abundance, 4th ed. HarperCollins, New York.

Rankin, M. A., ed. 1985. Migration: mechanisms and adaptive significance. Marine Science Institute, Port Aransas, TX.

Swingland, I. R. and P. J. Greenwood. 1983. The ecology of animal movement. Clarendon Press, Oxford, England.

CHAPTER

24

Populations and Life History

Populations are groups of organisms of the same species, present at the same place and time. Their boundaries may be natural, or they may be defined arbitrarily by an investigator. A central problem for mammalian ecologists for decades has been to understand and predict the population dynamics of mammals. Predicting outbreaks of pest species or declines of endangered species relates directly to our quality of life. In this chapter, we consider processes that affect populations of mammals; that is, how populations grow and how they are regulated in nature. We also deal with life history tactics of individuals, exploring tradeoffs that mammals make as they allocate resources to reproduction versus survival.

POPULATION PROCESSES

Populations of mammals increase via births, or **natality;** they decrease via deaths, or **mortality;** or they change in numbers via movements. One-way movement into a population is **immigration,** whereas one-way movement out of a population is **emigration.** The term *migration* refers to movements (usually round-trip) that are repeated each year, shown most spectacularly by some species of bats, ungulates, cetaceans, and pinnipeds (see chapter 23).

Rate of Increase

Populations of all organisms have a great capacity for increase, perhaps best illustrated by the human population of the world that is presently increasing at nearly 2% per year and doubling in less than 40 years. The English economist Thomas Malthus was perhaps the first to realize that populations had the potential to increase exponentially, but the means to support them did not (Malthus 1798). Arithmetical growth takes place by adding a constant amount during each time interval, but **exponential** growth occurs in proportion to the number already present. Populations increase exponentially because as each individual matures, it begins to breed along with the rest of the population. Similarly, in a savings account in which the interest is compounded, the interest begins earning interest as does the principle. The more often the interest is compounded, the faster the savings grow. When breeding occurs continuously, the population grows as if interest were compounded instantaneously, requiring the use of calculus. The difference between arithmetical and exponential growth can be seen clearly in figure 24.1. Darwin (1859) wrote, in *On the Origin of Species,* "There is no exception to the rule that every organic being naturally increases at so high a rate, that, if not destroyed, the earth would soon be covered by the progeny of a single pair." Darwin pointed out that even one of the slowest breeding animals, the elephant, would increase from a single pair to nearly 19 million in just 750 years if unchecked.

The rate of growth of a population undergoing exponential growth at a particular instant in time is a differential equation:

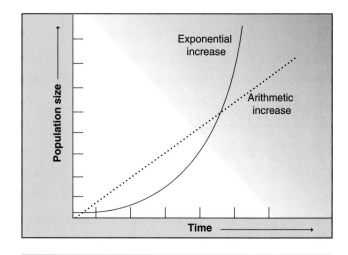

Figure 24.1   Arithmetic and exponential growth curves. The exponential curve increases more slowly at first but then accelerates past the arithmetic curve, which increases at a steady, incremental pace throughout.

$$\frac{dN}{dt} = rN \qquad (1)$$

where N = the number of individuals in the population; t = the time interval; and r = the growth rate per individual, or the intrinsic rate of natural increase. r can be calculated as the individual birth rate minus the individual death rate (sometimes written as $r = b - d$). If we assume a closed population, then immigration and emigration are zero.

In words, this formula can be read as

The rate of change in population size	=	The contribution of each individual to population growth	×	The number of individuals present in the population

If we plot the number of individuals against time, a J-shaped curve results during exponential growth (see figure 24.1). For example, reindeer (*Rangifer tarandus*) were introduced on several of the Pribilof Islands in the Bering Sea off Alaska. Four males and 21 females were released on 106-km^2 St. Paul Island in 1911. In the absence of hunting or predators, numbers increased rapidly, reaching a peak of about 2000 in 1938 (Scheffer 1951; figure 24.2). The population then crashed to only eight in 1950, as the habitat was overgrazed. Smaller species of mammals, with shorter generation times, can have even higher growth rates. Under laboratory conditions, vole populations can double in as little as 79 days (Leslie and Ranson 1940)!

Life Tables

Not all individuals in a population are identical, of course; they differ in sex, age, and size. These differences affect natality, mortality, and movement rates. Mortality is sum-

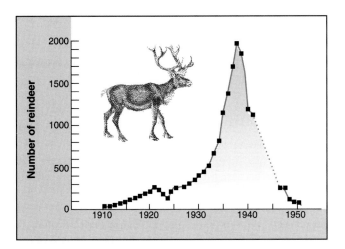

Figure 24.2 **Population growth of a herd of reindeer.**
Reindeer were introduced on one of the Pribilof Islands, Bering Sea, in 1911 and increased exponentially until 1938, after which numbers declined rapidly.

Source: Data from V.B. Scheffer, "The Rise and Fall of a Reindeer Herd"
in Sci. Mon. *73:356-362, 1951.*

marized in populations by means of **life tables,** originally developed by human demographers to provide data on life expectancy for life insurance companies. In his classic study of wolves (*Canis lupus*) of Mount McKinley, Alaska, Murie (1944) collected 608 Dall sheep (*Ovis dalli*) skulls and estimated the age at death based on the size of their horns. For instance, 121 sheep were less than 1 year old at death, 7 were between 1 and 2 years, and so forth. The oldest sheep were between 13 and 14 years. From these data, Deevey (1947) constructed a life table, shown in table

24.1. Note that mortality was high in the first year, declined to low levels from about 2 to 8 years, then increased once more. From life tables can be computed the expectation of further life (e_x), the amount of time an individual can expect to live once it survives to a particular age (the fifth column in table 24.1).

The data for life tables are usually collected in two different ways. **Static,** sometimes called vertical, life tables are generated from a cross section of the population at a specific time. Individuals are assigned an age group, and mortality rates are recorded for each age group. These are used by life insurance actuaries to calculate insurance risks for humans. The sheep life table (see table 24.1) is a type of static life table in which the age at death was observed from skulls. Static life tables require the assumption that the population is stationary and that the birth and death rates of each group are constant, both rather unlikely circumstances. **Cohort,** sometimes called horizontal, life tables are generated by following a group, or cohort, of individuals of similar age from birth to death; these data are much harder to obtain, especially for mammals that are often highly mobile and long-lived. An example of a cohort life table for a relatively short-lived species comes from white-footed mice (*Peromyscus leucopus*), few individuals of which lived more than 1 year (Goundie and Vessey 1986). In this population, spring-born individuals lived an average of just 10 weeks after weaning. Cohort life tables for a long-lived species come from a provisioned population of rhesus monkeys (*Macaca mulatta*) on islands off the coast of Puerto Rico, where a cohort was followed for 18 years (Meikle and Vessey 1988). Another island example comes from red deer on the Isle of Rhum, Scotland, where calves born in 1957 were followed for 9 years, by which time 92% had died

Table 24.1.   Life table for the Dall mountain sheep constructed from the age at death of 608 sheep in Mount McKinley (now Denali) National Park

Age Interval (Years) (x)	Number Dying During Age Interval (d_x)	Number Surviving at Beginning of Age Interval (n_x)	Number Surviving as a Proportion of Newborn (l_x)	Expectation of Further Life (e_x)
0–1	121	608	1.000	7.1
1–2	7	487	0.801	7.7
2–3	8	480	0.789	6.8
3–4	7	472	0.776	5.9
4–5	18	465	0.764	5.0
5–6	28	447	0.734	4.2
6–7	29	419	0.688	3.4
7–8	42	390	0.640	2.6
8–9	80	348	0.571	1.9
9–10	114	268	0.439	1.3
10–11	95	154	0.252	0.9
11–12	55	59	0.096	0.6
12–13	2	4	0.006	1.2
13–14	2	2	0.003	0.7
14–15	0	0	0.000	0.0

Source: *Based on data in Murie (1944), quoted by Deevey (1947).*

(Lowe 1969). Cohort life tables require the assumption that the age class followed is representative of the entire population; this assumption is rather unlikely because mortality rates are apt to change over time.

Survivorship Curves

From life tables, one can also construct **survivorship curves** to show graphically the pattern of mortality across different age groups. The number of survivors (l_x, the fourth column in table 24.1) is plotted against age. Often the number of survivors is expressed as a proportion of the total, as in table 24.1. It is best to plot the $\log_{10}$ of survivors, so that a straight line is obtained when mortality rates are constant across all age categories. Thus, if there were 1000 individuals to begin with and the mortality rate was 50% per year, an arithmetic plot would show a sharply declining curve, but a log plot of the same data would show a straight line. Different types of survivorship curves are illustrated in figure 24.3. Murie's sheep generally show a type I curve; once they survive the rather high mortality rate in the first year, the rate is nearly zero until 8 or 9 years, when it rises sharply and the curve breaks steeply down. Most other large mammals also show type I survivorship curves (figure 24.4). White-footed mice, and probably many other small mammals, show a type II curve, in which mortality rates are more or less constant at all ages (Schug et al. 1991), similar to the warthog curve in figure 24.4. Type III curves, in which mortality rates are highest at early stages of development, are characteristic of invertebrates and fish but probably are not seen in mammals because of the larger investment female mammals make in their offspring through gestation and lactation (chapter 21).

Age Structure

Birth rates as well as death rates vary with age, because very young and very old individuals do not breed in most species of mammals. The **age structure** of a population is determined in a static fashion by calculating the proportion of the total population made up of individuals of various ages. This information can then be graphed, as shown in figure 24.5. For humans with a type I survivorship curve, distributions with triangular, "pinched" shapes indicate a high potential for increase, because there are many individuals just reaching reproductive age and few old individuals. Much of the human population lives in less-developed countries with age structures similar to those shown in figure 24.5C. Stationary populations and those with slow growth rates, typical of developed countries, tend to be rectangular in shape, with more even percentages in each age group. Short-lived mammals with type II survivorship curves tend to have triangular distributions even when stationary because mortality rates are typically high and independent of age.

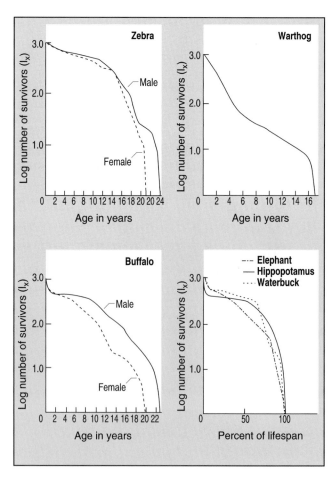

Figure 24.4 Survivorship curves of African ungulates. After the first few years, most resemble type I curves, with the exception of the warthog, which resembles a type II curve.

Source: Data from E.R. Pianka, Evolutionary Ecology, *5th edition, 1994, HarperCollins, New York, NY.*

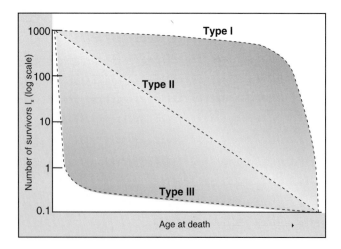

Figure 24.3 **Hypothetical survivorship curves.** Type I represents low mortality rates early in life, with most individuals dying at an old age. Type II represents constant mortality rates at all ages. Type III represents high mortality rates early in life. Most mammal species show type I or type II curves.

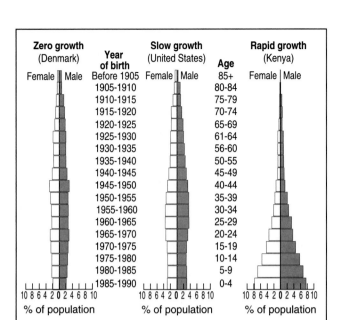

Figure 24.5 Age-structure diagrams for human populations. (A) Zero growth is shown for Denmark, (B) slow growth for the United States, and (C) rapid growth for Kenya.

Source: Data from U.S. Census Bureau, The United Nations, and Population Reference Bureau.

It is also useful to add to the life table the number of offspring produced by an average female of age x during that age period; this is called **fecundity** and is written as m_x (table 24.2). Multiplying this column times l_x gives the realized fecundity for each age class. Summing the $l_x m_x$ column gives the net reproductive rate R_0 for the population. **Reproductive value,** which is the sum of an individual's current reproductive output and its expected future output at age x, can also be estimated. A female's reproductive value increases with age once she passes through the juvenile stage, peaks early in adulthood, then declines in old age.

The ratio of males to females and the type of mating system can affect population growth. As we saw in chapter 21, many species of mammals have a polygynous mating system in which a small number of males do most of the mating. Although the sex ratio at birth tends to be 50:50 (chapter 21), males typically suffer higher mortality rates than do females. This leads to female-biased sex ratios among adult breeders. For this reason, most measures of reproductive rates are calculated for females rather than for both sexes.

LIFE HISTORY TRAITS

Life history traits, including size at birth, litter size, age at maturity, and degree of parental care, directly influence the life table schedules of fecundity and survival. Natural selection recognizes only one currency: successful offspring. Although all organisms have presumably been selected to maximize their own lifetime reproductive success, mammals vary widely in the relative amounts of energy they expend on activities that enhance fecundity versus those that enhance survival. Much of life history theory examines tradeoffs between reproduction and survival.

Semelparity Versus Iteroparity

As we saw in chapter 21, some organisms, such as annual plants, insects, and the Pacific salmon, reproduce only once during their entire lifetimes. These "big bang" breeders are called semelparous (from the Latin *semel* [once] and *pario* [to beget]). They exert a huge effort during their reproductive episode and then die. One genus of mammal, the marsupial mouse (*Antechinus* spp.) from Australia, is semelparous in that the males reach sexual maturity, mate during a brief period, then physically decline and die shortly thereafter (Cockburn et al. 1985; figure 24.6). Semelparity seems to be favored when extensive preparation for breeding is necessary or favorable environmental conditions are ephemeral or

Table 24.2. Hypothetical life table

Age (x)	Survivorship (l_x)	Fecundity (m_x)	Realized Fecundity ($l_x m_x$)	Expectation of Life (e_x)	Reproductive Value (v_x)
0	1.0	0.0	0.00	3.40	1.00
1	0.8	0.2	0.16	3.00	1.25
2	0.6	0.3	0.18	2.67	1.40
3	0.4	1.0	0.40	2.50	1.65
4	0.4	0.6	0.24	1.50	0.65
5	0.2	0.1	0.02	1.00	0.10
6	0.0	0.0	0.00	0.00	0.00
Sums		2.2	1.00 (R_0)		

Compare columns with table 24.1. Fecundity (m_x) has been added so that net reproductive rate (R_0) and reproductive value (v_x) can be calculated.

Figure 24.6   Marsupial mouse of the Genus *Antechinus*, from Australia.   Males of this genus are semelparous, or "big bang" breeders. After mating, males experience rapid senescence and die.

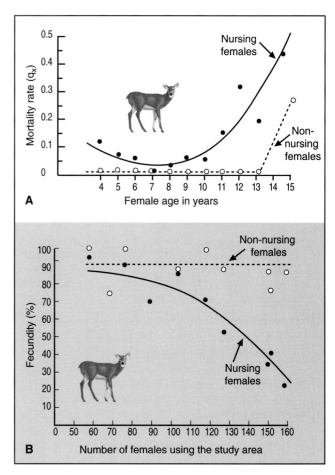

Figure 24.7   The cost of reproduction in female red deer and the effect of density on fecundity.   (A) Nursing females have a higher rate of mortality at all ages. (B) Nonnursing females have a higher probability of calving in the next year than nursing females at high densities.

Source: Data after Clutton-Brock et al., "The Costs of Reproduction to Red Deer Hinds" in J. Anim. Ecol. *52:367–383, 1983.*

uncertain. Most, if not all, other mammal species are iteroparous (from the Latin *itero* [to repeat]), that is, able to breed more than once during their lifetimes.

Reproductive Effort

Reproductive effort is the energy expended and risk taken to produce current offspring. It is measured in terms of the cost to future reproduction. Among iteroparous species, high fecundity early in life is correlated with decreased fecundity later in life, presumably because the cost of early reproduction reduces later survival and reproduction. In long-lived species with type I survivorship curves, such as many ungulates (see figure 24.4), reproductive effort is likely to increase with age as reproductive value declines. In other words, older animals, with little time left to live, should put maximum effort into current breeding. Long-lived species of mammals are expected to show a gradual increase in effort with age, as measured by the number or mass of offspring. Short-lived species, especially those with high extrinsic mortality rates (i.e., not related to breeding effort), should put maximum effort into early reproduction, because they are not likely to live long enough to suffer any negative consequences of high reproductive effort early in life.

Although the increased energy expenditure and risk associated with current reproduction have been documented in many species of mammals, only a few studies have shown that these costs reduce later survival and reproductive success. Lactation is much more costly for mammals than is gestation; lactation doubles the energy requirements of a mouse (Millar 1978), and it delays ovulation in most species examined (Short 1983). Bighorn sheep ewes suffer a cost of

reproduction in having higher parasite loads when lactating than when they are not (Festa-Bianchet 1989). Among red deer (*Cervus elaphus*), in years when winter food was low, females that successfully reared an offspring had lower survival rates than those that failed to get pregnant or lost a young before weaning (Clutton-Brock et al. 1989; figure 24.7). Parental care after weaning can also be costly. In red-necked wallabies (*Macropus rufogriseus*), mothers that associate closely with their offspring are more likely to lose their next infant (Johnson 1986).

Senescence

Inspection of the life table shown in table 24.2 indicates that fecundity increases until the fourth age interval and then declines. Most mammals experience **senescence,** that is, their fecundity gradually decreases and their mortality rate increases, resulting from deterioration in physiological function with age. Assuming that survival is advantageous to an individual of any age, we need to discover what causes senescence and why natural selection does not eliminate it.

The "rate-of-living" theory of aging links senescence to metabolic rate (Austad and Fischer 1991). According to this theory, aging is viewed as an inevitable consequence of physiological processes in which biochemical errors and toxic metabolic by-products accumulate. Life span should be negatively related to the rate at which physiological processes occur. Thus, organisms with low metabolic rates should live longer than those with high metabolic rates (Austad and Fischer 1991).

A second, evolutionary theory of aging argues that selection acts more strongly on traits expressed at an early age, when most individuals in the population are still alive. Selection gets weaker as organisms age because fewer individuals are around on which it can act. Also, some genes enhance fitness at an early age, and they would be selected for but might act incidentally to reduce fitness later on (Stearns 1992). According to this theory, species that suffer high extrinsic rates of mortality, as from predation or other environmental hazards, should senesce more quickly, as should those that invest heavily in reproduction at an early age. Thus, senescence would not be expected to begin until after sexual maturity, and it should show up earlier in populations with higher mortality rates. There is little evidence of senescence in most natural populations of mammals because few individuals live long enough to show it. Predation, disease, starvation, or other factors tend to kill individuals well short of their potential life spans. The situation is different in zoos or other captive populations, where unlimited food is available, diseases are controlled, and predators are absent; there, senescence is seen frequently.

One way to test between these theories is to compare groups of mammals that have different metabolic rates. Marsupials have basal metabolic rates about 70% that of similar-sized eutherian mammals. Yet life spans of captive marsupials are only about 80% of comparable eutherians, the opposite of the situation that would be predicted by the rate-of-living theory (Austad and Fischer 1991). Bats, on the other hand, have metabolic rates comparable to those of other eutherians, yet relative to their body size, are the longest-lived group of mammals. Why do bats live so long? Austad and Fischer (1991) argue that this difference is attributable to flight; bats are protected from predation and can roost in protected sites. They emphasize that birds are also relatively long-lived, as are species of gliding mammals (squirrels, marsupials, and dermopterans). These examples therefore tend to support an evolutionary theory of aging that links life span to levels of environmental hazard.

r- and *K*-Selection

The evolution of life history traits is linked to the growth rates of populations and their environments. In temperate and arctic regions, populations are frequently reduced by density-independent factors, mainly climate. Populations in such areas often have high intrinsic rates of natural increase (r from equation 1), including early maturity, high fecundity, and little parental care of offspring. Such opportunistic species are said to be *r*-selected. Conversely, populations of tropical species living in more constant climates often have low intrinsic rates of increase, including later maturation, low fecundity, and much parental care of offspring. Such populations are said to be *K*-selected because they remain at the carrying capacity (see chapter 21). Pianka (1970) lists a variety of life history traits associated with these types of species in addition to those mentioned earlier (table 24.3).

Table 24.3.   Some of the correlates of *r-* and *K*-selection

	r-Selection	*K*-Selection
Climate	Variable or unpredictable; uncertain	Fairly constant or predictable; more certain
Mortality	Often catastrophic, nondirected, density independent	More directed, density dependent
Survivorship	Often type III	Usually types I and II
Population size	Variable in time, nonequilibrium; usually well below carrying capacity of environment; unsaturated communities or portions thereof; ecologic vacuums; recolonization each year	Fairly constant in time, equilibrium; at or near carrying capacity of the environment; saturated communities; no recolonization necessary
Intra- and interspecific competition	Variable, often lax	Usually keen
Selection favors	1. Rapid development 2. High maximal rate of increase, r_{max} 3. Early reproduction 4. Small body size 5. Single reproduction 6. Many small offspring	1. Slower development 2. Greater competitive ability 3. Delayed reproduction 4. Larger body size 5. Repeated reproduction 6. Fewer, larger progeny
Length of life	Short, usually less than a year	Longer, usually more than a year
Leads to	Productivity	Efficiency
Stage in succession	Early	Late, climax

Source: *From E.R. Pianka, "On* r- *and* K-*selection" in* American Naturalist, *104:592–597, 1970. Copyright © 1970 University of Chicago Press, Chicago, IL. Reprinted with permission.*

In the eastern United States, two common small mammal species are the white-footed mouse and the meadow vole (*Microtus pennsylvanicus*). The white-footed mouse is widely distributed in relatively stable woodland habitats. Individuals mature at about 60 days, and the litter size is four to five. The meadow vole, on the other hand, inhabits fields, often those cleared by humans for farming. Fields typically undergo succession to woods unless maintained by fire or other disturbances and so are relatively temporary habitats. *Microtus* has litter sizes of 15 or more and reaches sexual maturity in as little as 30 days, giving it a high intrinsic rate of increase. Relative to *Peromyscus*, *Microtus* would thus be considered *r*-selected.

A consistent relationship between environmental fluctuations and life history traits has been difficult to demonstrate. Thus, many apparently *r*-selected species occur in tropical climates, and many *K*-selected species occur in temperate and arctic regions. Furthermore, many species tend to be intermediate between these extremes, making them difficult to classify. In spite of these problems, the terms *r*- and *K*-selection continue to be widely used by ecologists studying life history patterns.

Bet-Hedging

Another approach to understanding the relationship between life history traits and environmental conditions is to consider how an organism should expend reproductive effort when faced with unpredictable outcomes. When survival of offspring varies widely from year to year and is unpredictable, parents might be selected that have smaller litter sizes and spread breeding across several seasons, a strategy referred to as bet-hedging (Stearns 1976).

The predictions of bet-hedging contradict those of *r*- and *K*-selection in some cases. In unstable, unpredictable environments, *r*-selection, which favors high reproductive rates, should be important. Bet-hedging theory, however, suggests that the opposite would be favored: low reproductive rates, greater parental care, and spreading reproductive effort across many breeding seasons. Additional studies are needed to explain these contradictions.

POPULATION GROWTH AND REGULATION

Limiting Factors

Although a host of factors, such as food, water, and shelter, must be present for any population of mammals to survive, typically only a single factor limits the growth of a population at any one time. This idea originated with Liebig (1847), a plant physiologist who studied the effects that nutrients, such as nitrogen, had on plant growth. The axiom is referred to as Liebig's **law of the minimum.** Extending Liebig's idea to plant and animal populations, one factor will be the limiting factor. We can test which factor limits a population's growth by manipulating the amount of one factor

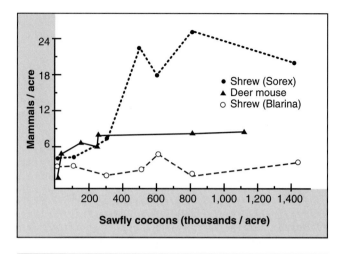

Figure 24.8 **Numerical response of small mammals to increasing food supply.** Populations of the shrew (*Sorex*) and deer mouse (*Peromyscus*) increase in size, up to a point, as food supply increases. Shrews of the Genus *Blarina,* however, show no response to increased food supply (sawfly cocoons), so their numbers must be limited by some other resource.

Source: Data from C.S. Holling, "The Components of Predation as Revealed by a Study of Small Mammal Predation of the European Pine Sawfly" in Can. Entomol. *91:293–320, 1959.*

while holding others constant and recording changes in the population. If food is the limiting factor, an increase in the population will result if the food supply is augmented.

A number of studies have shown that adding food leads to increased reproduction and to higher population size. For example, Holling (1959) studied the population response of some small mammals to different amounts of a staple food, cocoons of the European pine sawfly (*Neodiprion sertifer*). Shrews of the Genus *Sorex* increased more or less linearly with increasing food supply up to a population of 25 shrews per acre; then the population leveled off (figure 24.8). At that point, some resource other than sawflies became limiting. Likewise, deer mice increased in numbers up to about eight per acre, then remained constant despite the food supply (Holling 1959). The abundance of sawflies seemed to be unrelated to the population density of another genus of shrew (*Blarina*). Thus, some other factor was limiting the population of that species. Although Liebig's law explains how populations are limited in some instances, factors may interact in complex ways to regulate numbers, as we shall see later in this chapter.

Logistic Growth

As we saw with the reindeer example in figure 24.2, populations of mammals or anything else do not increase indefinitely. As density rises, the presence of other organisms reduces the birth rate, increases the death rate, or triggers emigration, and population growth slows or stops. One model for such growth, the **logistic equation,** yields a sigmoid, or S-shaped, curve. Numbers increase slowly at first, then in-

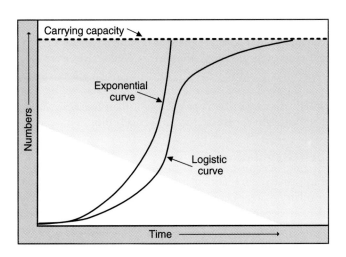

Figure 24.9   **Comparison of exponential and logistic population growth curves.**   The rate of logistic growth declines gradually to zero as the carrying capacity is approached.

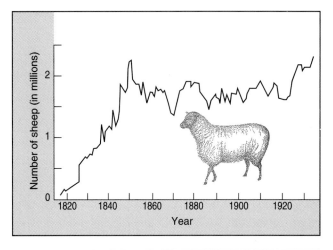

Figure 24.10   **Logistic growth of sheep population.** Number of sheep on the island of Tasmania following their introduction in the early 1800s.

Source: Data from J. Davidson, "On the Growth of the Sheep Population in Tasmania" Trans. Roy. Soc. South Aus. 62:342-346, 1938.

crease rapidly, as shown by equation (1). Then, however, the growth rate begins to slow as numbers approach an upper limit, or asymptote. The upper limit is often called the carrying capacity, or the equilibrium density. The carrying capacity is typically set by the amount of food or some other factor that is limiting the population at that time (figure 24.9).

We can modify equation (1) to give the equation for logistic growth:

$$\frac{dN}{dt} = rN\left(1 - \frac{N}{K}\right) \tag{2}$$

where K = the carrying capacity.

The term in parentheses is a density-dependent term that ranges from 0 to 1. When N is very small in relation to K, N/K is close to 0, so the term in parentheses is close to 1. When the term in parentheses is 1, equation (2) becomes the same as equation (1), and the population increases exponentially. As N approaches K, N/K approaches 1, and the term in parentheses approaches 1 minus 1, which is 0. At that point, the whole right hand part of the equation goes to 0 and so the growth rate, shown on the left side, is also 0, and population growth stops. If N should exceed K, the growth rate would be negative, and the population would decline.

Laboratory populations of mice often show logistic growth when a few pairs are used to start a colony and food, water, and nesting material is supplied freely (Terman 1973). Some natural populations remain at equilibrium for long periods of time, presumably at the carrying capacity. For instance, sheep were introduced on the island of Tasmania in the early 1800s. Numbers showed a logistic growth pattern, reaching an asymptote of about 2 million and remaining there for many years (Davidson 1938; figure 24.10). Often, however, populations show an exponential increase followed by rapid decline, as in the reindeer herd shown in figure 24.2. Many populations seem to fluctuate, increasing and decreasing without any

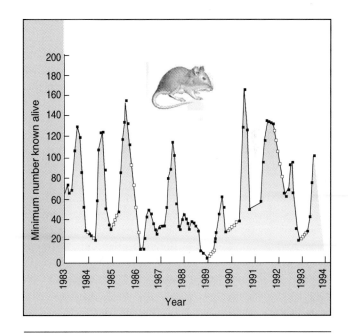

Figure 24.11   **Fluctuations of white-footed mice in a small, 2-hectare woodlot in northwest Ohio.**   Open blocks denote missing data estimates.

Source: Unpublished data from S.H. Vessey.

apparent pattern. Short-lived mammals that breed seasonally, such as mice, may show annual fluctuations, with year-to-year variation in the summer peaks and winter troughs, as seen for white-footed mice (figure 24.11). These populations tend to increase exponentially in spring and early summer, often crashing in autumn.

Density-Independent and Density-Dependent Factors

The factors in nature that keep populations from increasing indefinitely are traditionally divided into two types: those that are density-independent and those that are density-dependent. The type depends on whether the factor's effect on the population is or is not a function of population density. Density-independent factors are such things as storms, fires, floods, or other climatic conditions that destroy individuals without regard to how many are present. Density-dependent factors include food supply, shelter, predators, competitors, parasites, and disease. Their effect on the population is stronger in crowded than in sparse populations, and they can be expressed mathematically by the $1 - K/N$ term in equation (2).

If we define population regulation as the maintenance of numbers within certain limits and not simply as random fluctuations, we must look to density-dependent factors as regulatory agents. Only when a factor increases its negative effect on population growth as population increases can an equilibrium of numbers be maintained (Krebs 1994). Mammals, as warm-blooded organisms, are buffered against weather relative to most other taxa and thus may be less influenced by density-independent factors. The best examples of the role of density dependence from natural populations of mammals are managed game animals such as white-tailed deer (*Odocoileus virginianus*). There is good evidence that survival and reproduction of deer depend on the quality and quantity of the food supply. On good range, most females get pregnant and produce twins, whereas on poor range, pregnancy rates decline and twinning is rare (table 24.4). In areas where predators have been extirpated and hunting is not allowed, deer herds increase, range deteriorates, and reproduction then declines, a pattern that supports the notion of density-dependent regulation. These changes are reversed when hunting resumes (Cheatum and Severinghaus 1950).

Red deer on the Isle of Rhum, Scotland, show a number of density-dependent changes in fecundity and juvenile mortality (Clutton-Brock et al. 1985). The density of females on the Rhum study site was negatively correlated with fecundity and survival for both mothers and offspring (see

figure 24.7). Especially important was winter mortality of calves, which was very high in years of high density. Presumably, forage was scarce in those years. Male calves were especially affected, with over 70% mortality when female density was highest. As noted previously, males seem to be more sensitive to their mother's condition than females, perhaps because males grow faster in this sexually dimorphic species.

Self-Regulation

Intraspecific competition has received much attention as a regulatory factor because it is the only one that is "perfectly" density-dependent. In other words, the intensity of competition among members of the same population is likely to vary directly with the size of the population; as the population increases, so does competition for the limiting resource. Unsuccessful competitors may leave the area, die, or have lower reproductive success. Other biotic regulatory factors, such as predation and interspecific competition, include other species of organisms that, in turn, are affected by other agents; thus, these biotic regulators are not perfectly density-dependent. For example, predation involves a complex relationship between predator and prey. Although wolves eat caribou, they are not necessarily the regulators of the caribou population. Because the wolf population is affected by factors other than the numbers of caribou, such as diseases or the abundance of alternative prey species, it will not be able to track the caribou population exactly. An outbreak of disease might reduce the size of the wolf pack and allow the caribou to "escape" control and begin to destroy grazing land. Intraspecific competition among the caribou, however, varies directly with the population density and the limiting resource.

Intraspecific competition has been implicated in a number of hypotheses of self-regulation of populations. Self-regulation refers to an intrinsic tendency of populations to adjust birth, death, and movement rates so that numbers remain at or below the carrying capacity. In this way, resources are not exhausted and starvation is minimized. The impetus for proposing such hypotheses stems from the observation of naturalists that populations of mammals typically stop increasing before food resources are exhausted and that starving animals are rarely seen in nature.

Table 24.4.　Reproductive parameters of white-tailed deer in five regions of New York, 1939–1949

Region*	Percent of Females Pregnant	Embryos per Female	Corpora Lutea per Ovary†
Western (best range)	94	1.71	1.97
Catskill periphery	92	1.48	1.72
Catskill central	87	1.37	1.72
Adirondack periphery	86	1.29	1.71
Adirondack central (worst range)	79	1.06	1.11

Source: Data from E. L. Chaetum and C. W. Severinghaus "Variations in Fertility of White-Tailed Deer Related to Range Conditions" in Trans. North Am. Wildl. Conf., *15:170–189, 1950.*

*Arranged by decreasing suitability of range.
†Note the decline in fecundity with decreasing quality of range.

Behavioral-Physiological Factors

During cyclical peaks in abundance, snowshoe hares are seen under practically every bush; during these population peaks, a disturbance as slight as a hand clap was reported to be enough to send them into convulsions and coma, followed by death, apparently due to hypoglycemic shock (Green et al. 1939). Christian (1950) proposed that mammalian populations could be regulated by shock disease caused by exhaustion of the adrenal gland, following prolonged psychological stress from agonistic interactions at high population levels. This negative-feedback loop, which depends on intraspecific competition, is perfectly density-dependent. The idea grew from Selye's (1950) work on the **general adaptation syndrome** (GAS) in which nonspecific stressors, such as heat, cold, or defeat in a fight, produce a specific physiological response. Adrenocorticotrophic hormone (ACTH) released from the anterior pituitary, under control of the hypothalamus, stimulates production of glucocorticoids by the adrenal gland. These hormones, such as cortisone, function mainly to elevate blood glucose to prepare the body for fight or flight.

The phenomenon of death due to adrenal exhaustion turned out to be an extreme case, and Christian (1978) subsequently modified his hypothesis after a series of laboratory experiments. There is, in fact, a rise in adrenocortical output in response to increasing population density. These hormones, along with corticotropin-releasing factor, ACTH, and some others still being explored, seem to reduce direct fitness in numerous ways. For example, the body's two main defense mechanisms—the immune and the inflammatory responses—are inhibited by these hormones. Such changes obviously increase the likelihood of morbidity or mortality. At the same time, growth and sexual maturation are inhibited by increased adrenocortical output, as are spermatogenesis, ovulation, and lactation (Rivier et al. 1986). Some of these effects on reproduction persist even into subsequent generations, in spite of a reduction in population density (figure 24.12). Field data supporting these findings came from studies of a variety of mammals such as Sika deer (*Cervus nippon*), Norway rats (*Rattus norvegicus*), house mice (*Mus musculus*), and woodchucks (*Marmota monax*;

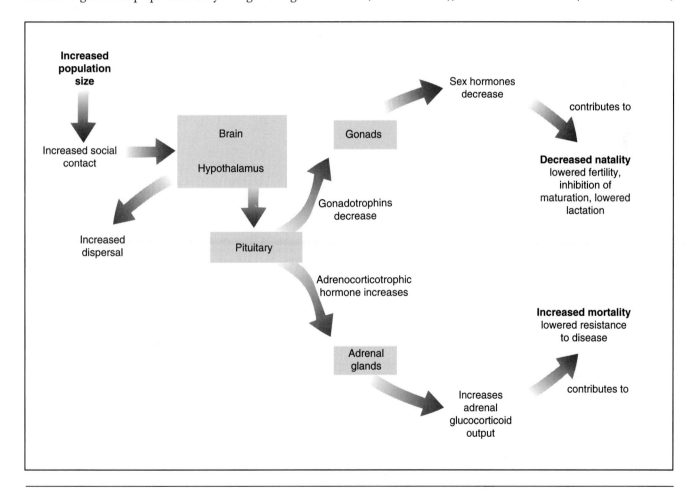

Figure 24.12   **Christian's model of population regulation in mammals.**   As the population increases (*upper left*), social contacts increase, activating the pituitary and adrenal glands and inhibiting the production of sex hormones. The population then declines because of the decrease in the birth rate and the increase in the death rate. As the population declines, further social contacts are reduced, and the pituitary hormone outputs are reversed, allowing the population to increase once again.

Source: Data from J.J. Christian, Populations of Small Mammals Under Natural Conditions, *D.P. Snyder, ed., 1978, The Pymatuning Symposia in Ecology, University of Pittsburg Press, Pittsburgh.*

Christian 1978). There was no evidence, however, of increased adrenal gland size with population density in lemmings (Krebs 1963) or meadow voles (To and Tamarin 1977). Part of the problem is that the stressor is not density itself but the agonistic behavior associated with competition for limited resources. Thus, in laboratory mice, as little as 2 minutes per day exposure to a trained "fighter" mouse produces a pronounced stress response, lowering the body's defense against parasites (Patterson and Vessey 1973). Clearly, this negative- feedback loop works under some conditions, but its generality in natural systems remains to be demonstrated.

Work with pheromones has augmented some of these findings. Substances released in the urine produce effects on conspecifics without a physical encounter. In some species of mice, the smell of strange male urine blocks pregnancy by preventing implantation, a phenomenon referred to as the Bruce effect (Bruce 1966); in crowded or unstable populations, the birth rate could thus be lowered. House mice avoid the urine of a mouse recently defeated by another mouse, and urine from stressed mice produces an adrenocortical response in naive mice (Bronson 1971, 1979).

Females grouped together produce substances voided in urine that delay sexual maturation in other females (Drickamer 1974). Six volatile organic compounds produced under the influence of the adrenal gland have been isolated by Novotny and colleagues (1986). Various combinations of these substances, when added to the urine of adrenalectomized females or to plain water, restore the delay effect.

The same delay in the onset of estrus has been produced in the laboratory by exposing young mice to urine of females taken from high-density populations in the field (Massey and Vandenbergh 1980). In an experimental study, field populations were increased by adding new mice to simulate a population explosion (Coppola and Vandenbergh 1987). Urine from resident females before the explosion had little delay effect on maturation of young mice in the laboratory, but after the explosion, urine from these same females produced a pronounced delay. In contrast, odor from mature males produces the opposite effect, accelerating the sexual maturation of young females (Vandenbergh 1969). Although most of these effects are best viewed as adaptations of individuals to gain reproductive advantage, they may act incidentally to regulate population size.

Behavioral-Genetic Factors

The ideas examined so far are complementary; they are all based on the belief that the behavior and physiology of animals change as a result of increased interaction rates associated with high population density. A competing hypothesis suggests that natural selection favors different genotypes at high and low population levels. While working with voles (*Microtus* spp.) that have 3- to 4-year population cycles, Chitty (1960) noticed that populations continued to decline even under seemingly favorable environmental conditions. Voles from declining populations were highly aggressive,

intolerant, and bred poorly; voles from increasing populations were mutually tolerant and rapid breeders. He postulated that a change took place in the quality of animals in a declining population. Through natural selection, the proportion of aggressive individuals increased, and, even though they could compete in crowded conditions, their reproductive rates were low, and the population declined (figure 24.13). It seems unlikely that a change in gene frequency could take place over a span of only a few years, but dispersal of animals of one genotype can produce a rapid change in the gene frequency in the rest of the population. Myers and Krebs (1971) reported that the frequency of one allele in the blood serum of voles was significantly different in dispersers than in residents. The behavioral changes reported in vole populations have not been linked to specific genes, however; nor have other genetic studies substantiated the findings of Myers and Krebs. Evidence from the laboratory shows that the age of puberty can be shifted in house mice after only a few generations of artificial selection (Drickamer 1981). The low birth rates observed in peak populations in the wild could be due, in part, to selection for late-maturing individuals.

Tamarin (1980) emphasized the role of dispersal itself as a regulating mechanism. Figures 24.12 and 24.13 demonstrate that dispersal is a factor in most regulatory schemes. If dispersal is blocked by a fence or is blocked naturally, as on islands, regulation fails and high populations and depleted resources result—the **fence effect** (Krebs et al. 1973). Dispersers are likely to be at a disadvantage reproductively because they are probably in a suboptimal habitat; also, predation is higher on transient individuals (Metzgar 1967). Emigrating individuals, however, have the potential of colonizing new areas and may even be the founders of new species (Christian 1970).

Several other mechanisms have been proposed that incorporate both genetics and aggressive behavior. One is that individuals behave less aggressively toward kin than nonkin. As a population increases, the likelihood of nonrelatives' interacting increases, leading to an increase in aggression, followed by population decline (Charnov and Finerty 1980). A second idea links increased genetic heterozygosity with increased aggression. Several studies have shown that heterozygous individuals are more aggressive than homozygous individuals. As a population increases, so does dispersal and outbreeding, leading to an increase in heterozygosity. The resultant increase in aggressiveness could force a decline via increases in mortality rates and decreases in reproduction (Smith et al. 1978). So far, little hard evidence exists to support either of these hypotheses.

Evolution of Population Self-Regulation

Proponents of self-regulation have not resolved two important problems. First, most of the data demonstrating behavioral and physiological changes in response to increasing population size are from laboratory populations, in which densities are often unrealistically high and emigration is usu-

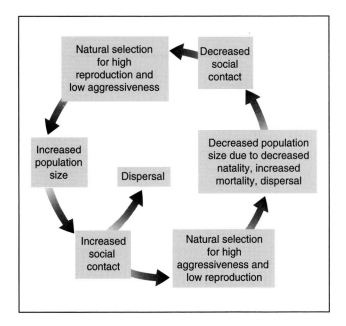

Figure 24.13 Chitty's model of population regulation in mammals. In response to increased population size and social contact, aggressive individuals with low reproductive rates are selected, leading to a decline in population. Modifications of the model emphasize the role of dispersal of certain phenotypes in causing the decline. This model predicts genetic changes in the population as different genotypes are favored at high versus low densities.

Source: Data from D. Chitty, "The Natural Selection of Self-regulatory Behavior in Animal Populations" in Proceedings of the Ecological Society of Australia, *2:51-78; and C. Krebs, "The lemming cycle at Baker Lake, Northwest Territories, during 1959-1962" in* Arctic Institute of North America Technical Paper No. 15, *1964.*

ally prevented. The results, therefore, cannot be extrapolated to natural populations. In natural populations, furthermore, many extrinsic factors come into play, and no single mechanism can be implicated. Second, if we accept the idea that such self-regulatory mechanisms exist, do they evolve specifically for that purpose, and, if so, what is the unit of selection?

We can probably agree that self-regulation is adaptive for a population or species that has the potential for destroying or using up its resources, but many of the mechanisms we have discussed seem to be maladaptive for the individuals involved. An animal that is genetically programmed to respond to increased fighting by cutting back on reproduction will lose out, genetically speaking, to another individual that can keep on reproducing. How could a system evolve in which an animal responds to aggression by reducing its defense mechanisms and therefore, by definition, becomes less fit?

One answer is by group selection, as Wynne-Edwards (1962, 1986) argued. Groups or populations that avoid overexploitation of the environment survive at higher rates and may later colonize the habitats left vacant by imprudent groups that became extinct. Although group selection is considered by most evolutionary biologists to be theoretically possible and several mechanisms for it have been proposed (Lewontin and Dunn 1960; Wilson 1980), the rate of

change in gene frequency would usually be too slow compared with the rate of change through individual selection. To maintain a trait that is adaptive for the group but not for the individual, group extinction would have to be more rapid than individual extinction—and that does not seem to be the case. More than three decades of study of diverse organisms have failed to produce strong support for the group selection hypothesis. Most population biologists, therefore, favor arguments based on individual selection and reject Wynne-Edward's (1962, 1986) hypothesis.

Arguments based on kin selection (see chapter 22) offer one way around the problem of group versus individual selection. We have seen that many social groups are composed of close relatives. Thus, behavioral and physiological traits that reduce direct fitness but increase indirect fitness, in this case by increasing the likelihood of survival of a kinship group, could be selected for. In other words, it might benefit an individual, evolutionarily speaking, to sacrifice its immediate reproductive output if the potential survival of the group as a whole, which contains its relatives, is enhanced.

A second, more widely accepted answer to the question of self-regulation is that what we are calling "mechanisms" are nothing more than the consequence of an individual's behaving in its own best interest. Reproductive restraint in the face of overcrowding may be the best individual strategy; a delay in maturation or the production of fewer young can save energy for a better time. The greatest genetic representation in the next generation is not synonymous with having the most offspring. Litter size in small mammals has evolved to produce the number of offspring that is most likely to survive and reproduce; too large a litter could result in survival of few or no young.

Similarly, at times of high density, aggressive individuals that are intolerant of others would be selected for if limiting factors were operating (Chitty's hypothesis). The most successful individuals at times of high density might be those that reduce the fitness of others through competition. We should keep in mind that fitness is a relative term that refers to increasing one's genetic complement in the next generation relative to that of other individuals.

The inhibition of the body's defense systems as a result of increased adrenocortical output at high population is more difficult to explain (Christian's hypothesis). Although it seems counterintuitive, such a response can be adaptive to the individual under some circumstances. The pituitary-adrenal response is part of the general adaptation syndrome. Only animals that are being attacked and defeated show this response, whereas dominant animals, even in crowded situations, respond little and are like individuals at low densities. The general adaptation syndrome enables an animal under stress to mobilize energy reserves just to survive. Other systems, such as the reproductive system, are temporarily shut down. The production of offspring at such times would likely be a waste of energy. Low rank or exclusion from a territory suitable for breeding may be a temporary situation. In fact, many territory and dominance systems are age-graded, and younger animals, especially males, attain prime territories or high rank only if they can live long enough.

The preceding arguments suggest that self-regulatory mechanisms do not evolve, as such, for the purpose of controlling populations. Most biologists who test hypotheses about the functions and origins of behavioral patterns assume that those patterns evolved by means of natural selection's acting at the lowest level consistent with the evidence. This level is usually that of the individual and its offspring (Williams 1966). Arguments for selection at higher levels (group or population) are invoked only if lower-level selection cannot explain the observed traits.

Field Studies of Population Regulation

Descriptive and Correlational Studies

Given the large number of factors that might act singly or in combination to regulate mammal populations, it is perhaps not surprising that much disagreement exists about their importance in different populations. Food and predators have received much attention; the role of interspecific competition is discussed in the next chapter. Descriptive studies, sometimes spanning many years, occasionally demonstrate correlations among various factors and population density. For instance, the abundance of mast (primarily acorns) is positively correlated with winter survival, winter breeding, and subsequent population densities of white-footed mice (Ostfeld et al. 1996; Wolff 1996). The mast appears to be a limiting factor for these mice, at least in some years. In correlational studies, however, it is difficult to show a cause-and-effect relationship. In this example, some other variable, also correlated with abundance of mast, might actually be the limiting factor.

The role of predators in regulating prey populations has also been the focus of many correlational studies. For instance, in southern Sweden, field voles (*Microtus agrestis*) and wood mice (*Apodemus sylvaticus*) are heavily preyed on by generalist predators: common buzzards (*Buteo buteo*), red foxes (*Vulpes vulpes*), and domestic cats (*Felis* [*catus*] *sylvestris;* Erlinge et al. 1983). Analysis of pellets, scats, and prey remains for two years showed that predators consumed as many mice and voles as were produced during each year! Predators can show two types of responses to changes in the availability of prey. A **numerical response** means increases in the numbers of predators (see figure 24.8). This, of course, takes time because the generation time of predators usually exceeds that of prey, and it is common for changes in predator numbers to lag behind their prey. Predators may also show a **functional response,** changing the numbers of a particular prey item in their diets. In the Swedish study, predators showed a functional response, switching from eating other prey to eating mice and voles as numbers of these two species increased, and thus exerted a density-dependent effect on mouse and vole populations. These researchers conclude that generalist predators are effective regulators of small mammals in southern Sweden and that they keep the vole populations from cycling. Vole populations do cycle every 3 to 4 years farther north, where these generalist predators are lacking.

Field Experiments

Cause-and-effect relationships can be best revealed by experimental manipulations. One of the simplest experiments is to add food to a population and monitor changes in birth, death, and movement rates as well as changes in home range size. Many studies have been done on populations of mammals, with mixed results. Typical responses to supplemental food are decreased home range size, increased body mass, and increased breeding activity (Boutin 1990). Food addition, however, usually does not prevent major declines in populations, such as in voles that undergo multiannual fluctuations, indicating that other factors must be involved.

Other field experiments have manipulated predators. European rabbits (*Oryctolagus cuniculus*), introduced into Australia, have become serious pests there, sparking interest in the possible role of predators in keeping rabbit numbers in check. In New South Wales, rabbits have several predators, including red foxes and domestic cats. Predators were removed from some plots and their stomach contents analyzed for rabbit remains (Pech et al. 1992). When rabbits were at low density, predators kept numbers in check, but if predators were removed, rabbits increased in numbers, as one might expect. When predators were allowed back into those areas, however, rabbit populations continued to increase. In other words, the rabbits "escaped" control by the predators. A different predator-prey system in Chile suggests a somewhat similar effect of predators on their mammalian prey. The degu (*Octodon degus*) is a medium-sized caviomorph rodent, weighing about 150 g. Its main predator is the culpeo fox (*Pseudalopex culpaeus*). On study plots where foxes and aerial predators such as owls were excluded, degus tended to achieve higher densities than on control plots, where predators were allowed access (Meserve et al. 1996). Survival rates of degus and other species of small mammals also were higher on predator-free plots.

Several attempts have been made to study the combined effects of food and predators in what is called a factorial design. At least four plots are needed: one with added food, one with predators excluded, one with both added food and predators excluded, and one left alone as a control. Each plot is usually replicated at least once, for a minimum of eight. For rodent studies, various types of fences and nets are used to exclude predators. In one such study, prairie vole (*Microtus ochrogaster*) densities increased in pens where high-quality food was added and in those where predators were excluded (Desy and Batzli 1989). Highest vole densities were achieved in those pens with both added food and predators excluded. In this study, the effects of increased food and reduced predation were additive (figure 24.14).

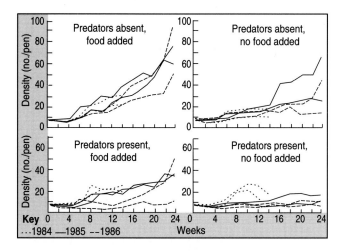

Figure 24.14   The role of food and predators in regulating vole populations.   Population density of voles in four treatments (two replicates each year).

Source: Data from E.A. Desy and G.O. Batzli, "Effects of Food Availability and Predation on Prairie Vole Demography: A Field Experiment" in Ecology, 70:411–421, 1989.

Food operated mostly by influencing reproduction, whereas predators affected survival of adults and young. For an example in which the effects of food and predators were nonadditive, see the study by Krebs and colleagues (1995) on snowshoe hares discussed in the following section.

Cycles

We have seen that many populations of mammals fluctuate rather widely, presumably in response to environmental changes. True cycles, those with constant periods, are rare in nature, and none is known in the Southern Hemisphere or the tropics (Sinclair and Gosline 1997). Mammals are unusual in having a number of species that cycle, ranging from 3 to 4 years in smaller species such as certain voles (*Microtus* spp.) and lemmings (*Lemmus* spp.) to 10 years or more in larger species such as lynx (*Lynx canadensis*) and snowshoe hares (*Lepus americanus*). The existence of cycles has fascinated and puzzled mammalogists and ecologists ever since Charles Elton (1924) introduced the subject. Cycle amplitude and length tend to increase with latitude (Hanski et al. 1991). Both the causes of these cycles and their apparent synchrony in time and space demand explanation.

As an example of a cyclic species, brown lemmings (*Lemmus sibiricus*) at Barrow, Alaska, have cycles of about 4 years (figure 24.15). During one of the increases, breeding starts in the fall, and by spring, large numbers of lemmings are present, many of which fall prey to snowy owls (*Nyctea scandiaca*), arctic foxes (*Alopex lagopus*), weasels (*Mustela* spp.), and other predators. During the summer, the population crashes to a low level, where it remains for 1 to 3 years.

Many agents, including extrinsic factors such as changes in food quantity or quality, parasites, disease, and predators, and intrinsic factors such as physiological stress from crowding and changes in gene frequency have been proposed to explain these cycles (Pianka 1994). Many researchers have focused on one or two factors in an effort to explain cycles. For instance, Hanski and colleagues (1991) have argued that small, specialist predators such as weasels, which are major predators at high latitudes in Scandinavia, increase the amplitude of the cycles. These authors contend that larger, generalist predators, which are more important at lower latitudes, tend to stabilize rodent populations.

Others have argued that single-factor explanations are not likely to be universal and that multifactorial explanations are needed to explain multiannual fluctuations. In these models, extrinsic and intrinsic factors act synergistically and sequentially to produce cycles (Lidicker 1988; figure 24.16). A problem with these models is that they are complex and difficult to test. Note the large number of factors in Lidicker's model for the California vole (*Microtus californicus*).

Most spectacular is the snowshoe hare cycle in the boreal forests and tundra of North America. Among the estimates of abundance of hares and their mammalian predators are the numbers of furs brought in by trappers to the Hudson Bay Company (MacLulich 1957). Each cycle is about 10 years in length, and the lynx and hare populations are highly synchronized, with the peaks in lynx numbers following those of the hare by a year or two (figure 24.17). One hypothesis for the cause of these cycles examines time delays in the interaction of three **trophic** (feeding) levels: the winter food supply of hares, the number of hares, and the number of predators (primarily lynx and great horned owls, *Bubo virginianus*). A

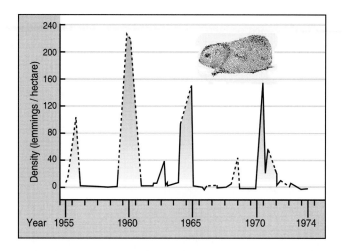

Figure 24.15   Lemming population cycles.   Estimated lemming densities in the coastal tundra at Barrow, Alaska, for a 20-year period.

Source: Data from E.O. Batzli, et al., "The Herbivore Based Trophic System" in An Arctic Ecosystem: The Coastal Tundra at Barrow, Alaska, J. Brown, et al., (eds.), Dowden, Hutchinson and Ross, Stroudsburg, PA, 1980.

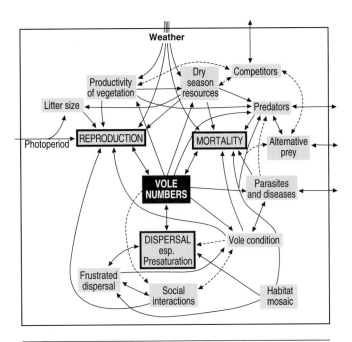

Figure 24.16   **Factors affecting vole populations.**   Schematic representation of factors known (*solid arrows*) or suspected (*dashed arrows*) to influence numbers of California voles.

Source: Data from W.J. Lidicker, Jr., "Solving the Enigma of Microtine 'Cycles'" in Journal of Mammalogy, *69:225-235, 1988.*

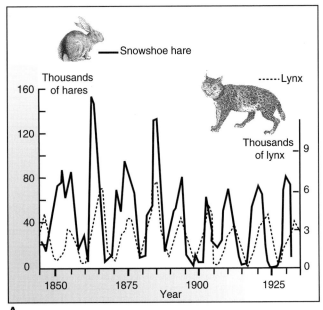

A

B

Figure 24.17   **Population cycles of the lynx and the snowshoe hare.**   (A) Comparison of population cycles of both species in the Hudson's Bay region of Canada, as indicated by fur returns to the Hudson's Bay Company. (B) The snowshoe hare well camouflaged by its winter pelage.

Source: (A) Data after MacLulich, 1937 in R.E. Ricklefs, Ecology, *4th edition, 1990, W.H. Freeman & Company, New York.*

second hypothesis includes only two trophic levels: hares and their predators.

Mathematical models and laboratory experiments demonstrate that cycles can occur in the absence of any regular environmental fluctuations. The logistic equation (equation 2) produces population cycles for certain values of *r*, the intrinsic rate of natural increase. When the response of population growth to density is time-delayed, either damped or increasing oscillations result. This result has been confirmed in laboratory populations of invertebrates, but field tests are lacking. Mathematical models of predator-prey interactions also predict stable cycles under certain conditions, perhaps explaining the lynx-hare cycles. Keith (1983) concludes, however, that food shortage during the winter initiates the decline in hare density, with predators playing a secondary role. Lynx density seems to depend on hare density, so the lynx is food-limited, but whether hares are limited only by predators or by both food and predators is unclear (Keith 1987).

In an experiment designed to test the role of food and predators on hare density, supplemental food and mammalian predator abundance were manipulated on 1-km² plots in the Yukon (Krebs et al. 1995). Mammalian predators were excluded from some plots by means of electric fences. Hare density during the cyclic peak and decline doubled when predators were excluded and tripled when food was added. When predator exclusion was combined with food addition, hare density increased 11-fold. Thus, both food supply and predation seem important in population cycles of

hares, and they seem to interact in a synergistic, or nonadditive, way. These results tend to support the three-trophic level hypothesis.

Whatever the immediate causes of hare cycles, the observation that cycles frequently are synchronized in time and space also requires an explanation. Could some continent-wide synchronizer be affecting climate and thus altering the hare's food supply over a wide region? One candidate suggested early on were peaks in sunspot activity that have about an 11-year mean. The role of sunspots was later discounted because the mean of the hare cycles was closer to 10 years (Keith 1963). This theory has been resurrected, however, based on new analyses (Sinclair et al. 1993). These researchers found a correlation between dark marks observed in tree rings from logged spruce trees and the abundance of hares in the Yukon. The dark marks represent stunted growth resulting from browsing by hares on the apical shoots of spruce. In peak hare years, after the more preferred foods of birch and willow are already consumed, hares feed on the less preferred spruce. Dark marks also correlate with the fur records from trapping between 1751 and 1983 in other parts of Canada. Both dark marks in tree rings and hare numbers are correlated in turn with sunspot activity, all with a 10-year periodicity. Sinclair and colleagues (1993) further suggest that solar variability is somehow connected with weather patterns, which directly or indirectly (e.g., through precipitation and food supply) affect hare numbers.

If a global synchronizer such as sunspot cycles is operating in the hare cycle, cycles might be expected to be synchronized among boreal forests within similar climatic regions of both the Old and New World. One study of the Finnish mountain hare (*Lepus timidus*) found that hare density cycles in Finland and Canada were not synchronized (Ranta et al. 1997), suggesting either that sunspots are not involved or that the effect of solar activity on weather is region-specific (Sinclair and Gosline 1997).

The topic of population cycles continues to be vexing. After decades of work by mammalian ecologists from many countries, including some studies of voles spanning more than 25 years, we know much about what are *not* the causes of cycles. A more complete understanding of their causes will await additional statistical tests of large data sets and further experiments.

Summary

Populations increase via natality and immigration and decrease via mortality and emigration. All species of mammals are capable of exponential increase in numbers, resulting in a J-shaped growth curve; however, they rarely do so in nature because of the operation of limiting factors (food, predators, disease, and others).

Life tables are used to compute expectation of further life, survivorship curves, net reproductive rate, and reproductive value. Most species of large mammals show type I survivorship curves, with low mortality rates until old age. Small mammals more often show constant mortality rates at all ages, giving a type II curve.

Life history traits affect fecundity and survival as reflected in the life table schedules. Most mammals are iteroparous, breeding repeatedly during their lives. Reproductive effort, the energy put into current reproduction that reduces future survival and reproduction, is predicted to increase with age in long-lived species. In short-lived species subject to heavy rates of predation, reproductive effort is high at sexual maturity. Senescence, a decline in physiological function with age, seems to be a consequence of natural selection acting more strongly on traits favoring survival and reproduction early in life than on traits favoring survival to old age.

Species adapted for severe, fluctuating climates are predicted to show *r*-selected traits, with early maturation and large litter sizes. Those adapted to stable habitats are *K*-selected, with larger body size and slower reproductive rates. Bet-hedging theory predicts that individuals should reproduce at lower rates for a longer period when survival of offspring is low and unpredictable.

Some populations of mammals show logistic growth, with densities gradually approaching the limit that can be supported by the environment. Density-dependent factors, whose negative effects on growth increase with density, are likely to be more important than density-independent factors in regulating such populations.

Several mechanisms have been proposed by which populations of mammals could regulate their own numbers, by means of either behavioral-physiological responses to increased density or behavioral-genetic mechanisms, in which the genetic structure of the population changes as a function of density. Although some evidence exists for both types of responses in natural populations of mammals, none shows that self-regulation is an evolved trait. The reproductive restraint seen in dense populations is more likely to be an adaptive response of individuals in the population to crowded conditions.

A number of studies have attempted to correlate food supply and predators with population growth, suggesting that both are important. A smaller number of field experiments have tested for interactions among food, predators, and competitors in regulating population density.

At high latitudes, densities of a few species cycle, typically every 10 years for larger mammals such as the lynx and hare and every 3 to 4 years for smaller mammals such as voles and lemmings. Both food and predators have been implicated in hare cycles, and there is some evidence that cycles are synchronized by large-scale climate fluctuations linked to solar activity.

Discussion Questions

1. Statements suggesting that many factors contribute to regulating animal populations seem to run counter to Liebig's law of the minimum. Is there any way to resolve this apparent contradiction?

2. Studies of litter size in white-footed mice have shown that litters of six produce the most surviving offspring, yet the average litter size is closer to five. Why would a female mouse produce less than the optimum number of offspring in a particular litter? For a discussion of this topic, see Morris (1992).

3. Understanding factors affecting the size of populations is a central problem in ecology and mammalogy. Why has social behavior been implicated in so many of the theories of population regulation?

4. The data in the following table refer to mice that were exposed 2 min/day either to trained fighter mice, to trained nonfighter mice, or to no mice for days 1 through 14. On day 8, each mouse was fed equal numbers of tapeworm eggs. On day 22, the mice were killed, their adrenal and thymus glands weighed, and the intestinal tapeworms counted and weighed. Adrenal gland weight is positively related to adrenocortical hormone secretion. The thymus is involved in the immune response. Discuss how these data may relate to population regulation and behavior.

Effect of fighter mice exposure on mean host weights, organ weights, and tapeworm number

| | Groups Exposed To | | |
	No Mice	Trained Nonfighter Mice	Trained Fighter Mice
Group size	10	10	7
Adrenal weight (mg)	4.39	4.42	5.88*
Adrenal weight (mg)/body weight (gm) $\times$ 100	13.1	13.1	16.6*
Thymus weight (mg)	37.0‡	35.0	21.9*
Body weight (gm)	33.7	34.0	35.8
Worm number/mouse	11.6§	15.4§	75.3†
Worm weight/mouse weight (mg)	6.37	4.04	82.28†

Source: *Data from Patterson, M.A. and S.H. Vessey, 1973. Tapeworm (*Hymenolepis nana*) Infection in Male Albino Mice: Effect of Fighting Among the Hosts.* Journal of Mammalogy. *54:784–786.*

Significantly different from other two groups, p < .01.
†*Significantly different from other two groups, p < .001.*
‡*Mean based on nine weights.*
§*Three of these mice had no worms.*

Suggested Readings

Krebs, C. J. 1994. Ecology: the experimental analysis of distribution and abundance, 4th ed. HarperCollins, New York.

Pianka, E. R. 1994. Evolutionary ecology, 5th ed. HarperCollins, New York.

Ricklefs, R. E. 1990. Ecology, 3d ed. W. H. Freeman, New York.

Roff, D. 1992. The evolution of life history: theory and analysis. Chapman & Hall, London.

Stearns, S. C. 1992. The evolution of life histories. Oxford Univ. Press, New York.

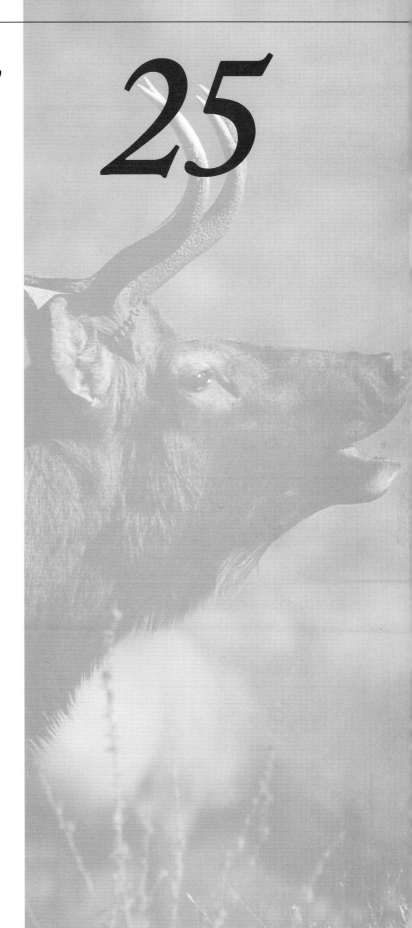

CHAPTER

25

Community Ecology

THE ECOLOGICAL NICHE

SPECIES INTERACTIONS AND COMMUNITY STRUCTURE

COMMUNITY FUNCTION

COMMUNITY PATTERNS

In this chapter, we focus on the interactions among species and the effect those interactions have on both the living and nonliving features of their environment—the subject of **community ecology.** Interactions among species affect the abundances of populations (chapter 24) and contribute to natural selection among phenotypes, thus influencing the evolution of coexisting species. These interactions ultimately explain why there are so many different species of mammals. Populations of different species interacting in an area at the same time make up a **biological community.** A more inclusive term is **ecosystem,** which includes the biotic (living) components (i.e., the community) plus the abiotic (nonliving) components. Communities and ecosystems are not simply random assemblages of species but seem to show similar patterns of assemblages across different continents. Understanding these so-called assembly rules is a central problem in ecology, and the study of mammalian communities has played a significant role in this research. **Community structure** encompasses patterns of species composition and abundance, temporal changes in communities, and relationships among locally coexisting species. At the simplest level, community ecology can be studied by looking at interactions between two species, which we do first in this chapter. Then we explore more complex and larger-scale patterns of mammalian diversity. Our goal is to understand the way single-species populations are integrated into larger biological levels of organization.

Mammalogists have played important roles in our understanding of the interactions among populations. One of the first North American community ecologists was C. Hart Merriam, a mammalogist who developed the concept of life zones. Working in the San Francisco Mountains of northern Arizona, Merriam (1894) concluded that temperature defined floral and faunal zones, forming elevational bands up the sides of mountains and latitudinal bands from the equator to the North and South Poles. Early researchers viewed communities as interrelated units responding collectively to abiotic factors (Mares and Cameron 1994). The degree to which communities can be considered discrete units rather than random assemblages of populations has been hotly debated by ecologists, as we shall see later in this chapter.

THE ECOLOGICAL NICHE

Each species of mammal occupies one or more habitats, has certain physical and chemical environmental tolerances, and performs a specific functional role in the community. Joseph Grinnell, who worked with birds and mammals in the early 1900s, was one of the first ecologists to use the term *niche*, by which he meant the habitat a species occupies as a function of its physiological and behavioral attributes (Grinnell and Swarth 1913). The term was defined somewhat differently by Elton (1927) to include an organism's functional role in the community in terms of its trophic (feeding) level (Mares and Cameron 1994). Today, most ecologists include both the

distributional and functional components, considering a niche to be the total of adaptations of a species to a particular environment (Pianka 1994). We can never know all the parameters necessary to completely define a niche, so we emphasize those factors that potentially limit distribution and abundance and those for which organisms compete.

Most studies of mammals provide information potentially useful in determining niche attributes of individual species. Laboratory studies can determine physiological tolerances of a species to such conditions as heat, cold, and moisture (see chapter 8). Measuring its niche in the field is more difficult, however, because biotic and abiotic factors often interact to produce unpredictable results. Competitors, predators, and symbionts affect the distribution, population dynamics, and niche of most species.

SPECIES INTERACTIONS AND COMMUNITY STRUCTURE

Interspecific Competition

We have seen that individuals of the same species may compete for limited resources such as food or shelter, leading to density-dependent changes in population growth (see equation 2 in chapter 24). Different species may also compete for those resources, affecting rates of population growth of both species. Equation 2 from chapter 24 can be modified to include the effect of competing species 2 on species 1 by adding a term called the coefficient of competition (a_{12}).

$$\frac{dN}{dt} = r_1 N_1 \left(1 - \frac{N_1}{K_1} - \frac{a_{12}N_2}{K_1} \right)$$

where the subscripts 1 and 2 refer to values for species 1 and species 2. Notice that the density-dependent term in parentheses now contains a term including the number of individuals of species 2. When $a = 1$, the interspecific competitor, species 2, has the same effect on population growth of species 1 as another member of species 1, and interspecific competition is intense. As a declines, approaching zero, the last term of the equation approaches zero, and the effect of species 2 on species 1 can be ignored. A similar equation can be written to describe the effect of species 1 on species 2, where the coefficient of competition is a_{21}. This equation is too simple to actually model competition in nature, but it serves as a starting place to develop hypotheses and design experiments to further understand this important ecological process.

Interspecific competition also affects the habitat and food preferences of different species. In the absence of competition, a species occupies its **fundamental niche,** the full range of conditions and resources in which the species can maintain a viable population. In the presence of competing species, however, the species may be restricted to a narrower **realized niche,** where some habitats or resources are not available because competitors occupy them (Hutchinson

1957). If the competitors are strong enough, a realized niche may become so small that local extinction results. This line of reasoning has led to the **competitive exclusion principle:** if two competing species coexist in a stable environment, they do so as a result of differentiation of their realized niches (Ricklefs 1990). If no differentiation takes place, then one species will exclude the other. Although this principle makes intuitive sense and has been widely accepted by ecologists, it is difficult to confirm experimentally.

One way to visualize interspecific competition is to look at only one niche component at a time, picking the one that is limiting for the competitors. Consider several species of desert rodents feeding on different-sized seeds from various plants. It is relatively easy to determine which types of seeds are stored for later consumption because seeds can be collected from the cheek pouches of trapped animals. The feeding niches of the rodents can be inspected by plotting seed size on one axis and the utilization (consumption) rate on the other. In the hypothetical example shown in figure 25.1, each of the three species shown has a roughly normal, or bell-shaped, utilization curve with a mean, or average. Niche width (w), measured by the standard deviation around the mean, and the distance between utilization peaks (d) indicate the amount of interspecific competition. When d is large and w is small, little overlap occurs, and the coefficient of interspecific competition is small (figure 25.1A). When d is small and w is large, overlap is great, and the coefficient is

large. In this case, we might expect d to increase or w to decrease, leading to reduced competition. Another outcome is that one species would be eliminated from the area. If possible, more than one niche dimension needs to be examined because coexisting species may overlap extensively in one niche dimension but only slightly in another (Schoener 1974).

How different do two species need to be to coexist in the same general habitat? The ratio d/w is a measure of niche overlap, and when this ratio is low (i.e., less than 1) coexistence is theoretically possible. However, differences in carrying capacity among the competing species, environmental fluctuations, and the importance of other niche parameters may all affect the outcome.

One way to indirectly evaluate a niche is to assess morphological traits related to important functions, such as feeding. The number of species packed into a niche space can be estimated by comparing morphological distances between nearest neighbors. Fenton (1972) used several dimensions to define morphological space among different species of bats. One was the ratio of ear length to forearm length; the higher the ratio, the larger the ears and the more important echolocation is presumed to be in the foraging behavior. A second was the ratio of third and fifth digits of the wing bones, yielding an index of wing shape. A high ratio on this dimension indicates a long, thin wing shape, whereas a low ratio indicates a short, broad wing shape (called aspect ratio—see chapter 12, figure 12.4). The former favors high-speed flight, whereas the latter favors low-speed flight and greater maneuverability. Plotting both ratios together provides an insight into the niche space occupied by different species (figure 25.2). In the temperate zone communities in Ontario, bat species are small insectivores, and the points are clustered fairly closely. In the tropics, bats fill many other roles, such as fruit-, nectar-, fish-, and even bat-eaters, and the scatter of the points reflects the wide range of wing and ear shapes. Other studies of bats (Findley and Black 1983; Schum 1984) have confirmed that external morphology and diet are correlated and that variation in morphology increases with the number of species present.

The role of interspecific competition in determining niche parameters and the species composition of communities has been controversial. Although competition is assumed to be a driving force, it is possible that competition played a major role in adaptation of the species in the past, but is hardly observable today. Another problem is that competition may occur only during occasional "crunches," when resources become limiting for a brief period (Wiens 1977) or when species newly come together. At other times, a particular resource may not be limiting. Nevertheless, these crunches may be crucial in determining niche breadth and the distribution of species across the landscape.

Several factors provide indirect evidence of the importance of competition in natural communities of mammals. One is the distributions of species, both in the presence and

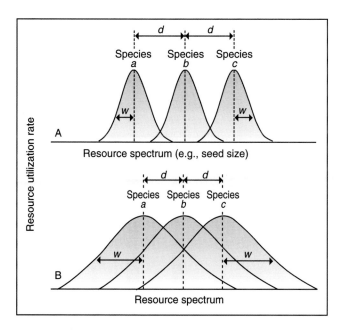

Figure 25.1   **Resource utilization curves for three species coexisting along a one-dimensional resource spectrum.**   The term *d* is the distance between adjacent curve peaks; *w* is the standard deviation of the curves. (A) Narrow niches with little overlap (*d* > *w*), i.e., relatively little interspecific competition; (B) broad niches with great overlap (*d* < *w*), i.e., relatively intense interspecific competition.

Source: Data from M. Begon, et al., Ecology, *1986, Sinauer Associates, Sunderland, MA.*

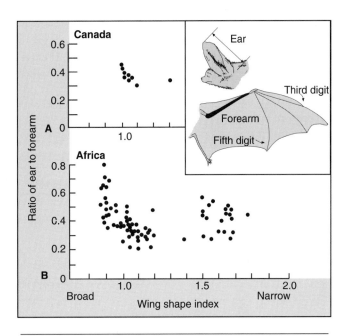

Figure 25.2   Niche space of insectivorous bats of (A) southeastern Ontario, Canada and (B) Cameroon, Africa. The horizontal axis is the ratio of the lengths of the third and fifth digits of the hand and the vertical axis is the ratio of ear length to forearm length.

Source: Data from M.B. Fenton, "The Structure of Aerial-feeding Bat Faunas as Indicated by Ears and Wing Elements" in Can. Journal Zoology, *50:287-296, 1972.*

the absence of presumed competitors. For instance, two species of chipmunk—*Eutamias dorsalis* and *E. umbrinus,* are found on isolated mountaintops in the Great Basin area of Nevada. When each species is present alone, it occupies a wide range of elevations, but when together, one species occupies lower elevations and the other higher (Hall 1948; figure 25.3). The assumption is that in the presence of competition, each species reduces niche overlap. Conversely, each species shows **competitive release,** expanding its range, in the absence of competitors.

Introductions by humans provide indirect evidence of competition, especially when mainland species are introduced on islands. The result is often extinction of the island species, presumably due to competitive exclusion. Thus, introduction of the placental canids, the dingo (*Canis lupus dingo*) and red fox (*Vulpes vulpes*), into Australia was a factor in the extinction of the marsupial fauna such as the Tasmanian wolf (*Thylacinus cynocephalus*) and the Tasmanian devil (*Sarcophilus harrisii*). Based on fossil evidence, the disapearance of both the Tasmanian wolf and devil from southern Australia coincided with the arrival of the dingo to that region. Both marsupial species survived on Tasmania because the island was cut off from the mainland before the dingo got to the southwestern mainland of Australia.

In the United States, introduced species of deer have had a negative effect on native species. Sika deer (*Cervus nippon*), native to Japan and the East Asian mainland, were introduced into Maryland's eastern shore in 1916. During the 1970s and 1980s, the proportion of white-tailed deer

(*Odocoileus virginianus*) harvested declined sharply, from 75% to 35%, and that of sika deer showed a corresponding increase, from 25% to 65% (Feldhamer and Armstrong 1993). Such negative correlations do not, however, demonstrate a cause-and-effect relationship.

Character Displacement

Other observational evidence for competition comes from comparisons of morphology and behavior of individuals in the presence or absence of close competitors. When the ranges of two similar species overlap (i.e., their distributions are **sympatric**), the species tend to differ more from each other than where their ranges do not overlap (i.e., their distributions are **allopatric**). This shifting of traits in areas where both species are present has been referred to as **character displacement.** In northern Europe, the pygmy shrew (*Sorex minutus*) and the common shrew (*S. araneus*) overlap extensively. Allopatric populations, however, occur on islands off the coasts of England and Sweden. Pygmy shrews have significantly smaller lower jaws in sympatric populations than in allopatric populations (Malmquist 1985; figure 25.4). In this study, however, jaw measurements of the larger, common shrew did not shift. Skull size is related to prey size, that is, individuals with larger skulls can eat larger prey than those with smaller skulls. The inference is that in zones of overlap, one or both of the species diverge in prey size selection, leading to reduced competition and coexistence of both species.

Although usually restricted to closely related species, character displacement may occur among more distantly related taxa on a community-wide basis. Three sympatric small cat species in Israel are strongly sexually dimorphic, and the sexes can be treated separately for comparison (Dayan et al.

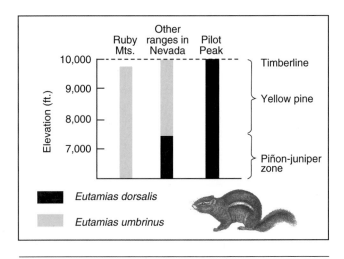

Figure 25.3   The distribution of two species of chipmunk in the Great Basin area of Nevada.   Note that where only one species is present, each occupies the entire slope, but if both are present, one occupies higher elevations and the other lower elevations.

Source: Data from E.R. Hall, Mammals of Nevada, *1948, University of California Press, Berkeley, CA.*

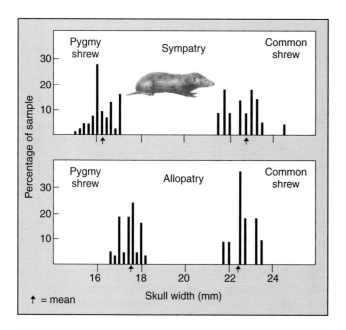

Figure 25.4   Character displacement in shrews.   The pygmy shrew and the common shrew are allopatric on islands off the coast of Sweden but sympatric on the mainland. Note that the measure of skull width is smaller for the pygmy shrew in sympatry.

Source: Data from M.G. Malmquist, "Character Displacement and Biogeography of the Pygmy Shrew in Northern Europe" in Ecology, 66:372-377, 1985.

1990). The average canine diameters for the six "morpho-species" were remarkably evenly spaced, consistent with the idea that niche partitioning of prey size occurs among species and between sexes. These cats are exclusively carnivorous, and canine teeth are used to kill prey. The canines are well adapted to wedge between the vertebrae of the prey's neck, severing the spinal cord or hind brain. Larger prey would clearly require thicker canine teeth, making canine diameter a good predictor of the size of prey captured. Dayan and colleagues (1990) argue that the evenly spaced sizes indicate character divergence to reduce competition for prey. Although competition for food is the presumed explanation, it is also possible that sexual selection has played a role in canine size. Males of these species may use canines for display or fighting with other males, as do many polygynous species of primates (see chapter 21). Thus, canine size may be affected by male–male competition for access to females as well as by size of prey.

Character displacement may also exist among mustelids. McNab (1971) hypothesized that prey size combined with character displacement governs body size for three species of North American weasels (*Mustela* spp.); sympatric species of weasels were forced by competition to maintain minimum size differences. Harvey and Ralls (1985), who examined condylobasal measurements of skull length in North American weasels, found no evidence of character displacement, however, concluding that most effects were due to latitudinal gradients in body size. In a still more recent study using the diameter of the upper canine, almost no overlap occurred among different sympatric species

(Dayan et al. 1989). Size ratios were nearly evenly spaced, as was true in the cat example. Equal size ratios between species adjacent in a size ranking were considered by Dayan and colleagues as evidence for community-wide character displacement as a result of competition, supporting McNab's (1971) original hypothesis.

Limiting Similarity

One of the ways two or more species can coexist is to evolve different body sizes, thus essentially dividing the area into different feeding niches and avoiding competition. A ratio in length of 1:1.3 was suggested for the minimum difference between two closely related competitors to permit coexistence (Hutchinson 1959), prompting a number of investigators to compare body size at the community level. Numerous species of seed-eating rodents occupy the deserts of the southwestern United States. When Brown (1975) compared the rodent communities in the Sonoran Desert with those of the Mojave and Great Basin deserts, he noticed that the distributions of body sizes of different species were remarkably similar (figure 25.5). Brown also observed that species of similar size tended to "replace" one another in different habitats. Thus, the 7-g pocket mouse (*Perognathus flavus*) in the Sonoran was replaced by the 7-g *P. longimembris* in the Great Basin, and the 11-g harvest mouse (*Reithrodontomys megalotis*) was replaced by the 12-g kangaroo mouse (*Microdipodops pallidus*). In addition, when the same species was present in both deserts, it was displaced in size, filling a gap (e.g., the deer mouse, *Peromyscus maniculatus*, was 24 g in the Sonoran Desert but only 18 g in the Great Basin).

Removal Experiments

Observational studies can provide only indirect, correlational evidence of the importance of competition in structuring ecological communities. One way to provide direct evidence of interspecific competition is to remove one of the competitors and monitor the response of the other species. In one such study, three sympatric species of small mammals—the white-footed mouse (*Peromyscus leucopus*), the golden mouse (*Ochrotomys nuttalli*), and the northern short-tailed shrew (*Blarina brevicauda*)—were live-trapped, marked, and released in a pine plantation in Tennessee (Seagle 1985). A number of vegetation characteristics were recorded at each capture site to determine the microhabitat preferences for each species. When white-footed mice were removed from one grid, golden mice demonstrated competitive release by shifting their microhabitat preference from open forest to areas with more fallen logs, denser canopy development, and denser understory.

Four species of chipmunks (*Eutamias* spp.) are found along an altitudinal-vegetational gradient on the eastern slope of the Sierra Nevada, California. The chipmunk species are contiguously allopatric, that is, their ranges are adjacent but not overlapping, and each is restricted to its own vegetational community along the gradient (figure 25.6). Chappell

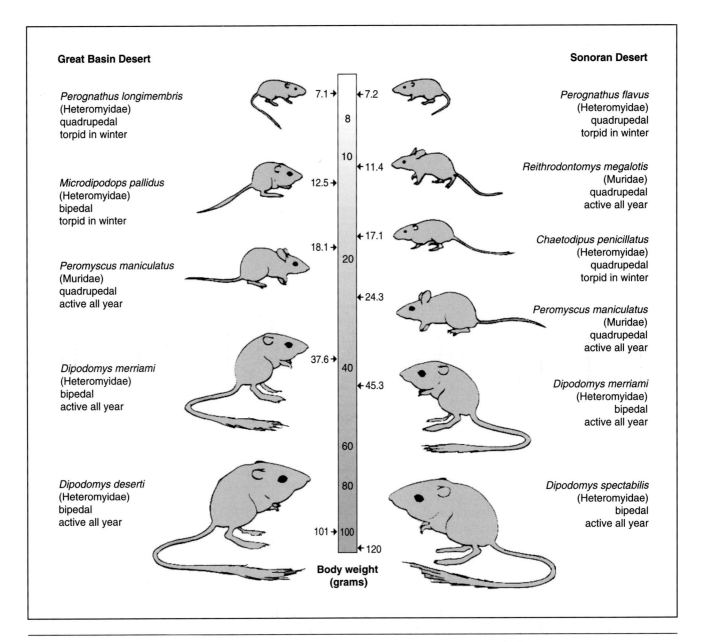

Figure 25.5 Convergence in structure between a six-species community from the Sonoran Desert and a five-species community from the Great Basin Desert. Numbers are average weights (grams). Note the similarities in body size, form, taxonomic affinity, and other characteristics between species occupying similar positions in each community. Also notice the displacement in body size in *Peromyscus maniculatus* and *Dipodomys merriami* (both are larger in the Sonoran Desert) to compensate for the different numbers and sizes of coexisting species.

Source: Data from J.H. Brown, "Geographical Ecology of Desert Rodents" in Ecology and Evolution of Communities, *M.L. Cody and J.M. Diamond, eds., 1975, Belknap Press of Harvard University Press, Cambridge, MA.*

(1978) conducted a series of observations and experiments to reveal the physiological and behavioral factors that might produce this distribution. For example, captures of *E. minimus* increased in wooded areas at higher elevations, where *E. amoenus* was removed. Behavioral observations confirmed that *E. amoenus* was dominant to *E. minimus*, suggesting that competitive exclusion was responsible for the failure of *E. minimus* to colonize the wooded zones. Where *E. minimus* had been removed, however, *E. amoenus* failed to invade. The failure of the socially dominant *E. amoenus* to move down

slope into the sagebrush zone in the absence of *E. minimus* is probably due to physiological constraints: *E. amoenus* is not adapted to tolerate the hot, dry conditions at lower elevations. In this example, both behavioral and physiological factors affect the distribution of species.

Not all removal studies provide clear-cut results. Along the Mediterranean coast of Israel is a narrow strip of sand dunes, of recent origin, perhaps only a few hundred years old. Two species of seed-eating gerbilline rodents occupy the area, and they show considerable overlap in diet and microhabitat

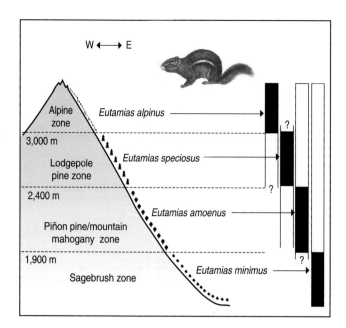

Figure 25.6 **Distribution of four species of chipmunk on the eastern slope of the Sierra Nevada, California.** Dark portions of bars denote realized niches; entire bars show fundamental niches.

*Source: Data from M.A. Chappell, "Behavioral Factors in the Altitudinal Zonation of Chipmunks (*Eutamias*)" in Ecology, 59:565-579, 1978.*

preference (Abramsky and Sellah 1982). The larger species, *Meriones tristrami*, colonized the dunes from the north. It can live on sand as well as other soil types. The smaller species, *Gerbillus allenbyi*, colonized from the south, and it is found only on sand. When alone, *M. tristrami* occupies sand. Where the two species are sympatric, however, *G. allenbyi* occupies sand and *M. tristrami* occupies soil types other than sand. The exclusion of *M. tristrami* from the habitat it prefers when alone (sand) has been interpreted as resulting from interspecific competition. When *G. allenbyi* was removed from an area containing both species, it was expected that *M. tristrami* would expand its habitat and occupy the sand areas, showing competitive release. No such effect was observed, however, suggesting that competition was not responsible for the different substrates occupied. One interpretation of these results is that these species did, in fact, compete in the past and that habitat selection is under genetic control. Thus, genetic changes in the population of *M. tristrami* in areas of overlap with *G. allenbyi* might have led to its failure to occupy sand in the absence of the other species. This phenomenon has been referred to as "the ghost of competition past."

In some cases, two very similar species are found in the same habitat. In the Appalachian Mountains, two species of the genus *Peromyscus* frequently coexist. White-footed mice (*P. leucopus*) and deer mice (*P. maniculatus*) are so similar in appearance in some regions that molecular techniques are needed to tell the two species apart with certainty. They eat the same foods (Wolff et al. 1985) and defend territories against members of their own species and the other species (Wolff et al. 1983). The two species may occupy different

microhabitats, because *P. maniculatus* shows more arboreal activity than does *P. leucopus* (Harney and Dueser 1987). Based on reciprocal removal of each species on different plots, neither species seemed to encroach on the other's microhabitat, however, suggesting little in the way of competitive interactions (Harney and Dueser 1987). By checking live traps in the middle of the night, Drickamer (1987) captured *P. leucopus* more often before 1:00 AM and *P. maniculatus* more often after 1:00 AM. Using live traps with timers on them, Bruseo and Barry (1995) confirmed that *P. leucopus* becomes active earlier each night than does *P. maniculatus*. Thus, these two similar species may reduce intraspecific competition and coexist by occupying slightly different microhabitats and by being active at different times of night.

Competition Among Different Phyla

Competition is not necessarily limited to members of the same taxonomic group. Rodents and ants compete with one another as they prey on plant seeds when seeds are in short supply. These complex relationships were explored by means of a long-term field experiment in the Chihuahuan Desert in southeastern Arizona (Brown et al. 1986). Twenty-four plots were fenced off as part of a 10-year study. Each plot was 50 m by 50 m, and three general manipulations were performed: (1) exclusion of some or all rodent species using different-sized gates in the fences; (2) exclusion of ant species by means of poisoning colonies; and (3) the addition of seeds. Rodent densities increased in the plots with added food, suggesting that seeds were a limiting factor. Excluding large granivorous rodents (kangaroo rats, *Dipodomys* spp.) led to increases in small granivorous rodents, suggesting competition among rodents for seeds. When all rodents were excluded, large-seeded plants were favored because the rodents preferred eating the larger seeds; small-seeded plants then declined, as did the ants that fed on them. Thus, although rodents and ants seemed to be competing for seeds, they actually had a mutualistic relationship. By feeding on seeds of different plants, both large- and small-seeded plants coexisted, as did the ants and rodents that feed on them (figure 25.7).

Predation

In chapter 24, we discussed predation from the perspective of prey population dynamics and whether predators can regulate the numbers of prey. Here we take a broader perspective and consider the role of predators in structuring communities. The term *predator* is used in its broadest sense to include herbivores "preying" on plants and rodents "preying" on seeds.

Keystone Predators

Some species appear to play critical roles in the community, such that "not all species are created equal." Although mammals typically account for only a small fraction of the biomass and energy flow in most ecosystems, they sometimes regulate the structure and dynamics of the entire community.

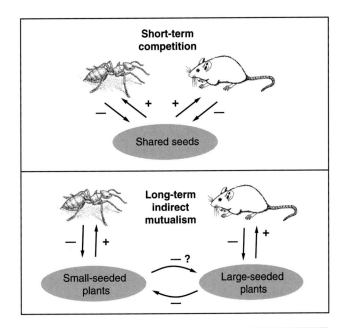

Figure 25.7   Interactions between granivorous rodents and ants.   In the short term, the two taxa compete if they overlap in their feeding on limited seeds, but in the longer term, the two taxa can have indirect mutualistic effects on each other if they feed differentially on different plant species that also compete with each other. Plus signs refer to positive interactions; negative signs to negative interactions.

Keystone predators are those that exert a controlling force in the community, affecting abundance and distribution of many other species and actually increasing the species diversity in the community.

For example, sea otters (*Enhydra lutris*) prey extensively on sea urchins along the coast of the western United States. Fur traders decimated otter populations from 1741 to 1911, except for a few remnant populations in Alaska and central California. Since being protected, sea otter populations have become reestablished over much of their original range. In areas where sea otters are present, sea urchins are sparse and algae communities (mainly kelp) thrive. In areas where sea otters are absent, sea urchins become abundant and eliminate much of the kelp and associated marine species (Estes et al. 1978).

When sea urchins were experimentally removed from plots on Torch Bay, Alaska, where no otters were present, a complex association of several species of kelp developed. Similar effects were seen when otters were transplanted into areas with high densities of sea urchins. The sea urchins were quickly depleted by the otters, and the kelp community developed several years later (Duggins 1980; table 25.1). Sea otters, therefore, exert a profound effect on community structure and fit our definition of keystone predators.

Herbivores can also act as keystone species. The African elephant (*Loxodonta africana*) prefers forest edge, woodland, and brushland. It feeds on a diet of browse (twigs, shoots, and leaves), supplemented by grass. While foraging on bark, elephants destroy the shrub understory and girdle trees, all of which leads to loss of woodland and the creation of grassland (figure 25.8). Once forest canopy is opened up, fire accelerates the process. In large parts of sub-Saharan Africa, the vegetation has shifted from woodland to grassland as a result of elephants. This change ultimately works to the disadvantage of the elephants, however, because they need woody species for browse, but it works to the advantage of grazing ungulates (Laws 1970).

More than one species may play a keystone role in the same community. Species occupying a particular trophic level can be subdivided into **guilds,** or groups of species that exploit a common resource base in a similar fashion. In the Chihuahuan Desert, several species of kangaroo rats form a

Table 25.1.   Sea otters as a keystone species.   Changes in sea urchin and kelp densities in response to the reintroduction of sea otters.

	Mean Density (no./m² ± SD)*		
	Torch Bay (no otters) N = 80	Deer Harbor (otters present < 2 y) N = 24	Surge Bay (otters present < 10 y) N = 80
Urchins			
Strongylocentrotus franciscanus	6±7 (79%)	0.03±0.03 (4%)	0
S. purpuratus	4±13 (31%)	0.08±0.06 (8%)	0
S. droebachiensis	6±14 (65%)	0.2±0.1 (12%)	0
Kelps			
Annuals	3±7 (33%)	10±5 (100%)	2±5 (28%)
Laminaria groenlandica	0.8±5 (6%)	0.3±0.6 (21%)	46±26 (99%)
Samples with no kelp	(64%)	(0%)	(0%)

Source: Data from D.O. Duggins, "Kelp Beds and Sea Otters: An Experimental Approach" in Ecology, *61:448, 1980.*

*Numbers in parentheses are the percent of these quadrats in which the species or group were observed. N = number of 1.0 m² quadrats.

Figure 25.8 **Elephants as a keystone species.** By destroying shrubs and trees, elephants promote grassland, which in turn favors grazing ungulates.

seed-eating guild (Brown and Heske 1990). When these rodents were excluded from this desert shrubland, a chain of events occurring over 12 years changed it to desert grassland. In the absence of kangaroo rats (*Dipodomys* spp.), annual plants with large seeds increased in number, raising the vegetative cover to the point where ground-feeding birds could no longer forage on grass seeds. This led to the increase in grasses. Both selective foraging by *Dipodomys* on large seeds and soil disturbance by foraging, caching, and burrowing tend to maintain the shrub desert. Of the 15 species of rodents that live in the Chihuahuan Desert, Brown and Heske (1990) considered the kangaroo rats to form a **keystone guild** of large-seed eaters.

Risk of Predation

Food partitioning is one way in which communities are structured, but risk of predation may also be important. Rodents in the Great Basin Desert in Nevada partition the microhabitat so that kangaroo rats and kangaroo mice (*Microdipodops*) forage in the open, whereas pocket mice (*Perognathus*) and deer mice (*Peromyscus*) forage near bushes. The former have hyperinflated auditory bullae, associated with high auditory acuity, and elongated hind legs, associated with bipedal locomotion. These adaptations aid in the detection of, and escape from, predators such as long-eared

owls (*Asio otus*), coyotes (*Canis latrans*), kit foxes (*Vulpes macrotis*), and gopher snakes (*Pituophis melanoleucus*). Dice (1945) showed that owls find prey more easily under moonlight than starlight. In a field experiment, illumination simulating moonlight was added to some grids by means of lanterns, and shadows were created at some sites by means of parachute canopies (Kotler 1989). In general, illumination led to reduced foraging in open microhabitats. Food in the form of bird seed was added to some bush sites and to other open sites. The bipedal species responded positively to additional food in both bush and open sites, whereas the quadripedal species responded only to food in the bush sites. Kotler (1989) concluded that both risk of predation and resource availability interacted to affect foraging behavior and habitat choice. Thus, both predators and distribution of resources need to be considered when trying to predict the species composition of a community.

Mutualism

Interactions involving mammals and members of other taxa can be mutualistic, in which both species benefit either directly or indirectly. Examples of direct mutualism, in which both species are in contact, include seed dispersal by rodents, pollination of many species of plants by bats (chapter 12), and digestion of cellulose by the endosymbionts within the rumens of their ungulate hosts (chapter 19). Mutualistic relationships between birds and mammals include one between the African honey guide (*Indicator indicator*) and the African honey badger (*Mellivora capensis*). The bird vocalizes and "leads" the badger to a bee's nest. The badger opens up the nest and feeds on honey, while the honey guide consumes wax (Vaughan 1986). Oxpeckers (*Buphaga* spp.) eat the ticks they remove from large African ungulates to the benefit of both bird and mammal. Unfortunately, as domestic cattle replace native species of ungulates, oxpeckers have declined in number, in part because cattle are treated with pesticides that then eliminate the oxpecker's food supply.

Indirect mutualism involves positive effects without direct contact between the species. An example is the ant-rodent mutualism in the Arizona desert discussed earlier (see figure 25.7). Commensal relationships, in which one species benefits and the other is more or less unaffected, include that between cattle egrets (*Bubulcus ibis*) and cattle (*Bos;* Heatwole 1965). Cattle egrets feed on insects stirred up by the cattle. Another example of commensalism is the moving substrate provided for barnacles attached to the whale's skin.

Plant-Herbivore Interactions

Herbivorous mammals, especially ungulates, are important components of many ecosystems, especially grassland communities. The effect of grazing on the plant community has been viewed by ecologists both as a predator-prey relationship, in which grazers benefit and plants are harmed, and as a

Figure 25.9   Migratory herd of wildebeests grazing in the Serengeti.   Grazing stimulates plant growth and improves forage for gazelles.

mutualistic one, in which numerous species of grazers and plants have coevolved adaptations in response to one another. The most spectacular concentration of grazing animals in the world occurs in the Serengeti-Masai-Mara ecosystem in Tanzania and Kenya, where hundreds of thousands of wildebeests (*Conochaetes taurinus*), Thomson's gazelles (*Gazella thomsonii*), zebras (*Equus burchelli*), and buffalos (*Syncerus caffer*) concentrate, along with lesser numbers of more than 20 other species (figure 25.9, see also figure 6.10). In the wet season, usually ending in May, huge numbers of migrating wildebeests pass through the area and reduce the aboveground green biomass by more than 80% and plant height by more than 50% (McNaughton 1976). Within several weeks, however, a dense lawn of green leaf forage appears in the heavily grazed areas, providing ideal grazing for Thomson's gazelles. Grazing by the wildebeests stimulates growth of senescent grasses, which provides forage of higher nutrient content and digestibility for the gazelles. This phenomenon has been called grazing facilitation, in which the feeding activity of one herbivore species increases the food supply for a second. No evidence exists that grazing ever increases plant production or fitness, however (Belsky 1986).

Community Assembly Rules

Most ecologists now agree that ecological communities are not random collections of species; certain combinations of species seem to occur more frequently than would be expected by chance. Competition, predation, and mutualism all play roles in structuring mammalian communities and may act in concert. Several researchers have developed rules for assembling communities based on the competitive use of resources. One such rule, suggested by Fox (1989), is that species entering a community are drawn from a different

group, or guild, until each group is represented, and then the series repeats. Communities lacking a representative from one group but having more than one from another group are less stable and more subject to invasion. Fox evaluated this rule for three groups of marsupials and for rodents in eastern Australia. When tested against a null model of chance assembly, significantly more sites fit the rule. One species from each of three sizes of shrews (small, medium, large) was usually present at a number of sites in western Kentucky and Tennessee, rather than several species from the same size group (Feldhamer et al. 1993). Other communities of shrews in North America seem to follow a similar rule (Fox and Kirkland 1992).

Findley (1989) asked the somewhat more general question of whether morphological differences between **syntopic** (present at the same time and place) species of rodents are greater than would be the case if the species were drawn randomly from the available pool of colonists. Using skull measurements of the rodents from a number of sites in New Mexico and Sonora, Mexico, he found that morphological differences in size, but not shape, among coexisting species were greater than expected by chance. He concluded that this nonrandom pattern may have resulted from interactions such as competition or from species-specific responses to resources that were differentially distributed.

Other investigators have failed to find evidence that competition for resources structures communities. Five feeding guilds of phyllostomid bats were identified from northeastern Brazil by means of fecal samples from field-caught specimens: foliage-gleaning insectivores, nectarivores, frugivores, sanguinivores, and omnivores (Willig and Moulton 1989). A number of morphological measurements were made from each species to see if community members were assorted by size. The observed species compositions were no different from what would be found if species were selected from the available species pool at random, suggesting that competition is not structuring the bat community. Similarly, in a comparison of bat communities from different regions of Venezuela, it was concluded that although competition is important at the population level, it was not necessarily important at the community level (Willig and Mares 1989). The failure to find a pattern could be due to the spatial heterogeneity of the habitats in tropical South America combined with the great dispersal abilities of bats.

COMMUNITY FUNCTION

Energy Flow and Community Metabolism

One way to view the interconnectedness of communities is to trace the flow of energy through different trophic levels. Although often presented as linear food chains for simplicity, most communities are really interconnected in a weblike fashion. **Energy,** defined as the ability to do work, enters the community as electromagnetic light energy from the sun.

Converted by photosynthetic plants into chemical energy (in the form of sugars), it is then available to animals. Under natural conditions, green plants convert less than 1% of the light energy available to them. In an old-field community in Michigan studied by Golley (1960), meadow voles (*Microtus pennsylvanicus*) consumed about 2% of the plant material available to them, and least weasels (*Mustela nivalis*) ate about 31% of the available voles (figure 25.10). In this linear food chain, so little energy was converted into weasels that a higher carnivore dependent on weasels for food could not be supported. Of course, many other organisms and energy pathways existed within this community, and most of the plants were eaten by herbivorous insects. A more complex food web (figure 25.11) depicts trophic interaction in the alpine tundra in the central Rocky Mountains. Actually measuring the flow of energy in such a system would be extremely difficult.

Pathways of energy flow through a desert scrub ecosystem were documented by Chew and Chew (1970). Thirteen species of small and medium-sized mammals were

present. Small mammals, both herbivores and granivores, played a small role in terms of energy, converting only 0.016% of the primary, aboveground production to mammal tissue that was then available to carnivores.

The production efficiency of small mammals was measured for populations in nine ecosystem types in Europe and North America (Grodzinski and French 1983). Productivity refers to the addition of new tissue in the form of growth of individuals plus new individuals from reproduction. Production was divided by respiration to give production efficiency. Efficiency was lowest for shrews (0.7%) and highest for herbivores (3.4%). Efficiencies were low for all these small mammals compared with poikilotherms because of the high respiratory cost of homeothermy in small mammals.

Although mammals typically do not account for a large proportion of energy flow in communities, their biomass may be quite high, resulting in a large "standing crop." Furthermore, the total effect of mammals on vegetation may be much greater than the amount assimilated. Voles (*Microtus arvalis*) in agricultural fields in Poland destroyed as much as 13 times more vegetation than they actually used for energy requirements (Grodzinski et al. 1977). Mammals affect the vegetation in many ways such as by cutting, trampling, burrowing, and nesting in it. Finally, because of their large size, long life, and high activity, some species of mammals play dominant roles in the community by influencing vegetation and other animals, as we have seen with elephants, sea otters, and other keystone predators.

Community Development

Ecological succession is the replacement of species in a habitat through a regular progression of stages leading ultimately to a stable state, the **climax** community. Species occupying early successional communities tend to be *r*-selected, with high dispersal rates, rapid growth rates, and high reproductive rates (chapter 24). Those in later stages are more likely to be *K*-selected, with lower dispersal rates, delayed maturation, and lower reproductive rates.

Old-field succession is often studied, in which fields are monitored for varying numbers of years since abandonment from agriculture. The diversity of species typically increases during succession, although it may decline somewhat at the climax. In one study in Minnesota, a census of small mammals was taken for 18 fields that ranged in age from 2 to 57 years since abandonment from agriculture (Huntly and Inouye 1987). The youngest fields were dominated by short-lived, introduced Eurasian plant species. In middle-aged and older fields, native species of prairie grasses dominated. Woody shrubs were found only in fields older than 50 years and were never common. Of the six species of small mammals trapped, white-footed mice were caught in fields of all ages, whereas meadow voles and masked shrews (*Sorex cinereus*) tended to be caught in older fields. Generally, the number of small mammal species increased with the age of the field; this increase was associated with a striking increase in plant nitrogen, a measure

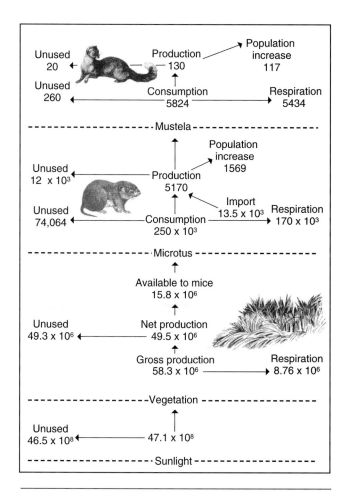

Figure 25.10   Energy flow diagram of a portion of a food web in an old-field community in southern Michigan. Numbers are in kilocalories per square meter.

Source: Data from F.B. Golley, "Energy Dynamics of a Food Chain of an Old-Field Community" in Ecological Monographs, *30:187-206, 1960.*

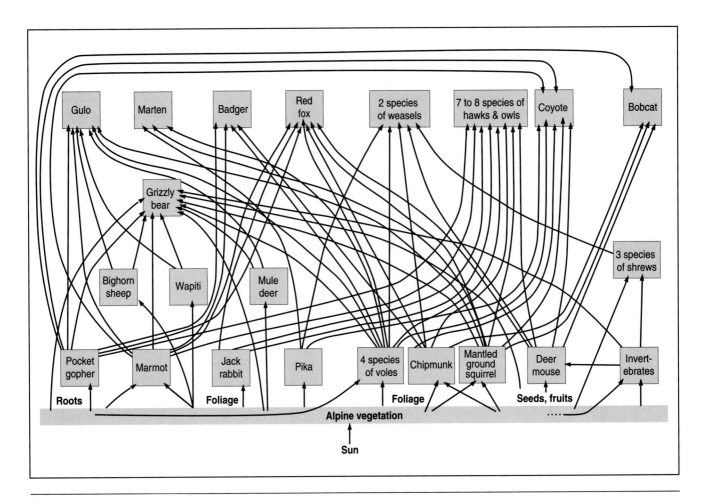

Figure 25.11   Food web in the alpine tundra community of the Beartooth Plateau.   Insectivorous and herbivorous birds are not included.

Source: Data from R. S. Hoffmann, "Terrestrial Vertebrates" in Arctic and Alpine Environments, *J.D. Ives and R.G. Barry, eds., 1974, Harper & Row.*

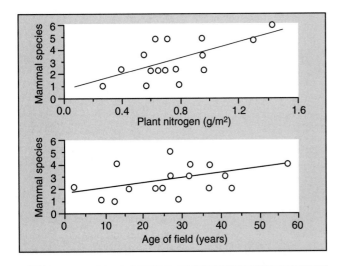

Figure 25.12   Effect of secondary succession on small mammal communities.   Total number of species of small mammals in fields as a function of field age (years since abandonment from agriculture) and nitrogen content of vegetation (g/m^2).

Source: Data from N. Huntly and R. S. Inouye, "Small Mammal Populations of an Old-field Chronosequence" in Journal of Mammalogy, *60:739–745, 1987.*

of primary productivity (figure 25.12). Thus, mammal communities change as the vegetation changes.

We assume that mammals respond to changes in vegetation, but they may affect the vegetation and thus the course of succession. Voles (*Microtus* spp.) can extensively influence vegetation, especially during peak years (Batzli and Pitelka 1970; Grodzinski et al. 1977). In the Serengeti, grazing by large herds of ungulates has a profound effect on the vegetation by increasing energy and nutrient flow rates (McNaughton 1985). McNaughton argued that mammals and plants both have coevolved traits, resulting in interdependence rather than a relationship in which animals gain and plants lose.

COMMUNITY PATTERNS

Island Biogeography

It has often been observed that island habitats have fewer species than do comparable mainland sites. MacArthur and Wilson (1967) developed a model that predicts a dynamic equilibrium of the number of species on islands. Although the species identity changes through time, the total number of species remains constant as new species colonize and resident species become extirpated. Immigration rates by new species are affected by island size; smaller islands are smaller targets and therefore have lower colonization rates. Extinction rates are higher on small islands, in part because population sizes are smaller. Also important is the distance of the island from the colonizing pool of species on the mainland; islands farther away have lower immigration rates (figure 25.13). Thus, small islands equilibrate at fewer species than large islands, and distant islands equilibrate at fewer species than near islands.

It makes sense that the number of species should increase as the area being sampled increases. Larger areas typically have more habitats, which reduces the chances that an individual species will become extirpated. The rela-

tionship, called the **species-area curve,** can be described by the equation

$$S = cA^z \qquad (1)$$

where S = number of species, c = a constant measuring the number of species per unit area, A = area of island, and $z = a$ constant measuring the slope of the line relating S and A.

The application of island biogeography theory to mammalian communities has met with mixed results. Compared with birds and insects, mammals are relatively poor dispersers across water. For instance, Lawlor (1986) compared isolated, oceanic islands with so-called landbridge islands, those that were connected to the mainland at the end of the Ice Age. The species-area curve for oceanic islands is nearly flat (z is low relative to landbridge and mainland areas). Oceanic islands have fewer species than predicted from equilibrium theory, probably because colonization rates are so low for mammals.

Island size and distance from the mainland were evaluated in the Thousand Islands region of the St. Lawrence River in New York (Lomolino 1986). For species such as the red fox (*Vulpes vulpes*) and raccoon (*Procyon lotor*) to be present, islands had to be above a critical size. For other species, such as deer mice, both island size and distance from the mainland were important factors in determining presence or absence of a species (figure 25.14).

The Great Basin of North America consists of a vast "sea" of sagebrush desert interspersed at irregular intervals by isolated mountain ranges. The upper slopes of these mountains are well vegetated, with cool, mesic (moist) conditions. The mammal species on these mountaintops (above approximately 2500 m) are derived from the boreal faunas of both the Sierra Nevada Mountains to the west and the Rocky Mountains to the east. As can be seen from figure 25.15, more species were found as the area of the mountaintop "island" increased, showing a linear relationship on log scale and approaching the number found in the saturated "mainland" areas (Brown 1971). The "islands" close to the "mainland," however, do not tend to have more species than those farther away, and the rate of colonization by new species was much lower than on true oceanic islands. These desert habitats were all connected at the end of the Pleistocene epoch, when rainfall was higher, and they probably shared the same species of mammals. After postglacial climate change took place, rainfall declined, mountaintops became isolated, and they gradually lost species of mammals. Because of the extreme isolation and relatively poor dispersal powers of mammals, immigration rates of new species were probably very low, as Lawlor (1986) also found.

This model has been applied to other favorable habitat "islands" in a "sea" of inhospitable habitat. In eastern Iowa, small patches of forest are surrounded by cornfields. These patches supported fewer species of mammals than did comparably sized areas within contiguous forest (Gottfried 1979; 1982). Smaller and more isolated patches had fewer species than larger and less isolated patches, as expected.

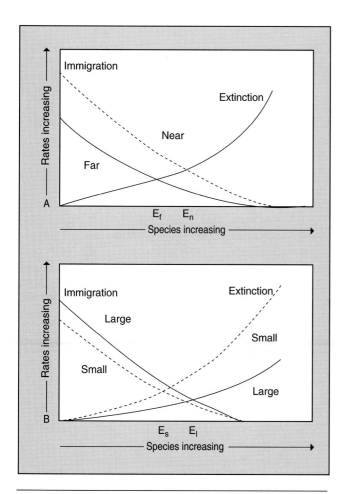

Figure 25.13   Equilibrium numbers of species on islands as a function of immigration and extinction rates.   In (A), an island far from a colonizing source or mainland should equilibrate with fewer species, E_f, than an otherwise identical near island, E_n. In (B) a large island equilibrates with more species, E_l, than a small one, E_s, at the same distance from the mainland. Equilibrium numbers are shown where immigration and extinction curves intersect.

Source: Data from Robert H. MacArthur and Edward O. Wilson, The Theory of Island Biogeography, *1967, Princeton University Press, Princeton, NJ.*

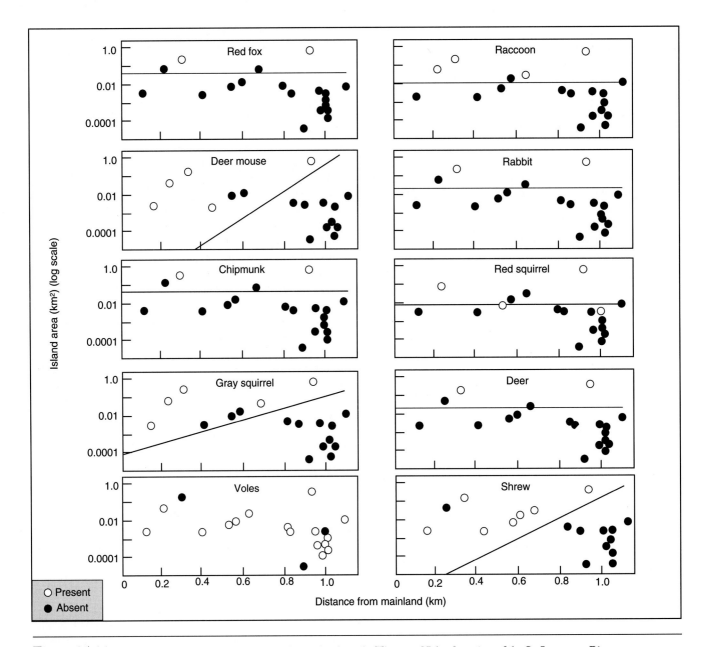

Figure 25.14   Occurrence of 10 species of mammals on islands in the Thousand Islands region of the St. Lawrence River in New York as a function of island size and isolation.   Many species occur only on islands above a certain critical size (*horizontal lines*). Other species, such as the deer mouse, are affected by both island size and distance from the mainland.

Source: Data from M.V. Lomolino, "Mammalian Community Structure in Islands" in Biological Journal of the Linnean Society, *28:1-21, 1986.*

Species Diversity

Because of the ever-increasing encroachment of humans into natural communities throughout the world, there is much interest today in **biodiversity.** Biodiversity has two main components: **species richness,** which is simply the number of species in an area, and **evenness,** the relative abundance of individuals within each species. Thus, we would say that a community made up of 10 species of mammals is more diverse than one with only 5. At the same time, comparing two communities with five species each and a total of 100 individuals, the one with abundances of 20, 20, 20,

20, and 20 would be considered more diverse than one with 92, 2, 2, 2, and 2. Two diversity indices often used by mammalian ecologists are the Shannon-Wiener index and Simpson's index. Both use richness and evenness to compare communities (Krebs 1989). In most communities, a few species are abundant and a large number are relatively rare. Large-scale patterns of species richness are often apparent; the diversity of mammals in the United States increases from east to west and from north to south. These patterns are discussed in more detail in the next chapter.

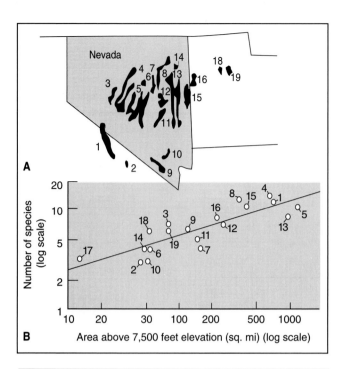

Figure 25.15   Island biogeography applied to mountaintops.
(A) Map of the Great Basin region of the western United States showing
the isolated mountain ranges between the Rocky Mountains on the east
and the Sierra Nevada on the west. (B) Species-area relationship for the
boreal mammal species. Numbers refer to sample areas on the map.

*Source: Data from J.H. Brown, "The Theory of Insular Biogeography and the
Distribution of Boreal Birds and Mammals" in* Great Basin Naturalist Memoirs,
2:209-227, 1978.

Landscape Ecology

A relatively new branch of ecology, called **landscape ecology,** is concerned with understanding population and community relationships within a large geographic region. Mammalogists have understood the influence of landscapes on animal distribution and abundance since the pioneering work of Merriam (1894), but only recently have methods been developed to quantify the effects of spatial scale on community structure. An important application of this approach is in the design of nature reserves, discussed more fully in chapter 29. Problems such as movements of individuals through habitat patches differing in size and shape, responses to habitat fragmentation, design of connecting corridors, and patterns of dispersal across habitat patches are modeled. This discipline explores the behavior of individuals, populations, and communities as they respond to different features of the environment at different levels of spatial scale (Forman and Godron 1986; Merriam and Lanoue 1990). New techniques include computer simulations based on fractal geometry, satellite imagery, and comuterized geographic information systems (GIS).

Landscape ecology has great potential for helping us predict the effects of habitat fragmentation, one of the major causes of species extinctions. Classical island biogeography

theory focuses on species richness and patterns of colonization and extinction as a function of habitat patch size and isolation, and it assumes that the surrounding matrix is homogeneous (e.g., water). For these and other reasons, it has met with mixed results when applied to terrestrial systems, as noted previously. Landscape ecology, on the other hand, incorporates information about how landscape patterns of many types influence reproduction and dispersal of local populations. Such studies can be done on natural landscapes or on experimental plots where the arrangement of habitat elements can be manipulated (Diffendorfer et al. 1995).

At the population level, it is important to understand how mammals living in isolated habitat fragments move through less suitable areas to get to more suitable ones (Merriam 1995). Studying how animals move about their home ranges and territories (chapters 20 and 23) is not sufficient; we must also study the spatial structure of the landscape mosaic and how dispersing animals move through it. Local interbreeding populations, called **demes,** can be connected to other demes via dispersal, forming **metapopulations** (Hanski 1996). It therefore becomes necessary to define the demographic unit, or population, under study. We know that immigration and emigration are important population forces (chapter 24), but we often have little idea of what the boundaries of a population actually are. For example, in a long-term study of white-footed mice in a small, isolated 2-hectare woodlot in Ohio, it had been assumed that the population was essentially defined by the boundaries of the woodlot. Live-trapping mammals in the surrounding agricultural fields, however, revealed that mice were making extensive use of cropland at certain times of the year (Cummings and Vessey 1994). In eastern Ontario, near the northern edge of its range, this forest-dwelling species has adapted to living in cornfields year-round, where densities were similar to those in wooded areas (Wegner and Merriam 1990). Densities were quite low compared with other studies, and the spatial scale over which the mice moved was unusually large.

Landscape ecology can also be viewed at the community level and above, where patches of the same community types are studied in a spatially explicit way (Lidicker 1995). Properties of landscapes that can be studied at this level include spatial configuration (the dispersion of patches), edge-to-area ratios (sizes and shapes of patches), and connectedness (links among patches of the same community type).

Macroecology

At a still larger scale is the study of **macroecology,** which explores patterns of body mass, population density, and geographic range at the scale of whole continents (Brown and Maurer 1989; Brown 1995). It involves the nonexperimental (i.e., nonmanipulative) investigation of relationships among populations, including patterns of abundance, body size, metabolic rates, geographic distribution, and diversity. Such large-scale studies require massive amounts of data on distributions, densities, and body sizes of representative species in each community before meaningful statistical patterns can be seen. Most of these data have already been collected for mammals, however, at least in temperate regions. The goal of macroecology is to identify and explain emergent statistical patterns in terms of general processes that can then be applied to unstudied areas. One statistical pattern already discussed is the species-area curve (see equation 1), which shows how the number of species increases with the area sampled (e.g., see figure 25.15). The species-area relationship can be studied at all spatial scales, from the smallest microhabitats to entire continents, and it has provided much insight into how communities are assembled.

Other patterns involve attributes of individual organisms. When the number of species of terrestrial mammals in North America is plotted against average body mass, one finds that there are many more small species than large ones, and a strong peak in mass occurs between 50 and 100 g (Brown and Maurer 1989; figure 25.16A). When smaller spatial scales are used, however, comparing deciduous forest, desert, or still smaller patches of relatively uniform habitat, the distributions become flatter, giving approximately equal numbers of species in each size category (figure 25.16B–E). Why are there so many small species, and why does the pattern shift as the scale changes? One possible explanation for this shift is interspecific competition, which reduces the number of similarly sized species in the same habitat. Large species could be relatively rare at the continental scale due to energy constraints. Smaller populations are more likely to go extinct. Finally, Brown and Maurer (1989) hypothesize that a greater number of small species exist at the continental scale because they have smaller geographic ranges and replace one another more frequently across the landscape than

do large species. These small species are specialized because they need high-quality food to maintain their relatively high metabolic rates.

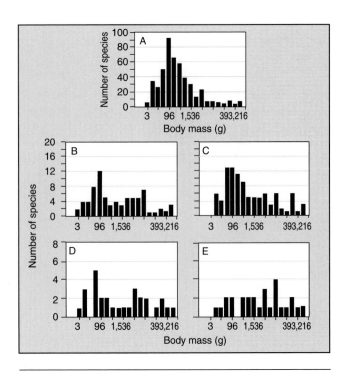

Figure 25.16   **Frequency distributions of body masses (log scale) among species of North American land mammals.**
(A) Distribution for the entire continent, for land mammals within biomes, (B) northern deciduous forest, and (C) desert; and for land mammals within small patches of relatively uniform habitat within each of these biomes—(D) Powdermill Reserve and (E) Rio Grande Bosque.

Source: Data from J.H. Brown and B.A. Maurer, "Macroecology: The Division of Food and Space Among Species on Continents" in Science, *243:1145-1150, 1990.*

Summary

Biological communities consist of interacting populations of organisms in a prescribed area. Communities can be studied at the simplest level by looking at interactions between two species and then expanding the analysis to include more complex interactions and large-scale patterns of distribution and abundance.

Central to the study of any species is the idea of ecological niche, an organism's habitat and functional role in the community. Interspecific competition affects both the distribution and abundance of species. The competitive exclusion principle states that complete competitors cannot coexist indefinitely.

Evidence that interspecific competition is an important force in structuring mammalian communities has been difficult to obtain. Distributions of species with and without presumed competitors sometimes show competitive release.

The divergence of traits sometimes seen in areas where close competitors coexist, called character displacement, is taken as evidence of competition. The best evidence of competition comes from removal experiments, in which species are experimentally removed and the response of the presumed competitor is monitored. Such experiments have produced mixed results.

Predators also may be important forces in structuring communities. Keystone predators are those that control the distribution and abundance of many other members of the community, often by limiting a particular species of prey. The existence of keystone predators is often discovered by removal experiments. Guilds are groups of species exploiting a common resource base in a similar fashion. Guilds may also act as keystone species complexes. Risk of predation may also act to structure communities; prey species may modify their foraging behavior and habitat choice to reduce predation risk.

Mutualistic relationships, in which both species benefit, are common but have not been well studied in mammals. Interactions among grazing ungulates and forage plants may be mutualistic in some systems, because some species of ungulates make more edible forage more available for others, and plant production increases.

Several attempts have been made to devise rules for assembling communities and to test the general hypothesis that communities are not simply random collections from the pool of available species. In some cases, such as small mammals in Australia, shrews in North America, and rodents in the deserts of North America, the composition is nonrandom, suggesting that competitive interactions have resulted in the inclusion of only those species sufficiently different from those already present to allow coexistence. Studies of other groups, however, especially Neotropical bats, fail to show anything but a random assortment of species based on size and shape.

Several studies have documented the flow of energy through the mammalian component of ecosystems. Herbivorous and granivorous mammals consume relatively little of the energy available to them and thus provide little to higher trophic levels. Some species of mammals play important roles by controlling other components of the ecosystem.

Mammals show changes in both species composition and abundance in response to ecological succession. In old-field succession, the number of species of small mammals tends to increase, possibly in response to higher quality food. Grazing ungulates influence succession by their effects on forage plants.

Studies of islands, varying in size and isolation from mainland habitats, have shown that communities consist of species in dynamic equilibria that result as new species colonize and resident species go extinct. Although some studies of mammals support the idea, islands that are highly isolated have fewer species than predicted, perhaps because mammals are relatively poor dispersers.

Landscape ecology is the study of the distribution of individuals, populations, and communities across different levels of spatial scale. This approach holds much promise for a better understanding of how components of communities respond to fragmented habitats and for the design of nature reserves. On a still larger scale, macroecology explores patterns of body mass, population density, and geographic range at a continental scale.

Discussion Questions

1. Character displacement is usually considered to be a result of competition among closely related species in sympatry. For what other reasons might species show character displacement?
2. Which species in figure 25.1B is most likely to be eliminated, and why?
3. Suppose that you conduct a removal experiment to test for competition between two sympatric species of mammals. What dependent measures would you use to demonstrate that competition is or is not occurring at your site? If you find no effects of removal of one species on the other, what conclusions can you draw?
4. Which of the following paragraphs best describes mammalian communities? Muster as much support as possible for your position. You may wish to consult an ecology textbook, such as Krebs (1994), Ricklefs (1990), or Begon and colleagues (1996).

   a. The distribution of individual species in space and time suggests that each responds to its own unique set of requirements independently of the effects on other species. Community boundaries are best understood as being arbitrarily defined units, more for the convenience of the investigator, than as highly integrated levels of organization.
   b. Communities include closely integrated species with complementary functional roles. Predictable patterns of species distribution and guild composition reflect the close coevolution of species in response to interspecific competition.

Suggested Readings

Brown, J. H. 1995. Macroecology. Univ. of Chicago Press, Chicago.

Lidicker, W. Z., Jr. 1995. Landscape approaches in mammalian ecology and conservation. Univ. of Minnesota Press, Minneapolis.

Morris, D. W., Z. Abramsky, B. J. Fox, and M. R. Willig (eds.) 1989. Patterns in the structure of mammalian communities. Texas Tech Univ. Press, Lubbock, TX.

Ricklefs, R. E. 1990. Ecology, 3d ed. W. H. Freeman, New York.

CHAPTER

26

Zoogeography

WHAT IS ZOOGEOGRAPHY?

FAUNAL REGIONS

Palearctic
Nearctic
Neotropical
Ethiopian
Oriental
Australian
Oceanic

HISTORICAL ZOOGEOGRAPHY OF MAMMALS

Plate Tectonics and Continental Drift
Glacial Periods
Refugia
Dispersal of Species and Centers of Origin
Extinction

ECOLOGICAL ZOOGEOGRAPHY OF MAMMALS

Present Mammalian Faunas
Some Principles of Ecological Zoogeography

WHAT IS ZOOGEOGRAPHY?

How do we explain the presence of marsupials in both Australia and South America? Why are there members of the Camelidae in North Africa and South America? What factors have lead to the present-day distribution of primates, extending from Japan to Africa and including South but not North America? **Biogeography** is the study of the patterns of distribution of organisms, including both extant and extinct species. **Zoogeography** is the study of these distributions in animals, including mammals.

Two main types of questions are used to investigate the distributions of animals. Examples of one type include: What was the ancestral distribution of lions (*Panthera leo*)? Where did the species originate? Where are its close living relatives? How has the distribution of lions been influenced by geological events such as continental drift? How have climate changes in the past, such as the periods of glaciation, influenced the distribution of lions? How is it that from living and fossil material we can determine that lions were or are living in a number of disjunct locations in the world? Together, these sorts of queries are generally termed **historical zoogeography.** Historical zoogeography of mammals attempts to explain, by reconstruction, the sequences of events involved in the origin, dispersal, and extinction of species. A subarea of historical zoogeography is **vicariance zoogeography.** Vicariance refers to situations in which a once widespread species becomes restricted or split into isolated geographic locations through the disappearance of the intervening populations or establishment of geographic barriers. For example, lions were once widespread throughout Africa and Asia; today, they are restricted to several disjunct populations scattered through Africa and a small population in northwest India (figure 26.1). We call their present zoogeographic pattern a **discontinuous distribution.**

A second type of investigation, **ecological zoogeography,** involves studying relationships between organisms and their environments. We discussed some aspects of ecological factors influencing mammal distributions in the previous chapter. Ungulates, for example deer and antelope, are among the most widely distributed mammals. What features of the environments of each species restrict them to their present ranges? How are they distributed with respect to climate, environmental features, and other living organisms, such as predators? Why are there more species of ungulates at tropical than at extreme northern and southern latitudes? What causes species replacement, whereby, as we move from a high mountain top to a valley or shoreline, we find a series of different ungulate species, each adapted for the zone in which it lives? Ultimately, an interaction of historical and ecological factors results in the observed distributions of animals and plants.

In this chapter, our purpose is to examine the distributions of mammals. Applying the principles of historical and ecological zoogeography, we discuss current and past distrib-

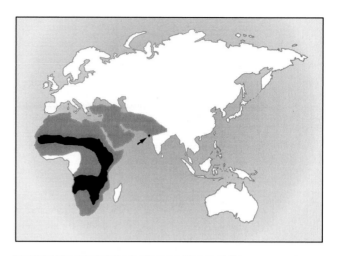

Figure 26.1   **Changes in distribution.**   The lion (*Panthera leo*) was once distributed throughout much of Africa and southwestern Asia, including the Arabian peninsula (*shaded*). Today, lions still inhabit many areas of Africa (*black area*), but their range has been dramatically reduced in Asia, now consisting of a small remnant population in the Gir Forest (*black dot with arrow*).

Source: Data from J. A. Burton and B. Pearson, Rare Mammals of the World, *1987, Collins (London).*

utions. Although we concentrate on mammals and the reasons for their distributions, mammals do not live in a vacuum. Much of the evidence concerning the distributions of mammals depends on knowledge of similar phenomena in other species. For example, in many locations, much of what we know about climate stems from in-depth studies of pollen, a discipline known as **palynology.** Pollen becomes preserved in bogs and other moist places. Because plants that are characteristic of a region at a given time reflect the existing climate, we can use our knowledge of plants' requirements to describe past climate conditions. That is, we can study ancient pollen grains to determine what plants were present then, and in turn, we can infer the prevailing climate at a particular location in the distant past. The climate and plants present at a location help to determine the kinds of mammals and other animals that might have lived there.

To understand the mammalian fauna of a region, we depend a great deal on fossil evidence. Unfortunately, this evidence is often incomplete, and for many regions of the world, it is almost nonexistent. Additional fossil evidence is necessary before some mammalian distribution patterns can be placed in historical perspective.

FAUNAL REGIONS

The land surface of the earth is generally divided into six faunal regions (figure 26.2), each with its own characteristic mammalian fauna. **Biomes** are broad ecosystems characterized by particular plant life, soil type, and climatic conditions. The major terrestrial biomes include tropical rain forest,

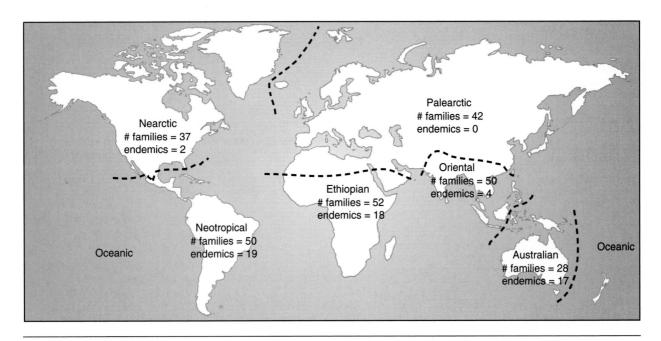

Figure 26.2   **Faunal regions.**   The land surfaces of the world are divided into six major faunal regions, based on major barriers separating these regions and past geological history. In addition, the islands of the oceans are sometimes designated as a seventh region, Oceania. Considerable differences occur in familial diversity and the number of endemic families in these different regions.

tropical deciduous forest, thorn forest, tropical savanna, desert, sclerophyllous woodland, subtropical evergreen forest, temperate deciduous forest, boreal forest, temperate rain forest, temperate grassland, and tundra (Brown and Gibson 1983). There are also marine and freshwater aquatic biomes. Various biome types occur within each of the major faunal zones. Characterizing mammalian fauna therefore involves both the major geographical regions of the world and the biomes represented within each region (Lydekker 1896; Darlington 1957; Brown and Gibson 1983; Vaughan 1986).

Palearctic

The Palearctic faunal region consists of the northern Old World, including Europe, Russia, central Asia, and China. This is the largest of the world's faunal regions and is separated from the Ethiopian region by deserts, from the Oriental region by the Himalayan Mountains, and from the Nearctic region by the Bering Straits. The total familial diversity for mammals in the Palearctic region is moderate relative to that in other faunal regions (see figure 26.2). There are no endemic families in the Palearctic region. The region shares about 50% of its families with the Nearctic region, primarily due to the Bering land bridge, which was open for dispersal for extended periods of time.

The mammalian fauna of this region represents a mix of Nearctic, Ethiopian, and Oriental faunas. The Families Cervidae, Bovidae, and Ursidae are shared with all of these other regions. The Families Suidae, Hyaenidae, and Viverridae are shared with the Oriental and Ethiopian regions. The Families Gliridae, Dipodidae, and Procaviidae occur in the

Ethiopian as well as the Palearctic regions, and Dipodidae is also present in the Nearctic region. The rodent Families Muridae and Sciuridae, and the carnivore Families Mustelidae, Canidae, and Felidae are relatively widespread and distributed continuously throughout the Palearctic region.

Nearctic

This region extends from the Arctic in northern Canada through the United States south to the central Mexican plateau and includes Greenland. It is separated from the Palearctic region by the Bering Sea and has been connected relatively recently to the Neotropical region via the Panamanian land bridge. Fewer families are present in the Nearctic relative to all other regions except the Australian region (see figure 26.2). This may be due, in part, to the relatively long isolation of this region, with fewer time periods for interchange of mammalian fauna, and to the cooler, temperate climate, which sometimes makes it more difficult for mammals to establish themselves. Approximately 50% of the fauna in the Nearctic region is shared with the Palearctic region and nearly the same proportion with the Neotropical region. More mammalian diversity occurs in the western and southern portions of this faunal region.

Three families—the Talpidae, Ochotonidae, and Castoridae—are shared only with the Palearctic region. Five families are shared only with the Neotropical region, including Didelphidae, Myrmecophagidae, Erethizontidae, Tayassuidae, and Dasypodidae. The two endemic families in the Nearctic region are the Aplodontidae and Antilocapridae, both of which occur in the western portion of the faunal region.

Neotropical

The Neotropical region extends from central Mexico southward to the southern tip of South America and generally includes the islands of the Caribbean. The only connection in recent times with the Nearctic region has been via the Panamanian land bridge. We'll examine the effects of this land bridge on faunas of the two regions shortly. The Neotropical region has the greatest number of endemics of any region (see figure 26.2). This situation can be attributed, at least in part, to the isolation of the Neotropical region and to its generally warm, favorable climate. Within the region, mammalian diversity increases from south to north.

Eleven of the families of mammals endemic to the Neotropical region are rodents: Abrocomidae, Agoutidae, Capromyidae, Caviidae, Chinchillidae, Dasyproctidae, Dinomyidae, Echimyidae, Hydrochaeridae, Myocastoridae, and Octodontidae. The other endemic families are three xenarthrans (Megalonychidae, Bradypodidae, Myrmecophagidae), two marsupials (Caenolestidae, Microbiotheriidae), two primates (Callitrichidae, Cebidae), and one insectivore (Solenodontidae).

Ethiopian

This region includes all of Africa, except that portion north of the Sahara Desert, as well as Madagascar. The region is connected to the Palearctic region via the Sahara. Much of the area is tropical. This region has the greatest familial diversity of mammals of any faunal area (see figure 26.2). The Ethiopian region shares just over 70% of its fauna with the Oriental region and nearly the same percentage with the Palearctic region.

Many of the endemic families are either rodents (Anomaluridae, Bathyergidae, Ctenodactylidae, Pedetidae, Petromuridae, Thryonomyidae) or primates (Cheirogaleidae, Daubentoniidae, Galagonidae, Indridae, Lemuridae, Megaladapidae). Five of the six primate groups are found only on Madagascar. In addition, two endemic families are insectivores (Chrysochloridae, Tenrecidae), two are artiodactyls (Giraffidae, Hippopotamidae), and the remaining two are elephant shrews (Macroscelididae) and the aardvark (Orycteropodidae).

Oriental

The Oriental region consists of the Indian subcontinent, the Malay Peninsula and Archipelago, the Philippine Islands, Sumatra, Java, and Borneo, and extends southward to Wallace's line. Wallace's line runs between Bali and Lombok Islands in the Malay Archipelago and divides the Oriental and Australian faunal regions. The Oriental region is isolated from the Palearctic region by deserts and mountains and from Australia by the sea. Much of this region is tropical. Nearly 75% of the species in this region are shared with the neighboring Palearctic region, whereas more than 60% of the mammalian fauna are shared with the Ethiopian region, and more than 50% are shared with the Nearctic region. A large number of different mammalian families is represented

in the Oriental fauna, but only four are endemic. A key reason for the diversity of mammalian families in this region is its central location with respect to other faunal regions. The warm climate also meant that mammals dispersing into this region could survive. The paucity of endemic families is probably related to the lack of isolation. The endemic families include Tupaiidae (tree shrews), Cynocephalidae (colugos), Tarsiidae (tarsiers), and Hylobatidae (gibbons).

Australian

This region includes the Australian continent plus New Guinea, Tasmania, Sulawesi, and those islands in the Malay Archipelago that are south of Wallace's line. This is the most isolated of any of the major faunal areas, sharing less than 20% of its mammalian fauna with any of the other regions. The only recent route of exchange involves crossing water. Two major groups of mammals have succeeded in crossing the barriers to reach the Australian region: bats and murid rodents. The 17 endemic families here are all monotremes or marsupials (see chapter 10 for detailed coverage of these families). The relative isolation of the Australian region provided an opportunity for adaptive radiation, particularly of the marsupials. Many forms of marsupials therefore occupy niches in Australia that are occupied by eutherian mammals in other faunal regions.

Oceanic

Though not historically considered a faunal region, the Oceanic division, recognized by some zoogeographers, consists of the many islands in the Pacific Ocean, including Polynesia, Micronesia, and those portions of Melanesia that are not part of the Australian region. The Oceanic region also sometimes includes the isolated islands of the Indian and Atlantic oceans. The mammalian fauna of these islands are greatly restricted by the inability of terrestrial animals to cross water. Marine mammals are not considered in our scheme here, and of course, they occupy a variety of locations in the oceans and seas of the world. A few species of bats that can fly between locales, and some rodents that probably rafted to the islands before human dispersal of other rodents occurred are the only mammals found on most of these islands. We have already considered islands as special cases with respect to mammalian distributions in chapter 25, and we'll examine this further later in this chapter and from a conservation viewpoint in chapter 29.

HISTORICAL ZOOGEOGRAPHY OF MAMMALS

Plate Tectonics and Continental Drift

The suggestion that land masses move over geological time first appeared in the scientific literature in the mid-1800s. The idea was formalized by Wegener (1912, 1915) in the early part of the twentieth century, but acceptance of this

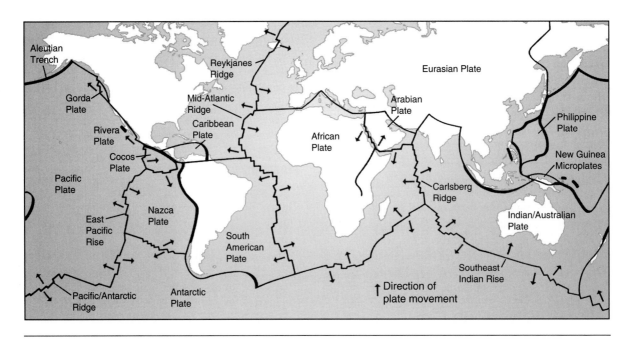

Figure 26.3   Tectonic plates.   Map of the surface of the world showing the locations and movement directions of the major tectonic plates. Arrows indicate the direction in which a plate is spreading. Oceanic trenches are indicated by thickened lines.

theory did not occur until the 1960s. **Plate tectonics** is the theory that the earth's crust, including the surfaces of continents and ocean floors, is made up of a series of "plates" (figure 26.3). At the ridges where plates join, volcanoes may form as magma pushes upward from the earth's mantle below. As plates collide, one moves under the other (subduction), causing earthquakes. These events may form oceanic islands, such as the Hawaiian Islands, and uplift mountains on the continents, such as the Andes in South America.

Continental drift is the movement, over geological time, of the large land masses of the earth's surface as a result of plate tectonics. Evidence for continental drift comes from several sources. The patterns of spreading observed along ridges in the floors of the oceans support the notion of continental drift (see figure 26.3). As rocks carrying iron-containing minerals cooled at the earth's surface, the particles became aligned along the lines of the prevailing magnetic field of the earth. Studying the orientation of such particles (paleogeomagnetism) today can reveal how far, and in what direction, the rocks were from the pole at the time they were formed. Compiling this sort of information also supports the theory that the continents were once a single land mass that has since broken up to form the current continents. Rock strata on different continents exhibit a general correspondence. Finally, in some instances, the outlines of the continents seem to fit together, such as the coastlines of West Africa and Brazil.

At one time (ca. 200 mya), a single large land mass existed—Pangea (figure 26.4; see also figures in chapter 4). This separated into two large land masses: Gondwanaland in the Southern Hemisphere and Laurasia in the Northern Hemisphere. Gondwanaland eventually formed the present-

day continents of South America, Australia, Africa, Antarctica, and part of Asia. Laurasia contained what today are North America, Europe, Greenland, and part of Asia. The processes of plate tectonics and continental drift continue today. Because of the relative newness of these geological theories, the discipline of historical zoogeography, including that of mammals, is in its infancy. Early mammalian evolution was occurring at a time when the continents were drifting apart; this accounts in some degree for why mammals are more diverse than reptiles (Kurten 1969; see chapter 4).

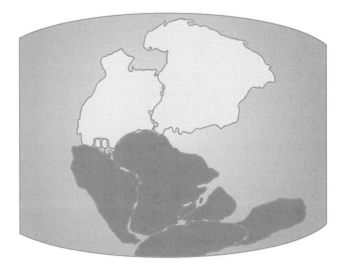

Figure 26.4   Continental movements.   A reconstruction of Pangea before it began to split apart. The northern regions that became Laurasia are shown in white. The southern regions that became Gondwanaland are shown shaded. Also see figure 4.11.

How does continental drift affect mammalian distributions? One of the questions we posed earlier concerned the presence of marsupials in South America and Australia. The evidence, consisting of both living forms and fossil material, suggests that these two land masses were both once part of Gondwanaland. Marsupials apparently originated in what is now North America and dispersed eastward and southward from there. Some ancestral marsupials were distributed throughout portions of the larger land mass and survived the isolation of the two continents in the Tertiary period; South America was then still linked with Australia. These marsupials then underwent radiations into a variety of forms in each location. As noted in chapter 10, the presence of the Virginia opossum (*Didelphis virginiana;* figure 26.5) in North America occurred later with formation of the Panamanian land bridge, 2 to 5 mya. This represented a reappearance for this family in North America; didelphids had been widespread during the Cretaceous period but died out 15 to 20 mya.

Other examples further demonstrate the roles of continental drift and changes in sea level in affecting the distribution patterns of mammals now and in the recent past. Hystricomorph rodents, another group with representation in Africa and South America, descended from a lineage that once inhabited Gondwanaland. The presence of Bactrian camels (*Camelus bactrianus*) and dromedaries (*C. dromedarius*) in North Africa and llama (*Lama glama*), alpaca (*L. pacos*), guanaco (*L. guanicoe*), and vicuña (*Vicugna vicugna*) in South America can be explained by the presence of a land bridge between North and South America some 2 to 5 mya through the movement of land masses. Camelids originally arose in North America and moved to Eurasia via the Bering land bridge. The Camelidae became extinct in North America by the late Pleistocene epoch.

Figure 26.5   Common opossum.   The only opossum found in North America, *Didelphis virginiana,* crossed the Panamanian isthmus. Fossil evidence indicates that other opossums existed in North America at an earlier time, but these became extinct.

Glacial Periods

The Pleistocene glaciations that affected large areas of the Northern Hemisphere played a significant role in the present distributions of some mammal species. The Pleistocene epoch has been characterized by pronounced fluctuations in temperature. Lower temperatures often result in the formation of glaciers, some of which cover enormous areas of land surface as continental ice sheets. To date, there have been four glacial periods—the first commencing about 600,000 years ago and the last ending some 12,000 years ago. Each episode of glaciation resulted in the displacement of the biota, including mammals, over large areas. During warmer periods, episodes of recolonization and resettlement occurred. In addition, sea levels have fluctuated, rising during warmer periods and decreasing during periods of glaciation. The lands exposed when sea levels are low, such as the Bering Sea bridge between North America and Asia, have also influenced mammalian distribution patterns.

What does knowledge of advancing and receding glaciers tell us about mammalian distributions? During periods when the glacial sheets were farthest south, the regions just beyond the ice sheets were often tundralike and inhabitable by mammals, with some areas of sand dunes. These regions extended into the south-central United States, including parts of Ohio, Indiana, and Illinois, and west through Missouri to the Great Plains. The Nebraska Sandhills are remnants of dry, sometimes sandy conditions produced by winds that resulted in sand accumulating in glacier-free areas (Darlington 1957; Pielou 1991). The most conspicuous large mammals in the tundra region were the mastodon (*Mastodon americanus*) and woolly mammoth (*Mammuthus primigenius*). Other genera and species of these proboscideans inhabited similar niches in Eurasia. In addition, musk ox (*Ovibos moschatus*) and caribou (*Rangifer tarandus*) lived in regions much farther south than their present ranges. These latter mammals survive today, though their present ranges are in the northern latitudes, where the tundra and conifer forests abound (Pielou 1991). Populations of musk ox and caribou were successful in shifting their ranges as the climate changed; they apparently followed the cool conditions and associated vegetation northward during the past 10,000 to 12,000 years. Other species, such as woodland musk ox (*Ovibos bombifrons*), shrub-ox (*O. cavifrons;* figure 26.6), and stag moose (*Cervalces* sp.), were not as adaptable and became extinct. Fossil evidence indicates that small mammals, for example the arctic shrew (*Sorex arcticus*) and the collared lemming (*Dicrostonyx torquatus*), also once lived in areas that were south of the glacial sheet. Their distributions have since shifted northward with climatic changes (Brown and Gibson 1983).

Refugia

Refugia (sing., refugium) are delimited geographical regions, often within a larger biome type, that help to preserve species diversity. They can be thought of as biogeographic arks (Lynch 1988). Several forms of refugia are recognized.

Figure 26.6   **Shrub-ox.**   The shrub-ox was one of a variety of herbivores that lived in the regions just south of the ice sheets during periods of glaciation. As the ice sheets receded, some of these species, such as musk ox, moved northward and survived, while others, like the shrub-ox, became extinct. Note the large horns, possibly for engaging in encounters with conspecifics, in the manner of today's bighorn sheep (*Ovis canadensis*).

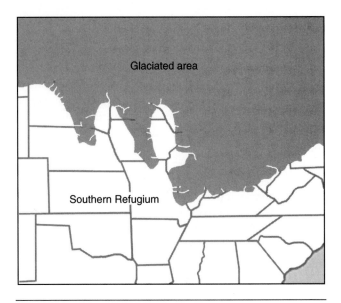

Figure 26.7   **Driftless region.**   An area in northwestern Illinois, eastcentral Iowa, and southwestern Wisconsin, known as the driftless plain, remained free of the ice sheets during recent glaciations. This region served as a refugium. Several plant and animal species, including a number of mammals, spread outward from this refugium as the glaciers receded.

Nunataks, refugia found within the ice sheets during periods of glaciation, were pockets of variable size that were not covered by the advancing glaciers. They therefore served as islands where fauna and flora could survive during a period of glaciation. Mountaintops that remained isolated when ice sheets were advancing also were potential refugia. Refugia served as a source of new populations to move into surrounding areas once the glaciers receded. The driftless area in southwestern Wisconsin, northwestern Illinois, and eastern Iowa was a large nunatak (figure 26.7). Knowledge of it aids us with the interpretation of present-day distributions of some animals and plants. For example, the least weasel (*Mustela nivalis*) and Franklin's ground squirrel (*Spermophilus franklinii*) likely survived in the driftless region during the most recent glaciation and then spread outward after the glacier receded.

Portions of the tropical rain forests around the world contain enormous species diversity. Historically, as glaciers covered much of the landscape at northern latitudes, the rain forests became very fragmented—islands of forest in large areas of open grasslands. These changes were due to climatic variation, including cooler temperatures and less rainfall in what are tropical regions today. Then, as glaciers receded, the pockets expanded to again cover large areas with continuous rain forest (Haffer 1969). Although Haffer's ideas concerning pockets of rain forest as refugia have not been universally accepted, they have been thought-provoking (Lynch 1988). According to this hypothesis, considerable differentiation occurred within the pockets of rain forest while they were isolated. When the forest again

became dominant and widespread, sufficient changes had occurred so that what had been single species before isolation were now no longer able to interbreed. Speciation events had occurred. An example involves the explosive radiation among sigmodontine rodents in South America (Hershkovitz 1962; Eisenberg 1981). In more recent times, some areas of tropical rain forest can be viewed as refugia due to humans removing large portions of the forest. The consequences of this deforestation for mammalian species diversity are examined in more detail in chapter 29.

Islands are sometimes considered refugia, particularly if they are located close to a larger land mass or other islands, where some form of interchange is possible. Madagascar is such an island refuge (Darlington 1957; Eisenberg 1981). Two groups of mammals—tenrecid insectivores and lemurid primates—have undergone extensive radiation on Madagascar, where they now occupy a wide variety of niches. Fossil evidence indicates that the ancestors of these groups lived on the African continent. In this sense, Madagascar has served as a refugium where, protected from potential mainland competitors, particular groups of mammals diversified.

Dispersal of Species and Centers of Origin

The term **dispersal** has several closely related meanings with respect to animals. Elsewhere (see chapter 23), we presented the idea of individuals or small groups of mammals leaving their natal location to disperse. Such dispersal movements occur within the lifetime of the individual and are sometimes referred to as **ecological dispersal.** We also use dispersal to refer to long-term movement patterns involving species in a historical zoogeographic sense. Thus, the expansion or shift of a species' range is referred to as a dispersal event. The two uses of the term *dispersal* are connected in that dispersal by individuals is the basis for an expansion or shift in a species' range. With respect to species dispersal, we usually recognize two major classes. **Passive dispersal** involves movements in which the dispersing organisms have no active role in the movement. Rafting and being carried by humans are two examples. **Active dispersal** involves an accumulation of ecological dispersal events in which individuals move by terrestrial locomotion or flight.

Active, long-term species dispersal movements, or **faunal interchanges** as Simpson (1940) referred to them, occur via several different routes. A **corridor route** provides minimal resistance to the passage of animals between two geographic locations. The present interconnection of Europe and Asia is a corridor that provides for almost complete interchange of mammals. Many mammalian taxa, down to the level of genera and even species, are distributed throughout Eurasia as a result of this corridor. This is one reason the Palearctic faunal region has no endemic families; species were able to move rather freely between this region and adjoining land masses.

A **filter route** allows only certain species to pass through. A good example is Beringia, the land bridge that existed at various times in the past between the Alaskan peninsula of North America and the Siberian portion of Asia. Only mammals that were adapted to the colder climate of the land bridge could successfully cross between continents. Several groups of rodents, for example, the voles (*Microtus* spp. and *Clethrionomys* spp.), likely used the Bering land bridge as a filter route to enter North America from Asia. Cervids and several carnivores crossed from Asia into North America via this route. Camelids crossed in the reverse direction, from North America into Eurasia. Another example of a filter is the Panamanian land bridge that formed between South and North America 2 to 5 mya. Some mammals have successfully crossed this filter in each direction, whereas others have been effectively blocked from expanding their ranges (figure 26.8). Other types of barriers that could serve as filters are geographic, for example the deserts or mountain ranges that separate faunal regions.

The most restrictive type of pathway for interchange is called a sweepstakes route. A **sweepstakes route** involves movement of animals by swimming, flying, rafting, or other means, so that the probability of their arriving by their method of transport at another destination, one that is inhabitable, is extremely limited, much like the odds of winning a lottery. One example of a sweepstakes route is the path between Australia and New Guinea. Several marsupial species have moved between Australia and New Guinea successfully. Examples are the cuscuses (*Phalanger*) and ring-tailed possums (*Pseudocheirus*). Because they can fly, bats have used the sweepstakes route for some of their dispersal movements. A number of bat species have made the journey from Asia to Australia. Only three species of bats, however, successfully traveled the additional sweepstakes route to New Zealand: the lobe-lipped bat (*Chalinolobus tuberculatus*) and the short-tailed bats (*Mystacina robusta* and *M. tuberculata*). *Mystacina robusta* has not been recorded since 1965 and is believed to have been driven to extinction by introduced rats invading its nesting areas (see chapter 29).

A **center of origin** for a genus or family is the location where a particular taxon arose (Darlington 1957; Müller 1973, 1974; Brown and Gibson 1983). Though somewhat hypothetical or problematic in a practical sense (Cain 1944), centers of origin are of interest to zoogeographers for two reasons. Because particular regions of the world have been centers of origin for more species than other locations, we

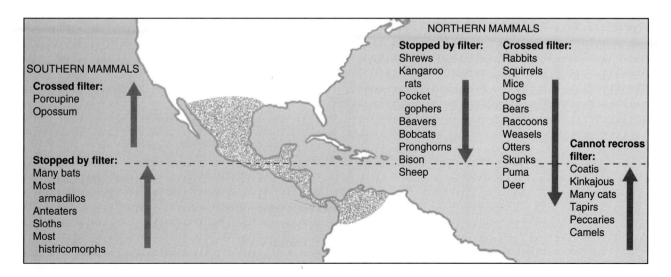

Figure 26.8   Filter routes.   The role of Central America and the Panamanian isthmus as a filter route. Some species have crossed through this filter, but others have been effectively blocked from increasing their geographic range.

Source: Data after Simpson, 1965, from T.A. Vaughan, Mammalogy, *3rd edition, 1986, W.B. Saunders Company, Philadelphia.*

can attempt to determine what conditions promote higher rates of evolutionary innovation. Second, centers of origin help us to determine species dispersal routes and how a particular assemblage of organisms arrived at its present composition. Three criteria are often used to define a center of origin (Davis and Golley 1963):

1. The earliest known fossil evidence for a group indicates the location of origin.

2. A fossil history of earlier progenitors of the group in question in a particular area is an indication of the place of origin.

3. The region where the highest diversity of a particular group occurs may indicate the place of origin.

How do these ideas apply to mammals? As noted in chapter 4, mammals evolved from reptiles over 200 mya. For the first 100 million years, mammals were a rare component of the faunal landscape. Eventual radiations involved more diversified forms of multituberculates, symmetrodonts, and triconodonts. During the Cretaceous period (ca. 135–100 mya), continental land masses began to separate, leading to the isolation of major mammalian stocks (see figure 4.12). The stage was then set for the major mammalian radiations that occurred from the Paleocene epoch to the middle of the Eocene epoch. These radiations involved a number of groups that underwent independent evolution on the various continental land masses (Eisenberg 1981).

We can use fossil history to discern patterns of origin for particular taxa and their possible spread to other locations. The routes between North America and Eurasia have been open throughout much of geological history, which has lead to a great similarity between the mammalian faunas of these two land masses. For example, shrews (Soricidae) and moles (Talpidae) originated in Europe in the Eocene epoch and spread to North America via a North Atlantic route in the Oligocene epoch. Cats (Felidae) originated in Asia and spread first to North America and eventually to Europe via the same North Atlantic connection. The Families Camelidae and Tapiridae appear to have originated in North America, spreading to Europe and eastward. The Panamanian land bridge formed relatively recently. New World porcupines (Erethizontidae) originated in South America, where they occur in a variety of forms. Just one of these species, *Erethizon dorsatum*, dispersed to North America via the Panamanian isthmus. Alternatively, it is possible that more than one form dispersed and this is the only survivor.

As an example of current work that combines ideas concerning refugia, species dispersal, and molecular techniques, consider the present distribution of the American pika (*Ochotona princeps;* Hafner and Sullivan 1995). Pikas live on rocky terrain in cool, mesic conditions. They are found in Asia and in isolated populations scattered throughout the mountains of western North America. Pikas originally entered North America from more northern source populations, originating with the Bering land bridge. Analy-

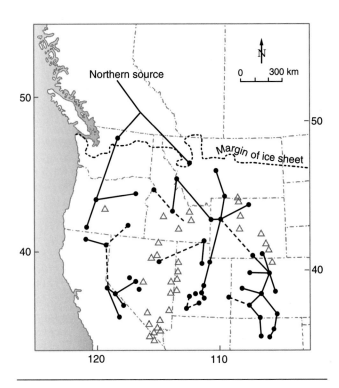

Figure 26.9   **Pika distribution.**   Pikas live in relatively isolated populations on mountains. Solid lines indicate the maximum elevation network connecting a series of sites (*dots*), suggesting how groups of populations can be related via dispersal. Connections are based on allozyme analyses from pikas in each population. Dashed lines indicate where dispersal corridors require an elevational displacement of more than 1000 m. Triangles indicate extralimital Wisconsin-age fossils of *O. princeps.*

*Source: Data from D.J. Hafner and R.M. Sullivan, "Historical and Ecological Biogeography of Nearctic Pikas (*Lagomorpha Ochotonidae*)" in* Journal of Mammalogy, *76:302-321, 1995.*

ses of allozyme patterns were used to examine possible degrees of genetic separation between the populations in different mountainous areas in western North America. During various glacial advances and retreats, pika populations became isolated in their montane refugia. Four major genetic groups of pikas diverged at some point prior to the Wisconsin glaciation, 120,000 years ago (figure 26.9). Genetic evidence suggests some degree of secondary contact between the isolated populations, but various geological and climatolological events (glaciers and lakes), as well as behavioral traits of the pikas (high philopatry), have resulted in the four distinct genetic units (see figure 26.9). Whether, given sufficient time, these will diverge to form separate species is an open question.

Extinction

Extinction, the loss of a species, is a natural process and the ultimate fate for all species. Extinction can be thought of as the evolutionary counterpoint of speciation, or the generation of additional diversity. Fossil history bears evidence that there are more extinct mammalian species than extant ones. It seems that a perpetual race is occurring. Each species is part of a complex community in which all of the organisms are evolving. Adaptations to changing environmental conditions or shifts involving interspecific competition for food or nest sites are examples of the ongoing evolutionary processes. Those species that can adapt over time remain in the race and may undergo further speciation (Van Valen 1973). Those that fail to adapt become extinct.

Several natural events may lead to extinction. Sometimes this process is gradual, and at other times, episodes of mass extinction occurred over relatively short periods (Marshall 1988). The factors involved in extinction include climatic shifts; changes in the ecology of a region; and cataclysmic events, such as an asteroid impact with the earth, volcanic eruptions, or major floods. Within the past several thousand years (but mostly in the past 100 years), humans have become a primary factor, causing the extinction of literally thousands of animal and plants species by destroying or contaminating habitats (Raup 1991). Mammals have not been spared in this process; we will discuss mammalian extinctions and endangered species in detail in the final chapter (see figure 29.4).

ECOLOGICAL ZOOGEOGRAPHY OF MAMMALS

This section covers two major topics. First, we explore a hypothesis for examining the present mammalian faunas of the different regions of the world. Second, we consider a series of specific topics relevant to the ecological zoogeography of mammals.

Present Mammalian Faunas

What sorts of mammalian faunas are characteristic of the different faunal regions (see figure 26.2)? We presented a great deal of information about the distributions of various groups of mammals in part III, where we examined each mammalian order and family, and a general distribution scheme for the faunal regions was presented earlier in this chapter. Let us examine a more dynamic approach to faunal regions that involves ecological zoogeography. The available terrestrial niches in various world regions were categorized by Eisenberg (1981) according to trophic position, or feeding ecology (16 types), and substrate used (8 types; table 26.1). Of 128 possible niches in a matrix cross-classifying these two sets of categories, mammals presently occupy about half. As Eisenberg notes, such classification schemes are, of necessity, based on modal tendencies for a particular species or genus with respect to diet and substrate preference. His matrix can be used to develop some useful insights concerning mammalian zoogeography.

Eisenberg (1981) divided the land masses into 11 different ecogeographic regions (table 26.2), a somewhat more refined classification than we used earlier. An examination of

Table 26.1.   Niche components.   Two major niche components of mammals are their feeding ecology and their substrate preference.*

Feeding Types	Substrate Types
1. Piscivore and squid eater	1. Fossorial
2. Carnivore	2. Semifossorial
3. Nectarivore	3. Aquatic
4. Gumivore	4. Semiaquatic
5. Crustacivore and clam-eater	5. Volant
6. Myrmecophage	6. Terrestrial
7. Aerial insectivore	7. Scansorial
8. Foliage-gleaning insectivore	8. Arboreal
9. Insectivore/omnivore	
10. Frugivore/omnivore	
11. Frugivore/granivore	
12. Frugivore/herbivore	
13. Herbivore/browser	
14. Herbivore/grazer	
15. Planktonivore	
16. Sanguivore	

Source: From J. F. Eisenberg, The Mammalian Radiations, *1981, University of Chicago Press, Chicago, IL.*
*Eisenberg (1981) provides the details concerning the constraints that shape the physiology and anatomy of mammals living in each of these feeding and substrate niches.

niche occupancy and nonoccupancy, according to feeding ecology and substrate type, reveals several interesting patterns. The long period during which tropical forest has been present in South America and Southeast Asia has resulted in more arboreal and volant species than occur in tropical Africa, with its shorter history of having a tropical climate. The longer presence of semiarid conditions in Africa, on the other hand, has resulted in a much greater occupancy of niches involving terrestrial herbivores that graze or browse for food. Proportionately many more fossorial and semifossorial mammals occur in Eurasia, North America, and the temperate regions of South America. Eisenberg (1981) postulates that this may be caused by problems with the capacity of mammals to dissipate heat, restricting their occupancy of these niches to higher elevations in tropical zones. The pattern of overall niche occupancy in Australia is intermediate among the faunal regions. The Australian pattern likely results from a combination of the arboreal evolutionary history of the marsupials combined with the xeric (dry) conditions that promote the presence of numerous grazing niches. Further study of this conceptual model should reveal a number of patterns and additional questions concerning mammalian distributions worthy of investigation.

Some Principles of Ecological Zoogeography

Endemism

No species of mammal is cosmopolitan, distributed in all habitats and regions of the world. Some mammalian species, including humans, are rather widespread, whereas others

Table 26.2 Niche occupancy. For each of 11 ecogeographic regions of the world, the table shows the number of genera of mammals according to the 16 feeding ecology and 8 substrate preference categories listed in table 26.1.

Region	Substrate Categories								Σ	Feeding Categories															
	1	2	3	4	5	6	7	8		1	2	3	4	5	6	7	8	9	10	11	12	13	14	15	16
Southeast Asia	3	3	5	4	42	44	24	34	159	9	10	2	—	—	1	22	2	22	27	31	14	13	4	—	—
Northern South America	2	7	4	14	69	66	19	38	219	12	9	9	1	—	8	27	9	14	48	41	19	14	6	—	3
Southern Africa	8	1	3	8	39	123	11	21	214	8	19	3	1	2	2	23	3	21	24	45	18	29	17	—	—
North America	10	19	11	5	16	51	14	2	128	10	8	2	—	1	—	11	3	16	13	24	15	16	8	—	—
North Asia	5	10	8	7	10	76	12	8	136	13	8	—	—	1	—	8	2	12	12	33	22	15	10	—	—
Australia	1	1	6	3	19	38	9	19	96	5	7	3	—	—	2	11	2	17	12	9	8	11	8	1	—
Europe	2	4	3	5	10	40	8	5	77	7	7	—	—	—	—	8	2	4	7	18	7	11	6	—	—
Middle East	1	5	—	2	17	50	6	7	88	2	9	—	—	—	—	13	3	8	9	21	10	8	5	—	—
India	1	—	3	2	22	45	9	8	90	5	6	—	—	—	1	15	1	8	15	16	6	8	8	—	1
Southern South America	2	6	7	6	5	35	8	1	70	6	4	—	1	—	1	5	—	9	11	7	14	6	6	1	1
Madagascar	—	1	—	1	15	14	7	13	51	1	3	—	—	—	—	12	—	14	6	4	5	4	—	—	—

Source: From J. F. Eisenberg, The Mammalian Radiations, *1981. Copyright © 1981 University of Chicago Press, Chicago, IL. Reprinted by permission.*
Note: Numbers refer to numbers of genera in each category. Figures are exclusive of the Cetacea.

have more limited distributions. Endemism (occurring nowhere else), or being endemic, refers to a taxon being restricted to a limited geographic area. This can happen through historical, physiological, or ecological constraints, or a combination of these. The area occupied by an endemic taxon can be small, such as an island of a few square kilometers, or an entire continent. Endemism can refer to a single species or some larger grouping, such as a genus, family, or order (see figure 26.2 and related discussion). For example, the Heteromyidae occupy a geographic region that encompasses the southwestern portion of North America and extends southward into northern South America (Brown and Gibson 1983). Within this overall distribution, each of the five genera occupy ranges of varying sizes. Kangaroo rats (*Dipodomys*) occur over much of the southwestern United States and southward to central Mexico. By contrast, kangaroo mice (*Microdipodops*) are found only in the Great Basin region of the American West. Because of their greater isolation, regions such as Australia, southern Africa, South America, and Madagascar have greater proportions of mammalian endemics than, for example, Europe and North America, which share considerable proportions of their fauna at the generic and family levels.

Mammals on Islands

The basic tenets of the theory of island biogeography were presented in chapter 25. We usually think of islands as land masses located in bodies of water. Isolated mountain ranges, small deserts, and other land features surrounded by differing habitat also can be thought of as "habitat islands" and can result in the same sorts of effects as occur on the more traditional islands (figure 26.10). Human fragmentation of habitats, occurring at an increasing rate around the world, is also creating numerous "habitat islands."

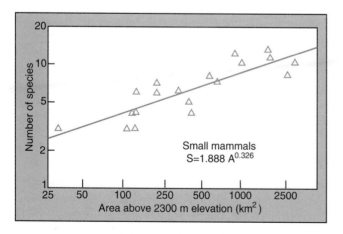

Figure 26.10   Island biogeography.   The graph depicts the species-area relationship between small terrestrial mammals inhabiting isolated mountain ranges (above 2300 m) in the Great Basin. The line is described by the equation: Number of species = 1.888 area$^{0.326}$.

Source: Data from J.H. Brown and A.C. Gibson, Biogeography, *1983, C.V. Mosby Company, St. Louis, MO.*

The ability of mammals to reach various islands through dispersal is a primary limiting factor. Few large mammals live on islands. Small mammals, rodents and bats in particular, have been the most successful colonizers of islands (Sondaar 1977). To see this best, we need to examine the distributions of mammals on islands prior to the spreading of mammals by humans. Rodents are the only mammals on the most remote of the Philippine Islands. Rodents and bats are the only terrestrial eutherian mammals to have reached Australia. Only rodents and bats managed to colonize the Galapagos Islands. Even in areas where some other mammals exist, such as the Greater Antilles in the Caribbean Sea, rodents and bats predominate. Much of the colonization of islands by rodents probably occurred via rafting, whereby the small mammals were able to survive transport on floating logs or other material for some days until a chance encounter occurred with some perhaps distant island.

Living on islands affects the characteristics of mammals. They are often subject to different selection pressures than are mammals inhabiting mainland areas. Interspecific competition is usually less on islands than on the mainland. Thus, niches of island-dwelling mammals may expand relative to their niche (or that of closely related species) on the mainland. Both morphology and behavior apparently can be affected in some mammals that live on islands. Body size is affected by island life in somewhat contradictory ways (Foster 1964). Some mammals on islands are smaller than their mainland counterparts. From fossil evidence, we know that the elephant, *Elaphus falconeri*, that once inhabited the island of Sicily was about a quarter the size of its mainland relatives (Sondaar 1977). Gray foxes (*Urocyon littoralis*) inhabiting the Channel Islands off the coast of California are significantly smaller in stature than the mainland form (*U. cinereoargenteus*). On the other hand, many rodents and marsupials living on islands are considerably larger than related mainland forms. Several factors probably interact in a complex manner to cause changes in body size. Many islands do not have the predators that are present in mainland locations. Lack of predation reduces the need for speed, which could lead to larger body size. Food supply may be relatively less consistent or abundant than on the mainland, which could select for smaller body size. Lack of competition with other species affects niche use and resulting body size. Where size is an advantage with predators present, it may not be selected for in the absence of predators.

As a more terrestrial example of the effects of "islands" on small mammals, Gottfried (1979) found that *Peromyscus leucopus* predominated in isolated woodlots (islands) of varying sizes, sometimes accompanied by one or more of the following species of small mammals: *Microtus pennsylvanicus, M. ochrogaster,* and *Blarina brevicauda* (table 26.3). Three additional small mammal species, however, were found in larger forest stands nearby: *P. maniculatus, Cryptotis parva,* and *Zapus hudsonius.* The *P. leucopus* inhabiting these more isolated woodlots therefore faced less competition than those inhabiting the larger forested areas.

Table 26.3   Woodlots as islands.   The number of species of small mammal inhabiting woodlots varies with both the distance from the source population and the size of the woodlot.

Isolation from Nearest Forest (km)	Woodlot Size (m²)	Number of Species
0.08	502	3
0.16	639	3
0.32	573	1
0.42	510	3
0.64	374	1
0.75	350	1
0.80	93	1
1.60	630	2
1.83	250	1
2.88	337	1

Source: Data from B. M. Gottfried, "Small mammal populations in woodlot islands" in American Midland Naturalist, *102:105-112, 1979.*

Many mammals, and birds even more, have been negatively affected by human introduction of rats and mice to islands. Native mammals and birds often have not previously dealt with these sorts of predators and are not adapted to respond to the new threat. Introductions of nonnative mammals to islands where there are no predators can have negative consequences. For example, eight different species of cervids, released in parts of New Zealand, have severely damaged much of the native habitat.

Convergence

In many instances when groups of mammals become isolated geographically, they diverge, meaning that they adapt to particular climates and geological and ecological situations. In contrast, convergence occurs when distantly related lineages inhabiting regions with similar ecological, geological, and climatic conditions evolve similar morphologies, life history patterns, and niche characteristics. Numerous examples exist of evolutionary convergence in mammals. The North American flying squirrels (*Glaucomys* spp.) are similar in numerous respects to the marsupial gliders (Family Petauridae) of Australia. The heteromyids (kangaroo rats and pocket mice) of western North American deserts and scrublands are convergent with the dipodids (jerboas) that inhabit similar arid regions in North Africa, the Middle East, and Central Asia. Faced with similar conditions, these rodent groups have independently evolved similar diets as seedeaters and a fossorial life style. They share many similarities in terms of social systems and life history traits (Genoways and Brown 1996).

The Sonoran and Great Basin deserts of western America have rodent communities that have a variety of species occupying different niches (Brown 1975; see chapter 25). Similarities have often evolved in species that are only distantly related. Environmental features at these locations play

critical roles in terms of the direction of evolution of morphology and behavior of the organisms that live there. Riddle (1995) used molecular techniques to study the relationships among these desert rodents and to examine the time frame for the evolution of these convergent forms. His central theme was that the initial patterns of divergence that lead to the isolated desert rodent communities resulted from the geological evolution of the western mountain ranges in the late Tertiary and early Quaternary periods. This contrasts with the previously held notion that the distributions and speciation resulted from more recent events surrounding the advancing and retreating glaciers during the Pleistocene epoch. We see an interesting mix of historical and ecological biogeography being used to analyze existing patterns of mammalian distribution within one portion of a single continent.

The habit of eating ants as a primary means of obtaining food offers a good example of convergence. Myrmecophagy occurs in distantly related groups, both taxonomically and geographically (figure 26.11). Convergence is evident among anteaters (Xenarthra: Myrmecophagidae), pangolins (Pholidota: Manidae), the aardwolf (*Proteles cristatus;* Carnivora: Hyaenidae), the aardvark (*Orycteropus afer;* Tubulidentata: Orycteropodidae), the numbat (*Myrmecobius fasciatus;* Dasyuromorphia: Myrmecobiidae), and echidnas (Monotremata: Tachyglossidae). These varied species inhabit the tropical and subtropical regions of Africa, Asia, and South America. Their convergence involves skulls that are rather long and cylindrical, a long rostrum, and powerful forelimbs with digging claws on one or more of the digits. As another example of convergence, consider those mammals that inhabit rock piles (Mares and Lacher 1987). The mammals that use this type of habitat include representatives of four orders: Diprotodontia (Families Petauridae and Macropodidae); Hyracoidea (Family Procaviidae); Rodentia (Families Sciuridae, Muridae, Caviidae, Chinchillidae, Petromuridae, and Ctenomyidae); and Lagomorpha (Families Ochotonidae and Leporidae). Convergence in this instance occurs with respect to a herbivorous diet; a tendency to climb trees that are associated with the rock pile habitat; padded feet with nail reduction, with some groups having a specialized grooming claw; the use of rock piles as lookout posts from which alarm calls or whistles can be made; and a significant proportion of the species having a social, mating system involving harem polygyny (see chapter 21).

Latitudinal Gradients

For mammals as a group, and for many of the orders and families of mammals, species diversity has generally been thought to increase along a gradient from the poles toward the equator. Within North and Central America, species diversity increases from Alaska and northern Canada to Mexico and southward (figure 26.12; Simpson 1964). The higher diversity of mammals in the American Southwest results from a number of rodent and bat species occupying diverse habitats in the rather arid landscape.

Figure 26.11   Convergence.   Convergence is evident in these unrelated mammalian lineages of anteaters. All have a long rostrum and sticky extensible tongue. (A) Echidna, (B) aardvark, (C) giant anteater, (D) scaly anteater, (E) numbat. Not to same scale.

Several differing explanations have been proposed to account for the consistent patterns of latitudinal gradients in most organisms, including mammals (see Brown and Gibson 1983; Kaufman 1995; Brown and Lomolino 1998). These include the higher productivity and stability of climatic conditions, and high habitat heterogeneity for more tropical habitats contrasted with harsher prevailing conditions at higher latitudes. Also, the presence of more species can lead to greater interspecific competition, which, in turn, can promote specialization for coexistence and thus add to species diversity. Although there is agreement that latitudinal gradients occur in mammals, considerable discussion surrounds the possible explanations for these gradients (Brown and Gibson 1983; Kaufmann 1995). Similar gradients of diversity occur with changes in elevation: the higher the elevation, the lower the diversity (e.g., Patterson et al. 1989). These elevational effects are likely due to decreasing temperature with increasing elevation, related changes in vegetation, and, for some mountainous areas, a pattern of greater scarcity of available water with higher elevation.

What sorts of latitudinal gradients exist for mammals? In North America, the Order Chiroptera is represented by a single species at the northern extreme (66°N latitude),

whereas there are about 80 species present at the southern reaches of this land mass (8°N latitude; Wilson 1974). Over the same latitudinal range, the diversity of quadrupedal land mammals (not including the bats) ranges from a low of 20 species in the Arctic to about 80 species in the tropics. Darlington (1957) compared the mammalian faunas of the temperate zone in eastern Asia with those in the tropics of the neighboring Oriental region. Within all three groups of mammals that he examined, more diversity existed in the tropical zone. For land mammals, except bats, rodents, and lagomorphs, 180 species occurred in the tropics versus 100 in the temperate zone. For rodents and lagomorphs, 135 species occurred in the tropics versus 109 in the temperate region. Lastly, for bats, 154 species occurred in the tropics versus only 43 species in the temperate zone. The general relationship between latitude and species diversity for bats was confirmed by Willig and Selcer (1989). For analyses involving molossids, phyllostomids, and all bats, the single best predictor of species diversity was latitude. For vespertilionids, however, biome richness was the best predictor of species diversity, and latitude was not a significant factor.

There are some exceptions to the general pattern for latitudinal gradients. For example, seals (Carnivora) and

Figure 26.12   Latitudinal gradients. Fewer mammalian species occur in the more northern latitudes of North and Central America, and more species occur at lower latitudes. Interesting exceptions are the deserts of the southwestern United States and northern Mexico, with higher diversity, and the Baja California peninsula, with much lower diversity. The numbers along the lines on the figure represent the number of different species. The lines thus demarcate regions of differing diversity.

Source: Data after Simpson, 1964; after Cook, 1969; in J.H. Brown and A.C. Gibson, Biogeography, 1983, C.V. Mosby Company, St. Louis, MO.

baleen whales (Mysticeti) reach peak diversities at high latitudes. These differences may relate to the diversity and quantities of food resources available in more northerly and southerly ocean waters. The amount of carbon dioxide dissolved in water increases as temperature decreases, so plankton densities can be extremely high at higher latitudes where the ocean waters are cooler. Thus, mammals that depend primarily (Mysticeti) or secondarily (seals) on the plankton can achieve higher levels of diversity in the higher latitudes.

Within the past decade, several researchers have provided evidence suggesting that our long-held conception of latitudinal gradients needs to be reexamined. In South America, the drylands, which are south of the tropics (at higher latitudes), support a greater species diversity of endemics than the lowland Amazon rain forest (Mares 1992). Ruggiero (1994) studied the geographic distributions of 536 South American mammals. The findings suggest that in addition to the general pattern of increased species diversity at lower latitude, other factors must be considered. These factors include broader climatic tolerances of some species inhabiting higher latitudes, resulting in broader species geographical ranges for those species; broader dispersal capabilities, as, for example, with bats; historical patterns of distribution for some mammals; the shape of a continental land mass, such as South America, which has more land area within the lower latitudes than at higher latitudes; and the presence of what are called accidental species. Accidental species are species that are not locally self-sustaining but persist as rare populations, maintained through immigration from other populations (sources).

Finally, Rosenzweig (1992) has noted that we probably know less about species diversity gradients than we thought we did. He points to experimental tests indicating that diversity may actually decline as productivity of the habitat increases. This runs counter to the accepted dogma. Rosenzweig feels that at least two errors have lead us to be too hasty in accepting latitudinal gradients as a primary or sole explanation for species diversity patterns. First, we failed to accept the fact that we could be wrong in spite of considerable evidence to the contrary. Second, a mixture of scales and biomes leads us down a murky path. The question of species diversity, and particularly of the role of latitudinal gradients, is only now receiving more careful attention from mammalogists and other biologists.

Summary

There are six major faunal regions of the world, each characterized by connections with other regions, varying degrees of endemism, and differing patterns of mammalian familial diversity. Zoogeography involves searching for answers to two types of questions. Historical zoogeography seeks to explain the sequences of events for the origin, dispersal, and extinction of species. Both extant and fossil evidence of mammals are important in these reconstructions, as are understanding the concepts of continental drift and plate tectonics. In more recent times, glaciations and the periods between them resulted in major climatic changes over large regions, leading to shifts in mammalian faunas. The ranges of caribou and musk oxen at more northern latitudes than in earlier times can be explained by patterns of glaciation; other species failed to shift their ranges as the climate changed. Refugia, including nunataks, consist of pockets of suitable habitat isolated by glaciers or other changes in habitat and climate. Mammals survived in refugia, then dispersed when the glaciers receded. Mammalian species dispersal occurs via corridors, filters, and sweepstakes routes. Corridor routes are the most expansive and have been used by many mammals, as, for example, between Asia and Europe. A filter route, such as the Isthmus of Panama, allows some species to pass through but not others. Sweepstakes routes are the most restrictive and involve traversing some major barrier. Because they can fly, bats have used the sweepstakes means of dispersal most frequently. Extinction—the loss of a species—occurs by both natural events and, recently, human activities.

Ecological zoogeography seeks to explain the present distributions of mammals using ecological, geological, and climatic information. Biomes are broad ecosystems characterized by particular plant and animal life, soil conditions, and climates. They provide a general model to map mammalian distributions. A more refined model involves the use of 128 possible niches resulting from a cross-classification of 16 feeding types and 8 substrate types. Examination of occupancy and nonoccupancy of these niches in 11 regions of the world using this matrix reveals that niches are disproportionately filled or open in different geographic regions. Past climatic conditions and the stability of vegetation patterns over time help to explain the degrees of niche occupancy.

Mammalian zoogeographers are also interested in

1. Endemism—the restriction of a particular taxon to a specific, sometimes relatively isolated, region. Endemism provides information on how long particular mammalian faunas have been isolated and the nature of the ecological constraints in these locales.

2. Convergence—the evolution of similar morphologies, life history patterns, and niche characteristics in distantly related organisms inhabiting regions with similar ecological, geological, and climatic conditions.

3. Latitudinal gradients—the increase in species diversity along a gradient from the poles toward the equator as is the case with many groups of mammals.

Discussion Questions

1. Using a field guide for mammals, select an order, such as Rodentia, that has numerous species that live in North America (or most other continents). Compile a rough table of the numbers of species in the order you selected that are found at 30°, 40°, and 50° N latitude. What factors might explain any differences in species richness across the latitudinal gradient?

2. Some families of mammals, elephants (Elephantidae) or pikas (Ochotonidae) for example, have relatively few living species. Other families, such as common bats (Vespertilionidae) or shrews (Soricidae), have many. What factors do you think help to explain these differences?

3. Select your favorite mammalian family. Determine the present distribution of the various genera and species of this family using published volumes on mammals (e.g., guide books, encyclopedias, etc.). Based on your knowledge of geography and climates, along with information gained about the niches of these species during your search for the distribution patterns, what can you say about the ecological factors that may be limiting the ranges of these species under present conditions? (A useful resource for the distributions and related niche information is *The Encyclopedia of Mammals*, edited by D. Macdonald (1984), or this textbook.)

4. As we learned in chapter 3 and elsewhere in this book, a variety of molecular techniques has been developed in recent years that can aid the mammalogist in certain ways. Describe several ways in which these molecular techniques might be applied to the study of mammalian zoogeography.

Suggested Readings

Brown, J. H. and M. V. Lomolino. 1998. Biogeography, 2d ed. Sinauer Assoc., Sunderland, MA.

Cox, C. B. and P. D. Moore. 1993. Biogeography: an ecological and evolutionary approach, 5th ed. Blackwell Scientific Pub., Oxford, England.

Eisenberg, J. F. 1981. The mammalian radiations. Univ. of Chicago Press, Chicago.

Myers, A. A. and P. S. Giller (eds.). 1988. Analytical biogeography. Chapman & Hall, London.

Pielou, E. C. 1991. After the ice age. Univ. of Chicago Press, Chicago.

Rickleffs, R. E. and D. Schluter (eds.). 1994. Species diversity in ecological communities: historical and geographical perspectives. Univ. of Chicago Press, Chicago.

PART FIVE

Special Topics

The final three chapters of this book, which make up part V, cover topics of increasing importance to mammalogists. In chapter 27, we discuss general characteristics of parasites and their effects on mammalian hosts. We also investigate host-parasite coevolution and specificity. Finally, we consider diseases that occur in mammals, including several of current interest, such as Lyme disease and hantaviruses, which are transmitted from nonhuman mammals to people.

The second special topic, discussed in chapter 28, is the domestication of mammals. The biological and cultural processes of mammalian domestication that began over 12,000 years ago continue today. Domestication has had a significant effect on human history. Nonetheless, relatively few species of mammals have been domesticated, and we will explore the reasons why.

The final chapter of the text deals with one of the most important aspects of current mammalogy—conservation. In chapter 29, we discuss various factors and their interactions that contribute to the decline of mammalian populations throughout many parts of the world, as well as potential solutions. A better understanding of these issues and how they affect mammalian fauna will ultimately lead to better and more effective conservation decisions.

Raccoon (*Procyon lotor*) populations are valuable as furbearers but may also carry rabies.

CHAPTER

27

Parasites and Diseases

Parasitism is a form of interaction, or **symbiosis** ("living together"), between two species; one species (the parasite) benefits at the expense of the second species (the host). Parasites are similar to predators in that they both benefit at the expense of another species. Unlike most parasites, however, predators directly remove individuals from the prey population. Parasites usually do not kill their hosts, although there are exceptions, and effects on prey populations often may be subtle and difficult to determine.

Parasites include a vast array of diverse life forms; there are more species of parasites in the world than nonparasites. Nonetheless, parasites share several general characteristics: they are smaller than their host, usually are physiologically dependent on the host, spend either their entire life (permanent parasites) or part of it (temporary parasites) in or on the host, and derive essential nutrients from the host. Parasites can be categorized into two general groups: microparasites and macroparasites. **Microparasites** include viruses, bacteria, and fungi—disease agents often not thought of as parasites—and protistans. They usually are microscopic and have rapid regeneration rates within their hosts. **Macroparasites** include the flatworms or platyhelminths (tapeworms and flukes); nematodes (roundworms); acanthocephalans (thorny-headed worms); and arthropods (ticks, fleas, lice, flies, and mites). They are larger than microparasites and have longer generation times; generally, most do not reproduce entirely within the host.

Parasites can be categorized in other ways as well. They are either **obligate,** meaning that they must spend at least part of their life cycle as a parasite, or **facultative,** meaning that they are organisms that are not normally parasitic but become so. Those that occur within the body of the host are termed **endoparasites,** whereas **ectoparasites** occur either on or embedded in the host's body surface. Janzen (1985) noted the following generalizations about animal ectoparasites, several of which also apply to endoparasites:

1. They are subject to almost no predation while on the host.

2. The chemical content of their food is relatively constant among mammalian hosts.

3. Close contact is common, and the host can chemically identify the parasite.

4. Hosts play a major role in the interhost movement of ectoparasites.

5. There are often large numbers of parasites on a host.

6. A large percentage of the parasite's life cycle is spent on or near the host.

In this chapter, we will discuss the effects of parasites on mammalian host populations, host-parasite coevolution, and the different types of parasites. We also will focus on mammalian parasites and the associated human **diseases** (clinical conditions that can be observed or measured) they may impart. More specifically, we will consider several **zoonoses** (from the Greek words *zoo,* meaning "animals," and *noses* meaning "diseases"), which for our purposes are defined as diseases transmitted from nonhuman mammals to people. Mammals sometimes are a primary **reservoir,** or source, of the infective organism of human zoonoses (Marcus 1992). Reservoirs may enable a disease to persist in an area at moderate to low levels, which is termed an **enzootic** phase. When various factors amplify the occurrence and distribution of an enzootic disease, it becomes **epizootic.** Arthropod parasites often serve as **vectors** (carriers) of viruses, bacteria, or other microparasites to humans and other mammalian species; these microparasites are the **etiological,** or causal, agents of the disease. We will concentrate on a few common or historically important zoonoses, as well as on some relatively recent diseases with which mammalogists should be familiar. Understanding mammalian parasites and diseases is important because of the risks they present in mammalian studies and the roles they play in the evolution and life histories of species (Childs 1995).

PARASITE COLLECTION

Mammalian ectoparasites can be collected from live or dead specimens in the field or laboratory. Snap-trapped small mammals can be placed in individual plastic bags that are then sealed to keep parasites with the host. Animals can be examined under a dissecting microscope for ectoparasites on or embedded in the skin or clinging to the fur, such as mites. The eyes, ears, lips, and genital area should be searched as well. Parasites can be removed with a needle. Another technique for removing ectoparasites from small mammals is to immerse them in water containing detergent. Whatever removal method is used, ectoparasites can be preserved in 70% alcohol. Certain ectoparasites, such as fleas and larger mites, may leave dead animals as the animals cool. This is less of a problem with live-trapped mammals.

Collection of endoparasites is through necropsy in a laboratory. Depending on the objectives of the study, all visceral organs, as well as the brain, should be examined. Stomach and intestinal contents can be emptied into a tray, or these organs can be slit open and the walls examined for embedded individuals. Parasitology laboratory manuals, for example Dailey (1996), should be consulted for details of specific techniques on examination, recovery, preparation, and identification of both micro- and macroparasites. Whitaker (1968) summarized methods of parasite collection and preservation for *Peromyscus* that are applicable to other small species. Keys to identification of parasite groups are available; however, identification to family level or below often is difficult and should be done by a specialist familiar with the group.

EFFECTS OF PARASITES ON HOST POPULATIONS

Direct Mortality Effects

Parasites and diseases occur in all mammalian species. Although parasites usually do not kill their **definitive hosts** (where the parasite reaches sexual maturity), they may or may not have an obvious adverse effect on the host. Yuill (1987) suggested three circumstances, however, in which parasitism can be a mortality factor (table 27.1). An intermediate host may also die if its death enhances transmission of the parasite to a definitive host. Also, mortality can occur when parasites associate with **accidental hosts,** that is, those in which they normally are not found. For example, rinderpest, a viral disease normally found in livestock, is highly pathogenic in wild African ungulates and has caused several epizootic outbreaks (Scott 1981). Likewise, *Baylisascaris procyonis,* a nematode normally found in raccoons (*Procyon lotor*), is pathogenic in woodrats (Genus *Neotoma*). Current population declines of woodrats in the northeastern United States may be related, at least in part, to *Baylisascaris* infection. White-tailed deer (*Odocoileus virginianus*) are the definitive host for meningeal worm (*Parelaphostrongylus tenuis*), another nematode. Although it has little effect on whitetails, in other cervids, especially moose (*Alces alces*), meningeal worm can cause fatal neurological disease (figure 27.1). The parasite is a serious management consideration in translocating white-tailed deer outside their native range (Comer et al. 1991; Samuel et al. 1992).

Biological Control

When a mammalian population reaches such a high density that it is considered a pest, an unusual pathogen can purposely be introduced as a **biological control** (using predators

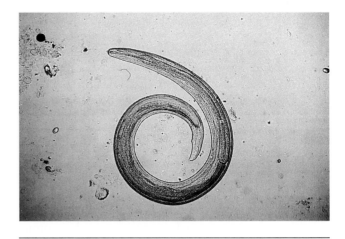

Figure 27.1   **Meningeal worm.**   This nematode is a serious mortality factor in moose and other cervids. The primary host, white-tailed deer, is not affected by the parasite.

or parasites to reduce host population density). As noted by Roberts and Janovy (1996), successful biological control agents have several characteristics:

1. High host-searching capacity
2. A limited but wide enough range of hosts to maintain the parasite population
3. A life cycle that is either shorter than or synchronized with the life cycle of the pest population
4. Ability to survive in all habitats occupied by the host
5. Easy replication so that large numbers are available to introduce
6. Ability for rapid control of the pest population

Biological control in mammals is rare and usually produces additional unforeseen problems (Spratt 1990). The European wild rabbit (*Oryctolagus cuniculus*) in Australia is one of the few examples of "successful" control. Introduced into Australia in the 1840s, wild rabbit populations increased dramatically. They soon devastated native flora and became a serious agricultural pest as well. In the 1950s, myxomatosis virus was introduced to control rabbit populations. Initially, control was successful as over 99% of infected rabbits died. Resistance to the virus developed in certain rabbits, however, and these individuals became the primary breeding nucleus (Yuill 1987). Also, attenuated, or nonlethal, strains of the virus developed, and eventually, rabbit populations once again reached pest proportions. Another attempt at control was made recently by introducing a different pathogen—rabbit hemorrhagic disease (RHD) virus. In some regions of Australia, rabbit populations have declined by 95%. It is hoped that rabbit control can be sustained with RHD virus in conjunction with other control methods, including fumigation, hunting, and destruction of warrens (Drollette 1997).

Table 27.1.   Circumstances in which parasites may act as mortality factors in mammalian host populations

Circumstance	Example
1. When death of the host facilitates transmission of the pathogen	Rabies (see text)
2. A "generalist" pathogen establishes transmission cycles involving many host species	The screwworm fly (*Cochliomyia hominivorax*) in deer
3. The pathogen "wanders" over a wide enough geographic area and long enough time period such that infected local host populations are not seriously threatened	The yellow fever virus in howler monkeys (Genus *Alouatta*) and marmosets (Genus *Callithrix*)

Source: From T.M. Yuill, "Diseases as Components of Mammalian Ecosystems: Mayhem and Subtlety" in Canadian Journal of Zoology, *65:1061–1066, 1987.*

Nonmortality Effects

Effects of parasitism on individual definitive hosts and mammalian populations may be subtle and inconspicuous. The impact a parasite has is a function of its numbers; the sex, age, and overall condition of the host; the season of the year; and other variables. Parasitism can increase energy costs and decrease energy gained in hosts (Yuill 1987). This may result in decreased movement or reproduction (Smith et al. 1993), reduced host growth, altered behavior, or reduced survival of offspring (Munger and Karasov 1994). All these factors can have negative consequences on host population dynamics, although Ostfeld and colleagues (1996) found that heavy tick infestation had no effect on the fitness of white-footed mice hosts. Many factors have received little attention by mammalogists, however, in part because they are very difficult to document in field studies. Grenfell and Gulland (1995) summarized observational, quantitative, and experimental studies done on the effects of micro- and macroparasites on mammalian reproduction and survival.

COEVOLUTION OF PARASITES AND MAMMALIAN HOSTS

Throughout the 200-million-year history of mammals, micro- and macroparasites have been adapting to mammals, while mammals in turn have been adapting as hosts. Host-parasite **coevolution** was defined by Kim (1985a:670) as "reciprocal evolutionary change in interacting species in a parasite community involving both the parasite species versus the host and the parasite species versus other parasites." Relationships (congruence) between host and parasite phylogenies may be difficult to assess, however, and may represent "coaccommodation" rather than coevolution (Brooks and McLennan 1991; Page 1993). Nonetheless, many mammal-parasite assemblages are believed to have coevolved (Waage 1979; Price 1980; Futuyma and Slatkin 1983), with the result that many parasites are closely associated with a particular group of mammals; that is, they show pronounced specificity.

Parasite Specificity

Most species of parasite probably descended from free-living ancestors. Through evolutionary time, they developed a close, often obligatory, association with hosts such that generalist parasites are relatively rare. As noted by Adamson and Caira (1994:S85), "Parasites are specialists, not just of hosts, but of microhabitat or tissue site. . . . Parasite specificity is not fundamentally different from specificity, or niche restriction, in free-living organisms, except that the niche of parasites is part of another organism."

Evolution can shape parasite specificity in three interacting ways (Adamson and Caira 1994). Initially, specificity is a vestige of the microhabitat and feeding mode of the par-asite's free-living ancestors and the manner in which the parasite-host relationship arose. A second factor is host ecology, especially important in passively transmitted parasites that have little effect on the physiological or immune systems of their host. These types of parasites (e.g., pinworm nematodes or trichomonadid protists) occur in hosts that are ecologically rather than phylogenetically similar. The third factor in host-parasite specificity can involve a number of coevolutionary adaptations, including

1. Synchrony with host phenology to maximize the transmission cycle of the parasite

2. Hatching or excystment cues, as in mammalian schistosomes

3. Migration cues, as occur in digeneans (trematode flatworms)

4. Specific cell surface receptors, which are important in a variety of protists, including *Giardia* (Lev et al. 1986), *Leishmania* (Alexander and Russell 1992), and *Plasmodium* (Braun Breton and Pereira da Silva 1993)

5. Ability to evade host immunological responses

6. Competitive exclusion, which reduces realized niches such that different parasites do not occur in the same host.

Adaptations of parasites include the means to find an individual host, attach to it, reproduce, and subsequently disperse either eggs, juveniles, or adults. Additional physiological and morphological adaptations involve the structure of mouth parts and the apparatus for feeding and digestion (Kim 1985b). Such a high degree of host specificity may evolve that a parasite may become species-specific, that is, not be able to survive on taxa other than its host.

A prime example of parasite-host specificity is the association of parasitic mites of the Family Myobiidae with their mammal hosts (Fain 1994). Two subfamilies of myobiids occur only on New World marsupials (chapter 10). The Subfamily Archemyobiinae occurs only on opossums (Family Didelphidae), with a single species of mite found on the monito del monte (Microbiotheriidae: *Dromiciops australis*), whereas the Subfamily Xenomyobiinae parasitizes only shrew opossums (Family Caenolestidae). A third myobiid subfamily, the Myobiinae, is divided into two tribes. The tribe Australomyobiini occurs only on Australian marsupials. More specifically, the Genus *Australomyobia* is found on marsupial "mice" (Family Dasyuridae), and the Genus *Acrobatobia* lives only on pygmy possums (Family Burramyidae). The tribe Myobiini includes all the genera and species of mites that parasitize eutherian mammals. For example, mammals within the Order Insectivora host 16 genera and 67 species of myobiid mites that are highly species-specific (Fain 1994). Likewise, the 21 genera and 230 species of myobiid mites that occur on bats are highly specific. Each genus of mite is generally restricted to one family or subfamily of bat, with many species of mite specialized for a single host genus. Numerous rodent species also are parasitized by myobiid mites (figure 27.2).

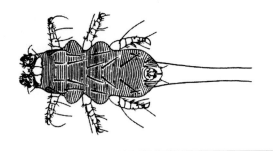

Figure 27.2   Host-parasite specificity.   The myobiid mite *Myobia musculinus* is a very common fur mite of house mice (*Mus musculus*).

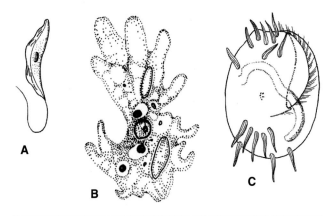

Figure 27.3   Protistans.   These parasites can be grouped by their means of locomotion: flagella, pseudopodia, or cilia. A single flagellum occurs in (A) *Trypanosoma*, the protozoan genus responsible for African sleeping sickness and Chagas' disease. (B) This *Amoeba* has pseudopodia ("false feet"). (C) Cilia occur in this free-living *Euplotes*.

Again, the ". . . concordance between the radiations of the Myobiidae and that of their rodent hosts is remarkable" (Fain 1994:1281). Coevolution also has been shown in pocket gophers (Rodentia: Geomyidae) and their associated chewing lice (Demastes and Hafner 1993; Reed and Hafner 1997). Studies of these types of relationships allow comparison of the rates of speciation and evolution in host and parasite (Hafner and Nadler 1990; Page 1993).

MAMMALIAN PARASITES AND DISEASES

Numerous species of parasites and diseases are associated with mammals. We will note examples of mammalian parasites from each major group, including parasites and diseases that affect humans. Space does not permit a discussion of most zoonoses; for more detailed information on zoonoses and other mammal-related diseases, see Acha and Szyfres (1989), Gorbach and colleagues (1992), and the *CRC Handbook Series on Zoonoses* (Beran 1994).

Protistans

Exclusive of the algae, this kingdom includes seven phyla and about 45,000 species of microscopic, single-celled, animal-like eukaryotes. Locomotion is by flagella, cilia, or pseudopodia (figure 27.3), and they reproduce asexually, sexually, or both. Thousands of species are free-living. Many other parasitic species cause no adverse effects for their hosts. For example, trypanosomes are widely distributed protistan parasites that usually are benign in mammals. A number of protistan species are pathogenic, however. For example, trypanosomiasis (*Trypanosoma evansi*) occurs in many mammals, including elephants, dogs, horses, camels, and deer, and is fatal if untreated. Nagana, caused by *Trypanosoma brucei brucei*, infects wild African ruminants but apparently causes no disease. It is pathogenic, however, in a variety of livestock species and has kept 4 million square miles of African grazing land out of livestock production. Toxoplasmosis (*Toxoplasma gondii*) is widespread in over 200 mammalian species, including humans, and can be fatal in

numerous species (Sanger 1971). Theileriosis (Genus *Theileria*) occurs primarily in numerous species of ruminant artiodactyls throughout the world, as well as rodents, primates, xenarthrans, and the aardvark (*Orycteropus afer*). *Theileria parva* is a tick-borne protist that infects red blood cells and is highly pathogenic in cattle. Babesiosis is another protistan disease carried by ticks, and it infects rodents, carnivores, and ungulates. Although babesiosis can be fatal, animals often do not show clinical disease. Mortality rates from *Babesia bigemina*, however, can approach 90% in cattle. Brucellosis (*Brucella* spp.) occurs in domestic and wild ungulates and a number of other species (Rementsova 1987), where it causes abortion. Leptospirosis (*Leptospira interrogans*) also causes abortion, but in many mammalian species, including humans, it can be fatal.

Some protistans are highly pathogenic to humans. A good example of the diseases they can cause is African sleeping sickness, which occurs throughout central Africa and is caused by two subspecies of *Trypanosoma brucei*: *T. b. rhodesiense* and *T. b. gambiense*. The vectors are several species of tsetse fly in the Genus *Glossina*. In people infected with *T. b. gambiense*, sleeping sickness progresses from headache and fever to muscular and neurological involvement, and eventually to coma and death. People infected with *T. b. rhodesiense* usually die before these symptoms develop. As many as 20,000 new cases of sleeping sickness occur annually, of which 50% prove fatal. Past epidemics have killed hundreds of thousands of people (Maguire and Hoff 1992). If not fatal, infection often results in permanent brain damage.

A closely related trypanosome protozoan, *T. cruzi*, causes American trypanosomiasis, or Chagas' disease. Reservoirs include over 100 mammalian species, particularly domestic cats and dogs, as well as bats and rodents. Several species within three genera of reduviid bugs ("assassin bugs") are the vectors. This disease affects 12 to 19 million people and is a leading cause of cardiovascular death in Central and

South America. In the United States, *T. cruzi* occurs predominately in the South, where it may be more prevalent than previously believed (Roberts and Janovy 1996).

Leishmaniasis is a complex of zoonotic diseases common to tropical and subtropical regions, and caused by some species in the Genus *Leishmania*. The reservoirs of these diseases, which are transmitted through the bite of sandflies (Family Psychodidae), are canids and rodents. Ashford and colleagues (1992) estimated at least 400,000 new cases of leishmaniasis annually worldwide. In humans, *Leishmania donovani* causes Dum Dum fever, also called kala-azar. After an incubation period of several weeks, respiratory or intestinal infection and hemorrhage occur. If untreated, kala-azar is usually fatal within 2 to 3 years.

Many campers and hikers are familiar with giardiasis. Found worldwide, it is the most prevalent protistan parasite in humans. *Giardia lamblia*, with reservoirs in beavers, dogs, and sheep, is picked up in water. It is highly contagious and can cause severe intestinal disorders ("beaver fever") but is not fatal. Giardiasis is common in developing countries with poor water and sewage treatment facilities, but it also occurs in industrialized nations. Most people with the parasite show no symptoms, however.

Finally, several species of the Genus *Plasmodium* infect mammals, including rodents, nonhuman primates, and humans. Species of *Plasmodium* cause several varieties of malaria. Worldwide, 1.5 billion people are exposed to the disease, and up to 3 million die each year.

Platyhelminths

Two classes within the Phylum Platyhelminthes are entirely parasitic and common in humans and other mammals. The Class Trematoda includes both visceral and blood flukes. The Class Cestoidea includes the tapeworms. Both groups have indirect life cycles. That is, adult parasites occur in the primary (definitive) host and produce eggs, which are usually expelled. The larvae enter one or more intermediate hosts and pass through several growth stages before once again entering a primary host. In mammals, the dwarf tapeworm (*Vampirolepis* [*Hymenolepis*] *nana*) and taenids such as *Taenia solium* are exceptions in that they can hatch and remain within the host.

Trematodes are usually dorsoventrally flattened, leaf-shaped parasites found in a wide variety of wild and domestic mammals throughout the world. Most trematodes require two intermediate hosts; at least one, a snail, is required for development. Hundreds of species of trematodes have been reported from numerous mammalian species. Many trematodes cause significant economic losses in livestock, such as the large liver flukes, *Fasciola hepatica* and *F. gigantica*, common in many species of wild and domestic herbivores. These flukes also can infect humans. Various species of *Paragonimus* infect mammals, primarily carnivores. Many are zoonotic and cause pulmonary disease in millions of people (Goldsmith et al. 1991). All trematodes

that parasitize humans belong to the Subclass Digenea. About 17 species of digenetic trematodes can infect the blood, liver, intestines, lungs, or brain of humans. For example, five different species of blood fluke cause schistosomiasis. This disease was known as early as 5000 years ago in ancient Egypt. Today, an estimated 300 million people are infected with schistosomiasis in parts of Africa, the Middle East, Southeast Asia, and South America, and a million die each year. The cycle of transmission to humans generally does not involve other mammals as intermediate hosts, although *Schistosoma mansoni* and *S. japonicum* occur in numerous mammalian species that may act as reservoirs.

Cestodes are the highly specialized, ribbonlike tapeworms found in the intestine of numerous mammalian species, including humans. Two orders of tapeworms infect people: Pseudophyllidea and Cyclophyllidea. Almost all species require two intermediate hosts to complete their life cycle. Juvenile tapeworms encyst in animals consumed as food by humans. A common pseudophyllidean tapeworm is *Diphyllobothrium latum* (figure 27.4). It occurs in numerous fish-eating carnivores, including canids, felids, mustelids, phocids, otariids, and ursids, as well as an estimated 9 million people worldwide (Hopkins 1992). Most tapeworms in wild mammals are cyclophyllideans. This order includes the Family Taeniidae, the most important medically for humans (Roberts and Janovy 1996). For example, infection from *Taeniarhynchus saginatus* results from eating undercooked beef. Usually, a single worm is present, which can attain a length of 20 m, although 3 to 5 m is typical. Likewise, consumption of undercooked pork may result in infection from *Taenia solium* (figure 27.5). Individual *T. solium* can survive in human hosts up to 25 years. Ingestion of eggs by people leads to cysticerosis, which can be fatal. The tapeworm Family Hymenolepididae contains many species that infect mammals. One, the dwarf tapeworm, is the most common human tapeworm, although pathogenicity is rare. It is unique in that an intermediate host is not necessary (Roberts and Janovy 1996).

Dogs and other carnivores are definitive hosts for the tapeworm *Echinococcus granulosus*. Herbivores, generally ruminants, serve as intermediate hosts. When infected, individual herbivores are more prone to predation by carnivores. The parasite thus "promotes" its own successful transmission. When eggs of *E. granulosus* are accidentally ingested by humans, they can cause a serious disease—hydatidosis. If growth of the juvenile parasite occurs in the central nervous system or heart, surgical removal may be necessary to prevent severe disability or death. Canids, cats, and rodents can also be infected with *E. multilocularis*. In humans, this parasite can cause alveolar hydatid disease, which also can be fatal.

Nematodes

These cylindrical, unsegmented worms are tapered at both ends. They usually have a direct life cycle, that is, no intermediate host is required. Generally called roundworms, they may be the most abundant animals on earth. Numerous

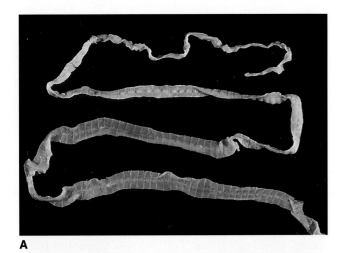

A

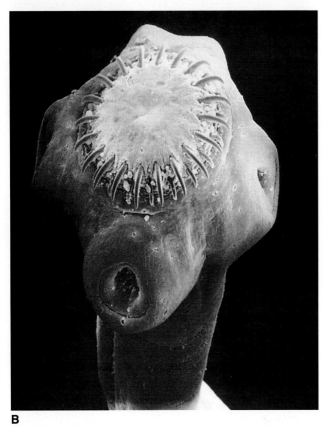

B

Figure 27.4   Tapeworms.   (A) This tapeworm may grow to be 10 m long. Tapeworms consist of a head, or scolex, (B) that attaches to the intestinal lining of the host; a germinal center where body segments, or proglottids, form; and the body segments themselves. Several different species of tapeworms occur in humans.

species of free-living nematodes, as well as thousands of parasitic species, infect many mammalian hosts. Nematode infection occurs most often in tropical and subtropical developing countries with poor sanitation facilities. "Some of the most dreaded, disfiguring, and debilitating diseases of humans are caused by nematodes" (Roberts and Janovy 1996:385).

Whipworms, in the Family Trichuridae (Order Trichurida), include about 70 described species in the Genus *Trichuris* that are found in mammals. One of these, *T. trichiura*, infects an estimated 750 million children worldwide (Hopkins 1992), although most cases are asymptomatic. Many mammalian species also harbor the liver nematode *Capillaria hepatica* (Family Capillariidae), although it usually is found in rodents. This nematode has been used as a biological control agent to reduce house mouse (*Mus musculus*) populations in Australia, apparently with little success (Singleton et al. 1995). Infections in humans are very rare but can be fatal.

A familiar human disease caused by a nematode parasite, and one of the most widespread, is trichinosis. Five species of *Trichinella* (Family Trichinellidae) are carried by wild carnivores and rats. Campbell (1988) recognized four types of infection cycles, one domestic and three wild (figure 27.6). People are most likely to become infected with trichinosis by eating undercooked pork carrying encysted larvae. Living juveniles may invade muscle tissue (figure 27.7), where they produce an immune response. An estimated 150,000 to 300,000 new cases of trichinosis occur each year in Europe and the United States (Bogitsh and Cheng 1990). Most cases are asymptomatic; no more than about 150 people yearly have heavy enough infections to produce clinical symptoms that are reported. In severe cases, however, death may result from respiratory or cerebral involvement or heart failure. Arctic explorers have died from eating infected polar bear meat.

The Order Strongylida includes numerous intestinal nematodes in domestic and wild mammals as well as humans. For example, two species of hookworms in the Family Ancylostomidae occur in over a billion people worldwide: *Necator americanus*, the most common, and *Ancylostoma duodenale*. Although most infected people are asymptomatic, heavy infections lead to malnutrition, heart involvement, and death in up to 60,000 people a year. Wild ruminants and domestic livestock also are parasitized by nematodes in the Family Trichostrongylidae. These include *Haemonchus contortus*, found in the abomasum, and several species of *Ostertagia* and *Trichostrongylus*, which can lead to extreme economic losses in livestock. Large intestinal roundworms (Family Ascarididae) such as *Ascaris lumbricoides* can reach 46 cm in length. It is estimated that 25% of the world's population harbors *A. lumbricoides* (Freedman 1992), with about 20,000 deaths yearly due to intestinal blockage. Lungworms, including *Dictyocaulus filaria*, *Protostrongylus rufescens*, and *Muellerius capillaris*, are found in a variety of ungulates, carnivores, and other mammals. Infection with these parasites may be seriously debilitating, often leading to pneumonia; heavy infections may be pathogenic. Lungworms can be a serious management consideration in big game species, including bighorn sheep (*Ovis canadensis*) (Arnett et al. 1993). *Toxocara canis* is another ascaridid found worldwide in canids. Humans are an accidental host in which juvenile nematodes "wander." This results in a disease called visceral larva migrans. Usually,

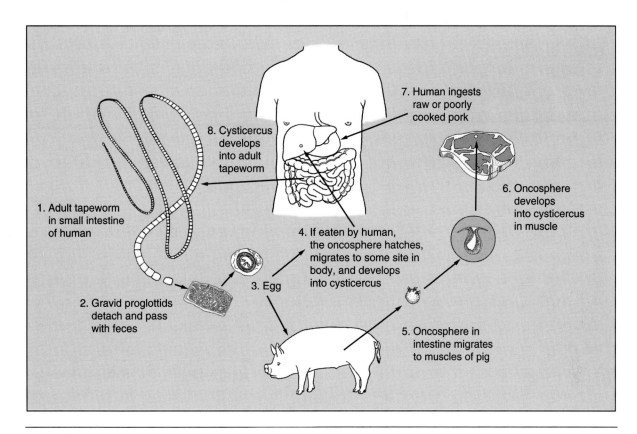

Figure 27.5   **The life cycle of *Taenia solium*.**   Individual parasites can persist 25 years in their human host.

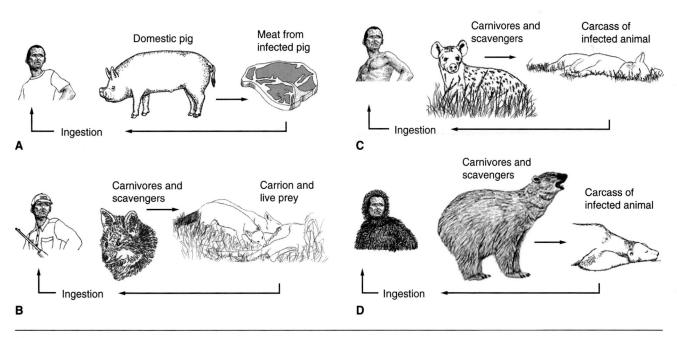

Figure 27.6   *Trichinella* **transmission.**   Domestic and sylvatic ("wild") zoonotic cycles involve humans as accidental hosts. (A) Domestic cycle of transmission; (B) temperate zone sylvatic cycle; (C) torrid zone sylvatic cycle; (D) frigid zone sylvatic cycle.

Figure 27.7   **Trichinosis.**   Juvenile *Trichinella spiralis* encysted in human muscle. This nematode parasite causes trichinosis.

symptoms are not serious, but depending on the number of parasites and the organ infected, death can result.

Seven nematode species, collectively referred to as fi-larial worms, infect the bloodstream of over 650 million people. They often cause debilitating diseases, including elephantiasis and river blindness, and result in up to 50,000 deaths per year. Filarial worms require an arthropod intermediate host. *Dirofilaria immitis*, heartworm, is a major nematode in dogs and other carnivores.

Acanthocephalans

Compared to parasitic platyhelminths and nematodes, the thorny-headed worms are less common in mammals. They are named for the hooks that cover the proboscis (figure 27.8),

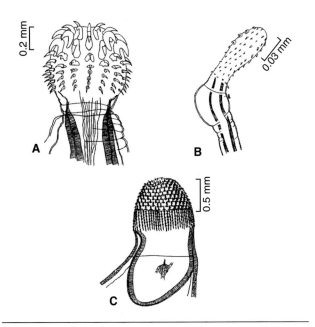

Figure 27.8   **Thorny-headed worms.**   A variety of different types of hooks occur on the proboscids of these representative acanthocephalens: (A) *Sphaerechinorhynchus serpenticola*, (B) *Pomphorhynchus yamagutii*, and (C) *Owilfordia olseni*.

which imbeds in the small intestine and holds the worm in place. Acanthocephalans have been reported from a wide range of mammalian species (DeGiusti 1971), but most often they are found in carnivores. Five species, including *Macra-canthorhynchus hirudinaceus*, have been reported in people (Schmidt 1971). They are of minor importance in human health, however. Pigs commonly carry *M. hirudinaceus*, and heavy infections can lead to death.

Arthropods

Arthropods include ticks, fleas, mites, flies, mosquitoes, and lice (figure 27.9). Arthropods are the largest phylum of animals, with over a million described species. They are **metameric** (segmented) and have a chitinous exoskeleton. They may serve as either intermediate or definitive hosts for protistans, platyhelminths, and nematodes. Although most species are of no medical importance, some arthropods are mammalian parasites that are among the most significant vectors in the transmission of zoonoses (table 27.2).

VECTOR-BORNE ZOONOSES

Plague

Plague has killed more people and has had a greater effect on human history than any other zoonotic disease (Twigg 1978; Lewis 1993). The etiological (infectious) agent is the bacterium *Yersinia pestis*, and fleas are the primary vector. There are more than 1500 species of fleas, most of which are probably capable of plague transmission (Poland et al. 1994). Female oriental rat fleas (*Xenopsylla cheopis*) are the best documented plague vectors. They transmit bacteria as they feed on an infected animal, often black rats (*Rattus rattus*) or Norway rats (*R. norvegicus*). In the United States, many other rodents carry plague, especially ground squirrels (*Spermophilus* spp. and *Cynomys* spp.) and the deer mouse (*Peromyscus maniculatus*; Gage et al. 1995). Over 200 mammalian species are known to be naturally infected with plague (Poland et al. 1994).

In about 12% of infected fleas, bacteria multiply rapidly and within 9 to 25 days fill their gut. These are called "blocked" fleas. When they try to feed on humans or other mammalian species, blood enters their gut and picks up bacteria. Because their gut is blocked, blood is regurgitated back into the wound, and plague bacteria enter the host (figure 27.10). Blocked fleas probably can transmit plague bacteria for about 14 days (Shrewsbury 1971) before they starve to death. Humans also may pick up the bacterium through flea fecal material rubbed into wounds.

There are three clinical types of plague: bubonic, septicemic, and pneumonic. Bubonic plague is the most common; bacteria concentrate in the lymph nodes of the armpits and groin, where, after a 1-to-8-day incubation period, they cause extreme swelling (figure 27.11). These swollen areas are called **buboes**—thus, the name bubonic plague. Because

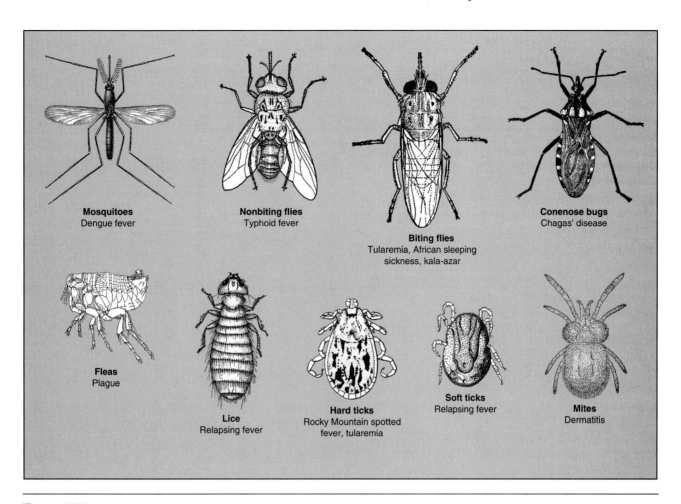

Figure 27.9 **Arthropods.** Different types of arthropods may serve as vectors, transmitting pathogens for human diseases.

Table 27.2. **Representative orders and genera of parasitic arthropods associated with mammals**

Order	Common name	Representative genera of mammalian parasites
	Class Insecta	
Dermaptera	Earwigs	*Hemimerus*
Mallophaga	Chewing lice	*Bovicola, Haematomyzus, Heterodoxus, Trichodectes*
Anoplura	Sucking lice	*Haematopinus, Pediculus*[1] *Phthirus*
Hemiptera	True bugs	*Cimex, Leptocimex, Rhodnius,*[2] *Triatoma*[2]
Coleoptera	Beetles	*Amblyopinus, Platypsyllus*
Siphonaptera	Fleas	*Ctenocephalides, Pulex, Tunga, Xenopsylla*[3]
Diptera	Flies, gnats, mosquitoes	*Aedes,*[4] *Anopheles,*[5] *Culex,*[6] *Cnephia, Cuterebra,*[7] *Glossina,*[8] *Psychoda, Simulium, Tabanus*
Lepidoptera	Moths	*Arcyophora, Calpe, Lobocraspis*
	Class Arachnida	
Ixodida	Hard ticks	*Amblyomma,*[9,10] *Argas, Boophilus, Dermacentor,*[9] *Haemaphysalis, Hyalomma, Ixodes,*[11] *Ornithodoros, Rhipicephalus*[9]
Mesostigmata	Mites	*Echinolaelaps, Halarachne, Liponyssus, Ornithonyssus, Pneumonyssus*
Prostigmata	Mites	*Cheyletiella,*[12] *Dermodex, Leptotrombidium, Psorergates, Trombicula*[13]
Astigmata[14]	Mites	*Chorioptes, Psoroptes, Sarcoptes*

Source: Taxonomy from Roberts, L. S. and J. Janovy. 1996. Foundations of Parasitology, *5th ed. Wm. C. Brown, Dubuque, IA.*

[1]—Transmits epidemic typhus, trench fever, and relapsing fever; [2]—vector for Chagas' Disease; [3]—vector for plague along with numerous other genera; [4]—transmits yellow fever and dengue fever; [5]—transmits malaria; [6]—transmits encephalitis; [7]—bot flies; [8]—transmits sleeping sickness; [9]—transmits Rocky Mountain spotted fever; [10]-transmits tularemia; [11]—transmits Lyme disease; [12]— causes mange; [13]—chiggers; [14]—causes mange

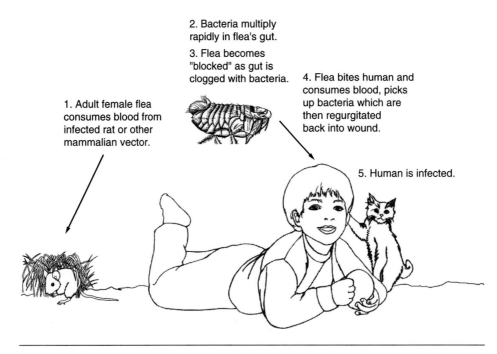

1. Adult female flea consumes blood from infected rat or other mammalian vector.

2. Bacteria multiply rapidly in flea's gut.

3. Flea becomes "blocked" as gut is clogged with bacteria.

4. Flea bites human and consumes blood, picks up bacteria which are then regurgitated back into wound.

5. Human is infected.

Figure 27.10   **Bubonic plague.**   The common vector-borne plague cycle involves "blocked" fleas and a mammalian reservoir.

of internal bleeding and necrosis of tissue under the skin, these areas turn black. Historically, bubonic plague is referred to as the "Black Death." Untreated, the human mortality rate for bubonic plague initially was about 75%. Today, 25% to 50% of untreated cases are fatal (Roberts and Janovy 1996). Of the three types of plague, however, bubonic is the least infectious.

Occasionally, plague bacteria do not concentrate in lymph nodes but instead infect the entire bloodstream. This results in septicemic plague. Pneumonic plague results when the bacteria in human hosts move to the lungs. The lungs fill with a frothy, bloody fluid—a hemorrhagic bronchiopneu-

monia. Droplet-borne bacteria are highly infectious and may be spread directly from person to person without the need of intermediate fleas. The human mortality rate for untreated septicemic or pneumonic plague is 100%, and following onset of symptoms, life expectancy is only 1 to 8 days.

In the last 1500 years, three plague **pandemics** (large-scale outbreaks over wide geographic areas) have been recorded. The first began in Arabia and spread to Egypt in 542 AD and to the Roman Empire and Europe between 558 and 664 (Twigg 1978; Acha and Szyfres 1989). An estimated 100 million people died. The second pandemic, the "Black Death" of Europe, probably began in Asia in the 1340s, spread across Europe, and reached England in 1348. Over 25 million people died as periodic **epidemics** (outbreaks affecting many people in an area) continued throughout Europe until the mid-1600s. Following this pandemic, it took 200 years for Europe to regain its 1348 population level (Twigg 1978). In the last plague pandemic, which broke out in 1894 and continued until the 1930s, 10 million people died throughout Southeast Asia, South Africa, and South America.

Plague is not just of historical interest, however. Several areas throughout the world continue to harbor the disease, with about 2000 cases reported each year. Plague is endemic in rodent populations in localized focal areas in western North America, southern Africa, the Middle East, China, and Southeast Asia (figure 27.12). Although deaths still occur from plague each year, it is treatable with a variety of antibiotics if properly diagnosed. However, given the ability of microorganisms to mutate rapidly and become immune to antibiotics, and the ability of people to travel anywhere in the world in a matter of hours, the potential exists for new outbreaks of plague.

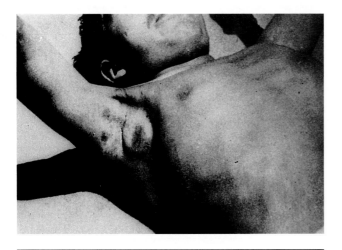

Figure 27.11.   **Plague symptoms**   A bubo (lymphadenitis) such as this typically occurs in people infected with bubonic plague.

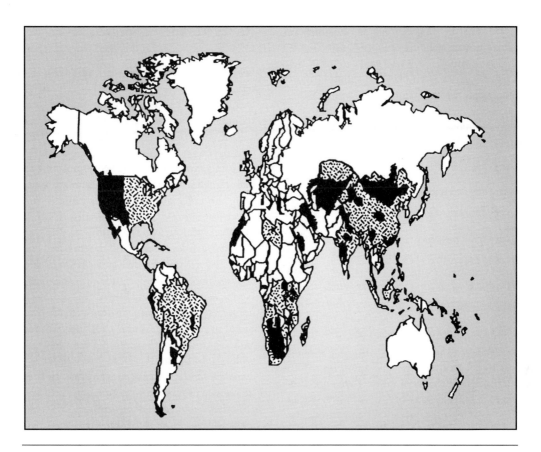

Figure 27.12   **Worldwide distribution of plague.**   Countries reporting plague (*gray*) and probable plague focal areas (*black*).

Source: Data from Centers for Disease Control.

Lyme Disease

Named for Lyme, Connecticut, where it was first recognized in the United States, this newly emerging disease is increasingly common (figure 27.13). The etiological agent is the spirochete (corkscrew-shaped) bacterium *Borrelia burgdorferi*. The vectors are ticks of the *Ixodes ricinus* complex, found throughout North America, Europe (where the disease was originally described), and Asia (Lane et al. 1991). The disease correlates closely with distribution of the deer tick (*I. scapularis*) in the northeastern and midwestern United States and the western black-legged tick (*I. pacificus*) in the western United States. Several other species of *Borrelia* carried by ticks or lice cause related diseases throughout Africa, Asia, and North America (Burgdorfer and Schwan 1991). Numerous mammalian species, primarily rodents, are reservoirs for Lyme disease throughout the world. The main reservoirs in central and eastern North America are white-footed mice (*Peromyscus leucopus*) and white-tailed deer (Ostfeld 1997). In the western United States, the primary reservoir is the dusky-footed woodrat (*Neotoma fuscipes*). In Europe, the main rodent reservoirs are the bank vole (*Clethrionomys glareolus*) and wood mice (*Apodemus* spp.). Rodents probably are not affected by the bacterium in terms of decreased reproduction or survival (Gage et al. 1995).

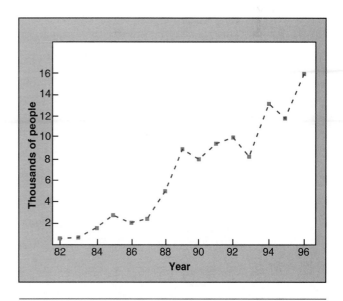

Figure 27.13   **Lyme disease.**   The number of cases of Lyme disease reported yearly in the United States is increasing.

Source: Data from National Center for Infectious Diseases.

Ticks likely transmit the spirochete to humans through saliva but only after 24 hours or more of attachment (Piesman et al. 1991; White 1993). Ticks pass through three developmental stages: larva, nymph, and adult (figure 27.14). They are most likely to transmit Lyme disease during the nymph stage. They actively feed during this developmental stage but are unlikely to be noticed because of their small size (about 1 mm). Although larval ticks also feed and are even smaller (0.5 mm), they rarely carry the infection. Adult ticks can also transmit Lyme disease. Because of their large size, however, they are much more likely to be noticed and removed in less than the 24 hours necessary to transmit infection.

Symptoms of Lyme disease may include fatigue, fever, muscle and joint pain, and a characteristic bull's-eye-shaped skin rash. This rash (erythema migrans), seen 3 to 30 days after infection, lasts 2 to 3 weeks. When diagnosed early, Lyme disease may be successfully treated by antibiotics. Unfortunately, positive diagnosis often is difficult because the diversity of symptoms varies from patient to patient. For example, a rash does not form in about 20% to 40% of cases (Barbour and Fish 1993). Conversely, a rash may result from an allergic reaction to the tick saliva rather than infection from Lyme disease. Clinically variable, Lyme disease can produce either acute or chronic disease. If untreated, the mortality rate is low (probably <5%), but infection can be disabling, leading to severe arthritis; nervous system involvement, including numbness, pain, or meningitis; and occasionally, cardiac arrhythmia (White 1993).

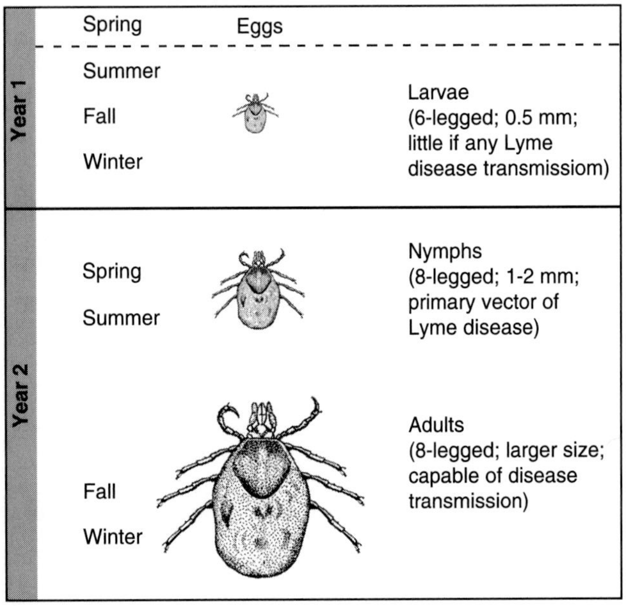

	Spring	Eggs	
Year 1	Summer		Larvae
	Fall		(6-legged; 0.5 mm; little if any Lyme disease transmissiom)
	Winter		
Year 2	Spring		Nymphs
	Summer		(8-legged; 1-2 mm; primary vector of Lyme disease)
	Fall		Adults
	Winter		(8-legged; larger size; capable of disease transmission)

Figure 27.14   Tick developmental stages.   Ticks have three developmental stages: larva, nymph, and adult. They are most likely to transmit Lyme disease during the nymph stage. Drawings illustrate relative size of each stage for *Dermacentor andersoni.*

Rocky Mountain Spotted Fever

Rocky Mountain spotted fever (RMSF) is another well-known tick-borne zoonosis caused by a rickettsial bacterium, *Rickettsia rickettsii.* Other rickettsial diseases in the United States and worldwide include rickettsialpox, murine typhus, louse-borne typhus, Q fever, and monocystic ehrlichiosis (Gage et al. 1995). *Rickettsia rickettsii* is transmitted transovarially; that is, female ticks pass the infection to their offspring. Nonetheless, the bacterium probably would not be maintained without the small mammals that serve as "amplifying" hosts (Gage et al. 1995) because *R. rickettsii* negatively affects the survival and reproduction of the ticks that carry it (McDade and Newhouse 1986).

Numerous species of rodents and lagomorphs serve as hosts for the ticks that carry RMSF, as do the opossum (*Didelphis virginiana*), carnivores, and deer (McDade and Newhouse 1986). The name of this disease actually is a misnomer, as it occurs from western Canada through the United States and Central America to Brazil (the Centers for Disease Control and Prevention frequently call it "tick-borne typhus"). In the southeastern United States, the tick *Dermacentor variabilis* is the primary vector. In the western United States, the usual vector is *D. andersoni.* In Central and South America, *Rhipicephalus sanguineus* and *Amblyomma cajennense* carry RMSF. Ticks must remain attached 10 to 20 hours to transmit infection. Clinical symptoms occur 2 to 14 days after inoculation and may include fever, headache, skin rash, and anorexia, with eventual vascular damage, kidney failure, and central nervous system involvement (Clements 1992). Antibiotics are successful in treating RMSF; left untreated, human mortality rates can approach 70%. About 800 cases of RMSF a year are reported in the United States.

Tularemia

This zoonosis is most commonly associated with lagomorphs, usually rabbits (*Sylvilagus* spp.), and is maintained primarily through the tick-lagomorph cycle. The causative agent, however, the bacterium *Francisella tularensis*, has been documented in over 100 mammalian species (Bell and Reilly 1981; Gage et al. 1995), including voles (*Microtus* spp. and *Clethrionomys* spp.), beavers (*Castor canadensis*), and muskrats (*Ondatra zibethicus*). Tularemia occurs worldwide in the Northern Hemisphere from above the Arctic Circle to 20° N latitude. Humans can contract the disease in several ways (figure 27.15): through direct contact with infected animals, from biting flies or ticks (primarily *Amblyomma americanum* but also other species), or from water contaminated by urine from infected animals. The incubation period to onset of symptoms is 3 to 5 days. Six types of tularemia have been described; all begin with fever, chills, muscle and joint pain, and malaise, and lead to respiratory involvement. The seriousness of the disease varies among individuals depending on the route of infection, bacterial load, and the particular strain of *F. tularensis* acquired

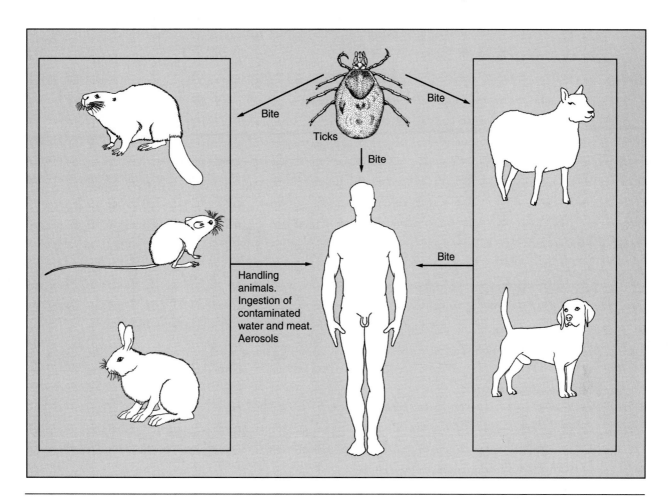

Figure 27.15 **Tularemia.** The several possible modes of transmission for tularemia from mammalian reservoirs to people.

(Hopla and Hopla 1994). Tularemia is successfully treated with antibiotics. In the United States, overall mortality rate of untreated cases is about 7%.

NONVECTOR ZOONOSES

The remaining zoonotic diseases we discuss do not involve arthropod vectors. The first, rabies, has an ancient history, whereas the hemorrhagic fevers and prion diseases have been recognized as a concern only relatively recently.

Rabies

Rabies is a disease of great antiquity. The Eshnunni Code of Mesopotamia (pre-2200 BC) called for fining the owner of a rabid dog that killed someone. The Greek philosopher Democritus was the first to describe cases of rabies in 500 BC. The causative agents of rabies and related diseases are distinct molecular strains of RNA-viruses in the Genus *Lyssavirus* (the Greek *lyssa* means "madness"), Family Rhabdoviridae. *Lyssavirus* occurs almost worldwide in a variety of mammalian hosts. Rabies virus is transmitted to humans most commonly through bite wounds or cuts and less commonly through mucous membranes or from inhalation. After inoculation, the virus infects nerve tissue. It eventually reaches the spinal cord and moves to the brain. Multiplying rapidly in the brain, the infection moves out along peripheral nerves throughout the body. In humans, symptoms generally appear 30 to 90 days after exposure, with fever, headache, and unusual tactile sensations. A number of other effects may follow, including apprehension, agitation, disorientation, hypersalivation, and paralysis. Once symptoms appear, death generally occurs in 100% of cases within a week or sooner (Fishbein 1991; Krebs et al. 1995).

Carnivores are the primary reservoir for the maintenance and transmission of rabies worldwide. Domestic dogs are the main carrier in developing countries; 22 types of rabies virus have been identified just from dogs. In North America, primary hosts are raccoons (*Procyon lotor*), skunks (most often the striped skunk, *Mephitis mephitis*), foxes (usually the red fox, *Vulpes vulpes*), and various species of bats (figure 27.16). Rabid vampire bats in Central and South America, principally *Desmodus rotundus* (see chapter 12), primarily affect the cattle industry, but they may also infect humans (Lopez et al. 1992). Rodents and lagomorphs also

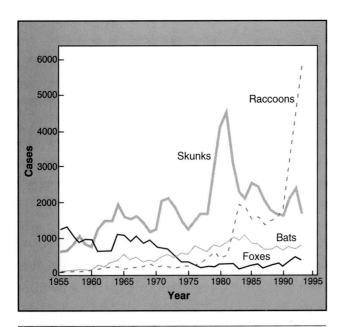

Figure 27.16   **Rabies in the United States.**   Among wild mammals in the United States, rabies occurs most commonly in raccoons. Geographic increases in rabies distribution can be caused by hunters releasing animals into new areas.

Source: Data from J.W. Krebs et al., "Rabies: Epidemiology, Prevention, and Future Research" in Journal of Mammalogy, *76:681-694, 1995.*

are capable of transmitting rabies but are unlikely to do so. They represent less than 1% of reported cases in the United States, almost all of which involve marmots (*Marmota monax;* Krebs et al. 1994).

Rabies infection in humans is relatively rare in developed countries. In the United States, only six people died from rabies in 1994. Controlling rabies, however, costs hundreds of millions of dollars a year (Smith 1996), including operating diagnostic laboratories in all 50 states. It remains a significant health problem in less developed countries (Krebs et al. 1995). Up to 50,000 people a year die of rabies in India. Similar numbers probably die in other Asian, African, and Latin American countries (Robinson 1992). Much research is directed toward developing vaccines to reduce the incidence of rabies in wildlife populations. Although untreated rabies in humans is almost always fatal, there have been rare exceptions. Brass (1994) documented three cases of people surviving rabies, but all suffered subsequent neurological impairment. Presumably, other mammalian carriers also succumb to infection. Brown and colleagues (1990) suggested, however, that a small percentage of individuals survive and become chronic carriers, a situation that potentially could have a significant effect on wild populations.

Hemorrhagic Fevers

Hemorrhagic fever actually encompasses a number of different viral diseases involving four families of viruses: flaviviruses, arenaviruses, bunyaviruses, and filoviruses. These

widely distributed viruses are often spread directly from rodent hosts to humans. The arenaviruses (Family Arenaviridae) include species that cause a variety of human diseases, including lymphocytic choriomeningitis, which is rarely serious; Lassa fever, which has a mortality rate of about 1%; and Argentine, Bolivian, and Venezuelan hemorrhagic fevers (Childs et al. 1995). Untreated, these last three diseases are fatal in 10% to 30% of cases.

Viruses in the Family Bunyaviridae include the Genus *Hantavirus.* In Europe and Asia, hantaviruses cause several zoonotic diseases, collectively referred to as hemorrhagic fever with renal syndrome, in up to 200,000 people a year. First recognized in the United States in 1993, hantavirus pulmonary syndrome (HPS) occurs primarily in the Southwest and secondarily in the Midwest and Southeast. By 1997, however, it was identified in 26 states, as well as Paraguay, Argentina, and southern Chile. Like all viral hemorrhagic fevers, symptoms of HPS include fever, muscle aches, headache, and cough. Progressing to severe lung involvement, HPS is fatal in over 50% of human cases. Rodents are the reservoir of the virus, primarily the deer mouse (*Peromyscus maniculatus*) and the cotton rat (*Sigmodon hispidus*). Nine different strains of hantavirus are currently recognized, each specific to a single rodent genus (Childs et al. 1995). The contagion is spread, often in dry, dusty areas, through infected rodent saliva or excreta inhaled as aerosols, or directly through broken skin (including bites) or contact with mucous membranes.

Hantavirus may be maintained and transmitted among wild rodent populations by individuals biting each other; arthropod vectors are not involved in transmission. Apparently, infection in rodents does not affect their survival or overall fitness. Unlike rabies, domestic pets are not known to carry hantavirus. Although few cases of HPS have been identified compared with other zoonoses, given the high mortality rates and the large number of researchers who work with wild rodents, mammalogists should be aware of the potential hazards and observe measures to minimize exposure (Mills et al. 1995).

Ebola hemorrhagic fever is caused by one or more of the filoviruses (McCormick and Fisher-Hoch 1994). Highly pathogenic, Ebola virus was first noted from outbreaks in 1976 in Zaire (Republic of Congo) and Sudan that resulted in over 340 deaths. Another outbreak occurred in Kikwit, Zaire, in 1995. The typical fever, headaches, and muscle aches begin 4 to 16 days after infection and progress to kidney and liver involvement. Eventually, patients may begin bleeding both internally and externally. The virus can be spread easily from person to person; the mortality rate is 50% to 90%. The natural reservoirs and hosts of the Ebola virus, as well as how it is spread, are unknown. Ebola-related viruses, including the Marburg virus from Africa and Reston virus from Asia, have been isolated from the Southeast Asian crab-eating monkey (*Macaca fascicularis*) and African green monkey (*Cercopithecus aethiops*).

Four distinct viral serotypes of the Genus *Filovirus* cause dengue hemorrhagic fever, or "breakbone fever," which does involve a vector. The virus is carried directly between humans by mosquitos of the Genus *Aedes*, especially *A. aegypti*, although wild monkeys can be involved as mammalian reservoirs. Hundreds of thousands of cases are reported yearly, with a mortality rate of about 5%.

Spongiform Encephalopathies

Finally, a group of transmissible, progressive, neurodegenerative diseases that afflict mammals (table 27.3) result from a very different method of infection from those discussed previously. One of these diseases is bovine spongiform encephalopathy (Bradley and Lowson 1992), or "mad cow" disease—one of several prion diseases. The infectious agents in each are **prions**—small, modified forms of cellular protein thought to be associated with synaptic function in neurons. Prion diseases, or spongiform encephalopathies, are characterized by large vacuoles (open areas) that occur in the cortex and cerebellum of the brain. These produce loss of motor control, dementia, paralysis, and eventually death.

In other infectious agents, genetic information is transmitted through nucleic acids. Besides being proteins, prions are unique disease agents because they appear to be

Table 27.3.   Prion diseases

Disease	Natural host
Scrapie	Sheep and goats
Transmissible mink encephalopathy (TME)	Mink
Chronic wasting disease (CWD)	Mule deer; elk
Bovine spongiform encephalopathy (BSE)*	Cattle
Kuru	Humans
Creutzfeldt-Jakob disease (CJD)	Humans
Gerstmann-Sträussler-Scheinker syndrome (GSS)	Humans
Fatal familial insomnia	Humans

Source: From S.B. Prusiner, "Prion Biology" in Prion Diseases of Humans and Animals, *S. Prusiner, J. Collinge, J. Powell, and B. Anderton (Eds.), 1992, Ellis Horwood, New York.*

*"Mad cow" disease.

both infectious and hereditary. Although transmission to humans is probably rare, evidence suggests that people can be infected by ingestion of prion-infected animal products. This is a very active area of current research (Prusiner et al. 1992; Prusiner 1993; Prusiner and Hsiao 1994; DeArmond and Prusiner 1995).

Summary

Parasitism is a symbiotic relationship between a parasite and a host. Parasites are relatively small, often microscopic; derive their nutrients from their host; and spend all or a part of their lifetime in or on the host. Depending on its size, life history characteristics, and host attachment, a given species of parasite is either a micro- or macroparasite and an endo- or ectoparasite. Microparasites include viruses, bacteria, fungi, and protistans. Macroparasites include platyhelminths, nematodes, acanthocephalans, and arthropods. Members of each of these groups are parasitic on mammals.

Many parasites do not kill their definitive host and may produce only subtle adverse effects. Parasites are more often pathological to intermediate and accidental hosts or to heavily infected definitive hosts. The influence of parasites on the dynamics of mammalian host populations is often subtle and inconspicuous. At the population level, a number of variables may affect overall mortality rate or reproductive potential of host populations. Parasites are rarely used for biological control of mammalian populations, but some attempts have been made.

Certain species of parasites are often closely associated with a particular mammalian taxon, resulting in a high degree of host specificity. These closely associated host-parasite assemblages may have coevolved; that is, reciprocal changes took place in the behavior, physiology, and morphology of both host and parasite.

Some parasitic species serve as vectors of disease-causing organisms. Organisms that cause disease are called etiological agents. Mammalian species serve as reservoirs for many diseases that can infect humans. Several zoonoses are discussed, representative of viral, bacterial, protistan, platyhelminth, and arthropod parasites. Many of these zoonotic diseases, including rabies and hemorrhagic fevers, have a high human mortality rate. Throughout history, for example, plague has killed hundreds of millions of people. Other important zoonotic diseases have been discovered relatively recently. These include Lyme disease, the Ebola and hantaviruses, and protein-based prion diseases, all fertile areas for basic and applied research.

Discussion Questions

1. Investigate current control methods for one of the zoonoses discussed in the text. Specifically, what efforts are possible for breaking the transmission cycle, enhancing host resistance, destroying the infectious agent, or destroying the vector?

2. What role does the mobility of mammalian hosts play in coevolutionary development of host-parasite assemblages? Why might host-specific relationships be more likely to develop in pocket gophers than in white-tailed deer?

3. We noted that relatively few studies have investigated the effects of parasitism on the dynamics of host populations. Discuss some of the practical difficulties in designing such a study.

4. Over 100 years ago, Robert Koch, a German medical bacteriologist, developed rules (a set of procedures) for identifying the causative agent of a disease. These procedures, Koch's Postulates, are still used today. Try to identify necessary steps for identifying the causative agent of a disease, then look up Koch's procedures in a microbiology textbook.

Suggested Readings

Brooks, D. R. 1979. Testing the context and extent of host-parasite coevolution. Syst. Zool. 28:299–307.

Davis, J. W., L. H. Karstad, and D. O. Trainer (eds.). 1981. Infectious diseases of wild mammals, 2d ed. Iowa State Univ. Press, Ames.

Ginsberg, H. S. (ed.). 1993. Ecology and environmental management of Lyme disease. Rutgers Univ. Press, New Brunswick, NJ.

May, R. M. and R. M. Anderson. 1990. Parasite-host coevolution. Parasitology (Suppl.)100:S89–S101.

Domestication and Domesticated Mammals

Most people have seen cows and horses grazing in fields and are familiar with barnyard animals such as goats and pigs. You may have a dog or cat as a pet. We realize that these species of mammals, living in close association with humans, are "different" in certain respects from wild species. In this chapter, we will discuss the domestication of mammals and the benefits to humans associated with it. Considering the number of mammalian species, relatively few have been **domesticated,** that is ". . . bred in captivity for purposes of economic profit to a human community that maintains complete mastery over its breeding, organization of territory, and food supply" (Clutton-Brock 1981:21). This definition is useful for our purposes; Bökönyi (1989) notes several other definitions used by researchers.

However domestication is defined, several questions arise concerning domesticated animals. Why domesticate at all? What was the process of domestication among early humans? How did it begin? Why did people choose to stop killing wild animals and begin to herd and propagate them? When did various species begin to be domesticated and where? Why were so few species of mammals "selected" for domestication? What behavioral or physiological characteristics lend a species to successful domestication? These are all interesting questions because of their effect on human history and because, in addition to mammalogy, they encompass anthropology, archaeology, and related disciplines. As we explore these questions, we will find that domestication includes both biological and cultural processes that continue today. Some mammalian species have been domesticated for thousands of years, and even today people are evaluating the potential of wild species for eventual domestication.

WHY DOMESTICATE MAMMALS?

From the dawn of human existence, our ancestors survived as hunters, gatherers, and scavengers. As the glaciers receded 10,000 to 12,000 years ago and the worldwide climate changed, humans began to make a transition from small groups of nomadic hunter-gatherers to a more sedentary life-style. This change allowed for the eventual establishment of discrete, larger human population centers. A major "cultural revolution" necessarily accompanied the transition from a nomadic to a sedentary life-style, however. This revolution involved the need for a sustainable food resource base to maintain permanent settlements (Reed 1984). Dependable, sustainable energy sources were provided by both cultivated food plants and domesticated animals. Thus, a major behavioral shift in humans gradually took place from simply hunting and killing animals to keeping and raising them. Eventually, domesticated animals provided numerous benefits to human society, economics, and culture.

Archaeological evidence suggests that agriculture, the cultivation of cereal grains (or "plant domestication"), began about 9000 years ago (Reed 1977, 1984). Several criteria were

necessary for agriculture to begin in an area (Sauer 1969). Primarily, food could not be chronically scarce; people had to have the time and resources to experiment with cultivation. They probably were already sedentary and had an alternative food source, such as fish. Whatever the circumstances, the rise of agriculture was preceded by the domestication of mammals. The first domesticated animals raised purely for food were sheep and goats as early as 10,000 to 11,000 years ago. This was a critical development because food storage in those days was a problem. Goats and sheep provided "live stock," or a "walking larder," (Clutton-Brock 1989) that could accompany small groups of people as they moved. One mammalian species was domesticated even earlier than sheep and goats, however. This was the dog, first domesticated about 12,000 years ago (figure 28.1), but not primarily to provide food (Davis and Valla 1978; Reed 1984; Morey 1994). Instead, dogs were valuable companions and probably assisted in hunting. They also scavenged human debris around camps and thus helped reduce intrusion by unwanted predators, warned of approaching predators, and, as they do today, no doubt provided mutual affection. Other mammalian species eventually were domesticated not only for food and milk, but also for a variety of other purposes. They provided hides for shelter or for use as carrying containers; wool or fur for clothing; dung for fuel; and transportation for people and products. They also were important in sport, war, or simply for prestige. Domestication of animals dictated the partial or complete separation of breeding stock from their wild ancestors. The process of domestication began human attempts to control nature. Substituting artificial selection for natural selection was an innovation in human history of the magnitude of controlling fire or developing tools (Davis 1987).

HOW DOMESTICATION BEGAN

The process of domestication began with Paleolithic people and continued over thousands of years. Domestication probably was not a conscious decision on the part of early humans. That is, initially people probably were unaware of what was happening as ephemeral associations between animals and people became more long-term and involved. Nonetheless, over a 5000-year period beginning about 10,000 years ago, humans changed from nomadic hunters and gatherers to sedentary farmers and herders in more densely populated areas. Domesticated mammals made this change possible and allowed people to become non-food-producing specialists, such as artisans, scholars, and soldiers (Diamond 1994). This revolution began with the development of dogs from wolves (see section on the dog in this chapter), as wolves and humans hunted the same prey and were necessarily in close association. Zeuner (1963) summarized the probable stages of mammal domestication, all of which can be easily visualized in the transition from wolves to dogs, but also apply to other species.

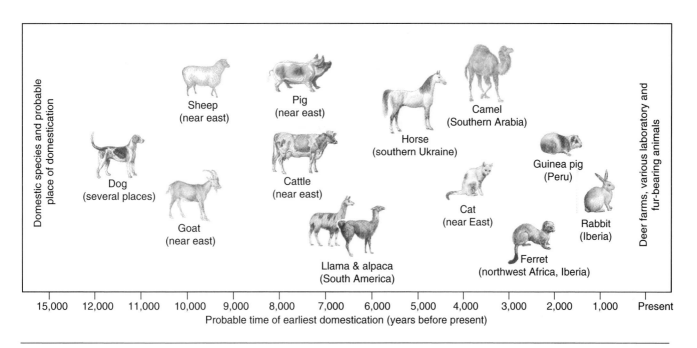

Figure 28.1   **Important domesticated mammals.**   The estimated times and places of origin of domesticated mammals are based on archaeological evidence.

Source: Data from D.F. Morey "The Early Evolution of the Domestic Dog" in American Scientist, *July-August, 1994, 82:336-247.*

Probable Stages of Mammal Domestication

1. Initially, the contacts between people and wild species were loose, with free breeding, uncontrolled by humans.

2. Through time, individual free-ranging animals were confined in and around human settlements, with breeding eventually occurring in captivity.

3. Breeding of confined individuals became more selective and organized by humans, probably with occasional crossing with wild individuals, to obtain certain desired behavioral and morphological characteristics.

4. Human economic considerations led to greater selectivity for various desirable properties, with the resultant formation of different "**breeds,**" that is, animals with a uniform, heritable appearance (Clutton-Brock 1992).

5. Because the wild ancestors of domesticated species were now competing for grazing resources or were predators on livestock, these wild species often were persecuted or exterminated.

Shared Behavioral and Physiological Characteristics

Regardless of the species that were domesticated or the exact process by which it happened, almost all domesticated mammals share certain behavioral and physiological characteristics that lend them to domestication. Because the

characteristics are rather specific, relatively few mammalian species have been successfully domesticated over many thousands of years. These characteristics were first summarized over 130 years ago by Galton (1865) and more recently by Clutton-Brock (1981). Specifically, we expect mammalian species suitable for domestication to be

1. Adaptable in terms of diet and environmental conditions

2. Highly social, with behavior based on a juvenile period with strong social bonding and eventual development of a dominance hierarchy

3. Easily maintained as a resource for products, as well as for food when necessary

4. Easily bred in captivity, because living offspring are one of the most important products

5. Closely herded, not adapted for instant flight, and thus easy for herders to tend, especially important for livestock such as sheep, goats, and pigs

These criteria necessarily preclude individuals that are strongly territorial and resist close association (cats are an exception). They also limit the potential number of domesticates. Archaeological evidence and the range overlap of ancestral species suggest that domestication of most mammals began in the Fertile Crescent area of the Middle East (figure 28.2) and possibly simultaneously in China and eastern Asia (Isaac 1970; Davis 1987).

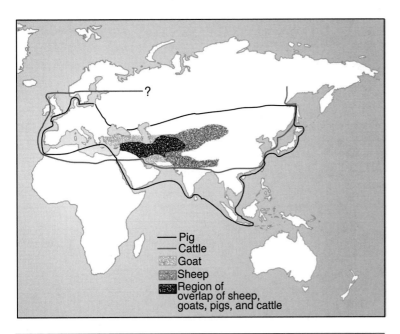

Figure 28.2   Areas of origin.   The hypothesized area of origin of domesticated sheep, goats, pigs, and cattle overlaps the region of early agriculture and the ranges of the ancestral species.

Source: Data from Erich Isaac, Geography of Domestication, *1970, Prentice-Hall, Inc., Englewood Cliffs, NJ.*

MAMMALS THAT HAVE BEEN DOMESTICATED

Relatively few mammalian species have been domesticated. Of the approximately 4600 species of extant mammals recognized today, less than 1% have been domesticated (table 28.1). Most mammalian species are small and of little use as domesticates. Nonetheless, of the almost 150 species of large herbivores (over 50 kg), only 5, all from Eurasia, have been domesticated and spread beyond their original range—horse, cow, sheep, goat, and pig (Diamond 1994).

Different stages in the development of domesticated mammals can be associated with evolving human social and economic development, including the concept of private ownership of animals. Zeuner (1963) grouped domesticated mammals into several temporal or functional stages as follows:

Temporal

Preagriculture	Dog, goat, sheep, reindeer
Early agriculture	Cattle (including water buffalo, yak, and banteng), pig

Functional

Transport and labor	Horse and ass, camels and llama, elephants
Destroy pests	Cats, ferrets
Experimental	Deer ranches, African bovids, fur farms

Rats, mice, and rabbits could be included within an additional functional category, namely medical research. These groups are useful as we briefly review the major groups of domesticated mammals. As noted in previous chapters, some species of domesticated mammals are no longer found in the wild, and their ancestry is uncertain. For other domesticated species, their wild ancestors are clearly evident.

Dog

The dog (*Canis* [*familiaris*] *lupus*) was the first domesticated mammal (see figure 28.1). Based on morphological, behavioral, and genetic criteria (Clutton-Brock and Jewell 1993), dogs clearly originated from wolves (*C. lupus*). They are an exception to the usual reasons mammals were domesticated, that is, as a source of food or raw materials. Instead, dogs originally provided a mutually beneficial hunting team with humans. Both wolves and Paleolithic people at the end of the Pleistocene epoch (about 10,000–12,000 years ago) were similar in many respects, and development of a mutually beneficial association is easy to envision.

Both wolves and early humans formed small hunting groups and stalked the same prey throughout similar environments around the world. As opportunistic scavengers, wolves and humans often hunted in the same areas. These wolves were tolerated by humans, and their pups may have been cared for and occasionally tamed. Soon their utility was evident, as they warned against predators or other enemies and helped to locate game animals (Isaac 1970). As generations passed, certain individual, tame wolves that had the requisite temperament and behavior (and were not eaten by humans during times of food shortage) began to breed and leave offspring with increasingly more submissive, placid dispositions. Through time and in different parts of the world, these lines of wolves eventually evolved into "dogs." With continued artificial selection, different breeds evolved. Several breeds are recognized from ancient Egypt and early Roman times (Olsen 1985; Clutton-Brock and Jewell 1993). Through continued artificial, selective breeding, many types of dogs exist today (figure 28.3); 134 breeds were officially recognized by the American Kennel Club in 1992. All are the same species and reflect the variety that artificial selection has produced over a 10,000-to-12,000-year period.

Goats and Sheep

Originating in central and western Asia, goats and sheep were the first mammals domesticated as livestock. They provided humans with a sustained resource base for food, milk, hides, and other products as early as 10,000 years ago. Domestication was favored by their environmental and feeding adaptability, social behavior, and ease of herding. Domesti-

Table 28.1.   Taxonomic grouping of domesticated mammals and their wild ancestors*

Domestic Form	Wild Ancestor
Lagomorphs	
Rabbit (*Oryctolagus cuniculus*)	European wild rabbit (*Oryctolagus cuniculus*)
Rodents	
Guinea pig (*Cavia porcellus*)	Cavy (*Cavia aperea*)
Laboratory mouse (*Mus* [*domesticus*] *musculus*)	House mouse (*Mus musculus*)
Laboratory rat (*Rattus norvegicus*)	Norway rat (*Rattus norvegicus*)
Chinchilla (*Chinchilla lanigera*)	Chinchilla (*Chinchilla lanigera*)
Golden hamster (*Mesocricetus auratus*)	Syrian hamster (*Mesocricetus auratus*)
Carnivores	
Dog (*Canis* [*familiaris*] *lupus*)	Wolf (*Canis lupus*)
Ranched fox (*Vulpes vulpes*)	Red fox (*Vulpes vulpes*)
Ferret (*Mustela furo*)	Polecat (*Mustela putorius*)
Ranched mink (*Mustela vison*)	Wild mink (*Mustela vison*)
Cat (*Felis* [*catus*] *sylvestris*)	Wild cat (*Felis sylvestris*)
Perissodactyls	
Horse (*Equus caballus*)	Wild horse (*Equus* [*ferus*] *hemionus*)
Donkey (*Equus asinus*)	Wild ass (*Equus* [*africanus*] *asinus*)
Artiodactyls	
Pig (*Sus scrofa*)	Wild boar (*Sus scrofa*)
Dromedary camel (*Camelus dromedarius*)	Unknown (*Camelus* sp.)
Bactrian camel (*Camelus bactrianus*)	Bactrian camel (*Camelus ferus*)
Llama (*Lama glama*)	Guanaco (*Lama guanicoe*)
Alpaca (*Lama pacos*)	Unknown (*Lama* sp.)
River buffalo (*Bubalus bubalis*)	Water buffalo (*Bubalus arnee*)
Domestic cattle and zebu (*Bos taurus*)	Aurochs (*Bos primigenius*)
Yak (*Bos grunniens*)	Wild yak (*Bos mutus*)
Bali cattle (*Bos javanicus*)	Banteng (*Bos javanicus*)
Domestic sheep (*Ovis aries*)	Asiatic mouflon (*Ovis orientalis*)
Domestic goat (*Capra hircus*)	Bezoar goat (*Capra aegagrus*)

Source: Data from J. Clutton-Brock, Domesticated Animals From Early Times, *1981, University of Texas Press, Austin, TX; G. B. Corbet and J. Clutton-Brock, "Appendix: Taxonomy and Nomenclature" in* Evolution of Domesticated Animals *(I. L. Mason, ed.), 1984, pp. 434-438, Longman, Inc., New York; and S. J. M. Davis,* The Archaeology of Animals, *1987, Yale University Press, New Haven, CT.*

*Species such as elephants and reindeer, which may be considered "exploited captives," are not included. There is little consensus among authorities on scientific names applied to domesticated mammals.

cation of early wild stock no doubt occurred either when young wild animals were reared by and imprinted on humans or when wild flocks were confined near water and eventually became habituated to people (Clutton-Brock 1981). Within 2000 years, several morphological changes were evident in domesticated stock, including smaller size with shorter limbs, changes in the shape of horns (figure 28.4), and hornless females. Another useful morphological change in domestic sheep involved the pelage; it no longer shed as occurred in wild species. Domestic sheep could be sheared, and their wool was not lost in the field.

Authorities do not agree on the systematics of sheep and goats. Grubb (1993) recognizes nine species of goats, several of which are considered subspecies by other authors. Domestic goats (*Capra hircus*) are derived from the ancestral bezoar goat (*C. aegagrus;* see table 28.1). Based on evidence from karyotypes, domestic sheep (*Ovis aries*) probably originated from early Asiatic mouflon (*O. orientalis*). Along with European mouflon (*O. musimon*), Asiatic mouflon now are **feral** (domesticated at one time but now wild).

Cattle

Economically, cattle are the most important livestock (table 28.2). The earliest domesticated cattle are known from 8000-year-old archaeological sites in southeastern Europe and western Asia. Domesticated cattle probably arose from the aurochs, or giant wild ox (*Bos primigenius*). Male wild oxen were very large, up to 2 m in shoulder height (figure 28.5), and reportedly very fierce. Widely distributed throughout much of the Northern Hemisphere, the aurochs became extinct in 1627, with the death of the last individuals in Poland. There has been much speculation as to the process of domestication of cattle from the wild progenitor. Given the size and fierceness of aurochsen (plural of *aurochs*, as in ox and oxen), they could not have been easy to capture or restrain. They probably were attracted to, and enticed to remain near, human settlements by humans making salt and water available. Because aurochsen would have trampled crops and attracted unwelcome predators, they eventually might have been driven into fenced areas where they were

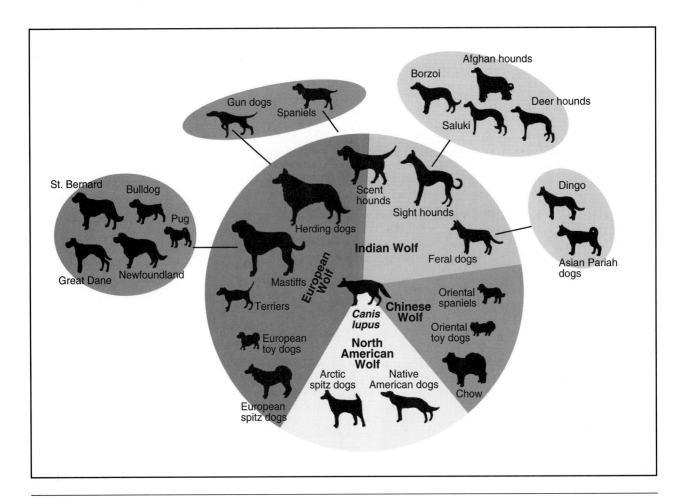

Figure 28.3 **Different dog breeds.** The numerous breeds of dog that occur in the world today (all descended from the wolf) are the products of thousands of years of artificial selection under human control.

Source: Data from J. Clutton-Brock and P. Jewel, "Origin and Domestication of the Dog" in Miller's Anatomy of the Dog, *3rd edition, 1993, W.B. Saunders, Philadelphia.*

easier to house and handle. Over time and through artificial selection resulting in smaller sizes (thus, easier to handle), distinctive breeds of "cattle" eventually emerged. Several breeds of cattle existed by the time of ancient Egypt, and today about 800 breeds are recognized.

Given thousands of years of domestication, interbreeding, and crossbreeding with wild species, it should not be surprising that the taxonomy of cattle is uncertain. They now are considered to encompass several species within the Family Bovidae, Subfamily Bovinae (see chapter 19). These are included in two groups: humpless cattle (*Bos taurus*), the most common species of domesticated mammals throughout the world; and humped cattle, which include the distinctive zebu (*B. [indicus] taurus*), with its large muscular hump, large hanging ears, and narrow face (figure 28.6). Zebu were considered by Grubb (1993) to be **conspecific** (the same species) with humpless cattle. There are several other species of cattle, none of which occurs beyond the range of its ancestral species. These include the domestic yak (*B. grunniens*), which today serves a variety of purposes in the high elevations of the Himalaya Mountains. Yaks interbreed with *B.*

taurus, although only the female offspring are fertile. Wild yaks (sometimes considered a distinct species, *B. mutus*) are now very rare. Cattle also include the Asian mithan, or gayal (*B. frontalis*), which is not truly domesticated. The mithan is used today, as it was historically, primarily for sacrificial purposes (Clutton-Brock 1981). The mithan is believed to be descended from the gaur (*B. [gaurus] frontalis*), the largest extant bovid. The banteng (*B. javanicus*) is another domesticated bovid bred in Malaysia, Sumatra, Borneo, and Java. There also are several species of domesticated water buffalo (Genus *Bubalus*). The domestic buffalo (*B. bubalis*) is the most common species of cattle in much of Asia.

Pig

The ancestor of domestic pigs (*Sus scrofa*) was the wild boar, and they are considered to be the same species. Wild boars were widely distributed throughout most of the Old World. They may have been domesticated first in the Middle East and western Asia, about 7000 to 8000 years ago (Reed 1969; Clutton-Brock 1981). Pigs are ideal for domestication be-

A   B

Figure 28.4 **Domestication changes morphology.** Among other characteristics, domestication results in changes in horn shape. (A) Domestic goats have horns that are either straight or twisted, whereas (B) their wild ancestors have large, scimitar-shaped horns.

cause of their diverse feeding habits. It is easy to envision the initial process of domestication. Pigs could have thrived by scavenging on discarded food remains in association with early human settlements, before they were eventually confined. Also, piglets are easily tamed and acclimated to humans. Historically, groups of domesticated pigs were allowed to roam loose, watched by a swineherd, or were housed in a "pig sty" or pen. Numerous breeds of pigs are now recognized (figure 28.7). Pigs have served as food and as sacrificial animals in many cultures. Later, pigs were considered unclean by Middle Eastern cultures, and religious sanctions were raised against eating pork. Such prohibitions must have developed only after pigs had been domesticated and eaten for thousands of years.

Table 28.2. **Estimated numbers of domesticated mammals worldwide***

Perissodactyla	Numbers
Horses	58,158,000
Asses	43,772,000
Mules	14,952,000
Artiodactyla	
Cattle	1,288,124,000
Sheep	1,086,661,000
Pigs	875,407,000
Goats	609,488,000
Buffalo	148,798,000
Camels	18,831,000

Source: Data from FAO Production Yearbook 1994, *Vol. 48, page 243. Food and Agriculture Organization of the United Nations, 1995.*

*These figures vary yearly.

Horse and Donkey

The horse (*Equus caballus*) was one of the last common livestock species to be domesticated (see figure 28.1), about 5000 years ago in the southern Ukraine (Bökönyi 1984). From this region, domesticated horses spread throughout Europe and Asia. As noted by Clutton-Brock (1981), this species is the least changed from the ancestral stock. Although originally used as a source of meat, the primary purpose of domesticated horses soon became carrying people and goods long distances as quickly as possible. Horses revolutionized transportation on land. For the first time in history, humans could move themselves and their goods to other places faster than their own legs could carry them.

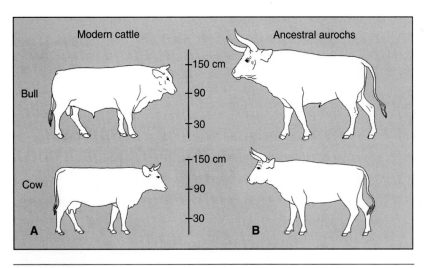

Figure 28.5   **Ancestors of cattle.**   Relative sizes of (A) representative modern cattle and (B) aurochsen. There are almost 800 breeds of cattle today, with much variation in size.
Source: Data from S.J.M. Davis, The Archaeology of Animals, *1987, Yale University Press.*

With the introduction of chariots around 2000 BC, horses also provided a revolutionary military advantage. Horses were already well adapted for these functions, so little artificial selection was necessary. Thus, it is very difficult to distinguish domestic from wild horse remains at archaeological sites, and the ancestor of the domestic horse is uncertain. Most authorities consider horses to be derived from the feral horse, or tarpan (*E.* [*ferus*] *hemionus*). Nevertheless, many breeds of horses are recognized today. The domestic horse has a long, flowing mane that is not shed. This is the result of a mutation not found in wild horses, which have a short, erect mane (figure 28.8) that is shed annually.

The origin and taxonomy of the donkey or domestic ass (*E. asinus*) also are uncertain and somewhat confusing. The only livestock species with an African ancestry, they probably were derived from the African wild ass (*E.* [*africanus*] *asinus*) at about the same time as the horse. Asian species and subspecies of wild equids also occur, including the onager, or kulan (*E. hemionus*), and the kiang (*E. kiang*), which have been interbred with each other for millennia. Distinguishing the remains of wild individuals from those of domesticated ones at archaeological sites is very difficult (see Clutton-Brock 1981), as is species designation of many living groups.

Camels, Llama, and Alpaca

As discussed in chapter 19, the taxonomy of the Family Camelidae is uncertain, because some of the domesticated species no longer exist in the wild. Camelids in the Old and New World always have been valued for their strength and endurance in harsh environments. As a result, as is true of horses, little artificial selection has taken place away from wild forms, and wild remains often cannot be differentiated from domesticated ones at archaeological sites. Little is known of the ancestors of domesticated camelids, due in part to early hybridization within both Old and New World groups (Stanley et al. 1994). The one-humped (dromedary) camel (*Camelus dromedarius*) may have been domesticated in southern Arabia about 5000 years ago, primarily as a pack animal and for riding (Bulliet 1975). Two-humped Bactrian

Figure 28.6   **Modern cattle.**   Cattle are often placed in two general categories: those with humps, such as this zebu, and those without humps, such as most breeds found in North America.

Figure 28.7   **Breeds of domestic pigs.**   More than 100 breeds of domestic pigs are recognized throughout the world today. The Yorkshire breed shown here is generally all white. Originally bred in England, it is still referred to there as the Large White.

A   B

Figure 28.8   **Horses.**   (A) Przewalski's horse (*Equus* [*przewalskii*] *caballus*), probably extinct in the wild, retains the characteristic short, erect mane of wild ancestral equids, unlike (B) the domestic horse.

camels (*C. bactrianus*), primarily draft and pack animals, may have been domesticated near present-day Iran about the same time.

Domestication of New World camelids, the llama (*Lama glama*) and alpaca (*L. pacos*), probably occurred earlier, 6000 to 7500 years ago, in high-elevation regions of the Andes Mountains (Pieres-Ferreira et al. 1976; Novoa and Wheeler 1984). Native llamas and alpacas flourished until the introduction of European livestock, which followed the Spanish conquest of the Inca Empire in 1532. Even today, llamas and alpacas are relegated to marginal lands unsuitable for introduced livestock.

Cat

Cats (*Felis* [*catus*] *sylvestris*) represent the exception to the general characteristics noted earlier that define a "suitable" domesticant. As carnivores, their diet is not particularly adaptable. Neither are they highly social, nor do they provide a resource for valuable products. Despite thousands of years of association with humans, cats are little changed from the ancestral wild cat (*F. sylvestris*) of Eurasia. Most domestic cats can easily revert to a feral state if necessary. It is unknown when cats were first domesticated; their association with humans may be very ancient. Initially, cats may have entered permanent human settlements in the Fertile Crescent of the Middle East by following their primary prey—rodents. Through time and continued association with people, cats may have progressed from a commensal, rodent-catching status to eventual domestication (Baldwin 1975). Again, it is impossible to assess whether remains at many early archaeological sites are wild or domestic cats; however, cats are known to have been fully domesticated by at least 3000 to 4000 years ago in ancient Egypt. The Egyptians considered cats to be sacred animals, and it was forbidden to harm them. When they died, they were mummified. Enor-

mous numbers of mummified cats have been excavated (one collection shipped to England weighed 19 tons!). Today, of course, cats compare with dogs as the most common household pets. There is no consensus on the number of recognized breeds among various cat associations in the United States; however, there are at least 7 breeds of longhair cats and 10 of shorthair.

Elephants and Reindeer

Other species of mammals, such as elephants (chapter 18) and reindeer, may be considered "exploited captives." These were defined by Clutton-Brock (1981:104) as ". . . mammals whose breeding remains more under the influence of natural rather than artificial selection" regardless of the length of association with humans. A degree of overlap and subjectivity certainly exists in these two categories. These species may be considered tamed in that their flight distance in response to humans is reduced to zero, but they are not truly domesticated.

Elephants

The difference between tamed and domesticated is especially evident in elephants. Because of their size, strength, and unique appearance, both the African (*Loxodonta africana*) and the Asian elephant (*Elephas maximus*) have been important economically, culturally, and historically for thousands of years: in the ivory trade, as beasts of burden, in zoos and circuses, for ceremonial purposes, and in early warfare. However, neither species has been bred in captivity for many generations nor subjected to artificial selection. This is partly because the reproductive behavior of elephants is difficult to control. As noted in chapter 18, females do not reach sexual maturity until they are 9 to 12 years old, gestation is 22 months with several years between estrous periods, and a single calf is produced that is dependent on its mother for a prolonged period. Also, as with horses and camels, artificial selection has not taken place in elephants because the characteristics of use to humans—size, strength and endurance—already exist naturally (Clutton-Brock 1981).

Humans have been taming elephants for about 4000 years, originally for use in warfare. Alexander the Great used elephants in his conquests; he first learned of them in his campaigns in India and then incorporated them into his armies. For several hundred years thereafter, and well before the famous march of Hannibal across the Alps beginning in 218 BC, elephants were used in numerous ancient battles (Scullard 1974). Elephants were used against cavalry, to trample infantry, and to break down walled encampments. Elephants can be unreliable and difficult to control, however, especially in the noise and confusion of a battlefield. They "... are usually finely balanced between fight and flight, and ... do not distinguish readily between friend and foe. In many ancient battles the elephants charged the enemy and then retreated through their own lines, causing havoc to friend and foe alike" (Douglas-Hamilton 1984:195).

The Indian elephant has been used as a beast of burden for several thousand years (Carrington 1959). Their principal function was in removing logs from forest tracts. Given the amount of food elephants consume and the fact they can work hard for only a few hours a day, they are relatively inefficient and expensive to maintain. Thus, their role in the timber industry today is decreasing because machinery is more efficient and because the decrease in wild populations makes it more difficult to replace tame individuals that are injured or die. The use of elephants as a domesticated mammal is rapidly becoming negligible.

Reindeer

Of all the species of cervids (deer), reindeer (*Rangifer tarandus*) are most likely to be associated with humans. Very gregarious and nonterritorial, they naturally form large herds. Nonetheless, they are considered exploited captives because they continue to live in semiwild conditions with little if any controlled breeding. Although reindeer may have been semidomesticated (or more properly "herded") very early, no direct evidence exists as to when this first occurred. Also, it is unknown whether they were first domesticated in one area or independently in several different areas. Nonetheless, the adaptations of reindeer that allow them to survive the extremes of harsh northern environments have been directly responsible for the habitation of northern regions of Scandinavia and Russia by humans, who depend greatly on reindeer. Reindeer provide meat and hides, and, secondarily, transportation. Their milk is not an important product, however. In the Western Hemisphere, domesticated reindeer have been introduced to the Seward Peninsula of Alaska, the lower Mackenzie Valley of northwest Canada, and parts of coastal Greenland (Skjenneberg 1984).

MORPHOLOGICAL EFFECTS OF DOMESTICATION

From the preceding discussion, we can see that of the common domesticated mammals, almost all exhibit distinct morphological, physiological, reproductive, and behavioral changes from their wild ancestors. Several generalizations are possible about these changes (Clutton-Brock 1992). The most obvious change is reduction in size. Most breeds of dog are smaller than ancestral wolves, and cattle are smaller than aurochsen. This trend may have resulted from intentional selection toward smaller individuals that were easier to handle or herd, or from dietary changes associated with captivity. Skulls are often shorter and dentition is reduced. Cranial capacity also is reduced in domesticated mammals, and they have relatively smaller brains than their wild counterparts. Changes in horn size, shape, and growth rate also have occurred; these especially are evident in the goat. Domesticated species also have a greater variety of coat colors and patterns. In wild counterparts, this variability might be maladaptive and reduce survival.

Many of these changes, viewed together, reflect **neoteny** (the retention and persistence of juvenile characteristics in adults). Neoteny results in several behavioral changes in response thresholds in domesticated species (Ratner and Boice 1975; Price 1984). Domesticants have slower reactions, reduced flight distance, and less perception of the immediate environment (Hemmer 1990). These characteristics lead to more docile and easily handled individuals. Domestication often leads to greater reproductive potential as well. Compared with wild species, domesticates usually have earlier puberty, larger litter sizes, and breed more often throughout the year (Setchell 1992). All these effects lead through time to distinctive breeds, whether in dogs, cats, pigs, or cattle. Although breeds are similar to subspecies in wild animals, the distinctiveness of breeds is maintained by human-influenced reproductive isolation, rather than by geographical isolation, as is the case in subspecies.

DOMESTICATION: HUMAN ARTIFACT OR EVOLUTIONARY PROCESS?

Domestication may be viewed as somewhat of an "artificial" process, against the "natural order" of things, and a condition "imposed" on certain species by humans. Budiansky (1992) has suggested, however, that domestication be viewed instead as an evolutionary process driven by natural selection. This process has resulted in a symbiosis between humans and certain other species; that is, a symbiosis similar to other mutually beneficial relationships between species, such as the one occurring between ants and aphids. Humans obviously benefit from domestic animals, but the domesticants also benefit. They receive a reliable food supply, shelter, protection from predation pressure, and increased reproductive potential. Domestication may be viewed as an inevitable evolutionary consequence of unpredictable, changing environmental conditions at the end of the Pleistocene epoch. These conditions brought humans and domesticants closer together and created a new ecological niche, namely, association with humans (Morey 1994). In this view, natural variation among individual animals in a population resulted in the physiological, morphological, and behavioral changes eventually seen in domesticated species. Thus, they retained more curiosity, a less developed species-specific sense of recognition, and care-soliciting behavior as the result of selection pressure rather than human preference.

Certainly prehistoric people worked toward domesticating species such as dogs, goats, and sheep. The question, however, is one of cause and effect. As noted by Morey (1994:346), "The issue lies with the presumption that the eventual result—highly modified animals under conscious human subjugation—explains the process that started those animals toward that end." Thus, we can consider that domesticated mammals selected us as much as we chose them. However it happened, placing domestication in the context of an evolutionary process, as opposed to considering it a strictly human sociocultural phenomenon, is an intriguing idea. In his landmark work *On the Origin of Species,* Charles Darwin included a chapter on domestication; artificial selection was one of the factors leading to his conclusion that nature selects the best suited variants, leading to change (evolution) through time.

CURRENT INITIATIVES

Although reindeer have been exploited for food and transportation for thousands of years, other species of cervids are just beginning to be considered for domestication. Deer farming—husbandry in penned conditions—to produce venison and antler velvet is a growing practice throughout the world (Hudson et al. 1989). Often, exotic deer species—most commonly red deer (*Cervus elaphus*), fallow deer (*Dama dama*), and sika deer (*C. nippon;* figure 28.9)—are preferred

Figure 28.9   **Deer farms.**   Thousands of deer farms exist throughout the United States and the world, such as this one in Ireland where Japanese sika deer are raised.

by game farmers. The increasing number of game farms and possible conflicts with native wildlife are a serious concern to wildlife managers (Feldhamer and Armstrong 1993; Wheaton et al. 1993). The potential for commercial exploitation of many other large ungulates, primarily bovids, is currently being explored in the United States and throughout the world. Likewise, debate continues in Australia about raising kangaroos as livestock. Foxes, mink (*Mustela vison*), and chinchillas are raised on farms for fur production. Likewise, several rodent species, including the capybara (*Hydrochoerus hydrochaeris*) and paca (Genus *Agouti;* see chapter 17), are being produced on farms as food sources (Robinson and Redford 1991). Various strains of rats and mice, important in medical research, occur only in laboratory colonies.

Today, there are billions of individual domesticated mammals throughout the world, as livestock (see table 28.2) and as pets. They play critical roles in our lives—socially, culturally, and economically. Unfortunately, the demands of livestock for pasture and grazing lands, and the needs of an increasing human population often mean less habitat and fewer resources are available for the wild mammals of the world. The extent to which wild mammals will remain a viable part of ecosystems depends on current and future conservation efforts—the subject of the next chapter.

Summary

All domestic mammals descended from wild ancestral species. This process has been occurring for thousands of years and for a variety of reasons—companionship, food, milk, fur, beasts of burden, sport, or prestige—and it continues today. Relatively few of the approximately 4600 mammalian species have been domesticated, however, because the behavioral and physiological attributes of species necessary for successful domestication are stringent. They must be social, have an adaptable diet, and be easily herded and bred under a variety of environmental conditions. Thus, only five species of large, herbivorous mammals have been domesticated and spread worldwide: horse, cow, goat, sheep, and pig. Once domesticated, species change morphologically and behaviorally from their wild ancestors in several ways. Body size often becomes smaller, pelage increases or decreases according to climate, the skull is shortened and dentition reduced, and submissive behavior patterns increase. Generally, domesticated species retain behavioral and morphological characteristics similar to juveniles, a condition referred to as neoteny. Neotenates have slower reaction times, reduced flight distance, and less environmental perception, all of which makes domesticated animals easier to handle and herd. It has been suggested that the initiation of domestication was not entirely a human innovation but an inevitable consequence of the evolutionary process.

Domestication is artificial selection for characters beneficial to humans and results in recognizable breeds. This is similar to evolution, including development of subspecies during natural selection of wild stock. Unlike subspecies, breeds are not necessarily geographically restricted. Domestication probably began 12,000 years ago between prehistoric humans and wolves, which led to domestic dogs. The first livestock species were sheep and goats. They were domesticated 10,000 years ago, before the development of agriculture. Sheep and goats provided the "walking larder" that helped make possible the major transition of humans from small groups of nomadic hunter-gatherers to permanent, sedentary, larger agricultural population centers. This in turn led to significant changes in the structure and function of human civilization. Domestication of many species was first centered in the Fertile Crescent area of the Middle East, but the process certainly arose independently in other regions of Eurasia and South America. The process of domesticating new species continues today, with several species of deer held on game farms, a situation that is not without controversy. Billions of individual livestock animals provide a critical economic base for much of human society today and have contributed to our social and cultural development for thousands of years.

Discussion Questions

1. Given the presence of pronghorn antelope, bighorn sheep, bison, and other large, social ungulates, speculate on why the native North Americans never domesticated any livestock species. Likewise, why was the kangaroo never domesticated in Australia?

2. Although livestock raising is now common in Africa, what factors might mitigate against livestock in Africa, especially several thousand years ago?

3. The question of private ownership of wildlife is central to deer farming. In addition to this, what other social, ethical, and biological questions and problems surround the commercialization of deer and other "wild" species?

Suggested Readings

Clutton-Brock, J. 1992. Horse power: a history of the horse and donkey in human societies. Harvard Univ. Press, Cambridge, MA.

Cole, H. H. and W. N. Garrett (eds.). 1980. Animal agriculture: the biology, husbandry, and use of domestic animals, 2d ed. W. H. Freeman, San Francisco.

Herre, W. and M. Röhrs. 1990. Animals in captivity: domestic mammals. Pp. 570–599 *in* Grzimek's encyclopedia of mammals, vol. 5. (S. P. Parker, ed.). McGraw-Hill, New York.

Loftus, R. T., D. E. MacHugh, D. G. Bradley, P. M. Sharp, and P. Cunningham. 1994. Evidence for two independent domestications of cattle. Proceedings of the National Academy of Science USA. 91:2757–2761.

Mason, I. L. (ed.). 1984. Evolution of domesticated animals. Longman, New York.

Ucko, P. J. and G. W. Dimbleby (eds.). 1969. Domestication and exploitation of plants and animals. Gerald Duckworth, London.

Wayne, R. K. 1993. Molecular evolution of the dog family. Trends Gen. 9:218–224.

CHAPTER

29

Conservation

NATURE OF THE PROBLEM

SOLUTIONS TO CONSERVATION PROBLEMS

CASE STUDIES

The gray wolf (*Canis lupus;* figure 29.1A) once roamed over most of North America. Its range has been reduced during the past two centuries so that today it is almost gone from the lower 48 states (figure 29.1B). Extirpation of the wolf over much of its range occurred primarily through hunting and trapping as wolves were considered dangerous and possible predators on livestock. Bison (*Bison bison*) numbered in excess of 60 million in North America about 1860. Three decades later, however, fewer than 200 were left in the wild, and shortly thereafter, the only remaining bison were in reserves. Again, human pressure was the major factor accounting for the dramatic decrease in numbers.

Madagascar (figure 29.2) is home to more species of prosimian primates than any other location on earth. During the past 50 to 75 years, however, a tremendous decline in numbers has occurred for most of these species. A number of the strepsirhine primates found only on Madagascar are on the verge of extinction (see chapter 13). Of the 32 species of primates that inhabit Madagascar, 17 (53%) are currently listed as endangered. The primary driving force behind this severe decline is **habitat destruction.** Many of the forests that were home to these primates have been cut down, so that the land could be used for agriculture and the wood used for fires and construction.

Many early mammalogy textbooks did not contain a separate chapter or much material at all on the subject of conservation (Cockrum 1962; Davis and Golley 1963; Gunderson 1976). Several of these books did contain information on trapping and sport hunting, however. More recently, a couple of editions of the text by Vaughan (1978, 1986) have a brief concluding chapter dealing with the effect of humans on mammals. In the last 20 years, however, interest in conservation biology has increased greatly. For an excellent in-depth treatment of conservation biology, see the recent book by Caughley and Gunn 1996.

Because we are mammals and some of our mammalian relatives, for example, wolves, elephants, primates, and pandas, are among the featured players in the current conservation drama, it seems logical that we conclude this text with a consideration of the plight of mammals in the modern world. In so doing, we will consider briefly the manner in which an integrated knowledge of the physiology, ecology, and behavior of mammals can be critical in making effective conservation decisions. What factors are responsible for the continuing decline in some mammal populations? Why are some species on the verge of extinction while others may actually be increasing their numbers? What are some of the solutions to these problems?

A

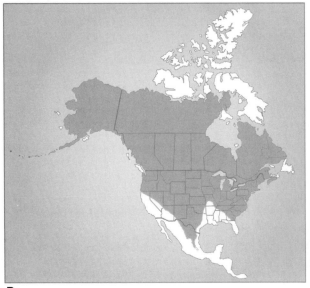

B

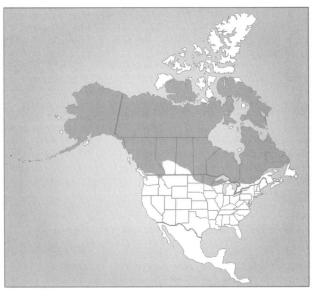

Figure 29.1   **Wolves.**   (A) The wolf is a predator and scavenger, living alone or in small packs that hunt as a unit. (B) Its range once encompassed much of North America but now is restricted to only the northern portion of the continent.

Figure 29.2 **Prosimian habitat loss.** Madagascar, an island of just over 587,000 km², is home to the richest prosimian primate fauna in the world. The loss of forested habitat, as shown here, has lead to sharp declines in the populations of many of these primates. The forested area, which once covered 62,000 km², has been reduced to 26,000 km².

NATURE OF THE PROBLEM

Conservation has a history almost as long as some of the problems it seeks to resolve. As humans have increased their standard of living, they have overused and degraded their environment. Today, some groups of humans still live in close association with their immediate surroundings. Their reverence for the land and all of its plant and animal resources, once true for all humans, has declined over the centuries. The land and associated resources have come to be considered expendable commodities, to be used up like any other commodity rather than being conserved for future generations (Leopold 1966).

In North America and other parts of the world, a conservation ethic has been reborn and has gained momentum during the past century (Owen 1971). Prior to that time, a few farsighted individuals, for example Thomas Jefferson, noted that we were despoiling our surroundings and needed to be concerned about the rate at which we were using up our natural resources. Establishment of Yellowstone National Park (1872) in the American West and the Adirondack Forest Preserve in upstate New York (1885) were among the first actions taken to preserve wilderness areas on a large scale.

The first modern event in conservation in North America was probably the White House Conference of 1908, organized and led by President Theodore Roosevelt. This conference resulted in the establishment of the National Conservation Commission, headed by Gifford Pinchot. The natural resources inventory, completed by this commission, resulted in the withdrawal of more than 200 million acres (80 million hectares) from further settlement. People like Roosevelt, Pinchot, and John Muir stand out for their commitment to conservation and their ability to balance the need for better resource management with the needs of increasing civilization. Together, they greatly increased the numbers of national parks and national forests.

President Franklin Roosevelt's establishment of the Works Progress Administration and Civilian Conservation Corps (CCC) resulted both in employment for millions during the Great Depression and afterward, and in a prodigious number of natural resource conservation and related endeavors in virtually every state in the country. Projects carried out by CCC members remain with us in many of our state and national parks, forests, and preserves. Also under Franklin Roosevelt, a Natural Resources Board was appointed. In 1934, the board completed the second full inventory of American natural resources. Other key agencies founded at this time include the Soil Conservation Service and the Tennessee Valley Authority, both of which continue to play roles in American conservation and resource management. The Wildlife Restoration Act of 1937 provided for exploration of conservation and resource problems, and for finding and implementing various means of remediating these problems.

The most recent wave of interest in conservation can be traced to the 1960s. This interest was spurred by such books as Rachel Carson's *Silent Spring* (1962) and Paul and Anne Ehrlich's *The Population Bomb* (1968). Legislative actions in the United States included the Clean Air Act (1964), National Wilderness Act (1964), Endangered Species Act (1973), Marine Mammal Protection Act (1972), and bills establishing numerous additional national parks, monuments, wilderness areas, and scenic rivers. The general awakening to such environmental concerns as the greenhouse effect, pollution, and diminishing supplies of nonrenewable natural resources that began in the 1960s led to the first Earth Day in 1970. The movement has continued to gain momentum. The need to maintain biodiversity on a global scale has become an acknowledged concern for people in many countries. Biodiversity is the summation of all of the living plants, animals, and other organisms that characterize a particular region, country, or the entire earth. With this brief background, we are now ready to explore in more detail some of the specific problems affecting mammals around the globe.

The list of factors that have led to declining numbers for many mammal species, and extinction for some, is lengthy, and the exact factors differ from one situation to another. To explain the decline in numbers for any particular species, its extirpation from certain regions, or even its extinction, it is necessary to understand the interactions of multiple factors.

Human Population

Early humans most likely lived in small subsistence groups, possibly consisting of one or several related families. Regional population density was quite low, but even then, exponential growth occurred. By about 1850, the total world population reached 1 billion people (figure 29.3). As of 1997, the population was in excess of 5.8 billion individuals, and it continues to increase exponentially. Predictions for the future vary, but figures of more than 6 billion by the start of the twenty-first century and more than 8 billion by 2020 are conservative estimates. The continued explosion of the human population worldwide results primarily from an unchecked birth rate and secondarily from somewhat lower death rates. Some areas, such as Europe, have approached a near stable population size. Other regions, including Latin America and much of Africa, have the highest annual growth rates of any of the world's regions, exceeding 2% per year.

The major consequence of the ever-increasing human population is a need for additional land and other resources. There is tremendous pressure to provide enough food, necessitating clearing more and more land for agriculture. Forests are cut down, to provide both land for growing crops and the wood for fuel and construction (Simmons 1981). Other critical habitats that are home to many mammals are transformed for human use, such as wetlands, which are

drained and filled. The accumulation of wastes, both human and those incidental to human activities (industrial and commercial), is an increasing problem. Armed conflicts that occur constantly around the world are, in large measure, a product of the growth of the human population and the competition for resources. The consequences for mammalian wildlife can be quite direct. The recent conflict in Rwanda resulted in the loss of some of the endangered mountain gorillas (*Gorilla gorilla beringei*). Several dozen species and subspecies of mammals have gone extinct since 1900 (figure 29.4), many due to direct or indirect human influences.

Habitat Destruction

Habitat destruction, caused primarily by human activities, is manifested in several forms. Probably the most publicized type of habitat destruction involves forests, particularly tropical rain forests. Over 40% of the original tropical rain forest of the world has been destroyed. This includes a 45% forest loss in Thailand between 1961 and 1985, all of the primary rain forest in Bangladesh, 85% of the forest in the Ivory Coast in West Africa, and more than 30% of the forest in Brazil (Lean et al. 1990). Much of this destruction has occurred since 1940. In addition, considerable forested areas in the temperate zones have been eliminated. Cutting of forests occurs for several reasons. One of these, provision of land for additional agriculture, has been mentioned. For example, vast areas of forest have been cut down in a number of regions of Africa and Latin America to grow crops and provide pastures for livestock (figure 29.5). Because most tropical soils are of relatively poor quality, a given area is productive for crops for only a few years. It is then abandoned, and people move on to repeat the process in another nearby locale.

Over time, vast areas of what were once forests of various types are destroyed. Most of the nutrients in these forest ecosystems are in the vegetation. When the area is logged or burned, the nutrients are lost, leaving an ecological desert. The secondary growth that replaces the original forest after the land is abandoned is a poor substitute for the original native forest. Leveling forests for agriculture does not occur only in less developed nations, however. In Australia, the need for grazing land for livestock is magnified by the low density of forage, and one cow can require up to 80 hectares/year to provide sufficient food. Many eucalyptus forests are leveled using ball-and-chain arrangements stretched between large bulldozers that move across the landscape.

A number of mammals are endangered or threatened as the result of deforestation. Endangered species are those likely to go extinct as a result of human activities and natural causes in all or a major portion of their range. Threatened (some listings use the term *vulnerable* instead) species are those that are likely to become endangered in the near future; they may be at risk for extinction. A variety of species of lemurs on Madagascar, where more than 80% of the original forest has been removed by humans, are endangered; black-and-white ruffed lemurs (*Varecia variegata*) and indris (*Indri indri*) are exam-

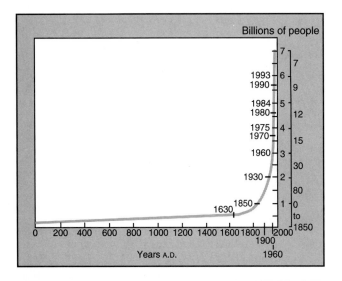

Figure 29.3   Human population growth.   The human population began to enter the exponential growth phase, as shown in this graph, even before the Industrial Revolution in the first half of the nineteenth century, when the total number of individuals was about 1 billion. Since then, the population has doubled almost three times.

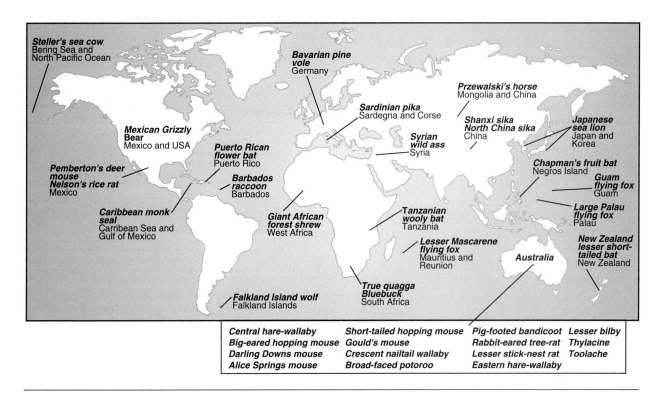

Figure 29.4   **Mammalian extinctions.**   The worldwide distribution of mammals that have become extinct in the past 400 years.

Source: Data from G. Caughley and A. Gunn, Conservation Biology in Theory and Practice, *1996, Blackwell Science Publications, Cambridge, MA.*

ples. Critically endangered are diademed sifakas (*Propithecus diadema*) and the aye-aye (*Daubentonia madagascarensis*). Western woolly lemurs (*Avahi occidentalis*) and ringtail lemurs (*Lemur catta*) are quite vulnerable due to habitat loss (Mittermeier 1987; Green and Sussman 1990). Similarly, some New World primates are on the verge of disappearing because of the loss of forest habitat. These include woolly spider monkeys (*Brachyteles arachnoides*) and the golden-lion tamarin (*Leontopithecus rosalia*), about which we will have more to say later in the chapter, when we deal with several specific case histories. Finally, among the primates, all three of the great apes, chimpanzees (Genus *Pan*), gorillas (Genus *Gorilla*), and orangutans (Genus *Pongo*) are endangered due primarily to the destruction of their forest habitats.

A   **B**

Figure 29.5   **Deforestation and agriculture.**   (A) In a number of the world's countries, slash-and-burn agriculture or similar practices result in deforestation of plots to provide land for grazing livestock or growing crops to support an increasing human population. Because of the generally poor quality of tropical soils, however, crops are successful for only a few years, and then the process is repeated on another plot. (B) In other regions, forests have been converted to rice paddies.

Nonprimate mammals are also threatened by deforestation. The giant pandas (*Ailuropoda melanoleuca*) are diminished throughout much of their range in China, due, in part, to destruction of their bamboo forest food resource (figure 29.6). The volcano rabbit (*Romerolagus diazi*), which inhabits the slopes of the volcanoes Popocatepetl and Ixtacihuatl and nearby ridges in Mexico, has diminished greatly in numbers due to hunting pressure and the clearing of its pine forest habitat for agriculture. Several subspecies of sika deer (*Cervus nippon*) are declining due to reduction of their forest habitats in Japan, Taiwan, and on the Asian mainland. The three South American tapirs (*Tapirus* spp.), the only living New World representatives of the Perissodactyla (of those that evolved in the New World and thus not including the horse and burro, which were introduced), live in a variety of forest habitats from lowlands to elevations above 3500 m. In spite of its generalized requirements, this species is diminishing rapidly over much of its range as its habitat is encroached on by the expanding human population. Subspecific populations of some mammals, such as the Florida panther (*Felis concolor coryi*), Asiatic lion (*Panthera leo persica*), and western giant eland (*Taurotragus derbianus derbianus*), are listed by various conservation agencies as threatened because of the loss of forest habitat.

Forests are not the only habitat altered by humans. Most other habitat types, including grasslands, prairies, savannas, swamps, marshes, and other wetlands, all have been usurped by humans for their own use, with negative consequences for certain mammal species. Much of the historical range of the American bison on the Great Plains is now used for agriculture. Though the great herds were killed for several reasons, their prairie habitat was turned into vast agricultural monocultures. Several species of prairie dogs (*Cynomys* spp.) likewise have been reduced. Their habitat has been taken for agriculture, and they have been poisoned or shot as pests

Figure 29.6   Habitat destruction.   Giant pandas (*Ailuropoda melanoleuca*) are declining precipitously throughout much of their range in China because of the loss of the bamboo forest habitat that is their home and source of food.

Figure 29.7   Value of captive herds.   Peré David's deer once roamed much of northeastern China but ceased to exist in the wild about 3500 years ago, when its swamp habitat was taken over for cultivation by local people. However, unlike some other animals that have been driven to extinction in such scenarios, these deer were maintained in herds in captivity. Although never domesticated, sizable herds of Peré David's deer live in many parks and reserves today.

(Miller et al. 1994). The drastic reduction of the black-footed ferret (*Mustela nigripes*) is directly related to declines of prairie dogs, their primary prey. In China, the extensive cultivation of wetlands for rice and other grain crops led to the virtual extinction in the wild of Pére David's deer (*Elaphurus davidianus*) over 3000 years ago (figure 29.7). Finally, the loss of relatively minor but nonetheless significant habitats such as mangroves in many tropical and subtropical regions has contributed to the reduction of numbers for species such as manatees (*Trichechus* spp.) and the dugong (*Dugong dugon;* Hartman 1979; see chapter 18).

Habitat Degradation

Habitat degradation may be due to (1) changes in habitats due to acid rain, introduction of toxic wastes, synthetic chemicals, oil spills, and so on; (2) introduction of predators and other alien or nonnative species, such as domestic dogs (*Canis familiaris*) and cats (*Felis* [*catus*] *sylvestris*), wild rats (*Rattus* spp.), or mongooses (*Herpestes* spp.) into particular habitats; and (3) the fragmentation of habitats, which can occur in a variety of ways. We will examine each of these briefly with a few mammalian examples.

The effects of acid rain have been most noticeable in the northeastern portion of the United States and neighboring areas of Canada, as well as in parts of Europe. While individual species may not have been threatened or extirpated, the habitat alterations associated with acid rain have undoubtedly had adverse effects on numerous mammals that depend on those habitats. These effects have yet to be adequately measured. The effects of industrial wastes and by-products, such as mercury, pesticides, and polychlorinated biphenyls (PCBs), on various habitats also remain largely unknown. Carnivores, higher on the food chain, are most likely to suffer from such pollution, because such dangerous chemicals are concentrated as they move through the food chain from producers to top-level carnivores. Populations of numerous species of bats worldwide have declined, and it is possible that increased exposure to chemicals has played a role in these declines. The effects of oil spills on populations of marine mammals such as sea otters and porpoises have been documented for each maritime accident, such as the *Exxon Valdez* disaster in Alaska (Bowyer et al. 1995). Although a species' existence does not appear to be threatened by such events, the ecological integrity of large regions is often affected for many years.

Another type of damaging effect results from the influences of some of the chemicals that we discharge into the environment. **Endocrine disrupters** are chemicals that mimic the effects of hormones or otherwise interfere with normal functioning of the endocrine system within an organism. Many of the livestock food additives and pesticides used in agriculture contain such synthetic chemical compounds. The negative effects of these chemicals with respect to endocrine functions in reproduction and development are only now being detected (Colburn et al. 1996). The effects include infertility, abnormal development, cancer, and others just now being discovered.

The introduction of predators, which has sometimes been included in the topic of biological invasions (Drake et al. 1989), has radically affected many mammalian populations and continues to pose problems for a number of species. In these instances, humans act as the dispersal agent, moving into a previously undisturbed area and bringing with them their domestic dogs and cats. On many Caribbean islands and in the Hawaiian Islands, humans introduced mongooses (Family Viverridae) in an attempt to control rodent populations; the rats had arrived earlier with the humans and their cargoes. One factor in the decline of the hutias (*Plagiodontia aedium* and *P. hylaeum*) on Hispaniola probably was the introduction of Burmese mongooses (*Herpestes javanicus*) for rodent control. Early aboriginal colonists brought the dingo (*Canis familiaris*) with them to Australia. One of the affects of the dingo was likely the extinction of the thylacine, or Tasmanian wolf (*Thylacinus cynocephalus*) on mainland Australia. Evidence also exists that humans, finding that Tasmanian wolves posed a threat to their sheep, aided in its demise, particularly on the island of Tasmania. The introduction of dogs and cats has played a critical role in the drastic decline in numbers of the numbat, or banded anteater (*Myrmecobius fasciatus*), in Australia; it is no longer found in New South Wales or South Australia.

A third type of habitat degradation is fragmentation. **Habitat fragmentation** is the process by which a previously continuous area of similar habitat is reduced in area and divided into smaller parcels. This process is best known because of its effect in eastern deciduous forests of the United States on migratory songbirds (Robinson et al. 1995). The amount of forested land in northeastern and midwestern United States has actually increased in the past 100 to 150 years. The effect of this additional forested land is diminished, however,

because much of it is fragmented and not in large, contiguous habitat. Fragmentation may in fact significantly benefit some mammalian populations. Clearly, the fragmented landscape of much of the central Atlantic states and Midwest has benefited the white-tailed deer (*Odocoileus virginianus*). Interspersing good cover with food resources has lead to sizable populations in most states. Not all fragmentation involves forests. Roadways have been tested as possible barriers for movements by small mammals (Oxley et al. 1974; Swihart and Slade 1984). Data for several rodent species indicate that construction of highways can be a barrier to genetic exchange. Morphological features diverged in populations of bank voles (*Clethrionomys glareolus*) from different sides of a highway (Sikorski and Bernshtein 1984). Furthermore, Oxley and colleagues (1974) reported that roads and highways clearly influenced the tendencies of white-footed mice (*Peromyscus leucopus*) and eastern chipmunks (*Tamias striatus*) to cross barriers during their daily movements (figure 29.8). Although the distances they moved should have lead to considerable road crossing, minimal numbers of both species actually ventured across the "barrier."

In addition to these forms of habitat degradation, much habitat is lost to grazing domestic livestock. Most affected are ungulates, such as buffalo and antelope, over much of the western United States, along with other mammals (Fleischner 1994). Many of these effects can be indirect in that the habitat is modified. The nature of the vegetation on the grasslands is altered, and riparian (riverbank) habitats, with their high levels of vegetative diversity, are trampled and perhaps irreparably altered. One example of small mammals adversely affected by various agricultural practices is the kangaroo rats (*Dipodomys* spp.): many of these species are now endangered, particularly in California (Price and Endo 1989; Price and Kelly 1994).

In a number of countries in Africa, humans graze their cattle on savannas or other similar grasslands, where the cattle compete with a variety of wild ungulates. A related problem is that cattle carry diseases such as rinderpest that can be transmitted to some of the wild ungulates, often decimating local or regional populations. Similar phenomena occur in areas of India and Southeast Asia.

A final, more specialized instance of habitat degradation concerns human activities in and around caves. Caves are a limited resource, and thus bats tend to be concentrated in them, putting the bats in peril should the caves be damaged or disturbances to the habitat be persistent. Many caves serve as roosting sites for bats, particularly during hibernation (see chapter 12). When humans repeatedly enter caves, disturb the bats, or despoil the habitat in some manner, bats may not return. Bat species whose habitat has been degraded in this manner and that have therefore suffered declines in North America include the gray bat (*Myotis grisescens*), the Indiana bat (*M. sodalis*), and Townsends's big-eared bat (*Plecotus townsendii*). Cave closure can also adversely affect bats. Furthermore, such closure may be a factor in the decline of woodrats (*Neotoma* spp.) in the eastern United States.

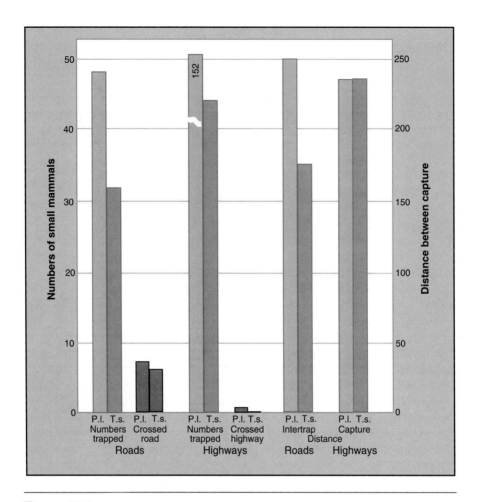

Figure 29.8   **Roads as possible barriers to genetic exchange.**   Relatively few white-footed mice (*Peromyscus leucopus* = P. l.) and eastern chipmunks (*Tamias striatus* = T. s.) cross country roads and highways. This is true even though the ranges of their movements, as determined from data on intertrap capture distances, indicate that they travel more than sufficient distances to cross such "barriers."

Source: Data from D. J. Oxley, M. B. Fenton, and G. R. Carmody, "The Effects of Roads on Small Mammals" in Journal of Applied Ecology, *1974, 11:51–59.*

Species Exploitation

Another major threat to mammals is the use of firearms to hunt them and traps to capture them. Some hunting and trapping are legitimate and certainly necessary for survival of native cultures and to provide some degree of control for populations of certain species, such as white-tailed deer, beaver, and raccoons (*Procyon lotor*). In some locations, mammals are killed by trapping, poisoning, or hunting because they are agricultural pests, competing in effect with humans for their crops (Burton and Pearson 1987). Regulated sport hunting is also a legitimate practice in many countries and occurs in many locales without adverse effect on mammal populations. In some nations, however, hunting occurs either illegally or with little regard to conservation and management of the mammalian prey. Examination of the *Red Data Book* (IUCN 1988) and related volumes on endangered mammals reveals that more than half the terrestrial nonrodent mammals in-

cluded in the list are there, in large measure, because of human hunting pressure. Examples of species whose numbers have declined due to hunting and related human activities include a variety of primates (blue monkeys, *Cercopithecus mitis*; common langur, *Semnopithecus entellus*), cervids (brow-antlered deer, *Cervus eldi*; several subspecies of sika deer, *Cervus nippon*), camelids (guanaco, *Lama guanicoe*; vicuña, *Vicugna vicugna*), and felids (leopard, *Panthera pardus*; tiger, *Panthera tigris*). Some species of mammals, such as sable antelope (*Hippotragus niger*) and North American beaver (*Castor canadensis*), have been killed or trapped for their horns or pelts. Species such as the sable antelope are endangered as a consequence of past hunting. Others, such as the beaver, were exploited for many decades (see chapter 2) but have recovered to such substantial numbers in most locations that they are considered pests. Trophy hunting, particularly for ungulates with impressive horns or antlers, has taken an additional toll on some species, particularly in Africa and Asia.

Some marine mammals have been devastated by hunting pressure. Most notable of these are the numerous species of whales that were killed in large numbers over many centuries (figure 29.9; see also figure 16.14). Both baleen and toothed whales have been hunted for a variety of purposes, most notably for their oil, which was used for fuel and in lamps until petroleum oil was discovered. Other whale products include meat for human and animal consumption, bone meal, spermaceti (a waxy solid used in ointments and cosmetics), ambergris (a waxy material from the intestines used as a fixative in perfumes), and extracts from the liver and endocrine glands. Such inventions as harpoon guns and steam-powered ships hastened the demise of many types of whales. Whaling continued at a steady pace with the use of factory ships and sonar devices until the last quarter of the twentieth century. International treaties now protect all whales from further hunting, although several nations persist in taking some whales each year.

Sea otters (*Enhydra lutris*) are another example of an overexploited sea mammal. Beginning in the late 1700s, these marine mammals were trapped in large numbers for their pelts. By the twentieth century, they had diminished to very low numbers. Recovery of sea otter populations in the northern Pacific Ocean was made possible via protection and reintroduction. In addition, numerous porpoises and dolphins have been ensnared in nets used for catching fish and other seafood, where they drown because they cannot reach the surface to breathe (see chapter 16).

Some mammal populations have declined, in part, because they make good pets. Especially affected have been some species of small South American primates, such as Goeldi's tamarin (*Callimico goeldii*), and related species of callitrichids. Some African primates are also kept as pets; Diana monkeys (*Cercopithecus diana*) are a good example. For several decades, numerous primates have been exported for use in medical testing and experiments. At the peak, some 30,000 to 40,000 wild rhesus macaques (*Macaca mulatta*) were exported annually from India, largely for use in testing the Salk polio vaccine. Primates are often ideal subjects for such purposes because of their close evolutionary relationship to humans. In the mid-1970s, the Indian government wisely put an end to these exports. We now rely instead on monkey farming for the production of some primates of medical importance.

A final, more subtle, process involves human domestication of wild mammals. Some stocks of formerly wild ungulates, probably the progenitors of stocks of domestic animals, have become extinct or nearly extinct because they were interbred with domestic forms over many generations. An example is the banteng (*Bos javanicus*) of southeast Asia, which interbred with domestic cattle.

Cultural/Religious Issues

In a number of cultures, particularly in Africa and Asia, ivory is believed to have special powers. Carved and decorated, it is valued as an art object in many human societies. The limited

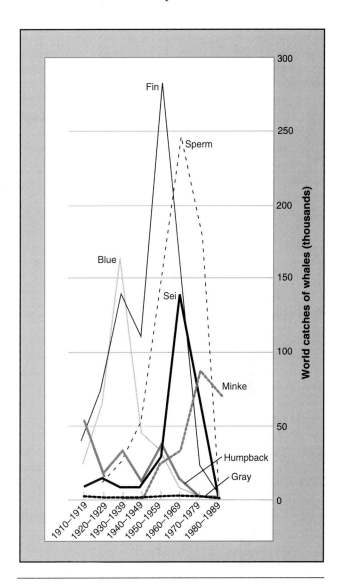

Figure 29.9   World catches of whales.   The rate of capture for many types of whales increased sharply in the early part of the twentieth century and continued to rise, with the use of newer technologies, until about the 1970s. Then a combination of virtual disappearance of some species and the implementation of whaling quotas followed by bans on hunting allowed some populations to recover.

supply of ivory, primarily from elephants (*Elaphus maximus* and *Loxodonta africana*) and walruses (*Odobenus rosmarus*), has put a premium on this commodity. The result has been extensive poaching of wild elephants (most often in Africa), primarily for the removal of the tusks (Eltringham 1979). Elephant feet are sometimes made into tables for the tourist trade and export. This problem has received much attention in the popular and scientific press in the past decade, although no clear solutions seem to be available (Cohn 1990; Milner-Gulland and Mace 1991). Similarly, the horns of rhinoceroses are valued for their use in traditional Asian medications and as an aphrodisiac. They are also prized as sheaths and handles for knives in some Middle Eastern cultures. Such practices lead to poaching of these animals throughout much of their range, to the point where fewer than 5000 black rhinoceroses (*Diceros bicornis*) now exist and the white rhinoceros (*Cratotherium simum*) is endangered. One practice initiated in the last few years to save rhinoceroses involves dehorning them, thus removing the prize for which they are sought by poachers. The efficacy of this practice is much debated, and possible behavioral side effects may result from it (Berger and Cunningham 1994; Berger et al. 1994; Rachlow and Berger 1997).

In North America, black bears (*Ursus americanus*), and sometimes other bear species, have been killed illegally to remove their gallbladders. This organ is thought to have medicinal or aphrodisiac powers by some people in the Near and Far East. These and other cultural practices involving particular mammals as delicacies or ceremonial foods have contributed to declining numbers for a variety of species.

Islands As Special Cases

Isolated in varying degrees from nearby land masses, islands of varying sizes—from small 1- to 2-hectare islets to larger islands, such as New Guinea—have, through geological time, tended to develop unique faunas and floras. The entire process has been described well in terms of island biogeography (MacArthur and Wilson 1967). The models developed by MacArthur and Wilson related the fauna and flora of islands to a series of factors. The size, age, and distance of an island from either a continent or other islands can influence the number of species present through differential rates of colonization and extinction (see chapter 25).

On islands, some of the factors affecting the decline of certain mammals are exacerbated. Mammals that are endemic to specific islands or groups of islands face special problems. Of the 83 documented cases of mammalian extinctions since the end of the fifteenth century, 51 (61.5%) have been of mammals living on islands (Ehrlich and Ehrlich 1981; Reid and Miller 1989).

The introduction of a predator or competitor can severely harm an endemic mammal species. We already mentioned the effects of mongooses introduced on Caribbean and Hawaiian islands for rodent control and of dingos accompanying humans as they migrated to Australia. Brown rats (*Rattus norvegicus*) and black rats (*R. rattus*), introduced to Madagascar possibly as long as 2000 years ago, decimated much of the native small rodent fauna (Goodman 1995). The introduction of the European rabbit (*Oryctolagus cunniculus*) to Australia has created intense competition with other small mammals. The solenodons on Hispaniola (*Solenodon paradoxus*) and Cuba (*S. cubanus*) have both been driven nearly extinct by a combination of predation from introduced dogs and cats and loss of their forested habitat. In New Zealand, several bat species, most notably the lobe-lipped bat (*Chalinolobus tuberculatus*) and short-tailed bat (*Mystacina tuberculata*), have been nearly eliminated by introduced predators that include dogs, cats, weasels (*Mustela* spp.), and rats (*Rattus* spp.). Since these island-based endemics do not have other source populations from which recolonization can occur, their demise is hastened if humans do not take special conservation measures.

Fragments of what were once larger habitats and the various parks and reserves of different sizes that we establish are considered by some to be ecological islands (Newmark 1995). The rates of extinction and recolonization in the various national parks in the western United States fit the predictions from models of island biogeography and related theory. These sorts of formulations—specifically regarding the sizes of the reserves and their possible connections via corridors—deserve careful consideration when making conservation plans for various mammals (Soulé and Wilcox 1980). New fields of study, such as landscape ecology, provide a broad framework within which particular conservation strategies can be evaluated.

Hybridization

One problem that we have only come to recognize fully in recent years concerns the possible extinction of species through hybridization (Avise and Hamrick 1996; Rhymer and Simberloff 1996). The problem can arise in at least three different ways. One way is through the introduction of an exotic species, either accidentally or deliberately (Drake et al. 1989). Exotic species are those that are not native to a particular locale or region, and their introduction can result in **introgression,** the mixing of the gene pools. The interbreeding of feral house cats with the wild cat (*Felis sylvestris*) in remote areas of Scotland and a similar problem involving the African wild cat (*F. libyca*) in southern Africa exemplify this problem (Stuart and Stuart 1991; Hubbard et al. 1992). Ongoing attempts to reintroduce the red wolf (*Canis rufus*) in eastern North America may be doomed, in part, because evidence from mitochondrial DNA indicates that red wolves already have significant genes from both gray wolves and coyotes (*C. latrans;* Wayne and Jenks 1991). Given the increasing coyote populations in most areas where red wolves could be released, the probability is high that introgression will occur wherever attempts are made to reestablish red wolves.

A second way occurs when habitat modification results in the possibility of two previously separated species meeting and interbreeding. This process can take any of the following forms. Local habitat change, such as clearing brush around natural or artificial ponds, has resulted in the hybridization of several species of tree frogs (Genus *Hyla;* Cade 1983). Regional habitat change, such as that involving clearing lands for agriculture or the reforestation process that occurs when such lands are abandoned, can apparently result in hybridization for some species. The ongoing mixing of the blue-winged warbler (*Vermivora pinus*) and golden-winged warbler (*V. chrysoptera*) in the eastern and central United States is an example (Gill 1980). Although clear examples have yet to be found for mammals, these problems are still a concern.

A third possibility involves our own conservation attempts and the introgression of gene pools that can result from such activities. When we attempt to save a species or subspecies by crossbreeding it with a closely related species or subspecies or when we attempt to enhance a stock of wild animals by introducing animals from a distant population, often of a different subspecies, two issues arise. First, outbreeding depression can occur when new genotypes produced by crossing stocks are inferior to the original native stock and when the new stock is at a disadvantage with respect to adaptation to local conditions. Mammalian examples of outcrossing for conservation reasons include the Florida panther (*Felis concolor coryi*), wisent (*Bison bonasus*), and wood bison (*Bison bison athabascae;* Fisher et al. 1969; Fergus 1991). In the first case, animals from the Texas subspecies (*Puma concolor cougar*) were introduced into Florida, and in the latter two cases, stocks were enhanced by introduction of American plains buffalo (*Bison bison americanus*). Second, we are, to varying degrees, transforming the original gene pool of a species or subspecies, thus opening up the myriad issues surrounding both the definition of a species and what exactly should be conserved. The magnitude of the possible negative effects is unclear in all cases, but what is certain is that introgression of gene pools occurs and the original species gene pool is altered forever.

SOLUTIONS TO CONSERVATION PROBLEMS

Just as a variety of factors contribute to the decline of many populations and species of mammals, so are there many solutions to the problem, some more general and others particular to specific situations. The single most important factor contributing to the demise of the other mammalian species is the growing human population. For a number of reasons, religious and cultural in nature, this unfortunately may be the most difficult problem to attack. Mammalian species conservation is most critical precisely where growth rates of the human population are at their highest levels—in Africa; Latin America; and southeast Asia, including all of the Malay Archipelago and the Philippine Islands. If even the more modest of the projections for human population growth are correct, then it seems likely that many dozens of threatened or endangered species will, in fact, become extinct before the middle of the next century. No programs we develop will have a chance to succeed without control of our own population.

To protect the remaining wildlife, several measures have received renewed emphasis. First, there is a continuing move to establish and maintain reserves, where native mammals, as well as the plants and animals that constitute their communities, have an opportunity to survive. National governments and nongovernmental agencies have aided in the establishment of new reserves as well as the addition of lands to existing reserves. This is a continuing process. In many

states within the United States and in a number of other countries, preservation of areas for hunting and trapping has provided a stimulus and funding for wildlife areas. In most instances, these reserves provide habitat for many species, not just those being hunted.

Second, in conjunction with the establishment of these reserves, national and international laws concerning hunting, trading, exporting, and other practices that directly affect endangered mammals must be enforced. Many countries have laws designed to protect endangered mammals. Human population pressure, however, drives a need for more food and space. Funding for enforcement and education of the populace concerning the laws is lacking, and corruption allows poaching and related activities to flourish in some countries. Many nations, including the United States, are feeling increasing pressure from development interests to permit exploration for and mining of minerals and other resources and other uses in areas that are designated reserves or wilderness preserves (figure 29.10). On top of this are efforts to develop agricultural land and natural areas for human habitation. Such activities invariably have negative consequences for wildlife, including mammals. Until these sorts of situations are changed, establishment of reserves will only protect mammals in a limited way.

A third major initiative, now being pursued in some countries, involves educating the local residents concerning the benefits to be gained from protecting organisms with which they share their environment. If residents can understand, for example, that large mammals living in their region are a potential economic resource through ecotourism, then there may be hope for saving some of these mammals. Ecotourism provides the funds needed to protect mammals and

Figure 29.10   Problems of development. Though designated as a national park, the Tauro National Park in Australia is used for cattle grazing. In other countries, reserves and parks have been degraded by the intrusion of development interests seeking oil and other minerals.

the habitat those mammals need to survive. Unfortunately, ecotourism sometimes has the negative consequence of overrunning an area with tourists: cheetahs, for example, are followed by hoards of tourists in minibuses, which interferes with their attempts to stalk and kill prey. The economics of conservation issues need to be examined further, but the basic principle that local people can benefit from conserving rather than killing and consuming the wildlife and the habitat must become firmly established.

Finally, a series of international treaties and agreements has potential benefits for many endangered mammals. The more than 80 national signatories to the Commission on International Trade in Endangered Species (CITES) have agreed to regulate the import and export of many animal species. Trade in products such as ivory and rare furs is regulated or banned altogether. In a similar manner, starting in 1946, treaties promulgated by the International Whaling Commission governing the hunting of whales eventually provided for the virtual cessation of whale hunting. Note, however, that many nations ceased whaling simply because the whale populations had already become so greatly diminished as to be economically unprofitable. A combination of agreements like these and cooperative multinational efforts may save some of the presently endangered species of mammals. For others, however, it is most likely already too late. Organizations such as the IUCN, CITES, and the IWC, as well as conservation groups such as the World Wildlife Fund and the Nature Conservancy provide a basis for obtaining the cooperation of all "shareholders" of the earth's riches that can lead to productive and successful conservation programs.

As the nations of the world become more aware of the problems, plans for conserving and managing resources are being developed. A Species Survival Commission (SSC) exists for each mammalian order. For some endangered species, Species Survival Plans (SSPs) have been created. These cover the most critically endangered species and involve breeding animals in captivity for reintroduction into their natural habitat (table 29.1). A desperation measure being employed with some mammals includes careful planning of a captive breeding program that will lead to the eventual reintroduction of some animals in the wild, with the hope that they will either re-establish an extirpated population or augment a critically low one, thus resulting in overall species survival. A similar program is presently underway to reintroduce gray wolves to habitat they formerly occupied in Yellowstone National Park in the western United States. In this instance, the animals are being captured in Canada and transplanted to the park. Such programs are generally expensive and labor-intensive (Fritts 1993; Mech 1995). Considerable discussion is taking place, even among knowledgeable conservation biologists, concerning whether it is more sensible and practical to invest in saving species that are on the verge of extinction via SSPs and reintroductions or to spend funds on mammals (or other animals) that are still living in the wild in sufficient numbers,

---

Table 29.1.   Species survival plan

A variety of factors are considered when a Species Survival Plan is put together for a particular mammal. The basic components of an SSP are listed with a synopsis of the procedure at each stage.

1. *Designation of species:* A critically endangered species is selected prior to extinction.
2. *Appointment of Species Coordinator:* An individual agrees to take charge of the SSP for the designated species and must then coordinate all activities relative to the SSP.
3. *Organization of Propagation Group:* The Propagation Group consists of representatives of the institutions that hold captive members of the species and have filed a Memorandum of Participation. It is the main advisory body for the SSP for the species in question.
4. *Appointment of Studbook Keeper:* One member of the Propagation Group is responsible for compiling this vital book of records on the captive members of the species.
5. *Compilation of Studbook:* This document is the collection of essential information on the identity, date and place of birth, date and cause of death, parents, and other characteristics (e.g., temperament) of each captive member of the species. It is becoming common practice to obtain DNA fingerprint information for many species. This book serves as the basis for all plans for captive breeding, and the information it contains is the heart of the process.
6. *Formulation of master plan:* Each individual in the captive population is evaluated with respect to its potential breeding status, relationship to others that might be used for breeding, health, and other characteristics. Each individual is assigned to one of five categories: (a) breed, (b) hold without breeding, (c) transfer to another location, (d) surplus, or (e) research. These decisions are made for the population at each participating institution and then in coordination with the other cooperating institutions. In this way, optimum use can be made of the existing genetic variability to attempt to retain that variability in subsequent generations.
7. *Review of master plan:* The master plan is reviewed by the Wildlife Conservation and Management Committee of the American Association of Zoological Parks and Aquariums.
8. *Update of master plan:* Each SSP requires an annual report summarizing all activities for the previous year, which is then published in the *International Zoo Yearbook.* These reports contain estimates of the existing genetic variability and the minimum viable population projected for the species.
9. *Compilation of Husbandry Handbook:* This handbook constitutes a record of information on how to care and manage a particular species that has an SSP. Recording detailed information on diet, behavior, diseases, and many other aspects of the species' biology can aid others working with the same or similar species. A thorough survey of all of the available literature and other information on the species should be included.

although threatened or endangered. Another current debate concerns whether we should be working on a species-by-species basis or instead working to preserve habitats. In the latter case, entire community structures may be preserved. A law to protect endangered or threatened habitats could be superior to the Endangered Species Act because it would protect all species, not just those currently listed as endangered. The recent book by Bowles and Whelan (1995) contains details concerning both the planning and applied aspects of recovery of threatened and endangered species.

An issue that is important to all efforts to save mammals is the notion of a **minimum viable population (MVP)**, or how many animals are needed in a population to prevent it from going extinct (Schonewald-Cox et al. 1983; Soulé 1987). Small populations are more likely to be lost than large ones (figure 29.11), as, for example, was the case for bighorn sheep (*Ovis canadensis;* Berger 1990). When a population is reduced below the MVP due to habitat loss, habitat degradation, or other human-related activities, loss of a particular population or extinction can occur rapidly (Soulé 1987). Related to the idea of an MVP is the concept of genetic variability. Genetic variability is critical for populations to retain the capacity to adapt to a changing environment (Avise and Hamrick 1996). If insufficient genetic variability remains, the probability of extinction increases. Thus, conservation biologists working with mammals must consider the sizes of existing populations when trying to determine if they are viable or susceptible to local extinction. Small populations can also suffer rapid declines due to short-term fluctuations in birth or death rates, or they may be affected by predation, food supplies, natural catastrophes, and other similar natural events.

Advances in DNA-related technologies are being applied to conservation problems to directly address the issue of genetic variation. By using available forms of DNA fingerprinting, it is possible to mate captive animals so as to maximize their genetic differences, aiding in maintaining overall heterozygosity of the species' gene pool. The potential for variation in phenotypic traits borne by those genes is thought by many to be a critical component of the process of successfully reintroducing animals to their native habitats from captive, often small, breeding populations.

CASE STUDIES

The following three examples illustrate some of the conservation problems faced by mammalogists and how they are being solved.

Arabian Oryx

Arabian oryx (*Oryx leucoryx;* figure 29.12) once roamed over virtually all of the Arabian peninsula. Hunting, enhanced by the use of motor vehicles (accompanying the development of oil reserves) and modern weapons technology (commencing with the active involvement of European powers in the region during World War I), reduced their numbers to a few, very small, scattered populations by the 1960s (Grimwood 1988). The last wild Arabian oryx was probably killed in 1972. Fortunately, some farsighted individuals had begun a program to save the species, commencing with the capture of some wild oryx in 1962. Captive populations already existed in Arabia. New captive breeding programs were established

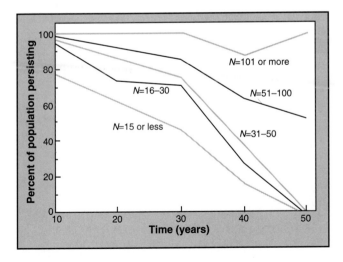

Figure 29.11   Minimum viable population size.
The relationship between original population size and the percentage of populations persisting over time. Most populations with about 100 bighorn sheep survived for 50 years, whereas smaller populations did not.

Source: Data from R.B. Primack, Essentials of Conservation Biology, *1993, Sinauer Associates, Sunderland, MA.*

Figure 29.12   Arabian oryx. These modest-sized ungulates, up to 1.4 m long and weighing about 55 kg, formerly inhabited much of the Arabian peninsula. Conservation efforts under way today have succeeded in reestablishing Arabian oryx populations at several locations.

first in the United States and then at several zoos in Europe. A critical problem, faced in the initial stages of this effort, was that few animals were available for breeding, perhaps only four or five! Additional stocks of Arabian oryx were discovered in private collections at several locations in the area of the Arabian peninsula as the project progressed. Because maintaining genetic diversity could be critical to the successful survival of such animals reintroduced into native habitat, a key goal of captive breeding programs like the one for the oryx is to maintain careful records on all known specimens. To accomplish this, a studbook (a record of all of the captive oryx, their locations, breeding status and, where known, their genealogy) was established.

By the 1970s, the breeding program was quite successful. The first reintroductions to the Middle East involved oryx established at the New Shaumari Reserve in Jordan, where they remain today. The primary site selected for reintroduction of Arabian oryx to the wild was at Yalooni in Oman. The site has rocky outcrops, temperatures that range from summer highs of 47°C to winter lows of 7°C, and minimal rainfall. Foggy seasons also occur, however, when moisture from the Arabian Sea is deposited on the landscape, providing for periods of relatively lush vegetation. In 1980, oryx were moved from captive breeding locations in the United States to small enclosures at the release site. Release into the wild occurred in stages. After confinement in small enclosures for several months, the oryx were then maintained in a larger 1-km² enclosure for a period of up to 2 years. Finally, in 1982, a herd of 10 animals was released into the wild. Herds set for release comprised specific numbers of males and females, adults and young, based on knowledge of the species and its reproductive patterns. Such knowledge was obtained through prior information on wild Arabian oryx and careful observation of the animals in captivity.

Education and local interest play vital roles in the success of such projects. The local people, the Harasis, act as wardens in protecting the oryx, as gamekeepers, in managing the stock, and as science aides in participating in scientific observations of the animals. Several subsequent releases have taken place, and the small herds have readily established territories. At this time, the project appears to be a success.

Whales

Numerous species of whales have been hunted for centuries to obtain a variety of products (see chapter 16). With the advent of modern power ships and technologies for tracking, killing, and processing whales, populations of many species plummeted (table 29.2). As is true for many conservation efforts, issues surrounding whale conservation become entangled in economic and political considerations. Whale hunting dates back thousands of years, when Basques first organized large-scale hunts. Initially, they exploited right whales (*Eubalaena glacialis*) in the Bay of Biscay. As they captured and used most of the whales available there, they moved farther and farther northward and out into the Atlantic Ocean. Whaling expanded during the seventeenth through nineteenth centuries, with major whaling fleets mounted by the Dutch, British, and Americans. Soon these fleets discovered the rich supply of whales in the Pacific Ocean. Later, into the twentieth century, Japan, Germany, Norway, Korea, Russia, and other nations joined in the hunt for whales.

Techniques used for hunting and harpooning whales and bringing them to the side of the ship for processing improved greatly during the nineteenth century. Herman Melville immortalized many aspects of this in *Moby Dick*. By the twentieth century, technological advances changed all of this. Large powered ships replaced sailing vessels, weapons such as explosive harpoons were used to kill the whales, and eventually large factory ships were launched, all of which resulted in a rapid increase in the number of whales taken. Up to 30,000 whales were taken each year beginning about 1930; at least 330,000 whales were captured and processed

Table 29.2.   Population estimates of several species of whales indicating the precipitous declines for some during the 20th century

Species (Common Name)	Estimated Population Sizes		
	Pre-1900	**1970**	**Present**
Balaenoptera musculus (blue whale)	200,000	5,000	10,000
B. acutorostrata (minke whale)	350,000	350,000	350,000
B. borealis (sei whale)	250,000	45,000	50,000
B. physalus (fin whale)	500,000	84,000	120,000
Eubalaena glacialis (right whale)	320,000	3,000	3,000
Physeter catodon (sperm whale)	2,000,000	487,000	1,000,000
Megaptera novaeagliae (humpback whale)	100,000	3,000	12,000
Balaena mysticetus (bowhead whale)	65,000	5,000	8,000

Sources: Data from P. B. Best, "Right Whales, Eubalaena australis *of Tristan da Cunha: a clue to the "Non-recovery" of Depleted Stock?" in* Biological Conservation, *1988, 46:23-51; S. D. Kraus, "Rates and Potential Causes of Mortality in North Atlantic Right Whale (*Eubalaena australis*)," in* Marine Mammal Science, *1990, 6:278-291; R. B. Primack,* Essentials of Conservation Biology, *1993, Sinauer Associates, Sunderland, MA.*
*Bans on whaling may have come just in time. Some species have already made a measurable recovery.

between 1910 and 1966. As populations of one species of whale were depleted, the factory ships and their crews moved to another species. By the early 1800s, right whales and bowhead whales (*Balaena mysticetus*) had been overharvested. Blue whales (*Balaenoptera musculus*), the largest mammals on earth, became scarce next, followed in order by fin (*B. physalus*), sei (*B. borealis*), and minke (*B. acutorostrata*) whales (Burton 1980; Burton and Pearson 1987). Some controls were attempted but proved generally ineffective, in large measure because whaling was so profitable. Whaling itself and those employed to use its products formed important components of the economies of the nations involved in the hunt. For political reasons, governments were hesitant to move forward with plans to slow down the slaughter.

The International Whaling Commission, formed in 1946, was largely ineffective. Its members tended to ignore the evidence provided by scientific advisers, who indicated that there would be dire consequences if restraint and long-term recovery plans were not initiated. Breakthroughs occurred in the early 1980s. In 1981, the commission voted to ban hunting sperm whales and the next year voted for a moratorium on all commercial whaling to take effect in 1985. Most nations complied, but several, most notably Russia, Norway, Peru, Norway, Korea, and Iceland, have all continued to do some whaling, though often under the guise of "scientific collecting." At this time, there is some pressure to resume commercial whaling, particularly of such species as minke whales, which apparently have survived in sufficient numbers or have undergone significant recovery during the moratorium. Additional chapters in the story of whale hunting remain to be written with the hopeful prospect that species whose stocks were so decimated may eventually recover.

Golden-Lion Tamarin

Only about 2% of the original Atlantic coastal forest in Brazil that is the home for *Leontopithecus rosalia* remains today. As the population dwindled to a few hundred wild tamarins and another 100 in zoos by about 1970, the first steps in the recovery for this species were taken. A national reserve, Poco das Antas Biological Reserve, was established in 1974. The year before, an official studbook for golden-lion tamarins was initiated, and the coordination of the recovery project was assigned to the National Zoological Park in Washington, D. C. (Kleiman et al. 1990). The cooperation that developed in the ensuing decades between scientists at zoological parks in several countries and the authorities and conservation biologists in Brazil can serve as a model for similar species preservation efforts.

Progress proceeded in several steps. First, the captive populations were not breeding well and were, in fact, declining. Information on behavior and reproductive physiology was collected. It was learned that in tamarin groups, only the dominant pairs generally breed; reproduction of other individuals is suppressed. In addition, female golden-lion tamarins are very aggressive toward one another. Initial steps thus involved forming appropriate breeding groups. This approach enhanced reproduction and the possibility that sufficient numbers could be produced to eventually plan for releases in native habitat. With the establishment of additional breeding centers, the captive population grew to nearly 400 individuals by the mid-1980s. A key feature of such programs is to maintain breeding while attempting to maximize the remaining genetic diversity. Any time we start with a limited stock of animals, as in this instance, only a portion of the total genetic diversity of the species will be represented. To foster diversity, a careful breeding program was established, using the information in the studbook. In this way, particular pairs could be bred to maintain the genetic heterozygosity. This often meant that tamarins were shipped between breeding centers at different zoos. The time needed for quarantine and adjustment under such circumstances adds to the logistical difficulties of such programs.

At the same time, additional research was taking place on the few tamarins remaining in the wild in Brazil. Steps were taken in Brazil to protect areas where these animals still remained. Furthermore, a campaign of public education concerning the tamarins heightened people's awareness of the need for their conservation and habitat preservation. In 1983, reintroductions to natural habitat in Brazil began with captive-bred golden-lion tamarins. Despite some losses and various setbacks, it appears that the program is succeeding overall in establishing new populations in former habitat and augmenting existing populations.

Summary

Conservation of mammals, something that has been with us in various forms for many centuries, has only really become an area of strong interest during the past 100 years. During that time, several waves of interest in conservation-related activities have occurred, most notably during the last 25 years. The passage of many new environmental laws, both in the United States and abroad, specifically including the Endangered Species Act, have helped to heighten our awareness of the fragility of our environment and the need for renewed vigilance.

The decline of a mammal species is likely due to a series of interrelated factors. The seemingly uncontrolled exponential growth of the human population is the single most critical factor in the decline of wild mammals and other animals and plants. Virtually all environmental problems can be traced directly or indirectly to human population growth. Habitat destruction—most commonly deforestation, but also the disappearance or degradation of savannas, marshes, and other habitats—poses a serious threat to a wide variety of species. Habitat degradation can take several forms, including the release of pollutants, introduction of predators, and fragmentation. Exploitation of some species through hunting has been a major factor in their decline. Finally, cultural practices, such as those associated with ivory or the use of particular animal parts as traditional medicines or aphrodisiacs, can lead to declining populations. Islands, with their endemic fauna and lack of source populations for recolonization, present special problems with respect to conservation of mammals.

Various solutions are being attempted to overcome these problems. Solutions must be found soon to control human population growth, or most other measures we might use for species conservation will be of little value. More reserves are being established, and existing reserves are being enlarged. Laws governing hunting, trapping, and trading in mammals and mammalian products need to be enforced. International cooperation in program development and financing is necessary to ensure the success of mammalian conservation plans. Examples of successful projects include the cessation of whale harvests and the reintroduction of captive-bred mammals such as the Arabian oryx and golden-lion tamarin. These provide us with different perspectives on how conservation biologists are approaching the issue of saving mammals from possible extinction.

Discussion Questions

1. Examine a week of issues of a major city newspaper and compile a list of articles and topics that relate to conservation. Given your knowledge of the conservation field from material in this and other courses, how balanced do you feel the coverage is of the various conservation issues? How many articles could have (but did not) refer to the human population increase as a serious problem? How many articles referred to the human population problem at all?

2. In addition to the topics we covered in this chapter, what other factors pose threats to mammals?

3. When reintroductions are done, such as the golden-lion tamarin in Brazil or the gray wolf in Yellowstone National Park, what measures must be taken to provide for the best possible success rate? Can you suggest some ways to accomplish these measures?

4. One of the biggest sources of conflict facing those who work on species conservation is economic development. From your own background and course work, what examples can you provide of ways in which corporations and conservationists have been able to cooperate, providing for the interests of both groups?

5. List the four or five most important conservation issues in the world today. How would you recommend solving these problems?

Suggested Readings

Hoyt, J. A. 1994. Animals in peril. Avery, Garden City Park, NY.

Leopold, A. 1966. A Sand County almanac. Oxford Univ. Press, New York.

Nash, R. 1967. Wilderness and the American mind. Yale Univ. Press, New Haven, CT.

Primack, R. B. 1993. Essentials of conservation biology. Sinauer Assoc., Sunderland, MA.

Soulé, M. E. 1987. Viable populations for conservation. Cambridge Univ. Press, New York.

Tudge, C. 1992. Last animals at the zoo. Oxford Univ. Press, Oxford, England.

GLOSSARY

A

ablation The destruction of tissue by electrical or chemical techniques.

abomasum The fourth and last chamber of the stomach of a ruminant; often called the "true stomach."

accidental host A host in which a parasite does not normally occur.

active dispersal Ecological dispersal events in which individuals move by terrestrial locomotion or flight.

adaptive Making an individual more fit to survive and reproduce in comparison with other individuals of the same species.

adaptive hypothermia A group of energy-conserving responses of mammals and birds characterized by the temporary abandonment of homeothermy.

adrenocortical hormones Hormones produced by the cortex of the adrenal gland.

age structure The proportion of individuals in a population in different age classes.

agenesis A tooth missing from the dental complement that is normally found in a species.

aggression Behavior that appears to be intended to inflict noxious stimulation or destruction on another organism.

agonistic behavior Behavior patterns used during conflict with a conspecific, including overt aggression, threats, and retreats.

agouti Hair pigments that exhibit mixtures or banding from pheomelanin and eumelanin.

albinism The condition in which all hairs are white because of an absence of pigment, caused by a genetic mutation.

allelopathic The direct inhibition of one species by another using noxious or toxic chemicals.

Allen's rule The biogeographic "rule" that states that extremities of endothermic animals are shorter in colder climates than those of animals of the same species found in warmer climates.

alliance A long-term association between two or more individuals in which they cooperate against a third party, as in dominance interactions.

allograft A piece of tissue or an organ transferred from one individual to another individual of the same species; a successful foreign transplant.

allopatric Occurring in different places, usually referring to geographical separation of populations. Compare to *sympatric*.

altricial Neonates that are born in a relatively undeveloped condition (eyes closed and with minimal fur present) and require prolonged parental care; as opposed to precocial.

altruistic Behavior that reduces personal fitness for the benefit of others.

alveolus The socket in a jaw bone for the root(s) of a tooth.

ambergris A form of excrement of sperm whales.

ambulatory locomotion Movement by walking; essentially plantigrade.

amphibious mammals Those that spend time in both terrestrial and aquatic habitats.

amphilestids A family of triconodont mammals that was extant for about 50 million years from the mid-Jurassic to the early Cretaceous periods.

ampullary glands Small, paired accessory reproductive glands in some male mammals that contribute their products to the semen.

anal sacs A paired glandular area in carnivores that produces secretory substances.

anestrus The nonbreeding, quiescent condition of the reproductive cycle of a mammal.

angle of attack The angle of the wings of a bat relative to the ground (i.e., the horizontal plane in the direction of movement).

angora Continuously growing, long, flowing hair that may or may not be shed.

anisogamy The condition in which the female gamete (ovum) is larger than the male gamete (sperm).

annual molt A rapid process each year during which most hairs are replaced in species living at temperate and northern latitudes.

Anomodontia A suborder of the Therapsida, these mammal-like reptiles were primarily herbivorous and were extinct by the late Triassic period.

anosmic Without the sense of smell.

antagonistic actions Having opposite effects, as hormones on target tissues.

anterior pituitary gland The anterior lobe of the pituitary gland.

antitragus A small fleshy process on the ventral margin of the pinnae of bats, often those that lack a tragus.

antlers Paired processes that are found only on the skull of cervids (deer), made entirely of bone, branched, and shed yearly.

apocrine sweat glands A type of sweat gland found on the palms of the hands and bottom of the feet of mammals; highly coiled structures located near hair follicles (sudoriferous glands).

apomorphic Characters that are derived or are of more recent origin.

appendicular skeleton The portion of the postcranial skeleton that consists of the pectoral and pelvic girdles, arms (forelimbs), and legs (hind limbs).

aquatic mammals Those that live most of the time in water but come onto land periodically for certain activities such as breeding and parturition.

Archaeoceti An extinct order of whales that had features intermediate between terrestrial mammals and fully marine species.

arousal The third or final stage of the cycle of dormancy. Characterized by shorter periods of dormancy and increases in body temperature; usually occurring in late winter.

articular A bone in the mandible of lower vertebrates and primitive mammals; becomes the malleus bone in the middle ear of modern mammals.

ascending ramus   The upward curving branch of the jaw that articulates with the base of the skull; the vertical portion of the dentary bone.

aspect ratio   The ratio of the length of a wing to its width; short, wide wings have a low aspect ratio.

association cortex   Regions of the cortex that are not specifically identifiable as sensory or motor cortex, where information is processed and integrated across the various sensory modalities.

astragalus   One of the ankle bones; in ungulates, its pulleylike surface limits motion to one plane.

atlas   One of the top two cervical vertebrae that articulate the skull and vertebral column and provide for head movement.

auditory bulla   The auditory (hearing) vessicle with the tympanic floor derived from only the petrosal plate and ectotympanic bone.

autocorrelation   As used in radiotelemetry, locations or fixes that are too close together in time, that is, dependent on the previous location.

awns   The most common guard hairs on mammals, having an expanded distal end with firm tips and a weak base. They exhibit definitive growth and usually lie in one direction, giving pelage a distinctive nap.

axial   The portion of the postcranial skeleton consisting of the spinal column and rib cage.

axis   One of the top two cervical vertebrae that articulate the skull and vertebral column and provide for head movement.

B

baculum   A penis bone found in certain mammalian orders (same as os penis).

baleen   The fringed plates of keratinized material that hang from the upper jaw of mysticete whales. Baleen grows throughout the life of an individual and is used to filter small marine organisms from the water for food.

Bateman gradient   The large variation in the reproductive success of males compared to females in relation to the number of mates they obtain.

behavioral ecology   The study of an animal's struggle for survival as it exploits resources and avoids predators, as well as how an animal's behavior contributes to its reproductive success.

Bergmann's rule   The biogeographic "rule" that races of warm-blooded species living in warmer climates are smaller than races from cooler climates.

bet-hedging   Spreading of risk, reducing the chance of catastrophic failure to survive and reproduce. Bet-hedging leads to more frequent but less intense bouts of reproduction.

bicornuate   Type of uterus in eutherian mammals that has a single cervix and the two uterine horns fused for a part of their length (found in insectivores, most bats, primitive primates, pangolins, some carnivores, elephants, manatees, dugongs, and most ungulates).

bifid   Divided into two equal parts.

bifurcated   Paired; with two corresponding halves.

binomial nomenclature   A system for naming all organisms in which there are two names, one for the genus and the other for the species.

biodiversity   The living plants, animals, and other organisms that characterize a particular region, country, or the entire earth.

biogeography   The study of the patterns of distribution of organisms, including both living and extinct species.

biological classification   The grouping of organisms into ordered categories according to their attributes, reflecting their similarities and consistent with their evolutionary descent.

biological communication   An action on the part of one organism (the sender) that alters the probability of occurrence of behavior patterns in another organism (the receiver) in a fashion adaptive to either one or both of the participants.

biological community   An association of interacting populations, usually defined by the nature of their interaction in the place in which they live.

biological control   Using a species of organism to reduce the population density of another species in an area through parasitism or predation.

biome   A broad ecosystem characterized by particular plant life, soil type, and climatic conditions.

bipartite   A type of uterus in eutherian mammals that is almost completely divided along the median line, with a single cervical opening into the vagina (found in whales and most carnivores).

blastomeres   Early cleavage cells.

blubber   A thick subcutaneous layer of fat below the dermis that prevents heat loss from the body core and provides energy, insulation, and bouyancy to whales, seals, and walruses.

body hair   The outer layer of hair or fur; also called guard hair.

Bowman's capsule   A part of the kidney; the invaginated distal portion of the uriniferous tubule, which contains the glomerulus (also called renal capsule).

brachiation   Using the forelimbs to swing from branch to branch.

brachyodont   Cheekteeth with low crowns found in certain omnivorous mammals such as primates, bears, pigs, and some rodents; as opposed to hypsodont.

bradycardia   A reduced heart rate associated with diving or torpor.

breasts   Milk-producing glands unique to mammals; see *mammae*.

breed   Individuals of the same species that have a uniform, heritable appearance.

bristles   Firm, generally long hairs that exhibit angora growth, such as in the manes of horses or lions. Bristles function in communication, augmenting or accentuating facial expressions or body postures.

brown adipose tissue   A special type of fat packed with mitochondria; the site of nonshivering thermogenesis in eutherian (placental) mammals.

Bruce effect   In mice, the effect of a strange male, or his odor, that causes a female to abort and become receptive.

buboes   Swollen lymph nodes.

bulbourethral glands   In male mammals, the small paired glands that secrete mucus into the urethra at the time of sperm discharge (also called Cowper's glands).

bunodont   Low-crowned teeth that have rounded, blunt cusps used primarily for crushing.

C

caching   The handling of food for the purpose of conserving it for future use; synonymous with food hoarding.

calcaneum   The heel bone; largest and most posterior of the ankle bones.

calcar   A process that extends medially from the ankle of bats and helps support the uropatagium.

calcarine fissure   A shallow groove or sulcus separating convolutions of the brain; located on the internal walls of the cerebral hemispheres.

callosities   Hardened, thick areas on the skin; for example, rough patches or outgrowths found on certain species of whales (same as excrescences) or ischial callosities in cercopithecid primates.

calyx   A small collecting area for the renal pyramids found within the mammalian kidney.

camber   Considered in cross section, the amount of curvature in a wing.

caniform   Doglike.

canine   Unicuspid tooth posterior to the incisors and anterior to the premolars. If present, there is never more than one canine tooth in each quadrant.

caniniform   Canine-shaped.

cannon bone   Metapodials that are fused to form a single long bone in many unguligrade species.

capacitation   The physiochemical changes in the spermatozoa that enable them to penetrate the protective covering of cells surrounding the oocyte.

cardiac muscle   Heart muscle.

carnassial   Bladelike, shearing cheekteeth found in most carnivores; the last upper premolar and the first lower molar in extant mammals. Most highly developed in felids and canids.

carnivorous (carnivory)   Consuming a diet primarily of animal material (characterizing meat-eating members of the Order Carnivora, marsupial dasyurids, and others).

carpal A group of bones of the forefoot distal to the radius and ulna that form the wrist.

carrion Animal matter after it is dead and often decaying.

carrying capacity (*K*) The number of individuals in a population that the resources can support; the equilibrium density reached in a logistic growth curve.

caudal The most posterior vertebrae; the number varies with tail length.

caudal undulation Vertical motion of the tail of a marine mammal to produce forward momentum.

cavernous sinus A network of small vessels immersed in cool venous blood located in the floor of the cranial cavity and important in heat exchange in certain carnivores and artiodactyls.

cellulolytic Cellulose-splitting enzymes.

cementum A layer of bony material that covers the roots of teeth and helps keep them in place. In many species, cementum annuli (rings) are evident that correlate with the age of the individual.

center of origin The location where a particular taxon arose.

cerebral cortex The upper main part of the brain, consisting of two hemispheres.

cervical The most anterior vertebrae, numbering seven in most mammals.

cervix The tip of the uterus that sometimes projects into the vagina.

character displacement Divergence in the characteristics of two otherwise similar species where their ranges overlap; thought to be caused by the effects of competition between the species in the area of overlap.

cheekteeth Dentition that is posterior to the canines, that is, premolars and molars.

chevron bones Found on the ventral part of the caudal vertebrae of marine mammals; sites for attachment of the muscles that depress the tail.

chorioallantoic placenta A type of placenta found in eutherians and peramelemorphians (bandicoots and bilbies) composed of two extra-embryonic membranes: an outer chorion and an inner vascularized allantois. Highly vascularized villi enhance nutrient exchange and mechanical connection to the uterine lining.

chorionic villi Fingerlike projections of capillaries from the outermost embryonic membrane that penetrate the endometrium; increase exchange between maternal and fetal systems.

choriovitelline placenta A type of "yolk-sac" placenta found in all marsupials except bandicoots; lacks villi and has a weak mechanical connection to the uterine lining.

circadian rhythms Activity patterns with a period of about 24 hours.

circannual rhythms Seasonal activity patterns with a period of about one year.

cladistics A process by which unweighted, nonmetric characters (traits) are used to organize organisms exclusively into taxa on the basis of joint descent from a common ancestor.

cladogram A simple representation of the branching pattern or phyletic lineage, which does not attempt to represent rates of evolutionary divergence.

clavicle The bone connecting the scapula and sternum.

claviculate Having a clavicle, the bone connecting the scapula and sternum.

claws These occur at the ends of the digits, grow continuously, and consist of two parts, a lower or ventral subunguis, which is continuous with the pad at the end of the digit, and an upper or dorsal unguis.

cleidoic Egg shells that are impermeable to air or nutrient exchange, as are found in birds.

climax community The end point of a successional sequence; a community that has reached a steady state.

climbing An often arboreal locomotion using the forelimbs and hind limbs to grasp branches and other objects to move about the habitat.

clitoris The small erectile body at the anterior angle of the female vulva; homologous to the penis in the male mammal.

cloaca A chamber into which the digestive, reproductive, and urinary systems empty and from which the products of these systems leave the body.

closed-rooted Teeth that do not grow throughout the life of an individual, as opposed to open-rooted teeth.

coagulating glands Anterior prostate glands of some mammals; upon ejaculation, the secretion of these glands, when mixed with the secretions of the seminal vesicles, sometimes forms a viscous substance that constitutes a "copulatory plug" left in the vagina.

cochlea The spirally coiled tubular cavity of the inner ear containing necessary organs for hearing.

coevolution In parasite-host relationships, reciprocal evolutionary changes that result in close association of certain species of parasites with certain host species.

cohort life table The age-specific survival and reproduction of a group of individuals of the same age, recruited into the population at the same time and followed from birth to death. Compare to *static life table.*

cold-blooded Pertaining to any animal whose body temperature remains close to that of ambient temperature. Includes all animals except birds and mammals; poikilothermy, ectothermy.

colostrum A special type of protein-rich mammalian milk secreted during the first few days before and after birth of young; contains antibodies that confer the mother's immunity to various diseases to the young.

commensal(ism) Different organisms living in close association with each other; one is benefited and the other is neither benefited nor harmed; in close association with humans.

community ecology The study of the interactions among species and the effect those interactions have on both the living and nonliving features of their environment.

community structure Patterns of species composition and abundance, temporal changes in communities, and relationships among locally coexisting species.

competition The attempt of two or more organisms (or species) to use the same limited resource.

competitive exclusion principle The hypothesis that two or more species cannot coexist on a single resource that is limiting to both.

competitive release The expansion of range of one species that sometimes occurs when a competing species is removed.

composite signal In communication, a signal that is made up of more than one signal.

concealed ovulation The absence of any behavioral or morphological changes associated with estrus.

conduction The movement of heat from regions of high temperature to regions of low temperature.

Condylarthra A diverse lineage of Paleocene herbivores, a generalized ancestral order, from which arose several orders including proboscideans, sirenians, cetaceans, perissodactyls, and artiodactyls.

condyle A rounded process at the end of a bone, providing for an articulation with a socket of another bone. Occipital condyles provide articulation between the skull and the vertebral column.

cones Retinal receptors for color vision.

conspecific Individuals or populations of the same species.

context In communication, stimuli other than the signal that are impinging on the receiver and might alter the meaning of a signal.

continental drift The movement over geological time of the large land masses of the earth's surface as a result of plate tectonics.

convection Movement of heat through a fluid (either liquid or gas) by mass transport in currents.

convergence (convergent) (1) The presence of a similar character for two taxa whose common ancestor lacked that character. (2) The evolution of similar morphologies by distantly related lineages that inhabit regions with similar ecological, geological, and climatic conditions.

coprophagy Feeding upon feces (as in shrews, lagomorphs, rodents); also called refection.

copulatory plug A plug of coagulated semen formed in the vagina after copulation; found only in certain species of mammals.

core area The area of heaviest use within the home range.

corpora albicans   The degenerated corpus luteum formed after the birth of the fetus or after the egg fails to implant in the uterus (also called "white bodies").

corpora cavernosa   A mass of spongy tissue surrounding the male urethra within the penis.

corpus callosum   The bundle of nerve fibers that integrates the left and right hemispheres of the brain in eutherian mammals.

corpus luteum   An endocrine structure that forms from the remnants of the ovarian follicle after ovulation. The corpus luteum ("yellow body") secretes progestins that support the uterine lining for blastocyst implantation.

corridor route   A faunal interchange where there is minimal resistance to the passage of animals between two geographic locations.

cortex   (1) The structure that surrounds the medulla and makes up most of the hair shaft. (2) The area of the kidney that contains the renal corpuscles, convoluted tubules, and blood vessels.

cortisol   An adrenal cortical hormone.

cosmopolitan   Essentially worldwide geographic distribution.

cotyledonary placenta   A type of chorioallantoic placenta in which the villi are grouped in well-shaped rosettes separated by stretches of smooth chorion.

countercurrent heat exchange   An arrangement of blood vessels that allows peripheral cooling particularly of appendages and at the same time maintains an adequate blood supply without excessive heat loss.

countershading   Having ventral body pelage that is more lightly colored than the dorsal surface.

cranium   The upper portion of the skull including the bones that surround the brain.

crenulations   A series of low humps or ridges on the back of gray whales.

crepuscular   Activity concentrated near sunrise and sunset.

crown   The portion of a tooth that projects above the gum, composed of enamel and dentine.

cryptic coloration   A pelage pattern that matches the general background color of the animal's habitat.

cup-shaped discoid placenta   The type of chorioallantoic placenta that is a variation on the discoidal arrangement of the placenta.

cursorial   A type of locomotion in which at least a portion of the time is spent running.

cusp   A projection or point on the chewing surface of a tooth.   Molariform teeth have several cusps, including a protocone, metacone, and so on.

cuticle   The thin, transparent, outer layer of the hair; forms a scalelike pattern on the surface.

cycle of dormancy   The three phases of dormancy: entrance, period of dormancy, and arousal.

Cynodontia   The diverse group of theriodont therapsid reptiles from which mammals evolved.

D

deceit   In communication, the sending of misleading information.

decidua   The uterine mucosa in contact with the trophoblast.

deciduous placenta   The type of placenta in which a portion of the uterine wall is torn away at parturition.

definitive hairs   Hairs that attain a particular length and are shed and replaced periodically.

definitive host   A host in which a parasite reaches sexual maturity.

Dehnel's phenomenon   The observation that reduction in the body size of the common shrew (*Sorex araneus*) in autumn is accompanied by shrinkage in the size of the skull.

delayed development   A condition found in some neotropical bats and characterized by a reduced growth rate of the embryo following implantation. Differs from delayed fertilization in that the blastocyst implants shortly after fertilization, but development is very slow.

delayed fertilization   An adaptation of certain species of hibernating bats where mating occurs in late summer or autumn. Following copulation, sperm floats free in the uterine tract during the winter. Ovulation then occurs in the spring with subsequent fertilization and implantation.

delayed implantation   The postponement of embedding of the blastocyst in the uterine epithelium for several days or months.

deme   A local population within which individuals theoretically mate more or less at random.

dendrogram   A treelike diagram of the relationships in a phylogeny.

dentary bone   The single bone of the lower jaw or mandible in mammals.

dentine   Hard, dense, calcareous (containing calcium carbonate), acellular material under the enamel in a tooth; mesodermal in origin.

dermis   The part of the skin that consists of connective tissue and is vascularized; located beneath the epidermis.

diaphragm   A muscular partition that separates the thoracic (chest) and abdominal cavities.

diastema   A gap between adjacent teeth, for example, between incisors and cheekteeth in rodents, lagomorphs, artiodactyls, and perissodactyls.

didactylous   Often used to refer to marsupial orders in which the digits are unfused, each with its own skin sheath; as opposed to syndactylous.

didelphous   Pertaining to the female reproductive tract of marsupials in which the uteri, oviducts, and vaginas are paired.

diestrus   The final stage of the estrous cycle; cornified cells are rare and some mucus may be present in a vaginal smear; progesterone levels increase, reach a peak, and then decline.

diffuse placenta   The type of chorioallantoic placenta in which the villi are expansive and distributed over the entire chorion.

digastric digestive system   See *foregut fermentation*.

digitigrade   A cursorial locomotion of running on one or more toes.

dilambdodont   Tooth cusps and associated ridges arranged in a W-shaped pattern.

dimorphic   Having more than one form, size, or appearance; usually referring to the difference between males and females of a species.

dioecious   Pertaining to an organism in which male and female reproductive organs occur in different individuals.

diphyletic   A group whose members are descended from two distinct lineages.

diphyodont   Two sets of teeth during a lifetime. In the typical mammalian pattern, deciduous teeth ("milk" or "baby" teeth) are followed by permanent counterparts.

diprotodont   Dentition in marsupials, specifically the Paucituberculata and Diprotodontia, in which the lower jaw is shortened and the single pair of lower incisors is elongated to meet the upper incisors; as opposed to polyprotodont.

discoidal placenta   The type of chorioallantoic placenta in which the villi are limited to one or two disc-shaped areas.

discontinuous distribution   A condition that occurs when a widespread species becomes restricted or split into isolated geographic locations.

discrete   In communication, signals that are all or none.

disease   The deviation from a normal, healthy state; illness with specific causes and symptoms.

dispersal   (1) Movements that occur within the lifetime of the individual, as, for example, when it leaves its natal site. (2) Long-term movement patterns involving species in a historical zoogeographic sense.

displays   Behavior patterns that convey messages from one individual to another.

disruptive coloration   Patterns of stripes or colors as part of the fur that stand out from the basic background fur pattern.

diurnal   Active primarily during daylight hours and quiescent at night.

DNA-DNA hybridization   Measures similarity between strands. The DNA is heated and the strands separate but remain intact. When the strands are allowed to cool, they collide with one another by chance, and complementary base pairs again link together. When DNA from a single species has been used, cooling results in homoduplexes. When DNA samples from two species are heated and cooled together, heteroduplexes are formed.

Strands are separated and tested for thermal stability by slowly raising the temperature and assessing the mixture for DNA strand disassociation.

DNA fingerprinting The DNA is cleaved by restriction enzymes that result in repetitive units of 16-64 base pairs in length. The fragments are run on gels and visualized by the Southern blot method. The band patterns produced in this manner are unique to each individual.

DNA sequencing The exact listing of the order of the four nucleotide bases (A [adenine], G [guanine], T [thymine], C [cytosine]) within a section of DNA.

Docodonta An order of late Jurassic mammals known only from the remains of complex tooth and jaw fragments.

domesticated Individuals or species that are bred in captivity to benefit a human community that controls breeding, territory, and food supply.

Doppler shift The apparent change in sound or light frequency caused by movement of the source or the receiver.

dormancy A period of inactivity in which an animal allows its body temperature to approximate ambient temperature.

drag As an object moves through a medium, the resistive force from friction with resulting loss of momentum. Friction results from resistance to movement of water or air and the size, shape, and speed of the object moving through it.

Dryolestidae A diverse family of early omnivorous mammals in the Order Eupantotheria that were extinct by the mid-Cretaceous period.

ductus deferens The tube that carries sperm from the epididymus to the cloaca or urethra in male mammals (also called vas deferens).

duplex A type of uterus in which the right and left parts are completely unfused and each has a distinct cervix; found in lagomorphs, rodents, aardvarks, and hyraxes.

durophagous Consuming a diet of shells, hulls, or other hard, often chitinous, materials.

E

eccrine sweat glands Sweat glands with separate ducts leading to the body surface, where water is forced outward leading to evaporative cooling. Found throughout the body, they are important as a means of evaporative cooling in mammals.

echolocate Emit high frequency sound pulses and gain information about the surrounding environment from the returning echoes.

ecological dispersal Movements that occur within the lifetime of the individual, as, for example, when it leaves its natal site.

ecological succession The replacement of populations in a community through a more or less regular series to a stable end point (see *climax community*).

ecological zoogeography Relationships between living organisms and their physical and biotic environment.

economic defendability A state in which the defense of a resource yields benefits that outweigh the costs of defending it.

ecosystem The interacting biotic (living) components (i.e., the community) plus the abiotic (nonliving) components in a defined area.

ectoparasites Parasites that occur on or embedded in the body surface of their host.

ectothermy The maintenance of body temperature primarily by sources outside (*ecto*) the body; cold-blooded, poikilothermy.

edentate Without teeth; true anteaters, pangolins, and the monotremes are all edentate.

Eimer organs The sensitive tactile organs located on the snouts of moles and desmans.

embryonic diapause A period of arrested development of an embryo at the stage of the blastocyst (70-100-cell stage); found in some kangaroos and wallabies.

emigration The movement of individuals out of a population.

enamel The outer portion on the crown of a tooth. Dense, acellular, and ectodermal in origin, it is the hardest, heaviest, most friction-resistant tissue in vertebrates.

endangered species Species that are likely to go extinct in all or a major portion of their range as a result of human activities and natural causes.

endemic (endemism) A native taxon that is restricted to a limited geographic area and nowhere else.

endocrine disrupters Chemical compounds that mimic the effects of hormones, facilitating or inhibiting processes normally regulated by the endocrine system; particularly important during prenatal and early postnatal development.

endocrine glands Specialized groups of cells that produce chemical substances that are released into the bloodstream.

endometrium The inner lining of the uterus in which blastocysts implant during gestation.

endoparasites Parasites that occur inside the body of their host.

endotheliochorial placenta The arrangement of the chorioallantoic placenta in which the chorion of the fetus is in direct contact with the maternal capillaries.

endothelioendothelial placenta The arrangement of the chorioallantoic placenta in which the maternal and fetal capillaries are next to each other with no connective tissue between them.

endothermy The maintenance of a relatively constant body temperature by means of heat produced from inside (*endo*) the body; emphasis is on the mechanism of body temperature regulation; homeothermy, warm-blooded.

energy The ability to do work.

entoconid One of the accessory cusps found in the lingual portion of the talonid of lower molars.

entrance In hibernation, the first stage of the cycle of dormancy; characterized by decrease in heart rate, reduction in oxygen consumption, and decrease in body temperature.

enucleate Without a nucleus. Red blood cells (erythrocytes) in adult mammals are enucleate.

enzootic A disease affecting animals only in certain areas, climates, or seasons.

epidemic A severe disease outbreak affecting many people, often over a widespread area.

epidermis The outer layer of the skin, which consists of three layers: the outer stratum corneum, the middle stratum granulosum, and the inner stratum germinativum.

epididymus A coiled duct that receives sperm from the seminiferous tubules of the testis and transmits them to the ductus deferens.

epipubic bones Paired bones extending anteriorly from the pelvic girdle. Seen in early reptiles, they occur in monotremes and almost all marsupials.

epitheliochorial placenta The arrangement of the chorioallantoic placenta typified by having six tissue layers, with the villi resting in pockets in the endometrium; the least modified placental condition.

epizootic Rapid, widespread disease; epidemic.

erythrocytes Red blood cells.

estivate (estivation) A period of several days or longer in summer during which a mammal allows its body temperature to approximate ambient temperature.

estradiol See *estrogen*.

estrogen Any of the C_{18} class of steroid hormones, so named because of their estrus-generating properties in female mammals; produced by developing ovarian follicles, under the control of follicle stimulating hormone. Biologically important estrogens include estradiol, estrone, and estriol.

estrous cycle A sequence of reproductive events, including hormonal, physiological, and behavioral, that typically occur at regular intervals in a female mammal; generally divided into four stages: proestrus, estrus, metestrus, and diestrus.

estrus The period during which female mammals will permit copulation (adj., *estrous*). Specifically, when ovulation occurs; detected via vaginal cytology or behavior. Also called "heat."

etiologic The specific causal agent in a disease.

eumelanin Pigment mixtures that provide shades of black and brown.

Eupantotheria An order of early mammals with tribosphenic teeth from which more advanced, therian lineages evolved by the mid-to late Cretaceous period.

eusocial (eusociality) A social system involving reproductive division of labor, that is, castes, and cooperative rearing of young by members of previous generations.

euthemorphic   An often square molar with four major cusps (protocone, hypocone, paracone, and metacone). Various modifications occur in different mammalian groups.

evaporation   The conversion of liquid into vapor.

evaporative cooling   Cooling due to absorption of heat when water changes state from a liquid to a vapor. Heat is absorbed from the surface at which the change of state occurs and is carried away with the water vapor produced.

evenness   The relative abundance of individuals within each species in a community.

excrescences   Hardened, thick areas on the skin; rough patches or outgrowths found on certain species of whales (same as callosities).

exotic   Nonnative; a species introduced to an area in which it does not occur naturally.

exploitation competition   A type of competition in which organisms passively use up resources; also called scramble competition. Contrast with **interference competition.**

exponential (growth)   In reference to rates of increase (or decrease) in population size in which the number present (N) is raised to a power; accelerating population growth as rate of change depends on the number of organisms present.

extinction   The loss of a species, which is often a natural process and the ultimate fate for all species.

F

facultative   An organism that is not dependent on establishing a parasitic relationship but can do so if the opportunity arises.

facultative delayed implantation   A form of delayed implantation in which the delay occurs because the female is nursing a large litter or faces extreme environmental conditions.

falcate   Curved or hooked.

faunal interchange   Active long-term species dispersal movements.

fecundity   The number of offspring produced during a unit of time.

feliform   Catlike.

female defense polygyny   A mating system in which males control access to females directly by competing with other males.

female reproductive system   Reproductive organs of females consisting of a pair of ovaries, a pair of oviducts, one or two enlarged uteri, a vagina, and a cervix.

fence effect   The tendency of populations to reach high densities when surrounded by a fence or natural barrier.

fenestrated   Describing an area of a skull with light, feathery, latticelike bone structure.

feral   Wild or free-ranging individuals or populations that were once domesticated.

fertilization   The penetration of an egg by a sperm with the subsequent combination of paternal and maternal DNA.

filiform   Thin and threadlike in shape.

filter route   A faunal interchange where only certain species move between land masses because of some type of barrier.

fimbriation (fimbriated)   Stiff fringe of hairs between the toes that aid species, such as shrews, in locomotion.

flagging behavior   Alarm signaling, as with the use of the tail or rump patch.

flehmen   A retraction of the upper lip exhibited soon after sniffing the anogenital region of another or while investigating freshly voided urine.

flight (powered)   Volant locomotion in the aerial environment through the use of wings that beat through the use of energy.

fluctuating asymmetry   Random deviations from bilateral symmetry in paired traits, such as horns.

fluke   The horizontal, dorsoventrally flattened distal end of a whale's tail.

focal animal sampling   The recording of all occurrences of specified behavior patterns or interactions of a selected individual or individuals during a period of prescribed length.

folivorous (folivory)   Consuming a diet of leaves and stems (characterizing koalas, gliders, ringtail possums, sloths, pandas, dermopterans, various megachiropterans, primates, and rodents).

follicle   A small cavity or pit; in the reproductive system of mammals a group of follicles enclose a single egg immediately under the surface of the ovary.

follicle-stimulating hormone (FSH)   A hormone produced by the anterior pituitary gland that stimulates development of ovarian follicles and secretion of estrogens; stimulates spermatogenesis; stimulates Leydig cell development and testosterone production in males.

food hoarding   The handling of food for the purpose of conserving it for future use; synonymous with caching or storing.

foramen magnum   An opening at the base of the skull that provides the pathway for the spinal cord to enter the brain.

foramina   Openings in bone; for example, in the facial region for the eyes and at the base of the skull for the spinal cord.

foregut fermentation   The digestive process characterized by mammals that possess a complex, multichambered stomach with cellulose-digesting microorganisms that enable them to derive nutrients from highly fibrous foods. Also called rumination; digastric digestive system.

fossorial   Digging under the ground surface to find food or create shelter.

frugivorous (frugivory)   Consuming a diet of fruit (characterizing pteropodid and phyllostomid bats, phalangerids, and primates such as indrids, lorisids, cercopithecids, colobines, pongids).

functional response   The change in the rate of consumption of prey by a predatory species as a result of a change in the density of its prey. Compare to **numerical response.**

fundamental niche   The full range of conditions and resources in which the species can maintain a viable population. Compare to **realized niche.**

fur   The most common underhair; consists of closely spaced, fine, short hairs.

fusiform   A cigar- or torpedo-shaped body form tapered at both ends.

G

gait   A pattern of regular oscillations of the legs in the course of forward movement.

gamete   A mature, haploid, functional sex cell (egg or sperm) capable of uniting with the alternate sex cell to form a zygote.

general adaptation syndrome (GAS)   The situation in which nonspecific stressors, such as heat, cold, or defeat in a fight, produce a specific physiological response.

geographic information systems (GIS)   Computer software programs used to store and manipulate geographic information.

gestation   The length of time from fertilization until birth of the fetus.

glans penis   The head or distal end of the penis.

gliding   The aerial locomotion involving the use of a membrane (patagium) to provide lift, but without any active power.

glissant   The gliding locomotion found in colugos, "flying" squirrels, and other species where patagia provide extended surface areas.

global positioning system (GPS)   The use of satellite-based radiosignals and a receiver to obtain the latitude and longitude at any location on the earth's surface.

Gloger's rule   The biogeographic "rule" that states, "Races in warm and humid areas are more heavily pigmented than those in cool and dry areas."

glomerulus   The minute, coiled mass of capillaries within a Bowman's capsule of the mammalian kidney.

graded   In communication, signals that are analog or varying continuously.

granivorous (granivory)   Consuming a diet of primarily fruits, nuts, and seeds.

graviportal   A mode of locomotion in large species in which the limbs are straight and pillarlike and adapted to support great mass. Species such as elephants and hippopotamuses have their legs directly under the body, sacrificing agility for support.

group selection   A selection that operates on two or more genetic lineages (groups) as units; broadly defined, this includes kin and interdemic selection.

guano   Bat fecal droppings that often accumulate in large amounts where colonies roost.

guard hair   The outer layer of hair or fur (overhair); comprised of three types, awns, bristles, and spines.

guilds   Groups of species that exploit a common resource base in a similar fashion.

gumivorous (gumivory) Consuming a diet of exudates from plants such as resins, sap, or gum (characterizing marmosets, mouse lemurs, petaurid gliders, and Leadbeater's possum).

H

habitat destruction A condition in which the expanding human population exerts strong negative effects on a variety of habitats in an effort to meet agricultural and industrial needs; habitats are altered such that some or all of the original fauna and flora can no longer exist in the community.

habitat fragmentation A condition in which the continuous area of similar habitat is reduced and divided into smaller sections because of roads, fields, and towns.

habitat selection The choosing of a place to live.

hair Cylindrical outgrowths from the epidermis composed of cornified epithelial cells; a unique feature of mammals.

hallux The first (most medial) digit of the pes (hind foot); the big toe in humans.

handicap In sexual selection, the hypothesis that apparently deleterious sexual ornaments possessed by males are attractive to females because they indicate that the males bearing them have such vigor that they can survive even with the handicap.

harmonics Integral multiples of a fundamental sound frequency.

heat See **estrus.**

heat load The sum of the environmental and metabolic heat gain.

Heimal threshold The depth of snow required to insulate the subnivean (below the snow) environment against fluctuating environmental temperatures.

hematocrit The number of red blood cells per unit volume of blood.

hemochorial placenta The arrangement of the chorioallantoic placenta that lacks maternal epithelium; villi are in direct contact with the maternal blood supply.

hemoendothelial placenta The arrangement of the chorioallantoic placenta in which the fetal capillaries are literally bathed in the maternal blood supply. This arrangement shows the greatest destruction of placental tissues and least separation of fetal and maternal bloodstreams.

heterodont Teeth that vary in form and function and generally include incisors, canines, premolars, and molars.

heterothermic (heterothermy) Animals that at times exhibit high and well-regulated body temperatures (homeothermic) and at other times exhibit body temperatures that are close to that of the environment (ectothermic).

hibernation A form of adaptive hypothermia characterized by profound dormancy in which the animal remains at a body temperature ranging from 2 to 5°C for periods of weeks during the winter season.

hindgut fermentation A digestive system in which food is completely digested in the stomach and passes to the large intestine and cecum, where microorganisms ferment the ingested cellulose; also called a monogastric system.

historical zoogeography Attempts to explain, by reconstruction, the sequences of events involved in the origin, dispersal, and extinction of species.

homeothermy The regulation of a constant body temperature by physiological means regardless of external temperature; endothermy, warm-blooded.

home range The area in which an animal spends most of its time engaged in normal activities.

homing The process of returning to a home range, nest site, or den.

homodont Teeth that do not vary in form and function; often peglike in structure, as in toothed whales and some xenarthrans.

hooves Large masses of keratin (the unguis) completely surrounding the subunguis; found in the ungulates (perissodactyls and artiodactyls); a specialized variation of claws.

hormones Chemical substances that are released into the bloodstream or into body fluids from endocrine glands and that affect target tissues.

horns Processes in bovids formed from an inner core of bone that extend from the frontal bone of the cranium, are covered by a sheath of keratinized material, and are derived from the epidermis.

hydric Wetland habitats; as opposed to xeric (very dry) areas.

hyoid apparatus A series of bones located in the throat region that are modified from ancestral gill arches in fish. These structures provide protection and support for the base of the tongue, trachea, larynx, and esophagus.

hypocone A cusp that is posterior to the protocone and lingual (toward the tongue) in upper molars. It is labial (toward the cheek) in lower molars (where it is called a hypoconid). The addition of this cusp often forms quadritubercular molars.

hypoconid One of the accessory cusps found in the labial portion of the talonid of lower molars.

hypoconulid One of the accessory cusps found in the posterior portion of the talonid of lower molars.

hypodermis The innermost layer of the integument, consisting of fatty tissue; the base of each hair follicle is located in this layer, along with vascular tissues, parts of sweat glands, and portions of the dermal sensory receptors.

hypothalamic-pituitary portal system The vascular connection between the hypothalamus and pituitary gland.

hypothalamus The part of the midbrain, located below the thalamus, that contains collections of neuron cell bodies (nuclei). Sensory input to the hypothalamus comes from other brain regions and from cells within the hypothalamus that monitor conditions in blood that passes through the region. The hypothalamus is the key brain region for body homeostasis; the mammalian "thermostat."

hypothermia A condition in which the temperature of the body is subnormal.

hypsodont Cheekteeth with high crowns, often with complex folding ridges; as opposed to brachyodont.

hystricognathous A mandible in certain rodents; in ventral view, the angular process is lateral to the alveolus of the incisor (as opposed to sciurognathus).

hystricomorph Rodents in which the infraorbital foramen is greatly enlarged.

I

ideal free distribution The distribution of individuals among resource patches of different quality that equalizes the net rate of gain of each individual. Assumes that organisms are free to move and have complete knowledge about patch quality.

imbricate Overlapping; as in fish scales, for example.

immigration The movement of individuals into a population.

implantation The attachment of the embryo to the uterine wall of the female mammal.

inbreeding depression The reduced reproductive success and survival of offspring from closely related parents compared to offspring of unrelated parents. It is caused by increased homozygosity of the inbred offspring and the resulting expression of deleterious recessive alleles.

incisors Usually unicuspid teeth anterior to the canines that are used for cutting or gnawing.

inclusive fitness The sum of an individual's direct and indirect fitness. Direct fitness is measured by reproductive success of one's own offspring (descendant relatives), and indirect fitness is measured by the reproductive success of one's nondescendant relatives.

incrassated Thickened or swollen.

incus The second of the three bones of the middle ear in mammals (ossicles); derived from the quadrate bone.

indicator models Models of sexual selection that assume that the trait favored by females in some way indicates male fitness.

induced ovulation Ovulation that occurs within a few hours following copulation; the act of copulation serves as a trigger for ovulation.

infanticide The killing of young.

infrasound Sound frequencies of less than 20 Hz.

insectivorous (insectivory) Consuming a diet of insects, other small arthropods, or worms.

insensible water loss   The mechanism by which water is lost by diffusion through the skin and from the surfaces of the respiratory tract. Also called transpirational water loss.

integument   The outer boundary layer between an animal and its environment; the skin.

interference competition   A form of competition in which organisms defend or otherwise control limited resources; also called contest competition. Contrast with *exploitative competition.*

interleukin   Any of several compounds that are produced by lymphocytes or monocytes and function especially in regulation of the immune system.

intersexual selection   The selection of characteristics of one sex (usually males) based on mate choices made by members of the other sex (usually females).

intrasexual selection   The selection of characteristics of the sexes based on competition among them (usually males) for access to members of the other sex (usually females).

introgression   The mixing of gene pools.

iteroparity (iteroparous)   The production of offspring by an organism in successive bouts. Compare to *semelparity (semelparous).*

K

K-selection   Selection favoring slow rates of reproduction and growth, characteristics that are adapted to stable, predictable habitats. Compare to *r-selection.*

karyotype   The characteristic number and shapes of the chromosomes of a species.

keratin   A tough, fibrous scleroprotein found in epidermal tissues (in hard structures such as hair and hooves, for example).

keratinized   Made of keratin.

keystone guild   A group of species exploiting a common resource and controlling the distribution and abundance of many other members of the community.

keystone predator   Species that control the distribution and abundance of many other members of the community, often by limiting a particular species of prey.

kidneys   Paired, bean-shaped structures in mammals located within the dorsal part of the abdominal cavity; the principal organ that regulates the volume and composition of the internal fluid environment.

kin selection   The selection of genes due to individual's assisting the survival and reproduction of nondescendant relatives who possess the same genes by common descent.

kinship   The possession of a common ancestor in the not-too-distant past.

krill   Small marine organisms fed on by baleen whales.

L

labia majora   Two large lateral folds of skin that border and cover the vulva area.

labia minora   Two lateral folds of skin that cover the vaginal opening and are largely covered by the labia majora.

lactation   The production of milk by mammary glands.

lactogenic hormone   A hormone, namely prolactin (PRL), produced by the anterior lobe of the pituitary gland that induces lactation and maintains the corpora lutea in a functioning state in mammals (originally called luteotropic hormone, LTH).

lactose   A 12-carbon sugar present in the milk of mammals.

lambdoidal crest   A bony ridge at the rear of the cranium.

laminae   Ridges on teeth that may have distinct cusps.

laminar flow   The smooth movement of air or water over a surface with a minimum of turbulence.

landscape ecology   The study of the distribution of individuals, populations, and communities across different levels of spatial scale.

laparoscopy   The use of fiber optic techniques to perform a laparotomy.

laparotomy   An internal examination of the female reproductive tract to determine the condition of the ovaries and uterus.

law of the minimum   The idea that only a single factor limits the growth of a population at any one time.

lek   An area used, usually consistently, for communal courtship displays.

lesion   Wound; an area of tissue destroyed via electrical or chemical means.

Leydig cells   The interstitial cells between the seminiferous tubules in the testes that produce androgens in response to luteinizing hormone secreted by the anterior pituitary gland.

life table   A summary by age of the survivorship and fecundity of individuals in a population.

lift   The upward force created as air moves over the top of a wing.

locomotion   A form of movement. Running, jumping, gliding, swimming, or flying are all types of locomotion that occur in various mammals.

logistic equation   The mathematical expression for a sigmoid (S-shaped) growth curve in which the rate of increase decreases in linear fashion as population size increases.

loop of Henle   A long, thin-walled kidney tubule present only in mammals and some birds. The concentrating ability of the mammalian kidney is closely related to the length of the loops of Henle and collecting ducts.

lophodont   An occlusal pattern in which the cusps of cheekteeth form a series of continuous, transverse ridges, or lophs, as in elephants.

lophs   Elongated ridges formed by the fusion of tooth cusps.

lower critical temperature   The temperature at which an animal must increase its metabolic rate to balance heat loss.

lumbar (vertebrae)   The lower back vertebrae that number from four to seven in mammals; sometimes partially or entirely fused.

lutenizing hormone (LH)   A hormone produced by the anterior pituitary gland that stimulates corpora lutea development and production of progesterone in females.

luteotropic hormone (LTH)   See *prolactin.*

lysozyme   A crystalline enzyme-like protein present in tears, saliva, milk, and many other animal fluids that is able to destroy bacteria by disintegration.

M

macroecology   The patterns of body mass, population density, and geographic range at a continental scale.

macroparasites   Parasites that are larger and have longer generation times than microparasites; they usually do not reproduce entirely within or on the host.

male dominance polygyny   A mating system in which males compete and acquire dominance ranks that influence their access to females, with higher-ranking males obtaining more mates.

male reproductive system   Reproductive organs of males consisting of paired testes, paired accessory glands, a duct system, and a copulatory organ.

malleus   The first of the three bones of the middle ear in mammals (ossicles), the "hammer" connects the tympanic membrane (ear drum) and the incus; derived from the reptilian articular bone.

mammae   Milk-producing glands unique to mammals (sing., *mamma*); see *breasts.*

mammalogy   The study of animals that constitute the class Mammalia, a taxonomic group of vertebrates in the Kingdom Animalia.

mammary glands (mammae)   Milk-producing, hormone-mediated glands, unique to mammals; similar to apocrine glands in development and structure.

mandible   The lower jaw; mobile and consists of a single dentary bone.

mandibular fossa   A part of the cranium with which the mandible (lower jaw) articulates.

manubrium   The long, lever-type arm of the malleus that attaches to the tympanic membrane (eardrum); also the anterior (uppermost) portion of the sternum.

manus   The forefoot; together, the carpals, metacarpals, and phalanges.

marine mammals   Mammals that spend their entire lives in the ocean and never come onto land.

marsupium   An external pouch formed by folds of skin in the abdominal wall. Found in many marsupials and in echidnas, the marsupium encloses mammary glands and serves as an incubation chamber.

masseter One of three main masticatory muscles of mammals that functions to close the mouth by raising the mandible. Pronounced in herbivorous mammals, the masseter aids with the horizontal movement of the jaw.

mass-specific metabolic rate The rate of energy necessary per gram of body mass. Refers to energy demands within the tissues of an animal.

maxilla One of a pair of large bones that form part of the upper jaw, carrying teeth; it also forms portions of the rostrum, hard palate, and zygomatic arch.

meaning In communication, how the recipient of a message interprets that message.

medulla (1) The central portion or shaft of the hair. (2) The internal area of the kidney divided into triangular wedges called renal pyramids.

melanism A condition in which an animal is generally all black, due to a genetic mutation.

meroblastic The type of egg cleavage in reptiles, birds, and monotremes in which only part of the cytoplasm is cleaved due to a large amount of yolk.

mesaxonic Having a weight-bearing axis of a limb pass through the third digit, as in perissodactyls.

mesial drift Molariform dentition that is replaced horizontally rather than vertically. As anterior teeth wear out, they move forward and are replaced from the rear of the jaw by posterior teeth; occurs in macropodids, elephants, and manatees.

message In communication, information about the state of the sender.

metabolic rate Energy expenditure measured in kilojoules per day.

metabolic water Water produced by aerobic catabolism of food. Also known as oxidation water.

metacone A cusp that is posterior to the protocone. It is labial (toward the cheek) in upper molars and lingual (toward the tongue) in lower molars (where it is called a metaconid).

metameric Segmented; repeated body units.

metapodials A general term for both the metacarpal and metatarsal bones.

metapopulation A set of local populations or demes linked together via dispersal.

metestrus The third stage of the estrous cycle: leucocytes appear among the cornified epithelial cells in a vaginal smear, corpora lutea are fully formed, and progesterone levels are high.

microfauna Symbiotic ciliated protozoans and bacteria in the forestomach or cecum of herbivores that break down cellulose and other plant materials.

microparasites Parasites that are microscopic and have rapid regeneration times generally within the host.

microsatellite DNA Tandem repeats of short sequences of DNA, usually multiples of two to four bases.

migration A persistent movement across different habitats in response to seasonal changes in resource availability and quality. In mammals, these typically are round-trip movements, and the individual returns to the same breeding and wintering areas each year.

minimum viable population The number of animals needed in a population to prevent it from going extinct within a given time period.

molar A nondeciduous cheektooth with multiple cusps that is posterior to the premolars.

molting The seasonal replacement of definitive hair and sometimes angora hair.

monestrous Having a single estrous period or heat each year.

monogamy (monogamous) A mating system in which a single male and female pair for some period of time and share in the rearing of offspring.

monogastric system See *hindgut fermentation.*

monophyletic Describing a group whose members are descended from a common ancestor.

monotypic Having only one member in the next lower taxon. For example, the aardvark is a monotypic order. It has only one family, one genus in that family, and only one species in the genus.

monozygotic polyembryony Reproductive process in some armadillos in which a single zygote splits into separate zygotes and forms several identical embryos all of the same sex.

morphometrics The measurement of the characteristics of organisms, including such features as the skeleton, pelage, antlers, or horns.

mortality Death, usually expressed as a rate.

multiparous Describing a female that has had several litters or young; often with evidence of placental scars of different ages.

Multituberculata An order of herbivorous mammals with large lower incisors and molariform teeth with numerous large cusps that extended for 120 million years from the late Jurassic period to the late Eocene epoch.

musk Secretions from scent glands found in mustelids and a variety of other mammalian species.

musth The reproductive period in male elephants.

mutualism (mutualistic) A mutually beneficial association between different kinds of organisms.

mycophagous (mycophagy) Consuming a diet of fungi (characterizes many sciurids, murids, and the marsupial Family Potoroidae).

mycorrhiza The mutualistic association of the mycelium of a fungus with the roots of a seed plant.

myoglobin A protein in the muscles that binds oxygen.

myometrium The thick muscular wall surrounding the highly vascular endometrium in mammals.

myomorph Rodents in which the infraorbital foramen is small to moderate in size.

myrmecophagous (myrmecophagy) Feeding primarily on colonial insects such as ants and termites. Many mammalian families are primarily or secondarily myrmecophagous.

Mysticeti A modern order of baleen whales; species that have two external nares and a symmetrical skull and do not echolocate.

N

nails A specialized variation of claws that evolved in primates to facilitate better gripping ability and precision in object manipulation by the hands and feet. Only the dorsal surface of the end of each digit is covered by the nail.

nares External nostrils or "blowholes" in whales.

natal dispersal More or less permanent movements from the natal site to a site where reproduction takes place.

natality Birth, usually expressed as a rate.

nectarivorous (nectarivory) (also *nectivorous, nectivory)* Consuming a diet of nectar; found in about six genera of bats and marsupial honey possums.

necton Larger marine organisms with movements independent of waves and currents.

neonates Newborn animals.

neotony The retention of juvenile characteristics in an adult.

nephrons Functional units of the mammalian kidney; consisting of the Bowman's capsule and a long, unbranched tubule running through the cortex and medulla and ending in the pelvis.

niche The role of an organism in an ecological community, involving its way of living and its relationships with other biotic and abiotic features of the environment.

nictating membrane A thin membrane that functions as a "third eyelid" in certain species.

nocturnal Exhibiting peak activity during hours of darkness and resting when there is daylight.

nondeciduous placenta A type of placenta that separates easily into embryonic and maternal tissue at parturition, resulting in little or no damage to the uterine wall.

nonshivering thermogenesis Means of heat production in mammals that does not involve muscle contraction.

norepinephrine A catecholamine found in sympathetic postganglionic neurons of mammals that stimulates production of heat by brown adipose tissue (also called noradrenaline).

nulliparous A female that has never given birth; shows no evidence of placental scars or pregnancy.

numerical response A change in the population size of a predatory species as a result of a change in the density of its prey. Compare to *functional response.*

numerical taxonomy   A system in which individuals are organized into taxa based on unweighted estimates of overall similarity.

nunatak   Refugia found within ice sheets during periods of glaciation; pockets of variable size that were not covered by the advancing glaciers.

O

obligate delayed implantation   A form of delayed implantation in which the delay occurs as a normal, consistent part of the reproductive cycle, as in armadillos.

obligate parasites   Organisms that must spend at least part of their life cycle as a parasite.

observability   (1) Extent to which habitat permits regular, direct observation. (2) When species are watched, individuals may be seen for different portions of the time period depending on age, sex, or dominance status.

occlusal   The surfaces of upper and lower teeth that contact each other during chewing. Occlusal surfaces of teeth have one or more cusps.

Odontoceti   A modern order of toothed whales; species with a single external nare, asymmetrical skull, and echolocation.

omasum   The muscular third chamber of the stomach of a ruminant.

omnivorous (omnivory)   Consuming both animal and vegetable food (characterizing most rodents, bears, raccoons, opossums, pigs, and humans).

open-rooted   Teeth that grow throughout the life of an individual, as opposed to closed-rooted teeth.

operational sex ratio   The number of reproductively active males and females in a population, expressed as a proportion.

organizational effects   Morphological, physiological, and behavioral differences in adults resulting from prenatal exposure to hormones that alter the developmental trajectories of various cells and tissues.

os baculum   see *os penis.*

os clitoris   A small bone present in the clitoris in some mammal species. Homologous to the baculum (*os penis*) in males.

os penis   A bone in the penises of certain mammals (also called a *baculum*).

os sacrum   Fused sacral vertebrae in mammals.

osmoregulation   The maintenance of proper internal salt and water concentrations; this function is performed principally by the kidneys in mammals.

ossicles (auditory)   The three bones (the malleus, incus, and stapes) of the middle ear in mammals that transmit sound waves from the tympanic membrane (eardrum) to the inner ear.

ossicones   Short, permanent, unbranched processes of bone that form the horns in giraffes.

outbreeding depression   A condition that occurs when new genotypes produced by crossing stocks are inferior to the original native stock; the new stock is at a disadvantage with respect to adaptation to local conditions, possibly because of the breaking up of coadapted gene complexes.

ovaries   The female gonads; the site of egg production and maturation (sing., *ovary*).

oviducts   The ducts that carry the eggs from the ovary to the uterus (also called Fallopian tubes).

oviparous   Able to reproduce by laying eggs, as in monotremes; unlike therian mammals, which are viviparous.

ovulation   The releasing of an egg by the ovary into the oviduct.

oxytocin   A hormone produced by the posterior pituitary gland that causes rhythmical contractions of the uterus during parturition and enhances milk "letdown."

P

pachyostotic   Describing bones that are very dense.

pacing   A gait in which both legs on the same side are raised together.

Paenungulata   A group within the generalized ancestral Order Condylarthra from which evolved elephants, dugongs, manatees, and hyraxes.

palmate   Flattened or weblike.

palynology   Studies of pollen preserved in bogs and other moist places. Because plants that are characteristic of a region at a given time reflect the existing climate, knowledge of flora requirements can be used to describe past climate conditions.

pandemic   A large-scale disease outbreak over a wide geographic area.

Pangaea   A large landmass formed by all the continents about 200 million years ago prior to their drifting apart.

panting   A method of cooling characterized by very rapid, shallow breathing that increases evaporation of water from the upper respiratory tract, as occurs in canids and small ungulates.

papillae   Small, protruding projections.

paracone   A cusp that is anterior to the protocone. It is labial (toward the cheek) in upper molars and lingual (toward the tongue) in lower molars (where it is called a paraconid).

paraxonic   Having a weight-bearing axis of a limb pass through the third and fourth digits, as in artiodactyls.

parental investment   Any investment in offspring that increases its chances of survival and reproduction at the expense of the parents' ability to invest in other offspring.

parental manipulation   The selective providing of care to some offspring at the expense of other offspring so as to maximize the parents' reproductive success.

parous   Describing female mammals that are pregnant or show evidence of previous pregnancies (e.g., possess placental scars).

parsimony (rule of)   The practice of adopting the simplest explanation for an observation consistent with the facts. In taxonomy, determining which cladogram best represents the evolution of a particular group. The tree with the fewest steps or branching points (character states) is generally accepted as the best representation of the phylogeny.

parturition   The process of giving birth in mammals.

passive dispersal   Movements in which the dispersing organisms have no active role.

patagium   A thin membrane of skin that often provides a gliding surface such as between the limbs of the colugo or flying squirrel; the wing membranes of bats are also patagia.

patella   Kneecap.

pectinate   A comblike structure with several prongs or projections in a row.

pectoral girdle   Bones of the shoulder region providing for articulation of the forelimbs; the scapula and clavicle or only the scapula form the shoulder joint in most mammals.

pedicel (pedicle)   (1) A short supporting stalk or stem. (2) In deer, the extension of the frontal bone on which the antlers occur.

pelage   All the hairs on an individual mammal.

pelagic   The open ocean; away from coastal areas.

pelvic girdle   The bones of the hip region, providing for articulation of the hind limbs, and consisting of the paired ilia, ischia, and pubic bones.

pelvis   A large cavity within the mammalian kidney; the renal pelvis empties into the ureter.

Pelycosauria   One of two orders within the reptilian Subclass Synapsida. Pelycosaurs had more primitive characteristics than the other order, the Therapsida.

penis   The male copulatory organ through which sperm are deposited in the female reproductive tract and urine leaves the body.

pentadactyl   Having five digits. The hands and feet of humans are pentadactyl as are those of insectivores.

Peramuridae   A family of Jurassic mammals that probably gave rise to the lineage of advanced therians, that is, mammals of metatherian-eutherian grade.

perineal swelling   The swelling and sometimes reddening of tissues in the anogenital region of some primates in estrus produced by the actions of estrogen.

phalanges   Bones of the fingers and toes; the distal-most bones in the manus and pes.

phenotype matching   A mechanism by which kin may recognize one another; individuals use as a reference kin whose phenotypes are learned by association.

pheomelanin (xanthophylls)   Pigment mixtures that produce various shades of red and yellow.

pheromones   Airborne chemical signals that elicit responses in other individuals, usually of the same species.

philopatric Living and breeding near the place of birth.

phylogenetics (phylogeny) The evolutionary history of various groups of living organisms.

phylogram A tree diagram attempting to represent the degree of genetic divergence among the taxa represented by the lengths of the branches and the angles between them.

piloerection Fluffing of the fur.

piloting The use of familiar landmarks to locate food or shelter.

pinnae External ears that surround the auditory meatus and channel soundwaves to the tympanic membranes (eardrums). Not found in many marine and fossorial mammals.

Pinnipedia (pinnipeds) Literally, "feather-footed"; aquatic carnivores that include the seals and walrus.

piscivorous (piscivory) Consuming a diet composed primarily of fish (characterizes bulldog bats [*Noctilio*]).

pituitary gland The master gland of the endocrine system; located below the hypothalamus.

placenta A highly vascularized endocrine organ developed during gestation from the embryonic chorion and the maternal uterine wall (endometrium). Connects to the umbilical cord through which nutrient and waste exchange occurs between mother and fetus.

placental scar A pigmented area on the uterine wall formed from prior attachment of a fetus.

plankton Floating plant and animal life in lakes and oceans; movements are primarily dependent on waves and currents.

plantigrade Walking on the soles of the hands and feet.

plate tectonics The theory that the earth's crust, including the surfaces of continents and the ocean floors, is made up of a series of geological plates.

plesiomorphic Characters that are primitive; those that are ancestral or appeared earlier.

pods Groups, schools, or herds of animals; specifically applied to whales.

poikilothermy Pertaining to animals whose body temperature is variable and fluctuates with that of the environment; includes all animals except birds and mammals; cold-blooded, ectothermy.

pollex The first (most medial) digit of the manus (forefoot); the thumb in humans.

polyandry A mating system in which females acquire more than one male as a mate.

polyestrous Pertaining to species that exhibit several periods of estrus or heat per year.

polygamy (polygamous) A mating system in which both males and females mate with several members of the opposite sex.

polygyny A mating system in which some males obtain more than one mate and females provide most of the care of offspring.

polygyny threshold The point at which a female will benefit more by joining an already mated male possessing a good territory rather than an unmated male on a poor territory.

polymerase chain reaction (PCR) A procedure for preparing large amounts of DNA from small amounts of sample.

polyprotodont Dentition in several orders of marsupials in which the lower jaw is equal in length to the upper jaw and the lower incisors are small and unspecialized; as opposed to diprotodont.

populations Groups of organisms of the same species, present at the same place and time.

postjuvenal molt A molt starting soon after weaning.

precocial Born in a relatively well-developed condition (eyes open, fully furred, and able to move immediately) and requiring minimal parental care, for example, snowshoe hares, deer, porcupines, and many bovids; as opposed to altricial.

prehensile Possessing digits or tail able to grasp branches and other objects.

premolar Cheekteeth that are anterior to the molars and posterior to the canines. Unlike molars, there are both deciduous and permanent premolars.

preputial glands Modified sebaceous glands that in males of some species contribute to the formation of the semen and in others secrete a scent used for "marking."

priming pheromones Chemical communication substances that produce generalized internal physiological responses, such as the production and release of hormones.

prions Small, modified proteins thought to be the disease agents in spongiform encephalopathies, including "mad cow" disease.

procumbent Projecting forward more or less horizontally, as teeth in shrews, horses, and prosimian primates.

proestrus The beginning stage of the estrous cycle when nucleated cells are present in a vaginal smear and when estrogen, progesterone, and lutenizing hormone levels reach their peak.

progesterone A steroid hormone produced in small quantities by the follicle and in larger quantities by the corpus luteum; promotes growth of the uterine lining and makes possible the implantation of the fertilized egg.

prolactin (PRL) A hormone produced by the anterior pituitary gland that has many actions relating to reproduction and water balance in mammals. For example, PRL promotes corpus luteum function in ovaries and stimulates milk production.

promiscuity A mating system in which there is no prolonged association between the sexes and in which multiple matings by both sexes occurs.

propatagium The anterior portion of a bat's wing that extends from the shoulder to the wrist.

prostaglandins Lipid-based hormones that communicate between cells over a short distance; involved in several aspects of reproductive function such as increased contractions of the uterus.

prostate gland A mass of muscle and glandular tissue surrounding the base of the urethra in male mammals; at the moment of sperm release it secretes an alkaline fluid that has a stimulating effect on the action of the sperm.

protein electrophoresis A method that uses the characteristic migration distance of various proteins in an electric field to identify and compare individuals; sometimes called allozyme analysis.

protein immunology The cross-reactivity of the homologous (original) antigen (protein) and a heterologous antigen (protein from a different, related species) to provide an estimate of the degree of genetic relationship between the two species.

protocone The primary cusp in a tribosphenic molar at the apex of the trigon. It is lingual in upper molars and labial in lower molars (where it is called the protoconid).

protrusible Capable of being turned inside out; bulging or jutting out.

pseudopregnancy Any period when there is a functional corpus luteum and buildup of the endometrial uterine layer in the absence of pregnancy (synonymous with luteal phase).

pterygoideus One of three main masticatory muscles of mammals that functions to close the mouth by raising the mandible. Important in stabilizing and controlling the movement of the jaw.

pulp cavity The part of the tooth below the gumline that contains nerves and blood vessels to maintain the dentine.

Q

quadrate A bone in the posterior part of the mandible of lower vertebrates; becomes the incus in modern mammals.

quadrituberular (quadrituberculate) Describing a square or rectangular cheektooth with four major cusps: protocone, paracone, metacone, and hypocone.

R

r-selection Selection favoring rapid rates of reproduction and growth, especially among species that specialize in colonizing short-lived, unstable habitats. Compare to *K-selection*.

radiation Energy transmitted as electromagnetic waves (e.g., ultraviolet, visible, and infrared).

radioimmunoassay (RIA) A method for assessing hormone levels on small blood samples involving a radioactively labeled antibody mixed with blood samples from an animal in a competitive binding assay.

radiotelemetry A method for determining the location and movements of an animal by using a transmitter affixed to the individual, the signals from which are monitored with an antenna and a receiver from known points in the study area.

random amplified polymorphic DNA (RAPDs) Involves the use of restriction enzymes with short primers in conjunction with PCR. The resulting DNA fragments are 200 to 2000 base pairs long and appear as a series of bands on a gel. The primers are not locus-specific, and thus the amplified loci are said to be anonymous.

ranging Movements that include forays outside the home range, usually in search of suitable habitat or mating opportunities.

realized niche The range of conditions and resources in which the species can maintain a viable population when in the presence of competitors or predators. Compare to *fundamental niche.*

reciprocal altruism The trading of altruistic acts by individuals at different times, that is, the payback to the altruist occurs some time after the receipt of the act.

refugium A delimited geographical region, often from the reduction of a larger range.

relaxin A hormone produced by the corpora lutea that acts to soften the ligaments of the pelvis, so it can spread and allow the fetus to pass through the birth canal.

renal corpuscle A unit of the mammalian kidney, located in the cortical portion of the mammalian kidney, that is composed of Bowman's capsule and the glomerulus.

renal papillae Narrow apices of the cortex of the mammalian kidney. Because the papillae are composed of long loops of Henle, their prominence gives an indication of the number and length of such loops.

renal pyramids Triangular wedges of the medulla of the mammalian kidney.

reproductive effort The energy expended and risk taken to reproduce, measured in terms of the decrease in ability of the organism to reproduce at a later time.

reproductive value The sum of an individual's current reproductive output and its expected future output at age *x.*

reservoir A source that maintains a disease agent in nature.

resource defense polygyny A mating system in which males control access to females indirectly by monopolizing resources needed by females.

restriction fragment-length polymorphisms (RFLPs) Fragments of DNA that have been isolated and cut with one or more restriction enzymes. They are placed on a gel for electrophoresis and stained to permit viewing of the fragments sorted by size.

rete mirabile A complex mass of intertwined capillaries specialized for exchange of heat or dissolved substances between countercurrent flowing blood (also called miraculous net, marvelous net, wonderful net).

reticulum The second of the four compartments of the stomach of ruminants. A blind-end sac with honeycomb partitions in its walls.

rhinarium An area of moist, hairless skin surrounding the nostrils.

ribs Bones attached to the thoracic vertebrae on the dorsal surface and, in most cases, to the sternum on the ventral surface. The rib cage, or thoracic basket, surrounds and protects the vital internal organs.

ritualized Describing behavior patterns that have become modified through evolution to serve as communication signals.

rods Retinal receptors for black-and-white vision.

root The portion of a tooth that is below the gum and fills the alveolus.

rorquals Literally, "tube throated;" large baleen whales with longitudinal grooves on their throats that allow for expansion as they fill with water during feeding.

rostrum The anterior portion of the face or cranium.

rumen The first and largest compartment of the four-part stomach of ruminants.

ruminant artiodactyl A member of the Order Artiodactyla that "chews its cud" or ruminates (e.g., cervids, bovids, antilocaprids, giraffids).

ruminate (rumination) To chew the cud; see *foregut fermentation.*

runaway selection Selection for ornaments (usually in males) that happens due to the genetic correlation and the resulting positive feedback relationship between the trait and the preference for the trait.

rut The mating season in cervids and other artiodactyls.

S

sacculated A stomach with more than one chamber and symbiotic microorganisms for cellulose digestion present in the first chamber(s). The stomach of certain herbivores, whales, and marsupials is sacculated.

sacral (vertebrae) In most mammals, the sacral vertebrae are fused to form the *os sacrum,* to which the pelvic girdle attaches.

sagittal crest The bony midline ridge on the top of the cranium formed by the temporal ridges.

salivary amylase A potent digestive enzyme produced by the salivary glands.

saltatorial locomotion Jumping and ricocheting. Jumping involves the use of all four feet; ricocheting involves propulsion provided only by the two hind limbs.

sanguinivorous (sanguinivory) Feeding on a diet of blood (e.g., vampire bats).

scaling Structural and functional consequences of a change in size or in scale among animals.

scan sampling The recording of the current activity of all or selected members of a group at predetermined intervals.

scapula A part of the pectoral girdle; the shoulder blade.

scent glands Modified sweat or sebaceous glands that produce substances used for a wide variety of functions in mammals.

schizodactylous Grasping digits in which the first two most medial oppose the remaining three.

sciurognathous Describing the mandible in certain rodents; in ventral view, the angular process is in line with the alveolus of the incisor (as opposed to hystricognathous).

sciuromorph Rodents in which the infraorbital foramen is relatively small.

scramble polygyny A mating system in which males actively search for mates without overt competition.

scrotum A bag or pouch of skin in the pelvic region of many male mammals that contains the testicles.

seasonal molt The change of the pelage more than once each year.

sebaceous glands Structures associated with hair follicles that secrete oils to keep the hair moist and waterproof.

sectorial Cutting or shearing teeth.

selenodont A cusp pattern in molariform teeth of goats, sheep, cows, and deer in which the lophs form cresent-shaped ridges or "half-moons" on the grinding surface.

sella A median projection of the nose leaf of horseshoe bats in the Subfamily Rhinolophinae.

semantic Of or relating to the meaning of signals; specifically used to denote the use of different alarm signals to warn about different predators.

semelparity (semelparous) The production of offspring by an organism once in its life. Compare to *iteroparity (iteroparous).*

semen A product of the male reproductive system including sperm and the secretions of various glands associated with the reproductive tract (also called seminal fluid).

semibrachiators An animal that moves by swinging from branch to branch.

seminal vesicles The swollen portion of a male reproductive duct in which sperm are stored and that secretes a fluid useful in the transmission of sperm during copulation (also called vesicular glands).

seminiferous tubules The long, convoluted tubules of vertebrate testes in which sperm cells are produced and undergo various stages of maturation or spermatogenesis.

senescence The gradual deterioration of function in an organism with age, leading to increased probability of death.

sensory channels The physical modality used for signaling, for example, odor or vision.

sequencing Determing the exact order of the four nucleotide bases (A [adenine], G [guanine], T [thymine], C [cytosine]) within a section of DNA.

Sertoli cells Cells that line the seminiferous tubules and that surround the developing sperm, which they nourish.

set point A "reference" temperature in the hypothalamus; analogous to a thermostatic control.

sexual dichromatism The presence of distinctly different pelage colors in males and females.

sexual dimorphism A difference in the sexes in form, such as size; males are often larger than females, although the opposite occurs in some species.

sexual selection Selection in relation to mating; composed of competition among members of one sex (usually males) for access to the other sex and choice of members of one sex by members of the other sex (usually females).

shaft The central structure of a hair, comprised of the inner medulla, surrounded by the cortex, and covered by a thin outer cuticle.

signal The physical form in which a message is coded for transmission through the environment.

signaling pheromones Airborne chemical signals that produce an immediate motor response, such as the initiation of a mounting sequence.

simplex A type of uterus in eutherian mammals in which all separation between the uterine horns is lacking; the single uterus opens into the vagina through one cervix (found in some bats, higher primates, xenarthrans).

smooth muscle An involuntary muscle.

society A group of individuals belonging to the same species and organized in a cooperative manner. Usually assumed to extend beyond sexual behavior and parental care of offspring.

sound window The use of frequencies for communication that are transmitted through the environment with little loss of strength (attenuation).

species–area curve The equation describing the increase in number of species as a function of the area sampled.

species richness The number of species in an area.

sperm competition A situation in which one male's sperm fertilize a disproportionate number of eggs when a female copulates with more than one male.

spermaceti An organ found in the head of certain species of toothed whales. It contains a waxy liquid that may function in diving physiology and echolocation.

spermatogenesis A series of cell divisions and chromosome and cytoplasmic changes involved in the production of functional spermatozoa, beginning with the undifferentiated germinal epithelium.

spines Stiff, enlarged guard hairs that exhibit definitive growth.

spontaneous ovulation Ovulation that occurs without copulation.

stapes The "stirrup" is the last of the three middle ear bones (ossicles) found in mammals. In other vertebrates, this is the only ossicle (the columella) in the middle ear.

state of dormancy The second stage of the cycle of dormancy characterized by leveling off of body temperature; usually occurs in early winter.

static life table A life table generated from a cross section of the population at a specific time. Compare to **cohort life table.**

station keeping Local movements of an animal within its home range, as it acquires resources or marks and defends its territory.

striated muscle A voluntary or skeletal muscle.

subunguis The lower or ventral portion of the claw, which is continuous with the pad at the end of the digit.

supernumerary An additional tooth or teeth in a position where they do not normally occur in a species.

survivorship curve (l_x) The proportion of newborn individuals alive at age x, plotted against age.

sweepstakes route A faunal interchange in which the movement of animals by swimming, flying, rafting, or other means is such that the probability of arriving at another habitable destination is extremely limited.

symbiosis Two species living together in which one benefits and the other may benefit (mutualism), be unaffected (commensalism), or harmed (parasitism).

Symmetrodonta An early order of therian mammals with tribosphenic teeth that include small carnivores or insectivores from the late Triassic period.

sympatric Occurring in the same place; usually referring to areas of overlap in species distributions. Compare to **allopatric.**

Synapsida One of four subclasses of reptiles, this is the subclass from which mammals evolved.

synaptomorphy A sharing of a derived trait by two or more taxa.

syndactylous Having digits in which the skeletal elements of the second and third toes are fused and share a common skin sheath, as in the marsupial Orders Peramelemorphia and Diprotodontia; as opposed to didactylous.

syndesmochorial placenta The arrangement of the chorioallantoic placenta that possesses one less layer than the epitheliochorial condition.

syntax The information provided by the sequence in which signals are transmitted.

syntopic Being present at the same time and place.

T

talonid The basin or heel in lower molariform teeth posterior to the trigonid that occludes with the protocone of the upper molar.

tapetum lucidum A reflective layer lying outside the receptor layer of the retina. Causes the eye shine when light strikes the retina at night. This structure aids in night vision by reflecting light that has passed through the receptor layer back toward the retina.

taxonomic key An arrangement of the traits of a group of organisms into a series of hierarchical, dichotomous choices.

taxonomy A description of species and the process of classifying them into groups that reflect their phylogenetic history.

telescoped skulls Compressed posterior bones and elongated anterior bones in the cranium of modern whales, with associated movement of the nares to the top of the skull.

temporalis One of three main masticatory muscles of mammals that functions to close the mouth by raising the mandible. Pronounced in carnivorous mammals; assists in holding the jaws closed and aids in the vertical chewing action.

territory An area occupied exclusively and defended by an animal or group of animals.

testosterone A steroid hormone secreted by the testes, especially in higher vertebrates; responsible for the development and maintenance of sexual characteristics and the normal production of sperm.

Therapsida One of two orders within the reptilian Subclass Synapsida. These mammal-like reptiles eventually gave rise to mammals.

Theriodontia One of two suborders within the Order Therapsida, the mammal-like reptiles. Primarily carnivorous, theriodonts encompassed several diverse lineages.

thermal conductance The heat loss from the skin to the outside environment.

thermal windows Bare or sparsely furred areas of certain mammals that reside in regions characterized by intense solar radiation and high air temperatures (e.g., guanacos and many desert antelopes). They function as sites through which some of the heat gained from solar radiation can be lost by convection and conduction.

thermogenin A mitochondrial protein responsible for heat production by brown adipose tissue due to uncoupling oxidative phosphorylation.

thermoneutral zone A range in environmental temperatures within which the metabolic rate of an animal is minimal.

thoracic (vertebrae) Articulating with the ribs; from 12 to 15 pairs in mammals.

threatened species Those that are likely to become endangered in the near future (some listings use the term *vulnerable* instead).

tine A point or projection on an antler.

torpor A form of adaptive hypothermia or dormancy in which body temperature, heart rate, and respiration are not lowered as drastically as in hibernation.

total metabolic rate The total quantity of energy necessary to meet energy demands of an animal.

tradition A behavior pattern that is passed from one generation to the next through the process of learning.

tragus A projection from the lower margin of the pinnae of many microchiropteran bats that functions in echolocation.

transpirational water loss The mechanism by which water is lost by diffusion through the skin and from the surfaces of the respiratory tract. Also called insensible water loss.

tribosphenic   Molars with three main cusps (the trigon) arranged in a triangular pattern. Cusp patterns of many modern mammalian groups are derived from this pattern.

Triconodonta   An order of small, carnivorous mammals characterized by molars that had three cusps in a row. The lineage extended for 120 million years until the late Cretaceous period.

trigon(id)   The three cusps (protocone, paracone, and metacone) of a tribosphenic molar.   The suffix *-id* is applied to the mandibular dentition; main cusps are the protoconid, paraconid, and metaconid.

Tritylodonts   A lineage of rodentlike reptiles that existed for about 50 million years from the late Triassic to the mid-Jurassic periods.

trophic   Pertaining to food or nutrition.

trophoblast   The outer layer of the blastocyst in mammals; attaches the ovum to the uterine wall and supplies nutrition to the embryo as part of the placenta.

truth in advertising   In sexual selection, the hypothesis that a male's ornaments or behavior are reliable indicators of his overall genetic fitness.

turbinal (turbinate) bones   Found within the nasal area, they increase the surface area for reception of chemical cues and secrete mucus to aid in filtering small particles from incoming air.

tympanoperiotic   The auditory bullae and middle ear apparatus of whales; not fused to the skull so that the direction of incoming sound waves can be determined.

tympanum   A membrane (ear drum) at the interior end of the external auditory meatus; connects to the ossicles of the middle ear.

U

ultradian rhythms   Activity rhythms with a period of less than 24 hours.

ultrasound   Sound frequencies greater than 20,000 kHz.

unguiculate   Having nails or claws instead of hooves.

unguis   The upper or dorsal portion of the claw, which is a scalelike plate that surrounds the subunguis.

ungulates   Mammals with hooves; perissodactyls and artiodactyls.

unguligrade   A running locomotion with only the hooves (tips of the digits) on the ground; characteristic of ungulates.

unicuspid   Teeth with a single cusp. Canine teeth are unicuspid, as are premolars in many species.

upper critical temperature   The temperature at which an animal must dissipate heat to maintain a stable internal temperature.

ureters   Ducts within the mammalian kidney that drain the renal pelvis.

urethra   The tube through which urine is expelled from the urinary bladder.

urinary bladder   The organ that stores urine in mammals.

urogenital sinus   A common chamber for the reception of products from the reproductive and urinary systems. In mammals, it is found in monotremes and marsupials.

uropatagium   The membrane between the hind legs of bats that encloses the tail; also called the interfemoral membrane.

uteri   In female mammals a muscular expansion of the reproductive tract in which the embryo and fetus develop; opens externally through the vagina (sing., *uterus*).

V

vagina   The part of the female reproductive tract that receives the male penis during copulation.

vaginal smear technique   A procedure for monitoring different stages of the estrous cycle by observing changes in the types of cells lining the vaginal canal.

valvular   Describing nostrils or ears that can be closed when an animal is under water.

variable number tandem repeats (VNTRs)   Individual loci where alleles are composed of tandem repeats that vary in terms of the number of core units.

vectors   Any agent or carrier that transmits a disease organism.

velli   Very short, fine hairs sometimes referred to as "down" or "fuzz."

velvet   Haired and highly vascularized skin covering growing antlers.

vestigial   Reduced; remnant; atrophied.

vibrissae   Long, stiff hairs with extensive enervation at the base of the follicle that are found on all mammals except humans.

vicariance zoogeography   A study of the state in which a once widespread species becomes restricted or split into isolated geographic locations through the disappearance of the intervening populations or establishment of geographic barriers.

villi   Fingerlike projections of capillaries from the outermost embryonic membrane that penetrate the endometrium; increases exchange between maternal and fetal systems (same as chorionic villi).

viviparous   Able to give birth to live young. Therian mammals are viviparous; prototherians are oviparous.

volant   Having powered flight.

W

warm-blooded   Characterized by having a constant body temperature, independent of environmental temperature.   Typified by birds and mammals only; endothermy, homeothermy.

wavelength   The distance from one peak to the next in a sound (or light) wave.

white adipose tissue   The major fatty tissue of mammals that functions in body insulation, mechanical support, buoyancy, and as an energy reserve.

wing loading   In bats (and birds), the body mass divided by the total surface area of the wings.

winter lethargy   A period of winter dormancy in which body temperature of the animal decreases only about 5 to 6°C from euthermy, as in black bears (*Ursus americanus*).

wool   Underhair that is long, soft, and usually curly.

X

xanthophylls   Pigments that produce mixtures of red and yellow (pheomelanin).

Z

zalambdodont   Tooth cusps that form a V-shape.

Zeitgeber   "Time giver"; environmental cues that serve to set and adjust biological clocks.

zona pellucida   A noncellular layer surrounding the zygote.

zonary placenta   The type of chorioallantoic placenta in which the villi occupy a girdlelike band about the middle of the chorionic sac.

zoogeography   The study of distributions of animals, including mammals.

zoonoses   Diseases transmitted from vertebrate animals (nonhuman mammals) to people.

zooplankton   Animal material including both plankton and necton fed on by baleen whales.

zygomatic arch   Bony structure that surrounds and protects the eye, and serves as a place of attachment for jaw muscles.

zygote   A diploid cell resulting from the union of the male and female gametes.

REFERENCES

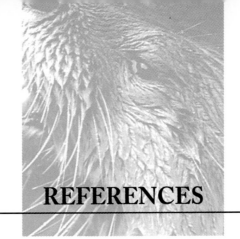

Abraham, G. E., F. S. Manlimos, and R. Gazara. 1977. Radioimmunoassay of steroids. Pp. 591–999 *in* Handbook of radioimmunoassay (G. E. Abraham, ed.). Marcel Dekker, New York.

Abramsky, Z. and C. Sellah. 1982. Competition and the role of habitat selection in *Gerbillus allenbyi* and *Meriones tristrami:* a removal experiment. Ecology. 63:1242–1247.

Acha, P. N. and B. Szyfres. 1989. Zoonoses and communicable diseases common to man and animals, 2d ed. Pan Am. Health Org., Sci. Publ. No. 503. Washington, D.C.

Adams, G. P., P. G. Griffin, and O. J. Ginther. 1989. In situ morphologic dynamics of ovaries, uterus, and cervix in llamas. Biol. Reprod. 41:551–558.

Adams, G. P., C. Plotka, C. Asa, and O. J. Ginther. 1991. Feasibility of characterizing reproductive events in large nondomestic species by transrectal ultrasonic imaging. Zoo Biol. 10:247–259.

Adams, L. and S. Hane. 1972. Adrenal gland size as an index of adrenocortical secretion rate in the California ground squirrel. J. Wild. Dis. 8:19–23.

Adamson, M. L. and J. N. Caira. 1994. Evolutionary factors influencing the nature of parasite specificity. Parasitol. 109:S85–S95.

Adcock, E. W., F. Teasdale, C. S. August, S. Cox, G. Meschia, F. C. Battaglia, and M. A. Naughton. 1973. Human chorionic gonadotropin: its possible role in maternal lymphocyte suppression. Science. 181:845–847.

Albrecht, E. D. and G. J. Pepe. 1990. Placental steroid hormone biosynthesis in primate pregnancy. Endocrin. Rev. 11:124–150.

Alexander, J. and D. G. Russell. 1992. The interaction of *Leishmania* species with macrophages. Adv. Parasitol. 31:175–254.

Alexander, R. D. 1974. The evolution of social behavior. Ann. Rev. Ecol. Syst. 5:325–383.

Alexander, R. M. and H. C. Bennet-Clark. 1977. Storage of elastic strain energy in muscle and other tissues. Nature. 265:114–117.

Allen, E. G. 1938. The habits and life history of the eastern chipmunk (*Tamias striatus lysteri*). Bull. N. Y. State Mus. 314:1–122.

Allsopp, H. 1960. The manatee: ecology and use for weed control. Nature. 188:762.

Altenbach, J. S. 1979. Locomotor morphology of the vampire bat, *Desmodus rotundus.* Spec. Pub., Am. Soc. Mammal. 6:1–137.

Altmann, J. 1974. Observational study of behavior: sampling methods. Behaviour 48:227–265.

Altmann, J., G. Hausfater, and S. A. Altmann. 1988. Determinants of reproductive success in savannah baboons, *Papio cynocephalus.* Pp. 403–418 *in* Reproductive success (T. H. Clutton-Brock, ed.). Univ. of Chicago Press, Chicago.

Altringham, J. D. 1996. Bats: biology and behaviour. Oxford Univ. Press, Oxford, England.

Altschuler, E. M., R. B. Nagle, E. J. Braun, S. L. Lindstedt, and P. H. Krutzsch. 1979. Morphological study of the desert heteromyid kidney with emphasis on the genus *Perognathus.* Anat. Record, 194:461–468.

Anderson, L. T. 1973. An analysis of habitat preference in mice as a function of prior experience. Behaviour. 47:302–339.

Anderson, P. K., G. E. Heinsohn, P. H. Whitney, and J. P. Huang. 1977. *Mus musculus* and *Peromyscus maniculatus:* homing ability in relation to habitat utilization. Can. J. Zool. 55:169–182.

Anderson, R. M. 1965. Methods of collecting and preserving vertebrate animals, 4th ed., rev. Bull. Nat. Mus. Canada 69:1–199.

Andersson, M. 1994. Sexual selection. Princeton Univ. Press, Princeton, NJ.

Andrews, R. V. and R. W. Belknap. 1986. Bioenergetic benefits of huddling by deer mice (*Peromyscus maniculatus*). Comp. Biochem. Physiol. 85A:775–778.

Angerbjörn, A. and J. E. C. Flux. 1995. Lepus timidus. Mamm. Species 495:1–11.

Aplin, K. P. and M. Archer. 1987. Recent advances in marsupial systematics with a new syncretic classification. Pp. xv–lxxii *in*

Possums and opossums: studies in evolution (M. Archer, ed.). Royal Zoological Society, New South Wales, Sydney.

Appolonio, M., M. Festa-Bianchet, F. Mari, S. Mattioli, and B. Sarno. 1992. To lek or not to lek: mating strategies of male fallow deer. Behav. Ecol. 3:25–31.

Aranoff, S. 1993. Geographic information systems: a management perspective. WDL Publications, Ottawa, Ontario.

Arcese, P. 1994. Harem size and horn symmetry in oribi. Anim. Behav. 48:1485–1488.

Archer, J. 1988. The behavioural biology of aggression. Cambridge Univ. Press, New York.

Archer, M. 1978. The status of Australian dasyurids, thylacinids and myrmecobiids. Pp. 29–43 *in* The status of endangered Australian wildlife (M. J. Tyler, ed.). Royal Zoological Society, South Australia.

Archer, M. and A. A. Bartholomai. 1978. Tertiary mammals of Australia: a syntopic review. Alcheringa. 2:1–19.

Archer, M., T. F. Flannery, A. Ritchie, and R. E. Molnar. 1985. First Mesozoic mammal from Australia-an early Cretaceous monotreme. Nature. 318:363–366.

Archer, M. and J. A. W. Kirsch. 1977. The case for the Thylacomyidae and Myrmecobiidae, Gill 1872, or why are marsupial families so extended? Proceedings of the Linnean Society of New South Wales, 102:18–25.

Archer, M., P. Murray, S. J. Hand, and H. Godthelp. 1992. Reconsideration of monotreme relationships based on the skull and dentition of the Miocene *Obdurodon dicksoni* (Ornithorhynchidae) from Riversleigh, Queensland, Australia. Pp. 75–94 *in* Mammalian phylogeny (F. Szalay, M. Novacek, and M. McKenna, eds.). Springer-Verlag, New York.

Archer, M., M. D. Plane, and N. S. Pledge. 1978. Additional evidence for interpreting the Miocene *Obdurodon insignis,* Woodburne and Tedford 1975, to be a fossil platypus (Ornithorhynchidae: Monotremata) and a reconsideration of the status of

511

Ornithorhynchus agilis De Vis, 1895. Aust. Zool. 20:9–27.

Arletazz, R. 1996. Feeding behaviour and foraging strategy of free-living mouse-eared bats, *Myotis myotis* and *Myotis blythii.* Anim. Behav. 51:1–11.

Arlton, A. V. 1936. An ecological study of the mole. J. Mammal. 17:349–371.

Armitage, K. B., J. C. Melcher, and J. M. Ward, Jr. 1990. Oxygen consumption and body temperature in yellow-bellied marmot populations from montane-mesic and lowland-xeric environments. J. Comp. Physiol, B. Biochem. Syst. Environ. Physiol. 160:491–502.

Armstrong, R. B. 1981. Recruitment of muscles and fibers within muscles in running animals. Symp. Zool. Soc. London. 48:289–304.

Arnett, E. B., L. R. Irby, and J. G. Cook. 1993. Sex- and age-specific lungworm infection in Rocky Mountain bighorn sheep during winter. J. Wildl. Dis. 29:90–93.

Arnold, W. 1988. Social thermoregulation during hibernation. J. Comp. Physiol. B. Biochem. Syst. Environ. Physiol. 158:151–156.

Arroyo-Cabrales, J., R. R. Hollander, and J. Knox Jones, Jr. 1987. Choeronycteris mexicana. Mamm. Species. 291:1–5.

Ashford, R. W., P. Desjeux, and P. deRaadt. 1992. Estimation of population at risk of infection and number of cases of leishmaniasis. Parasitol. Today. 8:104–105.

Ashton, D. G. 1978. Marking zoo animals for identification, Pp. 24–34 *in* Animal marking, (B. Stonehouse, ed.). University Park Press, Baltimore.

Au, W. W. L. 1993. The sonar of dolphins. Springer-Verlag, New York.

Audubon, J. J. and J. Bachman. 1846–1854. The viviparous quadrupeds of North America, 3 vols. V. G. Audubon, New York.

Augee, M. L. 1978. Monotremes and the evolution of homeothermy. Aust. Zool. 20:111–119.

August, P. V., S. G. Ayvazian, and J. G. T. Anderson. 1989. Magnetic orientation in a small mammal, *Peromyscus leucopus.* J. Mammal. 70:1–9.

Austad, S. N. and K. E. Fischer. 1991. Mammalian aging, metabolism, and ecology: evidence from the bats and marsupials. J. Gerontol. 46:B47–53.

Austad, S. N. and M. E. Sunquist. 1987. Sex ratio manipulation in the common opossum. Nature. 324:58–60.

Austin, C. R. and R. V. Short (eds.). 1972a. Reproduction in mammals, book 1. Germ cells and fertilization. Cambridge Univ. Press, Cambridge, England.

Austin, C. R. and R. V. Short (eds.). 1972b. Reproduction in mammals, book 2. Embryonic and fetal development. Cambridge Univ. Press, Cambridge, England.

Austin, C. R. and R. V. Short (eds.). 1972c. Reproduction in mammals, book 4 , Reproductive patterns. Cambridge Univ. Press, Cambridge, England.

Austin, C. R. and R. V. Short 1984. Hormonal control of reproduction, 2d ed. Cambridge Univ. Press, New York.

Avise, J. C. 1994. Molecular markers, natural history and evolution. Chapman & Hall, New York.

Avise, J. C. and J. L. Hamrick (eds). 1996. Conservation genetics: case histories from nature. Chapman and Hall, New York.

Axelrod, R. and W. D. Hamilton. 1981. The evolution of cooperation. Science. 211:1390–1396.

Ayala, F. J. 1986. On the virtues and pitfalls of the molecular evolutionary clock. J. Heredity. 77:226–235.

Baar, S. L. and E. D. Fleharty. 1976. A model of the daily energy budget and energy flow through a population of the white-footed mouse. Acta Theriol. 21:179–193.

Bachman, G. C. 1994. Food restriction effects on the body composition of free-living ground squirrels, *Spermophilus beldingi.* Physiol. Zool. 67:756–770.

Bailey, V. 1924. Breeding, feeding, and other life habits of meadow mice (*Microtus*). J. Agri. Res. 27:523–535.

Bailey, W. J., J. L. Slightom, and M. Goodman. 1992. Rejection of the "flying primate" hypothesis by phylogenetic evidence from the ε-globin gene. Science. 256:86–89.

Baird, S. F. 1859. General report on North American mammals. J. B. Lippincott, Philadelphia.

Baker, A. N. 1985. Pygmy right whale, *Caperea marginata* (Gray, 1846). Pp. 345–354 *in* Handbook of marine mammals vol. 3: the sirenians and baleen whales (S. H. Ridgway and R. Harrison, eds.). Academic Press, New York.

Baker, C. S., S. R. Palumbi, R. H. Lambertsen, M. T. Weinrich, J. Calambokidis, and S. J. O'Brien. 1990. Influence of seasonal migration on geographic distribution of mitochondrial DNA haplotypes in humpback whale. Nature. 34:238–240.

Baker, M. A. 1979. A brain-cooling system in mammals. Sci. Am. 240:114–123.

Baker, R. J., C. S. Hood, and R. L. Honeycutt. 1989. Phylogenetic relationships and classification of the higher categories of the New World bat family Phyllostomidae. Syst. Zool. 38:228–238.

Baker, R. J., M. J. Novacek, and N. B. Simmons. 1991. On the monophyly of bats. Syst. Zool. 40:216–231.

Baker, R. R. 1987. Human navigation and magnetoreception: the Manchester experiments do replicate. Anim. Behav. 35:691–704.

Baker, R. R. and M. A. Bellis. 1988. 'Kamikaze' sperm in mammals? Anim. Behav. 36:936–938.

Bakker, R. T. 1971. Dinosaur physiology and the origin of mammals. Evolution. 25:636–658.

Balasingh, J., J. Koilraj, and T. H. Kunz. 1995. Tent construction by the short-nosed fruit bat, *Cynopterus sphinx* (Chiroptera: Pteropodidae) in southern India. Ethology. 100:210–229.

Balcomb, K. C., III, and M. A. Bigg. 1986. Population biology of three resident killer whale pods in Puget Sound and off southern Vancouver Island. Zoo Biol. Monogr. 1:85–95.

Baldi, R., C. Campganga, S. Pedraza, and B. J. LeBouef. 1996. Social effects of space availability on the breeding behaviour of elephant seals in Patagonia. Anim. Behav. 51:717–724.

Baldwin, B. H., B. C. Tennant, T. J. Reimers, R. G. Gowan, and P. W. Concannon. 1985. Circannual changes in serum testosterone concentrations of adult and yearling woodchucks (*Marmota monax*). Biol. Reprod. 32:804–812.

Baldwin, J. A. 1975. Notes and speculations on the domestication of the cat in Egypt. Anthropos. 70:428–448.

Balmford, A. P. 1991. Mate choice on leks. Trends Ecol. Evol. 6:87–92.

Barash, D. P. 1974. The evolution of marmot societies: a general theory. Science. 185:415–420.

Barbour, A. G. and D. Fish. 1993. The biological and social phenomenon of Lyme disease. Science. 260:1610–1616.

Barbour, R. W. and W. H. Davis. 1969. Bats of America. Univ. of Kentucky Press, Lexington.

Barclay, R. M. R. and R. M. Brigham. 1994. Constraints on optimal foraging: a field test of prey discrimination by echolocating insectivorous bats. Anim. Behav. 48:1013–1021.

Barkley, M. S. and B. D. Goldman. 1977. The effects of castration and silastic implants of testosterone on intermale aggression in the mouse. Horm. Behav. 9:32–48.

Barnes, B. M. 1989. Freeze avoidance in a mammal: body temperatures below 0°C in an arctic hibernator. Science. 244:1593–1595.

Barnosky, A. D. 1985. Taphonomy and herd structure of the extinct Irish elk, *Megaloceros giganteus.* Science 228:340–344.

Barry, R. E. Jr. and E. N. Francq. 1980. Orientation to landmarks within the preferred habitat by *Peromyscus leucopus.* J. Mammal. 61:292–303.

Bartholomew, G. A. 1982. Body temperature and energy metabolism. Pp. 333–406 *in* Animal physiology: principles and adaptations (M. S. Gordon, G. A. Bartholomew, A. D. Grinnell, C. B. Jørgensen, and F. N. White, eds.), 4th ed. Macmillan, New York.

Bartholomew, G. A., W. R. Dawson, and R. C. Lasiewski. 1970. Thermoregulation and heterothermy in some of the smaller flying foxes (Megachiroptera) of New Guinea. Z. Vergl. Physiol. 70:196–209.

Bateman, A. J. 1948. Intra-sexual selection in *Drosophila.* Heredity. 2:349–368.

Bateman, J. A. 1959. Laboratory studies of the golden mole and mole-rat. Afr. Wildl. 13:65–71.

Batzli, G. O. 1985. Nutrition. Pp. 779–811 *in* Biology of New World *Microtus* (R. H.

Tamarin, ed.). Spec. Pub. Am. Soc. Mammal. No. 8.

Batzli, G. O. 1994. Special feature: mammal-plant interactions. J. Mammal. 75:813–815.

Batzli, G. O. and I. D. Hume. 1994. Foraging and digestion in herbivores. Pp. 313–314 *in* The digestive system in mammals: food, form and function (D. J. Chivers and P. Langer, eds.). Cambridge Univ. Press, Cambridge, England.

Batzli, G. O. and F. A. Pitelka. 1970. Influence of meadow mouse populations on California grassland. Ecology. 51:1027–1039.

Baverstock, P. and B. Green. 1975. Water recycling in lactation. Science. 187:657–658.

Baverstock, P. R., M. Krieg, J. Birrel, and G. M. McKay. 1990. Albumin immunologic relationships of Australian marsupials 2. The Pseudocheiridae. Aust. J. Zool. 38:519–526.

Bazin, R. C. and R. A. MacArthur. 1992. Thermal benefits of huddling in the muskrat (*Ondatra zibethicus*). J. Mammal. 73:559–564.

Beatley, J. C. 1969. Dependence of desert rodents on winter annuals and precipitation. Ecology. 50:721–724.

Bedford, J. M., J. C. Rodger, and W. G. Breed. 1984. Why so many mammalian spermatozoa-a clue from marsupials? Proceedings of the Royal Society of London, Series B, 221:221–233.

Beer, J. R. 1961. Winter home ranges of the red-backed mouse and white-footed mouse. J. Mammal. 42:174–180.

Begon, M., J. L. Harper, and C. R. Townsend. 1996. Ecology: individuals, populations and communities. 3rd ed. Blackwell Science, Cambridge, MA.

Beidleman, R. G., and W. A. Weber. 1958. Analysis of a pika hay pile. J. Mammal. 39:599–600.

Bekoff, M. 1977. Social communication in canids: evidence for the evolution of a stereotyped mammalian display. Science. 197:1097–1099.

Belk, M. C. and M. H. Smith. 1996. Pelage coloration in oldfield mice (*Peromyscus polionotus*): antipredator adaptation? J. Mammal. 77:882–890.

Bell, J. F. and J. R. Reilly. 1981. Tularemia. Pp. 213–231 *in* Infectious diseases of wild mammals, 2d ed. (J. W. Davis, L. H. Karstad, and D. O. Trainer, eds.). Iowa State Univ. Press, Ames.

Bell, R. H. V. 1971. A grazing ecosystem in the Serengeti. Sci. Am. 225:86–93.

Bell, W. J. 1991. Searching behavior. Chapman and Hall, New York.

Belovsky, G. E. 1978. Diet optimization in a generalist herbivore: the moose. Theoret. Pop. Biol. 14:105–134.

Belovsky, G. E. 1984. Herbivorous optimal foraging: a comparative test of three models. Am. Nat. 124:97–115.

Belovsky, G. E. and O. J. Schmitz. 1994. Plant defenses and optimal foraging by mammalian herbivores. J. Mammal. 75:816–832.

Belsky, A. J. 1986. Does herbivory benefit plants? A review of the evidence. Am. Nat. 127:870–892.

Benjaminsen, T. and I. Christensen. 1979. The natural history of the bottlenose whale *Hyperoodon ampullatus*. Pp. 143–164 *in* Behavior of marine animals vol. 3: cetaceans (H. E. Winn and B. L. Olla, eds.). Plenum Press, New York.

Bennett, A. F. and J. A. Ruben. 1979. Endothermy and activity in vertebrates. Science. 206:649–654.

Bennett, D. K. 1980. Stripes do not a zebra make, part I: a cladistic analysis of *Equus*. Syst. Zool. 29:272–287.

Bennett, S., L. J. Alexander, R. H. Crozier, and A. G. Mackinlay. 1988. Are megabats flying primates? Contrary evidence from a mitochondrial DNA sequence. Aust. J. Biol. Sci. 41:327–332.

Benson, S. B. 1933. Concealing coloration among desert rodents of the southwestern United States. Univ. Calif. Publ. Zool. 40:1–70.

Beran, G. W. (editor-in-chief). 1994. CRC handbook series on zoonoses, 2d ed. CRC Press, Boca Raton, FL.

Berger, J. 1986. Wild horses of the Great Basin: social competition and population size. Univ. of Chicago Press, Chicago.

Berger, J. 1990. Persistence of different-sized populations: an empirical assessment of rapid extinctions in bighorn sheep. Conserv. Biol. 4:91–98.

Berger, J. and C. Cunningham. 1994. Phenotypic alterations, evolutionarily significant structures, and rhino conservation. Conserv. Biol. 8:833–840.

Berger, J., C. Cunningham, and A. A. Gawuseb. 1994. The uncertainty of data and dehorning black rhinos. Conserv. Biol. 8:1149–1152.

Berger, P. J., N. C. Negus, and C. N. Rowsemitt. 1987. Effect of 6-methoxybenzoxazolinone on sex ratio and breeding performance in *Microtus montanus*. Biol. Reprod. 36:255–260.

Berndtson, W. E. 1977. Methods for quantifying mammalian spermatogenesis: a review. J. Anim. Sci. 44:818–833.

Bernstein, I. S. and L. G. Sharpe. 1966. Social roles in a rhesus monkey group. Behaviour. 26:91–104.

Berry, R. J. 1970. The natural history of the house mouse. Field Stud. 3:219–262.

Berta, A. 1991. New Enaliarctos (Pinnipedimorpha) from the Oligocene and Miocene of Oregon and the role of "Enaliarctids" in pinniped phylogeny. Smithsonian Contrib. Paleobiol. 69:1–33.

Bertram, B. C. R. 1976. Kin selection in lions and in evolution. Pp. 281–301 *in* Growing points in ethology (P. P. G. Bateson and R. A. Hinde, eds.). Cambridge Univ. Press, New York.

Bertram, B. C. R. and J. M. King. 1976. Lion and leopard immobilization using C1-744. East Afr. Wild. J. 14:237–239.

Bewick, T. 1804. A general history of quadrupeds. G. and R. Waite, New York.

Bhat, H. R. and T. H. Kunz. 1995. Altered flower/fruit clusters of the kitul palm used as roosts by the short-nosed fruit bat, *Cynopterus sphinx* (Chiroptera: Pteropodidae). J. Zool. Lond. 235:597–604.

Bick, Y. A. E. and W. D. Jackson. 1967. DNA content of monotremes. Nature. 215:192–193.

Biewener, A., R. M. Alexander, and N. Heglund. 1981. Elastic energy storage in the hopping of kangaroo rats (*Dipodomys spectabilis*). J. Zool. London. 195:369–383.

Biggers, J. D. 1966. Reproduction in male marsupials. Symp. Zool. Soc. London. 15:251–280.

Birdsall, D. A. and D. Nash. 1973. Occurrence of successful multiple-insemination of females in natural populations of deer mice (*Peromyscus maniculatus*). Evolution. 27:106–110.

Birney, E. C. and J. R. Choate (eds.). 1994. Seventy-five years of mammalogy 1919–1994. Special Pub. No. 11 Am. Soc. Mammal.

Bjornhag, G. 1994. Adaptations in the large intestine allow small animals to eat fibrous foods. Pp. 287–309 *in* The digestive system in mammals: food, form and function (D. J. Chivers and P. Langer, eds.) Cambridge Univ. Press, Cambridge, England.

Black, H. L. 1974. A north temperate bat community: structure and prey populations. J. Mammal. 55:138–157

Blair, W. F. 1951. Population structure, social behavior, and environmental relations in a natural population of the beach mouse (*Peromyscus polionotus leucocephalus*.) Contr. Lab. Vert. Biol. Univ. Mich. 48:1–47.

Bloemendal, H. 1977. The vertebrate eye lens. Science. 197:127–138.

Bodmer, R. E. and G. B. Rabb. 1992. Okapia johnstoni. Mammal. Species No. 422:1–8.

Bogitsh, B. J. and T. C. Cheng. 1990. Human parasitology. Saunders College Publ., Chicago.

Bökönyi, S. 1984. Horse. Pp. 162–173 *in* Evolution of domesticated animals (I. L. Mason, ed.). Longman, New York.

Bökönyi, S. 1989. Definitions of animal domestication. Pp. 22–27 *in* The walking larder: patterns of domestication, pastoralism, and predation (J. Clutton-Brock, ed.). Unwin Hyman, London.

Bolkovic, M. L., S. M. Caziani, and J. J. Protomastro. 1995. Food habits of the three-banded armadillo (Xenarthra: Dasypodidae) in the dry Chaco, Argentina. J. Mammal. 76:1199–1204.

Bonaparte, J. F. 1990. New late Cretaceous mammals from the Los Alamitos Formation, northern Patagonia. Natl. Geogr. Res. 6:63–93.

Bonner, N. 1989. Whales of the world. Facts on File Publ., New York.

Bonner, T. I., R. Heinemann, and G. J. Todaro. 1980. Evolution of DNA sequence has been retarded in Malagasy lemurs. Nature. 286:420–423.

Bonner, W. N. 1990. The natural history of seals. Facts on File Publ., New York.

Bookhout, T. A. (ed.). 1996. Research and management techniques for wildlife and habitats, 5th ed., rev. Wildlife Society, Bethesda, MD.

Boonstra, R., C. J. Krebs, S. Boutin, and J. M. Eadie. 1994. Finding mammals using far infra-red thermal imaging. J. Mammal. 75:1063–1068.

Bourliére, F. 1970. The natural history of mammals. Alfred A. Knopf, New York.

Boutin, S. 1990. Food supplementation with terrestrial vertebrates: patterns, problems, and the future. Can. J. Zool. 68:203–220.

Bovet, J. 1992. Mammals. Pp. 321–361 *in* Animal homing (F. Papi, ed.). Chapman and Hall, New York.

Bowen, W. D., O. T. Oftedal, and D. J. Boness. 1985. Birth to weaning in 4 days: remarkable growth in the hooded seal, *Cystophora cristata*. Can. J. Zool. 63:2841–2846.

Bowles, M. L. and C. J. Whelan. 1995. Restoration of endangered species. Cambridge Univ. Press, New York.

Bown, T. M. and M. J. Kraus. 1979. Origin of the tribosphenic molar and metatherian and eutherian dental formulae. Pp. 172–181 *in* Mesozoic mammals: the first two-thirds of mammalian history (J. A. Lillegraven, Z. Kielan-Jaworowska, and W.A. Clemens, eds.). Univ. of California Press, Berkeley.

Bowyer, R. T., J. W. Testa, and J. B. Faro. 1995. Habitat selection and home ranges of river otters in a marine environment: effects of the Exxon Valdez oil spill. J. Mammal. 76:1–11.

Bozinovic, F. and J. F. Merritt. 1992. Summer and winter thermal conductance of *Blarina brevicauda* (Mammalia: Insectivora: Soricidae) inhabiting the Appalachian Mountains. Ann. Carn. Mus. 61:33–37.

Bradbury, J. W. 1977. Lek mating behavior in the hammer-headed bat. Zeit. Tierpsychol. 45:225–255.

Bradley, R. and R. C. Lowson. 1992. Bovine spongiform encephalopathy: the history, scientific, political and social issues. Pp. 285–299 *in* Prion diseases of humans and animals (S. Prusiner, J. Collinge, J. Powell, and B. Anderton, eds.). Ellis Horwood, New York.

Bradshaw, G. V. R. 1961. Le cycle des reproduction des *Macrotus californicus* (Chiroptera, Phyllostomatidae). Mammalia. 25:117–119.

Bradshaw, G. V. R. 1962. Reproductive cycle of the California leaf-nosed bat, *Macrotus californicus*. Science. 136:645–646.

Bradshaw, R. H., R. F. Parrott, J. A. Goode, D. M. Lloyd, R. G. Rodway, and D. M. Broom. 1996. Behavioral and hormonal responses of pigs during transport-effect of mixing and duration of journey. Anim. Sci. 62:547–554.

Brain, P. F. and A. E. Poole. 1976. The role of endocrines in isolation-induced intermale fighting in albino laboratory mice: II. Sex steroid influences in aggressive mice. Aggr. Behav. 2:55–76.

Brandon, R. N. and R. M. Burian, ed. 1984. Genes, organisms, populations. MIT Press, Cambridge, MA.

Brass, D. A. 1994. Rabies in bats: natural history and public health implications. Livia Press, Ridgefield, CT.

Braun Breton, C. and L. H. Pereira da Silva. 1993. Malaria proteases and red blood cell invasion. Parasitol. Today. 9:92–96.

Breeden, S. and K. Breeden. 1967. Animals in eastern Australia. Australasian Publ., Sydney.

Brigham, R. M. and P. Trayhurn. 1994. Brown fat in birds? A test for the mammalian BAT-specific mitochondrial uncoupling protein in common poorwills. Condor. 96:208–211.

Britten, R. J. 1986. Rates of DNA sequence evolution differ between taxonomic groups. Science 231:1393–1398.

Britten, R. J., D. E. Graham, and B. R. Neufield. 1974. Analysis of repeating DNA sequences and a speculation of the origins of evolutionary novelty. Methods Enzymol. 29:363–418.

Bronner, G. N. 1995. Cytogenetic properties of nine species of golden moles (Insectivora: Chrysochloridae). J. Mammal. 76:957–971.

Bronson, F. H. 1971. Rodent pheromones. Biol. Reprod. 4:344–357.

Bronson, F. H. 1979. The reproductive ecology of the house mouse. Q. Rev. Biol. 54:265–299.

Bronson, F. H. and B. E. Eleftheriou. 1964. Chronic physiological effects of fighting in mice. Gen. Comp. Endocrinol. 4:9–14.

Brooke, A. P. 1994. Diet of the fishing bat, *Noctilio leporinus* (Chiroptera: Noctilionidae). J. Mammal. 75:212–218.

Brooks, D. R. and D. A. McLennan. 1991. Phylogeny, ecology, and behavior. Univ. of Chicago Press, Chicago.

Brower, J. E. and T. J. Cade. 1966. Ecology and physiology of *Napaeozapus insignis* (Miller) and other woodland mice. Ecology. 47:46–63.

Brown, C. H. and P. M. Waser. 1984. Hearing and communication in blue monkeys (*Cercopithecus mitis*). Anim. Behav. 32:66–75.

Brown, C. L., C. E. Rupprecht, and W. M. Tzilkowski. 1990. Adult raccoon survival in an enzootic rabies area of Pennsylvania. J. Wildl. Dis. 26:346–350.

Brown, G. 1993. The great bear almanac. Lyons and Burford, New York.

Brown, J. H. 1971. Mammals on mountaintops: nonequilibrium insular biogeography. Am. Nat. 105:467–478.

Brown, J. H. 1975. Geographical ecology of desert rodents. Pp. 315–341 *in* Ecology and evolution of communities (M. L. Cody and J. M. Diamond, eds.). Belknap Press of Harvard Univ. Press, Cambridge, MA.

Brown, J. H. 1995. Macroecology. Univ. of Chicago Press, Chicago.

Brown, J. H. and G. A. Bartholomew. 1969. Peroidicity and energetics of torpor in the kangaroo mouse, *Microdipodops pallidus*. Ecology. 50:705–709.

Brown, J. H. and D. W. Davidson. 1977. Competition between seed-eating rodents and ants in desert ecosystems. Science. 196:880–882.

Brown, J. H., D. W. Davidson, J. C. Munger, and R. S. Inouye. 1986. Experimental community ecology: the desert granivore system. Pp. 41–61 *in* Community ecology (J. Diamond and T. J. Case, eds.). Harper and Row, New York.

Brown, J. H. and A. C. Gibson. 1983. Biogeography. C. V. Mosby, St. Louis.

Brown, J. H. and E. J. Heske. 1990. Control of a desert-grassland transition by a keystone rodent guild. Science. 250:1705–1708.

Brown, J. H. and R. C. Lasiewski. 1972. Metabolism of weasels: the cost of being long and thin. Ecology. 53:939–943.

Brown, J. H. and M. V. Lomolino. 1998. Biogeography, 2d ed. Sinauer Assoc., Sunderland, MA.

Brown, J. H. and B. A. Maurer. 1989. Macroecology: the division of food and space among species on continents. Science. 243:1145–1150.

Brown, J. H., O. J. Reichman, and D. W. Davidson. 1979. Granivory in desert ecosystems. Ann. Rev. Ecol. Syst. 10:201–227.

Brown, J. H. and D. E. Wilson. 1994. Natural history and evolutionary ecology. Pp. 377–397 *in* Seventy-five years of mammalogy 1919–1994 (E. C. Birney and J. R. Choate, eds.). Special Pub. No. 11 American Society of Mammalogists.

Brown, J. L. 1964. The evolution of diversity in avian territorial systems. Wilson Bull. 76:160–169.

Brown, J. L. and A. Eklund. 1994. Kin recognition and the major histocompatibility complex: an integrative review. Am. Nat. 143:435–461.

Brown, L., R. W. Shumaker, and J. F. Downhower. 1995. Do primates experience sperm competition? Am. Nat. 146:302–306.

Brown, P. T., T. W. Brown, and A. D. Grinnell. 1983. Echolocation, development, and vocal communication in the lesser bulldog bat, *Noctilio albiventris*. Behav. Ecol. Sociobiol. 13:287–298.

Brown, R. D. (ed.). 1983. Antler development in Cervidae. Caesar Kleberg Wildlife Research Institute, Kingsville, TX.

Brown, R. E. and D. W. Macdonald. 1985. Social odours in mammals, 2 vols. Clarendon Press, Oxford, England.

Brown, S. G. 1978. Whale marking techniques. Pp. 71–80 *in* Animal marking (B. Stonehouse, ed.). University Park Press, Baltimore.

Bruce, H. M. 1966. Smell as an exteroceptive factor. J. Anim. Sci. Suppl. 25:83–89.

Bruns Stockrahm, D.M., B. J. Dickerson, S. L. Adolf, and R. W. Seabloom. 1996. Aging black-tailed prairie dogs by weight of eye lenses. J. Mammal. 77:874–881.

Bruseo, J. A. and R. E. Barry, Jr. 1995. Temporal activity of syntopic *Peromyscus* in the central Appalachians. J. Mammal. 76:78–82.

Bryant, P. J. 1995. Dating remains of gray whales from the eastern North Atlantic. J. Mammal. 76:857–861.

Buchler, E. R. and S. B. Childs. 1982. Use of post-sunset glow as an orientation cue by the big brown bat (*Eptesicus fuscus*). J. Mammal. 63:243–247.

Budiansky, S. 1992. The covenant of the wild: why animals choose domestication. W. Morrow, New York.

Buechner, H. K. and H. D. Roth. 1974. The lek system in Uganda kob. Am. Zool. 14:145–162.

Buffon, G. L. L. 1858. Buffon's natural history of man, the globe, and of quadrupeds. Hurst, New York.

Bulliet, R. W. 1975. The camel and the wheel. Harvard Univ. Press, Cambridge, MA.

Burda, H., S. Marhold, T. Westenberger, R. Wiltschko, and W. Wiltschko. 1990. Magnetic compass orientation in the subterranean rodent *Cryptomys hottentotus* (Bathyergidae). Experientia. 46:528–530.

Burgdorfer, W. and T. G. Schwan. 1991. Borrelia. Pp. 560–566 in Manual of clinical microbiology, 5th ed. (A. Balows, W. J. Hauslei, K. L. Herrmann, H. D. Isenberg, and H. J. Shadomy, eds.). American Society of Microbiology, Washington, D.C.

Burt, W. H. 1960. Bacula of North American mammals. Misc. Pub. Mus. Zool., Univ. of Mich. 113:1–75.

Burt, W. H. and R. P. Grossenheider. 1980. A field guide to mammals of North America north of Mexico, 3d ed. Macmillan, Boston.

Burton, J. A. and B. Pearson. 1987. Collins guide to the rare mammals of the world. Collins, London.

Burton, R. 1980. The life and death of whales. Andre Deutsch, Worcester, England.

Bush, M. 1996. Methods of capture, handling, and anesthesia. Pp. 25–40 in Wild mammals in captivity. (D. G. Kleiman, M. E. Allen, K. V. Thompson, and S. Lumpkin, eds.). Univ. of Chicago Press, Chicago.

Butcher, E. O. 1951. Development of the piliary system and the replacement of hair in mammals. An. N. Y. Acad. Sci. 53:508–516.

Butler, P. M. 1972. The problem of insectivore classification. Pp. 253–265 in Studies in vertebrate evolution (K. A. Joysey and T. S. Kemp, eds.). Winchester Press, New York.

Butler, P. M. 1978. Insectivora and Chiroptera. Pp. 56–68 in Evolution of African mammals (V. J. Maglio and H. B. S. Cooke, eds.). Harvard Univ. Press, Cambridge, MA.

Butler, P. M. 1992. Tribosphenic molars in the Cretaceous. Pp. 125–138 in Structure, function and evolution of teeth (P. Smith and E. Tchernov, eds.). Freund Publishing House, London.

Cabanac, M. 1986. Keeping a cool head. New Physiol. Sci. 1:41–44.

Cade, T. J. 1983. Hybridization and gene exchange among birds in relation to conservation. Pp. 288–309 in Genetics and conservation, (C. M. Schoenwald-Cox, S. M. Chambers, B. MacBryde, and L. Thomas, eds.). Benjamin Cummings, Menlo Park, CA.

Cain, S. A. 1944. Foundations of plant geography. Harper, New York.

Calder, W. A. 1984. Size, function, and life history. Harvard Univ. Press, Cambridge.

Caldwell, D. K. and M. C. Caldwell. 1989. Pygmy sperm whale *Kogia breviceps* (de Blainville, 1838): Dwarf sperm whale *Kogia simus* Owen, 1866. Pp. 235–260 in Handbook of marine mammals, vol. 4: river dolphins and the larger toothed whales. (S. H. Ridgway and R. Harrison, eds.). Academic Press, New York.

Camhi, J. M. 1984. Neuroethology. Sinauer Assoc., Sunderland, MA.

Campbell, C. B. G. 1974. On the phyletic relationships of the tree shrews. Mammal Rev. 4:125–143.

Campbell, W. C. 1988. Trichinosis revisited-another look at modes of transmission. Parasitol. Today. 4:83–86.

Carl, E. A. 1971. Population control in arctic ground squirrels. Ecology. 52:395–413.

Carleton, M. D. 1973. A survey of gross stomach morphology in New World Cricetinae (Rodentia, Muroidea), with comments on functional interpretations. Misc. Publ. Mus. Zool., Univ. of Michigan. 146:1–43.

Carleton, M. D. 1984. Introduction to rodents. Pp. 255–265 in Orders and families of recent mammals of the world (S. Anderson and J. K. Jones, Jr., eds.). John Wiley and Sons, New York.

Carleton, M. D. 1985. Macroanatomy. Pp. 116–175 in Biology of New World *Microtus* (R. H. Tamarin, ed.). Spec. Publ., American Society of Mammalogists. No. 8.

Carleton, M. D. and G. G. Musser. 1984. Muroid rodents. Pp. 289–446 in Orders and families of recent mammals of the world (S. Anderson and J. K. Jones, Jr., eds.). John Wiley and Sons, New York.

Caro, T. M. 1994. Cheetahs of the Serengeti plains. Univ. of Chicago Press, Chicago.

Carrick, F. N. and R. L. Hughes. 1978. Reproduction in male monotremes. Aust. Zool. 20:211–231.

Carrington, R. 1959. Elephants: A short account of their natural history, evolution, and influence on mankind. Basic Books, New York.

Carroll, R. L. 1988. Vertebrate paleontology and evolution. W. H. Freeman, New York.

Carson, R. 1962. Silent spring. Fawcett, Boston.

Cartmill, M. 1972. Arboreal adaptations and the origin of the Order Primates. Pp. 97–122 in Biology of the Primates (R. H. Tuttle, ed.). Aldine-Atherton, Chicago.

Cartmill, M. 1985. Climbing. Pp. 73–88 in Functional vertebrate morphology. (M. Hildebrand, D. M. Bramble, K. F. Liem, and D. B. Wake, eds.). Harvard Univ. Press, Cambridge.

Cartwright, T. 1974. The plasminogen activator of vampire bat saliva. Blood. 43:317–326.

Casey, T. M. 1981. Nest insulation: energy savings to brown lemmings using a winter nest. Oecologia. 50:199–204.

Catania, K. C. and J. H. Kaas. 1996. The unusual nose and brain of the star-nosed mole. BioScience. 46:578–586.

Catesby, M. 1748. The natural history of Carolina, Florida and the Bahama Islands, 3 vols. B. White, London.

Caughley, G. and A. Gunn. 1996. Conservation biology in theory and practice. Blackwell Science Publ., Cambridge, MA.

Cawthorn, J. M. 1994. A live-trapping study of two syntopic species of *Sorex*, *S. cinereus* and *S. fumeus*, in southwestern Pennsylvania, Pp. 39–43 in Advances in the biology of shrews (J. F. Merritt, G. L. Kirkland, Jr., and R. K. Rose, eds.). Carnegie Museum of Natural History (Special Pub. No. 18), Pittsburgh.

Chakraborty, R. and K. K. Kidd. 1991. The utility of DNA typing in forensic work. Science. 254:1735–1739.

Chaline, J., P. Mein, and F. Petter. 1977. Les grandes lignes d'une classification évolutive des Muroidea. Mammalia. 41:245–252.

Champoux, M. and S. J. Suomi. 1994. Behavioral and adrenocortical responses of rhesus macaque mothers to infant separation in an unfamiliar environment. Primates. 35:191–202.

Chan, L. K. 1995. Extrinsic lingual musculature of two pangolins (Pholidota: Manidae). J. Mammal. 76:472–480.

Chapman, J. A. and G. A. Feldhamer (eds.). 1982. Wild mammals of North America: biology, management, and economics. Johns Hopkins Univ. Press, Baltimore.

Chappell, M. A. 1978. Behavioral factors in the altitudinal zonation of chipmunks (*Eutamias*). Ecology. 59:565–579.

Chappell, M. A. and G. A. Bartholomew. 1981. Standard operative temperatures and thermal energetics of the antelope ground squirrel *Ammospermophilus leucurus*. Physiol. Zool. 54:81–93.

Charnov, E. and J. Finerty. 1980. Vole population cycles: a case for kin-selection? Oecologia. 45:1–2.

Charnov, E. L. 1976. Optimal foraging: the marginal value theorem. Theoret. Pop. Biol. 9:129–136.

Cheatum, E. L. 1949. The use of corpora lutea for determining ovulation incidence and variations of fertility in white-tailed deer. Cornell Vet. 39:282–291.

Cheatum, E. L. and C. W. Severinghaus. 1950. Variations in fertility of white-tailed deer related to range conditions. Transactions of the North American Wildlife Conference. 15:170–189.

Cheney, D. L. and R. M. Seyfarth. 1983. Nonrandom dispersal in free-ranging vervet monkeys: social and genetic consequences. Am. Nat. 122:392–412.

Chew, R. M. and A. E. Chew. 1970. Energy relationships of the mammals of a desert scrub (*Larrea tridentata*) community. Ecol. Monogr. 40:1–21.

Chew, R. M. and A. E. Dammann. 1961. Evaporative water loss of small vertebrates, as measured with an infrared analyzer. Science. 133:384–385.

Childs, J. E. 1995. Special feature: zoonoses. J. Mammal. 76:663.

Childs, J. E., J. N. Mills, and G. E. Glass. 1995. Rodent-borne hemorrhagic fever viruses: a special risk for mammalogists? J. Mammal. 76:664–680.

Chittenden, H. 1954. The American fur trade, 2 vols. Stanford Univ. Press, Stanford, CA.

Chitty, D. 1960. Population processes in the vole and their relevance to general theory. Can. J. Zool. 38:99–113.

Chitty, D. and M. Shorten. 1946. Techniques for the study of the Norway rat *Rattus norvegicus*. J. Mammal. 27:63–78.

Chopra, S. R. K. and R. N. Vasishat. 1979. Sivalik fossil tree shrew from Haritalyangar, India. Nature. 281:214–215.

Christian, J. J. 1950. The adreno-pituitary system and population cycles in mammals. J. Mammal. 31:247–259.

Christian, J. J. 1963. Endocrine adaptive mechanisms and the physiologic regulation of population growth. Pp. 189–353 *in* Physiological mammalogy, vol. I: mammalian populations (W. V. Mayer and R. C. Van Gelder, eds.). Academic Press, New York.

Christian, J. J. 1970. Social subordination, population density and mammalian evolution. Science. 168:84–90.

Christian, J. J. 1978. Neurobehavioral endocrine regulation of small mammal populations. Pp. 143–158 *in* Populations of small mammals under natural conditions (D. P. Snyder, ed.). Pymatuning Lab Ecology, Univ. of Pittsburgh, Spec. Publ., Ser. 5.

Christian, J. J. and D. E. Davis. 1964. Endocrines, behavior, and population: social and endocrine factors are integrated in the regulation of growth of mammalian populations. Science. 146:1550–1560.

Chuan-Kuen, L., R. W. Wilson, M. R. Dawson, and L. Krishtalka. 1987. The origin of rodents and lagomorphs. Pp. 97–108 *in* Current mammalogy (H. H. Genoways, ed). Plenum Press, New York.

Churchfield, S. 1982. The influence of temperature on the activity and food consumption of the common shrew. Acta Theriol. 27:295–304.

Churchfield, S. 1990. The natural history of shrews. Cornell Univ. Press, Ithaca, New York.

Cifelli, R. L. 1993. Early Cretaceous mammal from North America and the evolution of marsupial dental characters. Proceedings of the National Academy of Science, 90:9413–9416.

Clapham, P. J. and C. A. Mayo. 1987. Reproduction and recruitment of individually identified humpback whales, *Megaptera novaeangliae*, observed in Massachusetts Bay, 1979–1985. Can. J. Zool. 65:2853–2863.

Claridge, A. W. and S. J. Cork. 1994. Nutritional value of hypogeal fungal sporocarps for the long-nosed potoroo (*Potorous tridactylus*), a forest-dwelling mycophagous marsupial. Aust. J. Zool. 42:701–710.

Clark, A. B. 1978. Sex ratio and local resource competition in a prosimian primate. Science. 201:163–165.

Clark, T. W. and M. R. Stromberg. 1987. Mammals of Wyoming. Univ. of Kansas Press, Lawrence.

Clark, W. E .L. 1959. History of the primates, 5th ed. Univ. of Chicago Press, Chicago.

Clarke, M. R. 1978a. Structure and proportions of the spermaceti organ in the sperm whale. J. Mar. Biol. Assoc. U.K. 58:1–17.

Clarke, M. R. 1978b. Bouyancy control as a function of the spermaceti organ in the sperm whale. J. Mar. Biol. Assoc. U.K. 58:27–71.

Clements, M. L. 1992. Rocky Mountain spotted fever. Pp. 1304–1312 *in* Infectious diseases (S. L. Gorbach, J. G. Bartlett, and N. R. Blacklow, eds.). W. B. Saunders, Philadelphia.

Clutton-Brock, J. 1981. Domesticated animals from early times. Univ. of Texas Press, Austin.

Clutton-Brock, J. (ed.). 1989. The walking larder: patterns of domestication, pastoralism, and predation. Unwin Hyman, London.

Clutton-Brock, J. 1992. The process of domestication. Mammal Rev. 22:79–85.

Clutton-Brock, J. and P. Jewell. 1993. Origin and domestication of the dog. Pp. 21–31 *in* Miller's anatomy of the dog, 3d ed. (H. E. Evans, ed.). W. B. Saunders, Philadelphia.

Clutton-Brock, T. H. 1988. Reproductive success. Univ. of Chicago Press, Chicago.

Clutton-Brock, T. H., S. D. Albon, and F. E. Guinness. 1984. Maternal dominance, breeding success and birth sex ratios in red deer. Nature. 308:358–360.

Clutton-Brock, T. H., S. D. Albon, and F. E. Guinness. 1989. Fitness costs of gestation and lactation in wild mammals. Nature. 337:260–262.

Clutton-Brock, T. H., F. E. Guinness, and S. D. Albon. 1982. Red deer: behavior and ecology of two sexes. Univ. of Chicago Press, Chicago.

Clutton-Brock, T. H., M. Major, and F. E. Guinness. 1985. Population regulation in male and female red deer. J. Anim. Ecol. 54:831–846.

Coburn, D. K. and F. Geiser. 1996. Daily torpor and energy saving a subtropical blossom-bat, *Syconycteris australis* (Megachiroptera). Pp. 39–45 *in* Adaptations to the cold: tenth international hibernation symposium (F. Geiser, A. J. Hulbert, and S. C. Nicol, eds.). Univ. of New England Press, Armidale, NSW, Australia.

Cockburn, A., M. P. Scott, and D. J. Scotts. 1985. Inbreeding avoidance and male-biased natal dispersal in *Antechinus* spp. (Marsupialia: Dasyuridae). Anim. Behav. 33:908–915.

Cockrum, E. L. 1962. Introduction to mammalogy. Ronald Press, New York.

Cockrum, E. L. 1969. Migration in the guano bat, *Tadarida brasiliensis*. Univ. of Kansas Mus. Nat. Hist. Misc. Publ. No. 51:303–336.

Cohn, J. P. 1990. Elephants: remarkable and endangered. Bioscience. 40:10–14.

Colburn, T., D. Dumanoski, and J. P. Myers. 1996. Our stolen future. Plume Penguin Books, New York.

Cole, A. J. (ed.). 1969. Numerical taxonomy: proceedings of the colloquium in numerical taxonomy held in the University of St. Andrew, September 1968. Academic Press, London.

Coles, R. W. 1969. Thermoregulatory function of the beaver tail. Am. Zool. 9:203a.

Coles, R. W. 1970. Pharyngeal and lingual adaptations in the beaver. J. Mammal. 51:424–425.

Coley, P. D. and J. A. Barone. 1996. Herbivory and plant defenses in tropical forests. Ann. Rev. Ecol. Syst. 27:305–335.

Comer, J. A., W. R. Davidson, A. K. Prestwood, and V. F. Nettles. 1991. An update on the distribution of *Parelaphostrongylus tenuis* in the southeastern United States. J. Wildl. Dis. 27:348–354.

Conner, D. A. 1983. Seasonal changes in activity patterns and the adaptive value of haying in pikas (*Ochotona princeps*). Can. J. Zool. 61:411–416.

Connor, R. C., R. A. Smolker, and A. F. Richards. 1992. Dolphin alliances and coalitions. Pp. 415–443 *in* Coalitions and alliances in humans and other animals (A. H. Harcourt and F. B. M. de Waal, eds.). Oxford Univ. Press, New York.

Constantine, D. G. 1967. Bat rabies in the southwestern United States, Public Health Rep. 82:867–888.

Constantine, D. G. 1970. Bats in relation to the health, welfare, and economy of man. Pp. 319–449 *in* Biology of bats, vol. II (W. A. Wimsatt, ed.). Academic Press, New York.

Contreras, L. C., J. C. Torres-Mura, A. E. Sportorno, and L. I. Walker. 1994. Chromosomes of *Octomys mimax* and *Octodontomys gliroides* and relationships of octodontid rodents. J. Mammal. 75:768–774.

Cook, J. A. and T. L. Yates. 1994. Systematic relationships of the Bolivian tuco-tucos, genus *Ctenomys* (Rodentia: Ctenomyidae). J. Mammal. 75:583–599.

Cooke, J. A. and C. R. Terman. 1977. Influence of displacement distance and vision on homing behavior of the white-footed mouse (*Peromyscus leucopus noveboracensis*). J. Mammal. 58:58–66.

Coombs, M. C. 1983. Large mammalian clawed herbivores: A comparative study. Transactions of the American Philosophical Society. 73:1–96.

Cooper, S. M. 1991. Optimal hunting group size: the need for lions to defend their kills against loss to spotted hyaenas. Afr. J. Ecol. 29:130–136.

Coppola, D. M. and J. G. Vandenbergh. 1987. Induction of a puberty-regulating chemosignal in wild mouse populations. J. Mammal. 68:86–91.

Corbet, G. B. 1978. The mammals of the Palaearctic Region: a taxonomic review. British Museum of Natural History, London.

Corbet, G. B. 1988. The mammals of the Palearctic region: a taxonomic review. Cornell Univ. Press, Ithaca, NY.

Corbet, G. B. and J. E. Hill. 1991. A world list of mammalian species, 3d ed. British Museum of Natural History Publ., London.

Corbett, L. K. 1995. The dingo in Australia and Asia. New South Wales Univ. Press, Sydney.

Cork, S. J. and G. J. Kenagy. 1989a. Rates of gut passage and retention of hypogeous fungal spores in two forest-dwelling rodents. J. Mammal. 70:512–519.

Cork, S. J. and G. J. Kenagy. 1989b Nutritional value of hypogeous fungus for a forest-dwelling ground squirrel. Ecology. 70:577–586.

Cossins, A. R. and K. Bowler. 1987. Temperature biology of animals. Chapman and Hall, New York.

Cott, H. B. 1966. Adaptive coloration in animals. Methuen, London.

Coues, E. 1877. Monographs of North American Rodentia. No. 1-Muridae. Smithsonian Institution, Washington, D. C.

Couffer, J. 1992. Bat bomb: World War II's other secret weapon. Univ. of Texas Press, Austin.

Cowles, C. J., R. L. Kirkpatrick, and J. O. Newell. 1977. Ovarian follicular changes in gray squirrels as affected by season, age, and reproductive state. J. Mammal. 58:67–73.

Cox, C. R. and B. J. Le Boeuf. 1977. Female incitation of male competition: a mechanism in sexual selection. Am. Nat. 111:317–335.

Cox, F. E. G. 1979. Ecological importance of small mammals as reservoirs of disease. Pp. 213–238 *in* Ecology of small mammals (D. M. Stoddart. ed.). Chapman and Hall, London.

Craighead, D. J. and J. J. Craighead. 1987. Tracking caribou using satellite telemetry. Nat. Geog. Res. 3:462–479.

Crandall, L. S. 1964. The management of wild mammals in captivity. Univ. of Chicago Press, Chicago.

Cranford, J. A. 1978. Hibernation in the western jumping mouse (*Zapus princeps*). J. Mammal. 59:496–509.

Creel, S. and N. M. Creel. 1995. Communal hunting and pack size in African wild dogs, *Lycaon pictus*. Anim. Behav. 50:1325–1339.

Creel, S., N. Creel, D. E. Wildt, and S. L. Monfort. 1992. Behavioural and endocrine mechanisms of reproductive succession in Serengeti dwarf mongooses. Anim. Behav. 43:231–246.

Crockett, C. M. 1996. Data collection in the zoo setting, emphasizing behavior, Pp. 545–565 *in* Wild mammals in captivity (D. G. Kleiman, M. E. Allen, K. V. Thompson, and S. Lumpkin, eds.). Univ. of Chicago Press, Chicago.

Crockett, C. M., C. L. Bowers, G. P, Sackett, and D. M. Bowden. 1993. Urinary cortisol responses of longtailed macaques to five cage sizes, tethering, sedation and room change. Am. J. Primatol. 30:55–74

Crompton, A. W. 1972. The evolution of the jaw articulation of cynodonts. Pp. 231–251 *in* Studies in vertebrate evolution (K. A. Joysey and T. S. Kemp, eds.). Winchester Press, New York.

Crompton, A. W. and F. A. Jenkins, Jr. 1979. Origin of mammals. Pp. 59–73 *in* Mesozoic mammals: the first two-thirds of mammalian history (J. A. Lillegraven, Z. Kielan-Jaworowska, and W. A. Clemens, eds.). Univ. of California Press, Berkeley.

Crook, J. H. and S. S. Gartlan. 1966. Evolution of primate societies. Nature. 210:1200–1203.

Csada, R. 1996. Cardioderma cor. Mammal. Species No. 519:1–4.

Cumming, D. H. M., R. F. DuToit, and S. N. Stuart. 1991. African elephants and rhinos: status survey and conservation action plan. IUCN/SSC African Elephant and Rhino Species Group. Gland, Switzerland.

Cummings, J. R. and S. H. Vessey. 1994. Agricultural influences on movement patterns of white-footed mice (*Peromyscus leucopus*). Am. Mid. Nat. 132:209–218.

Cummings, W. C. 1985. Right whales *Eubalaena glacialis* (Müller, 1776) and *Eubalaena australis* (Desmoulins, 1822). Pp. 275–304 *in* Handbook of marine mammals, vol. 3: the sirenians and baleen whales. (S. H. Ridgway and R. Harrison, eds.). Academic Press, New York.

Curtis, H. 1975. Biology, 2d ed. Worth Publ., New York.

Daan, S. and S. Slopsema. 1978. Short-term rhythms in foraging behaviour of the common vole, *Microtus arvalis*. J. Comp. Physiol. A127:215–227.

Dagg, A. I. and J. B. Foster. 1982. The giraffe: biology, behavior, and ecology. R. E. Krieger, Huntington, NY.

Dailey, M. D. 1996. Meyer, Olsen and Schmidt's essentials of parasitology, 6th ed. Wm. C. Brown, Dubuque, IA.

Dalquest, W. W. 1950. The genera of the chiropteran family Natalidae. J. Mammal. 31:436–443.

Danilkin, A. A. 1995. Capreolus pygargus. Mamm. Species. 512:1–7.

Darlington, P. J., Jr. 1957. Zoogeography: the geographical distribution of animals. John Wiley and Sons, New York.

Darwin, C. 1871. The descent of man, and selection in relation to sex. D. Appleton, New York.

Darwin, C. R. 1859. On the origin of species. Dent, London.

Davidson, J. 1938. On the growth of the sheep population in Tasmania. Transactions of the Royal Society of South Australia. 62:342–346.

Davis, D. E. 1991. Seasonal change in irradiance: a *Zeitgeber* for circannual rhythms in ground squirrels. Comp. Biochem. Physiol. 98A:241–243.

Davis, D. E. and E. P. Finnie. 1975. Entrainment of circannual rhythm in weight of woodchucks. J. Mammal. 56:199–203.

Davis, D. E. and F. B. Golley. 1963. Principles in mammalogy. Reinhold Publishing, New York.

Davis, S. J. M. 1987. The archaeology of animals. Yale Univ. Press, New Haven, CT.

Davis, S. J. M. and F. R. Valla. 1978. Evidence for domestication of the dog 12,000 years ago in the Natufian of Israel. Nature. 276:608–610.

Davis, W. B. 1973. Geographic variation in the fishing bat, *Noctilio leporinus*. J. Mammal. 54:862–874.

Davis, W. B. 1976. Geographic variation in the lesser noctilio, *Noctilio albiventris* (Chiroptera). J. Mammal. 57:687–707.

Davis, W. H. and H. B. Hitchcock. 1965. Biology and migration of the bat, *Myotis lucifugus*, in New England. J. Mammal. 46:296–313.

Dawkins, R. and T. R. Carlisle. 1976. Parental investment, mate desertion and a fallacy. Nature. 262:131–133.

Dawkins, R. and J. R. Krebs. 1978. Animal signals: information or manipulation? Pp. 282–309 *in* Behavioural ecology: an evolutionary approach (J. R. Krebs and N. B. Davies, eds.). Sinauer Assoc., Sunderland, MA.

Dawson, M. R. and L. Krishtalka. 1984. Fossil history of the families of recent mammals. Pp. 11–57 *in* Orders and families of recent mammals of the world (S. Anderson and J. K. Knox, Jr., eds.). John Wiley & Sons, New York.

Dawson, T. J. 1983. Monotremes and marsupials: the other mammals. Edward Arnold, London.

Dawson, T. J. 1995. Kangaroos: biology of the largest marsupials. Comstock Publ. Assoc., Ithaca, NY.

Dayan, T., D. Simberloff, E. Tchernov, and Y. Yom-Tov. 1989. Inter- and intraspecific character displacement in mustelids. Ecology. 70:1526–1539.

Dayan, T., D. Simberloff, E. Tchernov, and Y. Yom-Tov. 1990. Feline canines: community-wide character displacement among the small cats of Israel. Am. Nat. 136:39–60.

Dearing, M. D. 1997. The function of hay piles of pikas (*Ochotona princeps*). J. Mammal. 78:1156–1163.

DeArmond, S. J. and S. B. Prusiner. 1995. Etiology and pathogenesis of prion diseases. Am. J. Pathol. 146:785–811.

DeBlase, A. F. and R. E. Martin. 1981. A manual of mammalogy with keys to the families of the world, 2d ed. William C. Brown, Dubuque, IA.

Deevey, E. S., Jr. 1947. Life tables for natural populations of animals. Q. Rev. Biol. 22:283–314.

DeGiusti, D. L. 1971. Acanthocephala. Pp. 140–157 *in* Parasitic diseases of wild mammals (J. W. Davis and R. C. Anderson, eds.). Iowa State Univ. Press, Ames.

Dehnel, A. 1949. Studies on the genus *Sorex*. Ann. Univ. M. Curie-Sklodowkka, Sect. C4:17–102.

de Jong, W. W. and Goodman, M. 1988. Anthropoid affinities of *Tarsius* supported by a-crystallin sequences. J. Hum. Evol. 17:575–582.

DelGiudice, G. D., U. S. Seal, and L. D. Mech. 1987. Effects of feeding and fasting on wolf blood and urine characteristics. J. Wild Mgmt. 51:1–10.

DeLong, R. L., G. L. Kooyman, W. G. Gilmartin, and T. R. Loughlin. 1984. Hawaiian monk seal diving behavior. Acta Zool. Fenn. 172:129–131.

DeLong, R. L. and B. S. Stewart. 1991. Diving patterns of northern elephant seal bulls. Mar. Mammal Sci. 7:369–384.

DeLong, R. L., B. S. Stewart, and R. D. Hill. 1992. Documenting migrations of northern elephant seals using daylength. Mar. Mammal Sci. 8:155–159.

Delpietro, H. A., N. Marchevsky, and E. Simonetti. 1993. Relative population densities and predation of the common vampire bat (*Desmodus rotundus*) in natural and cattle-raising areas in north-east Argentina. Prev. Vet. Med. 14:13–20

Demastes, J. W. and M. S. Hafner. 1993. Cospeciation of pocket gophers (*Geomys*) and their chewing lice (*Geomydoecus*). J. Mammal. 74:521–530.

de Meijere, J. C. 1894. Über die Haare der Säugetiere besonders über ihre Anordung. Gegenhaur's Morphol. Jahrb. 21:312–424.

Demers, M. N. 1997. Fundamentals of geographic information systems. John Wiley & Sons, New York.

Dene, H., M. Goodman, and W. Prychodko. 1976. An immunological examination of the systematics of Tupaioidea. J. Mammal. 59:697–706.

Dene, H., M. Goodman, W. Prychodko, and G. Matsuda. 1980. Molecular evidence for the affinities of Tupaiidae. Pp. 269–291 *in* Comparative biology and evolutionary relationships of tree shrews (W. P. Luckett, ed.). Plenum Press, New York.

Desy, E. A. and G. O. Batzli. 1989. Effects of food availability and predation on prairie vole demography: a field experiment. Ecology. 70:411–421.

de Waal, F. B. M. 1992. Coalitions as part of reciprocal relations in the Arnhem chimpanzee colony. Pp. 233–258 *in* Coalitions and alliances in humans and other animals (A. H. Harcourt and F. B. M. de Waal, eds.). Oxford Univ. Press, New York.

Dewsbury, D. A. 1982. Ejaculate cost and male choice. Am. Nat. 119:601–610.

Dewsbury, D. A. 1984. Sperm competition in muroid rodents. Pp. 547–571 *in* Sperm competition and the evolution of animal mating systems (R. L. Smith, ed.). Academic Press, New York.

Dewsbury, D. A. 1988. A test of the role of copulatory plugs in sperm competition in deer mice (*Peromyscus maniculatus*). J. Mammal. 69:854–857.

Diamond, J. 1994. Zebras and the Anna Karenina Principle: what can Tolstoy teach us about our failure to domesticate certain animals? Nat. Hist. 103:4,6,8–10.

Dice, L. R. 1939. Variation in the cactus mouse, *Peromyscus eremicus*. Contr. Lab. Vert. Biol. Univ. Mich. 8:1–27.

Dice, L. R. 1945. Minimum intensities of illumination under which owls can find dead prey by sight. Am. Nat. 79:385–416.

Diffendorfer, J. E., N. A. Slade, M. S. Gaines, and R. D. Holt. 1995. Population dynamics of small mammals in fragmented and continuous old-field habitat. Pp. 175–199 *in* Landscape approaches in mammalian ecology and conservation (W. Z. Lidicker, Jr., ed.). Univ. of Minnesota Press, Minneapolis.

Dilkes, D. W. and R. R. Reisz. 1996. First record of a basal synapsid ("mammal-like reptile") in Gondwana. Proc. R. Soc. Lond. 263:1165–1170.

Dinerstein, E. 1991. Sexual dimorphism in the greater one-horned rhinoceros (*Rhinoceros unicornis*). J. Mammal. 72:450–457.

Dinerstein, E. and L. Price. 1991. Demography and habitat use by greater one-horned rhinoceros in Nepal. J. Wildl. Manage. 55:401–411.

Dingle, H. 1980. Ecology and evolution of migration. Pp. 1–101 *in* Animal migration, orientation, and navigation (S. A. Gauthreaux, Jr. ed.). Academic Press, New York.

Dingle, H. 1996. Migration: the biology of life on the move. Oxford Univ. Press, New York.

Dobson, F. S. 1982. Competition for mates and predominant juvenile male dispersal in mammals. Anim. Behav. 30:1183–1192.

Domning, D. P. 1978. Sirenia. Pp. 573–581 *in* Evolution of African mammals (V. J. Maglio and H. B. S. Cooke, eds.). Harvard Univ. Press, Cambridge, MA.

Domning, D. P., G. S. Morgan, and C. E. Ray. 1982. North American Eocene sea cows (Mammalia: Sirenia). Smithsonian Contrib. Paleontol. 52:1–69.

Dötsch, C. and W. Koenigswald. 1978. Zur rotfärbung von soricidenzähnen [On the reddish coloring of soricid teeth]. Z. Säugetierk. 43:65–70.

Douglas-Hamilton, I. 1984. African elephant. Pp. 193–198 *in* Evolution of domesticated animals (I. L. Mason, ed.). Longman, New York.

Douglas-Hamilton, I. 1987. African elephants: population trends and their causes. Oryx. 21:11–24.

Drake, J. A., H. A. Mooney, F. diCastri, R. H. Groves, F. J. Kruger, M. Rejmánek, and M. Williamson. 1989. Biological invasions. John Wiley and Sons, New York.

Drickamer, L. C. 1974. Sexual maturation of female house mice: social inhibition. Dev. Psychobiol. 7:257–265.

Drickamer, L. C. 1974. Social rank, observability, and sexual behaviour of rhesus monkeys (*Macaca mulatta*). J. Reprod. Fert. 37:117–120.

Drickamer, L. C. 1981. Selection for age of sexual maturation in mice and the consequences for population regulation. Behav. Neur. Biol. 31:82–89.

Drickamer, L. C. 1987. Influence of time of day on captures of two species of *Peromyscus* in a New England deciduous forest. J. Mammal. 68:702–703.

Drickamer, L. C. 1989. Pheromones: behavioral and biochemical aspects. Pp. 269–348 *in* Advances in comparative and environmental physiology (J. Balthazart, ed.). Springer-Verlag, Berlin.

Drickamer, L. C. 1996. Intra-uterine position and anogenital distance in house mice: consequences under field conditions. Anim. Behav. 51:925–934.

Drickamer, L. C. and B. Shiro. 1984. Effects of adrenalectomy with hormone replacement therapy on the presence of sexual maturation-delaying chemosignal in the urine of grouped female mice. Endocrinology. 115:255–260.

Drickamer, L. C. and B. M. Vestal. 1973. Patterns of reproduction in a laboratory colony of *Peromyscus*. J. Mammal. 54:523–528.

Drollette, D. 1997. Wide use of rabbit virus is good news for native species. Science. 275:154.

Dugatkin, L. A. 1997. Cooperation among animals. An evolutionary perspective. Oxford Univ. Press, New York.

Duggins, D. O. 1980. Kelp beds and sea otters: an experimental approach. Ecology. 61:447–453.

Dyck, A. P. and R. A. MacArthur. 1993. Seasonal variation in the microclimate and gas composition of beaver lodges in a boreal environment. J. Mammal. 74:180–188.

Dyhrepoulsen, P., H. H. Smedegaard, H. H. Roed, and E. Korsgaard. 1994. Equine hoof function investigated by pressure transducers inside the hoof and accelerometers mounted on the first phalanx. Equine Vet. J. 26:362–366.

Eberhard, I. H., J. McNamara, R. J. Pearse, and I. A. Southwell. 1975. Ingestion and excretion of *Eucalyptus punctata* D.C. and its essential oil by the koala, *Phascolarctos cinereus* (Goldfuss). Aust. J. Zool. 23:169–179.

Ecke, D. H. and A. R. Kinney. 1956. Aging meadow mice, *Microtus californicus*, by observation of molt progression. J. Mammal. 37:249–254.

Economos, A. C. 1981. The largest land mammal. J. Theoretical Biol. 89:211–215.

Ehrlich, P. and A. Ehrlich. 1981. Extinction. Random House, New York.

Ehrlich, P. R. and A. H. Ehrlich. 1968. The population bomb. Amereon, Mattituck, NY.

Eisenberg, J. F. 1963. The behavior of heteromyid rodents. Univ. Cal. Publ. Zool. 69:111–114.

Eisenberg, J. F. 1975. Tenrecs and solenodons in captivity. Internl. Zoo Yearb. 15:6–12

Eisenberg, J. F. 1978. The evolution of arboreal herbivores in the Class Mammalia. Pp. 135–152 *in* The ecology of arboreal folivores (G. G. Montgomery, ed.), Smithsonian Institution Press, Washington, D.C.

Eisenberg, J. F. 1981. The mammalian radiations, an analysis of trends in evolution, adaptation, and behavior. Univ. of Chicago Press, Chicago.

Eisenberg, J. F., N. Muckenhirn, and R. Rudran. 1972. The relationship between ecology and social structure in primates. Science. 176:863–874.

Elgmork, K. 1982. Caching behavior of brown bears (*Ursus arctos*). J. Mammal. 63:607–612.

Ellegren, H. 1991. Fingerprinting birds' DNA with a synthetic polynucleotide probe (TG)$_n$. Auk. 108:956–958.

Elton, C. 1924. Periodic fluctuations in the numbers of animals: their causes and effects. Brit. J. Exptl. Biol. 2:119–163.

Elton, C. S. 1927. Animal ecology. Sidgwick and Jackson, London.

Eltringham, S. K. 1978. Methods of capturing wild animals for purposes of marking them. Pp. 13–23 *in* Animal marking (B. Stonehouse, ed.). University Park Press, Baltimore.

Eltringham, S. K. 1979. The ecology and conservation of large African mammals. University Park Press, Baltimore.

Emerson, S. B. 1985. Jumping and leaping. Pp. 58–72 *in* Functional vertebrate morphology (M. Hildebrand, D. M. Bramble, K. F. Liem, and D. B. Wake, eds.). Harvard Univ. Press, Cambridge.

Emlen, S. T. 1982. The evolution of helping. I. An ecological constraints model. Am. Nat. 119:29–39.

Emlen, S. T. 1984. Cooperative breeding in birds and mammals. Pp. 305–339 *in* Behavioural ecology: an evolutionary approach, 2d ed. (J. R. Krebs and N. B. Davies, eds.). Sinauer Assoc., Sunderland, MA.

Emlen, S. T. and L. W. Oring. 1977. Ecology, sexual selection, and the evolution of mating systems. Science. 197:215–223.

Enders, A. C. (ed). 1963. Delayed implantation. Univ. of Chicago Press, Chicago.

Engelmann, G. F. 1985. The phylogeny of the Xenarthra. Pp. 51–64 *in* The evolution and ecology of armadillos, sloths, and vermilinguas (G. G. Montgomery, ed.). Smithsonian Institution Press, Washington, D. C.

Erlinge, S., G. Göransson, L. Hansson, G. Högstedt, O. Liberg, I. N. Nilsson, T. Nilsson, T. von Schantz, and M. Sylvén. 1983. Predation as a regulating factor on small rodent populations in southern Sweden. Oikos. 40:36–52.

Estes, J. A., N. A. Smith, and J. F. Palmisano. 1978. Sea otter predation and community organization in the western Aleutian Islands, Alaska. Ecology. 59:822–833.

Estes, R. D. 1972. The role of the vomeronasal organ in mammalian reproduction. Mammalia. 36:315–341.

Etheridge, K., G. B. Rathbun, J. A. Powell, and H. I. Kochman. 1985. Consumption of aquatic plants by the West Indian manatee. J. Aquatic Plant Manage. 23:21–25.

Etienne, A. S., R. Maurer, and F. Saucy. 1988. Limitations in the assessment of path dependent information. Behaviour. 106:81–111.

Evans, P. G. H. 1987. The natural history of whales and dolphins. Facts on File Publ., New York.

Ewer, R. F. 1968. Ethology of mammals. Plenum Press, New York.

Ewer, R. F. 1973. The carnivores. Weidenfeld and Nicolson, London.

Fagen, R. 1981. Animal play behavior. Oxford Univ. Press, New York.

Fain, A. 1994. Adaptation, specificity and host-parasite coevolution in mites (Acari). Int. J. Parasitol. 24:1273–1283.

Farlow, J. O. 1987. Speculations about the diet and digestive physiology of herbivorous dinosaurs. Paleobiology. 13:60–72.

Feldhamer, G. A. and W. E. Armstrong. 1993. Interspecific competition between four exotic species and native artiodactyls in the United States. Transactions of the North American Wildlife Natural Resources Conference. 58:468–478.

Feldhamer, G. A., R. S. Klann, A. S. Gerard, and A. C. Driskell. 1993. Habitat partitioning, body size, and timing of parturition in pygmy shrews and associated soricids. J. Mammal. 74:403–411.

Feldhamer, G. A. and J. A. Rochelle. 1982. Mountain beaver, *Aplodontia rufa*. Pp. 167–175 *in* Wild mammals of North America: biology, management, and economics (J. A. Chapman and G. A. Feldhamer, eds.). Johns Hopkins Univ. Press, Baltimore.

Felsenstein, J. 1983. Parsimony in systematics: biological and statistical issues. Ann. Rev. Ecol. Syst. 14:313–333.

Fenton, M. B. 1972. The structure of aerial-feeding bat faunas as indicated by ears and wing elements. Can. J. Zool. 50:287–296.

Fenton, M. B. 1984a. Echolocation: Implications for ecology and evolution of bats. Q. Rev. Biol. 59:33–53.

Fenton, M. B. 1984b. Sperm competition? The case of vespertilionid and rhinolophid bats. Pp. 573–587 *in* Sperm competition and the evolution of animal mating systems (R. L. Smith, ed.). Academic Press, New York.

Fenton, M. B. 1985. Communication in the Chiroptera. Indiana Univ. Press, Bloomington.

Fenton, M. B. 1995. Constraint and flexibility-bats as predators, bats as prey. Pp. 277–289 *in* Ecology, evolution and behaviour of bats (P. A. Racey and S. M. Swift, eds.). Oxford Univ. Press, Oxford, England.

Fenton, M. B., D. Audet, D. C. Dunning, J. Long, C. B. Merriman, D. Pearl, D. M. Syme, B. Adkins, S. Pedersen, and T. Wohlgenant. 1993. Activity patterns and roost selection by *Noctilio albiventris* (Chiroptera: Noctilionidae) in Costa Rica. J. Mammal. 74:607–613.

Fenton, M. B. and D. W. Thomas. 1985. Migrations and dispersal of bats (Chiroptera). Pp. 409–424 *in* Migration: mechanisms and adaptive significance (M. A. Rankin, ed.). Marine Science Institute, Port Aransas, TX.

Fenton, M. B., D. W. Thomas, and R. Sasseen. 1981. *Nycteris grandis* (Nycteridae): an African carnivorous bat. J. Zool. 194:461–465.

Fergus, C. 1991. The Florida panther verges on extinction. Science. 251:1178–1180.

Ferraris, J. D. and S. R. Palumbi. 1996. Molecular zoology: advances, strategies, and protocols. Wiley-Liss, New York.

Ferron, J. 1996. How do woodchucks (*Marmota monax*) cope with harsh winter conditions? J. Mammal. 77:412–416.

Festa-Bianchet, M. 1989. Individual differences, parasites and the costs of reproduction for bighorn ewes (*Ovis canadensis*). J. Anim. Ecol. 58:785–795.

Findley, J. S. 1989. Morphological patterns in rodent communities of southwestern North America. Pp. 253–263 *in* Patterns in the structure of mammalian communities (D. W. Morris, Z. Abramsky, B. J. Fox and M. R. Willig, eds.). Texas Tech Univ. Press, Lubbock, TX.

Findley, J. S. 1993. Bats: a community perspective. Cambridge Univ. Press, Cambridge, England.

Findley, J. S. and H. Black. 1983. Morphological and dietary structuring of a Zambian insectivorous bat community. Ecology. 64:625–630.

Findley, J. S., E. H. Studier, and D. E. Wilson. 1972. Morphologic properties of bat wings. J. Mammal. 53:429–444.

Findley, J. S. and D. E. Wilson. 1974. Observations on the Neotropical disk-winged bat, *Thyroptera tricolor* Spix. J. Mammal. 55:562–571.

Findley, J. S. and D. E. Wilson. 1982. Ecological significance of chiropteran morphology. Pp. 243–260 *in* Ecology of bats. (T. H. Kunz, ed.). Plenum, New York.

Fish, F. E. 1992. Aquatic locomotion. Pp. 34–63 *in* Mammalian energetics: interdisciplinary views of metabolism and reproduction (T. E. Tomasi and T. H. Horton, eds.). Cornell Univ. Press, Ithaca, NY.

Fish, F. E. and C. A. Hui. 1991. Dolphin swimming-a review. Mammal Rev. 21:181–195.

Fishbein, D. B. 1991. Rabies in humans. Pp. 519–549 *in* The natural history of rabies (G. M. Baer, ed.). CRC Press, Boca Raton, FL.

Fisher, J., N. Simon, and J. Vincent. 1969. Wildlife in danger. Verlag, New York.

Fisher, R. A. 1958. Genetical theory of natural selection. Dover Publ., New York.

Flannery, T. 1995. Mammals of New Guinea. Cornell University Press, Ithaca, NY.

Flannery, T. F. 1987. An historic record of the New Zealand greater short–tailed bat, *Mystacina robusta* (Microchiroptera: Mystacinidae) from the South Island, New Zealand. Aust. Mammal. 10:45–46.

Fleagle, J. G. 1988. Primate adaptation and evolution. Academic Press, New York.

Fleischner, T. L. 1994. Ecological costs of livestock grazing in western North America. Conserv. Biol. 8:629–644.

Fleming, T. H. 1970. Notes on the rodent faunas of two Panamanian forests. J. Mammal. 51:473–490.

Fleming, T. H. 1971. *Artibeus jamaicensis:* delayed embryonic development in a neotropical bat. Science. 171:402–404.

Fleming, T. H. 1974. The population ecology of two species of Costa Rican heteromyid rodents. Ecology. 55:493–510.

Fleming, T. H. 1982. Foraging strategies of plant-visiting bats. Pp. 287–325 *in* Ecology of bats (T. H. Kunz, ed.). Plenum, New York.

Fleming, T. H. 1993. Plant-visiting bats. Am. Sci. 81:460–467.

Fleming, T. H. and A. Estrada (eds.). 1993. Frugivory and seed dispersal: ecological and evolutionary aspects. Adv. Veg. Sci. 15:1–392.

Fleming, T. H. and V. J. Sosa. 1994. Effects of nectarivorous and frugivorous mammals on reproductive success of plants. J. Mammal. 75:845–851.

Flower, W. H. and R. Lydekker. 1891. An introduction to the study of mammals living and extinct. Adam and Charles Black, London.

Flux, J. E. 1970. Colour change of mountain hares (*Lepus timidus scoticus*) in northeast Scotland. J. Zool. 162:345–358.

Flynn, L. J., N. A. Neff, and R. H. Tedford. 1988. Phylogeny of the Carnivora. Pp. 73–115 *in* The phylogeny and classification of the tetrapods, vol. I (Mammals) (M. J. Benton. ed.). Clarendon Press, Oxford.

Focardi, S., P. Marcellini, and R. Montanaro. 1996. Do ungulates exhibit a food density threshold? A field study of optimal foraging and movement patterns. J. Anim. Ecol. 65:606–620.

Fogel, R. and J. M. Trappe. 1978. Fungus consumption (Mycoghagy) by small animals. Northwest Sci. 52:1–31.

Foley, W. J., W. V. Engelhardt, and P. Charles-Dominique. 1995. The passage of digesta, particle size, and in vitro fermentation rate in the three-toed sloth *Bradypus tridactylus* (Edentata: Bradypodidae). J. Zool. Lond. 236:681–696.

Foley, W. J. and C. McArthur. 1994. The effects and costs of allelochemical for mammalian herbivores: an ecological perspective. Pp. 370–391 *in* The digestive system in mammals: food, form and function. Cambridge Univ. Press, New York.

Folk, G. E. Jr. and M. A. Folk. 1980. Physiology of large mammals by implanted radio capsules. Pp. 33–43 *in* A handbook of biotelemetry and radio tracking (C. J. Amlaner Jr. and D. W. Macdonald, eds.). Pergamon Press, Oxford, England.

Fooden, J. 1972. Breakup of Pangea and isolation of relict mammals in Australia, South America, and Madagascar. Science. 175:894–898.

Fordyce, R. E. 1980. Whale evolution and Oligocene southern ocean environments. Paleogeogr. Paleoclimatol. Paleoecol. 31:319–336.

Forman, R. T. and M. Godron. 1986. Landscape ecology. Wiley, New York.

Forsman, K. A. and M. G. Malmquist. 1988. Evidence for echolocation in the common shrew, *Sorex araneus*. J. Zool. Lond. 216:655–662.

Fortelius, M. and J. Kappelman. 1993. The largest land mammal ever imagined. Zool. J. Linnean Soc. 108:85–101.

Fossey, D. 1983. Gorillas in the mist. Houghton Mifflin, Boston.

Fossey, D. and J. Harcourt. 1977. Feeding ecology of free-ranging mountain gorilla (*Gorilla gorilla beringei*). Pp. 415–447 *in* Primate ecology: studies of feeding and ranging behaviour in lemurs, monkeys and apes. Academic Press, London.

Foster, J. B. 1964. Evolution of mammals on islands. Nature. 202:234–235.

Fox, B. J. 1989. Small-mammal community pattern in Australian heathland: a taxonomically-based rule for species assembly. Pp. 91–103 *in* Patterns in the structure of mammalian communities (D. W. Morris, Z. Abramsky, B. J. Fox and M. R. Willig, eds.). Texas Tech Univ. Press, Lubbock, TX.

Fox, B. J. and G. L. Kirkland, Jr. 1992. North American soricid communities follow an Australian small mammal assembly rule. J. Mammal. 73:491–503.

Francis, C. M., E. L. P. Anthony, J. A. Brunton, and T. H. Kunz. 1994. Lactation in male fruit bats. Nature. 367:691–692.

Frank, L. G., S. E. Glickman, and P. Licht. 1991. Fatal sibling aggression, precotial development, and androgens in neonatal spotted hyenas. Science. 252:702–704.

Fraser, D. 1990. Behavioral perspectives on piglet survival. J. Reprod. Fertil. Suppl. 40:355–370.

Fraser, D. and B. K. Thompson. 1991. Armed sibling rivalry by domestic piglets. Behav. Ecol. Sociobiol. 29:1–15.

Fraser, F. C. and P. E. Purves. 1960. Hearing in cetaceans. Bull. Br. Mus. Nat. Hist. 7:1–140.

Freedman, D. O. 1992. Intestinal nematodes. Pp. 2003–2008 *in* Infectious diseases (S. L. Gorbach, J. G. Bartlett, and N. R. Blacklow, eds.). W. B. Saunders, Philadelphia.

Freeland, W. J. and D. H. Janzen. 1974. Strategies in herbivory by mammals: the role of plant secondary compounds. Am. Nat. 108:269–289.

Freeman, P. W. 1979. Specialized insectory: beetle-eating and moth-eating molossid bats. J. Mammal. 60:467–479.

Freeman, P. W. 1981. Correspondence of food habits and morphology in insectivorous bats. J. Mammal. 62:166–177.

Freeman, P. W. 1988. Frugivorous and animalivorous bats (Microchiroptera): dental and cranial adaptations. Biol. J. Linn. Soc. 33:249–272.

French, A. R. 1992. Mammalian dormancy. Pp. 105–121 *in* Mammalian energetics (T. E. Tomasi and T. H. Horton, eds.), Comstock Publ. Assoc., Ithaca, New York.

French, A. R. 1993. Physiological ecology of the Heteromyidae: economics of energy and water utilization. Pp. 509–538 *in* Biology of the Heteromyidae (H. H. Genoways and J. H. Brown, eds.). Spec. Pub. American Society of Mammalogists. 10:1–719.

French, N. R., B. G. Maza, H. O. Hill, A. P. Aschwanden, and H. W. Kaaz. 1974. A population study of irradiated desert rodents. Ecol. Monogr. 44:45–72.

Fretwell, S. D. 1972. Populations in a seasonal environment. Princeton Univ. Press, Princeton, NJ.

Fretwell, S. D. and H. L. Lucas. 1970. On territorial behaviour and other factors influencing habitat distribution in birds. I. Theoretical development. Acta Biotheoretica. 19:16–36.

Friend, J. A. 1989. Myrmecobiidae. Pp. 583–590 *in* Fauna of Australia: Mammalia, vol. IB (D. W. Walton and B. J. Richardson, eds.). Australian Government Publ. Service, Canberra.

Friend, J. A. 1995. Numbat, *Myrmecobius fasciatus*. Pp. 160–162 *in* The mammals of Australia (R. Strahan, ed.). Reed Books, Chatswood, Australia.

Fries, R., A. Eggen, and G. Stranzinger. 1990. The bovine genome contains polymorphic microsatellites. Genomics. 8:403–406.

Fritts, S. H. 1993. Controlling wolves in the greater Yellowstone area. *In* Ecological issues in reintroducing wolves into Yellowstone National Park (R. S. Cook, ed.). National Park Service (Scientific Monograph NPS/NRYELL/NRSM-93/22), Washington, D.C.

Frost, H. C., W. B. Krohn, and C. R. Wallace. 1997. Age-specific reproductive characteristics in fishers. J. Mammal. 78:598–612.

Fujita, M. S. and M. D. Tuttle. 1991. Flying foxes (Chiroptera: Pteropodidae): threatened animals of key ecological and economic importance. Conserv. Biol. 5:455–463.

Fullard, J. H., J. A. Simmons, and P. A. Saillant. 1994. Jamming bat echolocation: the dogbane tiger moth *Cynia tenera* times its clicks to the terminal attack calls of the big brown bat *Eptesicus fuscus*. J. Exp. Biol. 194:285–298.

Fuller, M. R. 1987. Applications and considerations for wildlife telemetry. J. Raptor. Res. 21:126–128.

Fuller, W. A., L. L. Stebbins, and G. R. Dyke. 1969. Overwintering of small mammals near Great Slave Lake, northern Canada. Arctic. 22:34–55.

Futuyma, D. J. and M. Slatkin (eds.). 1983. Coevolution. Sinauer Assoc., Sunderland, MA.

Gage, K. L., R. S. Ostfeld, and J. G. Olson. 1995. Nonviral vector-borne zoonoses associated with mammals in the United States. J. Mammal. 76:695–715.

Gaisler, J. 1979. Ecology of bats. Pp. 281–342 *in* Ecology of small mammals (D. M. Stoddart, ed.). Chapman and Hall, London.

Galbreath, G. J. 1982. Armadillo, *Dasypus novemcinctus*. Pp. 71–79 *in* Wild mammals of North America: biology, management, and economics (J. A. Chapman and G. A. Feldhamer, eds.). Johns Hopkins Univ. Press, Baltimore.

Galton, F. 1865. The first steps toward the domestication of animals. Transactions of the Ethnological Society of London. 3:122–138.

Gardner, A. L. 1977. Feeding habits. Pp. 293–350 *in* Biology of bats of the New World Family Phyllostomatidae, part II. (R. J. Baker, J. Knox Jones, Jr., and D. C. Carter, eds.). Spec. Pub. Museum of Texas Tech Univ., No. 13, Lubbock.

Gardner, A. L. 1982. Virginia opossum, *Didelphis virginiana*. Pp. 3–36 *in* Wild mammals of North America: biology, management, and economics (J. A. Chapman and G. A. Feldhamer, eds.). Johns Hopkins Univ. Press, Baltimore.

Gartlan, J. S. and T. T. Struhsaker. 1972. Polyspecific associations and niche separation of rain-forest anthropoids in Cameroon, West Africa. J. Zool. (London). 168:221–266.

Gaudin, T. J. and A. A. Biewener. 1992. The functional morphology of xenarthrous vertebrae in the armadillo *Dasypus novemcinctus* (Mammalia: Xenarthra). J. Morphol. 214:63–81.

Gauthier-Pilters, H. 1974. The behaviour and ecology of camels in the Sahara, with special reference to nomadism and water management. Pp. 542–551 *in* The behaviour of ungulates and its relation to management (V. Geist and F. Walther, eds.). IUCN Publ. No. 24. Morges, Switzerland.

Gauthier-Pilters, H. and A. Dagg. 1981. The camel, its evolution, ecology, behavior, and relations to man. Univ. of Chicago Press, Chicago.

Gebczynska, Z. and M. Gebczynski. 1971. Insulating properties of the nest and social temperature regulation in *Clethrionomys glareolus* (Schreber). Acta Zool. Fennici. 8:104–108.

Geiser, F. and T. Ruf. 1995. Hibernation versus daily torpor in mammals and birds: physiological variables and classification of torpor patterns. Physiol. Zool. 68:935–966.

Geist, V. 1966. Validity of horn segment counts in aging bighorn sheep. J. Wildl. Manage. 30:634–635.

Geist, V. 1971. Mountain sheep: a study in behavior and evolution. Univ. of Chicago Press, Chicago.

Geist, V. 1987. Bergmann's rule is invalid. Can. J. Zool. 65:1035–1038.

Geist, V. 1991. Bones of contention revisited: did antlers enlarge with sexual selection as a consequence of neonatal security strategies? Appl. Anim. Behav. Sci. 29:453–469.

Geist, V. and P. T. Bromley. 1978. Why deer shed antlers. Z. Säugetierk. 43:223–231.

Geluso, K. N. 1978. Urine concentrating ability and renal structure of insectivorous bats. J. Mammal. 59:312–323.

Genoud, M. 1988. Energetic strategies of shrews: ecological constraints and evolutionary implications. Mammal Rev. 18:173–193.

Genoways, H. H. and J. H. Brown (eds.). 1996. Biology of the Heteromyidae. Spec. Publ. No. 10, American Society of Mammalogists.

Gentry, A. W. 1990. Evolution and dispersal of African Bovidae. Pp. 195–227 *in* Horns, pronghorns, and antlers: evolution, morphology, physiology, and social significance (G. A. Bubenik and A. B. Bubenik, eds.). Springer-Verlag, New York.

Gentry, R. L. and G. L. Kooyman. 1986. Methods of dive analysis. Pp. 28–40 *in* Fur seals: maternal strategies on land and at sea (R. L. Gentry and G. L. Kooyman, eds.). Princeton Univ. Press, Princeton, NJ.

Gerard, A. S. and G. A. Feldhamer. 1990. A comparison of two survey methods for shrews: pitfalls and discarded bottles. Am. Mid. Nat. 124:191–194.

Getz, L. L. and C. S. Carter. 1980. Social organization in *Microtus ochrogaster* populations. The Biologist. 62:56–69.

Gibbs, H. L., P. J. Weatherhead, P. T. Boag, B. N. White, L. M. Tabak, and D. J. Hoysak. 1990. Realized reproductive success of polygynous red-winged blackbirds revealed by DNA markers. Science. 250:1394–1397.

Giles, R. H. 1971. Wildlife management techniques, 3d ed., rev. Wildlife Society, Washington, D.C.

Gill, F. B. 1980. Historical aspects of hybridization between blue-winged and golden-winged warblers. Auk. 97:1–18.

Gillette, D. D. 1991. *Seismosaurus halli* gen. et sp. nov., a new sauropod dinosaur from the Morrison Formation (Upper Jurassic/Lower Cretaceous) of New Mexico, USA. J. Vert. Paleontol. 11:417–433.

Gillette, M. U., S. J. DeMarco, J. M. Ding, E. A. Gallman, L. E. Faiman, C. Liu, A. J. McArthur, M. Medanic, D. Richard, T. K. Tcheng, and E. T. Weber. 1993. The organization of the suprachiasmatic pacemaker of the rat and its regulation by neurotransmitters and modulators. J. Biol. Rhythms. 8:S53–S58 (suppl.).

Gilmore, D. P. 1969. Seasonally reproductive periodicity in the male Australian brush-tailed possum. J. Zool. 157:75–98.

Gingerich, P. D., B. H. Smith, and E. L. Simons. 1990. Hind limbs of Eocene *Basilosaurus*: evidence of feet in whales. Science. 249:154–156.

Gingerich, P. D., N. A. Wells, D. E. Russell, S. M. I. Shah. 1983. Origin of whales in epicontinental remnant seas: new evidence from the early Eocene of Pakistan. Science. 220:403–406.

Girardier, L. and M. J. Stock (eds.). 1983. Mammalian thermogenesis. Chapman and Hall, New York.

Gittleman, J. L. 1989. The comparative approach in ethology: aims and limitations. Persp. Ethol. 8:55–83.

Glander, K. E. 1977. Poison in a monkey's Garden of Eden. Nat. Hist. 86 (March):35–41.

Glander, K. E. 1982. The impact of plant secondary compounds on primate feeding behavior. Yearb. Phys. Anthro. 25:1–18.

Glantz, S. A. 1992. Primer of biostatistics, 3d ed. McGraw-Hill, New York.

Glaser, H. and S. Lustick. 1975. Energetics and nesting behavior of the northern white-footed mouse, *Peromyscus leucopus noveboracensis*. Physiol. Zool. 48:105–113.

Glass, J. D., U. E. Hauser, W. Randolph, S. Ferriera, and M. A. Rea. 1993. Suprachiasmatic nucleus neurochemistry in the conscious brain: correlation with circadian activity rhythms. J. Biol. Rhythms. 8:S47–S52 (suppl.).

Gleeson, S. K., A. B. Clark, and L. A. Dugatkin. 1994. Monozygotic twinning: an evolutionary hypothesis. Proceedings of the National Academy of Science. 91:11363–11367.

Godin, A. J. 1977. Wild mammals of New England. Johns Hopkins Univ. Press, Baltimore.

Goepfert, M. C. and L. T. Wasserthal. 1995. Notes on echolocation calls, food and roosting behaviour of the Old World sucker-footed bat, *Myzopoda aurita* (Chiroptera: Myzopodidae). Z. Saeugetierk. 60:1–8.

Goldingay, R. L. and R. P. Kavangah. 1991. The yellow-bellied glider: a review of its ecology and management considerations. Pp. 365–375 *in* Conservation of Australia's forest fauna (D. Lunney, ed.). Royal Zool. Soc., Sydney, Australia.

Goldsmith, R., D. Bunnag, and T. Bunnag. 1991. Lung fluke infections: paragonimiasis. Pp. 827–831 *in* Hunter's tropical medicine, 7th ed. (G. T. Strickland, ed.). W. B. Saunders, Philadelphia.

Goldspink, G. 1981. The use of muscles during flying, swimming, and running from the point of view of energy saving. Symp. Zool. Soc. London. 48:219–238.

Goldsworthy, S. D. 1995. Differential expenditure of maternal resources in Antarctic fur seals, *Arctocephalus gazella*, at Heard Island, southern Indian Ocean. Behav. Ecol. 6:218–228.

Golley, F. B. 1960. Energy dynamics of a food chain of an old-field community. Ecol. Monogr. 30:187–206.

Golley, F. B. 1961. Effect of trapping on adrenal activity in *Sigmodon*. J. Wild. Mgmt. 25:331–333.

Goodall, J. 1986. The chimpanzees of Gombe. Harvard Univ. Press, Cambridge, MA.

Goodman, M., D. A. Tagle, D. H. A. Fitch, W. Bailey, J. Czelusniak, B. F. Koop, P. Benson, and J. L. Slighton. 1990. Primate evolution at the DNA level and a classification of hominids. J. Mol. Evol. 30:260–266.

Goodman, S. M. 1995. *Rattus* on Madagascar and the dilemma of protecting the endemic rodent fauna. Conserv. Biol. 9:450–453.

Gorbach, S. L., J. G. Bartlett, and N. R. Blacklow (eds.). 1992. Infectious diseases. W. B. Saunders, Philadelphia.

Gordon, G. and A. J. Hulbert. 1989. Peramelidae. Pp. 603–624 *in* Fauna of Australia: Mammalia, vol. IB (D. W. Walton and B. J. Richardson, eds.). Australian Government Publ. Service, Canberra.

Gordon, M. S. 1982. Animal physiology: principles and adaptations, 4th ed. Macmillan, New York.

Gorman, M .L. and R. D. Stone. 1990. The natural history of moles. Comstock Publ. Assoc., Ithaca, NY.

Gosling, L. M. and M. Petrie. 1990. Lekking in topi: a consequence of satellite behaviour by small males at hotspots. Anim. Behav. 40:272–287.

Gottfried, B. M. 1979. Small mammal populations in woodlot islands. Am. Mid. Nat. 102:105–112.

Gottfried, B. M. 1982. A seasonal analysis of small mammal populations on woodlot islands. Can. J. Zool. 60:1660–1664.

Gould, E. 1983. Mechanisms of mammalian auditory communication. Pp. 265–342 *in* Advances in the study of mammalian behavior (J. F. Eisenberg and D. G. Kleiman, eds.). Spec. Pub. No. 7. American Society of Mammalogists.

Gould, E., W. McShea, and T. Grand. 1993. Function of the star in the star-nosed mole, *Condylura cristata*. J. Mammal. 74:108–116.

Goundie, T. R. and S. H. Vessey. 1986. Survival and dispersal of young white-footed mice born in nest boxes. J. Mammal. 67:53–60.

Grant, P. R. 1975. Population performance of *Microtus pennsylvanicus* confined to woodland habitat and a model of habitat occupancy. Can. J. Zool. 53:1447–1465.

Grant, T. R. 1989. Ornithorhynchidae. Pp. 436–450 *in* Fauna of Australia: Mammalia, vol. IB (D. W. Walton and B. J. Richardson, eds.). Australian Government Publ. Service, Canberra.

Green, B. 1984. Composition of milk and energetics of growth in marsupials. Pp. 369–387 *in* Physiological strategies in lactation (M. Peaker, R. G. Vernon, and C. H. Knight, eds.). Academic Press, New York.

Green, G. M .G. and R. W. Sussman. 1990. Deforestation history of the eastern rain forests of Madagascar from satellite images. Science. 248:212–215.

Green, R. G., C. L. Larson, and J. F. Bell. 1939. Shock disease as the cause of the periodic decimation of the snowshoe hare. Am. J. Hyg. 30:83–102.

Greenbaum, I. F. and C. J. Phillips. 1974. Comparative anatomy of and general histology of tongues of long-nosed bats (*Leptonycteris sanborni* and *L. nivalis*) with reference to infestation of oral mites. J. Mammal. 55:489–504.

Greenhall, A. M. 1972. The biting and feeding habits of the vampire bat, *Desmodus rotundus*. J. Zool. (Lond.). 168:451–461.

Greenhall, A. M., G. Joermann, and U. Schmidt. 1983. Desmodus rotundus. Mamm. Species. 202:1–6.

Greenhall, A.M., U. Schmidt, and G. Joermann. 1984. Diphylla ecaudata. Mamm. Species. 227:1–3.

Greenhall, A. M. and W. A. Schutt, Jr. 1996. Diaemus youngi. Mamm. Species. 533:1–7.

Greenwood, P. J. 1980. Mating systems, philopatry, and dispersal in birds and mammals. Anim. Behav. 28:1140–1162.

Greenwood, P. J. 1983. Mating systems and the evolutionary consequences of dispersal. Pp. 116–131 *in* The ecology of animal movement (I. R. Swingland and P. J. Greenwood, eds.). Clarendon Press, Oxford, England.

Grenfell, B. T. and F. M. D. Gulland. 1995. Ecological impact of parasitism on wildlife host populations. Parasitol. 111:S3–S14.

Griffin, D. R. 1958. Listening in the dark. Yale Univ. Press, New Haven, CT.

Griffin, D. R. 1970. Migrations and homing of bats. Pp. 233–265 *in* Biology of bats (W. A. Wimsatt, ed.). Academic Press, New York.

Griffiths, M. 1978. The biology of the monotremes. Academic Press, New York.

Griffiths, M. 1989. Tachyglossidae. Pp. 407–435 *in* Fauna of Australia: Mammalia, vol. IB (D. W. Walton and B. J. Richardson, eds.). Australian Government Publ. Service, Canberra.

Griffiths, M., M. A. Elliott, R. M. C. Leckie, and G. I. Schoefl. 1973. Observations on the comparative anatomy and ultrastructure of mammary glands and on the fatty acids of the triglycerides in platypus and echidna milk fats. J. Zool. 169:255–279.

Griffiths, M., R. T. Wells, and D. J. Barrie. 1991. Observations on the skulls of fossil and extant echidnas (Monotremata: Tachyglossidae). Aust. Mammal. 14:87–101.

Grigg, G. C., L. A. Beard, and M. L. Augee. 1989. Hibernation in a monotreme, the echidna *Tachyglossus aculeatus*. Comp. Biochem. Physiol. 92A:609–612.

Grigg, G. C., L. A. Beard, T. R. Grant, and M. L. Augee. 1992. Body temperature and diurnal activity pattern in the platypus, *Ornithorhynchus anatinus*, during winter. Australian J. Zool. 40:135–142.

Grimwood, I. 1988. "Operation oryx." The start of it all. Pp. 1–8 *in* The conservation and biology of desert antelopes, (A. Dixon and D. Jones, eds.). Helm, London.

Grinnell, J. and H. S. Swarth. 1913. An account of the birds and mammals of the San Jacinto area of southern California, with remarks upon the behavior of geographic races on the margins of their habitats. Univ. Calif. Pub. Zool. 10:197–406.

Grodzinski, W. and N. R. French. 1983. Production efficiency in small mammal populations. Oecologia. 56:41–49.

Grodzinski, W., M. Makomaska, R. Tertil, and J. Weiner. 1977. Bioenergetics and total impact of vole populations. Oikos. 29:494–510.

Grojean, R. E., J. A. Suusa, and M. C. Henry. 1980. Utilization of solar radiation by polar animals: an optical model for pelts. Appl. Optics. 19:339–346.

Gromko, M. H., D. G. Gilbert, and R. C. Richmond. 1984. Sperm transfer and use in the multiple mating system of *Drosophila*. Pp. 371–426 *in* Sperm competition and the evolution of animal mating systems (R. L. Smith, ed.). Academic Press, New York.

Groves, C. P. 1989. A theory of human and primate evolution. Oxford Univ. Press, New York.

Groves, C. P. 1993. Order Diprotodontia. Pp. 45–62 *in* Mammal species of the world: a taxonomic and geographic reference, 2d ed. (D. E. Wilson and D. M. Reeder, eds.). Smithsonian Institution Press, Washington, D.C.

Groves, C. P. and T. F. Flannery. 1990. Revision of the families and genera of bandicoots. Pp. 1–11 *in* Bandicoots and bilbies (J. H. Seebeck, P. R. Brown, R. W. Wallis, and C. M. Kemper, eds.). Surrey Beatty and Sons, Sydney.

Groves, C. P. and P. Grubb. 1987. Relationships of living deer. Pp. 21–59 *in* Biology and management of the Cervidae (C. M. Wemmer, ed.). Smithsonian Institution Press, Washington, DC.

Groves, C. P. and D. P. Willoughby. 1981. Studies on the taxonomy and phylogeny of the genus Equus. I. Subgeneric classification of the Recent species. Mammalia. 45:321–354.

Grubb, P. 1993. Order Artiodactyla. Pp. 377–414 *in* Mammal species of the world: a taxonomic and geographic reference, 2d ed. (D. E. Wilson and D. M. Reeder, eds.). Smithsonian Institution Press, Washington, DC.

Grubb, P. 1993a. Order Perissodactyla. Pp. 369–372 *in* Mammal species of the world: a taxonomic and geographic reference, 2d ed. (D. E. Wilson and D. M. Reeder, eds.). Smithsonian Institution Press, Washington, DC.

Grubb, P. 1993b. Order Artiodactyla. Pp. 377–414 *in* Mammal species of the world: a taxonomic and geographic reference, 2d ed. (D. E. Wilson and D. M. Reeder, eds.). Smithsonian Institution Press, Washington, DC.

Guarch-Delmonte, J. M. 1984. Evidencias de la existencia de *Geocapromys* y *Heteropsomys* (Mammalia: Rodentia) en Cuba. Misc. Zool. (Havana). 18:1.

Guinee, L. N., K. Chu, and E. M. Dorsey. 1983. Changes over time in the songs of known individual humpback whales (*Megaptera novaeangliae*). Pp. 59–80 *in* Communication and behavior of whales (R. Payne, ed.). Westview Press, Boulder, CO.

Gunderson, H. L. 1976. Mammalogy. McGraw-Hill, New York.

Gunn, A. 1982. Muskox (*Ovibos moschatus*). Pp. 1021–1035 *in* Wild mammals of North

America: biology, management, and economics. (J. A. Chapman and G. A. Feldhamer, eds.). Johns Hopkins Univ. Press, Baltimore.

Guthrie, R. D. 1971. A new theory of mammalian rump patch evolution. Behaviour. 38:132–145.

Gwinner, E. 1986. Circannual rhythms. Springer-Verlag, Berlin.

Gwynne, M. D. and R. H. V. Bell. 1968. Selection of vegetation components by grazing ungulates in the Serengeti National Park. Nature. 220:390–393.

Habersetzer, J., G. Richter, and G. Storch. 1994. Paleoecology of early middle Miocene bats from Messel, FRG: aspects of flight, feeding and echolocation. Hist. Biol. 8:235–260.

Hadley, M. E. 1988. Endocrinology. Prentice-Hall, Englewood Cliffs, NJ.

Haffer, J. 1969. Speciation in Amazonian forest birds. Science. 165:131–137.

Hafner, D. J. and R. M. Sullivan. 1995. Historical and ecological biogeography of Nearctic pikas (Lagomorpha: Ochotonidae). J. Mammal. 76:302–321.

Hafner, M. S., W. L. Gannon, J. Salazar-Bravo, and S. T. Alvarez-Casteñeda. 1997. Mammal collections in the Western Hemisphere. American Society of Mammalogists, Provo, UT.

Hafner, M. S. and S. A. Nadler. 1990. Cospeciation in host-parasite assemblages: comparative analysis of rates of evolution and timing of cospeciation. Syst. Zool. 39:192–204.

Haigh, G. R. 1979. Sun-compass orientation in the thirteen-lined ground squirrel, *Spermophilus tridecemlineatus*. J. Mammal. 60:629–632.

Hain, J. H. W., G. R. Carter, S. D. Kraus, C. A. Mayo, and H. E. Winn. 1982. Feeding behavior of the humpback whale, *Megaptera novaeangliae*, in the western North Atlantic. Fish. Bull. 80:99–108.

Hainsworth, F. R., 1981. Animal physiology, adaptations in function. Addison-Wesley, Reading, MA.

Haley, M. P., C. J. Deutsch, and B. J. Le Boeuf. 1994. Size, dominance and copulatory success in male northern elephant seals, *Mirounga angustirostris*. Anim. Behav. 48:1249–1260.

Hall, E. R. 1948. Mammals of Nevada. Univ. of California Press, Berkeley.

Hall, E. R. 1951. American weasels. Univ. of Kansas Pub., Mus. Nat. Hist. 4:1–466.

Hall, E. R. 1981. The mammals of North America, 2d ed., 2 vols. Wiley-Interscience, New York.

Hall, E. R. and K. R. Kelson. 1959. The mammals of North America, 2 vols. Ronald Press, New York.

Halle, S. 1995. Diel pattern of locomotor activity in populations of root voles, *Microtus oeconomus*. J. Biol. Rhythms. 10:211–224.

Halls, L. K. (ed.) 1984. White-tailed deer ecology and management. Stackpole Books, Harrisburg, PA.

Hamilton, W. D. 1963. The evolution of altruistic behavior. Am. Nat. 97:354–356.

Hamilton, W. D. 1964. The genetical evolution of social behavior I, II. J. Theoret. Biol. 7:1–52.

Hamilton, W. D. 1971. Geometry for the selfish herd. J. Theoret. Biol. 31:295–311.

Hamilton, W. D. and R. M. May. 1977. Dispersal in stable habitats. Nature. 269:578–581.

Hamilton, W. D. and M. Zuk. 1984. Heritable true fitness and bright birds: a role for parasites? Science. 218:384–387.

Hamilton, W. J., Jr. 1955. Mammalogy in North America. Pp. 661–688 *in* A century of progress in the natural sciences. California Academy of Sciences, San Francisco.

Hammond, K. A. and B. A. Wunder. 1991. Effects of food quality and energy needs: changes in gut morphology and capacity of *Microtus ochrogaster*. J. Mammal. 64:541–567.

Hammond, P. S., S. A. Mizroch, and G. P. Donovan (eds.). 1990. Individual recognition of cetaceans: use of photo-identification and other techniques to estimate population and pod characteristics. Sci. Rep. Whales Res. Inst. 29:59–85.

Handley, C. O., Jr., D. E. Wilson, and A. L. Gardner (eds.). 1991. Demography and natural history of the common fruit bat, *Artibeus jamaicensis*, on Barro Colorado Island, Panama. Smithsonian Contr. Zool. 511.

Hanken, J. and P. W. Sherman. 1981. Multiple paternity in Belding's ground squirrel litters. Science. 212:351–353.

Hansen, E. W. 1966. The development of maternal and infant behavior in the rhesus monkey. Behaviour. 27:107–149.

Hanski, I., L. Hansson, and H. Henttonen. 1991. Specialist predators, generalist predators, and the microtine rodent cycle. J. Anim. Ecol. 60:353–367.

Hanski, L. 1996. Metapopulation ecology. Pp. 13–43 *in* Population dynamics in ecological space and time (O. E. Rhodes, Jr., R. K. Chesser and M. H. Smith, eds.). Univ. of Chicago Press, Chicago.

Harcourt, A. H. 1989. Deformed sperm are probably not adaptive. Anim. Behav. 37:863–864.

Harcourt, A. H., P. H. Harvey, S. G. Larson, and R. V. Short. 1981. Testis weight, body weight and breeding system in primates. Nature. 293:55–57.

Harder, J. D. and A. Woolf. 1976. Changes in plasma levels of oestrone and oestradiol during pregnancy and parturition in white-tailed deer. J. Reprod. Fert. 47:161–163.

Harney, B. A. and R. D. Dueser. 1987. Vertical stratification of activity of two *Peromyscus* species: an experimental analysis. Ecology. 68:1084–1091.

Harper, S. J. and G. O. Batzli. 1996. Monitoring use of runways by voles with passive integrated transponders. J. Mammal. 77:364–369.

Harrington, M. E., D. M. Nance, and B. Rusak. 1985. Neuropeptide Y immunoreactivity in the hamster geniculo-suprachiasmatic tract. Brain Res. Bull. 15:465–472.

Harris, R. H. 1959. Small vertebrate skeletons. Museums J. 58:223–224.

Harris, V. T. 1952. An experimental study of habitat selection by prairie and forest races of the deer mouse, *Peromyscus maniculatus*. Contrib. Lab. Vert. Biol., Univ. Mich. 56:1–53.

Harrop, C. J. F. and I. D. Hume. 1980. Digestive tract and digestive function in monotremes and nonmacropod marsupials. Pp. 63–77 *in* Comparative physiology: primitive mammals (K. Schmidt-Nielsen, L. Bolis, and C. R. Taylor, eds.). Cambridge Univ. Press, Cambridge, England.

Hart, J. A. and T. B. Hart. 1988. A summary report on the behaviour, ecology, and conservation of the okapi (*Okapia johnstoni*) in Zaire. Acta Zool. Pathol. Antverpiensia. 80:19–28.

Hart, J. S. 1971. Rodents. Pp. 1–149 *in* Comparative physiology of thermoregulation (G. C. Whittow, ed.) Academic Press, New York.

Harthoorn, A. M. 1976. The chemical capture of animals. Balliére Tindall, London.

Hartman, D. S. 1979. Ecology and behavior of the manatee (*Trichechus manatus*) in Florida. Spec. Publ. No. 5, American Society of Mammalogists.

Harvey, P. H. and M. D. Pagel. 1991. The comparative method in evolutionary biology. Oxford Univ. Press, New York.

Harvey, P. H. and K. Ralls. 1985. Homage to the null weasel. Pp. 155–171 *in* Evolution: essays in honour of John Maynard Smith (P. J. Greenwood, P. H. Harvey and M. Slatkin, eds.). Cambridge Univ. Press, Cambridge, England.

Hasson, O. 1991. Pursuit-deterrent signals: communication between prey and predator. Trends Ecol. Evol. 6:325–329.

Hausfater, G. and S. B. Hrdy. 1984. Infanticide: comparative and evolutionary perspectives. Aldine, New York.

Hausfater, G. and D. F. Watson. 1976. Social and reproductive correlates of parasite ova emissions by baboons. Nature. 262:688–689.

Hawkey, C. M. 1966. Plasminogen activator in the saliva of the vampire bat, *Desmodus rotundus*. Nature. 211:434–435.

Hawkins, R. E., L. D. Martoglio, and G. G. Montgomery. 1968. Cannon-netting deer. J. Wild. Mgmt. 32:191–195.

Hay, K. A. and A. W. Mansfield. 1989. Narwhal *Monodon monoceros* Linnaeus, 1758. Pp. 145–176 *in* Handbook of marine mammals, vol. 4: river dolphins and the larger toothed whales (S. H. Ridgway and R. Harrison, eds.). Academic Press, New York.

Hayes, J. P. and T. Garland, Jr. 1995. The evolution of endothermy: testing the aerobic capacity model. Evolution. 49:836–847.

Hayssen, V. 1993. Empirical and theoretical constraints on the evolution of lactation. J. Dairy Sci. 76:3213–3233.

Hayssen, V. and T. H. Kunz. 1996. Allometry of litter mass in bats: maternal size, wing morphology, and phylogeny. J. Mammal. 77:476–490.

Hayssen, V., Ari van Tienhoven, and Ans van Tienhoven. 1993. Asdell's patterns of mammalian reproduction. Comstock Publ. Assoc., Ithaca, NY.

Hayward, J. S. and P. A. Lisson. 1992. Evolution of brown fat: its absence in marsupials and monotremes. Can. J. Zool. 70:171–179.

Hayward, J. S. and C. P. Lyman. 1967. Nonshivering heat production during arousal from hibernation and evidence for the contribution of brown fat. Pp. 346–355 *in* Mammalian hibernation III (K. C. Fisher, A. R. Dawe, C. P. Lyman, E. Schönbaum, and F. E. South, eds.). Oliver & Boyd, Edinburgh.

Heath, M. E. 1992. Manis temminckii. Mammal. Species No. 415:1–5.

Heath, M. E. 1995. Manis crassicaudata. Mammal. Species No. 513:1–4.

Heatwole, H. 1965. Some aspects of the association of cattle egrets with cattle. Anim. Behav. 13:79–83.

Hedricks, C. and M. K. McClintock. 1990. Timing of insemination is correlated with the secondary sex ratio of Norway rats. Physiol. Behav. 48:625–632.

Heideman, P. D. 1988. The timing of reproduction in the fruit bat *Haplonycteris fischeri* (Pteropodidae): geographic variation and delayed development. J. Zool. 215:577–595.

Heinsohn, R. and C. Packer. 1995. Complex cooperative strategies in group-territorial African lions. Science. 269:1260–1262.

Heldmaier, G. 1971. Relationship between non-shivering thermogenesis and body size. Pp. 73–80 *in* Non-shivering thermogenesis (L. Jansky, ed.). Swets and Zeitlinger, Amsterdam.

Heller, H. C. and T. L. Poulson. 1970. Circannian rhythms. II. Endogenous and exogenous factors controlling reproduction and hibernation in chipmunks (*Eutamias*) and ground squirrels (*Spermophilus*). Comp. Biochem. Physiol. 33:357–383.

Hellgren, E. C., D. R. Synatzske, P. W. Oldenburg, and F. S. Guthery. 1995. Demography of a collared peccary population in south Texas. J. Wildl. Manage. 59:153–163.

Hellgren, E. C., M. R. Vaughan, R. L. Kirkpatrick, and P. F. Scanlon. 1990. Serial changes in metabolic correlates of hibernation in female black bears. J. Mammal. 71:291–300.

Hemmer, H. 1990. Domestication: the decline of environmental appreciation. Cambridge Univ. Press, Cambridge, England.

Hensel, R. J., W. A. Troyer, and A. W. Erickson. 1969. Reproduction in the female brown bear. J. Wild Mgmt. 33:357–365.

Henshaw, J. 1997. Status of the saola (*Pseudoryx nghetinhensis*). Oryx. 31:89–91.

Henshaw, R. E., L. S. Underwood, and T. M. Casey. 1972. Peripheral thermoregulation: foot temperature in two arctic canines. Science. 175:988–990.

Hensley, A. P., and K. T. Wilkins. 1988. Leptonycteris nivalis. Mamm. Species. 307:1–4.

Herbert, J. 1972. Initial observations on pinealectomized ferrets kept for long periods in either daylight or artificial illumination. J. Endocrinol. 55:591–597.

Herman, L. M. and W. N. Tavolga. 1980. The communication systems of cetaceans. Pp. 149–209 *in* Cetacean behavior (L. M. Herman, ed.). Wiley Interscience, New York.

Hershkovitz, P. 1962. Evolution of Neotropical cricitine rodents (Muridae) with special reference to the phyllotine group. Fieldiana. 46:1–524.

Hershkovitz, P. 1971. Basic crown patterns and cusp homologies of mammalian teeth. Pp. 95–150 *in* Dental morphology and evolution (A. A. Dahlberg, ed.). Univ. of Chicago Press, Chicago.

Heyning, J. E. 1989. Comparative facial anatomy of beaked whales (Ziiphidae) and a systematic revision among the families of extant Odontoceti. Nat. Hist. Mus. Los Angeles County No. 405:1–64.

Heyning, J. E. and J. G. Mead. 1996. Suction feeding in beaked whales: morphological and observational evidence. Nat. Hist. Mus. Los Angeles County No. 464:1–12.

Hickman, C. P., Jr., L. S. Roberts, and A. Larson. 1997. Integrated principles of zoology, 9th ed. Wm. C. Brown, Dubuque, IA.

Hildebrand, M. 1968. Anatomical preparations. Univ. of Calif. Press, Berkeley.

Hildebrand, M. 1980. The adaptive significance of tetrapod gait selection. Am. Zool. 20:255–267.

Hildebrand, M. 1985a. Walking and running. Pp. 89–109 *in* Functional vertebrate morphology (M. Hildebrand, D. M. Bramble, K. F. Liem, and D. B. Wake, eds.). Harvard Univ. Press, Cambridge.

Hildebrand, M. 1985b. Digging of quadrupeds. Pp. 38–57 *in* Functional vertebrate morphology (M. Hildebrand, D. M. Bramble, K. F. Liem, and D. B. Wake, eds.). Harvard Univ. Press, Cambridge.

Hildebrand, M. 1995. Analysis of vertebrate structure, 4th ed., John Wiley & Sons, New York.

Hildebrand, M., D. M. Bramble, K. F. Liem, and D. B. Wake (eds.). 1985. Functional vertebrate morphology. Harvard Univ. Press, Cambridge.

Hill, J. E. 1974. A new family, genus, and species of bat (Mammalia: Chiroptera) from Thailand. Bull. Br. Mus. Nat. Hist. Zool. 27:301–336.

Hill, J. E. and S. D. Smith. 1984. Bats: A natural history. British Museum, London.

Hill, J. E. and S. E. Smith. 1981. Craseonycteris thonglongyai. Mammal. Species No. 160:1–4.

Hill, R. W. 1983. Thermal physiology and energetics of *Peromyscus;* ontogeny, body

temperature, metabolism, insulation, and microclimatology. J. Mammal. 64:19–37.

Hill, R. W., D. P. Christian, and J. H. Veghte. 1980. Pinna temperature in exercising jackrabbits. J. Mammal. 61:30–38.

Hill, R. W. and G. A. Wyse. 1989. Animal physiology, 2d ed. Harper & Row, New York.

Himms-Hagen, J. 1985. Brown adipose tissue metabolism and thermogenesis. Ann. Rev. Nutrition. 5:69–94.

Hinds, D. S., and R. E. MacMillen. 1985. Scaling of energy metabolism and evaporative water loss in heteromyid rodents. Physiol. Zool. 58:282–298.

Hoffmesiter, D. F. 1969. The first fifty years of the American Society of Mammalogists. J. Mammal. 50:794–802.

Hogg, J. T. 1988. Copulatory tactics in relation to sperm competition in Rocky Mountain bighorn sheep. Behav. Ecol. Sociobiol. 22:49–59.

Holekamp, K. E. and P. W. Sherman. 1989. Why male ground squirrels disperse. Am. Sci. 77:232–239.

Holekamp, K. E., L. Smale, R. Berg, and S. M. Cooper. 1997. Hunting rates and hunting success in the spotted hyena (*Crocuta crocuta*). J. Zool. Lond. 242:1–15.

Holling, C. S. 1959. The components of predation as revealed by a study of small mammal predation of the European pine sawfly. Can. Entomol. 91:293–320.

Holmes, W. G. 1988. Kinship and development of social preferences. Pp. 389–413 *in* Developmental psychobiology and behavioral ecology (E. M. Blass, ed.). Plenum Press, New York.

Holmes, W. G. 1995. The ontogeny of littermate preferences in juvenile golden-mantled ground squirrels: effects of rearing and relatedness. Anim. Behav. 50:309–322.

Holmes, W. G. and P. W. Sherman. 1983. Kin recognition in animals. Am. Sci. 71:46–55.

Honeycutt, R. L. 1992. Naked mole-rats. Am. Sci. 80:43–53.

Hoogland, J. L. 1979a. Aggression, ectoparasitism, and other possible costs of prairie dog (Sciuridae: *Cynomys* spp.) coloniality. Behaviour 69:1–35.

Hoogland, J. L. 1979b. The effect of colony size on individual alertness of prairie dogs (Sciuridae: *Cynomys* spp.). Anim. Behav. 27:394–407.

Hoogland, J. L. 1982. Prairie dogs avoid extreme inbreeding. Science. 215:1639–1641.

Hopkins, D. R. 1992. Homing in on helminths. Am. J. Tropical Med. Hyg. 46:626–634.

Hopla, C. E. and A. K. Hopla. 1994. Tularemia. Pp. 113–126 *in* Handbook of zoonoses. Section A: bacterial, rickettsial, chlamydial, and mycotic diseases, 2d ed. (C. W. Beran and J. H. Steele, eds.). CRC Press, Boca Raton, FL.

Hopson, J. A. 1995. Patterns of evolution in the manus and pes of non-mammalian therapsids. J. Vertebr. Paleontol. 15:615–639.

Horner, B. E., J. M. Taylor, A. V. Linzey, and G. R. Michener. 1996. Women in mammalogy (1940–1994): personal perspectives. J. Mammal. 77:655–674.

Hoss, M., A. Dilling, A. Current, and S. Paabo. 1996. Molecular phylogeny of the extinct ground sloth *Mylodon darwinii.* Proceedings of the National Academy of Science. 93:181–185.

Hotton, N. III, P. D. MacLean, J. J. Roth, and E. C. Roth (eds.). 1986. The ecology and biology of mammal-like reptiles. Smithsonian Institution Press, Washington, D.C.

Hovland, N. and H. P. Andreassen. 1995. Fluorescent powder as dye in bait for studying foraging areas in small mammals. Acta Theriol. 40:315–320.

Hoyt, J. A. 1994. Animals in peril. Avery, Garden City Park, NY.

Hoyt, R. A. and R. J. Baker. 1980. Natalus major. Mammal. Species No. 130:1–3.

Hrdy, S. B. 1977a. Infanticide as a primate reproductive strategy. Am. Sci. 65:40–49.

Hrdy, S. B. 1977b. Langurs of Abu: female and male strategies of reproduction. Harvard Univ. Press, Cambridge, MA.

Hubbard, A. L., S. McOrist, T. W. Jones, R. Boid, R. Scott, and N. Easterbee. 1992. Is survival of European wild cats *Felis silvestris* in Britain threatened by interbreeding with domestic cats? Biol. Conserv. 61:203–208.

Huck, U. W., J. Seger, and R. D. Lisk. 1990. Litter sex ratios in the golden hamster vary with time of mating and litter size and are not binomially distributed. Behav. Ecol. Sociobiol. 26:99–109.

Hudson, J. W. 1973. Torpidity in mammals. Pp. 97–165 *in* Comparative physiology of thermoregulation (G. C. Whittow, ed.). Academic Press, New York.

Hudson, J. W. 1978. Shallow, daily torpor: a thermoregulatory adaptation. Pp. 67–108 *in* Strategies in cold: natural torpidity and thermogenesis (L. C. H. Wang and J. W. Hudson, eds.). Academic Press, New York.

Hudson, R. J., K. R. Drew, and L. M. Baskin. 1989. Wildlife production systems: economic utilization of wild ungulates. Cambridge Univ. Press, Cambridge, England.

Hudson, R. J. and R. G. White. 1985. Bioenergetics of wild herbivores. CRC Press, Boca Raton, FL.

Hughes, J. J., D. Ward, and M. R. Perrin. 1994. Predation risk and competition affect habitat selection and activity of Namib desert gerbils. Ecology. 75:1397–1405.

Hui, C. A. 1994. Lack of association between magnetic patterns and the distribution of free-ranging dolphins. J. Mammal. 75:399–405.

Hume, I .D. 1994. Gut morphology, body size and digestive performance in rodents. Pp. 315–323 *in* The digestive system in mammals: food, form and function (D. J. Chivers and P. Langer, eds.). Cambridge Univ. Press, Cambridge, England.

Huntly, N. and R. S. Inouye. 1987. Small mammal populations of an old-field chronosequence: successional patterns and associations with vegetation. J. Mammal. 68:739–745.

Hutchinson, G. E. 1957. Concluding remarks. Cold Spring Harbor Symp. Quant. Biol. 22:415–427.

Hutchinson, G. E. 1959. Homage to Santa Rosalia or why are there so many kinds of animals? Am. Nat. 93:145–159.

Hyvärinen, H. 1969. On the seasonal changes in the skeleton of the common shrew (*Sorex araneus* L.) and their physiological background. Aquilo Ser. Zool. 7:1–32.

Hyvärinen, H. 1994. Brown fat and the wintering of shrews. Pp. 259–266 *in* Advances in the biology of shrews (J. F. Merritt, G. L. Kikland, Jr., and R. K. Rose, eds.). Spec. Pub., Carnegie Mus. Nat. Hist. No. 18.

Ilse, L. M. and E. C. Hellgren. 1995. Spatial use and group dynamics of sympatric collared peccaries and feral hogs in southern Texas. J. Mammal. 76:993–1002.

Inns, R. W. 1982. Seasonal changes in the accessory reproductive system and plasma testosterone levels of the male tamar wallaby, *Macropus eugenii,* in the wild. J. Reprod. Fert. 66:675–680.

International Union for the Conservation of Nature. 1988. Red list of threatened animals. IUCN, Cambridge, England.

Irving, L. 1972. Arctic life of birds and mammals, including man. Springer-Verlag, Berlin.

Isaac, E. 1970. Geography of domestication. Prentice-Hall, Englewood Cliffs, NJ.

IUCN red data book, 3d ed. 1985. IUCN, Cambridge.

Jacobs, L. F. 1992. Memory for cache locations in Merriam's kangaroo rats. J. Mammal. 43:585–593.

Jacobs, L. L. 1980. Siwalik fossil tree shrews. Pp. 205–216 *in* Comparative biology and evolutionary relationships of tree shrews (W. P. Luckett, ed.). Plenum Press, New York.

Jacquot, J. J. and S. H. Vessey. 1994. Non-offspring nursing in the white-footed mouse, *Peromyscus leucopus.* Anim. Behav. 48:1238–1240.

James, W. H. 1996. Evidence that mammalian sex ratios at birth are partially controlled by parental hormone levels at the time of conception. J. Theor. Biol. 180:271–286.

Janis, C. M. and K. M. Scott. 1987. The interrelationships of higher ruminant families with special emphasis on the members of the Cervoidea. Am. Mus. Novitates. 2893:1–85.

Jansky, L. 1973. Non-shivering thermogenesis and its thermoregulatory significance. Biol. Rev. 48:85–132.

Janzen, D. H. 1985. Coevolution as a process: what parasites of animals and plants do not have in common. Pp. 83–99 *in* Coevolution of parasitic arthropods and mammals (K. C. Kim, ed.). John Wiley and Sons, New York.

Jarman, P. J. 1974. The social organization of antelope in relation to their ecology. Behaviour. 48:215–267.

Jarvis, J. U. M. 1981. Eusociality in a mammal: cooperative breeding in naked mole-rat colonies. Science. 212:571–573.

Jarvis, J. U. M., M. J. O'Riain, N. C. Bennett, and P. W. Sherman. 1994. Mammalian eusociality: a family affair. Trends Ecol. Evol. 9:47–51.

Jefferson, T. A., S. Leatherwood, and M. A. Webber. 1993. Marine mammals of the world. United Nations Environment Programme, FAO, Rome.

Jeffreys, A. J. 1987. Highly variable minisatellites and DNA fingerprints. Biochem. Soc. Trans. 15:309–317.

Jenkins, P. D. 1987. Catalogue of Primates in the British Museum (Natural History) and elsewhere in the British Isles. Part 4: Suborder Strepsirhini, including the subfossil Madagascan lemurs and family Tarsiidae. British Museum of Natural History, London.

Jenness, R. and E. H. Studier. 1976. Lactation and milk. Pp. 201–218 *in* Biology of bats of the New World family Phyllostomatidae, Part I. (R. J. Baker, J. K. Jones, Jr., and D. C. Carter, eds.). Spec. Pub., Mus. Texas Tech Univ., No. 10.

Jepson, G. L. 1966. Early Eocene bat from Wyoming. Science. 154:1333–1339.

Jepson, G. L. 1970. Bat origins and evolution. Pp. 1–64 *in* Biology of bats, vol. I (W. A. Wimsatt, ed.). Academic Press, New York.

Jerison, H. J. 1973. Evolution of the brain and intelligence. Academic Press, New York.

Jike, L., G. O. Batzli, and L. L. Getz. 1988. Home ranges of prairie voles as determined by radiotracking and by powdertracking. J. Mammal. 69:183–186.

Jiménez, J. A., K. A. Hughes, G. Alaks, L. Graham, and R. C. Lacy. 1994. An experimental study of inbreeding depression in a natural habitat. Science. 266:271–273.

Jinling, L., B. S. Rubidge, and C. Zhengwu. 1996. A primitive anteosaurid dinocephalian from China: implications for the distribution of earliest therapsid faunas. S. Afr. J. Sci. 92:252–253.

Johnson, C. N. 1986. Philopatry, reproductive success of females and maternal investment in the red-necked wallaby. Behav. Ecol. Sociobiol. 19:143–150.

Johnson, C. N. 1994. Mycophagy and spore dispersal by a rat-kangaroo: consumption of ectomycorrhizal taxa in relation to their abundance. Funct. Ecol. 8:464–468.

Johnson, E. 1984. Seasonal adaptive coat changes in mammals. Acta Zool. Fennica. 171:7–12.

Johnson, E. and J. Hornby. 1980. Age and seasonal coat changes in long haired and normal fallow deer (*Dama dama*). J. Zool. 192:501–509.

Johnson, K. A. 1995. Marsupial mole, Notoryctes typhlops. Pp. 409–411 *in* The mammals of Australia (R. Strahan, ed.). Reed Books, Chatswood, Australia.

Jones, F. W. 1968. The mammals of South Australia, parts I–III. Government Printer, Adelaide.

Jones, G. 1990. Prey selection by the greater horseshoe bat (*Rhinolophus ferrumequinum*): optimal foraging by echolocation. J. Anim. Ecol. 59:587–602.

Jones, J. K., Jr., D. M. Armstrong, and J. R. Choate. 1985. Guide to mammals of the Plains States. Univ. of Nebraska Press, Lincoln.

Jones, J. K. Jr. and D. C. Carter. 1979. Systematic and distributional notes. Pp. 7–11 *in* Biology of bats of the New World family Phyllostomatidae (R. J. Baker, J. K. Jones, Jr., and D. C. Carter, eds.), part III. Spec. Publ. No. 16, Museum of Texas Tech Univ., Lubbock, TX.

Jones, W. T. 1987. Dispersal patterns in kangaroo rats (*Dipodomys spectabilis*). Pp. 119–127 *in* Mammalian dispersal patterns. The effects of social structure on population genetics (B. D. Chepko-Sade and Z. T. Halpin, eds.). Univ. of Chicago Press, Chicago.

Joslin, J. K. 1977a. Rodent long distance orientation ("homing"). Adv. Ecol. Res. 10:63–89.

Joslin, J. K. 1977b. Visual cues used in orientation by white-footed mice, *Peromyscus leucopus:* a laboratory study. Am. Mid. Nat. 98:308–318.

Jung, K. Y., S. Crovella, and Y. Rumpler. 1992. Phylogenetic relationships among lemuriform species determined from restriction genomic DNA banding patterns. Folia Primat. 58:224–229.

Kalcounis, M. C. and R. M. Brigham. 1995. Intraspecific variation in wing loading affects habitat use by little brown bats (*Myotis lucifugus*). Can. J. Zool. 73:89–95.

Kam, M. and A.A. Degen. 1992. Effect of air temperature on energy and water balance of *Psammomys obesus*. J. Mammal. 73: 207–214.

Kantongol, C. B., F. Naftolin, and R. V. Short. 1971. Relationship between blood levels of luteinizing hormone, testosterone in bulls, and the effects of sexual stimulation. J. Endocrinol. 50:457–465.

Kaufman, D. M., D. W. Kaufman, and G. A. Kaufman. 1996. Women in the early years of the American Society of Mammalogists (1919–1949). J. Mammal. 77:642–654.

Kaufmann, D. M. 1995. Diversity of New World mammals: universality of the latitudinal gradients of species and bauplans. J. Mammal. 76:322–334.

Kay, R. F., C. Ross, and B. A. Williams. 1997. Anthropoid origins. Science. 275:797–804.

Kayser, C. 1965. Hibernation. Pp. 180–278 *in* Physiological mammalogy, vol. 2., (W. V. Mayer and R. G. van Gelder, eds.). Academic Press, New York.

Keane, B. 1990. The effect of relatedness on reproductive success and mate choice in the white-footed mice *Peromyscus leucopus*. Anim. Behav. 39:264–273.

Keith, L. B. 1963. Wildlife's ten-year cycle. Univ. of Wisconsin Press, Madison.

Keith, L. B. 1983. Role of food in hare populations. Oikos. 40:385–395.

Keith, L. B. 1987. Dynamics of snowshoe hare populations. Curr. Mammal. 2:119–195.

Kemp, N., M. Dilger, N. Burgess, and C. VanDung. 1997. The saola (*Pseudoryx nghetinhensis*) in Vietnam: new information on distribution and habitat preferences, and conservation needs. Oryx. 31:37–44.

Kemp, S., A. C. Hardy, and N. A. Mackintosh. 1929. Discovery investigations. Objects, equipment and methods. Discovery Rep. 1:141–232.

Kenagy, G. J. 1972. Saltbush leaves: excision of hypersaline tissue by a kangaroo rat. Science. 178:1094–1096.

Kenagy, G. J. and G. A. Bartholomew. 1985. Seasonal reproductive patterns in five coexisting California desert rodent species. Ecol. Monogr. 55:371–397.

Kenagy, G. J. and D. F. Hoyt. 1980. Reingestion of feces in rodents and its daily rhythmicity. Oecologia. 44:403–409.

Kenagy, G. J. and D. F. Hoyt. 1989. Speed and time-energy budget for locomotion in golden-mantled ground squirrels. Ecology. 70:1834–1839.

Kenagy, G. J., D. Masman, S. M. Sharbaugh, and K. A. Nagy. 1990. Energy expenditure during lactation in relation to litter size in free-living golden-mantled ground squirrels. J. Anim. Ecol. 59:73–88.

Kenagy, G. J., S. M. Sharbaugh, and K. A. Nagy. 1989a. Annual cycle of energy and time expenditure in a golden-mantled ground squirrel population. Oecologia. 78:269–282.

Kenagy, G. J., R. D. Stevenson, and D. Masman. 1989b. Energy requirements for lactation and postnatal growth in captive golden-mantled ground squirrels. Physiol. Zool. 62:470–487.

Kenagy, G. J. and S. C. Trombulak. 1986. Size and function of mammalian testes in relation to body size. J. Mammal. 67:1–22.

Kennelly, J. J. and B. E. Johns. 1976. The estrous cycle of coyotes. J. Wildl. Mgmt. 40:272–277.

Kenward, R. 1987. Wildlife radio tagging. Academic Press, San Diego.

Kermack, D. M. and K. A. Kermack. 1984. The evolution of mammalian characters. Kapitan Szabo Publishing, Washington, D.C.

Keys, P. L. and M. C. Wiltbank. 1988. Endocrine regulation of the corpus luteum. Ann. Rev. Physiol. 50:465–482.

Khateeb, A. and E. Johnson. 1971. Seasonal changes of pelage in the vole (*Microtus agrestis*), I. Correlation with changes in the endocrine glands. Gen. Comp. Endocrinol.16:217–228.

Kim, K. C. 1985a. Parasitism and coevolution: epilogue. Pp. 661–682 *in* Coevolution of parasitic arthropods and mammals (K. C. Kim, ed.). John Wiley and Sons, New York.

Kim, K. C. 1985b. Evolutionary relationships of parasitic arthropods and mammals. Pp. 3–82 *in* Coevolution of parasitic arthropods and mammals (K. C. Kim, ed.). John Wiley and Sons, New York.

Kimura, K. A. and T. A. Uchida. 1983. Ultrastructural observations of delayed implantation in the Japanese long-fingered bat, *Miniopterus schreibersii fulginosis.* J. Reprod. Fertil. 69:187–193.

King, C. 1983. Mustela erminea. Mammal. Spec. 195:1–8.

King, C. 1984 The origin and adaptive advantages of delayed implantation in *Mustela erminea.* Oikos. 42:126–128.

King, C. M. 1989. The advantages and disadvantages of small size to weasels, *Mustela* species. Pp. 302–334 *in* Carnivore behavior, ecology, and evolution (J. L. Gittleman, ed.). Comstock Publ. Assoc., Ithaca, NY.

King, C. M. 1990. The natural history of weasels and stoats. Comstock Publ. Assoc., Ithaca, NY.

King, J. A., D. Maas, and R. G. Weisman. 1964. Geographic variation in nest size among species of *Peromyscus*. Evolution. 18:230–234.

King, J. E. 1983. Seals of the world, 2d ed. British Museum of Natural History, London.

Kingdon, J. 1971. East African mammals, vol. I. Academic Press, New York.

Kingdon, J. 1972. East African mammals: an atlas of evolution in Africa. vol. IIA. Univ. of Chicago Press, Chicago.

Kingdon, J. 1972. East African mammals: carnivores. Vol. IIIA. Univ. of Chicago Press, Chicago.

Kingdon, J. 1974. East African mammals, vol. I. Univ. of Chicago Press, Chicago.

Kingdon, J. 1979. East African mammals, vol. IIIB, large mammals. Univ. of Chicago Press, Chicago.

Kirby, L. T. 1990. DNA fingerprinting. Stockton Press, New York.

Kirkpatrick, R. L. 1980. Physiological indices in wildlife management. Pp. 99–127 *in* Wildlife management techniques manual, 4th ed. (S. D. Schemnitz, ed.). Wildlife Society, Washington, D.C.

Kirsch, J. A. W. 1977. The comparative serology of Marsupialia, and a classification of marsupials. Aust. J. Zool., Suppl. Ser. 52:1–152.

Kirsch, J. A. W., M. S. Springer, C. Krajewski, M. Archer, K. Aplin, and A. W. Dickerman. 1990. DNA/DNA hybridization studies of the carnivorous marsupials. I: The intergeneric relationships of bandicoots (Marsupialia: Perameloidea). J. Mol. Evol. 30:434–448.

Kitchen, D. W. and B. W. O'Gara. 1982. Pronghorn, *Antilocapra americana*. Pp. 960–971*in*Wild mammals of North America: biology, management and economics (J. A. Chapman and G. A. Feldhamer, eds.). Johns Hopkins Univ. Press, Baltimore.

Kleiber, M. 1932. Body size and metabolism. Hilgardia. 6:315–353.

Kleiber, M. 1961. The fire of life. John Wiley & Sons, New York.

Kleiman, D. G. 1976. International conference on the biology and conservation of the Callitrichidae. Primates. 17:119–123.

Kleiman, D. G. 1977. Monogamy in mammals. Qt. Rev. Biol. 52:39–69.

Kleiman, D. G., M. E. Allen, K. V. Thompson, and S. Lumpkin (eds.). 1996. Wild animals In captivity. Univ. of Chicago Press, Chicago.

Kleiman, D. G., B. B. Beck, A. J. Baker, J. D. Ballou, L. Dietz, and J. M. Dietz. 1990. The conservation program for the golden-lion tamarin, *Leontopithecus rosalia.* Endangered Species Update. 8:82–84.

Kleinebeckel, D. and F. W. Klussmann. 1990. Shivering. Pp. 235–253 *in* Thermoregulation: physiology and biochemistry (E. Schanbaum and P. Lomax, eds.), Pergamon Press, New York.

Kliman, R. M. and G. R. Lynch. 1992. Evidence for genetic variation in the occurrence of the photoresponse of the Djungarian hamster, *Phodopus sungorus.* J. Biol. Rhythms. 7:161–173.

Klinowska, M. 1985. Cetacean stranding sites relate to geomagnetic topography. Aquatic Mammals. 1:27–32.

Kodric-Brown, A. and J. H. Brown. 1984. Truth in advertising: the kinds of traits favored by sexual selection. Am. Nat. 124:309–323.

Koehler, C. E. and P. R. K. Richardson. 1990. Proteles cristatus. Mamm. Species. 363:1–6.

Koeln, G. T., L. M. Cowardin, and L. L. Strong. 1996. Geographic Information Systems. Pp. 540–566 *in* Research and management techniques for wildlife and habitats, 5th ed., rev. (T. A. Bookhout, ed.). Wildlife Society, Bethesda, MD.

Kohler-Rollefson, I. U. 1991. Camelus dromedarius. Mamm. Species. 1–8 .

Koop, B. F., M. Goodman, P. Zu, J. L. Chan, and J. L. Slighton. 1986. Primate gamma-globulin DNA sequences and man's place among the great apes. Nature. 319:234–238.

Koop, B. F., D. A. Tagle, M. Goodman, and J. L. Slighton. 1989. A molecular view of primate phylogeny and important systematic and evolutionary questions. Mol. Biol. Evol. 6:580–612.

Koopman, K. F. 1984. Bats. Pp. 145–186 *in* Orders and families of recent mammals of the world (S. Anderson and J. K. Jones, Jr., eds.). John Wiley and Sons, New York.

Koopman, K. F. 1993. Order Chiroptera. Pp. 137–241 *in* Mammal species of the world: a taxonomic and geographic reference, 2d ed. (D. E. Wilson and D. M. Reeder, eds.). Smithsonian Institution Press, Washington, D. C.

Kooyman, G. L. 1963. Milk analysis of the kangaroo rat, *Dipodomys merriami.* Science. 147:1467–1468.

Kooyman, G. L. 1981. Weddell seal: consummate diver. Cambridge Univ. Press, London.

Kooyman, G. L., R. L. Gentry, and D. L. Urquhart. 1976. Northern fur seal diving behavior: a new approach to its study. Science. 193:411–413.

Koteja, P. 1996. The usefulness of a new TOBEC instrument (ACAN) for investigating body composition in small mammals. Acta Theriol. 41:107–112.

Kotler, B. P. 1989. Temporal variation in the structure of a desert rodent community. Pp. 127–139 *in* Patterns in the structure of mammalian communities (D. W. Morris, Z. Abramsky, B. J. Fox and M. R. Willig, eds.). Texas Tech Univ. Press, Lubbock, TX.

Kowalski, K. 1976. Mammals: an outline of theriology. Panstwowe Wydawnictwo Naukowe, Warsaw (National Technical Information Service translation for the Smithsonian Institution and the National Science Foundation).

Krajewski, C., A. C. Driskell, P. R. Baverstock, and M. J. Braun. 1992. Phylogenetic relationships of the thylacine (Mammalia: Thylacinidae) among dasyuroid marsupials: evidence from cytochrome b DNA sequences. Proceedings of the Royal Society of London, Series B, 250:19–27.

Krapp, F. 1965. Schädel under Kaumuskulatur von Spalax leucodon (Nordmann, 1840). Z. Wissensch. Zool. 173:1–71.

Krebs, C. J. 1963. Lemming cycle at Baker Lake, Canada during 1959–62. Science. 140:674–676.

Krebs, C. J. 1989. Ecological methodology. Harper and Row, New York.

Krebs, C. J. 1994. Ecology: the experimental analysis of distribution and abundance, 4th ed. HarperCollins, New York.

Krebs, C. J. 1996. Population cycles revisited. J. Mammal. 77:8–24.

Krebs, C. J., S. Boutin, R. Boonstra, A. R. E. Sinclair, J. N. M. Smith, M. R. T. Dale, and R. Turkington. 1995. Impact of food and predation on the snowshoe hare cycle. Science. 269:1112–1115.

Krebs, J. R. and R. Dawkins. 1984. Animal signals: mind-reading and manipulation. Pp. 380–402 *in* Behavioural ecology: an evolutionary approach (J. R. Krebs and N. B. Davies, eds.). Sinauer Assoc., Sunderland, MA.

Krebs, C. J., M. S. Gaines, B. L. Keller, J. H. Myers, and R. H. Tamarin. 1973. Population cycles in small rodents. Science. 179:34–41.

Krebs, J. W., T. W. Strine, J. S. Smith, C. E. Rupprecht, and J. E. Childs. 1994. Rabies surveillance in the United States during 1993. J. Am. Vet. Med. Assoc. 205:1695–1709.

Krebs, J. W., M. L. Wilson, and J. E. Childs. 1995. Rabies-epidemiology, prevention, and future research. J. Mammal. 76:681–694.

Kripke, D. F. 1974. Ultradian rhythms in sleep and wakefulness. Adv. Sleep Res. 1:305–325.

Kruuk, H. 1972. The spotted hyena: a study of predation and social behavior. Univ. of Chicago Press, Chicago.

Kruuk, H. 1986. Interactions between Felidae and their prey species: a review. Pp. 353–374 *in* Cats of the world: biology, conservation, and management (S. D. Miller and D. D. Everett, eds.). National Wildlife Federation, Washington, D.C.

Kruuk, H. and W. A. Sands. 1972. The aardwolf (*Proteles cristatus* Sparrman) 1783 as a predator of termites. E. Afr. Wildl. J. 10:211–227.

Kulzer, E. 1969. Das Verhalten von *Eidolon helvum* (Kerr) in Gefangenschaft. Z. Säugetierk. 34:129–148.

Kummer, H. 1968. Social organization of hamadryas baboons. A field study. Univ. of Chicago Press, Chicago.

Kunz, T. H. 1981. Ecology of bats. Plenum Press, New York.

Kunz, T. H.(ed.). 1982. Ecology of bats. Plenum Press, New York.

Kunz, T. H. 1988. Ecological and behavioral methods for the study of bats. Smithsonian Institution Press, Washington, D.C.

Kunz, T. H. and C. A. Diaz. 1995. Folivory in fruit-eating bats, with new evidence from *Artibeus jamaicensis* (Chiroptera: Phyllostomidae). Biotropica. 27:106–120.

Kunz, T. H., M. S. Fujita, A. P. Brooke, and G. F. McCracken. 1994. Convergence in tent architecture and tent-making behavior among neotropical and paleotropical bats. J. Mammal. Evol. 2:57–78.

Kunz, T. H. and K. A. Ingalls. 1994. Folivory in bats: an adaptation derived from frugivory. Funct. Ecol. 8:665–668.

Kunz, T. H. and G. F. McCracken. 1996. Tents and harems: apparent defence of foliage roosts by tent-making bats. J. Tropical Ecol. 12:121–137.

Kunz, T. H. and E. D. Pierson. 1994. Bats of the world: an introduction. Pp.1–46 *in* Walker's bats of the world (R. Nowak, ed.). Johns Hopkins Univ. Press, Baltimore.

Kunz, T. H. and S. K. Robson. 1996. Postnatal growth and development in the Mexican free-tailed bat (*Tadarida brasiliensis mexicana*): birth size, growth rates, and age estimation. J. Mammal. 76:769–783.

Kunz, T. H. and A. A. Stern. 1995. Maternal investment and post-natal growth in bats. Pp. 123–138 *in* Ecology, evolution and behaviour of bats (P. A. Racey and S. M. Swift, eds.). Proceedings of the 67th Symposium of the zoological society of London, November 26–27, 1993.

Kunz, T. H., J. O. Whitaker, Jr., and M. D. Wadanoli. 1995. Dietary energetics of the insectivorous Mexican free-tailed bat (*Tadarida brasiliensis*) during pregnancy and lactation. Oecologia. 101:407–415.

Kurta, A. 1995. Mammals of the Great Lakes region. Univ. of Michigan Press, Ann Arbor.

Kurta, A., G. P. Bell, K. A. Nagy, and T. H. Kunz. 1990. Energetics and water flux of free-ranging big brown bats (*Eptesicus fuscus*) during pregnancy and lactation. J. Mammal. 71:59–65.

Kurta, A. and T. H. Kunz. 1987. Size of bats at birth and maternal investment during pregnancy. Symp. Zool. Soc. Lond. 57:79–106.

Kurten, B. 1969. Continental drift and evolution. Sci. Am. 220:54–63.

Kurten, L. and U. Schmidt. 1982. Thermoreception in the common vampire bat (*Desmodus rotundus*). J. Comp. Physiol. 146:223–228.

Lambertsen, R., N. Ulrich, and J. Straley. 1995. Frontomandibular stay of Balaenopteridae: a mechanism for momentum recapture during feeding. J. Mammal. 76:877–899.

Lancia, R. A., J. D. Nichols, and K. H. Pollock. 1996. Pp. 215–253 in Research and management techniques for wildlife and habitats, 5th ed., rev. (T. A. Bookhout, ed.). Wildlife Society, Bethesda, MD.

Lane, R. S., J. Piesman, and W. Burgdorfer. 1991. Lyme borreliosis: relation of its causative agent to its vectors and hosts in North America and Europe. Annu. Rev. Entomol. 36:587–609.

Lane-Petter, W. 1978. Identification of laboratory animals. Pp. 35–40 in Animal marking, (B. Stonehouse, ed.). University Park Press, Baltimore.

Langbauer, W. R., Jr., K. B. Payne, R. A. Charif, L. Rappaport, and F. Osborn. 1991. African elephants respond to distant playbacks of low-frequency conspecific calls. J. Exp. Biol. 157:35–46.

Langer, P. 1996. Comparative anatomy of the stomach of the Cetacea: ontogenetic changes involving gastric proportions-mesentaries-arteries. Z. Säugetierk. 61:140–154.

Langevin, P. and R. M. R. Barclay. 1991. Hypsignathus monstrosus. Mammal. Species No. 357:1–4

Lanyon, L. E. 1981. Locomotor loading and functional adaptations in limb bones. Symp. Zool. Soc. London. 48:305–330.

Lavigne, D. M., C. D. Bernholz, and K. Ronald. 1977. Functional aspects of pinniped vision. Pp. 135–173 in Functional anatomy of marine mammals, vol. III (R. J. Harrison, ed.). Academic Press, London.

Lavigne, D. M. and K. M. Kovacs. 1988. Harps and hoods: ice-breeding seals of the Northwest Atlantic. Univ. of Waterloo Press, Ontario.

Law, B. S. 1992. Physiological factors affecting pollen use by the Queensland blossom bat (*Syconycteris australis*). Funct. Ecol. 6:257–264.

Law, B. S. 1993. Roosting and foraging ecology of the Queensland blossom-bat (*Syconycteris australis*) in northeastern New South Wales: flexibility in response to seasonal variation. Wildl. Res. 20:419–431.

Lawlor, T. E. 1973. Aerodynamic characteristics of some Neotropical bats. J. Mammal. 54:71–78.

Lawlor, T. E. 1986. Comparative biogeography of mammals on islands. Biol. J. Linnean Soc. 28:99–125.

Lawrence, B. 1945. Brief comparison of short-tailed shrew and reptile poison. J. Mammal. 26:393–396.

Laws, R. M. 1967. Occurrence of placental scars in the uterus of the African elephant (*Loxodonta africanus*). J. Reprod. Fert. 14:445–449.

Laws, R. M. 1970. Elephants as agents of habitat and landscape change in East Africa. Oikos. 21:1–15.

Laws, R. M. 1974. Behaviour, dynamics and management of elephant populations. Pp. 513–529 in The behaviour of ungulates and its relation to management (V. Geist and F. Walther, eds.). IUCN Publ. new series No. 24, Morges, Switzerland.

Layne, J. N. 1969. Nest-building behavior in three species of deer mice, *Peromyscus*. Behaviour. 35:288–303.

Leach, W. J. 1961. Functional anatomy: mammalian and comparative, 3d ed. McGraw-Hill, New York.

Lean, G., D. Hinrichsen and A. Markham. 1990. Atlas of the environment. Prentice-Hall, New York.

Leatherwood, S., R. Reeves, and L. Foster. 1983. Sierra Club handbook of whales and dolphins. Sierra Club Books, San Francisco.

Le Boeuf, B. J. 1974. Male-male competition and reproductive success in elephant seals. Am. Zool. 14:163–176.

Le Boeuf, B. J. and J. Reiter. 1988. Lifetime reproductive success in northern elephant seals. Pp. 344–362 in Reproductive success (T. H. Clutton-Brock, ed.). Univ. of Chicago Press, Chicago.

Lee, A. K. and A. Cockburn. 1985. Evolutionary ecology of marsupials. Cambridge Univ. Press, New York.

Lehner, P. N. 1996. Handbook of ethological methods, 2d ed. Cambridge Univ. Press, New York.

Lemen, C. A. and P. W. Freeman. 1985. Tracking mammals with fluorescent pigments: a new technique. J. Mammal. 66:134–136.

Leopold, A. 1966. A Sand County almanac. Oxford Univ. Press, New York:.

Leslie, P. H. and R. M. Ranson. 1940. The mortality, fertility and rate of natural increase of the vole (*Microtus agrestis*) as observed in the laboratory. J. Anim. Ecol. 9:27–52.

Leutenegger, W. 1973. Maternal-fetal weight relationships in primates. Folia Primat. 20:280–293.

Leutenegger, W. 1976. Metric variability in the anterior dentition of African colobines. Am. J. Phys. Anthropol. 45:45–52.

Lev, B., H. Ward, G. T. Keusch, and M. E. A. Perreira. 1986. Lectin activation in *Giardia lamblia* by host protease: a novel host-parasite interaction. Science. 232:71–73.

Lewis, M. and W. Clark. (E. Coues, ed.). 1979. The history of the Lewis and Clark expedition, 3 vols. Dover Publications, New York.

Lewis, P. O. and A. A. Snow. 1992. Deterministic paternity exclusion using RAPD markers. Mol. Ecol. 1:155–160.

Lewis, R. E. 1993. Fleas (Siphonaptera). Pp. 529–575 in Medical insects and arachnids (R. P. Lane and R. W. Crosskey, eds.). Chapman and Hall, London.

Lewontin, R. C. and L. C. Dunn. 1960. The evolutionary dynamics of a polymorphism in the house mouse. Genetics. 45:705–722.

Ley, W. 1968. Dawn of zoology. Prentice-Hall, Englewood Cliffs, NJ.

Lidicker, W. Z., Jr. 1973. Regulation of numbers in an island population of the California vole, a problem in community dynamics. Ecol. Monogr. 43:271–302.

Lidicker, W. Z., Jr. 1988. Solving the enigma of microtine "cycles." J. Mammal. 69:225–235.

Lidicker, W. Z., Jr. 1995. The landscape concept: something old, something new. Pp. 3–19 in Landscape approaches in mammalian ecology and conservation (W. Z. Lidicker, Jr., ed.). Univ. of Minnesota Press, Minneapolis.

Liebig, J. 1847. Chemistry application to agriculture and physiology, 4th ed. Taylor and Walton, London.

Lillegraven, J. A. 1975. Biological considerations of the marsupial-placental dichotomy. Evolution. 29:707–722.

Lillegraven, J. A. 1979. Introduction. Pp. 1–6 in Mesozoic mammals: the first two-thirds of mammalian history (J. A. Lillegraven, Z. Kielan-Jaworowska, and W. A. Clemens, eds.). Univ. of California Press, Berkeley.

Lillegraven, J. A. 1987. The origin of eutherian mammals. Biol. J. Linn. Soc. 32:281–336.

Lillywhite, H. B. and B. R. Stein. 1987. Surface sculpturing and water retention of elephant skin. J. Zool. Lond. 211:727–734.

Lindstedt, S. L 1980. Energetics and water economy of the smallest desert mammal. Physiol. Zool. 53:82–97.

Lindstedt, S. L. and M. C. Boyce. 1985. Seasonality, fasting endurance, and body size in mammals. Am. Nat. 125:873–878.

Linn, I. J. 1978. Radioactive techniques for small mammal marking, Pp. 177–191 in Animal marking (B. Stonehouse, ed.). University Park Press, Baltimore.

Lombardi, J. R. and J. G. Vandenbergh. 1977. Pheromonally induced sexual maturation in females: regulation by the social environment of the male. Science. 196:545–546.

Lomolino, M. V. 1986. Mammalian community structure on islands: the importance of immigration, extinction and interactive effects. Biol. J. Linn. Soc. 28:1–21.

Longland, W. S. and C. Clements. 1995. Use of fluorescent pigments in studies of seed caching by rodents. J. Mammal. 76:1260–1266.

Lopez, A., P. Miranda, E. Tejada, and D. B. Fishbein. 1992. Outbreak of human rabies in the Peruvian jungle. Lancet. 339:408–411.

Louwman, J. W. W. 1973. Breeding the tailless tenrec *Tenrec ecaudatus* at Wassenaar Zoo. Intern. Zoo Yearb. 13:125–126.

Low, B. S. 1990. Marriage systems and pathogen stress in human societies. Am. Zool. 30:325–339.

Lowe, V. P. W. 1969. Population dynamics of the red deer (*Cervus elaphus* L.) on Rhum. J. Anim. Ecol. 38:425–457.

Lowery, G. H., Jr. 1974. The mammals of Louisiana and its adjacent waters. Louisiana State Univ. Press, Baton Rouge.

Luckett, W. P. 1975. Ontogeny of the fetal membranes and placenta: their bearing on primate phylogeny. Pp. 157–182 *in* Phylogeny of the primates (W. P. Luckett and F. S. Szalay, eds.). Plenum Press, New York.

Luckett, W. P. 1980. Monophyletic or diphyletic origins of Anthropoidea and Hystriscognathi: evidence of the fetal membranes. Pp. 347–368 *in* Evolutionary biology of the New World monkeys and continental drift (R. L. Ciochon and A. B. Chiarelli, eds.). Plenum Press, New York.

Luckett, W. P. (ed.). 1980. Comparative biology and evolutionary relationships of tree shrews. Plenum Press, New York.

Luckett, W. P. 1994. Suprafamilial relationships within Marsupialia: resolution and discordance from multidisciplinary data. J. Mammal. Evol. 2:225–283.

Luckett, W. P. and F. S. Szalay. 1975. Phylogeny of the primates. Plenum Press, New York.

Luckett, W. P. and P. A. Woolley. 1996. Ontogeny and homology of the dentition in dasyurid marsupials: development in *Sminthopsis virginiae*. J. Mammal. Evol. 3:327–364.

Lydekker, R. A. 1896. A geographical history of mammals. Cambridge Univ. Press, Cambridge, England.

Lyman, C. P., J. S. Willis, A. Malan, and L. C. H. Wang. 1982. Hibernation and torpor in mammals and birds. Academic Press, New York.

Lynch, G. R. 1973. Seasonal change in the thermogenesis, organ weights, and body composition in the white-footed mouse, *Peromyscus leucopus*. Oecologia. 13:363–367.

Lynch, J. D. 1988. Refugia. Pp. 301–342 *in* Analytical biogeography, (A. A. Myers and P. S. Giller, eds.). Chapman and Hall, New York.

Lynch, M. and B. Milligan. 1994. Analysis of population genetic structure with RAPD markers. Mol. Ecol. 3:91–100.

M'Closkey, R. T. 1975. Habitat dimensions of white-footed mice, *Peromyscus leucopus*. Am. Midl. Nat. 93:158–167.

M'Closkey, R. T. 1978. Niche separation and assembly in four species of Sonoran Desert rodents. Amer. Nat. 112:683–694.

MacArthur, R. A. and M. Aleksiuk. 1979. Seasonal microenvironments of the muskrat (*Ondatra zibethicus*) in a northern marsh. J. Mammal. 60:146–154.

MacArthur, R. H. and E. R. Pianka. 1966. On the optimal use of a patchy environment. Am. Nat. 100:603–609.

MacArthur, R. H. and E. O. Wilson. 1967. The theory of island biogeography. Princeton Univ. Press, Princeton, NJ.

Macdonald, D. 1976. Food caching by red foxes and some other carnivores. Z. Tierphysiol. 42:170–185.

Macdonald, D. 1984. The encyclopedia of mammals. Facts on File Publications, New York.

Macdonald, D. (ed.) 1984. The encyclopedia of mammals. Facts on File Publications, New York.

MacLulich, D. A. 1957. The place of chance in population processes. J. Wildl. Manage. 21:293–299.

MacMillen, R. E. 1965. Aestivation in the cactus mouse, *Peromyscus eremicus*. Comp. Biochem. Physiol. 16:227–248.

MacMillen, R. E., and T. Garland, Jr. 1989. Adaptive physiology. Pp. 143–168 *in* Advances in the study of *Peromsycus* (G. L. Kirkland, Jr., and J. N. Layne, eds.). Texas Tech Univ. Press, Lubbock, TX.

MacMillen, R. E. and A. K. Lee. 1969. Water metabolism of Australian hopping mice. Comp. Biochem. Physiol. 28:493–514.

MacMillen, R. E. and A. K. Lee. 1970. Energy metabolism and pulmocutaneous water loss of Australian hopping mice. Comp. Biochem. Physiol., 35:355–369.

Madison, D. M. 1985. Activity rhythms and spacing. Pp. 373–419 *in* Biology of New World *Microtus* (R. H. Tamarin, ed.). Am. Soc. Mammal. Spec. Pub. No. 8.

Madison, D. M., R. W. FitzGerald, and W. J. McShea. 1984a. Dynamics of social nesting in overwintering meadow voles (*Microtus pennsylvanicus*): possible consequences for population cycling. Behav. Ecol. Sociobiol. 15:9–17.

Madison, D. M., J. P. Hill, and P. E. Gleason. 1984b. Seasonality in the nesting behavior of *Peromyscus leucopus*. Am. Midl. Nat. 112:201–204.

Magnanini, A., A. F. Coimbra-Filho, R. A. Mittermeier, and A. Aldright. 1975. The Tijuca Bank of lion marmosets, *Leontopithecus rosalia*: a progress report. Int. Zoo Yearb. 15:284–287.

Maguire, J. H. and R. Hoff. 1992. Trypanosoma. Pp. 1984–1991 *in* Infectious diseases (S. L. Gorbach, J. G. Bartlett, and N. R. Blacklow, eds.). W. B. Saunders, Philadelphia.

Mahboubi, M., R. Ameur, J. Y. Crochet, and J. J. Jaeger. 1984. Earliest known proboscidean from early Eocene of northwest Africa. Nature. 308:543–544.

Maher, C. R. 1991. Activity budgets and mating system of male pronghorn antelope at Sheldon National Wildlife Refuge, Nevada. J. Mammal. 72:739–744.

Mahoney, R. 1966. Techniques for the preparation of vertebrate skeletons. Pp. 327–351 *in* Laboratory techniques in zoology. Butterworth, Washington, D.C.

Maier, W., J. Van Den Heever, and F. Durand. 1996. New therapsid specimens and the origin of the secondary hard and soft palate of mammals. J. Zool. Syst. Evol. Res. 34:9–19.

Malmquist, M. G. 1985. Character displacement and biogeography of the pygmy shrew in Northern Europe. Ecology. 66:372–377.

Malthus, R. T. 1798. An essay on the principle of population as it affects the future improvement of society. Johnson, London.

Mammal collections in the Western Hemisphere. American Society of Mammalogists, Provo, UT.

Manger, P. R. and J. D. Pettigrew. 1995. Electroreception and the feeding behaviour of platypus (*Ornithorhynchus anatinus*: Monotremata: Mammalia). Philosophical Transactions of the Royal Society of London, Series B, 347:359–381.

Manning, J. T. and A. T. Chamberlain. 1993. Fluctuating asymmetry, sexual selection and canine teeth in primates. Proceedings of the Royal Society of London B. 251:83–87.

Manning, J. T. and A. T. Chamberlain. 1994. Fluctuating asymmetry in gorilla canines: a sensitive indicator of environmental stress. Proceedings of the Royal Society of London B. 255:189–193.

Mansergh, I. M. and L. S. Broome. 1994. The mountain pygmy-possum of the Australian Alps. New South Wales Univ. Press, Sydney.

Marchand, P. J. 1996. Life in the cold: an introduction to winter ecology, 3d ed. Univ. Press of New England, Hanover, NH.

Marcus, L. C. 1992. Infections acquired from animals. Pp. 1267–1269 *in* Infectious diseases (S. L. Gorbach, J. G. Bartlett, and N. R. Blacklow, eds.). W. B. Saunders, Philadelphia.

Mares, M. A. 1992. Neotropical mammals and the myth of Amazonian biodiversity. Science. 255:976–979.

Mares, M. A. and G. N. Cameron. 1994. Communities and ecosystems. Pp. 348–376 *in* Seventy-five years of mammalogy (E. C. Birney and J. R. Choate, eds.). Spec. Publ. No. 11, American Society of Mammalogists.

Mares, M. A. and T. E. Lacher, Jr. 1987. Ecological, morphological, and behavioral convergence in rock-dwelling mammals. Curr. Mammal. 2:307–348.

Marimuthu, G., J. Habersetzer, and D. Leippert. 1995. Active acoustic gleaning from the water surface by the Indian false vampire bat, *Megaderma lyra*. Ethology. 99:61–74.

Marler, P. 1968. Aggregation and dispersal: Two functions in primate communication. Pp. 420–438 *in* Primates: studies in adaptation and variability (P. C. Jay, ed.). Holt, Rinehart and Winston, New York.

Marler, P. 1973. A comparison of vocalizations of red-tailed monkeys and blue monkeys, *Cercopithecus ascanius* and *C. mitis*, in Uganda. Zeit. Tierpsychol. 33:223–247.

Marshall, A .G. 1985. Old World phytophagous bats (Megachiroptera) and their food plants: a survey. Zool. J. Linn. Soc. 83:351–369.

Marshall, L. G. 1980. Marsupial paleobiogeography. Pp. 345–386 *in* Aspects of vertebrate history (L. L. Jacobs, ed.). Museum of Northern Arizona Press, Flagstaff.

Marshall, L. G. 1984. Monotremes and marsupials. Pp. 59–115 *in* Orders and families of recent mammals of the world. (S. Anderson and J. K. Jones, Jr., eds.). John Wiley & Sons, New York.

Marshall, L. G. 1988. Extinction. Pp. 219–254 *in* A. A. Myers and P. S. Giller (eds.). Analytical biogeography. Chapman and Hall, New York.

Marshall, L. G., J. A. Case, and M. O. Woodburne. 1990. Phylogenetic relationships of the families of marsupials. Pp. 433–506 *in* Current mammalogy (H. H. Genoways, ed.). Plenum Press, New York.

Marshall, L. G., T. Sempere, and R. F. Butler. 1997. Chronostratigraphy of the mammal-bearing Paleocene of South America. J. South Am. Earth Sci. 10:49–70.

Marshall, L. G., S. D. Webb, J. J. Sepkoski, and D. M. Raup. 1982. Mammalian evolution and the Great American Interchange. Science. 215:1351–1357.

Martan, J. and B. A. Shepherd. 1976. The role of the copulatory plug in reproduction of the guinea pig. J. Exp. Zool. 196:79–84.

Martensson, L. 1982. The pregnant rabbit, guinea pig, sheep and rhesus monkey as models in reproductive physiology. Europ. J. Obstet. Reprod. Biol. 18:169–182.

Martin, I. G. 1981. Venom of the short-tailed shrew (*Blarina brevicauda*) as an insect immobilizing agent. J. Mammal. 62:189–192.

Martin, I. G. 1983. Daily activity of short-tailed shrew (*Blarina brevicauda*) in simulated natural conditions. Am. Midl. Nat. 109:136–144.

Martin, L. D. 1989. Fossil history of the terrestrial Carnivora. Pp. 536–568 *in* Carnivore behavior, ecology, and evolution (J. L. Gittleman, ed.). Cornell Univ. Press, Ithaca, N.Y.

Martin, R. A., M. Florentini, and F. Connors. 1980. Social facilitation of reduced oxygen consumption in *Mus musculus* and *Meriones unguiculatus*. Comp. Biochem. Physiol. 65A:519–522.

Martin, R. D. 1979. Phylogenetic aspects of prosimian behavior. Pp. 45–77 *in* The study of Prosimian behavior (G. A. Doyle and D. Martin, eds.). Academic Press, New York.

Martin, R. D. 1990. Primate origins and evolution. Princeton Univ. Press, Princeton, NJ.

Martin, R. D., A. F. Dixson, and E. J. Wickings (eds.). 1992. Paternity in primates: genetics tests and theories. S. Karger, Basel, Switzerland.

Maser, C. and Z. Maser. 1987. Notes on mycophagy in four species of mice in the genus *Peromyscus*. Great Basin Nat. 47:308–313.

Maser, C., J. M. Trappe, and R. A. Nussbaum. 1978. Fungal-small mammals interrelationships with emphasis on Oregon coniferous forests. Ecology. 59:799–809.

Maser, C., Z. Maser, J. W. Witt, and G. Hunt. 1986. The northern flying squirrel: a mycophagist in southwestern Oregon. Can. J. Zool. 64:2086–2089.

Maser, Z., C. Maser, and J. M. Trappe. 1985. Food habits of the northern flying squirrel (*Glaucomys sabrinus*) in Oregon. Can. J. Zool. 63:1084–1088.

Massey, A. 1977. Agonistic aids and kinship in a group of pigtail macaques. Behav. Ecol. Sociobiol. 2:31–40.

Massey, A. and J. G. Vandenbergh. 1980. Puberty delay by a urinary cue from female house mice in feral populations. Science. 209:821–822.

Matthews, L. H. 1978. The natural history of the whale. Weidenfeld and Nicolson, London.

Mattson, D. J., B. M. Blanchard, and R. R. Knight. 1991. Food habits of Yellowstone grizzly bears, 1977–1987. Can. J. Zool. 69:1619–1629.

Maxim, P. E., D. M. Bowden, and G. P. Sackett. 1976. Ultradian rhythms of solitary and social behavior in Rhesus monkeys. Physiol. Behav. 17:337–344.

May, M. 1991. Aerial defense tactics of flying insects. Am. Sci. 79:316–328.

Maylon, C. and S. Healy. 1994. Fluctuating asymmetry in antlers of fallow deer, *Dama dama*, indicates dominance. Anim. Behav. 48:248–250.

Maynard Smith, J. and R. Savage. 1956. Some locomotor adaptations in mammals. J. Linn. Soc. Zool. 42:603–622.

Mayr, E. 1970. Populations, species, and evolution. Belknap Press of Harvard Univ. Press, Cambridge.

Mayr, E. and P. D. Ashlock. 1991. Principles of systematic zoology. McGraw-Hill, New York.

McBee, R. H. 1971. Significance of intestinal microflora in herbivory. Ann. Rev. Ecol. Syst. 2:165–176.

McCarty, R. 1975. Onychomys torridus. Mamm. Species. 59:1–5.

McCarty, R. 1978. Onychomys leucogaster. Mamm. Species. 87:1–6.

McCormick, J. B. and S. P. Fisher-Hoch. 1994. Zoonoses caused by Filoviridae. Pp. 375–383 *in* Handbook of zoonoses. Section B: viral, 2d ed. (C. W. Beran and J. H. Steele, eds.). CRC Press, Boca Raton, FL.

McCracken, G. F. 1984. Communal nursing in Mexican free-tailed bat maternity colonies. Science. 223:1090–1091.

McCracken, G. F. and J. W. Bradbury. 1977. Paternity and genetic heterogeneity in the polygynous bat, *Phyllostomus hastatus*. Science 198:303–306.

McCullough, D. R. 1985. Long range movements of large terrestrial mammals. Pp. 444–465 *in* Migration: mechanisms and adaptive significance (M. A. Rankin, ed.). Marine Science Institute, Port Aransas, TX.

McDade, J. E. and V. F. Newhouse. 1986. Natural history of *Rickettsia rickettsii*. Annu. Rev. Microbiol. 40:287–309.

McFarland, W. N. and W. A. Wimsatt. 1969. Renal function and its relation to the ecology of the vampire bat, *Desmodus rotundus*. Comp. Biochem. Physiol. 28:985–1006.

McLaren, S. B. and J. K. Braun (eds.). 1993. GIS applications in mammalogy. Univ. of Oklahoma Press, Norman.

McManus, J. J. 1974. Didelphis virginiana. Mammal. Spec. 40:1–6.

McNab, B. K. 1963. Bioenergetics and the determination of home range size. Am. Nat. 97:133–140.

McNab, B. K. 1971. On the ecological significance of Bergmann's rule. Ecology. 52:845–854.

McNab, B. K. 1974. The energetics of endotherms. Ohio J. Sci. 74:370–380.

McNab, B. K. 1979. Climatic adaptation in the energetics of heteromyid rodents. Comp. Biochem. Physiol. 62A:813–820.

McNab, B. K. 1980. On estimating thermal conductance in endotherms. Physiol. Zool. 53:145–156.

McNab, B. K. 1985. Energetics, population biology, and distribution of Xenarthrans, living and extinct. Pp. 219–232 *in* The evolution and ecology of armadillos, sloths, and vermilinguas (G. G. Montgomery, ed.). Smithsonian Institution Press, Washington, D.C.

McNab, B. K. 1988. Complications inherent in scaling the basal rate of metabolism in mammals. Quart. Rev. Biol. 63:25–54.

McNab, B. K. 1989. Basal rate of metabolism, body size, and food habits in the Order Carnivora. Pp. 335–354 *in* Carnivore behavior, ecology, and evolution (J. L. Gittleman, ed.). Comstock Publ. Assoc., Ithaca, NY.

McNab, B. K. 1995. Energy expenditure and conservation in frugivorous and mixed-diet carnivorans. J. Mammal. 76:206–222.

McNab, B. K. and F. J. Bonaccorso. 1995. The energetics of pteropodid bats. Pp. 111–122 *in* Ecology, evolution and behaviour of bats (P. A. Racey and S. M. Swift, eds.). Zool. Soc. London, Clarendon Press, London.

McNab, B. K. and P. Morrison. 1963. Body temperature and metabolism in subspecies of *Peromyscus* from arid and mesic environments. Ecol. Monogr. 33:63–82.

McNair, J. N. 1982. Optimal giving-up times and the marginal value theorem. Am. Nat. 119:511–529.

McNaughton, S. J. 1976. Serengeti migratory wildebeest: facilitation of energy flow by grazing. Science. 171:92–94.

McNaughton, S. J. 1985. Ecology of a grazing ecosystem: the Serengeti. Ecol. Monogr. 55:259–294.

McQuade, D. B., E. H. Williams, and H. B. Eichenbaum. 1986. Cues used for localizing food by the grey squirrel (*Sciurus carolinensis*). Ethology. 72:22–30.

Mead, J. G. and R. L. Brownell, Jr. 1993. Order Cetacea. Pp. 349–364 *in* Mammal species of the world: a taxonomic and geographic reference, 2d ed. (D. E. Wilson and D. M. Reeder, eds.). Smithsonian Institution Press, Washington, D.C.

Mead, R. A. 1968. Reproduction in western forms of the spotted skunk (Genus *Spilogale*). J. Mammal. 49:373–390.

Mead, R. A. 1989. The physiology and evolution of delayed implantation in carnivores. Pp. 437–464 *in* Carnivore behavior, ecology, and evolution (J. L. Gittleman, ed.). Comstock Publ. Assoc., Ithaca, NY.

Mearns, E. A. 1907. Mammals of the Mexican boundary of the United States. Bull. U.S. Nat. Mus. 56:1–530.

Mech, L. D. 1970. The wolf: the ecology and behavior of an endangered species. Natural History Press, Garden City, NY.

Mech, L. D. 1995. The challenge and opportunity of recovering wolf populations. Conserv. Biol. 9:270–278.

Mech, L. D., U. S. Seal, and G. D. DelGiudice. 1987. Use of urine in snow to indicate condition of wolves. J. Wildl. Mgmt. 51:10–13.

Mech, L. D. and F. J. Turkowski, 1966. Twenty-three raccoons in one winter den. J. Mammal. 47:529–530.

Meikle, D. B. and S. H. Vessey. 1981. Nepotism among rhesus monkey brothers. Nature. 294:160–161.

Meikle, D. B. and S. H. Vessey. 1988. Maternal dominance rank and lifetime survivorship of male and female rhesus monkeys. Behav. Ecol. Sociobiol. 22:379–383.

Mengak, M. T. and D. C. Guynn Jr. 1987. Pitfalls and snap traps for sampling small mammals and herptofauna. Am. Midl. Nat. 118:284–288.

Merriam, C. H. 1890. Results of a biological survey of the San Francisco mountain region and desert of the Little Colorado, Arizona. North Am. Fauna. 3:1–136.

Merriam, C. H. 1894. Laws of temperature control of the geographical distribution of terrestrial animals and plants. Natl. Geog. 6:229–238.

Merriam, G. 1995. Movement in spatially divided populations: responses to landscape structure. Pp. 64–77 *in* Landscape approaches in mammalian ecology and conservation (W. Z. Lidicker, Jr., ed.). Univ. of Minnesota Press, Minneapolis.

Merriam, G. and A. Lanoue. 1990. Corridor use by small mammals: field measurement for three experimental types of *Peromyscus leucopus.* Landscape Ecol. 4:123–132.

Merritt, J. F. (ed.) 1984. Winter ecology of small mammals. Spec. Pub., Carnegie Mus. Nat. Hist., No. 10.

Merritt, J. F. 1986. Winter survival adaptations of the short-tailed shrew (*Blarina brevicauda*) in an Appalachian montane forest. J. Mammal. 67:450–464.

Merritt, J. F. 1987. Guide to the mammals of Pennsylvania. Univ. of Pittsburgh Press, Pittsburgh.

Merritt, J. F. 1995. Seasonal thermogenesis and changes in body mass of masked shrews, *Sorex cinereus.* J. Mammal. 76: 1020–1035.

Merritt, J. F., G. L. Kirkland, Jr., and R. K. Rose. 1994. Advances in the biology of shrews. Carnegie Museum of Natural History, Spec. Publ. No. 18. Pittsburgh.

Merritt, J. F. and J. M. Merritt. 1978. Population ecology and energy relationships of *Clethrionomys gapperi* in a Colorado subalpine forest. J. Mammal. 59: 576–598.

Merritt, J. F. and S. H. Vessey. In press. Shrews-Small insectivores with polyphasic patterns. (In press) *in* Activity patterns in small

mammals (S. Halle and N. C. Stenseth, eds.). Springer Verlag, Berlin.

Meserve, P. L. 1976. Food relationships of a rodent fauna in a California coastal scrub community. J. Mammal. 57:300–319.

Meserve, P. L., J. R. Gutiérrez, J. A. Yunger, L. C. Contreras, and F. M. Jaksic. 1996. Role of biotic interactions in a small mammal assemblage in semiarid Chile. Ecology. 77:133–148.

Messier, F. 1985. Solitary living and extraterritorial movements of wolves in relation to social status and prey abundance. Can. J. Zool. 63:239–245.

Metzgar, L. 1967. An experimental comparison of screech owl predation on resident and transient white-footed mice (*Peromyscus leucopus*). J. Mammal. 48:387–391.

Mikesic, D. G. and L. C. Drickamer. 1992a. Effects of radiotransmitters and fluorescent powders on activity of wild house mice (*Mus musculus*). J. Mammal. 73:663–667.

Mikesic, D. G. and L. C. Drickamer. 1992b. Factors affecting home-range size in house mice (*Mus musculus domesticus*) living in outdoor enclosures. Am. Midl. Nat. 127:31–40.

Miles, A. E. W. and C. Grigson. 1990. Colyer's variations and diseases of the teeth of animals. Cambridge Univ. Press, New York.

Milinkovitch, M. C. 1992. DNA–DNA hybridization support for ungulate ancestry of Cetacea. J. Evol. Biol. 5:149–160.

Millar, J. S. 1971. Breeding of the pika in relationship to the environment. Ph.D. diss., University of Alberta, Edmonton.

Millar, J. S. 1978. Energetics of reproduction in *Peromyscus leucopus:* the cost of lactation. Ecology. 59:1055–1061.

Millar, J. S. and F. C. Zwickel. 1972. Characteristics and ecological significance of hay piles of pikas. Mammalia. 36:657–667.

Miller, B., D. Biggins, L. Hanebury, and A. Vargas. 1994. Reintroduction of the black-footed ferret (*Mustela nigripes*). Pp. 455–464 *in* Creative conservation: interactive management of wild and captive animals (P. J. S. Olney, G. M. Mace, and A. T. C. Feistner, eds.). Chapman and Hall, London.

Miller, B., G. Ceballos, and R. Reading. 1994. The prairie dog and biotic diversity. Conserv. Biol. 8:677–681.

Miller, G. S., Jr. and J. W. Gidley. 1934. Mammals and how they are studied, Part II, in Warm-blooded vertebrates. Smithsonian Scientific Series, vol. 9. Smithsonian Institution, Washington, D. C.

Mills, J. N., T. L. Yates, J. E. Childs, R. P. Parmenter, T. G. Ksiazek, P. E. Rollin, and C. J. Peters. 1995. Guidelines for working with rodents potentially infected with hantavirus. J. Mammal. 76:716–722.

Mills, M. G. L. 1985. Related spotted hyaenas forage together but do not cooperate in rearing young. Nature. 316:61–62.

Mills, M. G. L. 1996. Methodological advances in capture, census, and food-habits studies of

large African carnivores. Pp. 223–266 *in* Carnivore behavior, ecology and evolution, vol. 2, J. Gittleman, (ed.). Cornell Univ. Press, Ithaca, NY.

Milner-Gulland, E. J. and R. Mace. 1991. The impact of the ivory trade on the African elephant *Loxodonta africana* population as assessed by data from the trade. Biol. Conserv. 55:215–229.

Milton, K. 1985. Mating patterns of woolly spider monkeys, *Brachyteles arachnoides:* implications for female choice. Behav. Ecol. Sociobiol. 17:53–59.

Miththapala, S., J. Seidensticker, and S. J. O'Brien. 1996. Phylogenetic subspecies recognition in leopards (*Panthera pardus*): molecular genetic variation. Cons. Biol. 10:1115–1132.

Mittermeier, R. A. and D. L. Cheney. 1986. Conservation of primates and their habitats. Pp. 477–490 *in* Primate societies (B. B. Smuts, D. L. Cheney, R. M. Seyfarth, R. W. Wrangham, and T. T. Struhsaker, eds.). Univ. of Chicago Press, Chicago.

Mittermeier, R. A., I. Tattersall, W. R. Konstant, D. M. Meyers, and R. B. Mast. 1994. Lemurs of Madagascar. Conservation International, Washington, D.C.

Mivart, St. G. J. 1873. On *Lepilemur* and *Cheirogaleus* and on the zoological rank of Lemuroidea. Proceedings of the Zoological Society of London. 1873:484–510.

Mladenoff, D. J., T. A. Sickley, R. G. Haight, and A. P. Wydeven. 1995. A regional landscape analysis and prediction of favorable gray wolf habitat in the northern Great Lakes Region. Cons. Biol. 9:279–294.

Moehlman, P. D. 1979. Jackal helpers and pup survival. Nature. 277:382–383.

Moehlman, P. D. 1983. Socioecology of silverbacked and golden jackals, *Canis mesomelas* and *C. aureus.* Pp. 423–453 *in* Recent advances in the study of mammalian behavior (J. F. Eisenberg and D. G. Kleiman, eds.). Spec. Publ. No. 7, American Society of Mammalogists.

Moehlman, P. D. 1986. Ecology of cooperation in canids. Pp. 64–86 *in* Ecological aspects of social evolution: birds and mammals (D. I. Rubenstein and R. W. Wrangham, eds.). Princeton Univ. Press, Princeton, NJ.

Møller, A. P. 1988. Ejaculate quality, testes size and sperm competition in primates. J. Hum. Evol. 17:479–488.

Møller, A. P., J. J. Cuervo, J. J. Soler, and C. Zamora-Muñoz. 1996. Horn asymmetry and fitness in gemsbok, *Oryx g. gazella.* Behav. Ecol. 7:247–253.

Montgomery, G. G. 1978. The ecology of arboreal folivores. Smithsonian Institution Press, Washington, D.C.

Moore, J. and R. Ali. 1984. Are dispersal and inbreeding avoidance related? Anim. Behav. 32:94–112.

Moore, R. Y. 1973. Retinohypothalamic projection in mammals: a comparative study. Brain Res. 49:403–409.

Moore, R. Y. 1982. Organization and function of a central nervous system circadian oscillator: the suprachiasmatic nucleus. Fed. Proc. 42:2783–2789.

Morales, J. C. and J. W. Bickham. 1995. Molecular systematics of the genus *Lasiurus* (Chiroptera: Vespertilionidae) based on restriction-site maps of the mitochondrial ribosomal genes. J. Mammal. 76:730–749.

Morell, V. 1996. New mammals discovered by biology's new explorers. Science. 273:1491.

Morey, D. F. 1994. The early evolution of the domestic dog. Am. Sci. 82:336–347.

Morgan, L. H. 1868. The American beaver and his works. J. B. Lippincott, Philadelphia.

Morin, P. A., J. J. Moore, R. Chakraborty, L. Jin, J. Goodall, and D. S. Woodruff. 1994. Kin selection, social structure, gene flow, and the evolution of chimpanzees. Science. 265:1193–1201.

Morris, D. W. 1992. Optimum brood size: tests of alternative hypotheses. Evolution. 46:1848–1861.

Morris, R. J. 1986. The acoustic faculty of dolphins. Pp. 369–399 *in* Research on dolphins (M. M. Bryden and R. Harrison, eds.). Clarendon Press, Oxford.

Morrison, P. 1960. Some interrelations between weight and hibernation function. Bull. Mus. Zool. 124:75–91.

Morrison, P. and B. K. McNab. 1967. Temperature regulation in some Brazilian phyllostomid bats. Comp. Biochem. Physiol. 21:207–221.

Morrison, P., F. A. Ryser, and A. R. Dawe. 1959. Studies on the physiology of the masked shrew *Sorex cinereus*. Physiol. Zool. 32:256–271.

Morrison, P. R. 1948. Oxygen consumption in several small wild mammals. J. Cell. Comp. Physiol. 31:69–96.

Morrison, P. R. 1965. Body temperatures in some Australian mammals. Aust. J. Zool. 13:173–187.

Moss, C. J. 1983. Oestrous behaviour and female choice in the African elephant. Behaviour. 86:167–196.

Mossing, T. 1975. Measuring small mammal locomotory activity with passage counters. Oikos. 26:237–239.

Mouchaty, S., J. A. Cook, and G. F. Shields. 1995. Phylogenetic analysis of northern hair seals based on nucleotide sequences of the mitochondrial cytochrome b gene. J. Mammal. 76:1178–1185.

Moyer, K. E. 1976. Psychobiology of aggression. Harper & Row, New York.

Muchlinski, A. E. 1980. Duration of hibernation bouts in *Zapus hudsonius*. Comp. Biochem. Physiol. 67A:287–289

Mueller, H. C. 1966. Homing and distance-orientation in bats. Z. Tierpsychol. 23:403–421.

Muizon, C. de 1994. A new carnivorous marsupial from the Paleocene of Bolivia and the problem of marsupial monophyly. Nature. 370:208–211.

Müller, P. 1973. Dispersal centres of terrestrial vertebrates in the Neotropical realm. Junk, The Hague.

Müller, P. 1974. Aspects of zoogeography. Junk, The Hague.

Muller-Schwarze, D. 1971. Pheromones in black-tailed deer. Anim. Behav. 19:141–152.

Mullican, T. R. 1988. Radiotelemetry and fluorescent pigments: a comparison of techniques. J. Wildl. Mgmt. 52:627–631.

Mullis, K., F. Faloona, S. Scharf, R. Saiki, G. Horn, and H. Erlich. 1986. Specific enzymatic amplification of DNA in vitro: the polymerase chain reaction. Cold Spring Harb. Symp. Quant. Biol. 51:263–273.

Mumford, R. E. and J. O. Whitaker, Jr. 1982. Mammals of Indiana. Indiana Univ. Press, Bloomington.

Munger, J. C. and W. H. Karasov. 1994. Costs of botfly infection in white-footed mice: energy and mass flow. Can. J. Zool. 72:166–173.

Murie, O. 1944. The wolves of Mount McKinley. U. S. Dept. Interior Nat. Park Serv., Fauna Ser. No. 5, Washington, D.C.

Murie, O. 1954. A field guide to animal tracks. Macmillan, New York.

Murie, O. J. and A. Murie. 1931. Travels of *Peromyscus*. J. Mammal. 12:200–209.

Murphy, B. P., K. V. Miller and R. L. Marchinton. 1994. Sources of reproductive chemosignals in female white-tailed deer. J. Mammal. 75:781–786.

Murtagh, C. E. 1977. A unique cytogenetic system in monotremes. Chromosoma. 65:37–57.

Musser, G. G. and M. Dagosto. 1987. The identity of *Tarsius pumilus,* a pygmy species endemic to the montane mossy forests of central Sulawesi. Am. Mus. Novitates. 2867:1–53.

Mutere, F. A. 1965. Delayed implantation in an equitorial fruit bat. Nature. 207:780.

Muul, I. 1968. Behavioral and physiological influences on the distribution of the flying squirrel, *Glaucomys volans*. Misc. Pub., Mus. Zool., Univ. of Mich. 124:1–66.

Myers, A. A. and P. S. Giller. 1988. Analytical biogeography. Chapman and Hall, London.

Myers, J. and C. Krebs. 1971. Genetic, behavioral, and reproductive attributes of dispersing field voles *Microtus pennsylvanicus* and *Microtus ochrogaster*. Ecol. Monogr. 41:53–78.

Nagel, A. 1977. Torpor in the European white-toothed shrews. Experientia. 33:1455–1458.

Nagy, K. A. 1987. Field metabolic rate and food requirement scaling in mammals and birds. Ecol. Monogr. 57:111–128.

Nagy, K. A. 1989. Field bioenergetics: accuracy of models and methods. Physiol. Zool. 62:237–252.

Nagy, K. A. and M. H. Knight. 1994. Energy, water, and food use by springbok antelope (*Antodorcas marsupialis*) in the Kalahari Desert. J. Mammal. 75:860–872.

Nagy, K. E., C. Meienberger, S. D. Bradshaw, and R. D. Woller. 1995. Field metabolic rate of a small Australian mammal, the honey possum (*Tarsipes rostratus*). J. Mammal. 76:862–866.

Nagy, T. R. 1993. Effects of photoperiod history and temperature on male collared lemmings, *Dicrostonyx groenlandicus*. J. Mammal. 74: 990–998.

Napier, J. R. and P. H. Napier. 1967. A handbook of living primates. Academic Press, New York.

Nash, L. T., S. K. Bearder, and T. R. Olson. 1989. Synopsis of *Galago* species characteristics. Int. J. Primat. 10:57–80.

Nash, R. 1967. Wilderness and the American mind. Yale Univ. Press, New Haven, CT.

Neal, E. 1986. The natural history of badgers. Christopher Helm, London.

Negus, N. C. and P. J. Berger. 1977. Experimental triggering of reproduction in a natural population of *Microtus montanus*. Science. 196:1230–1231.

Nelson, J. F. and R. M. Chew. 1977. Factors affecting seed reserves in the soil of a Mojave Desert ecosystem, Rock Valley, Nye County, Nevada. Am. Midl. Nat. 97:300–320.

Nelson, R. J. 1995. An introduction to behavioral endocrinology. Sinauer Assoc., Sunderland, MA.

Neuweiler, G. 1990. Auditory adaptations for prey capture in echolocating bats. Physiol. Rev. 70:615–641.

Newmark, W. D. 1995. Extinction of mammal populations in western North American national parks. Conserv. Biol. 9:512–526.

Nichols, J. D. and J. E. Hines. 1984. Effects of permanent trap response in capture probability of Jolly-Seber capture-recapture model estimates. J. Wildl. Mgmt. 48:289–294.

Nicol, S. C. and N. A. Andersen. 1993. The physiology of hibernation in an egg-laying mammal, the echidna. Pp. 56–64 *in* Life in the cold: ecological, physiological, and molecular mechanisms (C. Carey, G. L. Florant, B. A. Wunder, and B. Horwitz, eds.). Westview Press, Boulder, CO.

Nicoll, M. E. 1983. Mechanisms and consequences of large litter production in *Tenrec ecaudatus* (Insectivora: Tenrecidae). Ann. Mus. R. Afr. Centr. 237:219–226.

Noll-Banholzer, U. 1979. Body temperature, oxygen consumption, evaporative water loss and heart rate in the fennec. Comp. Biochem. Physiol. 62A:585–592.

Norberg, U. M. 1985. Flying, gliding, and soaring. Pp. 129–158, *in* Functional vertebrate morphology (M. Hildebrand, D. M. Bramble, K. F. Liem, and D. B. Wake, eds.). Harvard Univ. Press, Cambridge.

Norberg, U. M. 1990. Vertebrate flight: mechanisms, physiology, morphology, ecology, and evolution. Springer-Verlag, New York.

Nordenskiöld, E. 1928. The history of biology. Tudor Publishing, New York.

Norris, K. S. 1968. The evolution of acoustic mechanisms in odontocete cetaceans. Pp.

297–324 *in* Evolution and environment (E. T. Drake, ed.). Yale Univ. Press, New Haven.

Norris, K. S. and G. W. Harvey. 1972. A theory for the function of the spermaceti organ of the sperm whale (*Physeter catodon* L.). NASA Spec. Publ. No. 262:397–417.

Norris, K. S. and G. W. Harvey. 1974. Sound transmission in the porpoise head. J. Acoustic Soc. Am. 56:659–664.

Novacek, M. J. 1985. Evidence for echolocation in the oldest known bats. Nature 315:140–141.

Novick, A. 1977. Acoustic orientation. Pp. 77–289 *in* Biology of bats, vol. III (W. A. Wimsatt, ed.). Academic Press, New York.

Novick, A. and D. R. Griffin. 1961. Laryngeal mechanisms in bats for the production of sounds. J. Exp. Zool. 148:125–145.

Novoa, C. and J. C. Wheeler. 1984. Llama and alpaca. Pp. 116–128 *in*Evolution of domesticated animals (I. L. Mason, ed.). Longman, New York.

Novotny, M., B. Jemiolo, S. Harvey, D. Wiesler, and A. Marchlewska-Koj. 1986. Adrenal-mediated endogenous metabolites inhibit puberty in female mice. Science. 231:722–725.

Nowak, R. M. 1991. Walker's mammals of the world, 5th ed., 2 vols. Johns Hopkins Univ. Press, Baltimore.

Nowak, R. M. 1994. Walker's bats of the world. Johns Hopkins Univ. Press, Baltimore.

O'Brien, S. J. and J. A. M. Graves. 1990. Geneticists converge on divergent mammals: an overview of comparative mammalian genetics. Pp. 5–12 *in* Mammals from pouches and eggs (J. A. Marshall Graves, R. M. Hope, and D. W. Cooper, eds.). CSIRO, Melbourne, Australia.

Obrist, M. K., M. B. Fenton, J. L. Eger, and P. A. Schlegel. 1993. What ears do for bats: a comparative study of pinna sound pressure transformation in chiroptera. J. Exp. Biol. 180:119–152.

O'Connor, B. M. 1988. Host associations and coevolutionary relationships of astigmatid mite parasites of New World primates. I. Families Psoroptidae and Audycoptidae. Fieldiana. 39:245–260.

Odell, D. K. 1982. West Indian manatee, *Trichechus manatus*. Pp. 828–837 *in* Wild mammals of North America: biology, management, and economics (J. A. Chapman and G. A. Feldhamer, eds.). Johns Hopkins Univ. Press, Baltimore.

Oftedal, O. T. 1984. Milk composition, milk yield and energy output at peak lactation: a comparative review. Symp. Zool. Soc. London. 52:33–85.

Oftedal, O. T., D. J. Boness, and R. A. Tedman. 1987. The behavior, physiology, and anatomy of lactation in the Pinnipedia. Pp. 175–234 *in* Current mammalogy (H. H. Genoways, ed.). Plenum Press, New York.

O'Gara, B. W. and G. Matson. 1975. Growth and casting of horns by pronghorns and exfoliation of horns by bovids. J. Mammal. 56:829–846.

Oleyar, C. M. and B. S. McGinnes. 1974. Field evaluation of diethylstilbestrol for suppressing reproduction in foxes. J. Wildl. Mgmt. 38:101–106.

Oliphant. L. W. 1983. First observations of brown fat in birds. Condor. 85:350–354.

Olsen, S. J. 1985. Origins of the domestic dog: the fossil record. Univ. of Arizona Press, Tucson.

Olson, J. M., W. R. Dawson, and J. J. Camilliere. 1988. Fat from black-capped chickadees: avian adipose tissue? Condor. 90:529–537.

Orcutt, E. E. 1940. Studies on the muscles of the head, neck, and pectoral appendages of *Geomys bursarius*. J. Mammal. 21:37–52.

Orians, G. H. 1969. On the evolution of mating systems in birds and mammals. Am. Nat. 103:589–603.

Øritsland, T. 1977. Food consumption of seals in the Antarctic pack ice. Pp. 749–768, *in* Adaptations within Antarctic ecosystems. (G. A. Llano, ed.). Smithsonian Institution Press, Washington, D.C.

Orr, R. T. 1970. Animals in migration. Macmillan, New York.

Ortmann, S., J. Schmid, J. U. Ganzhorn, and G. Heldmaier. 1996. Body temperature and torpor in a Malagasy small primate, the mouse lemur. Pp. 55–61 *in* Adaptations to the cold: tenth international hibernation symposium (F. Geiser, A. J. Hulbert, and S. C. Nicol, eds). Univ. of New England Press, Armidale, NSW, Australia.

Osgood, D. W. 1980. Temperature sensitive telemetry applied to studies of small mammal activity patterns. Pp. 525–528 *in* A handbook of radiotelemetry and radio tracking (C. J. Amlaner and D. W. Macdonald, eds.). Pergamon Press, Oxford, England.

Osgood, W. H. 1909. Revision of mice of the American genus *Peromyscus*. North Am. Fauna. 28:1–285.

Ostfeld, R. S. 1997. The ecology of Lyme-disease risk. Am. Sci. 85:338–346.

Ostfeld, R. S., C. G. Jones, and J. O. Wolff. 1996. Of mice and mast. BioScience. 46:323–330.

Ostfeld, R. S., M. C. Miller, and K. R. Hazler. 1996. Causes and consequences of tick (*Ixodes scapularis*) burdens on white-footed mice (*Peromyscus leucopus*). J. Mammal. 77:266–273.

Owen, O. S. 1971. Natural resource conservation: an ecological approach. Macmillan, New York.

Owen-Smith, N. 1975. The social ethology of the white rhinoceros *Ceratotherium simum* (Burchell 1817). Z. Tierpsychol. 38:337–384.

Owen-Smith, N. 1993. Assessing the constraints for optimal diet models. Evol. Ecol. 7:530–531.

Owen-Smith, R. N. 1988. Megaherbivores: the influence of very large body size on ecology. Cambridge Univ. Press, New York.

Oxley, D. J., M. B. Fenton, and G. R. Carmody. 1974. The effects of roads on populations of small mammals. J. Appl. Ecol. 11:51–59.

Packer, C. 1977. Reciprocal altruism in *Papio anubis*. Nature. 265:441–443.

Packer, C. 1986. The ecology of sociality in felids. Pp. 429–451 *in* Ecological aspects of social evolution: birds and mammals (D. I. Rubenstein and R. W. Wrangham, eds.). Princeton Univ. Press, Princeton, NJ.

Packer, C., D. A. Collins, A. Sindimwo, and J. Goodall. 1995. Reproductive constraints on aggressive competition in female baboons. Nature. 373:60–63.

Packer, C., L. Herbst, A. E. Pusey, D. J. Bygott, J. P. Hanby, S. J. Cairns, and M. Borgerhoff-Mulder. 1988. Reproductive success in lions. Pp. 363–383 *in* Reproductive success (T. H. Clutton-Brock, ed.). Univ. of Chicago Press, Chicago.

Packer, C., S. Lewis, and A. Pusey. 1992. A comparative analysis of non-offspring nursing. Anim. Behav. 43:265–282.

Packer, C. and A. E. Pusey. 1982. Cooperation and competition within coalitions of male lions: kin selection or game theory? Nature. 296:740–742.

Packer, C., D. Scheel, and A. E. Pusey. 1990. Why lions form groups: food is not enough. Am. Nat. 136:1–19.

Padilla, M. and R. C. Dowler. 1994. Tapirus terrestris. Mammal. Species No. 481:1–8.

Page, R. D. M. 1993. Parasites, phylogeny and cospeciation. Int. J. Parasitol. 23:499–506.

Palo, R. T. and C. T. Robins (eds.) 1991. Plant defenses against mammalian herbivory. CRC Press, Boca Raton, FL.

Pan American Health Organization. 1978. The armadillo as an experimental model in biomedical research. World Health Organization Scientific Publ. No. 366. Washington, D.C.

Paradiso, J. L. and R. M. Nowak. 1982. Wolves, *Canis lupus* and allies. Pp. 460–474 *in* Wild mammals of North America: biology, management, and economics (J. A. Chapman and G. A. Feldhamer, eds.). Johns Hopkins Univ. Press, Baltimore.

Parker, G. A. 1970. Sperm competition and its evolutionary consequences in the insects. Biol. Rev. 45:525–568.

Parmagiano, S. and F. S. vom Saal. 1994. Infanticide and parental care. Harwood Academic, Langhorne, PA.

Parsons, L. M. and C. R. Terman. 1978. Influence of vision and olfaction on the homing ability of the white-footed mouse (*Peromyscus leucopus noveboracensis*). J. Mammal. 59:761–771.

Pascual, R., M. Archer, E. O. Jaureguizar, J. L. Prado, H. Godthelp, and S. J. Hand. 1992. The first non-Australian monotreme: an early Paleocene South American platypus (Monotremata: Ornithorhynchidae). Pp. 1–14 *in* Platypus and echidnas (M. L. Augee, ed.). Royal Zoological Society, New South Wales, Sydney.

Pastor, J. B. Dewey, and D. P. Christian. 1996. Carbon and nutrient mineralization and fungal spore composition of fecal pellets from voles in Minnesota. Ecography. 19:52–61.

Patterson, B. 1965. The fossil elephant shrews (Family Macroscelidae). Bull. Mus. Comp. Zool. Harvard. 133:295–335.

Patterson, B. 1975. The fossil aardvarks (Mammalia: Tubulidentata). Bull. Mus. Comp. Zool. 147:185–237.

Patterson, B. D., P. L. Meserve, and B. K. Lang. 1989. Distribution and abundance of small mammals along an elevational transect in temperate rainforests of Chile. J. Mammal. 70:67–78.

Patterson, M. A. and S. H. Vessey. 1973. Tapeworm (*Hymenolepis nana*) infection in male albino mice: effect of fighting among the hosts. J. Mammal. 54:784–786.

Payne, K., P. Tyack, and R. Payne. 1983. Progressive changes in the songs of humpback whales (*Megaptera novaeangliae*): a detailed analysis of two seasons in Hawaii. Pp. 9–57 *in* Communication and behavior of whales (R. Payne, ed.). Westview Press, Boulder, CO.

Payne, K. B., W. R. Langbauer Jr., and E. M. Thomas. 1986. Infrasonic calls of the Asian elephant (*Elephas maximus*). Behav. Ecol. Sociobiol. 18:297–301.

Payne, R. S. and S. McVay. 1971. Songs of humpback whales. Science. 173:585–597.

Pech, R. P., A. R. E. Sinclair, A. E. Newsome, and P. C. Catling. 1992. Limits to predator regulation of rabbits in Australia: evidence from predator-removal experiments. Oecologia. 89:102–112.

Pedler, C. and R. Tilley. 1969. The retina of a fruit bat (*Pteropus giganteus* Brünnich). Vision Res. 9:909–922.

Pelton, M. R. 1982. Black bear. Pp. 504–514 *in* Wild mammals of North America, biology, management, and economics (J. A. Chapman and G. A. Feldhamer, eds.). Johns Hopkins Univ. Press, Baltimore.

Perry, J. W. 1972. The ovarian cycle of mammals. Hafner, New York.

Petter, J.-J. 1962a. Ecological and behavioural studies of Madagascan lemurs in the field. Ann. N. Y. Acad. Sci. 102:267–281

Petter, J.-J. 1962b. Remarques sur l'ecologie et l'ethologie comparées des Lémuriens Malagaches. Mem. Mus. Nat. Hist. 27:1–146.

Petter, J.-J. 1972. Order of primates: Suborder of lemurs. Pp. 683–702 *in* Biogeography and ecology of Madagascar (R. Battastini and G. Richard-Vindard, eds.). W. Junk, The Hague.

Petter, J.-J. 1977. The aye-aye. Pp. 38–57 *in* Primate conservation (H. S. H. Prince Ranier III and G .H. Bourne, eds.). Academic Press, New York.

Petter, J.-J. and A. Petter. 1967. The aye-aye of Madagascar. Pp. 195–205 *in* Social communication among primates (S .A. Altmann, ed.). Univ. of Chicago Press, Chicago.

Petter, J.-J. and A. Petter-Rousseaux. 1979. Classification of the prosimians. Pp. 1–44, *in* The study of prosimian behavior (G. A. Doyle and R. D. Martins, eds.). Academic Press, London.

Pettigrew, J. D. 1986. Flying primates? Megabats have the advanced pathway from eye to midbrain. Science. 231:1304–1306.

Pettigrew, J. D. 1995. Flying primates: crashed or crashed through. Pp. 3–26 *in* Ecology, evolution and behaviour of bats (P. A. Racey and S. M. Swift, eds.). Oxford Univ. Press, Oxford, England.

Pettigrew, J. D., B. G. M. Jamieson, S. K. Robson, L. S. Hall, I. I. McAnally, and J. M. Cooper. 1989. Phylogenetic relations between microbats (Mammalia: Chiroptera and Primates). Philosophical Transactions of the Royal Society of London, Series B, Biol. Sci. 325:489–559.

Phelan, J. P. and R. H. Baker. 1992. Optimal foraging in *Peromyscus polionotus*: the influence of item-size and predation risk. Behaviour. 121:95–109.

Phillips, M. K. 1990. Measures of the value and success of a reintroduction project: red wolf reintroduction in Alligator River National Wildlife Refuge. Endangered Sp. Update. 8:24–26.

Pianka, E. R. 1970. On *r*- and *K*-selection. Am. Nat. 104:592–597.

Pianka, E. R. 1994. Evolutionary ecology, 5th ed. HarperCollins, New York.

Pielou, E. C. 1991. After the ice age. Univ. of Chicago Press, Chicago.

Pierce, S. S. and F. D. Vogt. 1993. Winter acclimatization in *Peromyscus maniculatus gracilis*, *P. leucopus noveboracensis*, and *P. l. leucopus*. J. Mammal. 74:665–677.

Pieres-Ferreira, J. W., E. Pieres-Ferreira, and P. Kaulicke. 1976. Preceramic animal utilization in the Central Peruvian Andes. Science. 194:483–490.

Piesman, J., G. O. Maupin, E. G. Campos, and C. M. Happ. 1991. Duration of adult female *Ixodes dammini* attachment and transmission of *Borrelia burgdorferi*, with description of a needle aspiration isolation method. J. Infect. Dis. 163:95–97.

Pilleri, G. 1990. Adaptation to water and the evolution of echolocation in the Cetacea. Ethology Ecol. Evol. 2:135–163.

Pivorunas, A. 1979. The feeding mechanisms of baleen whales. Am. Sci. 67:432–440.

Pleasants, J. M. 1989. Optimal foraging by nectarivores: a test of the marginal value theorem. Am. Nat. 134:51–71.

Poland, J. D., T. J. Quan, and A. M. Barnes. 1994. Plague. Pp. 93–112 *in* Handbook of zoonoses. Section A: bacterial, rickettsial, chlamydial, and mycotic diseases, 2d ed. (C. W. Beran and J. H. Steele, eds.). CRC Press, Boca Raton, FL.

Pond, C. M. 1978. Morphological aspects and the ecological and mechanical consequences of fat deposition in wild vertebrates. Pp. 519–570 *in* Annual review of ecology and systematics, vol. 9 (R. F. Johnston, P. W. Frank, and C. D. Michener, eds.), Annual Reviews, Palo Alto, CA.

Poole, J. H., K. Payne, W. R. Langbauer Jr., and C. J. Moss. 1988. The social contexts of some very low frequency calls of African elephants. Behav. Ecol. Sociobiol. 22:385–392.

Popper, A. N. 1980. Sound emission and detection by delphinids. Pp. 1–52 *in* Cetacean behavior (L. M. Herman, ed.). John Wiley and Sons, New York.

Popper, A. N. and R. R. Fay (eds.). 1995. Hearing in bats. Springer-Verlag, New York.

Porter, R. H., V. J. Tepper, and D. M. White. 1981. Experimental influences on the development of huddling preferences and "sibling" recognition in spiny mice. Devel. Psychobiol. 14:375–382.

Porter, W. P. and David M. Gates. 1969. Thermodynamic equilibria of animals with environment. Ecol. Monogr. 39:227–244.

Post, D. M. and O. J. Reichman. 1991. Effects of food perishability, distance, and competitors on caching behavior by eastern woodrats. J. Mammal. 72:513–517.

Post, D. M., O. J. Reichman, and D. E. Wooster. 1993. Characteristics and significance of the caches of eastern woodrats (*Neotoma floridana*). J. Mammal. 74:688–692.

Potts, W. K., C. J. Manning, and E. K. Wakeland. 1991. Mating patterns in seminatural populations of mice influenced by MHC genotype. Nature. 352:619–621.

Pough, F. H., J. B. Heiser, and W. N. McFarland. 1989. Vertebrate life, 3d ed. Macmillan New York.

Pough, F. H., J. B. Heiser, and W. N. McFarland. 1996. Vertebrate life, 4th ed. Macmillan, New York.

Powell, R. A. 1993. The fisher: life history, ecology, and behavior, 2d ed. Univ. of Minnesota Press, Minneapolis.

Prager, E. M. and A. C. Wilson. 1978. Construction of phylogenetic tress for proteins and nucleic acids: empirical evaluation of alternative matrix methods. J. Mol. Evol. 11:129–142.

Prager, E. M., A. C. Wilson, J. M. Lowenstein, and V. M. Sarich. 1980. Mammoth albumin. Science. 209:287–289.

Prentice, E. F., T. A. Flagg, C. S. McCutcheon, D. Brastow, and D. C. Cross. 1990. Equipment, methods and an automated data-entry station for PIT tagging. Am. Fish. Soc. Symp. 7:335–340.

Price, E. O. 1984. Behavioral aspects of animal domestication. Qt. Rev. Biol. 59:1–32.

Price, M. V. and P. R. Endo. 1989. Estimating the distribution and abundance of a cryptic species, *Dipodomys stephensi* (Rodentia: Heteromyidae), and implications for management. Conserv. Biol. 3:293–301.

Price, M. V. and P. A. Kelly. 1994. An age-structured demographic model for the endangered Stephens' kangaroo rat. Conserv. Biol. 8:810–821.

Price, P. W. 1980. Evolutionary biology of parasites. Princeton Univ. Press, Princeton, NJ.

Proske, U., A. Iggo, A. K. McIntyre, and J. E. Gregory. 1993. Electroreception in the platypus: a new mammalian sense. J. Comp. Physiol. A 173:708–710.

Pruitt, W. O., Jr. 1957. Observations of the bioclimate of some taiga mammals. Arctic. 10:131–138.

Prusiner, S. B. 1993. Genetic and infectious prion diseases. Arch. Neurol. 50:1129–1153.

Prusiner, S. B. and K. K. Hsiao. 1994. Human prion diseases. Ann. Neurol. 35:385–395.

Prusiner, S., J. Collinge, J. Powell, and B. Anderton (eds.). 1992. Prion diseases of humans and animals. Ellis Horwood, New York.

Pucek, Z. 1965. Seasonal and age changes in the weight of internal organs of shrews. Acta Theriol. 10:369–438.

Pulliam, H. R. and B. J. Danielson. 1991. Sources, sinks, and habitat selection: a landscape perspective on population dynamics. Am. Nat. 137:S50–S66.

Purves, P. E. 1967. Anatomical and experimental observations on the cetacean sonar system. Pp. 197–270 in Animal sonar systems: biology and bionics, vol. 1 (R. G. Busnel, ed.). Laboratoire de physiologie acoustique. Jouy-en-Josas, France.

Purves, P. E. and G. Pilleri. 1983. Echolocation in whales and dolphins. Academic Press, London.

Pusey, A. E. and C. Packer. 1987. The evolution of sex-biased dispersal in lions. Behaviour. 101:275–310.

Putman, R. 1988. The natural history of deer. Comstock Publ. Assoc., Ithaca, NY.

Puttick, G. M. and J. U. M. Jarvis. 1977. The functional anatomy of the neck and forelimbs of the cape golden mole, *Chrysochloris asiatica* (Liptophyla, Chrysochloridae). Zool. Afr. 12:445–458.

Pyke, G. H., H. R. Pulliam, and E. L. Charnov. 1977. Optimal foraging: a selective review of theory and tests. Q. Rev. Biol. 52:137–154.

Quilliam, T. A. 1966. The mole's sensory apparatus. J. Zool. 149:76–78.

Quinn, T. H. and J. J. Baumel. 1993. Chiropteran tendon locking mechanism. J. Morph. 216:197–208.

Qumsiyeh, M. B. and R. J. Baker. 1985. G- and C-banded karyotypes of the Rhinopomatidae (Microchiroptera). J. Mammal. 66:541–544.

Qumsiyeh, M. B. and J. K. Jones, Jr. 1986. Rhinopoma hardwickii and Rhinopoma muscatellum. Mammal. Species. 263:1–5.

Rabb, G. B. 1959. Toxic salivary glands in the primitive insectivore *Solenodon*. Chicago Acad. Sci. Nat. Hist. Mus. 170:171–173.

Racey, P. A. 1982. Ecology of bat reproduction. Pp. 57–104 in Ecology of bats (T. H. Kunz, ed.). Plenum, New York.

Racey, P. A. and S. M. Swift (eds.). 1995. Ecology, evolution and behaviour of bats. Symp. Zool. Soc. London. 67:1–421.

Rachlow, J. I. and J. Berger. 1997. Conservation implications of patterns of horn regeneration in dehorned white rhinos. Conserv. Biol. 11:84–91.

Rafael, M., F. Trillmich, and A. Honer. 1996. Energy allocation in reproducing and nonreproducing guinea pigs (*Cavia porcellus*) females and young under ad libitum conditions. J. Zool. 239:437–452.

Rainey, W. E., E. D. Pierson, T. Elmqvist, and P. A. Cox. 1995. The role of flying foxes (Pteropodidae) in oceanic island ecosystems of the Pacific. Pp. 47–62 in Ecology, evolution and behaviour of bats (P. A. Racey and S. M. Swift, eds.). Oxford Univ. Press, Oxford, England.

Ralls, K. 1971. Mammalian scent marking. Science. 171:443–449.

Ralls, K. 1976. Mammals in which females are larger than males. Qt. Rev. Biol. 51:245–276.

Ralls, K., K. Brugger, and J. Ballou. 1979. Inbreeding and juvenile mortality in small populations of ungulates. Science. 206:1101–1103.

Ramsey, E. M. 1982. The placenta: human and animal. Praeger, New York.

Randall, J. A. 1984. Territorial defense and advertisement by footdrumming in bannertail kangaroo rats (*Dipodomys spectabilis*) at high and low population densities. Behav. Ecol. Sociobiol. 16:11–20.

Randall, J. A. 1993. Behavioural adaptations of desert rodents (Heteromyidae). Anim. Behav. 45:263–287.

Randolph, J. C. 1973. Ecological energetics of a homeothermic predator, the short-tailed shrew. Ecology. 54:1166–1187.

Ransome, R. 1990. The natural history of hibernating bats. Christopher Helm, London.

Ranta, E., J. Lindström, V. Kaitala, H. Kokko, H. Lindén, and E. Helle. 1997. Solar activity and hare dynamics: a cross-continental comparison. Am. Nat. 149:765–775.

Ratner, S. C. and R. Boice. 1975. Effects of domestication on behaviour. Pp. 3–19 in The behaviour of domestic animals, 3d ed. (E. S. E. Hafez, ed.). Baillière Tindall, London.

Raup, D. M. 1991. Extinction: bad genes or bad luck? W. W. Norton, New York.

Rawlins, R. G., M. J. Kessler, and J. E. Turnquist. 1984. Reproductive performance, population dynamics and anthropometrics of the free-ranging Cayo Santiago rhesus macaques. J. Med. Primatol. 13:247–259.

Ray, C. E. 1976. Geography of phocid evolution. Syst. Zool. 25:391–406.

Rebar, C. E. 1995. Activity of *Dipodomys merriami* and *Chaetodipus intermedius* to locate resource distributions. J. Mammal. 76:437–447.

Reed, C. A. (ed.). 1977. Origins of agriculture. Mouton, Paris.

Reed, C. A. 1969. Animal domestication in the Near East. Pp. 361–380 in The domestication and exploitation of plants and animals (P. J. Ucko and G. W. Dimbleby, eds.). Gerald Duckworth, London.

Reed, C. A. 1984. The beginnings of animal domestication. Pp. 1–6 in Evolution of domesticated animals (I. L. Mason, ed.). Longman, New York.

Reed, D. L. and M. S. Hafner. 1997. Host specificity of chewing lice on pocket gophers: a potential mechanism for cospeciation. J. Mammal. 78:655–660.

Reeve, H. K., D. F. Westneat, W. A. Noon, P. W. Sherman, and C. F. Aquadro. 1990. DNA "fingerprinting" reveals high levels of inbreeding in colonies of the eusocial naked mole-rat. Proceedings of the National Academy of Science. 87:2496–2500.

Reichman, O. J. 1981. Factors influencing foraging in desert rodents. Pp. 195–213 in Foraging behavior; ecological, ethological, and psychological approaches (A. C. Kamil and T. D. Sargent, eds.). Garland STPM Press, New York.

Reichman, O. J. and D. Oberstein. 1977. Selection of seed distribution types by *Dipodomys merriami* and *Perognathus amplus*. Ecology. 58:636–643.

Reichman, O. J. and M. V. Price. 1993. Ecological aspects of heteromyid foraging. Pp. 539–574 in Biology of the Heteromyidae (H. H. Genoways and J. H. Brown, eds.). Spec. Pub. American Society of Mammalogists. No. 10.

Reichman, O. J. and S. C. Smith. 1990. Burrows and burrowing behavior of mammals. Pp. 197–244 in Current Mammalogy, vol. 2 (H. H. Genoways, ed.). Plenum, New York.

Reichman, O. J. and K. M. Van De Graaff. 1975. Association between ingestion of green vegetation and desert rodent reproduction. J. Mammal. 56:503–506.

Reid, W. V. and K. R. Miller. 1989. Keep options alive: the scientific basis for conserving biodiversity. World Resources Institute, Washington, D.C.

Reijnders, P., S. Brasseur, J. van der Toorn, P. van der Wolf, I. Boyd, J. Harwood, D. Lavigne, and L. Lowry. 1993. Seals, fur seals, sea lions, and walrus: status survey and conservation action plan. International Union for Conservation of Nature and Natural Resources, Gland, Switzerland.

Reiter, R. J. 1980. The pineal and its hormones in the control of reproduction in mammals. Endocrinol. Rev. 1:109–131.

Rementsova, M. M. 1987. Brucellosis in wild animals. Amerind Publ., New Delhi.

Renfree, M. B., E. M. Russell, and R. D. Wooller. 1984. Reproduction and life history of the honey possum, *Tarsipes rostratus*. Pp. 427–437 in Possums and gliders (A. P. Smith and I. D. Hume, eds.). Surrey Beatty and Sons, Sydney.

Repenning, C. A. and R. H. Tedford. 1977. Ontaroid seals of the Neogene. U.S. Dept. Inter., Geol. Surv. Prof. Paper 992:1–93.

Rewcastle, S. C. 1981. Stance and gait in tetrapods: an evolutionary scenario. Symp. Zool. Soc. London. 48:239–268.

Rhymer, J. M. and D. Simberloff. 1996. Extinction by hybridization and introgression. Ann. Rev. Ecol. Syst. 27:83–109.

Ribble, D. O. 1991. The monogamous mating system of *Peromyscus californicus* as revealed by DNA fingerprinting. Behav. Ecol. Sociobiol. 29:161–166.

Rice, C. G. and P. Kalk. 1996. Identification and marking techniques. Pp. 56–66 in, Wild mammals in captivity (D. G. Kleiman, M. E. Allen, K. V. Thompson, and S. Lumpkin, eds.). Univ. of Chicago Press, Chicago.

Rice, D. W. 1984. Cetaceans. Pp. 447–490 *in* Orders and families of recent mammals of the world (S. Anderson and J. K. Jones, Jr., eds.). John Wiley and Sons, New York.

Rice, D. W. and A. A. Wolman. 1971. The life history and ecology of the gray whale (*Eschrichtius robustus*). Spec. Publ. No. 3, American Society of Mammalogists.

Rice, W. R. 1987. Speciation via habitat specialization: the evolution of reproductive isolation as a correlated character. Evol. Ecol. 1:301–314.

Richard, P. B. 1973. Capture, transport and husbandry of the Pyrenian desman *Galemys pyrenaicus*. Int. Zoo Yearb. 13:175–177.

Richards, G. C. 1990. Rainforest bat conservation: unique problems in a unique environment. Aust. J. Zool. 26:44–46.

Richards, G. C. 1995. A review of ecological interactions of fruit bats in Australian ecosystems. Pp. 79–96 *in* Ecology, evolution and behaviour of bats (P. A. Racey and S. M. Swift, eds.). Oxford Univ. Press, Oxford, England.

Richardson, E. G. 1977. The biology and evolution of the reproductive cycle of *Miniopterus schreibersii* and *M. australis* (Chiroptera: Vespertilionidae). J. Zool. 183:353–375.

Richardson, J. 1829. Fauna boreali Americana, 4 vols. J. Murray, London.

Ricklefs, R. E. 1990. Ecology, 3d ed. W. H. Freeman, New York.

Riddle, B. R. 1995. Molecular biogeography in the pocket mice (*Perognathus* and *Chaetodipus*) and grasshopper mice (*Onychomys*): the late Cenozoic development of a North American aridlands rodent guild. J. Mammal 76:283–301.

Ride, W. D. L. 1970. A guide to the native mammals of Australia. Oxford Univ. Press, Melbourne.

Rivier, C., J. Rivier, and W. Vale. 1986. Stess-induced inhibition of reproductive functions: role of endogenous corticotropin-releasing factor. Science. 231:607–609.

Robbins, C. T. 1993. Wildlife feeding and nutrition, 2d ed. Academic Press, New York.

Roberts, L. S. and J. Janovy. 1996. Foundations of parasitology, 5th ed. Wm. C. Brown, Dubuque, IA.

Robinson, J. G. and K. H. Redford (eds.). 1991. Neotropical wildlife use and conservation. Univ. of Chicago Press, Chicago.

Robinson, M. F. 1996. A relationship between echolocation calls and noseleaf widths in bats of the genera *Rhinolophus* and *Hipposideros*. J. Zool. 239:389–393.

Robinson, P. 1992. Rabies. Pp. 1269–1277 *in* Infectious diseases (S. L. Gorbach, J. G. Bartlett, and N. R. Blacklow, eds.). W. B. Saunders, Philadelphia.

Robinson, S. K., F. R. Thompson III, T. M. Donovan, D. R. Whitehead, and J. Faarborg. 1995. Regional forest fragmentation and the nesting success of migratory birds. Science. 267:1987–1990.

Roby, D. D. 1991. A comparison of two noninvasive techniques to measure total body lipid in live birds. Auk. 108:509–518.

Roeder, K. D. and A. E. Treat. 1961. The detection and evasion of bats by moths. Am. Sci. 49:135–148.

Romer, A. S. and T. S. Parsons. 1977. The vertebrate body. W. B. Saunders, Philadelphia.

Rood, J. P. 1975. Population dynamics and food habits of the banded mongoose. East Afr. Wildl. J. 13:89–111.

Rood, J. P. 1980. Mating relationships and breeding suppression in the dwarf mongoose. Anim. Behav. 28:143–150.

Rood, J. P. 1983. The social system of the dwarf mongoose. Pp. 454–488 *in* Advances in the study of mammalian behavior (J. F. Eisenberg and D. G. Kleiman, eds.). Spec. Publ. No. 7, American Society of Mammalogists.

Rood, J. P. 1986. Ecology and social evolution in the mongoose. Pp. 131–152 *in* Ecological aspects of social evolution: birds and mammals (D. I. Rubenstein and R. W. Wrangham, eds.). Princeton Univ. Press, Princeton, NJ.

Rosenzweig, M. L. 1968. The strategy of body size in mammalian carnivores. Am. Midl. Nat. 80:299–315.

Rosenzweig, M. L. 1985. Some theoretical aspects of habitat selection. Pp. 517–540 *in* Habitat selection in birds (M. L. Cody, ed.). Academic Press, New York.

Rosenzweig, M. L. 1992. Species diversity gradients: we know more and less than we thought. J. Mammal. 73:715–730.

Rottman, S. J. and C. T. Snowden. 1972. Demonstration and analysis of an alarm pheromone in mice. J. Comp. Physiol. Psychol. 81:483–490.

Rowell, T. E. 1974. The concept of social dominance. Behav. Biol. 11:131–154.

Rubidge, B. S. 1994. *Australosyodon*, the first primitive anteosaurid dinocephalian from the Upper Permian of Gondwana. Palaeontology. 37:579–594.

Rudnai, J. 1973. The social life of the lion. Washington Square East, Wallingford, PA.

Ruggiero, A. 1994. Latitudinal correlates of the sizes of mammalian geographical ranges in South America. J. Biogeogr. 21:545–559.

Rumpler, Y. S. and B. Dutrillaux. 1986. Evolution chromosoinque des prosimiens. Mammalia. 50:82–107.

Rumpler, Y., S. Warter, J.-J. Petter, R. Albignac, and B. Dutrillaux. 1988. Chromosomal evolution of Malagasy lemurs. XI. Phylogenetic position of *Daubentonia madagascariensis*. Folia Primat. 50:124–129.

Rusak, B. and I. Zucker. 1979. Neural regulation of circadian rhythms. Physiol. Rev. 59:449–526.

Russell, E. M. 1982. Patterns of parental care and parental investment in marsupials. Biol. Rev. Cambridge Phil. Soc. 57:423–486.

Ryan, M. J. and M. D. Tuttle. 1983. The ability of the frog-eating bat to discriminate among novel and potentially poisonous frog species using acoustic cues. Anim. Behav. 31:827–833.

Ryder, M. L. 1973. Hair. Edward Arnold, London.

Ryder, O. A. (ed.). 1993. Rhinoceros biology and conservation. San Diego Zool. Soc., San Diego, CA.

Saarela, S., R. Hissa, A. Pyörnilä, R. Harjula, M. Ojanen, and M. Orell. 1989. Do birds possess brown adipose tissue? Comp. Biochem. Physiol. 92A:219–228.

Saarela, S., J. S. Keith, E. Hohtola, and P. Trayhurn. 1991. Is the "mammalian" brown fat-specific mitochondrial uncoupling protein present in adipose tissue of birds? Comp. Biochem. Physiol. 100B:45–49.

Sabol, B. M. and M. K. Hudson. 1995. Technique using thermal infrared-imaging for estimating populations of gray bats. J. Mammal. 76:1242–1248.

Sade, D. S. 1965. Some aspects of parent-offspring and sibling relations in a group of rhesus monkeys, with a discussion of grooming. Am. J. Phys. Anthro. 23:1–17.

Sadleir, R. M. F. S. 1973. The reproduction of vertebrates. Academic Press, New York.

Safar-Hermann, N., M. N. Ismail, H. S. Choi, E. Möstl, and E. Mamberg. 1987. Pregnancy determination in zoo animals by estrogen determination in feces. Zoo Biol. 6:189–193.

Samuel, M. D. and M. R. Fuller. 1996. Wildlife radiotelemetry. Pp. 370–418 *in* Research and management techniques for wildlife and habitats, 5th ed., rev. (T. A. Bookhout, ed.). Wildlife Society, Bethesda, MD.

Samuel, W. M., M. J. Pybus, D. A. Welch, and C. J. Wilke. 1992. Elk as a potential host for meningeal worm: implications for translocation. J. Wildl. Manage. 56:629–639.

Sanchez-Cordero, V. and T. H. Fleming. 1993. Ecology of tropical heteromyids. Pp. 596–617 *in* Biology of the Heteromyidae (H. H. Genoways and J. H. Brown, eds.). Spec. Pub. American Society of Mammalogists. No. 10.

Sandell, M. 1984. To have or not to have delayed implantation: the example of the weasel and the stoat. Oikos. 42:123–126.

Sanger, F., S. Nicklen, and A. R. Coulson. 1977. DNA sequencing with chain-terminating inhibitors. Proc. Natl. Acad. Sci. USA. 74:5463–5467.

Sanger, V. L. 1971. Toxoplasmosis. Pp. 326–334 *in* Parasitic diseases of wild mammals (J. W. Davis and R. C. Anderson, eds.). Iowa State Univ. Press, Ames.

Sapolsky, R. 1997. Testosterone rules. Discovery. 18:44–50.

Sapolsky, R. M. 1990. Adrenocortical function, social rank, and personality among wild baboons. Biol. Psychiatry. 28:862–885.

Sapolsky, R. M. 1991. Testicular function, social rank and personality among wild baboons. Psychoneuroendocrinology. 16:281–293.

Sarich, V., J. M. Lowenstein, and B. J. Richardson. 1982. Phylogenetic relationships of *Thylacinus cynocephalus*, Marsupialia, as reflected in comparative serology. Pp. 707–709 *in* Carnivorous marsupials (M. Archer, ed.). Royal Zoological Society, New South Wales, Sydney.

Sarich, V. M. and J. E. Cronin. 1976. Molecular systematics of the Primates. Pp. 141–170 *in* Molecular anthropology (M. Goodman and R. E. Tashian, eds.). Plenum Press, New York.

Sarich, V. M. and A. C. Wilson. 1967a. Immunological time scale for hominid evolution. Science. 158:1200–1203.

Sarich, V. M. and A. C. Wilson. 1967b. Rates of albumin evolution in primates. Proc. Nat. Acad. Sci. U.S.A. 58:142–148.

Sauer, C. O. 1969. Agricultural origins and dispersals: the domestication of animals and foodstuffs, 2d ed. MIT Press, Cambridge, MA.

Savage-Rumbaugh, E. S. 1986. Ape language. From conditioned response to symbol. Columbia Univ. Press, New York.

Savage-Rumbaugh, E. S. and K. E. Brakke. 1990. Animal language: methodological and interpretive issues. Pp. 313–343 *in* Interpretation and explanation in the study of animal behavior (M. Bekoff and D. Jamieson, eds.). Westview Press, Boulder, CO.

Sawyer, T. G., K. V. Miller, and R. L. Marchinton. 1994. Patterns of urination and rub-urination in female white-tailed deer. J. Mammal. 74:477–479.

Schaller, G. B. 1963. The mountain gorilla: ecology and behavior. Univ. of Chicago Press, Chicago.

Schaller, G. B. 1972. The Serengeti lion: a study of predator-prey relations. Univ. of Chicago Press, Chicago.

Schaller, G. B. 1993. The last panda. Univ. of Chicago Press, Chicago.

Schaller, G. B., T. Qitao, K. G. Johnson, W. Xiaoming, S. Heming, and H. Jinchu. 1989. The feeding ecology of giant pandas and Asiatic black bears in the Tangjiahe Reserve, China. Pp. 212–241 *in* Carnivore behavior, ecology, and evolution (J. L. Gittleman, ed.). Cornell Univ. Press, Ithaca, NY.

Schantz, V. S. 1943. Mrs. M. A. Maxwell, a pioneer mammalogist. J. Mammal. 24:464–466.

Scheel, D. and C. Packer. 1991. Group hunting behaviour of lions: a search for cooperation. Anim. Behav. 41:697–709.

Scheffer, V. B. 1951. The rise and fall of a reindeer herd. Sci. Mon. 73:356–362.

Scheffer, V. B. 1958. Seals, sea lions and walruses. Stanford Univ. Press, Stanford, CA.

Scheich, H., G. Langner, C. Tidemann, R. B. Coles, and A. Guppy. 1986. Electroreception and electrolocation in platypus. Nature. 319:401–402.

Schemnitz, S. D. 1996. Capturing and handling wild animals, Pp. 106–124 *in* Research and management techniques for wildlife and habitats, 5th ed., rev. (T. A. Bookhout, ed.). Wildlife Society, Bethesda, MD.

Schliemann, H. and B. Maas. 1978. *Myzopoda aurita.* Mammal. Species No. 116:1–2.

Schmidly, D. J., K. T. Wilkins, and J. N. Derr. 1993. Biogeography. Pp. 319–356 *in* Biology of the Heteromyidae (H. H. Genoways and J. H. Brown, eds.). Spec. Pub., Am. Soc Mammal, No. 10.

Schmidt, G. D. 1971. Acanthocephalan infections of man, with two new records. J. Parasitol. 57:582–584.

Schmidt-Nielsen, K. 1964. Desert animals: physiological problems of heat and water. Oxford Univ. Press, New York.

Schmidt-Nielsen, K. 1997. Animal physiology: adaptation and environment, 4th ed. Cambridge Univ. Press, New York.

Schmidt-Nielsen, K., W. L Bretz, and C. R. Taylor. 1970a. Panting in dogs: unidirectional air flow over evaporative surfaces. Science, 169:1102–1104.

Schmidt-Nielsen, K., F. R. Hainsworth, and D. E. Murrish. 1970b. Counter-current heat exchange in the respiratory passages: effect on water and heat balance. Resp. Physiol. 9:263–276.

Schmidt-Nielsen, K. and R. O'Dell. 1961. Structure and concentrating mechanism in the mammlian kidney. Am. J. Physiol. 200:1119–1124.

Schmidt-Nielsen, K., B. Schmidt-Nielsen, S. A. Jarnum, and T. R. Houpt. 1957. Body temperature of the camel and its relation to water economy. Am. J. Physiol. 188:103–112.

Schnabel, Z. E. 1938. The estimation of the total fish population in a lake. Am. Math. Mon. 45:348–352.

Schneider, H., M. P. C. Schneider, I. Sampaio, M. L. Harada, M. Stanhope, J. Czelusniak, and M. Goodman. 1993. Molecular phylogeny of the New World monkeys (Platyrrhini, Primates). Mol. Phylogen. Evol. 2:225–242.

Schnitzler, H. U., E. K. V. Kalko, I. Kaipf, and A. D. Grinnell. 1994. Fishing and echolocation behavior of the greater bulldog bat, *Noctilio leporinus,* in the field. Behav. Ecol. Sociobiol. 35:327–345.

Schoener, T. W. 1974. Resource partitioning in ecological communities. Science. 185:27–39.

Scholander, P. F. 1955. Evolution of climatic adaptation in homeotherms. Evolution. 9:15–26.

Scholander, P. F., V. Waters, R. Hock, and L. Irving. 1950. Body insulation of some arctic and tropical mammals and birds. Biol. Bull. 99:225–235.

Schonewald-Cox, C. M., S. M. Chambers, B. MacBryde, and L. Thomas (eds.). 1983. Genetics and conservation. Benjamin Cummings, Menlo Park, CA.

Schug, M. D., S. H. Vessey, and A. I. Korytko. 1991. Longevity and survival in a population of white-footed mice (*Peromyscus leucopus*). J. Mammal. 72:360–366.

Schum, M. 1984. Phenetic structure and species richness in North and Central American bat faunas. Ecology. 65:1315–1324.

Schwagmeyer, P. L. 1988. Scramble-competition polygyny in an asocial mammal: male mobility and mating success. Am. Nat. 131:885–892.

Schwartz, J. H. 1986. Primate systematics and a classification of the order. Pp. 1–41 *in* Comparative primate biology, vol. 1, Systematics, evolution, and anatomy (D. R. Swindler and J. Erwin, eds.). Alan R. Liss, New York.

Schwartz, J. H., I. Tatersall, and N. Eldridge. 1978. Phylogeny and classification of the primates revisited. Yearb. Phys. Anthrop. 21:95–133.

Scott, G. R. 1981. Rinderpest. Pp. 18–30 *in* Infectious diseases of wild mammals (J. W. Davis, L. H. Karstad, and D. O. Trainer, eds.). Iowa State Univ. Press, Ames.

Scott, J. P. 1966. Agonistic behavior of mice and rats: a review. Am. Zool. 6:683–701.

Scott, J. P. 1972. Animal behavior, 2d ed. Univ. of Chicago Press, Chicago.

Scullard, H. H. 1974. The elephant in the Greek and Roman world. Thames and Hudson, Cambridge, England.

Seagle, S. W. 1985. Competition and coexistence of small mammals in an east Tennessee pine plantation. Am. Mid. Nat. 114:272–282.

Sealander, J. A. 1952. The relationship of nest protection and huddling to survival of *Peromyscus* at low temperature. Ecology. 33:63–71.

Seber, G. A. F. 1965. A note on the multiple recapture census. Biometrika. 52:249–259.

Seber, G. A. F. 1982. The estimation of animal abundance and related parameters, 2d ed. Griffin, London.

Seebeck, J. H. and P. G. Johnston. 1980. *Potorous longipes* (Marsupialia: Macropodidae): a new species from eastern Victoria. Aust. J. Zool. 28:119–134.

Selander, R. K. 1970. Behavior and genetic variation in natural populations. Am. Zool. 10:53–66.

Selander, R. K. 1982. Phylogeny. Pp. 32–59 *in* Perspectives on evolution (R. Milkman, ed.). Sinauer Assoc., Sunderland, MA.

Selander, R. K., M. H. Smith, S. Y. Yang, W. E. Johnson, and J. B. Gentry. 1971. Biochemical polymorphism and systematics in the genus *Peromyscus*. I. Variation in the old-field mouse (*Peromyscus polionotus*). Studies in genetics VI. Univ. of Texas Pub. 7103:49–90.

Selye, H. 1950. The physiology and pathology of exposure to stress. Acta, Montreal.

Sempéré, A. J., V. E. Sokolov, and A. A. Danilkin. 1996. Capreolus capreolus. Mamm. Spec. 538:1–9.

Setchell, B. R. 1992. Domestication and reproduction. Anim. Repro. Sci. 28:195–202.

Seyfarth, R. M., D. L. Cheney, and P. Marler. 1980a. Monkey responses to three different alarm calls: evidence for predator classification and semantic communication. Science. 210:801–803.

Seyfarth, R. M., D. L. Cheney, and P. Marler. 1980b. Vervet monkey alarm calls: semantic communication in a free-ranging primate. Anim. Behav. 28:1070–1094.

Sharman, G. B. 1963. Delayed implantation in marsupials. Pp. 3–14 *in* Delayed implantation (A. C. Enders, ed.). Univ. of Chicago Press, Chicago.

Sharman, G. G. 1965. The effects of suckling on normal and delayed cycles of reproduction in the red kangaroo. Z. Saugetierk. 30:10–20.

Sheffield, S. R., R. P. Morgan II, G. A. Feldhamer, and D. M. Harman. 1985. Genetic variation in white-tailed deer (*Odocoileus virginianus*) populations in western Maryland. J. Mammal. 66:243–255.

Sheridan, M. and R. H. Tamarin. 1988. Space use, longevity, and reproductive success in meadow voles. Behav. Ecol. Sociobiol. 22:85–90.

Sherman, P. W. 1977. Nepotism and the evolution of alarm calls. Science. 197:1246–1253.

Sherman, P. W. 1981. Kinship, demography, and Belding's ground squirrel nepotism. Behav. Ecol. Sociobiol. 8:251–259.

Sherman, P. W., J. U. M. Jarvis, and R. D. Alexander, eds. 1991. The biology of the naked mole-rat. Princeton Univ. Press, Princeton, NJ.

Shields, W. M. 1982. Philopatry, inbreeding, and the evolution of sex. State Univ. of New York Press, Albany.

Shkolnik, A. and A. Borut. 1969. Temperature and water relations in two species of spiny mice (*Acomys*). J. Mammal. 50:245–255.

Short, R. V. 1983. The biological bases for the contraceptive effects of breast feeding. Pp. 27–39 *in* Advances in international maternal and child health (D. B. Jellife and E. F. B. Jellife, eds.). Oxford Univ. Press, Oxford, England.

Shrewsbury, J. F. D. 1971. A history of the bubonic plague in the British Isles. Cambridge Univ. Press, Cambridge, England.

Sibley, C. G. and J. E. Ahlquist. 1984. The phylogeny of homonoid primates as indicated by DNA-DNA hybridization. J. Mol. Evol. 20:2–15.

Siegel, S. and N. J. Castellan, Jr. 1988. Nonparametric statistics for the behavioral sciences, 2d ed. McGraw-Hill, New York.

Sikes, S. K. 1971. The natural history of the African elephant. Weidenfeld and Nicolson, London.

Sikorski, M. D. and A. D. Bernshtein. 1984. Geographical and intrapopulation divergence in *Clethrionomys glareolus*. Acta Theriol. 29:219–230.

Silk, J. B., D. L. Cheney, and R. M. Seyfarth. 1996. The form and function of post-conflict interactions between female baboons. Anim. Behav. 52:259–268.

Sillén-Tullberg, B. and A. P. Møller. 1993. The relationship between concealed ovulation and mating systems in anthropoid primates: a phylogenetic analysis. Am. Nat. 141:1–25.

Simmons, I. G. 1981. The ecology of natural resources. John Wiley and Sons, New York.

Simmons, J. A., D. J. Howell, and N. Suga. 1975. Information content of bat sonar echoes. Am. Sci. 63:204–215.

Simmons, N. B. 1995. Bat relationships and the origin of flight. Pp. 27–43 *in* Ecology, evolution and behaviour of bats (P. A. Racey and S. M. Swift, eds.). Oxford Univ. Press, Oxford, England.

Simmons, N. B. and T. H. Quinn. 1994. Evolution of the digital tendon locking mechanism in bats and dermopterans: a phylogenetic perspective. J. Mammal. Evol. 2:231–254.

Simons, E. L. 1972. Primate evolution. Macmillan, New York.

Simpson, C. D. 1984. Artiodactyls. Pp. 563–587 *in* Orders and families of recent mammals of the world (S. Anderson and J. K. Jones, Jr., eds.). John Wiley and Sons, New York.

Simpson, G. G. 1940. Mammals and land bridges. J. Wash. Acad. Sci. 30:137–163.

Simpson, G. G. 1945. The principles of classification and a classification of mammals. Bull. Am. Mus. Nat. Hist. 85:1–350.

Simpson, G. G. 1961. Principles of animal taxonomy. Columbia Univ. Press, New York.

Simpson, G. G. 1964. Species density of North American recent mammals. Syst. Zool. 13:57–73.

Sinclair, A. R. E. and J. M. Gosline. 1997. Solar activity and mammal cycles in the Northern Hemisphere. Am. Nat. 149:776–784.

Sinclair, A. R. E., J. M. Gosline, G. Holdsworth, C. J. Krebs, J. N. M. Smith, R. Boonstra, and M. Dale. 1993. Can the solar cycle and climate synchronize the snowshoe hare cycle in Canada? Am. Nat. 141:173–198.

Singer, C. 1959. A history of biology to about the year 1900. Abelard-Schuman, London.

Singleton, G. R., L. K. Chambers, and D. M. Spratt. 1995. An experimental field study to determine whether *Capillaria hepatica* (Nematoda) can limit house mouse populations in Eastern Australia. Wildl. Res. 22:31–53.

Skjenneberg, S. 1984. Reindeer. Pp. 128–138 *in* Evolution of domesticated animals (I. L. Mason, ed.). Longman, New York.

Slater, P. J. B. 1983. The study of communication. Pp. 9–42 *in* Animal behaviour (T. R. Halliday and P. J. B. Slater, eds.). W. H. Freeman, New York.

Slijper, E. J. 1979. Whales, 2d ed. Cornell Univ. Press, Ithaca, NY.

Smale, L., K. E. Holekamp, M. Weldele, L. G. Frank, and S. E. Glickman. 1995. Competition and cooperation between littermates in the spotted hyaena, *Crocuta crocuta*. Anim. Behav. 50:671–682.

Smith, A. L., G. R. Singleton, G. M. Hansen, and G. Shellam. 1993. A serologic survey for viruses and *Mycoplasma pulmonis* among wild house mice (*Mus domesticus*) in southeastern Australia. J. Wildl. Dis. 29:219–229.

Smith, A. T. 1987. Population structure of pikas: dispersal versus philopatry. Pp. 128–142 *in*

Mammalian dispersal patterns. The effects of social structure on population genetics (B. D. Chepko-Sade and Z. T. Halpin, eds.). Univ. of Chicago Press, Chicago.

Smith, A. T., N. A. Formozov, R. S. Hoffmann, Z. Changlin, and M. A. Erbajeva. 1990. The pikas. Pp. 14–60 *in* Rabbits, hares and pikas: status survey and conservation action plan (J. A. Chapman and J. E. C. Flux, eds.). IUCN/SSC Lagomorph Specialist Group. Gland, Switzerland.

Smith, C. C. and O. J. Reichman. 1984. The evolution of food caching by birds and mammals. Ann. Rev. Ecol. Syst. 15:329–351.

Smith, F. A. and D. M. Kaufman. 1996. A quantitative analysis of the contributions of female mammalogists from 1919 to 1994. J. Mammal. 77:613–628.

Smith, J. D. 1972. Systematics of the chiropteran family Mormoopidae. Univ. of Kansas Publ. Mus. Nat. Hist. 56:1–32.

Smith, J. S. 1996. New aspects of rabies with emphasis on epidemiology, diagnosis, and prevention of the disease in the United States. Clin. Microbiol. Rev. 9:166–176.

Smith, M., M. Manlove, and J. Joule. 1978. Spatial and temporal dynamics of the genetic organization of small mammal populations. Pp. 99–113 *in* Populations of small mammals under natural conditions (D. Snyder, ed.). Pymatuning Lab Ecology, Univ. of Pittsburgh, Spec. Publ., Ser. 5.

Smith, M. H., R. K. Selander, and W. E. Johnson. 1973. Biochemical polymorphism and systematics in the genus *Peromyscus*. III. Variation in the Florida deer mouse (*Peromyscus floridanus*), a Pleistocene relict. J. Mammal. 54:1–13.

Smith, P. and E. Tchernov (eds.). 1992. Structure, function and evolution of teeth. Freund Publishing House, London.

Smith, R. E. and B. A. Horwitz. 1969. Brown fat and thermogenesis. Physiol. Rev. 49:330–425.

Smith, W. J. 1984. Behavior of communicating, 2d ed. Harvard Univ. Press, Cambridge, MA.

Smuts, B. B. 1985. Sex and friendship in baboons. Aldine, New York.

Smythe, N. 1977. The function of mammalian alarm advertising: social signals or pursuit invitation? Am. Nat. 111:191–194.

Sober, E. 1984. The nature of selection. MIT Press, Cambridge, MA.

Soholt, L. F. 1973. Consumption of primary production by a population of kangaroo rats (*Dipodomys merriami*) in the Mojave desert. Ecol. Monogr. 43:357–376.

Sokal, R. R. and F. J. Rohlf. 1981. Biometry, 2d ed. W. H. Freeman, San Francisco.

Sondaar, P. Y. 1977. Insularity and its effect on mammal evolution. Pp. 671–707 *in* Major patterns of vertebrate evolution, (M. K. Hecht, P. C. Goody, and B. M. Hecht, eds.). Plenum Press, New York.

Soulé, M. E. 1987. Viable populations for conservation. Cambridge Univ. Press, Cambridge, England.

Soulé, M. E. and B. A. Wilcox (eds.). 1980. Conservation biology: an evolutionary-ecological perspective. Sinauer Assoc., Sunderland, MA.

Sparti, A. 1992. Thermogenic capacity of shrews (Mammalia, Soricidae) and its relationship with basal rate of metabolism. Physiol. Zool. 65:77–96.

Speakman, J. R. 1993. The evolution of echolocation for predation. Pp. 39–63 *in* Mammals as predators (N. Dunstone and M. L. Gorman, eds.). Proceedings of the Zoological Society of London, Clarendon Press, Oxford.

Speakman, J. R. 1995. Chiropteran nocturnality. Pp. 187–201 *in* Ecology, evolution and behaviour of bats (P. A. Racey and S. M. Swift, eds.). Oxford Univ. Press, Oxford, England.

Sperber, I. 1944. Studies on the mammalian kidney. Zool. Bidrag Fran Uppsala. 22:249–430.

Spratt, D. M. 1990. The role of helminths in the biological control of mammals. Int. J. Parasitol. 20:543–550.

Stains, H. J. 1984. Carnivores. Pp. 491–521 *in* Orders and families of recent mammals of the world (S. Anderson and J. K. Jones, Jr., eds.). John Wiley and Sons, New York.

Stallings, R. L., A. F. Ford, D. Nelson, D. C. Torney, C. E. Hildebrand, and R. K. Moyzis. 1991. Evolution and distribution of (GT)$_n$ repetitive sequences in mammalian genomes. Genomics. 10:807–815.

Stanley, H. F., M. Kadwell, and J. C. Wheeler. 1994. Molecular evolution of the family Camelidae: a mitochondrial DNA study. Proceedings of the Royal Society of London. 256:1–6.

Stapp, P. 1992. Energetic influences on the life history of *Glaucomys volans*. J. Mammal. 73:914–920.

Stapp, P., P. J. Pekins, and W. W. Mautz. 1991. Winter energy expenditure and the distribution of southern flying squirrels. Can. J. Zool. 69:2548–2555.

Start, A. N. 1972. Pollination of the baobab (*Adansonia digitata* L.) by the fruit bat *Rousettus aegyptiacus* Geoffroy. E. Afr. Wildl. J. 10:71–72.

Stearns, S. C. 1976. Life-history tactics: a review of the ideas. Q. Rev. Biol. 51:3–47.

Stearns, S. C. 1992. The evolution of life histories. Oxford Univ. Press, New York.

Stebbings, R. E. 1978. Marking bats. Pp. 81–94 *in* Animal marking (B. Stonehouse, ed.). University Park Press, Baltimore.

Stein, B. R. 1996. Women in mammalogy: the early years. J. Mammal. 77:629–641.

Stephens, D. W. and J. R. Krebs. 1986. Foraging theory. Princeton Univ. Press, Princeton, NJ.

Stevenson, R. D. 1986. Allen's rule in North American rabbits (*Sylvilagus*) and hares (*Lepus*) is an exception, not a rule. J. Mammal. 67:312–316.

Stewart, B. S. and R. L. DeLong. 1995. Double migrations of the northern elephant seal, *Mirounga angustirostris*. J. Mammal. 76:196–205.

Stonehouse, B. (ed.) 1978. Animal marking. University Park Press, Baltimore.

Storer, T. I. 1969. Mammalogy and the American Society of Mammalogists, 1919–1969. J. Mammal. 50:785–793.

Storer, T. I., F. C. Evans, and F. G. Palmer. 1944. Some rodent populations in the Sierra Nevada of California. Ecol. Monogr. 14:165–192.

Storey, K. B. and J. M. Storey. 1988. Freeze tolerance in animals. Physiol. Rev. 68:27–84.

Storrs, E. E. 1971. The nine-banded armadillo: a model for leprosy and other biomedical research. Int. J. Leprosy 39:703–714.

Stouffer, R. L. (ed.). 1987. The primate ovary. Plenum Press, New York.

Strahan, R. (ed.). 1995. The mammals of Australia. Reed Books, Chatsworth, Australia.

Strauss, J. F., III, F. Martinez, and M. Kiriakidou. 1996. Placental steroid hormone synthesis: unique features and unanswered questions. Biol. Reprod. 54:303–311.

Strickler, T. L. 1978. Functional osteology and myology of the shoulder in the Chiroptera. S. Karger, New York.

Strier, K. B. 1992. Faces in the forest. Oxford Univ. Press, New York.

Struhsaker, T. T. 1967. Auditory communication among vervet monkeys (*Cercopithecus aethiops*). Pp. 281–324 *in* Social communication among primates (S. A. Altmann, ed.). Univ. of Chicago Press, Chicago.

Stuart, C. and T. Stuart. 1988. Field guide to the mammals of Southern Africa. New Holland Publishers, London.

Stuart, C. and T. Stuart. 1991. The feral cat problem in southern Africa. Afr. Wildl. 45:13–15.

Stuart, M. D., K. B. Strier, and S. M. Pierberg. 1993. A coprological survey of parasites of wild muriquis, *Brachyteles arachnoides* and brown howling monkeys, *Alouatta fusca*. J. Helminth. Soc. Wash. 60:111–115.

Studier, E. H., S. H. Sevick, D. M. Ridley, and D. E. Wilson. 1994. Mineral and nitrogen concentrations in feces of some neotropical bats. J. Mammal. 75:674–680.

Stuenes, S. 1989. Taxonomy, habits, and relationships of the subfossil Madagascan hippopotami *Hippopotamus lemerlei* and *H. madagascariensis*. J. Vertebr. Paleontol. 9:241–268.

Sudre, J. 1979. Nouveaux mammiferes Eocene du Sahara occidental. Palaeovertebrata. 9:83–115.

Suga, N. 1990. Biosonar and neural computation in bats. Sci. Am. 262:34–41.

Sulkin, S. E. and R. Allen. 1974. Virus infections in bats. S. Karger, New York.

Surlykke, A., L. A. Miller, B. Mohl, B. B. Andersen, J. Christiansen-Dalsgaard, and M. B. Jorgensen. 1993. Echolocation in two very small bats from Thailand: *Craseonycteris thonglongyai* and *Myotis siligorensis*. Behav. Ecol. Sociobiol. 33:1–12.

Sutherland, W. J. 1996. Ecological census techniques. Cambridge Univ. Press, New York.

Suthers, R. A. 1970. Vision, olfaction, taste. Pp. 265–309 *in* Biology of bats, vol. II (W. A. Wimsatt, ed.). Academic Press, New York.

Svare, B. B. and M. A. Mann. 1983. Hormonal influences on maternal aggression. Pp. 91–104 *in* Hormones and aggressive behavior (B. Svare, ed.). Plenum, New York.

Swartz, S. M., M. B. Bennett, and D. R. Carrier. 1992. Wing bone stresses in free flying bats and the evolution of skeletal design for flight. Nature. 359:726–729.

Swihart, R. K. and N. A. Slade. 1984. Road crossing in *Sigmodon hispidus* and *Microtus ochrogaster*. J. Mammal. 65:357–360.

Swofford, D. L. and G. J. Olsen. 1990. Phylogeny reconstruction. Pp. 411–501 *in* Molecular systematics (D. M. Hillis and C. Moritz, eds.). Sinauer Assoc., Sunderland, MA.

Szalay, F. S. 1994. Evolutionary history of the marsupials and an analysis of osteological characters. Cambridge Univ. Press, Cambridge, England.

Szalay, F. S. and E. Delson. 1979. Evolutionary history of the primates. Academic Press, New York.

Taber, A. B., C. P. Doncaster, N. N. Neris, and F. H. Colman. 1993. Ranging behavior and population dynamics of the Chacoan peccary, *Catagonus wagneri*. J. Mammal. 74:443–454.

Taber, R. D. and I. McT. Cowan. 1969. Capturing and marking wild animals. Pp. 277–317 *in* Wildlife management techniques, 3d ed. (R. H. Giles, ed.). Wildlife Society, Washington, D.C.

Talmage, R. V. and G. D. Buchanan. 1954. The armadillo (*Dasypus novemcintus*): a review of its natural history, ecology, anatomy and reproductive physiology. Rice Inst. Pamphlet. 41:1–135.

Tamarin, R. H. 1980. Dispersal and population regulation in rodents. Pp. 117–133 *in* Biosocial mechanisms of population regulation (M. N. Cohen, R. S. Malpass and H. G. Klein, eds.). Yale Univ. Press, New Haven, CT.

Tattersall, I. 1972. Of lemurs and men. Nat. Hist. 81:32–43.

Tattersall, I. 1982. The primates of Madagascar. Columbia Univ. Press, New York.

Tautz, D. 1989. Hypervariability of simple sequences as a general source for polymorphic DNA markers. Nucleic Acid Res. 17:6463–6471.

Taylor, C. R. 1972. The desert gazelle: a paradox resolved. Pp. 215–227 *in* Comparative physiology of desert animals (G. M. O. Maloiy, ed.). Symp. Zool. Soc. London, 31.

Taylor, C. R. and C. P. Lyman. 1972. Heat storage in running antelopes: independence of brain and body temperatures. Am. J. Physiol. 222:114–117.

Taylor, J. M. 1984. The Oxford guide to mammals of Australia. Oxford Univ. Press, New York.

Taylor, R. J. 1992. Seasonal changes in the diet of the Tasmanian bettong (*Bettongia gaimardi*), a mycophagous marsupial. J. Mammal. 73:408–414.

Tedford, R. H., M. R. Banks, N. Kemp, I. McDougal, and F. L. Sutherland. 1975. Recognition of the oldest known fossil marsupials from Australia. Nature. 255:141–142.

Terman, C. R. 1973. Reproductive inhibition in asymptotic populations of prairie deermice. J. Reprod. Fertil., Suppl. 19:457–463.

Tevis, L. 1958. Interrelations between the harvester ant *Veromessor pergandei* (Mayr) and some desert ephemerals. Ecology. 39:695–704.

Thomas, D. W. 1983. The annual migrations of three species of West African fruit bats (Chiroptera: Pteropodidae). Can. J. Zool. 61:2266–2272.

Thomas, D. W. 1995. The physiological ecology of hibernation in vespertilionid bats. Pp. 233–244 in Ecology, evolution and behaviour of bats (P. A. Racey and S. M. Swift, eds.). Zool. Soc. London, Clarendon Press, London.

Thomas, T. R. and L. R. Irby. 1990. Habitat use and movement patterns by migrating mule deer in southeastern Idaho. Northw. Sci. 64:19–27.

Thompson, D'A. 1942. On growth and form: a new edition. Cambridge Univ. Press, Cambridge, England.

Thorpe, W. H. 1945. The evolutionary significance of habitat selection. J. Anim. Ecol. 14:67–70.

Thurber, J. M. and R. O. Peterson. 1993. Effects of population density and pack size on the foraging ecology of gray wolves. J. Mammal. 74:879–889.

To, L. P. and R. H. Tamarin. 1977. The relation of population density and adrenal gland weight in cycling and non-cycling voles (*Microtus*). Ecology. 58:928–934.

Toldt, K. 1935. Aufbau under natürliche Färbung des Haarkledes der Wildsäugetiere. Deuts. Gesch. Kleintier Pelztierzucht, Leipzig.

Tomasi, T. E. 1978. Function of venom in the short-tailed shrew, *Blarina brevicauda*. J. Mammal. 59:852–854.

Tomasi, T. E. 1979. Echolocation by the short-tailed shrew, *Blarina brevicauda*. J. Mammal. 60:751–759.

Tomlinson, R. F., H. W. Calkins, and D. F. Marble. 1976. Computer handling of geographical data. UNESCO Press, Paris.

Tonnessen, J. N. and A. O. Johnsen. 1982. The history of modern whaling. Univ. of California Press, Berkeley.

Trappe, J. M. and C. Maser. 1976. Germination of spores of *Glomus macrocarpus* (Endogonaceae) after passage through a rodent digestive tract. Mycologia. 68:433–436.

Trayhurn, P. and D. G. Nicholls (eds.). 1986. Brown adipose tissue. Edward Arnold, London.

Trivers, R. L. 1971. The evolution of reciprocal altruism. Quart. Rev. Biol. 46:35–57.

Trivers, R. L. 1972. Parental investment and sexual selection. Pp. 136–179 in Sexual selection and the descent of man, 1871–1971 (B. Campbell, ed.). Aldine, Chicago.

Trivers, R. L. 1974. Parent-offspring conflict. Am. Zool. 14:249–264.

Trivers, R. L. and D. E. Willard. 1973. Natural selection of parental ability to vary the sex ratio of offspring. Science. 179:90–92.

Tudge, C. 1992. Last animals at the zoo. Oxford Univ. Press, Oxford, England.

Turner, C. D. and J. T. Bagnara. 1976. General endocrinology. W. B. Saunders, Philadelphia.

Turner, D. C. 1975. The vampire bat: A field study in behavior and ecology. Johns Hopkins Univ. Press, Baltimore.

Turner, T. R., M. L. Weiss, and M. E. Pereira. 1992. DNA fingerprinting and paternity assessment in Old World monkeys and ringtailed lemurs. Pp. 96–112 in Paternity in primates: genetic tests and theories (R. D. Martin, A. F. Dixson, and E. J. Wickings, eds.). S. Karger, Basel, Switzerland.

Twigg, G. I. 1978. Marking animals by tissue removal. Pp. 109–118 in Animal marking (B. Stonehouse, ed.). University Park Press, Baltimore.

Twigg, G. I. 1978. The role of rodents in plague dissemination: a worldwide review. Mammal Rev. 8:77–110.

Tyndale-Biscoe, H. and M. Renfree. 1987. Reproductive physiology of marsupials. Cambridge Univ. Press, New York.

Ure, D. C. and C. Maser. 1982. Mycophagy of red-backed voles in Oregon and Washington. Can. J. Zool. 60:3307–3315.

Utzurrum, R. C. B. 1995. Feeding ecology of Philippine fruit bats: patterns of resource use and seed dispersal. Pp. 63–77 in Ecology, evolution and behaviour of bats (P. A. Racey and S. M. Swift, eds.). Oxford Univ. Press, Oxford, England.

Uzzell, T. and D. R. Pilbeam. 1971. Phyletic divergence dates of homonoid primates: a comparison of fossil and molecular data. Evolution. 25:615–635.

Valdés, A. M., M. Slatkin, and N. B. Freimer. 1993. Allele frequencies at microsatellite loci: the stepwise mutation model revisited. Genetics. 133:737–749.

Vandenbergh, J. G. 1967. Effect of the presence of a male on the sexual maturation of female mice. Endocrinology. 81:345–348.

Vandenbergh, J. G. 1969. Endocrine coordination in monkeys: male sexual responses to the female. Physiol. Behav. 4:261–264.

Vandenbergh, J. G. 1969. Male odor accelerates female sexual maturation in mice. Endocrinology. 84:658–660.

Vandenbergh, J. G. 1983. Pheromonal regulation of puberty. Pp. 95–112 in Pheromones and mammalian reproduction (J. G. Vandenbergh, ed.). Academic Press, New York.

Vandenbergh, J. G. and D. M. Coppola. 1986. The physiology and ecology of puberty modulation by primer pheromones. Adv. Stud. Behav. 16:71–108.

Vandenbergh, J. G. and L. C. Drickamer. 1974. Reproductive coordination among free-ranging rhesus monkeys. Physiol. Behav. 13:373–376.

Vander Wall, S. B. 1990. Food hoarding in animals. Univ. of Chicago Press, Chicago.

van Tienhoven, A. 1983. Reproductive physiology of vertebrates, 2d ed. Cornell Univ. Press, Ithaca, NY.

van Tulnen, P., T. J. Robinson, and G. A. Feldhamer. 1983. Chromosome banding and NOR location in sika deer. J. Heredity. 74:473–474.

Van Valen, L. 1973. A new evolutionary law. Evol. Theory. 1:1–33.

Van Valkenburgh, B. 1989. Carnivore dental adaptations and diet: a study of trophic diversity within guilds. Pp. 410–436 in Carnivore behavior, ecology, and evolution (J. L. Gittleman, ed.). Cornell Univ. Press, Ithaca, NY.

Vaughan, T. A. 1959. Functional morphology of three bats: Eumops, Myotis, and Macrotus. Univ. of Kansas Publ. Mus. Nat. Hist. 12:1–153

Vaughan, T. A. 1966. Morphology and flight characteristics of molossid bats. J. Mammal. 47:249–260.

Vaughan, T. A. 1967. Food habits of the northern pocket gopher on shortgrass prairie. Am. Midl. Nat. 77:176–189.

Vaughan, T. A. 1970. Flight patterns and aerodynamics. Pp. 195–216 in Biology of bats, vol. I. (W. A. Wimsatt, ed.), Academic Press, New York.

Vaughan, T. A. 1970a. The muscular system. Pp. 139–194 in Biology of bats, vol. I (W. A. Wimsatt, ed.). Academic Press, New York.

Vaughan, T. A. 1970b. The skeletal system. Pp. 97–138 in Biology of bats, vol. I (W. A. Wimsatt, ed.). Academic Press, New York.

Vaughan, T. A. 1970c. Flight patterns and aerodynamics. Pp. 195–216 in Biology of bats, vol. I (W. A. Wimsatt, ed.). Academic Press, New York.

Vaughan, T. A. 1978. Mammalogy, 2d ed. W. B. Saunders, Philadelphia.

Vaughan, T. A. 1986. Mammalogy, 3d ed. W. B. Saunders, Philadelphia.

Verme, L. J. and J. J. Ozoga. 1981. Sex ratio of white-tailed deer and the estrus cycle. J. Wildl. Manage. 45:710–715.

Vessey, S. H. 1971. Free-ranging rhesus monkeys: Behavioural effects of removal, separation and reintroduction of group members. Behaviour. 40:216–227.

Vessey, S. H. 1973. Night observations of free-ranging rhesus monkeys. Am. J. Phys. Anthrop. 38:613–620.

Vickery, W. L. and J. S. Millar. 1984. The energetics of huddling by endotherms. Oikos. 43:88–93.

Voelker, W. 1986. The natural history of living mammals. Plexus, Medford, OR.

Vogel, P. 1976. Energy consumption of European and African shrews. Acta Theriol. 21:195–206.

Vogel, S. 1994. Life in moving fluids: The physical biology of flow. Princeton Univ. Press, Princeton, NJ.

Vogt, F. D. and G. R. Lynch. 1982. Influence of ambient temperature, nest availability, huddling, and daily torpor on energy expenditure in the white-footed mouse *Peromyscus leucopus*. Physiol. Zool. 55:56–63.

Vogt, F. D., G. R. Lynch, and S. Smith. 1983. Radiotelemetric assessment of diel cycles in euthermic body temperature and torpor in a free-ranging small mammal inhabiting man-made nest sites. Oecologia. 60:313–315.

Vogt, J. L. 1984. Interactions between adult males and infants in prosimians and New World monkeys. Pp. 346–376 *in* Primate paternalism (D. M. Taub, ed.). Van Nostrand Reinhold, New York.

vom Saal, F. 1979. Prenatal exposure to androgen influences morphology and aggressive behavior of male and female mice. Horm. Behav. 12:1–11.

vom Saal, F. 1989. Sexual differentiation in litter bearing mammals: influence of sex of adjacent fetuses *in utero*. J. Anim. Sci. 67:1824–1840.

Waage, J. K. 1979. The evolution of insect/vertebrate associations. Biol. J. Linn. Soc. 12:187–224.

Wade-Smith, J. and B. J. Verts. 1982. Mephitis mephitis. Mamm. Species. 173:1–7.

Walker, M. M., J. L. Kirschvink, G. Ahmed, and A. E. Dizon. 1992. Evidence that fin whales respond to the geomagnetic field during migration. J. Exp. Biol. 171:67–78.

Walsberg, G. E. 1983. Coat color and solar heat gain in animals. BioScience. 33:88–91.

Walsberg, G. E. 1988. Evaluation of a nondestructive method for determining fat stores in small birds and mammals. Physiol. Zool. 61:153–159.

Walther, F. R. 1984. Communication and expression in hoofed mammals. Indiana Univ. Press, Bloomington.

Wang, L. C. H. 1978. Energetic and field aspects of mammalian torpor: the Richardson's ground squirrel. Pp. 109–145 *in* Strategies in cold: natural torpidity and thermogenesis (L. C. H. Wang and J. W. Hudson, eds.). Academic Press, New York.

Wang, L. C. H. 1979. Time patterns and metabolic rates of natural torpor in the Richardson's ground squirrel. Can. J. Zool. 57:149–155.

Wang, L. C. H., and M. W. Wolowyk. 1988. Torpor in mammals and birds. Can. J. Zool. 66:133–137.

Ward, S. J. 1990. Reproduction in the western pygmy possum, *Cercartetus concinnus* (Marsupialia: Burramyidae), with notes on reproduction of some other small possum species. Aust. J. Zool. 38:423–438.

Warren, W. S. and V. M. Cassone. 1995. The pineal gland: photoreception and coupling of behavioral, metabolic, and cardiovascular circadian outputs. J. Biol. Rhythms. 10:64–79.

Waser, P. M. 1975. Diurnal and nocturnal strategies in the bushbuck (*Tragelaphus scriptus*) (Pallas). East Afr. Wildl. J. 13:49–63.

Wasser, S. K., L. Risler, and R. A. Steiner. 1988. Excreted steroids in primate feces over the menstrual cycle and pregnancy. Biol. Reprod. 39:862–872.

Wasser, S. K., S. L. Montfort, and D. E. Wildt. 1991. Rapid extraction of faecel steroids for measuring reproductive cyclicity and early pregnancy in free-ranging yellow baboons (*Papio cynocephalus cynocephalus*). J. Reprod. Fert. 92:415–423.

Wayne, R. K. and S. M. Jenks. 1991. Mitochondrial DNA analysis implying extensive hybridization of the endangered red wolf *Canis rufus*. Nature. 351:565–568.

Webb, P. W. and R. W. Blake. 1985. Swimming. Pp. 110–128 *in* Functional vertebrate morphology (M. Hildebrand, D. M. Bramble, K. F. Liem, and D. B. Wake, eds.). Harvard Univ. Press, Cambridge.

Webb, S. D. 1985. The interrelationships of tree sloths and ground sloths. Pp. 105–112 *in* The evolution and ecology of armadillos, sloths, and vermilinguas (G. G. Montgomery, ed.). Smithsonian Institution Press, Washington, D.C.

Webb, S. D. 1985. Late Cenozoic mammal dispersals between the Americas. Pp. 357–386 *in* The great American biotic interchange (F. G. Stehli and S. D. Webb, eds.). Plenum Press, New York.

Webster, A. B. and R. J. Brooks. 1981. Daily movements and short activity periods of free-ranging meadow voles, *Microtus pennsylvanicus*. Oikos. 37:80–87.

Wecker, S. C. 1963. The role of early experience in habitat selection by the prairie deer mouse, *Peromyscus maniculatus bairdi*. Ecol. Monogr. 33:307–325.

Wegener, A. L. 1912. Die Entstehung der Kontinente. Geolog. Rundsch. 3:276–292.

Wegener, A. L. 1915. Die Entstehung der Kontinente und Ozeane. Vieweg, Braunschweig, Germany.

Wegner, J. and G. Merriam. 1990. Use of spatial elements in a farmland mosaic by a woodland rodent. Biol. Conserv. 54:263–274.

Weir, B. J. and I. W. Rowlands. 1973. Reproductive strategies of mammals. Pp. 139–163 *in* Annual review of ecological systems 4, (R. F. Johnston, P. W. Frank, and C. D. Michener, eds.), Annual Reviews, Palo Alto, CA.

Weir, B. S. 1996. Genetic data analysis II. Sinauer Assoc., Sunderland, MA.

Wells, R. T. 1978. Field observations of the hairy-nosed wombat (*Lasiorhinus latifrons* [Owen]). Aust. Wildl. Res. 5:299–303.

Wenstrup, J. J. and R. A. Suthers. 1984. Echolocation of moving targets by the fish-catching bat, *Noctilio leporinus*. J. Comp. Physiol., ser. A. 155:75–89.

West, S. D. and H. T. Dublin. 1984. Behavioral strategies of small mammals under winter conditions: solitary or social? Pp. 293–299 *in* Winter ecology of small mammals (J. F. Merritt, ed.). Spec. Pub., Carnegie Mus. Nat. Hist., No. 10.

Wetzel, R. M. 1985. The identification and distribution of recent Xenarthra (Edentata). Pp. 5–21 *in* The evolution and ecology of armadillos, sloths, and vermilinguas (G. G. Montgomery, ed.). Smithsonian Institution Press, Washington, D.C.

Wetzel, R. M., R. E. Dubos, R. L. Martin, and P. Myers. 1975. *Catagonus*, an "extinct" peccary, alive in Paraguay. Science. 189:379–381.

Wheaton, C., M. Pybus, and K. Blakely. 1993. Agency perspectives on private ownership of wildlife in the United States and Canada. Transactions of the North American Wildlife and Natural Resources Conference. 58:487–494.

Whitaker, J. O. Jr. 1963. Food of 120 *Peromyscus leucopus* from Ithaca, New York. J. Mammal. 44:418–419.

Whitaker, J. O. Jr. 1966. Food of *Mus musculus, Peromyscus maniculatus bairdi* and *Peromyscus leucopus* in Vigo County, Indiana. J. Mammal. 47:473–486.

Whitaker, J. O. Jr. 1968. Parasites. Pp. 254–311 *in* Biology of *Peromyscus* (Rodentia) (J. A. King, ed.). Spec. Publ. No. 2 American Society of Mammalogists.

Whitaker, J. O. Jr. 1994. Academic propinquity. Pp. 121–138 in Seventy-five years of mammalogy 1919–1994. (Birney, E. C. and J. R. Choate, eds.). Special Pub. No. 11 American Society of Mammalogists.

Whitaker, J. O., Jr. 1994. Food availability and opportunistic versus selective feeding in insectivorous bats. Bat Res. News. 35:75–77.

Whitaker, J. O., Jr. 1995. Food of the big brown bat *Eptesicus fuscus* from maternity colonies in Indiana and Illinois. Am. Midl. Nat. 134:346–360.

Whitaker, J. O. Jr., C. Neefus, and T. H. Kunz. 1996. Dietary variation in the Mexican free-tailed bat (*Tadarida brasiliensis mexicana*). J. Mammal. 77:716–724.

Whitaker, J. O., Jr. and L. J. Rissler. 1993. Do bats feed in winter? Am. Midl. Nat. 129:200–203.

White, D. J. 1993. Lyme disease surveillance and personal protection against ticks. Pp. 99–125 *in* Ecology and environmental management of Lyme disease (H. S. Ginsberg, ed.). Rutgers Univ. Press, New Brunswick, NJ.

White, G. C. and R. A. Garrott. 1990. Analysis of wildlife radio-tracking data. Academic Press, San Diego.

Wickler, S. J. 1980. Maximal thermogenic capacity and body temperatures of white-footed mice (*Peromyscus*) in summer and winter. Physiol. Zool. 53:338–346.

Wiens, J. A. 1976. Population responses to patchy environments. Ann Rev. Ecol. Syst. 7:81–120.

Wiens, J. A. 1977. On competition and variable environments. Am. Sci. 65:590–597.

Wildt, D. E., M. Bush, K. L. Goodrowe, C. Packer, A. E. Pusey, J. L. Brown, P. Joslin, and S. J. O'Brien. 1987. Reproductive and genetic consequences of founding isolated lion populations. Nature. 329:328–331.

Wiley, R. W. 1980. Neotoma floridana. Mamm. Species. 139:1–7.

Wilkinson, G. S. 1984. Reciprocal food sharing in the vampire bat. Nature. 308:181–184.

Wilkinson, G. S. 1985. The social organization of the common vampire bat. I. Pattern and cause of association. Behav. Ecol. Sociobiol. 17:111–121.

Wilkinson, G. S. 1987. Altruism and cooperation in bats. Pp. 299–323 *in* Recent advances in the study of bats (M. B. Fenton, P. A. Racey, and J. M. V. Rayner, eds.). Cambridge Univ. Press, New York.

Wilkinson, G. S. 1990. Food sharing in vampire bats. Sci. Am. 262:64–70.

Williams, E. S., E. T. Thorne, D. R. Kwiatkowski, and B. Oakleaf. 1992. Overcoming disease problems in the black-footed ferret recovery program. Trans. North Am. Wildl. Nat. Resourc. Conf. 57:474–485.

Williams, G. C. 1966. Adaptation and natural selection: a critique of some current evolutionary thought. Princeton Univ. Press, Princeton, NJ.

Williams, G. C. 1975. Sex and evolution. Princeton Univ. Press, Princeton, NJ.

Willig, M. R. and M. A. Mares. 1989. A comparison of bat assemblages from phytogeographic zones of Venezuela. Pp. 59–67 *in* Patterns in the structure of mammalian communities (D. W. Morris, Z. Abramsky, B. J. Fox and M. R. Willig, eds.). Texas Tech Univ. Press, Lubbock, TX.

Willig, M. R. and M. P. Moulton. 1989. The role of stochastic and deterministic processes in structuring Neotropical bat communities. J. Mammal. 70:323–329.

Willig, M. R. and K. W. Selcer. 1989. Bat species density gradients in the New World: a statistical assessment. J. Biogeogr. 16:189–195.

Willner, G. R. 1982. Nutria, *Myocastor coypus.* Pp. 1059–1076 *in* Wild mammals of North America: biology, management, and economics (J. A. Chapman and G. A. Feldhamer, eds.). Johns Hopkins Univ. Press, Baltimore.

Wilson, D. E., F. R. Cole, J. D. Nichols, R. Rudran, and M. S. Foster. 1996. Measuring and monitoring biological diversity: standard methods for mammals. Smithsonian Institution Press, Washington, D. C.

Wilson, D. E. and J. F. Eisenberg. 1990. Origin and applications of mammalogy in North America. Pp. 1–35 in Current mammalogy, vol. 2 (H. H. Genoways, ed.). Plenum, New York.

Wilson, D. E. and G. L. Graham. 1992. Pacific island flying foxes: proceedings of an international conservation conference. USDI, USFWS, Biol. Rep. 90(23).

Wilson, D. E. and D. M. Reeder (eds.). 1993. Mammal species of the world: a taxonomic and geographic reference, 2d ed. Smithsonian Institution Press, Washington, D. C.

Wilson, D. S. 1980. The natural selection of populations and communities. Benjamin/Cummings, Menlo Park, CA.

Wilson, E. O. 1975. Sociobiology: the new synthesis. Harvard Univ. Press, Cambridge, MA.

Wilson, J. W., III. 1974. Analytical zoogeography of North American mammals. Evolution. 28:124–140.

Wimsatt, W. A. 1945. Notes on breeding behavior, pregnancy, and parturition in some vespertilionid bats of the eastern United States. J. Mammal. 26:23–33.

Wimsatt, W. A. and A. L. Guerriere. 1962. Observations on the feeding capacities and excretory functions of captive vampire bats. J. Mammal. 43:17–27.

Wimsatt, W. A. and B. Villa-R. 1970. Locomotor adaptations in the disc-winged bat, *Thyroptera tricolor.* Am. J. Anat. 129:89–119.

Winn, H. E., T. J. Thompson, W. C. Cummings, J. Hain, J. Hudnall, H. Hays, and W. W. Steiner. 1981. Song of the humpback whale-population comparisons. Behav. Ecol. Sociobiol. 8:41–46.

Wishmeyer, D. L., G. D. Snowder, D. H. Clark, and N. E. Cockett. 1996 Prediction of live lamb chemical composition utilizing electromagnetic scanning (TOBEC). J. Anim. Sci. 74:1864–1872

Wolfe, J. L. 1970. Experiments on nest building behavior in *Peromyscus* (Rodentia: Cricetinae). Animal Behav. 18:613–615.

Wolfe, J. L. and S. A. Barnett. 1977. Effects of cold on nest-building by wild and domestic mice, *Mus musculus.* L. Biol. J. Linn. Soc. 9:73–85.

Wolff, J. O. 1980. Social organization of the taiga vole (*Microtus xanthognathus*). Biologist. 62:34–45.

Wolff, J. O. 1985. Maternal aggression as a deterrent to infanticide in *Peromyscus leucopus* and *P. maniculatus.* Anim. Behav. 33:117–123.

Wolff, J. O. 1989. Social behavior. Pp. 271–291 *in* Advances in the study of *Peromyscus* (G. L. Kirkland, Jr. and J. N. Layne, eds.). Texas Tech Univ. Press, Lubbock, TX.

Wolff, J. O. 1996. Population fluctuations of mast-eating rodents are correlated with production of acorns. J. Mammal. 77:850–856.

Wolff, J. O., R. D. Dueser, and K. S. Berry. 1985. Food habits of sympatric *Peromyscus leucopus* and *Peromyscus maniculatus.* J. Mammal. 66:795–798.

Wolff, J. O., M. H. Freeberg, and R. D. Dueser. 1983. Interspecific territoriality in two sympatric species of *Peromyscus* (Rodentia: Cricetidae). Behav. Ecol. Sociobiol. 12:237–242.

Wolff, J. O. and D. S. Durr. 1986. Winter nesting behavior of *Peromyscus leucopus* and *Peromyscus maniculatus.* J. Mammal. 67:409–412.

Wolff, J. O. and W. Z. Lidicker, Jr. 1981. Communal winter nesting and food sharing in taiga voles. Behav. Ecol. Sociobiol. 9:237–240.

Wolff, J. O., K. I. Lundy, and R. Baccus. 1988. Dispersal, inbreeding avoidance, and reproductive success in white-footed mice. Anim. Behav. 36:456–465.

Wood, A. E. 1965. Grades and clades among rodents. Evolution. 19:115–130.

Woodburne, M. O. and J. A. Case. 1996. Dispersal, vicariance, and the late Cretaceous to early Tertiary land mammal biogeography from South America to Australia. J. Mammal. Evol. 3:121–161.

Woodburne, M. O., B. J. MacFadden, J. A. Case, M. Springer, N. S. Pledge, J. D. Power, J. M. Woodburne, and K. Johnson. 1993. Land mammal biostratigraphy and magnetostratigraphy of the Etadunna Formation (late Oligocene) of South Australia. J. Vert. Paleontol. 13:132–164.

Woodburne, M. O. and R. H. Tedford. 1975. The first Tertiary monotreme from Australia. Am. Mus. Novitates. 2588:1–11.

Woolley, P. 1974. The pouch of *Planigale subtilissima* and other dasyurid marsupials. J. R. Soc. West. Aust. 57:11–15.

World Health Organization. 1984. Expert committee on rabies. Seventh report. WHO Tech. Report No. 709, Geneva.

Worthington, W. J., C. Moritz, L. Hall, and J. Toop. 1994. Extreme population structuring in the threatened ghost bat, *Macroderma gigas:* evidence from mitochondrial DNA. Proceedings of the Royal Society of London. 257:193–198.

Wozencraft, W. C. 1989a. The phylogeny of the Recent Carnivora. Pp. 495–535 *in* Carnivore behavior, ecology, and evolution (J. L. Gittleman, ed.). Cornell Univ. Press, Ithaca, NY.

Wozencraft, W. C. 1989b. Classification of the Recent Carnivora. Pp. 569–593 *in* Carnivore behavior, ecology, and evolution (J. L. Gittleman, ed.). Cornell Univ. Press, Ithaca, NY.

Wozencraft, W. C. 1993. Order Carnivora. Pp. 279–348 *in* Mammal species of the world, 2d ed. (D. E. Wilson and D. M. Reeder, eds.). Smithsonian Institution Press, Washington, D.C.

Wrabetz, M. J. 1980. Nest insulation: a method of evaluation. Can. J. Zool. 58:938–940.

Wright, P. L. and M. W. Coulter. 1967. Reproduction and growth in Maine fishers. J. Wild. Mgmt. 31:70–87.

Wroot, A. 1984. Hedgehogs. Pp. 751–757 *in* The encyclopedia of mammals (D. Macdonald, ed.). Facts on File Publ., New York.

Wunder, B. A. 1978. Implications of a conceptual model for the allocation of energy resources by small mammals. Pp. 68–75 *in* Populations of small mammals under natural conditions (D. P. Snyder, ed.). Pymatuning

Lab. Ecol., Univ. of Pittsburgh, Spec. Pub., Ser. 5.

Wunder, B. A. 1984. Strategies for, and environmental cueing mechanisms of, seasonal changes in thermoregulatory parameters of small mammals. Pp. 165–172 *in* Winter ecology of small mammals (J.F. Merritt, ed.). Spec. Pub., Carnegie Mus. Nat. Hist., No. 10.

Wunder, B. A. 1985. Energetics and thermoregulation. Pp. 812–844 *in* Biology of New World *Microtus* (R. H. Tamarin, ed.). Spec. Pub. Am. Soc. Mammal, No. 8.

Wunder, B. A., D. S. Dobkin, and R. D. Gettinger. 1977. Shifts of thermogenesis in the prairie vole (*Microtus ochrogaster*): strategies for survival in a seasonal environment. Oecologica. 29:11–26.

Wunder, B. A. and R. D. Gettinger. 1996. Effects of body mass and temperature acclimation on the nonshivering thermogenic response of small mammals. Pp. 131–139 *in* Adaptation to the cold: tenth international hibernation symposium (F. Geiser, A. J. Hulbert and S. C. Nicol, eds.). Univ. of New England Press, Armidale, NSW, Australia.

Würsig, B. 1988. The behavior of baleen whales. Sci. Am. 258:102–107.

Würsig, B. and M. Würsig. 1977. The photographic determination of group size, composition, and stability of coastal porpoises (*Tursiops truncatus*). Science. 198:755–756.

Wynne-Edwards, V. C. 1962. Animal dispersion in relation to social behavior. Oliver and Boyd, Edinburgh.

Wynne-Edwards, V. C. 1986. Evolution through group selection. Blackwell Scientific Publ., Boston.

Wyss, A. R. and J. J. Flynn. 1993. A phylogenetic analysis and definition of the Carnivora. Pp. 32–52 *in* Mammal phylogeny: placentals (F. S. Szalay, M. J. Novacek, and M. C. McKenna, eds.). Springer-Verlag, New York.

Xia, X. and J. S. Millar. 1991. Genetic evidence of promiscuity in *Peromyscus leucopus*. Behav. Ecol. Sociobiol. 28:171–178.

Ya-Ping, Z. and S. Li-Ming. 1993. Phylogenetic relationships of macaques inferred from restriction endonuclease analysis of mitochondrial DNA. Folia Primat. 60:7–17.

Yager, D. D. and M. L. May. 1990. Ultrasound-triggered, flight-gated evasive manoeuvres in the praying mantis, *Parasphendale agrionina*. II. Tethered flight. J. Exp. Biol. 152:41–58.

Yager, D. D., M. L. May, and M. B. Fenton. 1990. Ultrasound-triggered, flight-gated evasive escape manoeuvres in the praying mantis, *Parasphendale agrionina*. I. Free flight. J. Exp. Biol. 152:17–39.

Yagil, R. 1985. The desert camel. Karger, Basel.

Yamazaki, K., G. K. Beauchamp, C. J. Wysocki, J. Bard, L. Thomas, and E. A. Boyse. 1983. Recognition of H-2 types in relation to the blocking of pregnancy in mice. Science. 221:186–188.

Yamazaki, K., E. A. Boyse, V. Mike, H. T. Thaler, B. J. Mathieson, J. Abbott, J. Boyse, and Z. A. Zayas. 1976. Control of mating preferences in mice by genes in the major histocompatibility complex. J. Exp. Med. 144:1324–1335.

Yates, T. L. 1984. Insectivores, elephant shrews, tree shrews, and Dermopterans. Pp. 117–144 *in* Orders and families of recent mammals of the world (S. Anderson and J. K. Jones, Jr., eds.). John Wiley and Sons, New York.

Young, J. Z. 1957. The life of mammals. Oxford Univ. Press, New York.

Yousef, M. K., S. M. Horvath, and R. W. Bullard (eds.). 1972. Physiological adaptations, desert and mountain. Academic Press, New York.

Yuill, T. M. 1987. Diseases as components of mammalian ecosystems: mayhem and subtlety. Can. J. Zool. 65:1061–1066.

Zahavi, A. 1975. Mate selection—a selection for a handicap. J. Theor. Biol. 53:205–214.

Zar, J. H. 1996. Biostatistical analysis, 3d ed. Prentice-Hall, Upper Saddle River, NJ.

Zeuner, F. E. 1963. History of domesticated animals. Hutchinson, London.

Zeveloff, S. I. 1988. Mammals of the intermountain West. Univ. of Utah Press, Salt Lake City.

Zeveloff, S. I. and M. S. Boyce. 1980. Parental investment and mating systems in mammals. Evolution. 34:973–982.

Zima, J. and M. Macholán. 1995. B chromosomes in the wood mice (genus *Apodemus*). Acta Theriol. Suppl. 3:75–86.

Zucker, I. 1985. Pineal gland influences period of circannual rhythms of ground squirrels. Am. J. Physiol. 249:R111–R115.

Zucker, I., M. Boshes, and J. Dark. 1983. Suprachiasmatic nuclei influence circannual and circadian rhythms of ground squirrels. Am. J. Physiol. 244:R472–R480.

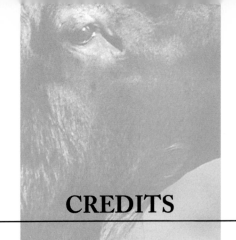

CREDITS

Photographs

Unit Openers(all)

Hal Korber

Chapter 1

Figure 1.1b: © Leonard Lee Rue III/Visuals Unlimited; **Figure 1.2:** James L. Amos/National Geographic Image Collection; **Figure 1.3:** © Smithsonian Institution, NMNH Chip Clark, Photographer.

Chapter 2

Figure 2.1: Hubertus Kanus/Photo Researchers, Inc.; **Figure 2.2:** Art Resource/Field Museum of Natural History, Chicago, USA; **Figure 2.3:** Special Collections, University of South Florida Library; **Figure 2.4,2.5:** Corbis-Bettmann; **Figure 2.6:** Independence National Historical Park Collection; **Figure 2.7(1):** photo by J.D. Haweeli/American Society Mammalogists; **Figure 2.7(2):** photo by P.V. August/American Society Mammalogists; **Figure 2.8:** © Francois Gohier/Photo Researchers, Inc.; **Figure 2.9:** Des Bartlett/Photo Researchers, Inc.; **Figure 2.10a,b & 2.10d,e:** American Society Mammalogists; **Figure 2.11:** © Hal H. Harrison/Photo Researchers, Inc.; **Figure 2.12:** © Leonard Lee Rue III/Visuals Unlimited.

Chapter 3

Figure 3.2: © Nicholas De Bore III/Bruce Coleman Inc.; **Figure 3.4:** Mark Fuller/The Wildlife Society; **Figure 3.6:** Jeanne & Stuart Altmann; **Figure 3.9:** From National Audubon Society/Photo Researchers, Inc.; **Figure 3.11:** Courtesy of J.S. Yoon, Bowling Green State University; **Figure 3.12:** R.P. Canham, "Serum Protein Variations and Selection in Fluctuating Populations of Cricetid Rodents," Ph.D. thesis, University of Alberta, 1969.

Chapter 5

Figure 5.1: © Mark Newman/Bruce Coleman, Inc.; **Figure 5.9a:** © Ken Lucas/Visuals Unlimited; **Figure 5.9b:** © Charlie Ott/Photo Researchers, Inc.; **Figure 5.9c:** © Verna R. Johnston, Photo Researchers, Inc.; **Figure 5.9d:** Science VU/Visuals Unlimited; **Figure 5.13:** © Bill Bachman 1986/Photo Researchers, Inc.; **Figure 5.14a:** Jen & Des Bartlett/Bruce Coleman, Inc.; **Figure 5.14b:** © Nigel J. Dennis/Photo Researchers, Inc.; **Figure 5.15:** Leonard Lee Rue III/From National Audubon Society/Photo Researchers, Inc.; **Figure 5.17:** © Leonard Lee Rue III/Visuals Unlimited; **Figure 5.18:** Science VU/Visuals Unlimited.

Chapter 6

Figure 6.4a: © Dwight Kuhn; **Figure 6.4b:** © Dr. Edwin Gould (National Zoological Perk)/Dwight Kuhn Photography; **Figure 6.5a:** Animals Animals © Studio Carlo Dani; **Figure 6.5b:** © Nancy Adams/Tom Stack & Assoc.; **Figure 6.7a:** © M.W. Larson/Bruce Coleman Inc.; **Figure 6.7b:** from Gunderson, *Mammalogy* fig 7-3 p. 160. McGraw-Hill Book Company; **Figure 6.13:** R. Van Nostrand from National Audubon Society/Photo Researchers, Inc.; **Figure 6.15a:** John Cooke © Oxford Scientific Films; **Figure 6.15b:** from *Analysis of Vertebrate Structure,* Milton Hildebrand © 1995. Reprinted by permission of John Wiley & Sons, Inc.; **Figure 6.16:** © John Alcock/Visuals Unlimited.

Chapter 7

Figure 7.7: Des Bartlett/Photo Researchers, Inc.; **Figure 7.8a:** © Leonard Lee Rue III/Photo Researchers, Inc.; **Figure 7.8b:** © Len Rue, Jr./Visuals Unlimited.

Chapter 8

Figure 8.7: Leonard Lee Rue III, from National Audubon Society/Photo Researchers, Inc.; **Figure 8.8:** Courtesy of P.F. Scholander, University of California, San Diego; **Figure 8.11:** © Leonard Lee Rue III/Visuals Unlimited; **Figure 8.12:** Carnegie Museum of Natural History; **Figure 8.13a,b:** Courtesy Robert MacArthur; **Figure 8.18b:** from Hayward and Lyman, 1967; **Figure 9.3:** from *Bacula of North American Mammals* by William Henry Burt (Ed) Dr. Jerry Smith; **Figure 9.18a:** Jen and Des Bartlett/Photo Researchers, Inc.

Chapter 9

Figure 9.16: © John D. Cunningham/Visuals Unlimited.

Chapter 10

Figure 10.16a: Picture taken by Peter L. Meserve in La Picada, Chile in 1984; **Figure 10.21a,b:** © Tom McHugh/Photo Researchers, Inc.; **Figure 10.27:** Turner & McKay (1987); Burramyidae. p. 652-664 in Walton, D.W. & Richardson, B.J. (eds). Fauna of Australia Volume 1B Mammalia. Australian Government Publishing Service, Canberra.

Chapter 11

Figure 11.2a: A.W. Ambler/Photo Researchers, Inc.; **Figure 11.2b:** © Irene Vandermolen/Visuals Unlimited; **Figure 11.3:** © Kim Taylor/Burce Coleman Inc.; **Figure 11.4a:** © Willaim J. Weber/Visuals Unlimited; **Figure 11.6a,b:** American Society Mammalogists; **Figure 11.16:** Photo by C.J. Phillips/American Society Mammalogists.

Chapter 12

Figure 12.11: © Meril D. Tuttle, Bat Conservation International; **Figure 12.13:** David L. Pearson/Visuals Unlimited; **Figure 12.17a:** Photo by P.V. August/American Society Mammalogists; **Figure 12.17b:** © Stephen Dalton/Animals Animals; **Figure 12.23:** © Charles E. Mohr/Photo Researchers, Inc.

Chapter 13

Figure 13.6: James Burke, *Life Magazine* © Time Inc.; **Figure 13.9:** © David Haring / Oxford Scientific Films; **Figure 13.10:** Rob Williams/Bruce Coleman, Inc.; **Figure 13.11:** A.W. Ambler/Photo Researchers, Inc.; **Figure 13.13:** © Sidney Bahrt/Photo Researchers, Inc.; **Figure 13.15:** © Nigel J. Dennis; **Figure 13.16:** Animals Animals © Doug Wechsler; **Figure 13.18:** Arthur W. Ambler/Photo Researchers, Inc.; **Figure 13.20:** © Norman Owen Tomalin/Bruce Coleman Inc.; **Figure 13.22:** Animals Animals © BatesLiddlehales; **Figure 13.23a:** Animals Animals © Rod Williams; **Figure 13.23b:** Animals Animals © Stewart D. Halperin; **Figure 13.23c:** Animals Animals © Pat Crowe.

Chapter 14

Figure 14.6(1): Jeanne White/Photo Researchers, Inc.; **Figure 14.6(2):** Animals Animals © Michael Fogden; **Figure 14.10a:** Roy Pinney/Photo Researchers, Inc.; **Figure 14.11a:** Gary Milburn/Tom Stack & Associates.

Chapter 15

Figure 15.2: Animals Animals © Joe McDonald; **Figure 15.3a:** Len Rue, Jr./Visuals Unlimited; **Figure 15.3b:** © Karl H. Maslowski/Photo Researchers, Inc.; **Figure 15.12:** Courtesy Kay Holekamp; **Figure 15.13:** © John Shaw/Bruce Coleman Inc.; **Figure 15.15, 15.18:** © Leonard Lee Rue III/Visuals Unlimited; **Figure 15.19:** © Don W. Fawcett/Visuals Unlimited.

Chapter 16

Figure 16.13: © M. DeMocker/Visuals Unlimited.

Chapter 17

Figure 17.6a: © Leonard Lee Rue III/Visuals Unlimited; **Figure 17.6b:** Science VU/Visuals Unlimited; **Figure 17.8:** © Karl H. & Stephen Maslowski/Visuals Unlimited; **Figure 17.9, 17.10a:** © Woodrow Goodpaster/Photo Researchers, Inc.; **Figure 17.11:** Animals Animals/Hope Sawyer Guyukmihci; **Figure 17.14a:** Animals Animals © Michael Dick; **Figure 17.14b:** © Patrick Morris/Oxford Scientific Films; **Figure 17.14c:** Jen and Des Bartlett/Photo Researchers, Inc.; **Figure 17.14d:** Joe Whittaker; **Figure 17.15:** © A.W. Ambler From National Audubon Society/Photo Researchers, Inc.; **Figure 17.16:** Jen and Des Bartlett/Photo Researchers, Inc.; **Figure 17.18a:** Animals Animals © Robert Maier; **Figure 17.18b:** © Leonard Lee Rue III/Visuals Unlimited; **Figure 17.19a:** © Tom McHugh/Photo Researchers, Inc.; **Figure 17.19b:** Jen and Des Bartlett/Photo Researchers, Inc.; **Figure 17.20:** Animals Animals © Patti Murray; **Figure 17.23:** © Leonard Lee Rue III/Visuals Unlimited; **Figure 17.26:** © Cleveland P. Hickman, Jr./Visuals Unlimited; **Figure 17.28a:** © Steve Maslowski/Visuals Unlimited; **Figure 17.28b:** © Leonard Lee Rue III/Visuals Unlimited.

Chapter 18

Figure 18.8: © Len Rue, Jr/Visuals Unlimited; **Figure 18.10:** © Leonard Lee Rue III/Visuals Unlimited; **Figure 18.11:** H.N. Hoeck, Constance; **Figure 18.16:** Daniel K. Odell.

Chapter 19

Figure 19.6a: © A.W. Ambler From National Audubon Society/Photo Researchers, Inc.; **Figure 19.6b:** © Leonard Lee Rue III/Visuals Unlimited; **Figure 19.11a:** Jen and Des Bartlett/Photo Researchers, Inc.; **Figure 19.13:** © A.W. Ambler/Photo Researchers, Inc.; **Figure 19.14:** © Mark N. Boulton/Photo Researchers, Inc.; **Figure 19.16:** © George Holton/Photo Researchers, Inc.; **Figure 19.19:** © Len Clifford/Visuals Unlimited.

Chapter 20

Figure 20.3: Courtesy Kay Holekamp; **Figure 20.5:** Animals Animals © Robert Maier; **Figure 20.7:** © Fiona Guinness; **Figure 20.8:** Irven DeVore/Anthro-Photo; **Figure 20.9:** Scott & Nancy Creel; **Figure 20.10:** Dr. Marc Bekoff; **Figure 20.12:** David Overcash/Bruce Coleman Inc.; **Figure 20.13:** Sarah Blaffer-Hardy/Anthro-Photo.

Chapter 21

Figure 21.3: Animals Animals © Leonard Rue III; **Figure 21.4:** New York Zoological Society; **Figure 21.5:** © Irene Vandermolen/Visuals Unlimited; **Figure 21.8:** Burney J. Le Boeuf; **Figure 21.11:** Dr. Laurence Frank/University of California, Berkeley; **Figure 21.13:** © Meril D. Tuttle, Bat Conservation International.

Chapter 22

Figure 22.1: © Craig Packer; **Figure 22.2:** Scott & Nancy Creel; **Figure 22.3:** Photo: Kelly C. Fletcher; **Figure 22.4:** L.T. Nash/Arizona State University; **Figure 22.5a:** © R.B. Wells/Anthro-Photo; **Figure 22.6:** © Raymond A. Mendez/Animals Animals; **Figure 22.7:** photo by George D. Lepp, courtesy of Paul Sherman; **Figure 22.9:** © Patricia D. Moehlman; **Figure 22.11:** photo by William A. Wimsatt, Courtesy of the Mammal Slide Library at Cornell University.

Chapter 23

Figure 23.1: Dave Wildt; **Figure 23.8:** © Len Rue, Jr./Visuals Unlimited.

Chapter 24

Figure 24.6: Animals Animals © Fritz Prenzel; **Figure 24.17b:** © Leonard Lee Rue III/Photo Researchers, Inc.

Chapter 25

Figure 25.8, 25.9: Leonard Lee Rue III/Visuals Unlimited.

Chapter 26

Figure 26.5: © Leonard Lee Rue III/Visuals Unlimited.

Chapter 27

Figure 27.1: Courtesy Roy C. Anderson; **Figure 27.4a:** Dwight Kuhn; **Figure 27.4b:** © G.Shih-R. Kessel/Visuals Unlimited; **Figure 27.7:** Bruce Iverson/Science Photo Library/Photo Researchers, Inc.; **Figure 27.11:** Science VU-FIP/Visuals Unlimited.

Chapter 28

Figure 28.4a: Animals Animals © Terence A. Gili; **Figure 28.4b:** © Tom McHugh/Photo Researchers, Inc.; **Figure 28.6:** © Eric & David Hosking; **Figure 28.7:** Judy Rains; **Figure 28.8a:** Animals Animals © Leonard Lee Rue III; **Figure 28.8b:** © Joan Iaconetti/Bruce Coleman Inc.

Chapter 29

Figure 29.1a: © Len Rue, Jr./Visuals Unlimited; **Figure 29.2:** © Frank Lambrecht/Visuals Unlimited; **Figure 29.5a:** © Asa C. Thoresen/Photo Researchers, Inc.; **Figure 29.5b:** Michael Freeman/Bruce Coleman Inc.; **Figure 29.6:** Photo by D.G. Huckaby/American Society Mammalogists; **Figure 29.7:** © Rod Williams/Bruce Coleman Inc.; **Figure 29.12:** Photo by A.H. Shoemaker/American Society Mammalogists.

Line Art

Figure 4.12: From Cleveland P. Hickman, Jr., et al., Integrated Principles of Zoology, 10th edition. Copyright © 1996 McGraw-Hill Companies, Inc., Dubuque, Iowa. All Rights Reserved. Reprinted by permission.

Figure 7.9: From D. M. Madison, "Activity Rhythme and Spacing" in Biology of New World Microtus, R. Tamarin (ed.), 1985, 8:373-419, 1985 American Society of Mammalogists. Reprinted by permission of the author.

Figure 8.23: From K. Schmidt-Nielsen, et al., "Counter-Current Heat Exchange in the Respiratory Passages: Effect on Water and Heat Balance" in Respiratory Physiology, 9:263-276, 1970.

Figure 9.17: After C.R. Austin and R.V. Short, eds., Reproduction in Mammals, vol. 4, Reproductive Patterns, 1972 in Cleveland P. Hickman, Jr., et al., Integrated Principles of Zoology, 10th edition. Copyright © 1996 McGraw-Hill Company, Inc., Dubuque, Iowa. All Rights Reserved. Reprinted by permission.

Figure 12.16: From J.E. Hill and J.D. Smith, Bats: A Natural History. Natural History Museum, London. Reprinted by permission.

Figure 13.24: From Cleveland P. Hickman, Jr., et al., Integrated Principles of Zoology, 10th edition. Copyright © 1996 McGraw-Hill Companies, Inc., Dubuque, Iowa. All Rights Reserved. Reprinted by permission.

Figure 17.22: From O.A. Reig, "Ecology of Spalacopus," in Journal of Mammalogy, 51(3). Copyright © American Society of Mammalogist. Reprinted by permission.

Figure 18.5: From V.J. Maglio, "Origin and Evolution of Elephantidae" in Trans. American Philosophical Society, Vol. 63, 1973. Reprinted by permission of American Philosophical Society, Bethesda, MD.

Figure 18.7: From H.F. Osborn, Proboscidea Volume II, American Museum Press. Reprinted by permission of American Museum of Natural History.

Figure 19.4: From Storer and Usinger, Elements of Zoology, 2nd edition. Copyright © 1961 California Academy of Sciences. Reprinted by permission.

Figure 19.5: From B.J. McFadden "Patterns of PhylogenyÖ." in Paleobiology, 11:245-257, 1985, Smithsonian Institution, Washington, D.C. Reprinted by permission.

Figure 20.2: From D. Muller-Schwarze, "Pheromones in Black-tailed Deer" in Animal Behavior, 19:141-152, 1971. Academic Press, London. Reprinted by permission.

Figure 24.16: From W.Z. Lidicker, "Solving the Enigma of Microtine Cycles" in Journal of Mammalogy, 1988, 69:225-235. Reprinted by permission of American Society of Mammalogists.

Figure 26.3: After W. Hamilton, U.S. Geological Survey as appeared in Carla W. Montgomery and David Dathe, Earth Then and Now, 3rd edition. Copyright © 1997 McGraw-Hill Company, Inc., Dubuque, Iowa. All Rights Reserved. Reprinted by permission.

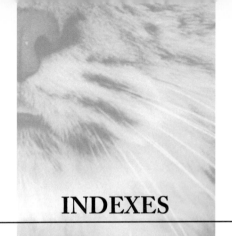

INDEXES

Page numbers followed by f and t refer to figures and tables, respectively.

SUBJECT INDEX

SCIENTIFIC NAMES